24th Innovative Manufacturing Engineering and Energy International Conference (IManEE 2020)

AA007524

IOP Conference Series: Materials Science and Engineering
Volume 1037

Athens, Greece
14 - 15 December 2020

Editors:

Angelos P. Markopoulos
Irene P. Koronaki
Laurentiu Slatineanu
Gheorghe Cretu
Oana Dodun
Andrei Marius Mihalache

ISBN: 978-1-7138-4562-1
ISSN: 1757-8981

Printed from e-media with permission by:

Curran Associates, Inc.
57 Morehouse Lane
Red Hook, NY 12571

Some format issues inherent in the e-media version may also appear in this print version.

Institute of Physics
Dirac House, Temple Back
Bristol BS1 6BE UK

Phone: 44 1 17 929 7481
Fax: 44 1 17 920 0979

techtracking@iop.org

Additional copies of this publication are available from:

Curran Associates, Inc.
57 Morehouse Lane
Red Hook, NY 12571 USA
Phone: 845-758-0400
Fax: 845-758-2633
Email: curran@proceedings.com
Web: www.proceedings.com

TABLE OF CONTENTS

IOP Conf. Series: Materials Science and Engineering 1037 (2021) 011001 doi:10.1088/1757-899X/1037/1/011001

24th Edition of Innovative Manufacturing Engineering and Energy International Conference (IManEE 2020)

1st Virtual Edition

14-15 December 2020

Edited by:

Angelos P. MARKOPOULOS
National Technical University of Athens, Greece

Irene P. KORONAKI
National Technical University of Athens, Greece

Laurenţiu SLATINEANU
"Gheorghe Asachi" Technical University of Iaşi, România

Gheorghe CRETU
"Gheorghe Asachi" Technical University of Iaşi, România

Oana DODUN
"Gheorghe Asachi" Technical University of Iaşi, România

Andrei Marius MIHALACHE
"Gheorghe Asachi" Technical University of Iaşi, România

IOP Conf. Series: Materials Science and Engineering 1037 (2021) 011001 doi:10.1088/1757-899X/1037/1/011001

Preface

The 24th edition of Innovative Manufacturing Engineering & Energy International Conference (IManEE 2020) is also the 1st Virtual edition of the Conference. Initially the conference was planned to take place in Athens, Greece, however, due to the fact that COVID-19 has significantly affected the travels between countries and in many countries total lockdowns have prohibited the organizing of events such as this, it was decided to go on-line, rather than postponing the event. Nevertheless, the School of Mechanical Engineering of the National Technical University of Athens (Greece) and the Department of Manufacturing Engineering of the "Gheorghe Asachi" Technical University of Iaşi (Romania) have joint forces to make this event come true and continue its long history.

Looking back at the time, the first conference of the series was organized in May 1996 in Iaşi and, since 1999, the conferences have been organized alternately, in Iaşi and Chişinău, by the collaboration of the Machine Manufacturing Departments from the "Gheorghe Asachi" Technical University of Iaşi (Romania) and by the Technical University of Moldova from Chisinau (Republic of Moldova). Considering the new trends and requests valid in the field of manufacturing engineering from the entire world, starting from 2013, the name of the conference was changed to "Innovative Manufacturing Engineering" and later to "Innovative Manufacturing Engineering and Energy". Over the last few years, other departments from universities in Romania (Braşov, Bucureşti, Cluj, Galaţi and Piteşti) or other countries (Serbia, Greece) have been involved in organizing the conference. This would be the first time for the conference to be hosted by the National Technical University of Athens and hopefully it will be organized in Athens in the near future. In the course of time, the conference has become a significant event with regular participants who meet to exchange ideas and make collaborations. The active scientific community involved in the conference considers this yearly gathering important and thus a conference, even in a virtual concept, was deemed necessary.

The conference has managed to draw the attention of many researchers, some new and some old friends, from 16 countries i.e., Bulgaria, Georgia, Germany, Greece, Israel, Italy, Nigeria, Poland, Portugal, Romania, Russia, Slovakia, Spain, Ukraine, the United Kingdom and the United States of America. In the conference, 72 papers were selected by the Scientific Committee to be presented, after single-blind review from at least two reviewers. All papers were presented on-line on Webex platform in about 15 minutes plus 5 minutes for Q&A, managed by the National Technical University of Athens, in Greece, divided in 10 sessions. Furthermore, 4 plenary speeches on interesting topics on manufacturing engineering and energy have enlightened the participants and given a floor for useful discussions. More specifically, the plenary sessions and the presenters, exceptional members of the academic community, were the following: Professor S. Karellas from the National Technical University of Athens, Greece, speaking on Carbon Capture and Utilization within the frame of the circular economy, Dr. A.M. Mihalache from Gheorghe Asachi Technical University of Iasi, Romania, with a lecture on 3D printing for small projects, Professor A. But from the Politehnica University of Timişoara, Romania, presenting Haas Technical Education Centers in Europe and the world and Professor A. Sadeh from HIT Holon Institute of Technology in Israel, with a speech on the success of innovative firms and the technological risk.

It is the hope of the organizers of IManEE 2020 Conference, despite the difficulties raised by the pandemic, that it was a good opportunity to exchange information of mutual interest in the fields of manufacturing technology and energy and for new collaborations among participants to be developed, that will lead to joint research projects for the benefit of all.

For its success, the Conference has been generously supported by the Section of Manufacturing Technology of the School of Mechanical Engineering, National Technical University of Athens. Special thanks are addressed to the organizers of the Conference, to the members of the Scientific Committee, to the reviewers and, last but not least, to all the participants; all colleagues that have wholeheartedly worked towards the success of IManEE 2020.

The organizing committee wishes all the participants to be successful and expresses conviction that there will be an opportunity to meet again, in person, in the near future, in the National Technical University of Athens, at another IManEE conference.

Assistant Professor Angelos P. Markopoulos,
National Technical University of Athens
School of Mechanical Engineering
Director of the Section of Manufacturing Technology
Athens, Greece

HONORARY COMMITTEE

Honorary President
Prof. Gheorghe NAGÎŢ, PhD, Head of Machine Manufacturing Department of the Machine Manufacturing and Industrial Management Faculty, Gheorghe Asachi Technical University of Iasi, Romania

Honorary Committee:
Prof. Gheorghe NAGÎŢ, PhD, Head of Machine Manufacturing Department of the Machine Manufacturing and Industrial Management Faculty, Gheorghe Asachi Technical University of Iasi, Romania

Prof. Laurenţiu SLĂTINEANU, PhD, Former Head of Machine Manufacturing Department of the Machine Manufacturing and Industrial Management Faculty, Gheorghe Asachi Technical University of Iasi, Romania

Prof. Oana DODUN, PhD, Former Head of Machine Manufacturing Department of the Machine Manufacturing and Industrial Management Faculty, Gheorghe Asachi Technical University of Iasi, Romania

CONFERENCE CHAIRS

Prof. Angelos P. MARKOPOULOS, PhD, National Technical University of Athens, School of Mechanical Engineering (Manufacturing)

Prof. Irene P. KORONAKI, PhD, National Technical University of Athens, School of Mechanical Engineering (Energy)

SCIENTIFIC COMMITTEE

Adrian TRIF, TU of Cluj-Napoca, România
Adriana MUNTEANU, Gheorghe Asachi TU of Iaşi, România
Alexander MIKHAYLOV, NTU Donetsk, Ukraine
Alexandra Banu, U Politehnica of Bucharest, România
Alexandru FILIP, Transilvania U of Braşov, România
Alexandru SOVER, Ansbach U of Applied Sciences, Germany
Alexandru EPUREANU, Dunarea de Jos U of Galaţi, România
Alexei TOCA, TU of Moldova, Chisinau, Republic of Moldova
Alin Daniel RIZEA, U of Piteşti, România
Alina MINEA, Gheorghe Asachi TU of Iaşi, România
Ancuţa PĂCURAR, TU of Cluj-Napoca, România
Andrei MIHALACHE, Gheorghe Asachi TU of Iaşi, România
Andrzej BUCHACZ, Silesian TU of Gliwice, Poland
Andrzej DZIERWA, Rzeszow University of Technology, Poland
Andrzej Wróbel, Silesian U of Technology, Gliwice, Poland
Angelos MARKOPOULOS, NTUA, Greece
Anna NOCIVIN, Ovidius U of Constanţa, România
António M. Gonçalves COELHO, U Nova de Lisboa, Portugal
António MOURÃO, U Nova de Lisboa, Portugal
Apostolos Korlos, TEI of Thessaloniki, Greece
Apostolos Tsagaris, Western Macedonia U of Appl Sciences, Greece
Arik SADEH, Holon Institute of Technology, Israel

Petru BERCE, TU of Cluj-Napoca, România
Péter TAMAS, University of Miskolc, Hungary
Petru DUŞA, Gheorghe Asachi TU of Iaşi, România
Radu IBANESCU, Gheorghe Asachi TU of Iaşi, România
Răzvan PĂCURAR, TU of Cluj-Napoca, România
Richard HERMAN, Polytechnic U of Timişoara, România
Roberto TETI, U Federico II of Naples, Italy
Romeo IONESCU, Stefan cel Mare U of Suceava, România
R.M. Chandima RATNAYAKE, University of Stavanger, Norway
Sergiu MAZURU, TU of Moldova, Chisinau, Republic of Moldova
Sevasti MITSI, Aristotle U of Thessaloniki, Greece
Sofiene DELLAGI, Université de Lorraine – Metz, France
Stanislaw LEGUTKO, Poznan U of Technology, Poland
Stergios Maropoulos, Western Macedonia U of Appl Scien. Greece
Tom SAVU, U Politehnica of Bucharest, România
Toufik BOUDOUH, U Technologie de Belfort-Montbéliard, France
Traian BUIDOS, U of Oradea, România
Traian MAZILU, U Politehnica of Bucharest, România
Valentin ZICHIL, Vasile Alecsandri U of Bacau, România
Valeriu DULGHERU, TU of Moldova, Chisinau, Rep. of Moldova
Vasile MERTICARU, Gheorghe Asachi TU of Iaşi, România
Vasilios DIMITRIOU, Hellenic Mediterranean University, Greece
Victoria RĂDULESCU, U Politehnica of Bucharest, România
Viorel COHAL, Gheorghe Asachi TU of Iaşi, România
Viorel PALEU, Gheorghe Asachi TU of Iasi, România
Viorel PĂUNOIU, Dunărea de Jos U of Galaţi, România
Virgil TEODOR, Dunarea de Jos U of Galaţi, România
Witold HABRAT, Rzeszow University of Technology, Poland

ORGANIZING COMMITTEE

Members of Organizing Committee from National Technical University of Athens:
Nikolaos GALANIS; Nikolaos KARKALOS; Panagiotis KARMIRIS-OBRATAŃSKI; Emmanouil Lazaros PAPAZOGLOU; Dimitrios SKONDRAS-GIOUSIOS.

Members of Organizing Committee from TU of Iasi:
Eugen AXINTE; Mihai BOCA; Constantin CĂRAUŞU; Viorel COHAL; Margareta COTEAŢĂ; Gheorghe CREŢU; Oana DODUN, Petru DUŞA; Laurenţiu GHENGHEA; Octavian LUPESCU; Vasile MERTICARU; Andrei MIHALACHE; Gavril MUSCĂ; Florin NEGOESCU; Gheorghe NAGIT; Marius RÎPANU; Lucian TĂBĂCARU.

KEYNOTE SPEAKERS

Sotirios KARELLAS, National Technical University of Athens, Greece
"Carbon Capture and Utilization: a step forward in the circular economy society"
Andrei Marius MIHALACHE, Gheorghe Asachi Tehnical University of Iasi, Romania
"3D printing. An alternative for small projects?"
Adrian BUT, Politehnica University of Timişoara, Romania
"HTEC presentation"
Arik SADEH, HIT Holon Institute of Technology, Israel
"The Success of Innovative Firms, Technological Risk, and Market Novelty"

IOP Conf. Series: Materials Science and Engineering 1037 (2021) 011001 doi:10.1088/1757-899X/1037/1/011001

SESSION CHAIRS

Session 1
Laurenţiu SLĂTINEANU, Gheorghe Asachi Tehnical University of Iasi, Romania
Pantelis G. NIKOLAKOPOULOS, University of Patras, Greece
Session 2
George-Christopher VOSNIAKOS, National Technical University of Athens, Greece
Witold HABRAT, Rzeszow University of Technology, Poland
Session 3
Katarina MONKOVA, Technical University of Kosice, Slovakia
Vasilis DIMITRIOU, Hellenic Mediterranean University, Greece
Session 4
Arik SADEH, HIT Holon Institute of Technology, Israel
John KECHAGIAS, University of Thessaly, Greece
Session 5
Panorios BENARDOS, National Technical University of Athens, Greece
Panagiotis KYRATSIS, University of Western Macedonia, Greece
Session 6
Irene P. KORONAKI, National Technical University of Athens, Greece
Yulia RAFIKOVA, Lomonosov Moscow State University, Russian Federation
Session 7
Mehrdad ASADI, Anglia Ruskin University, United Kingdom
Daniela SPASOVA, Technical University of Varna, Bulgaria
Session 8
Dimitrios TZETZIS, International Hellenic University, Greece
Nikolaos M. VAXEVANIDIS, School of Pedagogical and Technological Education, Greece
Session 9
Alexandru SOVER, Ansbach University of Applied Sciences, Germany
George ANTONAKOS, National Technical University of Athens, Greece
Session 10
Oana DODUN, Gheorghe Asachi Tehnical University of Iasi, Romania
Andrei Marius MIHALACHE, Gheorghe Asachi Tehnical University of Iasi, Romania

IOP Conf. Series: Materials Science and Engineering 1037 (2021) 011002 doi:10.1088/1757-899X/1037/1/011002

Peer review declaration

All papers published in this volume of *IOP Conference Series: Materials Science and Engineering* have been peer reviewed through processes administered by the Editors. Reviews were conducted by expert referees to the professional and scientific standards expected of a proceedings journal published by IOP Publishing.

- **Type of peer review: Single-blind**
  The reviewers (at least two for each paper) received the manuscript with all the details of the authors and a form, prepared for the conference, with certain questions and a field for comments. The authors, at the next step, received anonymous reviews with the following possible recommendations: accept as is/minor revisions/major revisions/reject. The paper was revised by the authors accordingly and then it was forwarded to the same reviewers for final recommendation. Final Decision was taken by the editors.

- **Conference submission management system:**
  The papers were submitted to an e-mail, handled by the editor. The handling editor managed the correspondence between reviewers and authors.

- **Number of submissions received:**
  76

- **Number of submissions sent for review:**
  76

- **Number of submissions accepted:**
  72

- **Acceptance Rate (Number of Submissions Accepted / Number of Submissions Received X 100):**
  94,7%

- **Average number of reviews per paper:**
  2

- **Total number of reviewers involved:**
  48

- **Any additional info on review process:**
  There was a pre-screening process of the abstracts by the editors, prior to the submission of the final papers.

- **Contact person for queries:**
  Associate Professor Angelos P. Markopoulos, National Technical University of Athens, Greece, amark@mail.ntua.gr

3D printing. An alternative for small projects?

A M Mihalache[1]**, G Nagit**[1]**, M I Ripanu**[1] **and L Slatineanu**[1]

[1]Gheorghe Asachi Technical University of Iasi, Department of Machine Manufacturing Technology, Bld. Dimitrie Mangeron 59A, Iasi, Romania

Corresponding author's e-mail: andrei.mihalache@tuiasi.ro

Abstract. 3D printing is one of the big top up-to-date technologies. Our aim is to show that up to a point 3D prints can successfully replace traditionally manufactured assemblies in the case of educational projects or lab experiments. We have considered a 3D printed simple-frame robotic arm upon which we have performed FEA analyses. They have shown promising results against its steel-based peer up to certain loads under the same working conditions in terms of safety factor and safety margin.

1. Introduction

3D printing techniques are various and address different domains. They became a viable alternative to classic manufacturing especially for small projects or prototypes. In this paper we are focusing on a comparation between two functional models of a robotic arm. One has the 1.0490 DIN [1] as the assigned manufacturing material and the other is using PLA (Polylactic acid) [2]. We aim to show by means of FEA that in case of small loads the PLA-based model can withstand multiple loading cycles. We are proofing the design by taking into consideration two parameters: *the safety factor* and *margin of safety*. Previous work included analyses that verified the structural integrity of spherical PLA based models against compression [3]. We wanted to find out also if 3D printing is accurate enough to consider in small projects or prototypes [4]. This wider research pool helped us formulate conclusions that did encourage us to move on with this new one.

Failure may be predicted using tests, instruments and mathematical models. Designs of all kinds are being implemented using a reliability index which stand on top of a safety factor. There are different opinions regarding the safety factor of a mechanical design. It generally describes the capacity of a system to exceed certain limits thus highlighting how much stronger a mechanical part or assembly is than it was intended to be under a certain load. Subsequently there is another parameter to take into account and that is the margin of safety. We may asses it as a measure of a system's structural capacity. It gives information about the system's excess capacity. If a mechanical part or assembly is subjected to a maximum load, the margin of safety informs us about how many more loads of the same force can it withstand before failing. The two can offer an insight for a specific mechanical part or assembly to be considered safe, overengineered or poorly designed.

G. Bolzon *et al.* have conducted tests which would highlight the behavior of the materials under uniaxial tensile loading. They found out "*that the indentation curves in different states are almost insensitive to the variations of the yield limits of the considered metals while they clearly reflect the change of the ultimate tensile strength*" [5]. This shows a possible failure if aging produces embrittlement or any other sort of treatment (*chemical, temperature dependent, etc*.).

Under a certain load we can obtain results in form of stress or strain. Mathematical equations have been proposed by Yu.G. Matvienko for plane stress and plane strain [6]:

$$SF_{FAD} = 1.44 \cdot SF_Y^{1.34} \quad (1)$$

$$SF_{FAD} = -0.064 + 1.1 \cdot SF_Y \quad (2)$$

Where SF_{FAD} represents the safety factor against fracture and SF_Y the safety factor against plastic collapse.

In case of failure assessment if we are to consider plane strain the two were found to be very similar. In case of plane stress, the author concluded "*that the probabilistic safety factors increase with the decrease of the probability of failure*" [6].

M. Chen *et al.* have presented results about the effect of process repairs taking into consideration the margin of safety. They considered a plastic pre-strain effect, and a thermal aging effect. They found out that the synergistic action of plastic pre-strain and thermal aging effect does not interfere in the evaluation. Failure still appears in plastic domain and "*as work hardening is induced by plastic pre-strain, the safety margin is increased after the repair process for the typical crack in the case study*" [7].

2. Theoretical considerations

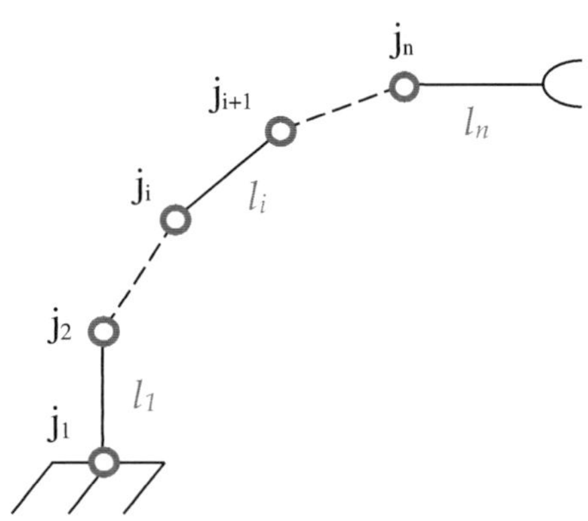

Figure 1. Decomposing a robotic arm into n elements.

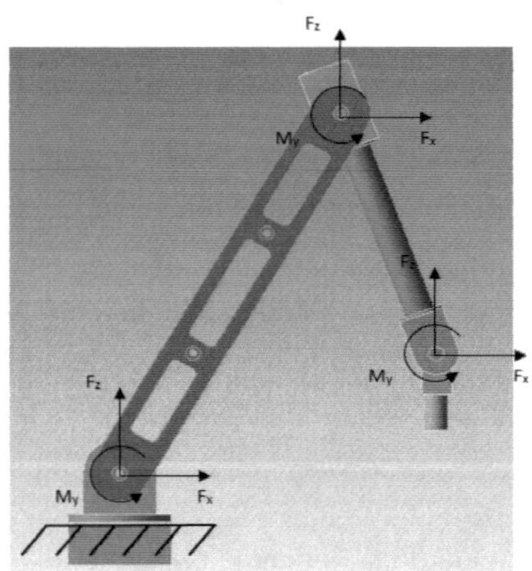

Figure 2. Identification of forces and moments for each joint of the considered robotic arm.

We have taken into account a generic robotic arm model. Mathematically, this model has to be decomposed into single components in order to be properly assessed. The decomposing produces n links: $l_1, l_2 \ldots, l_n$ and n joints: $j_1, j_2, \ldots, j_n$ (*see figure 1*). This method allows further decomposing into beam elements and joints. Regarding our model we may have a clear image of the forces that act upon each joint by applying external forces F_x and F_z and the moment M_y in the center of gravity for each joint (*see figure 2*).

Safety of the entire assembly depends much on its joints. If they fail the entire system would collapse. In this regard, L. Huo and L. Baron proposed a mathematical approach for joint-limits avoidance. They found out that by considering a weighting vector too small "*the redundant displacement may not be sufficient to avoid the joint limits*" and if too high it will produce high velocities in each joint [8].

Another element to be taken into account is the robotic arm's transmission ratio. Considering joint spaces S.H.H. Zargarbashi *et al.* proposed that it should be expressed as an angle between n-dimensional joint-torque and joint-rate-vectors making it as low as possible. This will act as a protection against overload of the joint-rate and joint-torque vectors [9].

In case of FEA analyses both the weighting vector and the transmission ratio were considered ideal.

IOP Conf. Series: Materials Science and Engineering 1037 (2021) 012001 doi:10.1088/1757-899X/1037/1/012001

3. Setup

By comparing safety factor and margin of safety values obtained under different loads for the same system we can predict how many loads that system will or will not withstand. For the chosen materials we have considered computing the safety factor both for yield and ultimate stress. Since yield will result in a value of the safety factor until the system starts to plastically deform, we have chosen only the evaluation of the safety factor by means of ultimate tensile strength since it gives us a value until failure occurs. Assessing the margin of safety, we may encounter three situations. If zero, the mechanical part or assembly will not accept additional loads before failure. If negative, failure will appear before maximum load is achieved. If one, the mechanical part or assembly will withstand one additional maximum load. We have considered three positions of the robotic arm: *horizontal*, *intermediate* and *vertical* (*see figure 3*). In each case we have applied a load under the form of a pulling force. Values of $10N$, $100N$ and $1000N$ were considered. This translates into approximately $1kgf$, $10kgf$ and $100kgf$ to be lifted by the robotic arm. The type of analysis was set to *Static Structural* and for each one we have set the stress tool to *Maximum Equivalent Stress* with the stress limit type set to *Tensile Ultimate per material*. We have then added the two materials using *Ansys Granta*.

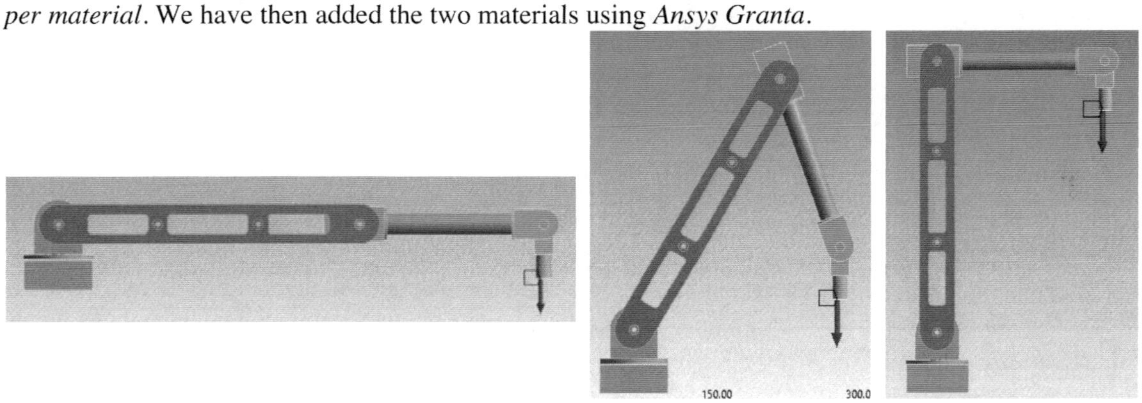

(a) (b) (c)

Figure 3. Considered positions of the robotic arm for FEA analyses.

4. Results and discussions

Table 1. Safety factor.

Force (N)	Horizontal 1.0490 DIN Min.	Horizontal 1.0490 DIN Max.	Horizontal PLA Min.	Horizontal PLA Max.	Intermediate 1.0490 DIN Min.	Intermediate 1.0490 DIN Max.	Intermediate PLA Min.	Intermediate PLA Max.	Vertical 1.0490 DIN Min.	Vertical 1.0490 DIN Max.	Vertical PLA Min.	Vertical PLA Max.
10*N*	1.2383e-004	2.4679	8.8699	15	3.6144e-004	13.858	15	15	15	15	4.2784	15
100*N*	1.2383e-005	0.2467	0.88697	15	3.6144e-005	1.3859	3.1229	15	14.275	15	0.43188	15
1000*N*	1.624e-006	2.4532e-002	1.9829e-002	15	3.6144e-006	0.1386	0.3123	15	1.5058	15	3.8802e-002	15

Positions of the robotic arm

Table 2. Margin of safety.

Force (N)	Horizontal 1.0490 DIN Min.	Horizontal 1.0490 DIN Max.	Horizontal PLA Min.	Horizontal PLA Max.	Intermediate 1.0490 DIN Min.	Intermediate 1.0490 DIN Max.	Intermediate PLA Min.	Intermediate PLA Max.	Vertical 1.0490 DIN Min.	Vertical 1.0490 DIN Max.	Vertical PLA Min.	Vertical PLA Max.
10*N*	-0.99988	1.4679	7.8699	14	-0.99964	12.858	14	14	14	14	3.2784	14
100*N*	-0.99999	-0.7533	-0.11303	14	-0.99996	0.38586	2.1229	14	13.275	14	-0.56812	14
1000*N*	-1	-0.97547	-0.98017	14	-1	-0.8614	-0.6877	14	0.50579	14	-0.9612	14

Positions of the robotic arm

Results have been plotted for both the safety factor and margin of safety for all considered positions of the robotic arm. We have decided to present where the minimum occurs for both the maximum and minimum values over time. Those values were then introduced in *Table 1* and *Table 2*.

To get a clear view of behaviors for each type of material-based model, graphical representations were obtained by stacking results obtained for all considered force values. This gave us the response of the assembly to an increase loading over time. Combining all positions, we may observe the differences of the assembly's response to a specific load value. By interpreting the charts, we have discovered similar behaviors that led to the conclusion that up to one-point PLA proves to be a valid alternative to its steel peer given the assembly's small dimensions.

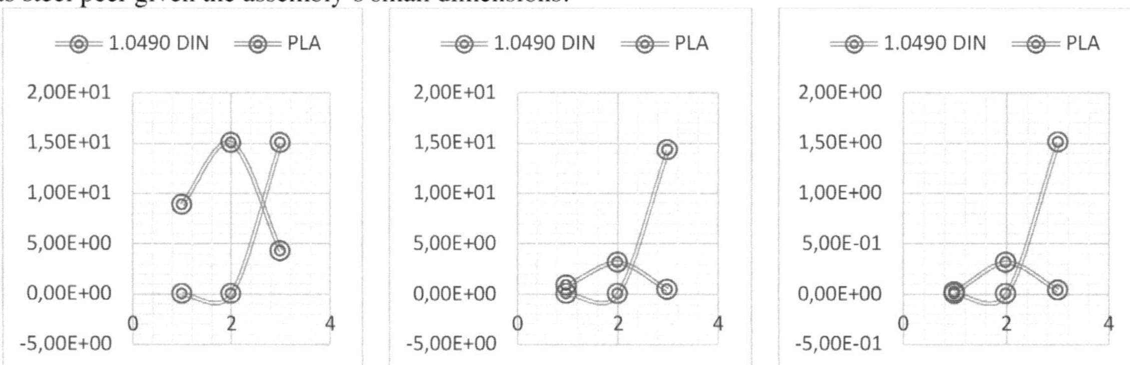

(a) (b) (c)

Figure 4. Safety factor comparative charts for both type of materials at minimum values.

We have considered both minimum and maximum values for each load combining results obtained for each position as follows: first the horizontal one, then the intermediate and finally the vertical one. In case of the safety factor, minimums show an interesting evolution. 1.0490 DIN proves as expected very reliable for all loads but reports only 1.5058 in the vertical position at 1000N load. PLA barely makes it over zero with 0.019829 at 1000N load in the horizontal positions (*see figure 4*).

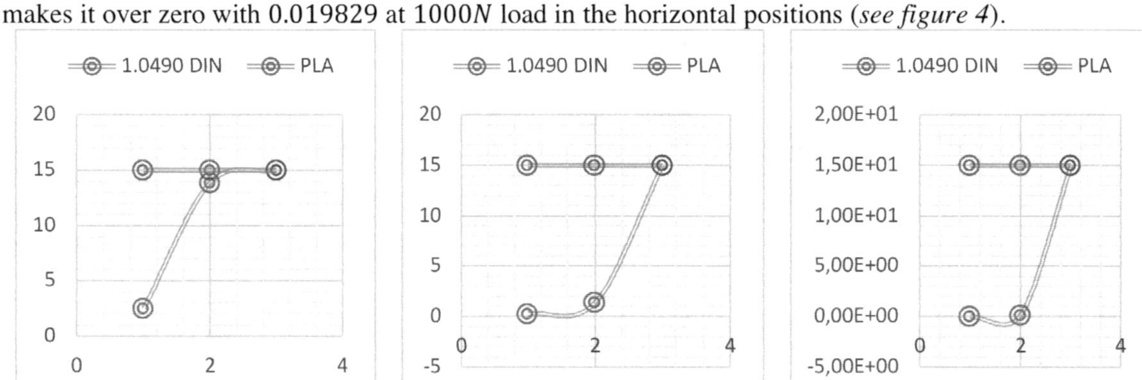

(a) (b) (c)

Figure 5. Safety factor comparative charts for both type of materials at maximum values.

Then we have compiled results for the margin of safety's maximum values. The graphical representation from the left corresponds to a 10N load, the middle one to 100N load and the right graphical representation corresponds to 1000N load. Except the 10N load, 1.0490 DIN reports values around 0.2 in the horizontal position as PLA reaches the upper limit. This means that PLA based assembly does not make it under these conditions and further refinement might be necessary. Although maximum reaches the upper limit, the average indicates that PLA proves valid for the 10N load in terms of safety factor requirements (*see figure 5*).

IOP Conf. Series: Materials Science and Engineering 1037 (2021) 012001 doi:10.1088/1757-899X/1037/1/012001

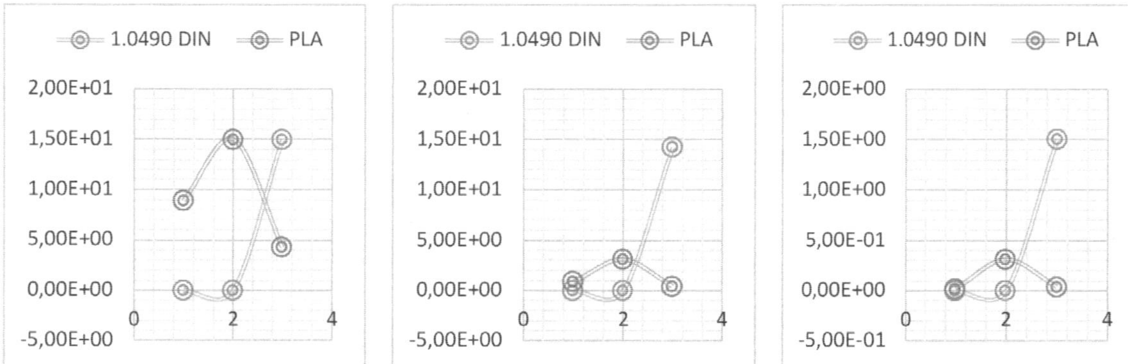

(a) (b) (c)

Figure 6. Margin of safety comparative charts for both type of materials at minimum values.

In case of margin of safety, the same working principle was applied. Charts were computed for minimum and maximum results for all loads by combining them for each position of the assembly. As mentioned above values below zero indicate a possible failure before the maximum load is achieved. 1.0490 DIN fails at $1000N$ load reaching -1 in the horizontal position of the assembly. So, even though the safety factor indicated that maximum load may be achieved this load cannot be sustained for multiple times at $1000N$ or even $100N$ for that matter. As well, for the minimums, PLA reports that it will support repetitive loading cycles only for a $10N$ load (*see figure 6*).

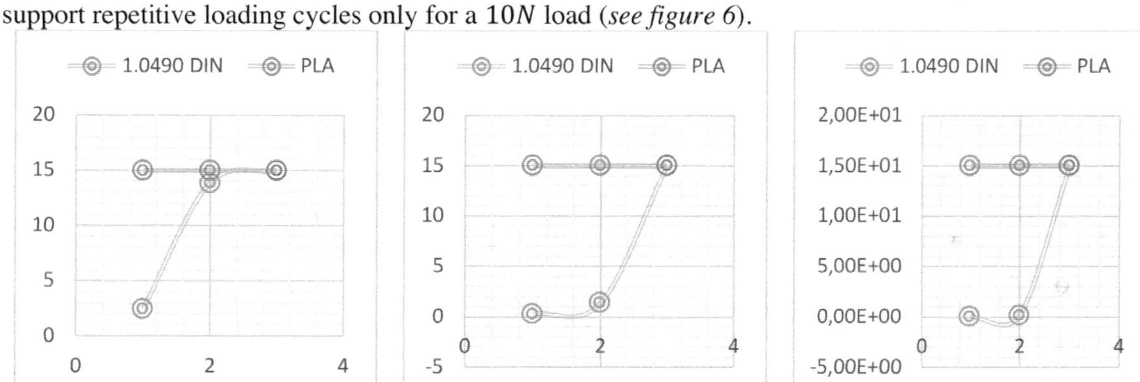

(a) (b) (c)

Figure 7. Margin of safety comparative charts for both type of materials at maximum values.

In case of the margin of safety maximums the horizontal position proves to be difficult for the 1.0490 DIN based assembly as it reaches only 1.4679 at $10N$ load meaning that it will support around 1.5 loading cycles at maximum load. As expected, PLA does not make it by quickly reaching the upper limit. Still, the average reports that the PLA based assembly almost withstands for at least 7 loading cycles. For loads above $10N$ PLA shows that failure may appear before even one loading cycle at maximum load will get to be finished (*see figure 7*).

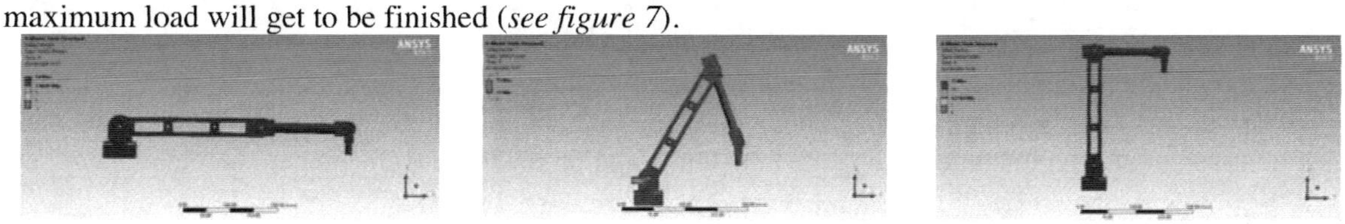

(a) (b) (c)

Figure 8. Print screens of PLA based model al $10N$ load.

Our PLA based model reported failure for one position. In horizontal position PLA based assembly reports 8.8699 for the safety factor and 7.8699 for the margin of safety. In vertical position PLA based assembly scores 4.2784 for the safety factor and 3.2784 for the margin of safety. Intermediate position reports that the assembly will not support a $10N$ load and paints it in red (*see figure 8*). By repetitive tests we found out that just about half of $10N$ load will report a "*safe*" design considering the two parameters.

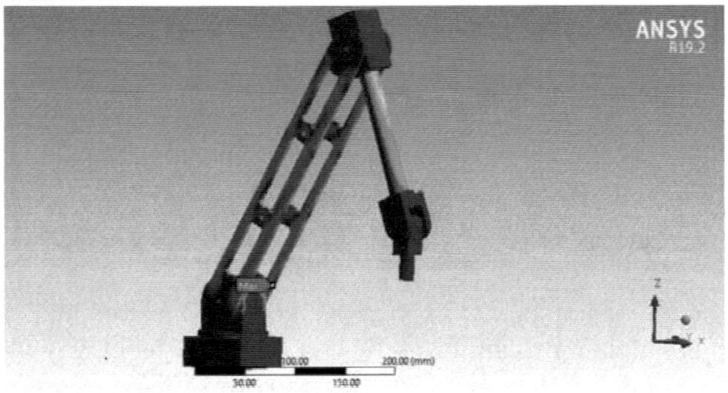

Figure 9. Print screen of PLA based model al $5N$ load.

The intermediate position proves to be the most demanding one. Because joints are seen as part of the assembly we can spot weak sections of the model and where it is more likely that it would fail. Pulling down with $0.5098581065kgf$ the beam sections tend to deform all length. That means that this particular design may be greatly improved if we choose for example a single filled middle section instead of two beams connected by bolts between them (*see figure 9*).

5. Conclusions

Does PLA prove a viable alternative against its steel peer in case of small projects or prototypes? In the case of our assembly the answer is *Yes* if small enough loads are used.

The proposed design was imposed by traditional manufacturing techniques where lighter models with the same pulling force are required. But keeping in mind that for smaller projects we need a structurally stronger model that would stand against its steel peers perhaps a fully rigid assembly would be more appropriate. Cylindrical components may prove highly adequate for this type of assembly much like it is the case for the upper arm of the current one.

Nevertheless, our design proved valid both in horizontal and vertical positions with lower loading cycles capacity. The intermediate position poses a threat to our design at maximum load thus making it interesting to conduct a transient structural type of analysis under *Rigid Dynamics* for the same model.

In this paper joints were considered as part of the assembly in order to find out which components of the assembly suffer the most at maximum load. But if joints are to be considered separately on their own, then that would change completely the results. Trying to evaluate the design we chose to overrule them and asses our assembly from the safety factor and margin of safety points of view only.

Future work will also try to take these analyses one step further by imposing a vibratory movement to the selected model thus being able to "*see*" some of its modal positions and further asses the opportunity of 3D printing for prototypes or small projects.

References

[1] https://www.totalmateria.com/subgroup.aspx?LN=EN&id1=379252&db=S, accesed at 30.03.2020
[2] https://www.totalmateria.com/TM_material_details.aspx?id1=91804&LN=EN, accesed at 30.03.2020
[3] Hrituc A *et al*. 2020 *Plastic Materials (in Romanian)* **57**(1) pp 13-20
[4] Hrituc A *et al*. 2020 *Macromolecular Symposia* **389**(1) 1900064
[5] Bolzon G *et al*. 2018 *Engineering Failure Analysis* **90** pp 434-439
[6] Matvienko Y G *et al*. 2020 *Engineering Failure Analysis* **117** 104814
[7] Chen M *et al*. 2019 *International Journal of Pressure Vessels and Piping* **172** pp 283-294
[8] Huo L, Baron L 2011 *Robotics and Computer-Integrated Manufacturing* **27** pp 367-376
[9] Zargarbashi S H H *et al*. 2012 *Mechanism and Machine Theory* **51** pp 74-86

Elaboration of the initial requirements in the design activities

L Slătineanu[1], B Oroian[1], I Condrea[1], C Botezatu[1], A Hrițuc[1], M A Boca[1], M Coteață[1] and O Dodun[1]

[1] Department of Machine Manufacturing Technology, "Gheorghe Asachi" Technical University of Iaşi, Blvd. D. Mangeron, 700050 Iaşi, Romania

Corresponding author's e-mail: hrituc.adelina3295@yahoo.com

Abstract. In solving the design problems, the initial requirements' clear statement significantly affects the design process's final result. Right from when the design problems are identified, the main requirements must be defined so that the final product meets the designed equipment requirements. Designers noted that design requirements do not have equal significance, and different solutions have been proposed to consider their weight in terms of end-use properties. The weighting can be achieved by simpler and operative methods and by more detailed methods of analysis. The paper approaches the problem of using the double input matrix method for weighting the functional requirements when discussing the problem of designing equipment to investigate the behavior of a computer component when the environment temperature is different compared to the normal temperature. The functional requirements assessment could provide an overview of the financial resources that could be invested in components to adequately meet the functional requirements. A first version of the tracked equipment was designed so that the previously established functional requirements are met.

1. Introduction

In machine building, *the design process* refers to the way of solving a technical problem. The researchers have distinct opinions concerning the stages of the design process.

Thus, when evaluating the design activity as an innovation process, 7 steps could be highlighted (identify the client need, researching the problem, brainstorming possible solutions, selecting a promising solution using the engineering analysis, achieving and testing a prototype, improving and redesigning/manufacturing the product) [1].

Other opinions could consider the four groups of activities within the so-called rational model: the pre-production design, the design during production, the post-production design, and the redesign [2].

The majority of opinions concerning the design process's definition considers an early step when based on the client's needs, the design process's goals are stated.

In such a stage, the client needs must be examined to define the functional requirements subsequently. Usually, the client's needs are expressed. The client desires a product or a process that allows him to achieve another product or process. In this stage, the designer must be clarified about what the client needs, and he must state the functional requirements so that the final product or process satisfies the client's desires.

On the other hand, when the final product or process includes known components and functions, there is *a routine design*, and when we have a new or at least improved product or process, there is *a creative design*. Over time, the evolution of human society showed the importance of creative design to improve the quality of life.

In such conditions, it is necessary to highlight the importance of establishing adequate functional requirements and even their selection that all the client needs be met - however, the product must be achieved quickly and at the lowest cost. One of the conditions to be met by the future product could refer to the product quality; we will understand that the future product must have such characteristics to satisfy the client's needs.

Within the so-called axiomatic approach of the design activity or *the axiomatic design*, the functional requirement must be defined according to the client's needs. It is necessary to highlight that there are functional requirements and, on the other hand, constraints or restrictions.

Professor Nam Suh considers that the functional requirements are a minim set of independent requirements that fully describe the functional needs of the product in the functional field. On the other hand, professor Nam Suh defines the constraints as bounds for acceptable solutions. There *are input constraints*, valid for the design specifications and *system constraints*, that refers to the conditions imposed by the system in which the future product must be integrated [3].

The requirements engineering concept was introduced to define, document, and maintain requirements during the engineering design activity.

Nuseibeh and Easterbrook proposed a roadmap to approach the requirements engineering [4]. They considered that the adequate establishment of the requirements would continue to play a significant role in generating a project's success or failure. They also formulated opinions concerning the future evolution of the research in the field of requirements engineering.

Martin et al. compared the requirements engineering process applied in two companies, considered the literature's models, and appreciated that no model corresponds to every project context [5]. They concluded that a requirements engineering process could combine linear and iterative structures, but the model could be appreciated as generally linear until the prototyping phase.

Socaciu and Blebea considered that a specifications document could clearly define the future product's requirements, and this document will be one of the documents for product development [6]. They showed that during software development, the requirements could be affected by changes due to the encounter of unforeseen situations. A general image concerning the so-called *field of requirements engineering* is addressed in the paper developed by these authors, including the various activities specific to this field.

Marcelino-Jesus et al. proposed a requirements engineering methodology to evaluate companies' technological innovation [7]. They appreciated that using such a methodology, an increase of acceptance of the companies' entrepreneurship initiative, and collaboration among enterprises are possible.

After establishing the functional requirements, a general defining of the product components and finding adequate solutions for them develops. Such an activity can be preceded or combined with an information activity to outline the best solutions to the addressed problem.

Due to the shorter time available, the information activity could be reduced or removed in routine design. However, if it is a creative design, the designer must resort to a selection process when determining each component of the final assembly or even the final assembly.

A problem of optimizing could be thus approached. Essentially, *the optimization* supposes the selection of certain versions among many available versions and using one or many selection criteria.

The problem addressed in this paper concerns the weighting of the functional requirements, starting from the fact that not all the functional requirements are of the same importance.

2. Use of methods of weighting the functional requirements

As mentioned above, not all the functional requirements are of the same importance. For this reason, the attention directed by the designer to the statement and the ways of achieving the functional requirements could not be of equal importance.

The need for weighting the functional requirements could be thus approached. The methods applied to weight the functional requirements could be the same or similar to those used in the optimization processes. It is known that the optimization process is connected with the field of decisional analysis.

Taking into consideration the circumstances in which a decision is established, it is possible to classify the methods of decision making:

- Methods applied in certitude conditions;
- Methods used in the risk conditions;
- Method applied in incertitude conditions.

Some methods from the first group of methods (applied in certitude conditions) could inclusively weigh the functional requirements.

Such methods could be the double input matrix method, the pairwise comparison matrix method, the analytic hierarchy process method, etc. [8].

3. Use of the double input method to weight the initial design requirements

One of the innovations resulted from the doctoral research was finding a solution to test a computer component's behavior when the working temperature is higher or lower than the normal temperature (for example, 20 °C).

As a component of the computer to be tested, the hard disk was selected. The choice was determined by the existence of specific software able to be used to evaluate the reading or the writing speed of the hard disk.

The initial idea was to use or to adapt a refrigerating box. Other functional requirements were formulated considering such an idea. A list of the main identified functional requirements found in the sequence of the conceptual design was the following:

1. Maintain a constant temperature inside the box;
2. Ensure a low temperature in the box;
3. Provide a working temperature higher than normal temperature (20 °C) in the box;
4. Provide easy insertion, extraction, and positioning of the computer component inside the box and quick connection of the computer component to the computer found out of the box;
5. Provide the possibility of relatively simple connection/disconnection of a heating device (with electrical resistance) found inside the box to a power supply source;
6. Ensure the operation of the cooling-heating subsystem until a preset temperature is reached;
7. Provide additional cooling to the cooling due to the cooling subsystem of the box, so when the latter does not provide sufficient cooling;
8. Ensure the highlighting of the values of temperature and humidity inside the box;
9. Ensure the box lid is sealed tightly.

If the design is deeper analyzed, other functional requirements could be identified.

In the next stage, the problem of weighting the identified functional requirements was formulated. If the double input matrix method is applied, a double input matrix must be considered, as the method name shows. This matrix's image could be seen in Table 1.

Such a table could also be used when establishing the value of the different equipment components to adequately allocate financial resources intended to be spent to obtain each equipment component.

As a result of information activity developed when identifying the functional requirements and including various results in the table, it was found that there are boxes able to refrigerate and heat the object found inside them.

In this way, some of the functional requirements are fulfilled by the same practical solution. In the axiomatic design, when some functional requirements are not independent, the design methodology (axiomatic design) requests actions to improve such a situation.

Also, during the short information activity, it was found that sometimes, the available refrigerating boxes do not ensure low enough temperature.

An additional solution to help decrease the box's temperature must be found to solve this new problem.

Table 1. Use of the double input matrix method to evaluate the functional requirements of the desired equipment.

Functional requirements	1. Constant temperature	2. Low temperature	3.Higher temperature	4. Easy component insertion	5.Simple connection/disconnection of the heating device	6. Reach a preset heating temperature	7. Additional cooling	8. Highlight the temperature and humidity e	9. Tightly closing	Proposed problem solving
1.Constant temperature	X	0.5	0.5	1	1	1	1	1	0.5	Insulating box
2. Low temperature	0.5	X	0.5	1	0	1	1	1	1	Refrigerating box
3. Higher temperature	0.5	0.5	X	1	0.5	1	1	0.5	0	Refrigerating/ heating box
4. Easy component insertion	0	0	0	X	0	0	1	1	0	Wireframe attached to the box cover
5.Simple connection/ disconnection of the heating device	0	1	0.5	1	X	1	1	1	0	Connection cables that cross the box lid
6. Reach a preset heating temperature	0	0	0	1	0	X	1	1	0	Temperature controller
7. Additional cooling	0	0	0	0	0	0	X	1	0	Passive cooling boxes
8. Highlight the temperature and humidity	0	0	0.5	0	0	0	0	X	0	Devices for measuring and highlighting the temperature and humidity, mounted in the lid of the cold box
9. Tightly closing	0.5	0	1	1	1	1	1	1	X	Sealing gaskets and device for pressing the lid on the refrigerator box
Sum S_i of marks	1.5	2.0	3.0	5.0	1.5	4.0	7.0	7.5	1.5	
Relative sum V_{ri} of marks	0.041	0.055	0.083	0.138	0.041	0.111	0.194	0.208	0.041	S=912
Order of the functional requirements taking into consideration the sum of their relative marks	8 (highlight the temperature and humidity) - 7 (Additional cooling) - 4 (Easy component insertion - 6 (reach a preset heating temperature - 3 (Higher temperature) - 2 (low temperature - 1 (constant temperature) - 5 (simple connection/disconnection of the heating device) - 9 (tightly closing)									

After including the double input matrix's functional requirements, each functional requirement's comparisons with the other functional requirements must be made. A simpler result of the comparisons activity could facilitate the inclusion in the table of scores (marks) of 1-0 when the first requirement is considered more important, 0-1 when the second requirement is evaluated as more important, and 0.5-0.5 when two requirements are considered of equal importance. There is still the possibility to consider a more detailed evaluation of the functional requirements, for example, giving marks between 1 and 10 to each functional requirement. However, in Table 1, the simpler evaluation method was preferred.

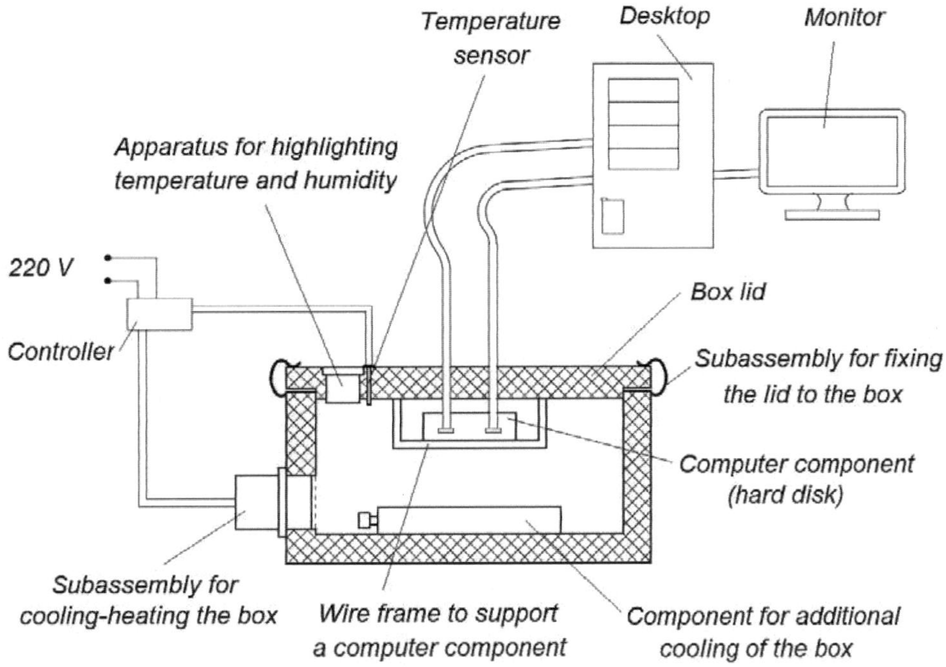

Figure 1. Schematic representation for equipment of investigation the behavior of the computer subsystems when the temperature is lower or higher than the normal working temperature

Both in the first column and the first line of the matrix, the main functional requirements identified were included; since sometimes the full name of the requirement is long enough, abbreviated forms of the proper requirements (mentioned above) were used.

The mark will be so included in the table that the sum corresponding to each functional requirement could be calculated and inscribed in the table's antepenultimate line along each vertical column. In the ultimate line, the relative value V_{ri} of each functional requirement could be introduced. This relative value could be determined by devising the sum S_i of the marks given to each functional requirement to the total number of the comparisons N_c:

$$V_{ri} = \frac{S_i}{N_c}, \tag{1}$$

where

$$N_c = \frac{n(n-1)}{2}, \tag{2}$$

where n is the number of functional requirements.

In the analyzed case, N_c=9·8/2=36.

A new order based on the evaluation of their significance could be identified. This order is highlighted in the ultimate line of Table 1, considering the sum of the relative sum V_{ri} of each functional requirement. Afterward, adequate solutions must be found to fulfill each functional requirement, and the first identified versions were included in the last column of Table 1.

It can notice that the most important functional requirement seems to be highlighting the temperature and humidity. The lowest weight was given to the necessity of tightly closing the refrigerating/heating box. A first version thought considering the proposed solution to fulfill the functional requirements can be seen in Figure 1.

4. Conclusions

The establishment of the design engineering requirements can exert a decisive influence on the evolution of product development. Over the years, the researchers formulated different opinions about solving the problems specific to elaborating on the design requirements. It has been noticed that not all the functional requirements are of equal significance. A weighting method could be used to order the functional requirements considering the sum of their relative marks. An application of the double input matrix method in the case of equipment for investigation of the computer subsystems behavior at the changing of the working temperature was developed. Adequate initial solutions for each functional requirement were identified. It noticed that while establishing the functional requirements, improved versions of requirements and solutions for their solving were found. In the future, there is the intention to apply more detailed analysis methods to weigh the functional requirements valid in the case of research or industrial equipment.

5. References

[1] Hands-on Activity: Design Step 1: Identify the Need 2016 https://www. teachengineering.org/activities/view/cub_creative_activity1 Accessed 28.03.2020

[2] Hiwase S D 2016 https://www.researchgate.net/ publication/305724231_Product_Design_-_Innovative_Trends_on_Design_Philosophies_Approaches_Processes_Techniques_Methods_Standards. Accessed 26.09.2020

[3] Suh N P 2001 *Axiomatic Design: Advances and Applications* (New York: Oxford University Press)

[4] Nuseibeh B, Easterbrook S 2000 *ICSE00: 22nd International Conference on Software Engineering* (Limerick, Ireland) pp 35–46

[5] Martin S, Aurum A, Jeffery R, Paech B 2002 *Proceedings of 7th Australian workshop process models in practice. AWRE'2002* (Melbourne, Australia) pp 141-155

[6] Socaciu L G, Blebea I 2010 *Proceedings of the 10th National Multidisciplinary Conference with international participation "Professor Dorin Pavel - the founder of Romanian hydropower"* (in Romanian) vol 17 (Sebeş, Romania: Publishing House AGIR) pp 47-54

[7] Marcelino-Jesus E, Sarraipa J, Agostinho, C, and Jardim-Goncalves, R A 2014 *Proceedings of the 21st ISPE Inc. International Conference on Concurrent Engineering, CE 2014* (Beijing, China: IOS Press BV) pp 577-586

[8] Slătineanu L 2019 *Fundamentals of scientific research* (in Romanian) (Iaşi – Romania: PIM Publishing House)

On machining Ti6Al4V ELI with EDM by using copper and graphite electrodes: a comparison study

P. Karmiris-Obratański[1,2], K. Zagórski[1], E.L. Papazoglou[2*], N.E. Karkalos[2] and A.P. Markopoulos[2]

[1] School of Mechanical Engineering and Robotics,
AGH University of Science and Technology, Cracow, Poland
[2]School of Mechanical Engineering,
National Technical University of Athens, Athens, Greece

Corresponding author's e-mail: mlpapazoglou@mail.ntua.gr

AbstractElectrical Discharge Machining (EDM) is a non-conventional machining process, in which, the material removal occurs by repetitive sparks. By utilizing EDM, conductive materials can be machined in complex shapes and geometries, with high dimensional accuracy, regardless of their mechanical properties. Titanium alloys, due to their unique inherent properties, find extensive use in the modern industrial environment. Nevertheless, they are still classified as hard-to-cut materials, facing significant difficulties in their machining with conventional methods. The current paper presents a comparative study regarding the machining of Ti6Al4V ELI with EDM, by using copper and graphite electrodes. The control parameters were the pulse-on current and time, while the machining performance was estimated in terms of Material Removal Rate (MRR) and Tool Wear Ratio (TWR). Additionally, the surface roughness was evaluated according to the Ra. For the aforementioned indexes, Analysis Of Variance (ANOVA) was performed.

1. Introduction

Electrical Discharge Machining (EDM) is a non-conventional machining process, which finds extensive use in modern industry. Conceptually, in EDM, a pulsed voltage is applied between the working electrode and the workpiece, in presence of dielectric fluid, forming a plasma channel. Extremely high temperatures are developed locally, ranging from 6000 up to 12000K, resulting in the melting and/or ablation of the workpiece material [1,2]. The main advantage of EDM is the capability to machine any conductive material, regardless its mechanical properties, with high dimensional accuracy, in complex shapes and geometries. Hence, EDM is rendered as a feasible alternative of conventional machining processes, for application in the fields of aerospace, automotive, mold and die industries [3]. Although EDM is a multi-parameter process, the pulse-on current (I_p), the pulse-on time (T_{on}) and the Duty Factor (DF) are the parameters that mostly affect the machining results, thus, research commonly focuses on them. Moreover, aiming in optimizing the machining efficiency, by reducing the costs and maximizing the productivity, different types (materials) of electrodes are tested and utilized. The machining performances are evaluated in terms of Material Removal Rate (MRR) and Tool Wear Ratio (TWR), while the machined Surface Quality (SQ) is usually evaluated by employing the Surface Roughness (SR) indexes of Ra and Rt [4].

Titanium and its alloys own exceptional properties, thus they are being widely used in a wide range of high-end applications in automotive, aerospace and biomedical industries. The main advantages of titanium alloys are the superior strength-to-weight ratio, the excellent corrosion and wear resistance, the good fatigue

property, and their high biocompatibility [5]. Nevertheless, due to their inherent properties of low thermal conductivity and high chemical reactivity, they are classified as difficult-to-cut materials, suffering from significantly low machinability. High tool wear, combined with the low material removal rates, along with vibrations and poor surface quality, are common difficulties and problems in machining of titanium alloys [6]. Hence, the research is mainly focused on utilizing non-conventional processes, like EDM, as feasible alternatives in machining of titanium alloys. Especially, the Ti6Al4V ELI (Extra Low Interstitial) is a titanium alloy of particular interest for researchers due to its properties. This alloy, also listed as Grade 23, has higher purity than the Grade 5 and in comparison with the common Ti6Al4V, it is an alpha-beta alloy, having lower amount of interstitials elements, such as iron, hydrogen and oxygen, resulting an improved corrosion resistance, at the expense of its endurance. Although the lower presence of oxygen increases the plasticity of titanium Grade 2, due to the low thermal conductivity its machinability remains low.
Due to the aforementioned reasons, it becomes evident that the machining of titanium alloys with EDM is of extreme interest, and thus, extensive research is carried out aiming in optimization of the process. Many studies aim to correlate the machining parameters of I_p, T_{on}, and DF with the MRR, TWR and SR. Specifically, Laxman et al. [7] carried out experiments regarding the machining of titanium alloy with EDM, for pulse-on currents and pulse-on times up to 21A and 500μs, respectively, by using copper working electrode. Similarly, Santos et al. [8] conducted research concerning the machining of Ti6Al64V with EDM, utilizing graphite electrode, and as control parameters the pulse-on current and time, the duty factor, and the machining polarity. Additionally to a mere correlation between machining parameters and results, Mhatre et al. [9] used the Grey Relational Approach to optimize the process in terms of MRR, TWR and Ra for two different electrode materials, copper and aluminum. Likewise, Verma et al. [10] employed the Taguchi method to optimize the machining of Ti6Al4V with EDM. Moreover, in order to render EDM of titanium alloys more efficient and feasible, the concept of utilizing electrodes from different materials has been investigated as well. Hascalik et al. [11] conducted a series of experiments concerning the machining of Ti6Al4V with EDM using graphite, copper and aluminum electrodes. The importance of the electrode's material on the machining results and performances was deduced. Similarly, Sivakumar et al. [12] studied the machining of Ti6Al4V with EDM using brass, copper, silver, tungsten, copper tungsten, silver tungsten, tungsten carbide, graphite and copper graphite electrodes, concluding that the most suitable one was the copper impregnated graphite electrode. Finally, the significant effect of the electrode's material on the MRR and TWR was investigated in the study of Bhaumik et al. [13], where cooper, brass, and zinc electrodes were tested.
The current paper presents a comparison study regarding the machining of Ti6Al4V-ELI with EDM, by utilizing graphite and copper electrodes. The experiments were carried out for high pulse-on currents, varying from 25 up to 65A, and for pulse-on times from 25 up to 200μs, thus covering a wide range of per pulse energies, something that has been rarely investigated in the relevant literature. The machining performance was estimated in terms of MRR and TWR, while the SR by the Ra. For all the aforementioned indexes, Analysis of Variance (ANOVA) was performed as well, in order to determine the importance of each process parameter to the MRR, TWR and surface quality when different types of electrodes are used.

2. Experimental Procedure

The experiments were carried out on an Agie-Charmilles Roboform 350 Sp Die Sink EDM. Slices of 10mm thickness cut from a Ti6Al4V ELI rod were used as workpieces, while kerosene was utilized as dielectric fluid, which properly channeled into the working tank through a flushing nozzle. Two different electrode materials were tested, copper and graphite, with the electrodes having nominal dimensions of and 19x14mm respectively. A full-scale experiment was conducted, with control parameters being the pulse-on current and time, ranging from 25 to 65A and from 25 to 200μs respectively. The close-circuit voltage was set constant at 30V and the duty factor at 50%.The workpieces' and electrodes' material properties are presented in Table 1, while in Table 2, the machining parameters are listed in detail.

Table 1. Material Properties

Material	Copper	Graphite	Ti6Al4V ELI
Density (g/cm^3)	8.94	1.77	4.45
Melting Point (°C)	1083	3300	1604 - 1660
Electrical Resistivity (μΩcm^{-1})	0.0192-0.0230	1400	178
Thermal Conductivity (W/mK)	300-355	168	6.7

Table 2. Experimental parameters

Machining Conditions	Level 1	Level 2	Level 3	Level 4
Discharge current I_p (A)	25	33	49	65
Pulse on-Time T_{on} (μs)	25	50	100	200
Duty Factor	0.5			
Dielectric	Synthetic Hydrocarbon Fluid			
Dielectric Flushing	Side Flushing with pressure			
Dielectric Flushing Pressure (MPa)	0.7			

The MRR and the TWR were calculated according to eq. (1) and eq. (2), respectively.

$$MRR = \frac{W_{st} - W_{fin}}{t_{mach}} \cdot \frac{1}{\rho_{work}} \tag{1}$$

$$TWR = \frac{El_{st} - El_{fin}}{W_{st} - W_{fin}} \cdot \frac{\rho_{elec}}{\rho_{work}} \tag{2}$$

where, MRR units are mm^3/min, TWR units are mm^3/mm^3, El_{st}, El_{fin} represent the workpiece weight before and after the machining, respectively, in gr, t_{mach} the machining time is measured in min, ϱ_{work}, ρ_{elec} represent the workpiece and electrode material density in gr/mm^{3}respectively, and W_{st}, W_{fin} the working electrode weight before and after machining respectively in gr. Finally, SR in terms of Ra was measured using a TOPO 01P contact profilometer, by employing the ISO 25178-2 standard for data analysis. The TOPO 01P profilometer is equipped with an induction measuring head, having a diamond cone-shaped tip with radius of 2μm and apex angle of 90°. The equipped software system has a confocal sensor with 130μm range, and vertical resolution of 8nm. According to the employed norm, 2.5ls was used with 8mm cutting length. In Figure 1 the experimental setup is depicted.

Figure 1. Experimental setup

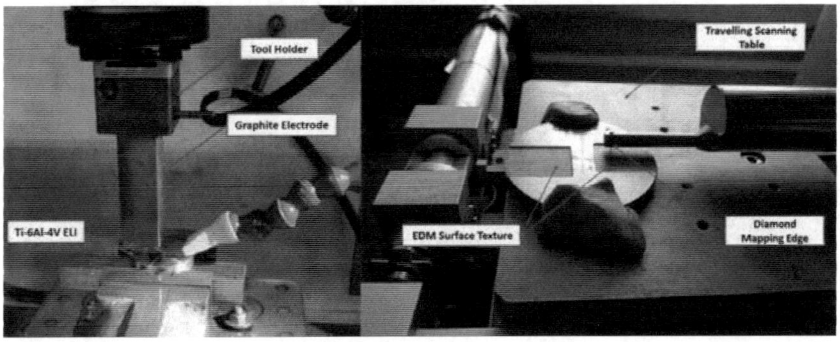

3. Results and discussion

In Table 3 the experimental results are presented.

Table 3. Experimental Results

I_p (A)	T_{on} (μs)	Graphite MRR (mm^3/min)	Graphite TWR	Graphite Ra (μm)	Copper MRR (mm^3/min)	Copper TWR	Copper Ra (μm)
25	25	1.077	2.022	13.92	5.483	0.064	8.307
33	25	0.696	3.160	10.204	6.729	0.062	9.243
49	25	1.588	3.134	11.544	9.888	0.137	9.400
65	25	3.633	1.617	14.312	9.644	0.511	12.523
25	50	0.979	2.343	16.025	2.502	0.089	14.517
33	50	1.316	3.187	10.268	6.738	0.041	11.207
49	50	1.938	3.376	12.249	10.635	0.058	13.116
65	50	3.056	2.142	11.996	16.829	0.129	12.873
25	100	1.443	1.626	17.55	1.855	0.066	12.445
33	100	2.651	1.575	17.342	2.346	0.060	16.456
49	100	4.068	1.771	14.036	11.273	0.022	18.303
65	100	4.591	1.265	14.289	21.235	0.031	17.494
25	200	1.229	1.654	15.242	1.484	0.054	14.741
33	200	2.372	1.290	15.791	2.351	0.046	16.632
49	200	4.779	1.255	15.619	4.865	0.037	21.510
65	200	6.483	1.267	16.206	9.651	0.027	22.197

From the experimental results of Table 3, and the respective plots in Figure 2, it is deduced that the MRR by utilizing copper electrode is significantly higher than by using graphite. Namely, by average, it is almost 300% higher, although for some sets of machining parameters the values are comparable. In specific, the greatest differences are observed for the lower pulse-on times i.e. 25 and 50μs, while for 100 and 200μs T_{on}, the differentiation still exists, but more limited. The interpretation of this behavior can be given by focusing on the primary mechanisms that take place during machining. The plasma channel melts and ablates material from both the workpiece and the electrode, while the dielectric fluid is decomposed in hydrogen and carbon. The free carbon is deposited on the electrode and workpiece surfaces, forming a protective layer, and hence, limiting the material removal. When a graphite electrode is used, there is a higher concentration of free carbon as the electrode wears out, thus, it can be reasonably concluded that the aforementioned protective mechanism becomes more intense, reducing the MRR. For higher pulse-on times, where the energy per pulse is increased, the effect of this mechanism seems to decline, thus, a more limited variation in MRR values is observed. Finally, by the ANOVA diagrams of Figure 2 the influence of I_p and T_{on} on MRR can be concluded. For both electrode materials increase in pulse-on current results in an increase in MRR; specifically, as the I_p increased from 25 to 65A the mean MRR increased by 276% and 406% for copper and graphite electrode respectively. On the other hand, the T_{on} has a different effect, depending on the electrode's material. For copper electrode, up to 100μs the T_{on} has a minor influence on the mean MRR, while a further increase up to 200μs results in a decrease in mean MRR. On the other hand, for the graphite electrode, the mean MRR is continuously increased as the pulse-on time increases.

Figure 2. MRR for Copper and Graphite electrode, and the ANOVA analysis

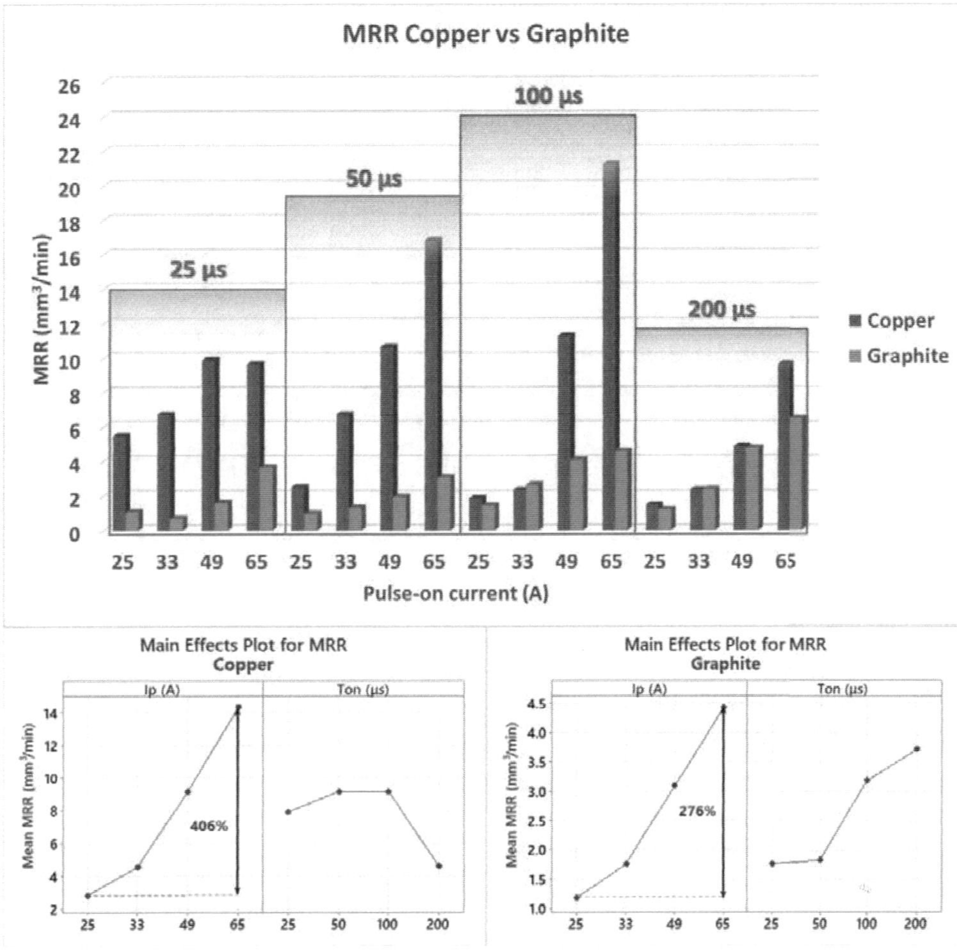

In Figure 3, the experimental results regarding the TWR are presented, along with the corresponding ANOVA. At this point, clarification is important to be done that the TWR, as indicated by eq. 2, is relevant to the ratio of the volumes that are removed from the electrode and the workpiece, and not to the masses. Thus, the comparison of TWR between different material electrodes having different densities is more representative. As the results of Table 3, and the plots of Figure 3 indicate, the TWR values when graphite electrode is utilized, are orders of magnitude higher than in the cases where the copper electrode is used. This occurs mainly due to the lower MRR when graphite electrode is employed, and to the higher wear rate of graphite. This is actually a loopback mechanism, where the high wear rate of graphite results in lower MRRs, and hence lower TWRs. Although graphite has almost three times higher sublimation point than copper, due to the intense machining conditions, i.e. high pulse-on currents and times, the sublimation temperature is reached, and subsequently, the wear of the graphite electrode becomes intense. Moreover, from the Main Effects Plots for TWR of Figure 3, it can be concluded that the pulse-on current has a fuzzy effect on TWR for both electrode materials, since increase in I_p does not monotonously affect the mean TWR. On the contrary, the pulse-on time seems to influence TWR more clearly; an increase in T_{on} above 50μs results in a decrease in mean TWR, a decrease that is coming from the greater increase of MRR comparatively with the increase of electrode's wear rate.

Figure 3. TWR for Copper and Graphite electrode, and the ANOVA analysis

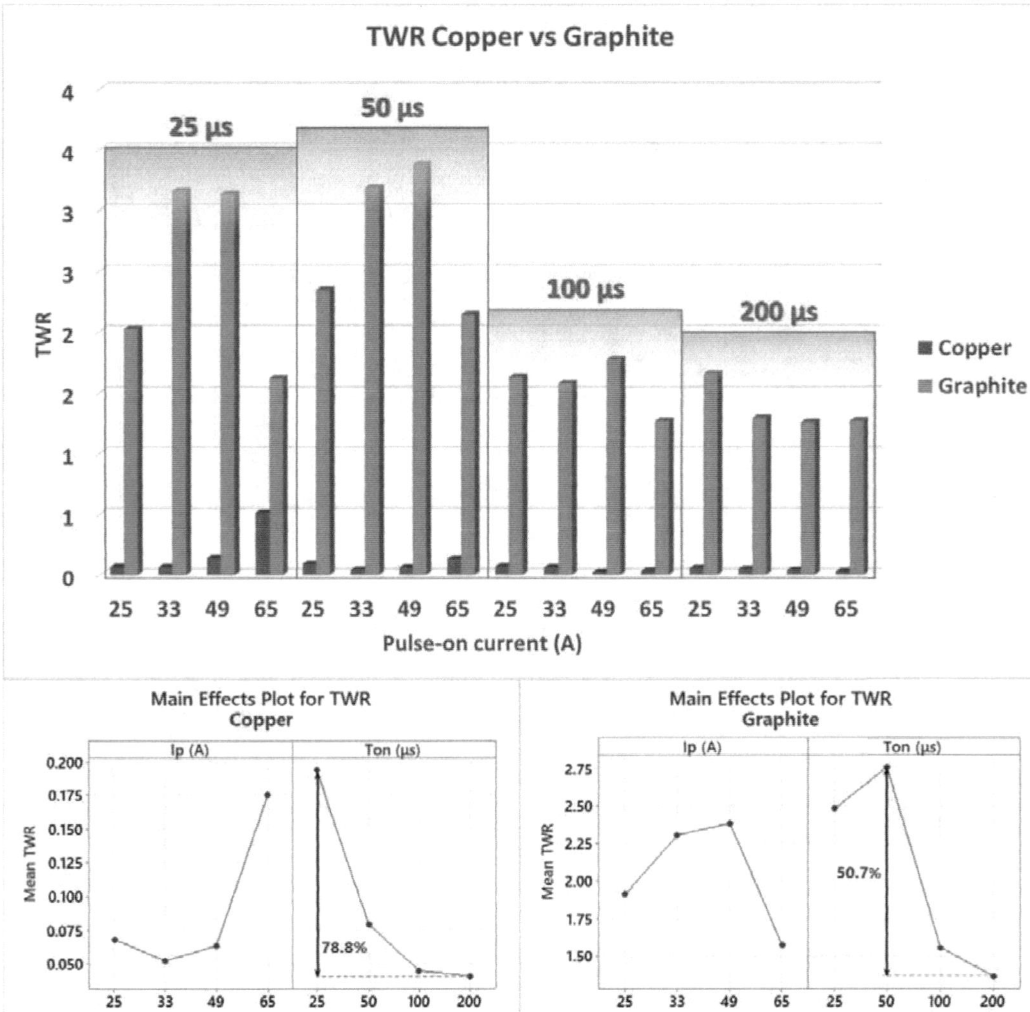

Finally, in Figure 4, the Ra values of the machined surfaces in each experimental case are plotted. Generally, no significant difference was found when different electrode materials were utilized, with Ra values being comparable for most of the sets of machining parameters. There were some exceptions, e.g. for 25A and 25μs the graphite electrode results in higher Ra than the copper one, while in contrast, for 49 and 65A at 200μs the copper electrode produces surfaces with higher Ra. By performing a Two-Sample T-Test, it is confirmed that Ra does not significantly differentiate based on the electrode material, since the T-Test's p-value is 0.82. From the Mean Effects Plot, it is deduced that increase in pulse-on time tends to lead in higher Ra. As the T_{on} was increased from 25 to 200μs, the mean Ra was increased by 90% and 25.7% for the copper and graphite electrode, respectively. On the other hand, the pulse-on current has a clear effect on Ra only when copper electrode is utilized, where the increase of I_p results in a respective increase in mean Ra. For the graphite electrode, a similar correlation cannot be established, since the increase in I_p has a vague effect on mean Ra.

Figure 4. Ra for Copper and Graphite electrode, and the ANOVA analysis

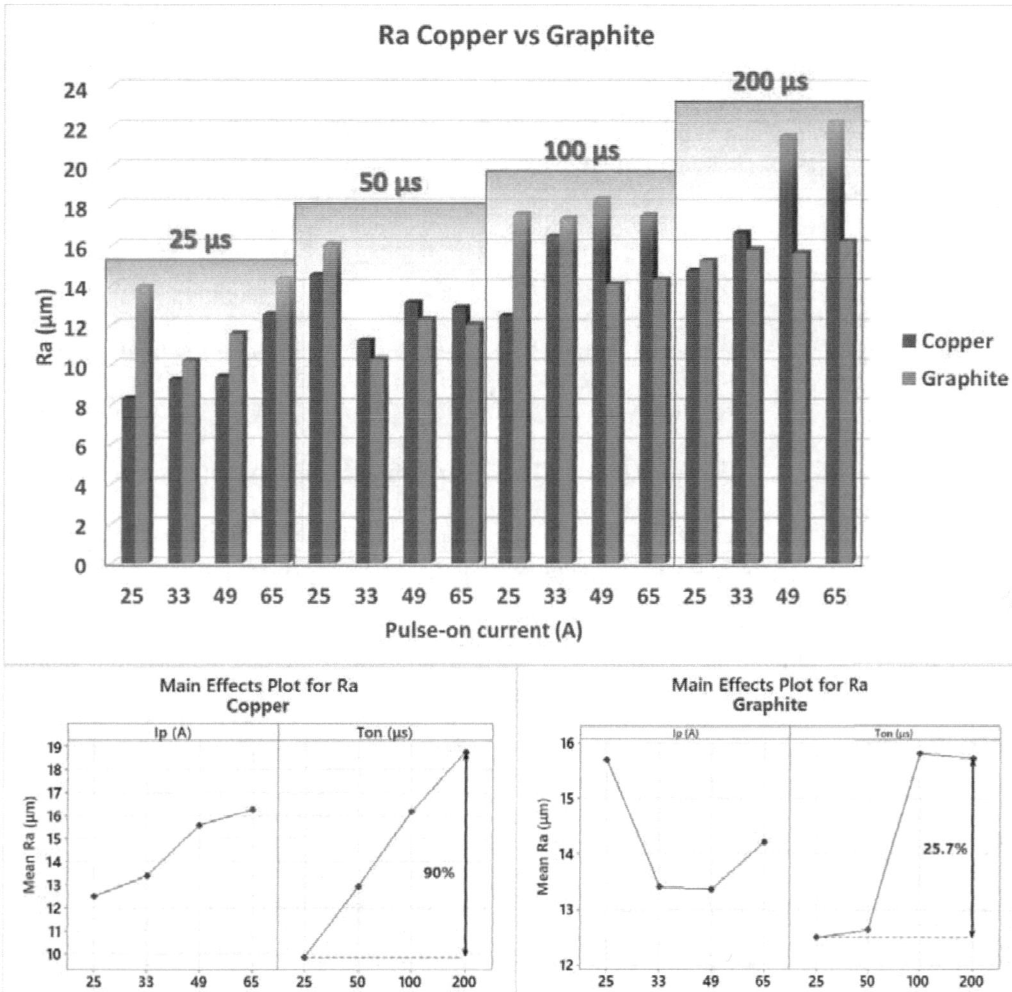

4. Conclusions

The current paper presents a comparison study concerning the machining of Ti6Al4V with EDM, by employing copper and graphite electrodes. Control parameters were the pulse-on current and time, while the machining performances were estimated in terms of MRR, TWR and Ra. For those performance indexes ANOVA was performed. The main conclusions, which were deduced, are:

- The MRR is significantly higher when copper electrodes are used rather than graphite ones for almost all the sets of parameters. Moreover, the pulse-on current strongly affects the MRR, with an increase in I_p resulting in an increase in mean MRR.
- The TWR for the graphite electrode is always orders of magnitude higher than that of copper electrode. For TWR, the pulse-on time has the major influence; namely, as the T_{on} increased above the 50μs the mean TWR was considerably decreased.
- There is no significant difference in Ra depending on the electrode material, while the mean Ra was increased for higher pulse-on times.

References

1. Jahan MP. Electrical Discharge Machining (EDM) Types, Technologies and applications. Jahan MP, editor. New York: Nova Science Publishers; 2015. 507 p.
2. Jameson EC. Electrical Discharge Machining. Michigan: Society of Manufacturing Engineers; 2001. 342 p.
3. Mohd Abbas N, Solomon DG, Fuad Bahari M. A review on current research trends in electrical discharge machining (EDM). Int J Mach Tools Manuf. 2007;47(7–8):1214–28.
4. Qudeiri JEA, Mourad AI, Ziout A, Abidi MH, Elkaseer A. Electric discharge machining of titanium and its alloys : review. 2018;
5. Manjaiah M, Narendranath S, Basavarajappa S. A review on machining of titanium based alloys using EDM and WEDM. Rev Adv Mater Sci. 2014;36(2):89–111.
6. Davim PJ. Machining of Titanium Alloys. Machining of Titanium Alloys. Springer International Publishing; 2014. 153 p.
7. Laxman J, Raj KG, Rao PV. EDM Process Parameters effects on Titanium Super Alloy. 2015;508–13.
8. Santos I, Polli ML, Daniel H. Influence of input parameters on the electrical discharge machining of titanium alloy (TI-6AL-4V). Int J Manuf Res [Internet]. 2015;10(3):286. Available from: http://www.inderscience.com/link.php?id=71626
9. Mhatre MS, Sapkal SU, Pawade RS. Electro Discharge Machining Characteristics of Ti-6Al-4V Alloy: A Grey Relational Optimization. Procedia Mater Sci [Internet]. 2014;5:2014–22. Available from: http://linkinghub.elsevier.com/retrieve/pii/S2211812814008992
10. Verma V, Sajeevan R. Multi Process Parameter Optimization of Diesinking EDM on Titanium Alloy (Ti6Al4V) Using Taguchi Approach. Mater Today Proc [Internet]. 2015;2(4–5):2581–7. Available from: http://dx.doi.org/10.1016/j.matpr.2015.07.212
11. Hasçalik A, Çaydaş U. Electrical discharge machining of titanium alloy (Ti-6Al-4V). Appl Surf Sci. 2007;253(22):9007–16.
12. Sivakumar KM, Gandhinathan R. Establishing Optimum Process Parameters for Machining Titanium Alloys (Ti6Al4V) In Spark Electric Discharge Machining. Int J Eng Adv Technol. 2013;2(4):201–4.
13. Bhaumik M, Maity K. Effect of different tool materials during EDM performance of titanium grade 6 alloy. Eng Sci Technol an Int J [Internet]. 2018;21(3):507–16. Available from: https://doi.org/10.1016/j.jestch.2018.04.018

The effect of different MQL supply strategies into the cutting zone on the tool wear when turning of Ti-6Al-4V alloy

Joanna Lisowicz[1], Witold Habrat[1], Krzysztof Krupa[1] and Piotr Janocha[2]

[1] Faculty of Mechanical Engineering and Aeronautics, Rzeszow University of Technology, 12 Al. Powstancow Warszawy street, 35-959 Rzeszów, Poland

[2]PROEKO OIL Sp. z o.o., Stanisława Sulimy 1 street, 82-300 Elbląg, Poland

Corresponding author's e-mail: j.lisowicz@prz.edu.pl

Abstract. Titanium alloys, especially Ti-6Al-4V alloy, are widely used in industry due to their good mechanical properties, such as high specific strength and excellent corrosion resistance. However, titanium alloys are also difficult-to-cut materials. Metalworking fluids are employed in the machining process to improve the machinability of titanium alloys. In recent years much attention has been paid to the use of the minimum quantity lubrication (MQL) techniques in the machining of the titanium alloys. In this study, the tool wear and its influence on the components of cutting forces have been investigated for different MQL supply strategies into the cutting zone - MQL via external nozzle and MQL via tool holder were investigated. The obtained results were compared with dry turning. It was concluded that the mechanism of tool wear is very similar for each cutting condition, however, it was also noted that the tool life is 28% and 68% longer for the MQL via external nozzle and MQL via tool holder, respectively, compared with the dry cutting conditions.

1. Introduction

Titanium alloys are often used in the aerospace industry, the chemical industry, and medical engineering primarily due to their two properties: high specific strength and excellent corrosion resistance [1]. However, the same properties make them also difficult-to-cut materials. The main problems in the machining of the titanium alloys are intensive tool wear and high cutting temperatures, which can reach beyond 1000°C at the tool-chip interface, due to low thermal conductivity and high chemical reactivity [2,3].

In order to improve the machinability of titanium alloys metalworking fluids are utilized in the machining process. The most common method of transfer of the heat generated during the cutting process is flood cooling. However, the conventional metalworking fluids are non-biodegradable and toxic in nature [4,5]. On the other hand, dry machining shortens the tool life especially in high-speed cutting [6]. Therefore minimum quantity lubrication (MQL) techniques are used to reduce friction at the tool-chip and tool-workpiece contact zones and results in lower cutting temperatures [5]. In the MQL technique, a small quantity of lubricant is supply into the tool-workpiece interface by a compressed air flow [7].

Raza et al. [8] compared the effect of six different cooling and lubrication strategies on the tool wear during turning of Ti-6Al-4V. The uncoated carbide cutting tool was used. The following cutting conditions were evaluated: dry machining, flood cooling, cryogenic machining (with liquid nitrogen), cooled air lubrication, vegetable oil MQL machining, and vegetable oil + cooled air / MQCL machining. It was found that the MQL and MQCL technique with the use of vegetable oil could be a sustainable alternative to synthetic cooling in terms of tool wear and surface roughness.

Rahim and Sasahara [7] investigated the tool wear and surface integrity during drilling of Ti-6Al-4V alloy in dry and MQL cutting conditions. For the MQL tests, the synthetic ester and palm oil were used. It was found that the tool life increased by 306% in comparison to dry cutting. It was also found that the implementation of the MQL technique improved the thrust force and torque and thus reduced power consumption.

Rahim and Sasahara [9] investigated the effect of palm oil as MQL lubricant in high-speed drilling of Ti-6Al-4V. The effects were compared to the air flow, MQL with synthetic ester and flood conditions. The MQL with palm oil and synthetic ester exhibited comparable performance with the flood conditions and much better comparing with dry cutting in the term of tool life. The cutting forces for MQL with palm oil were lower than that produced in MQL with synthetic ester and almost equal to the flood conditions.

Deiab et al. [10] analyzed the tool wear, energy consumption, and surface roughness during turning of Ti-6Al-4V under six different cutting conditions: dry machining, flood cooling, cooled air lubrication, vegetable oil MQL machining, cryogenic machining, and vegetable oil + cooled air / MQCL machining. The cold-pressed rapeseed oil was used for the MQL and MQCL cutting conditions. It was found that the vegetable oil as a lubricant in MQL and MQCL technique is a sustainable alternative to synthetic cooling in terms of tool wear, cutting energy consumption, and surface quality.

Khan and Maity [11] investigated the influence of cutting conditions on the cutting forces, tool wear, friction coefficient, chip morphology, and surface roughness during turning of commercially pure titanium (CP-Ti) grade II. The cutting conditions selected for the test were dry cutting, flood cooling, and MQL with vegetable oil. Under MQL machining the tool wear was lower by 57% and 34% comparing with dry and flood conditions, respectively. Also, the cutting force was lower for the MQL cutting in comparison to dry and flood cutting conditions (46% and 16%, respectively).

The aim of the study was investigation of the effect of different MQL supply strategies on the tool wear when turning of Ti-6Al-4V. MQL via external nozzle and MQL via tool holder were analyzed. Moreover, changes in the cutting force components as a result of blade wear were determined. The obtained results were referenced for dry turning.

2. Materials and methods

The Ti-6Al-4V alloy was used as the workpiece material in this study. The mechanical properties are given in the Table 1.

Table 1. Mechanical properties of Ti-6Al-4V alloy

Density, kg/m3	4430
Young’s modulus, GPa	114
Poisson’s ratio	0.33
Ultimate strength, MPa	1170
Yield strength, MPa	1100
Melting point, °C	1660
Thermal conductivity, W/(m·K)	6.7

The machining tests were conducted on the NEF 600 lathe. The cutting insert and tool holder used for the tests were VBGT160404-M3 HX and SVJBL2525M16 JET, respectively. Values of cutting parameters were as follow: cutting speed v_c = 120 m/min, feed f = 0.1 mm/rev and depth of cut a_p = 0.25 mm. Table 2 summarizes the experimental setup and cutting parameters.

Table 2. Experimental setup and cutting parameters.

	Component	Description
Experimental setup	Workpiece material	Ti-6Al-4V
	Machine	NEF 600 lathe
	Cutting tool	VBGT160404-M3 HX
	Tool holder	SVJBL2525M16 JET
Cutting parameters	Cutting speed, m/min	120
	Feed rate, mm/rev	0.1
	Depth of cut, mm	0.25

The tests were conducted in two different MQL supply strategies into the cutting zone: MQL via external nozzle (Figure 1), and MQL via tool holder (Figure 2), where the mist was supply on the rake face of the cutting tool. The obtained results were compared with dry cutting. The cutting fluid used in the MQL was a mixture of 46 % of eco-friendly vegetable oil, 46 % of diester, 4 % of extreme pressure (EP) / anti-wear additives, and 4% of antistatic agent. The volume flow of the cutting fluid was set up at 30 ml/ h and the air pressure was set up at 0,7 MPa.

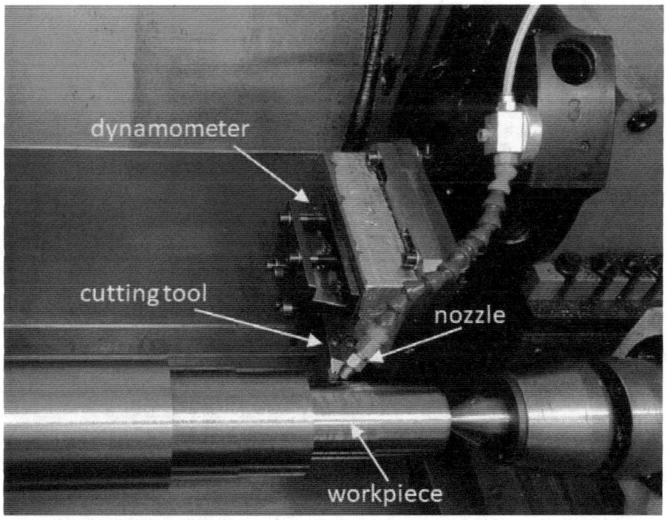

Figure 1. Configuration of the test stand for the cutting with MQL via external nozzle.

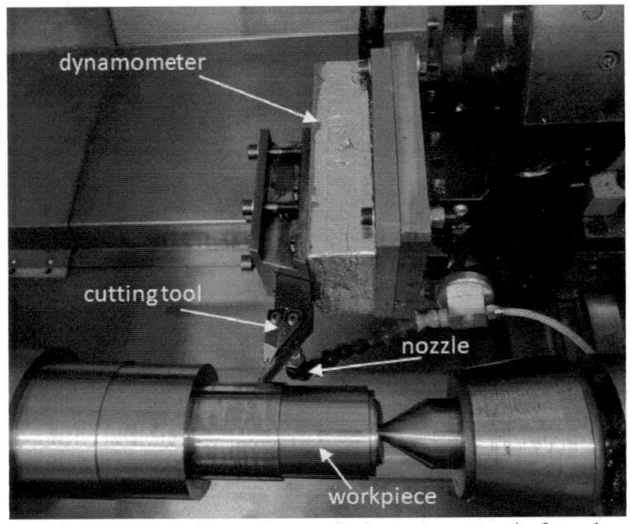

Figure 2. Configuration of the test stand for the cutting with the MQL via tool holder.

Kistler 9257B piezoelectric three-dimensional dynamometer was used to measure of the components of cutting forces. The dynamometer was mounted on the turret of the lathe machine by a VDI holder.

3. Results

3.1. Tool wear

The wear of the cutting tool was measured for specific volumes of material removed until the wear criterion was reached. The wear criterion was 0.2 mm of flank wear - *VB* indicator. In Table 3, the volumes of the removed material of the workpiece for reaching the tool wear criterion for different MQL supply strategies and dry cutting were presented.

Table 3. The effects of the different MQL supply strategies and dry cutting on tool wear.

	Volume of removed material, cm3
MQL via external nozzle	37.11
MQL via tool holder	48.63
Dry cutting	28.97

Figure 3 shows the tool wear in the function of the volume of removed material for MQL via external nozzle, and MQL via tool holder and dry cutting conditions.

The use of MQL cooling during finish turning of the Ti-6Al-4V alloy reduced the intensity of the cutting tool blade wear when compared to dry machining. The presence of the lubricant in the cutting zone causes a decrease of the coefficient of friction between tool-chip and tool-workpiece interfaces and therefore the tool life was extended. However the better results were achieved with the application of the MQL via tool holder. In this case, a more precise supply of the lubricant allowed more lubricant to go directly into the cutting zone.

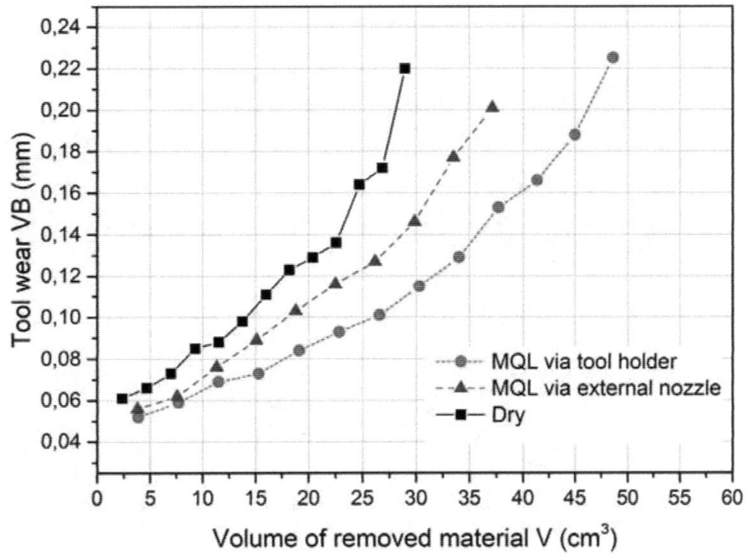

Figure 3. Tool wear VB in the function of volume of removed material

3.2. The influence of the tool wear on the cutting forces

Figures 4 and 6 show relationship of the cutting force in the function of the tool wear VB for MQL via external nozzle and MQL via tool holder cutting conditions, respectively. Figures 5 and 7 present the wear on the flank face respectively, for the same cooling strategies, for the points 1-4 marked at the graphs (Fig. 4, 6).

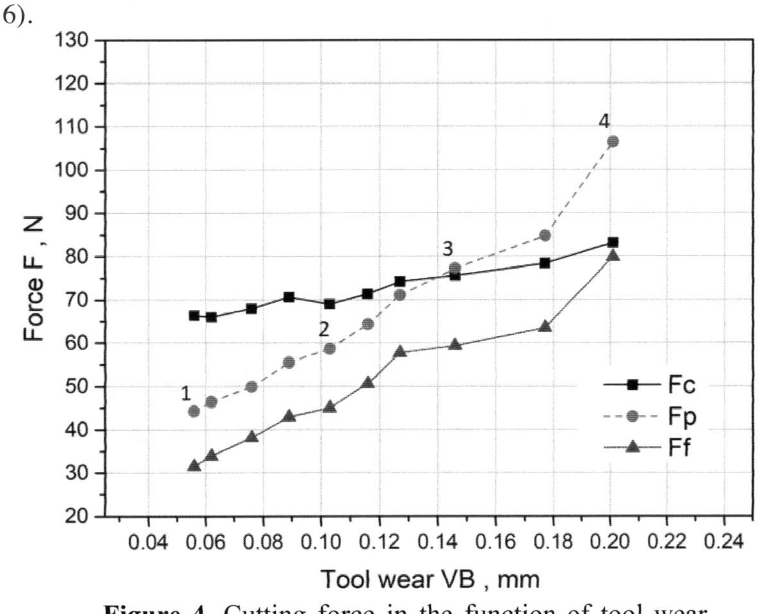

Figure 4. Cutting force in the function of tool wear VB for MQL via external nozzle.

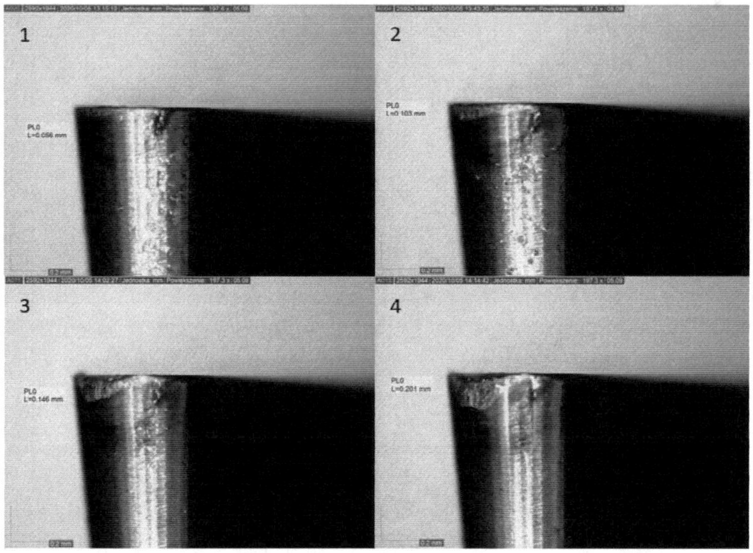

Figure 5. Photos of the cutting insert at different points of tool wear for MQL via external nozzle.

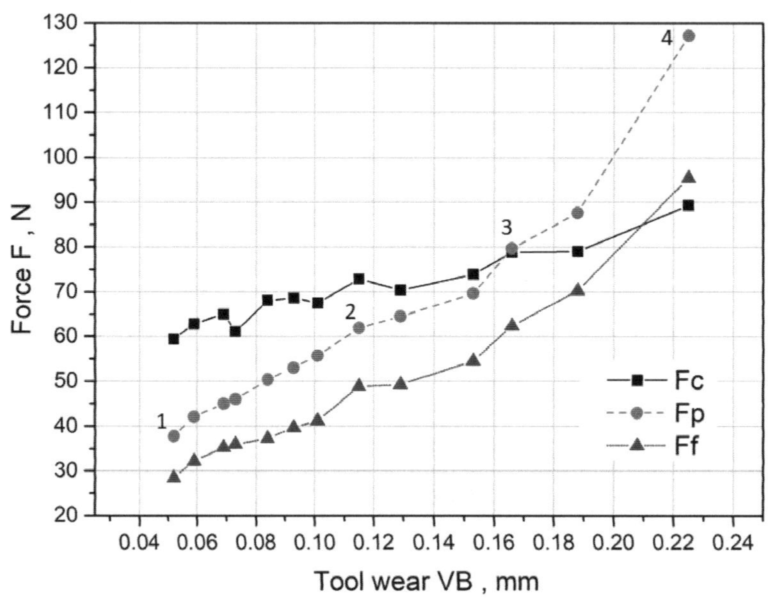

Figure 6. Cutting force in the function of tool wear VB for MQL via tool holder.

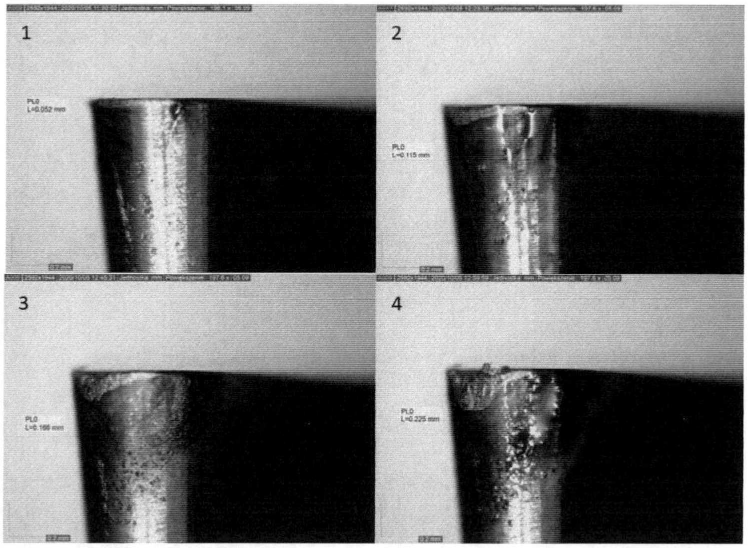

Figure 7. Photos of the cutting insert at different points of tool wear for MQL via tool holder.

In the case of dry machining, a similar relationship of the cutting force in the function of the tool wear VB was observed to that of MQL (Fig. 8). However, research has shown a greater tendency of the workpiece material to adhere to the rake face and flank face of the cutting insert (Fig. 9).

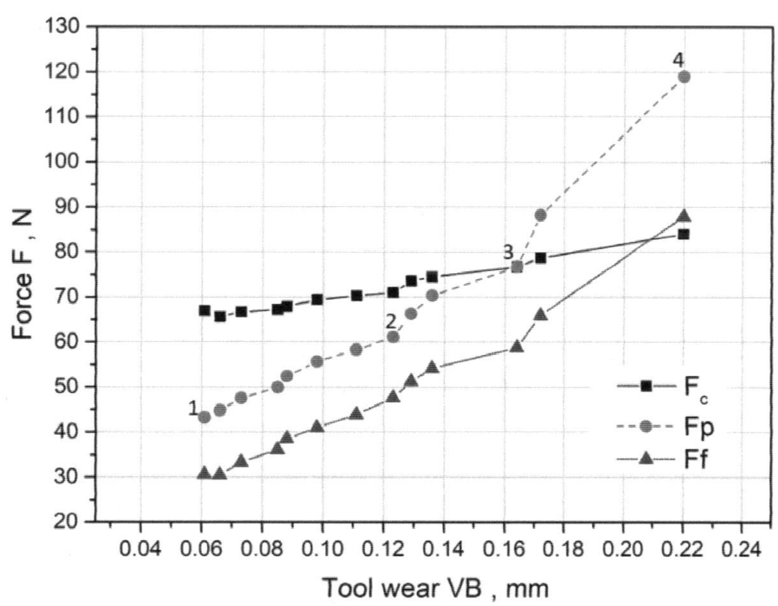

Figure 8. Cutting force in the function of tool wear VB for dry cutting conditions.

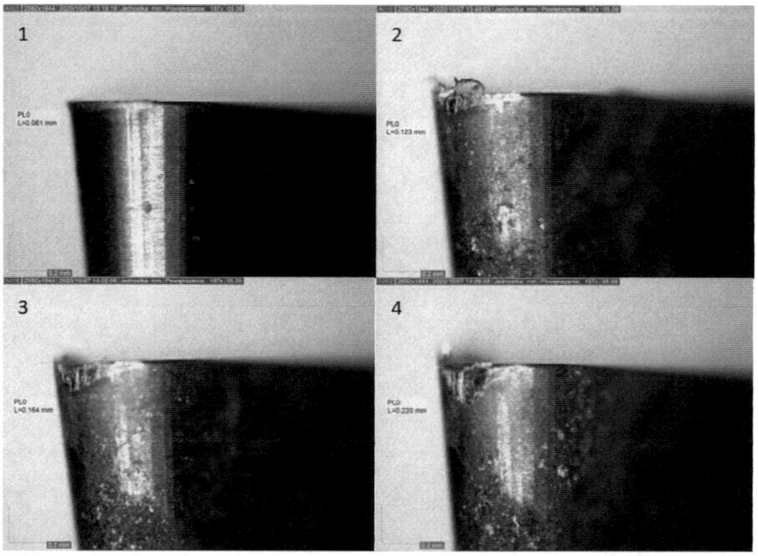

Figure 9. Photos of the cutting insert at different points of tool wear for dry cutting conditions.

It was observed that in the initial stage of machining, the main cutting force Fc is the largest component of the total cutting force. The progressive wear of the tool causes an intensive increase of the passive force Fp. With wear at the level of VB = 0.16 mm, the passive force becomes dominant, which is unfavorable due to the tendency of titanium alloys to strengthening For the cutting with the MQL via tool holder the cutting force components values are slightly smaller than for the dry machining and MQL via external nozzle. This is most likely due to the fact that directing the oil mist on the rake face under the chip formed reduce the contact surface of the chip to the cutting tool and therefore reduce the values of cutting force components. Moreover, for each tool wear value the cutting force components values were very similar for analyzed cutting condition.

4. Conclusions

The effect of different MQL supply strategies into the cutting zone on the tool wear and cutting forces has been studied when finish turning of Ti-6Al-4V. MQL via external nozzle and MQL via tool holder were analyzed and referenced with dry cutting. The following conclusions can be drawn:

- Volume of the removed material was 28% and 68% bigger for the MQL via external nozzle and MQL via tool holder, respectively, compared with the dry cutting conditions.
- At the basis of the measurments of the cutting forces and tool wear value it can be said that the machanism of tool wear is similar for all analyzed cutting conditions.
- Above the tool wear of about 0.16 mm the passive force becomes dominant components of the cutting force.

In conclusion, it can be said that the MQL via tool holder provides the best results from the tested strategies in the term of tool wear.

References

[1] Peters M, Hemptenmacher J, Kumpfert J and Leyens C 2003 Structure and Properties of Titanium and Titanium Alloys *Titanium and Titanium Alloys* ed M Peters and C Leyens (Weinheim: WILEY-VCH Verlag GmbH) pp 1–36

[2] Rahman M, Wong Y S and Zareena A R 2003 Machinability of titanium alloys *JSME Int. Journal, Ser. C Mech. Syst. Mach. Elem. Manuf.* **46** 107–15

[3] Namb M and Paulo D 2011 Influence of Coolant in Machinability of Titanium Alloy (Ti-6Al-4V) *J. Surf. Eng. Mater. Adv. Technol.* **01** 9–14
[4] Sokovic M and Mijanovic K 2001 Ecological aspects of the cutting fluids and its infuence on quantifiable parameters of the cutting processes *J. Mater. Process. Technol.* **109** 181–9
[5] Pervaiz S, Rashid A, Deiab I and Nicolescu C M 2016 An experimental investigation on effect of minimum quantity cooling lubrication (MQCL) in machining titanium alloy (Ti6Al4V) *Int. J. Adv. Manuf. Technol.* **87** 1371–86
[6] Park K H, Suhaimi M A, Yang G D, Lee D Y, Lee S W and Kwon P 2017 Milling of titanium alloy with cryogenic cooling and minimum quantity lubrication (MQL) *Int. J. Precis. Eng. Manuf.* **18** 5–14
[7] Rahim E A and Sasahara H 2011 Investigation of tool wear and surface integrity on MQL machining of Ti-6AL-4V using biodegradable oil *Proc. Inst. Mech. Eng. Part B J. Eng. Manuf.* **225** 1505–11
[8] Raza S W, Pervaiz S and Deiab I 2014 Tool wear patterns when turning of titanium alloy using sustainable lubrication strategies *Int. J. Precis. Eng. Manuf.* **15** 1979–85
[9] Abd Rahim E and Sasahara H 2017 Performance of palm oil as a biobased machining lubricant when drilling inconel 718 *MATEC Web Conf.* **101** 1–5
[10] Deiab I, Raza S W and Pervaiz S 2014 Analysis of lubrication strategies for sustainable machining during turning of titanium ti-6al-4v alloy *Procedia CIRP* **17** 766–71
[11] Khan A and Maity K 2018 Influence of cutting speed and cooling method on the machinability of commercially pure titanium (CP-Ti) grade II *J. Manuf. Process.* **31** 650–61

IOP Conf. Series: Materials Science and Engineering 1037 (2021) 012005 doi:10.1088/1757-899X/1037/1/012005

The finite element modelling of machining of NST 37-2 steel

A O Oluwajobi[1,*] and O E Egbebode[1]

[1] Department of Mechanical Engineering, Obafemi Awolowo University, Ile-Ife, Osun-State, 220005 Nigeria

*Corresponding author's e-mail: j.oluwajobi@gmail.com

Abstract. NST 37-2 has vast applications in building construction and in machine components production in Nigeria, but the effects of induced residual stress in the material, optimum cutting parameters, efficient process planning, chip formation process and surface roughness of the machined products are some of the challenges faced in the industry. In this study, the finite element simulation of orthogonal cutting of NST 37-2 steel was carried out, and the effect of geometry on chip morphology, residual stress distribution, strain energy was investigated to address informed predictability on the material. Different rake angles of -15°, 0°, 15°, and 30° were adopted for the tool geometry. In each case, the chip formed was continuous due to the plastic nature of the material. At negative and 0 rake angles, there were narrow bands of heavily strained chip, and as the rake angle increased, the chip tended to narrow due to low adhesion in the tool-work-piece region and decreasing work for chip folding. Uniform deformation was also observed with positive rake angle simulations. Residual stress was found to increase with negative rake angle. However, with positive rake angle, residual stress led to an increase in tensile strength due to the shearing action of the cutting tool in the cutting direction on the material layers below the cut plane. Cutting and thrust forces were observed to decrease with increasing rake angles. Adiabatic shear bands with high temperatures and thermoplastic shear instability were observed in machining with negative rake angles. High temperatures were generated in the shear zone due to large amount of work done by deformation of work material in machining with negative rake tools. The shear zone temperatures were much less in machining with positive rake tools. Temperatures on the rake face reduced with increasing rake angles, but not as significantly as the temperatures in the primary shear zone.

1. Introduction

NST 37-2 steel is one of the commercial carbon steel grades produced by Nigeria steel industry, which represents about 75 percent volume of production. This grade of steel is produced from diverse scrap metals and iron ore imported into the country from various uncoordinated sources like Malaysia, Brazil, China, etc. [1]. This steel is commonly used in building construction as structural members and machine components in production. NST 37-2 has vast applications as a substitute to foreign steels in Nigeria. There is increasing need for efficient machining and better manufacturing of this steel at low cost. The effects of induced residual stress in the work piece, optimum cutting parameters, efficient process planning, chip formation process and surface roughness of the machined products etc. are some of the challenges faced in the industry. The experimental approach to study machining process is expensive and time consuming, especially when a wide range of parameters are included, like tool geometry, materials, cutting conditions and so on. Because of these difficulties, alternative approaches have been developed, such as simulations where numerical methods are used. Among these numerical methods, the Finite Element Method (FEM) has proved to be useful and widely used [2]. Kovac et al. [3] presented a computer modelling of tool deformation and thermal load during turning operation using FEM. They found out that the deformation of cutting tools is greatly caused by cutting temperature and cutting forces. Ali et al [4] simulated the orthogonal milling of Ti6Ai4V alloy and predicted the cutting forces of high speed milling by varying the feed rate and analyzed the effect on surface roughness, feed cutting force and cutting forces. Residual stress and shear strain are induced into the work piece and these determine the performance and fatigue strength of the work piece. Zhang et al [5] studied the limiting shear stress at the tool chip interface and analyzed the effect of the coefficient of friction, cutting speed and tool rake angle. A similar approach of FEM was employed to study the effect of rake angle in the distribution of stress and strain energy of AISI 1045 steel [6]. It was discovered that the area of Von Mises Stress in the primary deformation zone decreases as the tool rake angle increases.

Fadare and Asafa [7], developed a predictive model to determine the optimum cutting parameters for turning NST 37.2 steel with uncoated carbide cutting inserts. They employed artificial neural network-based tool wear predictive model and a genetic algorithm-based optimization model. Multi-layer, feed-force, back - propagation network was used in the predictive model, while the maximum metal removal rate (MRR) was used as the objective function. Samples of NST 37.2 steel bars with 25 mm diameter and 400 mm lengths work piece and Sandvice Coromant uncoated carbide inserts with International Standard Organization (ISO) designation SNMA 12406 were used. Dry machining at different cutting conditions with cutting speed, feed rate and depth of cut ranging from 20.42-42.42 mm/min, 1.0-2.2 mm/rev and 0.2-0.8 mm, respectively were carried out. Eight passes of 50 mm length of cut were machined at each condition, the spindle power and tool wear (flank and nose) were measured during each cutting operation. They found out that the predictive model had acceptable accuracy and the optimum cutting parameters were obtained. Considerable work has been carried out on the NST 37-2 steel, but FEM investigation on the optimization of its machining parameters are nonexistent or scarce, hence this study.

1.1 Locally Rolled Steels in Nigeria

In Nigeria the steel industry comprises of Oshogbo rolling mills, Ajaokuta rolling mills, Jos and Kastina rolling mills. These rolling mills make use of hot rolling process principle to produce majorly, thirteen grades of steel in form of wire, ribble and round rods with Nigerian Standards (NST) tagged as a designation system. Table 1 shows the grades of locally rolled steel produced in Nigeria. Amongst these grades, NST 37-2 is the most commercially produced steel in Nigeria (about 75%). Due to its great mechanical properties; it is widely used in mechanical and in building constructions.

Table 1: The grades of steel produced in Nigeria

Billet Qualities	C	Si	Mn	P	S	Cu	Cr	Ni	Sn
NST 33	0.100	0.160	0.600	0.040	0.040	0.250	0.100	0.010	0.050
NST 34-LC	0.120	0.280	0.500	0.450	0.050	0.700	0.700	0.010	
NST 37-2	**0.200**	**0.370**	**0.650**	**0.040**	**0.050**	**0.700**	**0.700**	**0.010**	**0.050**
NST 37-3	0.220	0.250	0.650	0.040	0.080	0.060	0.700	0.010	0.050
NST 44-2	0.220	0.350	0.650	0.045	0.055	0.300	0.300	0.010	
NST 44-3	0.200	0.350	0.680	0.045	0.050	0.250	0.300	0.010	0.050

NST 50-2	0.370	0.370	0.800	0.045	0.045	0.300	0.300	0.010	
NST 52-3	0.380	0.370	0.750	0.045	0.050	0.300	0.300	0.010	
NST 60-2	0.490	0.370	1.800	0.045	0.045	0.300	0.300	0.010	
NST 60 MN	0.420	0.300	1.200	0.040	0.040	0.250	0.100	0.010	0.050
NST 65 MN	0.420	0.400	1.600	0.040	0.040	0.250	0.100	0.010	0.050
NST 70-2	0.490	0.400	1.200	0.045	0.045	0.300	0.100	0.010	
NST 70 MN	0.450	0.400	1.800	0.045	0.045	0.300	0.100	0.010	0.050

Source: (Osogbo Steel Rolling, 1982)

2. Methodology

For this study, the finite element modelling of machining of NST 37-2 was developed using the ABAQUS software. The following assumptions were made for the simulation, viz;

- The work piece material is a homogeneous polycrystalline, isotropic, and incompressible solid.
- There is constant friction at the tool- chip interface and at the tool-work piece interface.
- The machine tool is perfectly rigid and no influence of machine tool dynamics on machining is considered.
- Cutting is performed in air (dry machining) and no coolants are used. Adiabatic temperature boundary conditions were assumed [8].

2.1 Tool Modelling

The tool material was selected as uncoated carbide tool [9]. Thermal and mechanical properties of the tool with geometric variables of the tool are given in Table 2. Finite element mesh of the tool was modelled using 1120 nodes and 1057 elements. Iso-parametric quadrilateral elements with structured mesh smoothing algorithms were used for the analysis.

Table 2: Thermal and mechanical properties of uncoated carbide tool [9]

Properties	Amount
Density (Kg/m^3)	14900
Elastic Modulus, E (GPa)	650
Poisson's Ratio	0.25
Thermal Conductivity (W/m°C)	79
Specific heat (J/Kg°C)	138
Rake Angle, α (°)	0
Clearance Angle, c (°)	5
Tip Radius, rt (mm)	0.05

2.2 Work piece Modelling

The work material was considered as deformable. Accurate and reliable flow stress models are considered highly necessary to represent work piece material plastic behaviour under high-speed cutting conditions. The Von-Mises yield criteria combined with the rate dependent Johnson-Cook (J-C) plastic method were employed to the finite element models.

2.3 Steps and Boundary Conditions

The work piece is constrained against any movement at the bottom edges (in 2D) or surface (in 3D), while tool is advancing with a constant velocity by penetrating into work piece. A constant step-time is applied to solve the problem for all simulations. Surface to surface contact algorithm was employed in the model by using penalty mechanical constraint.

For the heat transfer calculations, the following assumptions were made:

i) The contact between the tool and the chip is thermally perfect. Hence a very large value of the interface heat transfer coefficient (h_{int}) was used and it was fixed to 1000 kW/m^2K. [10].

ii) The boundaries that are sufficiently away from the cutting zone remain at the room temperature (T_∞=20 °C)

iii) The chip and the tool loss heat due to heat convection (h_∞=20 W/m^2 °C) on the free surfaces on the work piece.

iv) Heat loss due to radiation is very small and it is neglected.

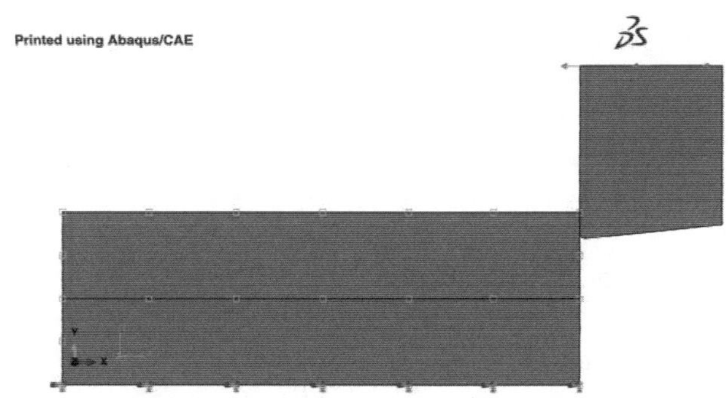

Figure 1: Displacement boundary conditions of the cutting system

2.4 Friction Model

A Coulomb's friction law was assumed to model the tool-chip and the tool-work piece contact zones. According to the Coulomb's law, the relative motion (slip) occurs at the contact point, when chip shear stress, τ is equal to or greater than the critical friction stress $\tau_t = \mu\sigma_n$. When τ is smaller than τ_t there is no relative motion and the contact point is in a state of stiff [11]. In this study, conventional Coulomb frictional law was used, by setting σ_n to infinity. Penalty method was used as a friction formulation.

2.5 Properties of Material and Simulation Approach

The physical properties of NST 37-2 steel are shown in Table 3. The material constants for Johnson-Cook material model and the damage model are presented in Table 4 and Table 5 respectively; which were obtained experimentally.

The finite element simulation of cutting process was carried out by using ABAQUS/Explicit. The 2D and 3D Arbitrary Lagrangian and Eulerian formulation were applied. Adiabatic temperature boundary conditions were assumed. The effect of any cooling materials was not investigated in this study. The machining operation was assumed to be performed in air without application of any liquid coolant.

The combination of cutting conditions used in the simulation are listed below (for the cutting length of 10 mm and depth of cut of 4.0 mm):

1. Cutting Speed range: 29.06 to 69.24 m/min at rake angle 15^0 and feed rate 1.0 mm/rev.
2. More simulations were run for 43.96 and 69.24 m/min cutting speed and with different feed rates (0.1, 0.2 and 0.8 mm/rev) to compare the trend of cutting force and thrust force with experimental results reported by [12].
3. More simulations were also run for 35.32, 59.34, 100.32, 169.56 m/min cutting speed and feed rate 1.65 mm/rev to compare the trend of rake face temperatures with experimental results reported by [13].

Table 3: Mechanical Properties of NST 37-2 steel

Properties	Values
Yield Strength (MN/m^2)	420.000
Elongation (%)	3.667
Reduction in Area (%)	15.050
Maximum Tensile Strength	683.449
Young Modulus (GPa)	28.565
Density (Kg/m^3)	8150

Table 4: Johnson-Cook material model constants for NST 37-2 steel

A (MPa)	B (MPa)	C	N	m	Room Temp	Melting Temp
340.00	683	0.062	0.11	0.42	20	1365

Table 5: Johnson-Cook Damage model constants for NST 37.2 steel

D1	D2	D3	D4	D5
0.026	0.853	-0.0388	0.3281	0.328

3. Results and Discussion

The effects of rake angle on chip morphology, residual stress distribution, rake face temperature, strain energy, surface finish and cutting force were investigated in the simulations. Different rake angles of -15°, 0°, 15°, and 30° were adopted for the tool geometry at cutting speed of 42.46 m/min and feed rate 0.10 mm/rev. In each case, the chip formed was continuous due to the plastic nature of NST 37-2 material. At negative and 0 rake angle, there were narrow bands of heavily strained chip, and as the rake angle increased, the chip tended to narrow due to low adhesion in the tool- work piece region and decreasing work for chip folding. The tip of the chip gets smoother as rake angle increases due to the transformation of the cutting action into pushing and squeezing force given a small rake angle which also leads to smoother surface finish of the work piece. Uniform deformation was also observed with positive rake angle simulations. Due to the inhomogeneous distribution of plastic deformation in machining process, the residual stress generated was computed, represented by Von Mises parameter on the machined surface in the simulations. Residual stress was found to increase with negative rake angle. However, with positive rake angle, residual stress led to an increase in tensile strength due to the shearing action of the cutting tool in the cutting direction on the material layers below the cut plane.

Tool forces are influenced by rake angles as tool-chip contact area decreases with increasing rake angles. Cutting and thrust forces were observed to decrease with increasing rake angles.

Adiabatic shear bands with high temperatures and thermoplastic shear instability were observed in machining with negative rake angles. High temperatures are generated in shear zone due to large amount of work done by deformation of work material in machining with negative rake tools. The shear zone temperatures are much less in machining with positive rake tools. Temperatures on the rake face reduce with increasing rake angles, but not as significantly as the temperatures in the primary shear zone.

Reduced area for frictional contact between tool and chip causes less heat generation at higher rake angles, which leads to lower temperatures on the rake face.

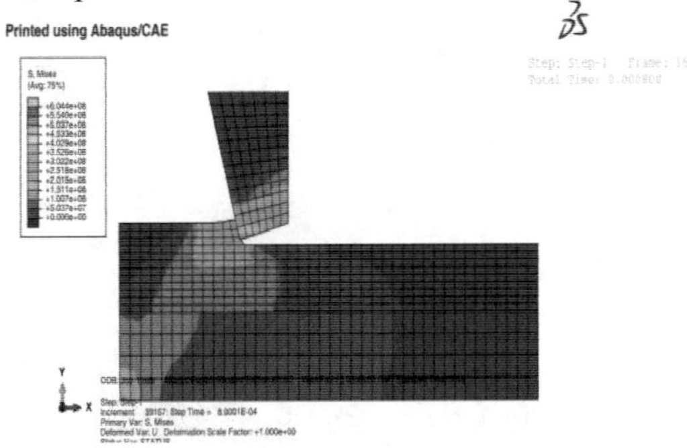

Figure 2: Effect of rake angle on chip formation and residual stress in simulation of machining NST 37.2 steel at cutting speed 42.46 m/min and feed rate of 1.0 mm/rev. at -15° rake angle. At rake angle -15°, residual stress is maximum (604 MPa)

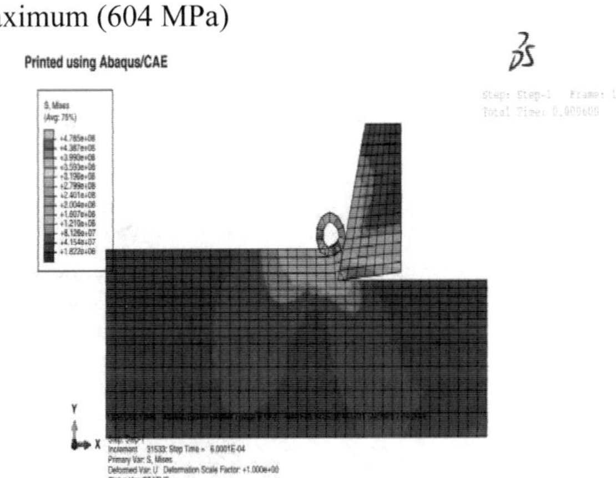

Figure 3: Effect of rake angle on chip formation and residual stress in simulation of machining NST 37.2 steel at cutting speed 42.46 m/min and feed rate of 1.0 mm/rev. At rake angle 15°, residual stress is 478 MPa

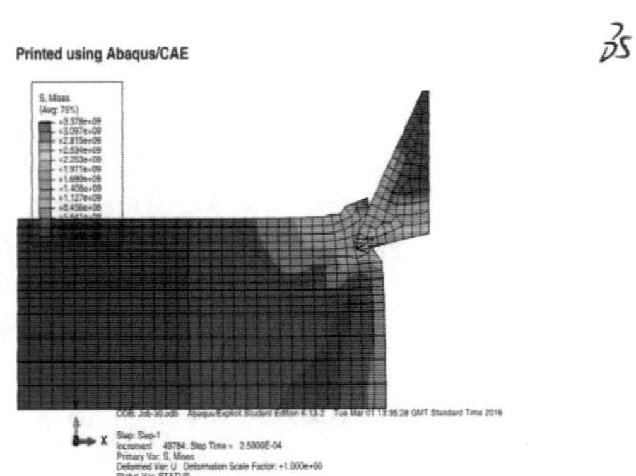

Figure 4: Effect of rake angle on chip formation and residual stress in simulation of machining NST 37.2 steel at cutting speed 42.46 m/min and feed rate of 1.0 mm/rev.
At rake angle 30°, residual stress is 338 MPa

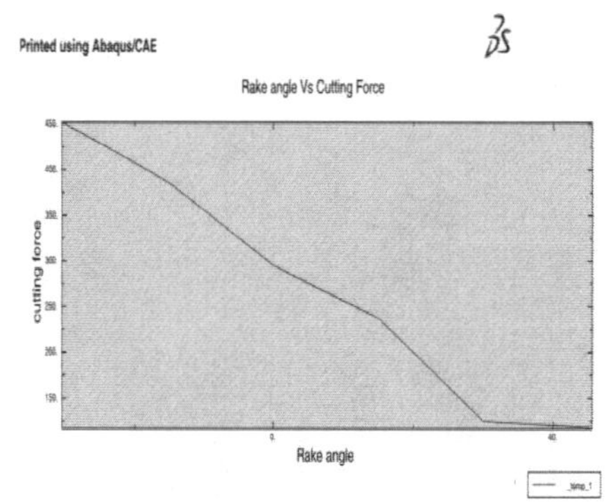

Figure 5: Effect of rake angles on cutting force of simulations of machining NST 37-2 steel at cutting speed 42.42 m/min and feed rate of 1.0 mm/rev.

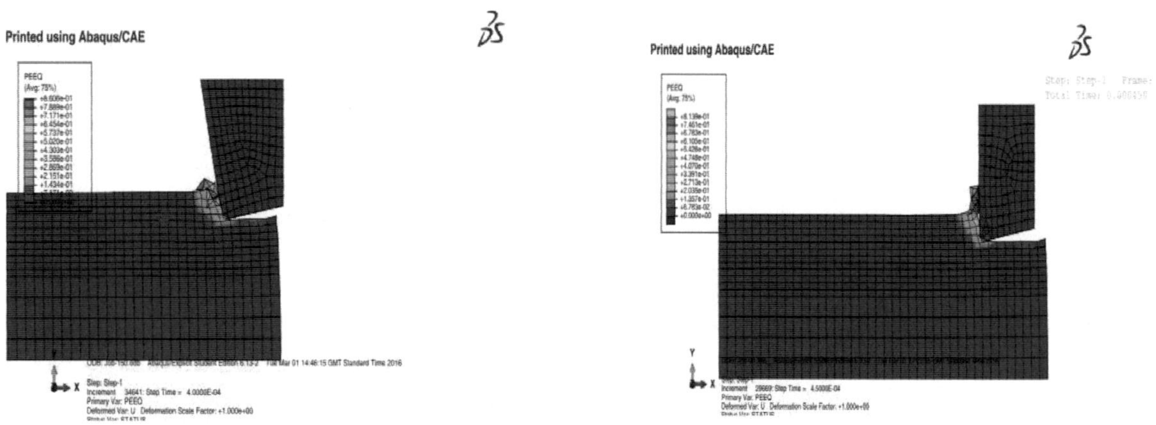

(a) (b)

Figure 6: Effect of rake angle on equivalent plastic strain (a) rake angle 15°, and (b) rake angle 30°, at 42.46 m/min cutting speed, 1.0 mm/rev feed rate and 0.8 mm depth of cut

The magnitude of the strain energy stored by the system decreases with increase in positive rake angles. However, the influence of negative rake angles affected energy stored in the system, this behaviour is certainly affected by the friction, cutting forces and cutting power which is usually bigger than positive rake angles. Based on the results of finite element simulations of machining NST 7-2 steel carried out in this study under different cutting conditions; cutting speed at 42.46 m/min, feed rate at 2.0 mm/rev and depth of cut at 0.8 mm were suggested as optimum cutting parameters for machining NST 37-2 steel. These were suggested because, at the integration progress of the results in the visualization module of the simulations, higher cutting speed above 42.46 m/min and feed rate above 2.0 mm/rev at 0.8 mm depth of cut were observed to lead to excess element distortion, and consequently, excess temperature, equivalent plastic strain and residual stress of the work piece which can pose adverse effects on the fatigue strength and the performance of the product in service. At data below these cutting parameters, chips were observed to be denser and tended to be discontinuous which are undesirable for better machining.

4. Conclusion

In this study, finite element simulation approach has been used to study the effect of machining conditions on chip formation, temperatures, equivalent plastic strain, residual stress distribution, cutting and thrust forces in machining of Nigeria locally rolled NST 37-2 steel by using a commercial software ABAQUS. The dynamics, explicit Arbitrary Lagrangian Eulerian method was utilized to develop the

FEM simulation model with a round edge uncoated carbide cutting tool. The Johnson-Cook work material model and a detailed friction model were also used, and work material flow around the round edge of the cutting tool was simulated in conjunction with an adaptive meshing scheme. The findings of this investigation are as follows:

- In this study optimal cutting parameters of 15° rake angle, 0.3 friction coefficient, 42.46 m/min cutting speed 2.0 mm/rev feed rate and 0.8 mm depth of cut are suggested.
- Johnson cook material constants for plastic model and damage model were provided and applied to represent material behaviour in the simulation, the chip formed throughout the simulation was continuous,
- Residual stress appeared larger with negative rake angle and also decreased as the positive rake angle increased. This is due to the shearing action of cutting tool.

References

[1] Malomo B O, Ibitoye S A amd Adekoya L O 2010, The study of elastic- plastic fatigue behaviour of Nigerian rolled NST 37-2 steel. *International Journal of Engineering Research in Africa*, ISSN 1663-4144, **3**: p. 28-41.

[2] Cenk Kiliçaslan 2009, Modelling and simulation of metal cutting by finite element Method. A Thesis Submitted to the Graduate School of Engineering and Sciences of İzmir Institute of Technology.

[3] Kovac P, Savkovic B, Serdar B and Sekalic M 2011, Modelling mechanical and thermal load of cutting tool. *ACTA T.C. -Bulletin of Engineering*: p. 78-80.

[4] Ali Moaz H, Khidir B A, Ansari M N M and Mohamed B 2013, FEM to predict the effect of feed rate on surface roughness for face milling of titanium alloy. *Journal of Housing and Building National Research Centre*. **7**: p. 263-269.

[5] Zhang Y C, Mabrouki T, Nehas D and Gong Y D 2011, Chip formation in orthogonal cutting considering interface limiting shear stress and damage evolution based on fracture energy approach, *Journal of Finite Elements Analysis and Design*. **47**: p. 850-863.

[6] Everton Ruggeri S A 2014, Finite element simulation of chip formation in machining process, State University of Santa Catarina UDESL CCT, ISSN 8219-710: p. 4-5.

[7] Fadare D A and Asafa T B 2010, Optimization of turning NST 37.2 steel with uncoated carbide cutting tools. *Journal of Nigerian Institution of Mechanical Engineers*. **2**(1): p. 31-40.

[8] Dhananjay Joshi 2004, Finite element simulation of machining a nickel-based superalloy – inconel 718. A MSc. Thesis Submitted to Graduate College of the Oklahoma State University: p. 45-48.

[9] Tounsi N, Vincenti J, Otho A and Elbestawi M A 2002, From the basics of orthogonal metal cutting toward the identification of the constitutive equation, *International. Journal of Machine Tools and Manufacture*, **42**: p. 1373-1383.

[10] Filice L, Micari F, Rizutti S and Umbrello D 2007. A critical analysis on the friction modelling in orthogonal machining. *International Journal of Machine Tools and Manufacturing*. **47**: p. 709-714.

[11] Shet C and Deng X 2000, Finite Element Analysis of the Orthogonal Metal Cutting Process. *Journal of Materials Processing Technology*, **105**: p. 95-109.

[12] Korka Z I, Miclosina C-O and Cojocam V 2013, An Experimental Study of the Cutting Forces in Metal Turning, *Journal of Universita II Eftimie Murgu Resita* Anul XX, **2**, ISSN 1453-7397.

[13] Sushil D. Ghodam (2014), Temperature Measurement of a Cutting Tool in Turning Process by Using Tool Work Thermocouple. *International Journal of Research in Engineering and Technology* ISSN: 2319-1163 ISSN: 2321-7308.

IOP Conf. Series: Materials Science and Engineering 1037 (2021) 012006 doi:10.1088/1757-899X/1037/1/012006

Research of design an educational modular 5 axis milling machine

F D Anania[1], A E Pena[2] R M Canarache[3]

[1]Associated professor, University Politehnica from Bucharest, Robots and Manufacturing Systems Department, Romania
[2] Lecturer, University Politehnica from Bucharest, Robots and Manufacturing Systems Department, Romania
[3] CEO INICAD DESIGN, Bucharest, Romania

Corresponding author's e-mai: Dorel.anania@upb.ro

Abstract. Industrial CNC machine tools are supplied as closed architecture systems. Their structure, kinematics and CNC controllers are designed to be used exclusively for their basic purpose, without any possibility of being changed. Reconfiguration is seen as the ability to add or remove modules from the machine, to be equipped with additional degrees of freedom (axes). For the concept to become a reality, it is necessary that the machine tool can be easily assembled and disassembled by an end-user and can adapt to changing projects or machining requirements. The approached of this research was to develop a 5-axis modular machine tool, with technological capabilities similar to industrial ones, at an affordable level of development costs and suitable to be used for teaching CAM techniques for future engineers

1. Introduction

The mechanical modules, from which a reconfigurable modular machine (RMM) is created, can be assembled in a "building" manner. This allows the mechanical modules to be assembled in several configurations, offering a variation of the degrees of freedom in processing and processing functions on a single platform.[1] The primary stage in the development of an RMM is the synthesis of a library with precompiled mechanical modules, which includes only the modules needed to provide the desired machine configuration.[2] The goal is to design a modular machine tool to represent three different possible architectures of 5x machine (Table-Table, Table-head and head-Head). By modular, we mean being able to disassemble the machine tool into different parts: at first glance each degree of freedom is at least divided into one part. Some intermediate parts are needed for easy assembly.

The main requirement of a machine tool is to provide a guide for continued accurate movement under certain condition: [5] ability of the structure of the bed to resist distortion effectively due to static and dynamic loads; stability of the motion of under load; wear resistance of guides; freedom from residual stresses; leveling; resistance to vibrations; freedom from slip-stick.

2. Desktop milling machine tool concept

When designing this machine, we placed great emphasis on the versatility of the parts we created. In other words, we've made sure that as many parts as possible are reusable from one configuration to another. For this, our parts are not necessarily optimized for an assembly, but optimized to have a place in several assemblies to reduce the production cost.

The design concept is for 5 axis NC milling machine in 3 basic configuration Table-Table (fig.1.a), Table-Head (fig.1.b) and Head-Head (fig.1.c)

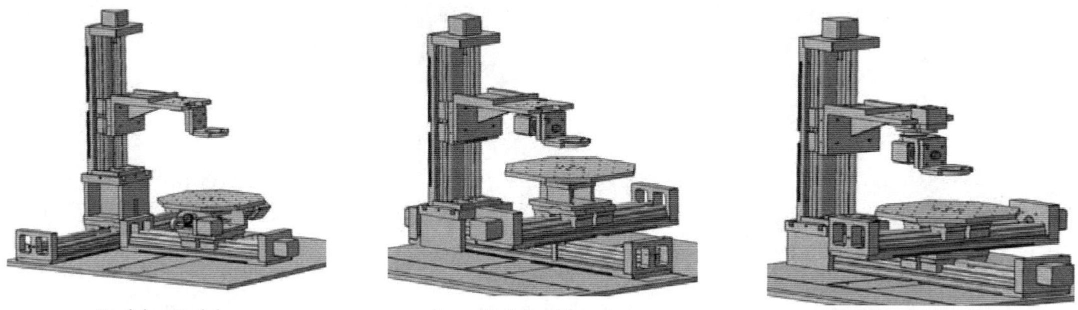

a. Table-Table b. Table-Head c. Head-Head

Fig.1 Basic Modular Machine tool architecture

According to the nomenclature of machine tools, the Table-Table assembly will be designated by the code (YCAOXZ), which means that the translations according to the rotations of axis X and Z as well as the translation of axis Y are located at the table level, while the X and Z axis translations are at the machining head.

According to the nomenclature of machine tools, the Table-Head assembly will be designated by the code (XYCOZB), that is to say that the translations along the X and Y axes and the rotation along the Z axis are located at the table level, and that the translation along the Z axis and the rotation along the Y axis are at the machining head. The two rotations are done with two separate motors, an SSPM HT-62 motor for the rotation of the table, and a stepper motor (same model as for the translations) for the rotation of the head.

According to the nomenclature of machine tools, the Head-Head assembly will be designated by the code (XYOZCA), that is to say that the translations along the X and Y axes are located at the table level, and that the translation along the Z axis and the rotations along the X and Z axes are at the level of the machining head.

3. Hypothesis and simplification of milling machine tool concept

An element of originality consists in the design of numerically controlled axes of modular translation and rotation, compact, completely independent that can be assembled in different architectures specific to machines with numerical control up to 5x NC used for educational purposes.

The modules of a machine must have standard interfaces that allow a wide variety of machine tool configurations. Reconfiguring a machine by replacing modules will require disassembling the machine and reassembling it as well as calibrating, setting set-up parameters and other operations.

The main criteria of the concept are the flexibility and versatility of the machines. Still some requirements like stability of the motion under loads, resistance to vibrations, machine accuracy (squareness of the axis, repeatability, and positioning) must be accomplished even if are not to the level of industrial machine.

For linear axis, a classic kinematic chain of direct drive has been created formed by step motor, leading screw, and nuts. For the guiding, linear bearings are used. All three linear axis were made in the same manner to simplify the construction and decrease the costs.

The assembly solution of the linear axis between them for construct the XYZ architecture is very simple, based on carriage shape which ensure the perpendicularity of the axis and fit position at the same time. In figure 2 a and b are marked the alignment surfaces of the carriage and the base of the next axis. In figure 3 a simple XY axis assembly is presented. For this axis, a physical prototype was made.

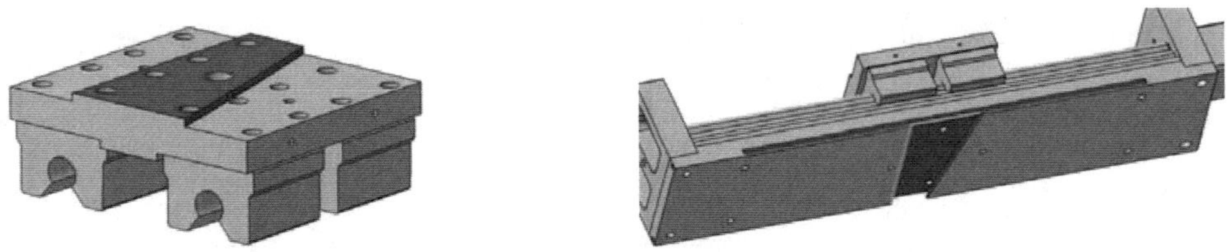

a. Carriage design b. Base design

Fig.2 Linear axis assembly mode design

The machine table has been designed in a specific shape to allow the mounting on linear axis as well on rotational C axis in any configuration In XYOZCA configuration the table is mounted over the X linear axis and will be aligned by the green surfaces (fig.3 a and b).

For XYCOZB the table will be centered on the yellow surface and aligned on the green surface (fig.3 a and b). The Z axis carriage design tries to respect the simple form and at the same time to ensure the stiffness of the machine. In figure 4 the Z axis assembly and components are specified. On Z carriage (4.a) can be mounted directly the B axis (fig.4b) or C and B assembly formed by B axis and C axis (fig.4.c) without the support used in XYCOZB machine configuration.

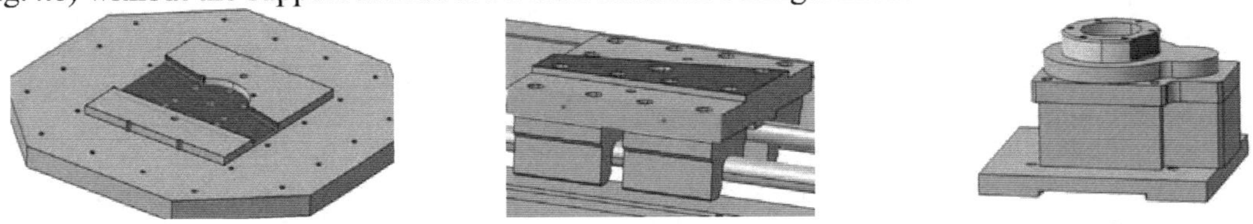

a. Machine table b. Linear axis c. C axis

Fig.3 Machine table design

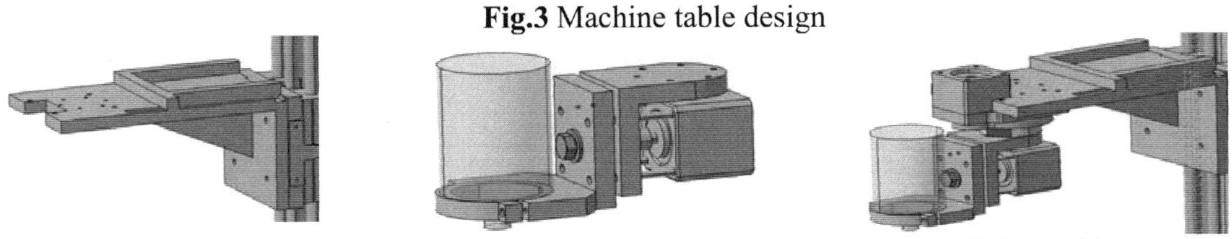

a. Z carriage b. B axis c. Z Assembly

Fig.4 Z axis design

4. Design and analysis

Due to basic configuration of the machine type on Z carriage the main spindle is mounted along with one (B) or two (C and B) rotational axis some verification and validation of the design must be done.

Finite element simulations were made for the XYCOZB configuration by take in account the possible load on the table and B axis flange in a conservative milling process (in polymers and composite material for prototyping). Some aluminum 5083 series has been used for all components with the fallowing characteristics: Density 265 g/cm^3; Modulus of Elasticity 72 GPa; Thermal Expansion 25 x10^-6 /K; Tensile Strength 270 - 345 MPa; Poisson's Ratios 0.334 used in FEM analysis.

To simulate the reaction of the XYCOZB configuration during its operation, the following conditions were imposed: an embedding on the underside of the assembly's support plate, a force of 300N imposed on the machining table, which represents the weight of the stock to be machined (the mass has been overestimated to observe the response of the assembly in extreme cases), and a force of 245N along the Z axis has been applied at the level of the machining head (overestimated again), to represent the forces imposed by the cutting process (Figure 5.a). First analysis was on a simple model design (initial design of components) (figure 5.b)

A maximum deformation of 0.67mm can be observed at the level of the machining head. This value represents a significant loss of precision during machining under the imposed conditions. For this, it was necessary to stiffen the "square" part (fig.9) to limit the bending phenomena that can be observed in the simulation, but also to prevent any torsion that could appear. For this, the rib of the square has been enlarged to better withstand bending forces and a bar has been added to limit the effects of torsion.

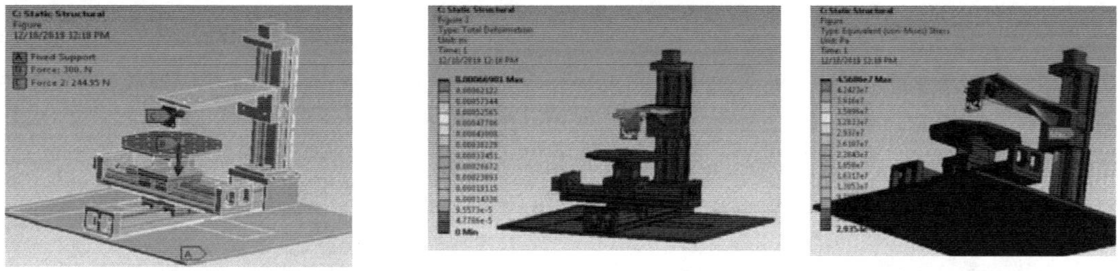

a. The FEM model b. Finite elements analyse c. Von-Mises stress

Fig.5. Finite elements analyse of static loads

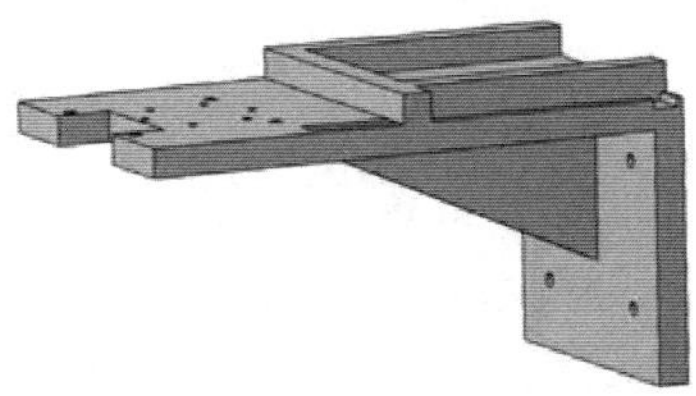

Fig.6 Redesigned of the "square" part

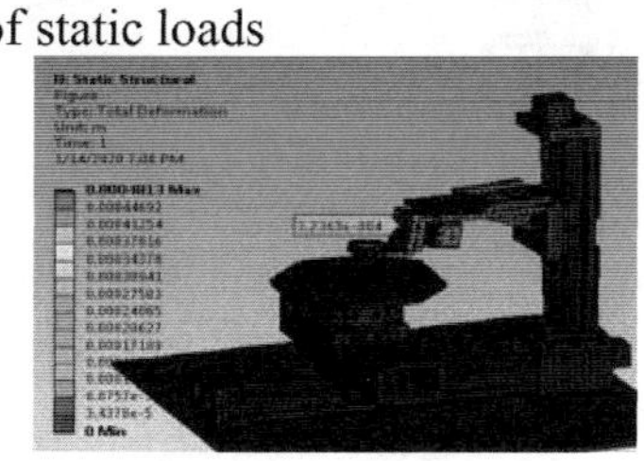

Fig.7 Total deformation for redesigned part

5. Conclusion

An simple to build educational machine tool up to 5x , in modular architecture can be used in higher education and technical school for better understanding of CAM software capabilities.

A modular system must have several characteristics: it must be modular, convertible, it can be customized, it can be integrated, and it can be diagnosed.

All parts were machined by milling from aluminum sheets (some of them of 20 mm thickness) with a conservative cutting regime in order not to induce a remanent stress in material which over time will affect the part accuracy.

Regarding the stresses, a maximum of 46 MPa can be observed (Figure 5) at the level of the Z axis slide connection. This value can be easily supported by steel and aluminum without impairing performance or performance. life of parts, no further action was taken

The design changes made to the square have reduced the deformation of the square to 0.2mm, but the addition of a "machining head holder" part results in deformations of the order of 0.48mm. This value should not be exceeded in practice, since the forces applied for finite element analyzes are greater than those applied during machine operation (fig.7).

References

[1]. Darabă, C. Dogariu, D. F. Anania , Aspects on the analysis with the finite elements method of the machine tools structures regarding remanufacturing, The International Conference Of The Carpathian Euro-Region Specialists In Industrial Systems 7th Edition (2008), Baia Mare, Romania.

[2]. F. D. Anania, C. Mohora, "Research Concerning FEA for a Machine Structure Assembly", Proceedings of the 1st WSEAS International Conference on VISUALIZATION, (2008).

[3]. Kuang -Hua Chang and Chienchih Chen, 3D Shape Engineering and Design Parameterization, Computer Aided Design& Applications, 8(5)(2011).

[4]. Andra Elena Pena, Florea Dorel Anania, Miron Zapciu - Research regarding software developing for machining parameters optimization calculus, Applied Mechanics and Materials, Vol.656 (2014), pp192-199, ISSN 1660-9336.

[5]. I. PETRESCU, I. PISCAN [a]I M. TICA, Static and dynamic analysis of portal milling machine using ANSYS workbench, Conference Proceedings of the Academy of Romanian Scientists, 2010

On the micro-modelling of surface roughness in pulsed laser machining

E Kaselouris[1,2], Y Orphanos[1,2], K Kosma[1,2], A Skoulakis[1,3], I Fitilis[1,3], M Bakarezos[1,2], N Papadogiannis[1,2], M Tatarakis[1,3] and V Dimitriou[1,2*]

[1]Hellenic Mediterranean University Research Centre, Institute of Plasma Physics & Lasers, Rethymnon, GR-74100, Greece
[2]Hellenic Mediterranean University, Department of Music Technology & Acoustics, Rethymnon, GR-74100, Greece
[3]Hellenic Mediterranean University, Department of Electronic Engineering, Chania, GR-73133, Greece

*Corresponding author's e-mail: dimvasi@hmu.gr

Abstract. The surface roughness and the evolution of its morphology of a pulsed laser irradiated aluminum workpiece, is investigated via numerical simulations and experimental measurements. A three-dimensional transient thermo-structural finite element model is developed to simulate the machining process. The developed finite element material model considers the effects of plastic strain, strain rate and temperature, along with a fracture model. For the experiments, a single laser pulse of 6 ns duration and 0.7 mJ of energy at 532 nm is employed as the heat source, and the surface roughness is measured using white light interferometry set-up and related experimental diagnostics. A step-like linear approximation is used to model the surface roughness. A satisfactory agreement between the experimental data and the simulation results is found. This preliminary study aims to contribute to a better understanding of the initial physical processes involved in pulsed laser machining considering the influence of surface roughness and is beneficial for industrial applications such as laser polishing, engraving and cutting.

1. Introduction

Laser machining is non-conventional machining manufacturing process of high scientific and technologic interest, where complex physical processes take place. Modeling of laser machining is essential to predict the heated matter's behavior and to better comprehend the fundamentals of the physical problem. The laser parameters, the thermal and optical properties of the material, and the surface morphology are critical factors that influence the laser-solid interaction and the subsequent heated matter's phase changes [1-7]. Surface roughness is a typically used criterion for characterization of surface quality in all machining processes. The knowledge of the fundamental basis of the laser machining process can be improved by the development of computational models of the matter–radiation interaction by considering the effects of the initial surface roughness. Finite element method (FEM) is a numerical method capable to model complex geometries when the domain changes, the solution lacks of smoothness or the desired precision varies on the whole domain [4,8].

In laser machining studies there is a lack of numerical models that consider the influence of the surface roughness by considering both mechanical and thermal properties of the heated matter. Conde et al [9,10] developed a thermal ablation finite element method (FEM) model that incorporates the initial target roughness to explain the target microstructure formation and the laser plume deflection. In laser shock peening, Ran et al [11] developed a FEM structural numerical model, to predict the surface roughness by applying the computed results as inputs to analytical equations and then compare them with experimental results. Hasser et al [12] in order to explore surface roughness effects in laser shock peening developed a structural FEM model and simulated roughness produced by displacing surface nodes using a statistical method.

In our study, a 3D FEM thermal-structural analysis is performed, accompanied with experimental measurements using an optical set-up to record the influence of the surface roughness and the evolution of its morphology under pulsed laser irradiation. An Al-6061 orthogonal part is properly machined for the generation of a periodic rough surface profile. This pre-definition of the type of the resulting roughness provides the ability of developing a simple geometry of a FEM micro-model, capable to carry out the simulation needs of our study. The FEM model considers the phase changes of matter and highlights the induced elastic and plastic effects using a material model that takes into account the effects of plastic strain, strain rate and temperature, along with a fracture model. For the optical measurements, a profilometry configuration based on a Michelson white light interferometer is used. A satisfactory agreement between the experimental data and the simulation results is found. This preliminary study aims to contribute to a better understanding of the initial physical processes involved in pulsed laser machining considering the influence of surface roughness and is beneficial for industrial applications such as laser polishing, engraving and cutting [13-15]. Furthermore, the determination of the influence of surface micro-defects on the laser-matter interaction processes provides crucial information for the manufacturing and response of the solid targets in inertial confinement fusion [16].

2. Experimental methodology

In this work an Al-6061 orthogonal part of 9 mm thickness is selected as the laser irradiated target of our study, due to its overall thermomechanical behavior and well-known physical properties. The 10 × 3 × 0.9 cm sized Al workpiece was dry milled with spindle rotation frequency of $n = 1910$ rpm and a feed rate of $f = 2292$ mm/min (feed per tooth $f_z = 0.4$ mm/z). The chosen tool-tip radius was $r = 0.8$ mm. The resulting rough periodic surface profile of the Al target is presented in figure 1b. The part of the surface studied under the present conditions of laser irradiation, shown in the next figures, covers the region from about 1200 μm to 1800 μm (mirrored on y-axis) along the x axis on figure 1b.

The optical geometry for the present experiments is shown in detail in figure 1a: The laser source used for the interaction with the Al samples is a pulsed laser source (6 ns pulse duration) emitting radiation of 0.5 to 1.2 mJ per pulse, at 532 nm. The output of the laser system is guided through metallic mirrors to a converging lens (f = 100 cm) and the beam is then focused onto the target surface. The Al targets, which are produced with a specific surface roughness, are hit with a single pulse of energy > 0.5 mJ; this value constitutes a threshold below which no detectable effect was induced by the laser on the target surface.

The profilometry set-up is presented in figure 1a. It is incorporated in the optical arrangement and is used for the characterization of the samples before and after they have been irradiated with the laser pulse. The profilometry configuration is based on a Michelson white light interferometer, described in detail in [4]. Briefly, a beam splitter separates the white light from the source, generating the reference and object beams, the latter being reflected by the sample surface. A metallic mirror placed on a lineal piezoelectric actuator is used for altering the optical path of the reference beam. The two beams are recombined on a CCD camera and a series of acquired interferograms for the different positions of the moving metallic mirror allows for the determination of the surface topology [4]. Using a roughness meter (Roughness gauge TESA Rugosurf 20) the same results for R_t were indicated, as shown for validity in figure 1d.

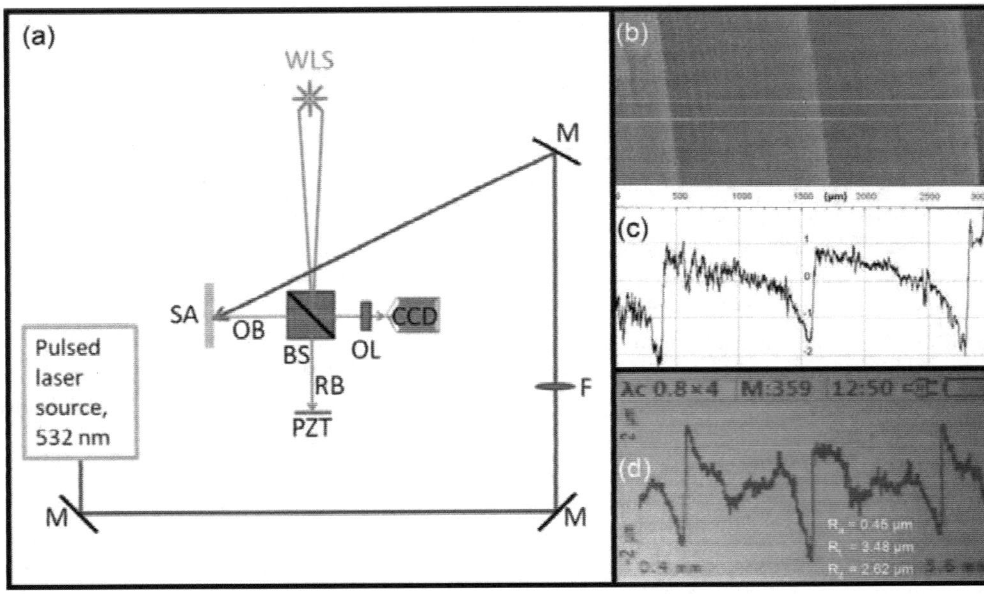

Figure 1. a) Experimental set-up for the target-laser interaction measurements: **M**-metallic mirror, **F**-focusing lens with focal length f=100 cm, **SA**-aluminum sample. Profilometry set-up: **WLS**-white light Source, **BS**-beam splitter, **RB**-reference beam, **OB**-object beam, **PZT**-piezoelectric transducer, **OL**-objective lens.b) 2D grayscale surface image of the SA as extracted from the profilometry measurements. c) The corresponding lineout of b) image. d) The periodic roughness profiles and the measured values of R_a, R_t, R_z.

3. Finite element modelling

The LS-DYNA software is used for the development of the FEM model [17]. A 3D coupled thermal and mechanical analysis is performed. The temperature distribution is found by solving the heat conduction equation:

$$\rho(T)C_p(T)\frac{\partial T(x,y,z,t)}{\partial t} - \nabla[k(T)\nabla T(x,y,z,t)] = Q(x,y,z,t) - L_i \quad (1)$$

where x, y and z are the space coordinates while ϱ, C_p, k are the mass density, specific heat at constant pressure and the thermal conductivity of the target material, respectively. The source term, Q, represents the laser energy absorbed by the sample, while L_i is the latent heat of melting and is equal to zero in the thermoelastic regime.

The conservative equations of mass, momentum and energy are also solved:

$$\frac{\partial \rho}{\partial t} + \rho\nabla \boldsymbol{v} = 0 \quad (2)$$

$$\rho\frac{\partial^2 U_i}{\partial^2 t} = \mu\frac{\partial^2 U_i}{\partial^2 k} + (\lambda+\mu)\frac{\partial}{\partial i}\left(\frac{\partial U_k}{\partial k}\right) - (3\lambda + 2\mu)\alpha_T\frac{\partial T}{\partial i} \quad (3)$$

$$\rho \frac{\partial E}{\partial t} = \sigma_{ij} \nabla \boldsymbol{v} \tag{4}$$

where U is the displacement, ϱ the density, λ and μ the *Lamé* coefficients, α_T the thermal expansion coefficient, E is the energy and σ_{ij} the stress tensor.

The source term $Q(x, y, z, t)$, has an elliptic spatial form and represents the absorbed laser energy per unit volume per unit time by the sample:

$$Q(x,y,z,t) = I_0(1-R)e^{-4ln2(\frac{t}{t_0})^2} e^{-[(X/r_a)^2+(Y/r_b)^2]} a_b e^{-a_b z} \tag{5}$$

$$X = (x - x_o)\cos\theta + (y - y_o)\sin\theta$$
$$Y = (x - x_o)\sin\theta - (y - y_o)\cos\theta \tag{6}$$

where I_0 is the incident laser intensity on target, R is the optical reflectivity of the sample, α_b is the optical absorption coefficient, t_0 is the laser pulse duration at full width at half maximum (FWHM), r_a the semi-major axis and r_b semi-minor axis of the laser spot, x_o, y_o the centre coordinates, and θ the rotation angle.

The 3D FEM model simulates the transient structural response of a homogeneous isotropic Al target. The dimensions of the solid target are 1.5 × 1.5 × 0.012 mm. A hexahedral, 3D-solid eight-node element is adopted for the transient analysis. A total of approximately 4.8 million elements is used for the model where surface roughness is not considered (model), while a total of approximately 4.2 million elements is used when surface roughness is modelled (R-model). Moreover, the beam spot has an elliptic shape. The major and minor axes of the elliptical laser spot are 160 μm and 75 μm long as indicated in the experiment. Since the laser spot defines the area of interest in our study, the sub-model of size 1.5 × 0.6 mm is considered, as shown in figure 3a (top-right) for the FEM models. A total of 960000 and of 840000 elements, were used for the FEM model (z = 12 μm, $0 \leq y \leq 600$) and the R-model (z = 12 μm, $0 \leq y \leq 400$ and z = 9 μm, $400 \leq y \leq 600$) simulations, respectively. The hydrodynamic and deviatoric behavior of the metallic target is considered by using simultaneously the analytical Grüneisen EOS coupled with the Johnson-Cook (J-C) strength material model, that considers plastic and fracture effects [18]. The temperature dependent thermomechanical properties and the J-C material model and failure parameters, as well as the Grüneisen EOS coefficient properties of the workpiece, are taken from literature [19-21].

4. Results and discussion

The laser fluence threshold for transition from the elastoplastic to the melting phase is computed to be 1.5 J/cm^2, a value that has been also validated by the experiments. The effect of the laser pulse on the surface roughness of the workpiece is demonstrated in figure 2 for the pulse energy of 0.7 mJ and a fluence of 2 J/cm^2, which is above the melting threshold of Al. Full and appropriately cropped images, tight around the elliptical shaped laser spot, are shown in figure 2. The corresponding 3D surface plots and lineouts of the Al target, prior and after the laser pulse hits the surface, are also demonstrated. To follow the experimental measurements alignment and the presentation of the results, the direction of the roughness pattern shown in figures 2 and 3 is mirrored on the y axis (compared to figure 1).

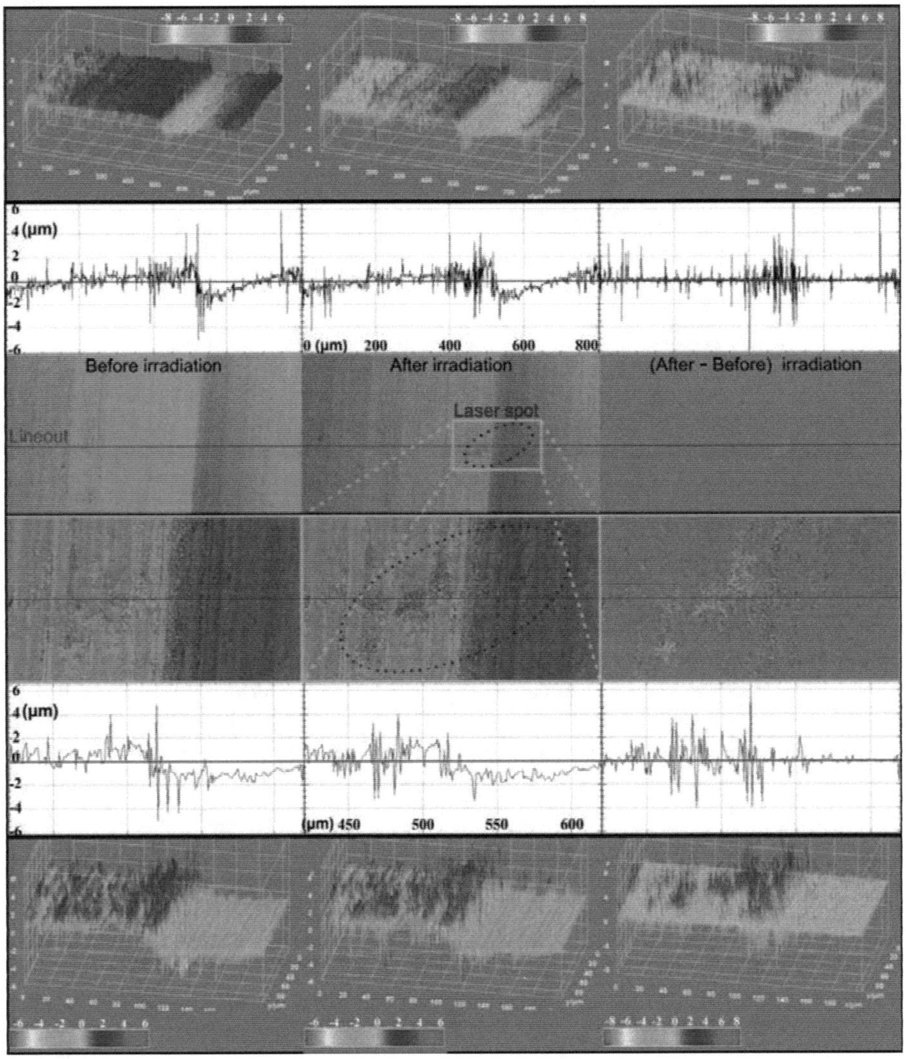

Figure 2. Top three rows: 3D surface plots and lineout graphs of the Al surface before (left) and after (middle) irradiation and of the subtracted images (right). **Bottom three rows:** same as before of the cropped images, demonstrating the vicinity of where the elliptical laser spot hits the surface of the workpiece.

The full and cropped images and the 3D and lineout images apart from the area clearly affected by the laser pulse, also reveal details on the height/depth of the affected peaks. According to the subtracted images (prior and after the laser pulse) some resulting laser-affected roughness peaks show a difference reaching up to 4 μm in height/depth, for one spike even up to 6 μm. Such spikes exist also outside the laser-affected region of the sample and are considered as artefacts (measurements noise). Most peaks inside the laser-affected region, influenced by the laser pulse, show differences in height/depth below 1 μm.

To model the surface roughness, a rough step-like linear approximation is carried out, as shown via the profile data from the original Al image in figure 3a. On the top-right is also presented the 3D full model by considering the step-like approximation of the roughness periodic profile. On the bottom of figure 3a the meshed sub-model of roughness (R-model) is shown. The FEM results of temperature,

Von Mises stresses and the plastic strain of the FEM model and the R-model are presented in figure 3b for 10 ns and 20 ns and for laser fluence of 2 J/cm^2.

Furthermore, the final nodal positions after the laser beam irradiation are presented in the graph on the bottom-right of the figure. At 10 ns the heated target is in the elastic regime, at 13 ns plastic effects are initiated and at 16 ns the surface temperature exceeds the melting point. At 20 ns the heated material is in the melting regime. The total fractured volume for the R-model is computed to be ~172 μm^3, while this value becomes ~163 μm^3 for the model without roughness. The maximum depth due to material fracture is 0.1 μm for both models. This value along with the maximum computed temperature of 850 ^{o}C, indicates that ablation phenomena do not take place for this laser fluence.

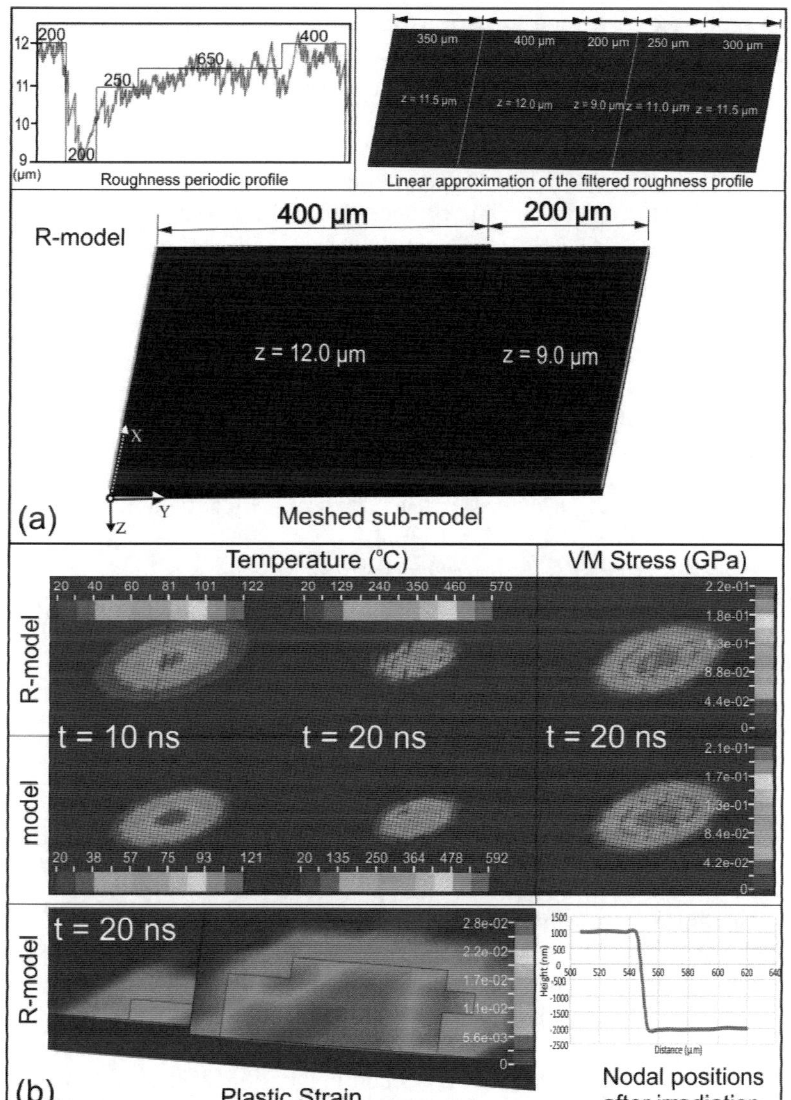

Figure 3. a) Top left: Image processing of the roughness profile data (figure 1c) in MATLAB (red) filtered using Gaussian smoothing (green). A rough linear approximation (black) is used to define the preliminary roughness FEM R-model (top right). The meshed sub-model of the R-model is presented at the bottom. b) FEM results of temperature, Von Mises stress and plastic strain distribution for the FEM model and the R-model at 10 ns and 20 ns for laser fluence of 2 J/cm^2 and the nodal positions after laser irradiation (bottom right).

5. Summary

The surface roughness and the effect of a single laser pulse on an Al 6061 workpiece is investigated via numerical simulations and experimental measurements. A single pulse of 6 ns duration an 0.7 mJ of energy (fluence of 2 J/cm^2, plastic regime) is used to hit the Al target and induce changes in the surface roughness, which are visualized by a white-light profilometry set up with sub-micrometer resolution. A FEM thermal-structural model is used to model the laser irradiated rough surface, with dimensions using a step-like linear approximation of the experimentally obtained periodic surface profile. The experimental and computational outcomes agree well. Combining the numerical models with the experimental techniques presented in this study constitutes a promising tool for an in-depth analysis of the influence of heat parameters on the surface roughness of different materials, below the ablation threshold, and can be further extended to various laser machining and micromachining applications. Future temperature measurement experiments are expected to provide additional insight on these issues. Higher spatial modelling of the micro-roughness characteristics will be further included in the models that will be developed for our future studies.

Acknowledgments
We acknowledge support of this work by the project "ELI - LASERLAB Europe Synergy, HiPER& IPERION-CH.gr" (MIS 5002735) which is implemented under the Action "Reinforcement of the Research and Innovation Infrastructure", funded by the Operational Programme "Competitiveness, Entrepreneurship and Innovation" (NSRF 2014-2020) and co-financed by Greece and the European Union (European Regional Development Fund). This work was also supported by computational time granted from the Greek Research & Technology Network (GRNET) in the National HPC facility—ARIS—under project ID pr009023— LaMIPlaS III.

References

[1] Kaselouris E, Nikolos I K, Orphanos Y, Bakarezos E, Papadogiannis N A, Tatarakis M, and Dimitriou V 2013 *J. Multiscal Model.* **5** 1330001
[2] Shirk M D, Molian P A 1998 *J. Laser Appl.* **10** 18
[3] Kaselouris E, Nikolos I K, Orphanos Y, Bakarezos M, Papadogiannis N A, Tatarakis M and Dimitriou V 2016 *Int. J. Damage Mech.* **25** 42
[4] Dimitriou V, Kaselouris E, Orphanos Y, Bakarezos M, Vainos N, Nikolos I K, Tatarakis M, Papadogiannis N A 2015 *Appl. Phys. A* **118** 739
[5] Kaselouris E et *al.* 2020 *IOP Conf. Ser.: Mater. Sci. Eng.* **916** 012050
[6] Kaselouris E *et al.* 2020 *Key Eng. Mater.* **827** 122
[7] Kaselouris E *et al.* 2017 *Key Eng. Mater.* **261** 339
[8] Efstathopoulos N et al. 2010 *Injury* **41** 259-265
[9] Conde J C, Lusquiños F, González P, Serra J, León B, Cultrera L, Guido D and Perrone A 2004 *Appl. Phys. A* **79** 1105
[10] Conde J C *et al.* 2004 *Thin Solid Films* **453–454** 323
[11] Ran Z, Yongkang Z, Guifang S, Xuting S, Pu L 2018 *Rare Metal Mat. Eng.* **47** 33
[12] Hasser P J, Malik S A, Langer K, Sprandlin T J 2013 Simulation of Surface Roughness Effects on Residual Stress in Laser Shock Peening, In Proceedings of the ASME 2013 International Manufacturing Science and Engineering Conference Collocated with the 41st North American Manufacturing Research Conference: American Society of Mechanical Engineers, Madison, WI, USA, 10–14 June 2013.
[13] Zhou J, Liao C, Shen H and Ding X 2019 *Surf. Coat. Tech.* **380** 125016
[14] Pandey A K, Dubey A K 2013 *J Mech Sci Technol* **27** 2115
[15] Adeniji D, Schoop J, Gunawardena S, Hanson C and Jahan M 2020 *J. Manuf. Mater. Process* **4** 59
[16] Hu S X, Fiksel G, Goncharov V N, Skupsky S, Meyerhofer D D and Smalyuk V A 2012 *Phys. Rev. Lett.* **108**, 195003

[17] Hallquist J O 2006 LS-DYNA Theory Manual (Livermore Software Technology Corporation, California)
[18] Kaselouris E et *al*. 2018 *Plasma Phys. Control. Fusion* **60** 014031
[19] Karimi M R, Sedighi M and Afshari D 2015 *Int. J. Adv. Manuf. Technol.* **77** 885
[20] Akram S, Jaffery S H I, Khan M, Fahad M, Mubashar A and Ali L 2018 *Adv. Mech. Eng.* **10** 1
[21] Steinberg D J 1991 *Equation of State and Strength Properties of Selected Materials* - Lawrence Livermore National Laboratory report UCRL-MA-106439

Rolling element bearings fault classification based on feature extraction from acceleration data and artificial neural networks

K Kotsanidis[1]**, P Benardos**[1*]

[1]National Technical University of Athens, School of Mechanical Engineering, Section of Manufacturing Technology, Heroon Polytechniou 9, Athens, GR15780, Greece

[*]Corresponding author's e-mail: pbenard@mail.ntua.gr

Abstract. This paper presents the development of an artificial neural network (ANN) model for rolling element bearings fault classification that uses features extracted from acceleration data collected during run-to-failure experiments. The presented approach initially employs a wavelet decomposition method for signal denoising and subsequently relies on a Fourier transform to analyse the acceleration signal in the frequency domain. Several features that correspond to the entire signal range as well as to specific frequency bands are then extracted and used as inputs in the ANN model, which is trained to identify three different operational states, namely, no fault, inner race fault and outer race fault. The developed ANN model is validated using experimental data from the publicly available dataset provided by the Center of Intelligent Maintenance Systems (IMS) of the University of Cincinnati. The results show that the trained ANN model has a classification accuracy of 90.2% in the training data and 100% in the test data.

1. Introduction

Predictive or condition-based maintenance (CBM) refers to a maintenance strategy where the relevant equipment is monitored during operation so as to identify patterns that are representative of evolving fault and failure mechanisms. Appropriate measures and actions can then be scheduled and performed only when they are necessary, resulting in significant cost savings and minimized equipment downtime compared to the, more common, preventive maintenance strategy [1]. Any effective implementation of CBM therefore requires a three step approach: i) design and deployment of an appropriate sensor and data collection system, ii) development of the required fault prediction and/or component reliability models based on the collected data and iii) design and deployment of an efficient decision making policy that dictates the most appropriate actions to be performed depending on the previous prediction models.

Modern manufacturing mainly uses equipment with rotating parts and components and among them, machine tools play a critical role in the global economy [2]. Focusing on cutting processes, one of the most important subsystems of a CNC machine tool is the main spindle assembly as it is responsible for providing the required power over a specific speed range while ensuring the desired precision [3]. The final performance characteristics of the spindle assembly depend on the individual performance of its components, especially the motor, shaft and bearings. For the latter, their role is to support and accurately locate the shaft, while reducing friction despite the static and dynamic loads and thermal conditions that they are subjected. While there are many types of bearings that are used in spindle assemblies, the most common type is rolling element bearings due to their low cost and ease of lubrication.

Bringing the concepts of CBM and machine tool spindles together, there have been studies that have identified spindle failures as one of the key sources of machine tool downtime [4] and similarly, bearing problems as one of the key sources of spindle failures [5]. The recommended approach for CBM implementation in these cases is to monitor mechanical vibrations. Initially, a baseline level that corresponds to normal operation is established, which is compared to any other subsequent measurement. As the vibration levels increase, the larger the deviation from this baseline becomes, which indicates a larger probability for failure to occur. There is an abundance of published work that studies the various vibration characteristics to evaluate the actual condition of bearings. Regardless of the specific approach that is adopted, a common initial step is followed, dealing with feature extraction from the collected experimental data. Typically extracted features include RMS, kurtosis, skewness and crest factor in the time domain and power peak, spectral area and kurtogram in the frequency domain [6-8]. In recent years, there is a trend of using data-driven, i.e. machine learning methods to fuse such features so as to develop the relevant predictive models [9-12]. The main advantages of these methods is that the developed models can be easily integrated in on-line monitoring and control schemes, they possess an inherent robustness when dealing with noisy data and can offer increasingly higher prediction accuracy as more data are collected over time [13].

This paper focuses on developing an artificial neural network (ANN) model that is able to classify the operational state of a rolling element bearing in one of three possible classes, namely without fault, with inner race fault or with outer race fault. To achieve this goal, the ANN model uses as inputs features that have been extracted from acceleration data collected during run-to-failure experiments. Section 2 presents the proposed approach, which involves signal denoising based on wavelet decomposition, frequency domain analysis through a Fourier transform, feature extraction and finally ANN model development. Section 3 presents the model validation procedure using a publicly available dataset and the obtained results and finally, Section 4 summarizes the obtained conclusions and offers suggestions for future work.

2. The proposed approach

The proposed approach can be implemented according to the flowchart shown in Figure 1.

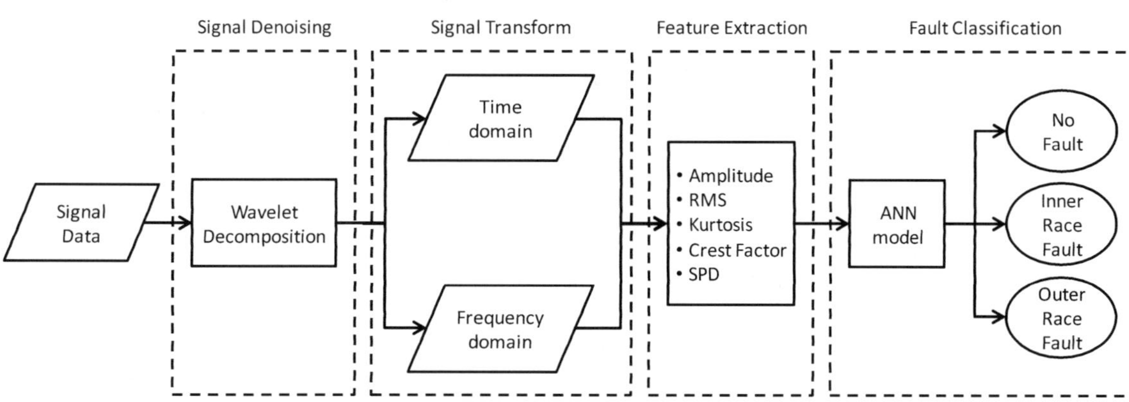

Figure 1. Flowchart of the proposed approach.

The approach is independent of the way that the necessary mechanical vibration signal is collected. Instead, it assumes that the relevant data are already available and focuses on how to process them so as to successfully develop the fault classification model. As can be seen, there are four main stages involving signal denoising, signal transform, feature extraction and fault classification respectively.

2.1. Signal denoising

In the majority of applications, vibration data are collected through appropriate sensors, e.g. accelerometers, that are installed on, or as close as possible to, the bearing housing. Any such experimental setup will also include additional mechanical components and parts that effectively act as significant noise sources therefore corrupting the measurements that characterize the bearing operational state. In the general case, the noisy signal can be represented by equation (1):

$$x(n) = s(n) + \sigma w(n) \tag{1}$$

where $x(n)$ is the noisy signal, $s(n)$ is the desired signal, $w(n)$ can be considered as Gaussian white noise and σ is a constant number indicating noise level. Therefore, the objective of the denoising process is to suppress $w(n)$ and recover $s(n)$.

Wavelet decomposition has been successfully applied for denoising vibration signals as shown in [14]. The method involves three steps:

- Signal decomposition that is accomplished by initially choosing a wavelet and level N and subsequently computing the wavelet decomposition of the $x(n)$ signal at level N
- Thresholding of detail coefficients, during which a threshold is selected for each level from 1 to N and is applied to the detail coefficients and
- Signal reconstruction that is accomplished by using the original approximation coefficients of level N and the modified detail coefficients of levels 1 to N.

The main advantage of this method is optimal noise reduction with simultaneous preservation of the signal. Moreover, it can also be implemented with computational efficiency in either software and/or hardware form as shown in [15] and for these reasons, it was adopted in this work.

2.2. Signal transform

The sensors that collect the vibration data essentially provide a series of amplitude measurements as a function of time. The analysis of such vibration data in the time domain is limited to only a few statistical parameters that mostly quantify the strength of the vibration profile such as amplitude, RMS, kurtosis and crest factor to name the most common (see Table 1).

Table 1. Time domain statistical parameters.

Parameter	Usage	Definition
RMS	Value increases gradually as fault develops	$RMS = \sqrt{\frac{1}{N}\sum_{i=1}^{N} x_i^2}$
Kurtosis	Ideally, for rolling element bearings without faults it should be equal to 3	$Ku = \frac{1}{N}\sum_{i=1}^{N}\left(\frac{x_i - \bar{x}}{\sigma}\right)^4$
Crest factor	Appropriate for impulsive signals	$CF = \frac{max(\lvert x_i \rvert)}{\sqrt{\left(1/N\right)\sum_{i=1}^{N} x_i^2}}$

However, it is also important to determine the frequency characteristics of the vibration and therefore the analysis also takes place in the frequency domain after a Fourier transform has been performed. In this way, the causes of the vibration can be identified by observing peaks at specific frequencies. In the case of rolling element bearings, this is highly desirable for two reasons. Firstly, because the evolution of faults in the bearing will generate characteristic failing frequencies that can be typically found in low ranges and which can be easily verified by observing the spectrum. Such examples are the ball pass

frequency outer (BPFO), ball pass frequency inner (BPFI) and ball spin frequency (BSF). Secondly, because through further evaluation of the obtained spectrum, specific frequency bands (close to the failing frequencies or to their harmonics) can be selected for feature extraction based on the assumption that these frequencies will contain the most useful information for fault detection.
To further enhance this assumption, the kurtogram method was also used to detect the presence of nonstationary transient faults and accurately indicate their location in the frequency bands [16].

2.3. Feature extraction

The main task of the third stage is to extract the appropriate features from both the time and frequency domains in an attempt to gather as much information as possible about the current vibration state of the rolling element bearing. Especially for the frequency domain and as previously described, some of these features will be extracted from the entire frequency range, while there will also be features specific to certain frequency bands that are identified through manual inspection of the spectrum and kurtogram.
For each of the identified frequency bands the features of Table 1 will be extracted, supplemented by the Power Spectral Density (PSD) value. The reason to include PSD in the approach is that in this way random vibrations can also be accounted for. This is possible since PSD is computed by multiplying each frequency bin in an FFT by its complex conjugate, which results in the real only spectrum of amplitude, and then normalizing this amplitude to the frequency bin width.
Therefore, the final set of features that will be extracted will depend on the number of identified frequency bands and in general will have the following composition (Figure 2).

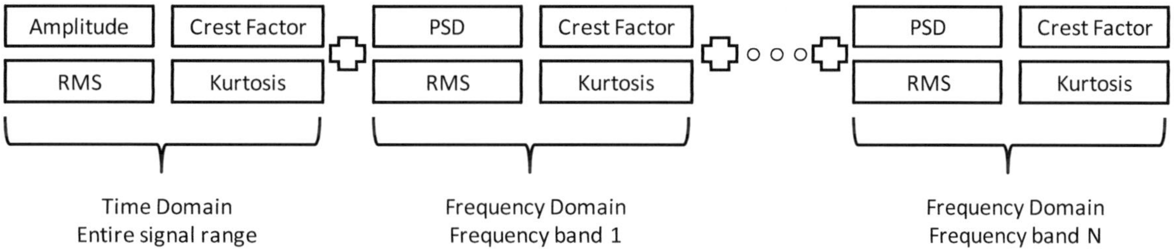

Figure 2. Set of extracted features.

2.4. Fault classification

The final stage of the proposed approach deals with using the extracted features to develop a fault classification model. As previously mentioned, machine learning methods have exhibited very good results in successfully developing accurate and consistent predictive models, see Section 1. In the proposed approach, a feedforward ANN is used to treat the fault prediction as a classification problem. The aim is to be able to classify the operational state of the rolling element bearing in one of three distinct classes, namely without fault or with an inner race fault or with an outer race fault. Therefore, the architecture of the ANN model comprises as many input neurons as the number of extracted features and three output neurons, effectively adopting the "1-of-C" coding scheme for its output in order to perform the actual classification task.
It has to be noted that determining the optimal number of hidden layers and hidden neurons is a case dependent problem. In the simplest implementation, a repetitive trial-and-error procedure can be followed, while more systematic approaches can be used as well.

The choice of a feedforward ANN model, combined with the choices made in the previous stages of the approach, ultimately aims at developing a rolling element bearing fault classification system with low computational requirements that can be deployed as a near real-time health monitoring system with low additional effort and cost. Once the ANN model has been trained, it can be straightforwardly deployed through an FPGA for an on-the-edge application.

3. Experimental validation and results

3.1. Validation

In order to validate the proposed approach a publically available dataset was used provided by the Center for Intelligent Maintenance Systems (IMS) of the University of Cincinnati with support from Rexnord Corp. [17]. The dataset comprises data collected from a specially designed test rig as shown in Figure 3. The test rig used an AC motor and belt to drive a shaft that was supported by four double row bearings. A constant radial load was applied by means of a spring mechanism and a constant rotational speed of 2000 RPM was selected. A set of accelerometers and thermocouples at each bearing was used to sample data at 20 kHz and the test rig was left to run until a failure in any of the bearings occurred. This experiment was ran three times; in the first run, the third bearing developed an inner race fault and the fourth bearing developed a rolling element fault, in the second run, the first bearing developed an outer race fault while in the third run, the third bearing developed an outer race fault.

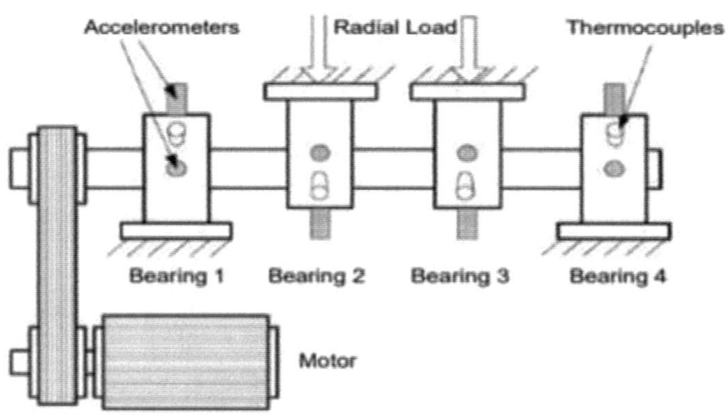

Figure 3. Test rig for IMS bearing dataset [14].

In this paper, data corresponding to the first and second runs were used so that a total of three operational states could be evaluated, namely a bearing with no fault, a bearing with inner race fault and a bearing with outer race fault. Furthermore, as the original data spanned a very large time interval of several days until the faults occurred, a uniform sampling procedure was employed. Thus, a total of 102 samples were selected, with 52 samples originating from the first experimental run and 50 samples originating from the second experimental run. Additionally, the 52 samples from the first experimental run were further divided so that 30 samples represent the no fault operational state and 22 samples represent the inner race fault operational state. Similarly, the 50 samples from the second experimental run were further divided so that 30 samples represent the no fault operational state and 20 samples represent the outer race fault operational state.

3.2. Results

As per the proposed approach, the raw time domain data were denoised using the described wavelet decomposition method. For the specific application, a level of N=5 was selected for the signal decomposition step along with Stein's unbiased risk estimate (SURE) thresholding selection rule for the wavelet coefficients. Figures 4 and 5 present the raw and denoised signals respectively.

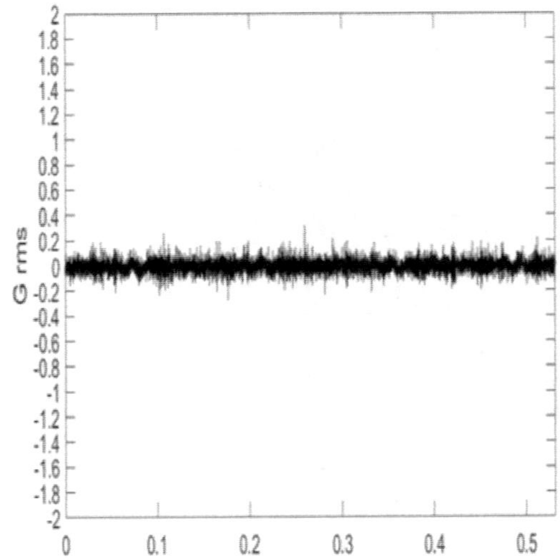

Figure 4. Raw acceleration signal in the time domain.

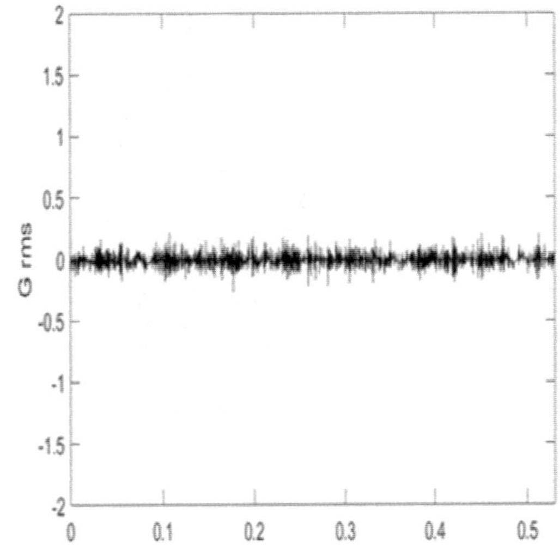

Figure 5. Denoised signal after wavelet decomposition.

Following with the signal transform stage, FFT was performed to obtain the signal in the frequency domain. Figure 6 shows the spectrum corresponding to a sample of the first experimental run for which the inner race fault has already occurred, resulting in the presence of characteristic failing frequencies and harmonics.

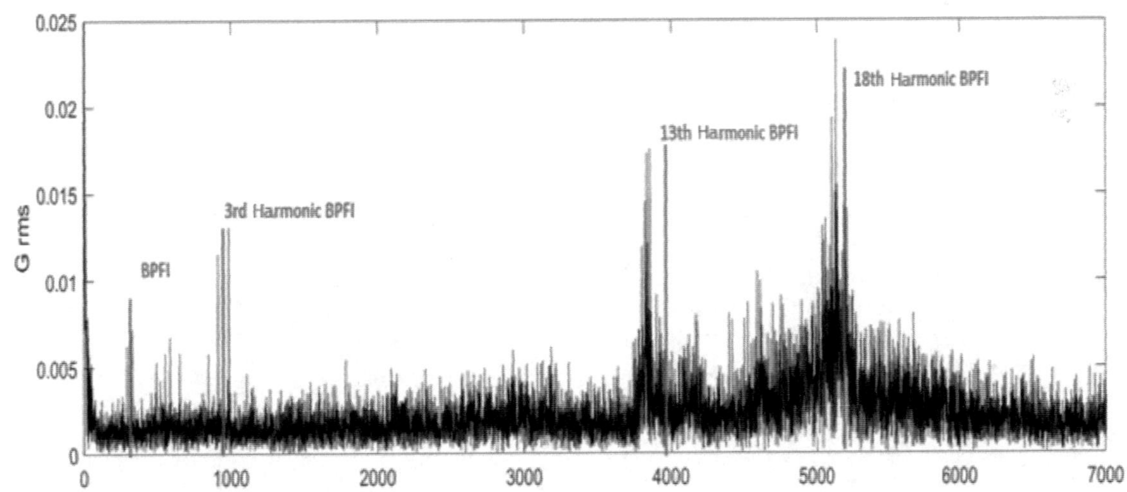

Figure 6. Frequency spectrum of acceleration signal corresponding to inner race fault.

As also discussed in section 2.2, kurtograms were calculated to assist in identifying desirable frequency bands for feature extraction. Figure 7 shows the kurtogram for the same sample as the one in Figure 6. As supported by both Figures 6 and 7, the frequency band between 4 and 6 kHz that contains the 13th and 18th harmonics of BPFI are good candidates for feature extraction.

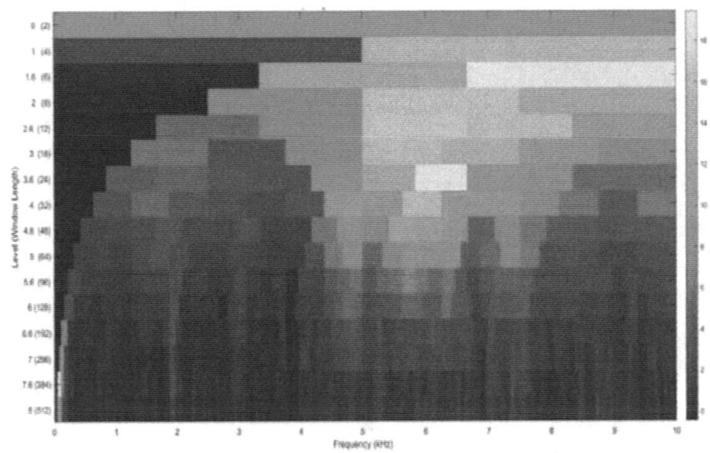

Figure 7. Kurtogram corresponding to inner race fault.

Furthermore, another two frequency bands between 100 to 500 Hz and between 500 Hz to 1 kHz were also selected resulting in a total of 20 extracted features to be used as inputs for the feedforward ANN model.

To develop and train the ANN model a repetitive trial-and-error procedure was selected using confusion matrices to evaluate the performance of each model. As is typical, the total 102 samples were divided to form three subsets, one used for training (approx. 80% or 82 samples), one used for early stopping of the training procedure (approx. 10% or 10 samples) and one to be used as unknown data (approx. 10% or 10 samples) so as to test the ANN's generalization ability. The ANN with the best performance had 2 hidden layers with 20 hidden neurons each for a 20x20x20x3 final architecture. The confusion matrices of the training and test subsets are shown in Figures 8 and 9 respectively. Class 1 represents the no fault state, class 2 the inner race fault and class 3 the outer race fault. It can be clearly seen that the ANN model has very high accuracy both in training (90.2%) and testing subsets (100%) and is therefore suitable for rolling element bearing fault classification.

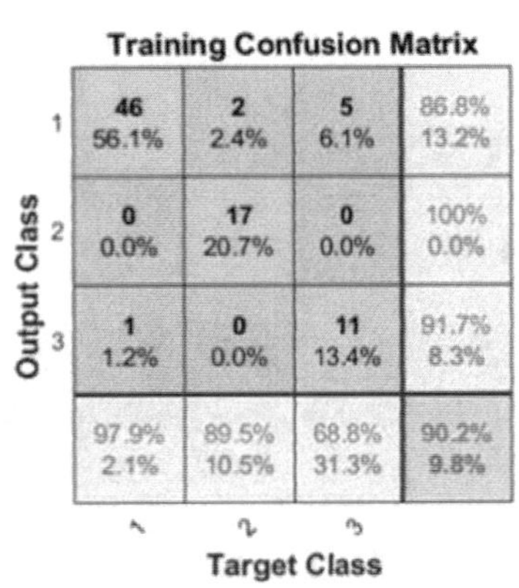

Figure 8. Training confusion matrix.

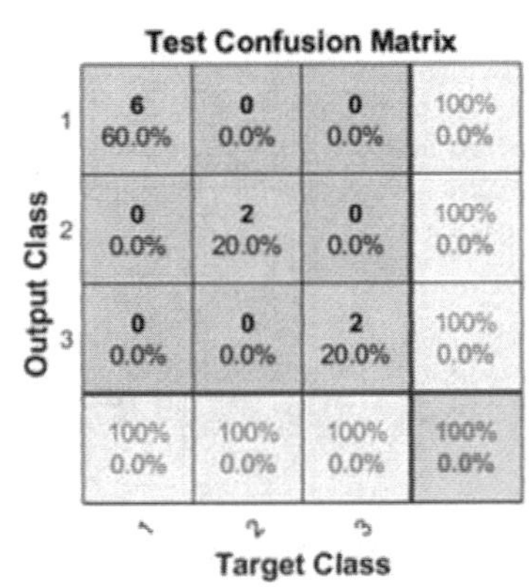

Figure 9. Testing confusion matrix.

4. Conclusions and future work

This paper presented the development of a rolling element bearing fault classification approach based on an ANN model. The ANN model uses as inputs features that are extracted from the time and frequency domains and correspond to the entire signal range as well as to specific frequency bands. In summary, the following conclusions can be drawn:

- The presented approach is generic and can be applied to any rolling element bearing fault classification problem.
- The extracted features represent the entire time domain as well as specific frequency bands that are manually identified by observing FFT obtained spectrums and kurtograms.
- The final ANN model is able to classify with very high accuracy (90.2% in training data and 100% in testing data) the different fault types of rolling element bearings.

In terms of future work, the two main directions that will be explored concern the investigation of dimensionality reduction techniques to decrease the number of ANN inputs without loss of classification performance and the automated kurtogram evaluation to systematically identify critical frequency bands.

References

[1] Simeu-Abazi Z, Iung B, Leger JB, Ly F, 2000. Maintenance in Manufacturing Systems. *IFAC Proceedings Volumes*, **33(17)**, pp 831-36.

[2] Gardner, 2015. International machine tool production and consumption (Metalcutting and Metal Forming). Available online at: http://www.gardnerweb.com/forecast/international.htm

[3] Abele E, Altintas Y, Brecher C, 2010. Machine tool spindles. *CIRP Annals*, **59**, pp 781-802.

[4] Fleischer J, Broos A, Schopp M, Wieser J, 2008. Sustainable design of machine tools through load-dependent interventions and adapted services. *Manufacturing Systems and Technologies for the New Frontier*, Springer London, pp 173-76.

[5] Neugebauer R, Fischer J, Praedicow M, 2011. Condition-based preventive maintenance of main spindles. *Production Enginnering*, **5**, pp 95-102.

[6] de Castelbajac C, Ritou M, Laporte S, Furet B, 2014. Monitoring of distributed defects on HSM spindle bearings, *Applied Acoustics*, **77**, pp 159-68.

[7] Vogl GW, Donmez MA, 2015. A defect-driven diagnostic method for machine tool spindles, *CIRP Annals*, **64**, pp 377-80.

[8] Nguyen PH, Kim J-M, 2015. Multifault diagnosis of rolling element bearings using a wavelet kurtogram and vector median-based feature analysis, *Shock and Vibration*, article ID 320508.

[9] Dong S, Yin S, Tang B, Chen L, Luo T, 2014. Bearing degradation process prediction based on the support vector machine and Markov model. *Shock and Vibration*, article ID 717465.

[10] Ahmad W, Khan SA, Kim J, 2017. A hybrid prognostics technique for rolling element bearings using adaptive predictive models, *IEEE Transactions on Industrial Electronics*, **65(2)**, pp 1577-84.

[11] Gunerkar RS, Jalan AK, Belgamwar SU, 2019. Fault diagnosis of rolling element bearing based on artificial neural network, *Journal of Mechanical Science and Technology*, **33**, pp 505-11.

[12] Hoang D-T, Kang H-J, 2019. Rolling element bearing fault diagnosis using convolutional neural network and vibration image, *Cognitive Systems Research*, **53**, pp 42-50.

[13] Rao BKN, Srinivasa Pai P, Nagabhushana TN, 2012. Failure diagnosis and prognosis of rolling element bearings using aritificial neural networks: a critical overview, *Journal of Physics: Conference Series*, **364**, 012023.

[14] Qiu H, Lee J, Lin J, Yu G, 2006. Wavelet filter-based weak signature detection method and its application on rolling element bearing prognostics, *Journal of Sound and Vibration*, **289(4-5)**, pp 1066-90.

[15] Parikh VN, Baraniecki AZ, 2016. Computationally efficient discrete wavelet transform model, *International Journal of Modelling and Simulation*, **16(3)**, pp 142-50.

[16] Antoni J, 2007. Fast computation of the kurtogramfor the detection of transient faults, *Mechanical Systems and Signal Processing*, **21(1)**, pp 108–24.

[17] Lee J, Qiu H, Yu G, Lin J, Rexnord Technical Services, 2007. Bearing data set. University of Cincinnati, Available online at: https://ti.arc.nasa.gov/tech/dash/groups/pcoe/prognostic-data-repository/.

FEM based investigation on thrust force and torque during Al7075-T6 drilling

Anastasios Tzotzis[1], Angelos P. Markopoulos[2], Nikolaos E. Karkalos[2], Dimitrios Tzetzis[3] and Panagiotis Kyratsis[4,*]

[1]University of Zaragoza, Department of Design and Manufacturing, C/Maria de Luna 3, 50018 Zaragoza, Spain
[2]National Technical University of Athens, School of Mechanical Engineering, Heroon Polytechniou 9, 15780 Zografou, Athens, Greece
[3]International Hellenic University, School of Science and Technology, 14km Thessaloniki - N. Moudania, 57001, Thermi, Greece
[4]University of Western Macedonia, Dept. of Product and Systems Design Engineering, Kila Kozani, GR50100, Greece

[*]Corresponding author's e-mail: pkyratsis@uowm.gr

Abstract. As modern industry advances, the demand for more time and cost effective machining is rising. In order to achieve high levels of standard during machining it is necessary to employ sophisticated techniques for precise prediction of various important parameters that relate to the machining processes. Such technique is the implementation of finite element modelling (FEM) which can become a valuable tool for researchers and industry engineers alike. In this work, the 3D modelling of Al7075-T6 drilling process with solid carbide tooling is being presented. DEFORM3D™ finite element analysis (FEA) software was utilized for simulating the drilling process based on frequently used cutting conditions; cutting speed of 100m/min and feed of 0.15mm/rev, 0.20mm/rev and 0.25mm/rev respectively. In order to approximate the complex phenomena that occur during drilling, the most critical factors were considered in the presented model such as the developed friction, heat transfer and damage interaction between the tool and the workpiece. Additionally, a validation of the generated results for thrust force and torque was performed by comparing the simulated results with experimental data. Three drilling experiments were carried out with the aid of a CNC machining center and a four component dynamometer in order to acquire the experimental values of thrust force and torque. Most of the simulations yielded results in accordance to the experimental ones with the agreement percentage reaching 95% in most cases for both the thrust force and torque, confirming the validity of the models and the accuracy of the simulated results.

1. Introduction

Finite element modelling (FEM) of cutting processes is a relatively new area of research with an increasing interest. Especially, the investigation of standard machining processes such as turning, drilling and milling seems to benefit from the employment of FEM and the recent advantages.

Maranhão and Davim [1] investigated the thermo-mechanical behavior when machining stainless steel AISI 316 to determine the influence of the friction coefficient in the tool-chip interface on several parameters such as cutting and feed forces, cutting temperature, plastic strain, plastic strain rate, maximum shear stress and residual stresses. In addition, authors performed a number of experiments for validation of the numerical results. Maurel-Pantel [2] realized simulations of shoulder milling operations on AISI 304L stainless steel using commercial software. Moreover, authors carried out a number of experiments in shoulder milling configurations in order to study the machinability of the material and also to validate the results obtained from the developed 3D FE model. Attanasio et al. [3] simulated the tool wear in drilling of Inconel 718 to study its influence on tool life, on final part quality and on cutting force and power consumption. Dou et al. [4] developed a prediction model based on a new constitutive model for Al6063 to predict thrust force and torque in drilling, which involves in the volume fraction of the material reinforcement. Additionally, authors established a 3D finite element (FE) model for simulating the drilling process of SiCp/Al6063 by using the new constitutive model. Nan et al. [5] studied the three-dimensional FE modeling for simulating the small-hole drilling process of AISI 1045. The verification of the model was performed via drilling experiments with 3mm diameter solid carbide drills at different rotational speeds and feeds. Xiang et al. [6] worked on a universal method that can be used to determine whether the constitutive formalism is suitable to describe the material constitutive behavior by measuring goodness-of-fit. A quantitative comparison of different fitting strategies on identifying Al6063/SiCp's material parameters was made in terms of performance evaluation including noise elimination, correlation, and reliability. Lastly, authors developed a 3D FE model in small-hole drilling of Al6063/SiCp composites to compare with the experimental observations in thrust force, torque, and chip morphology. Tzotzis et al. employed 3D FEM for developing prediction models for the generated cutting forces during turning of AISI-4140 [7] and drilling of Al7075-T6 [8] respectively.

In the present study, a 3D FEA was realized with DEFORM3D™ for investigation purposes of the Al7075-T6 behavior during drilling. Furthermore, validation of the numerical results was performed via experimental testing. With the utilization of a rotating dynamometer, the thrust force and the generated torque were determined and compared to the numerical values obtained from the 3D FE model. The comparison indicates that the FE modelling of drilling is feasible and yields accurate results.

2. Materials and Methods

2.1. *Overview of the experimental process*

In order to acquire the experimental results, three drilling experiments were conducted. A CNC machining center was employed for improved accuracy of the experiments. A rectangular piece (150mm x 130mm x 15mm) of Al7075-T6 alloy was chosen to serve as the workpiece, whereas a solid carbide drill from Kennametal (product code: B041A1000CPG - KC7325 grade), was used as the tool for the drilling experiments. The used tool contains 10% cobalt and is double coated. Moreover, the outer layer of the coating is a layer of TiN with 15 micron thickness and the inner layer of the coating is a multilayer of TiN/TiAlN with 3.5 micron thickness. In addition, Table 1 presents the most important geometric aspects of the used tool.

Table 1. Geometric parameters of B041A1000CPG drill.

Product code	Cutting diameter	Point angle	Helix angle	Chisel edge length	Point height
B041A1000CPG	10mm	140°	30°	0.35mm	1.71mm

A Kistler 9123C rotating four-component dynamometer was utilized to acquire both the thrust force and the produced torque during the experiments. The experimental values were monitored with Dynoware software (type 2825D-02) via a multichannel charge amplifier and a standard data acquisition system. Additionally, the sampling rate was set to approximately 10 kHz.

Figure 1 depicts the physical drill used in this study (a), the HAAS VF1 CNC machining center (b) and the setup of the tool – dynamometer – workpiece (c).

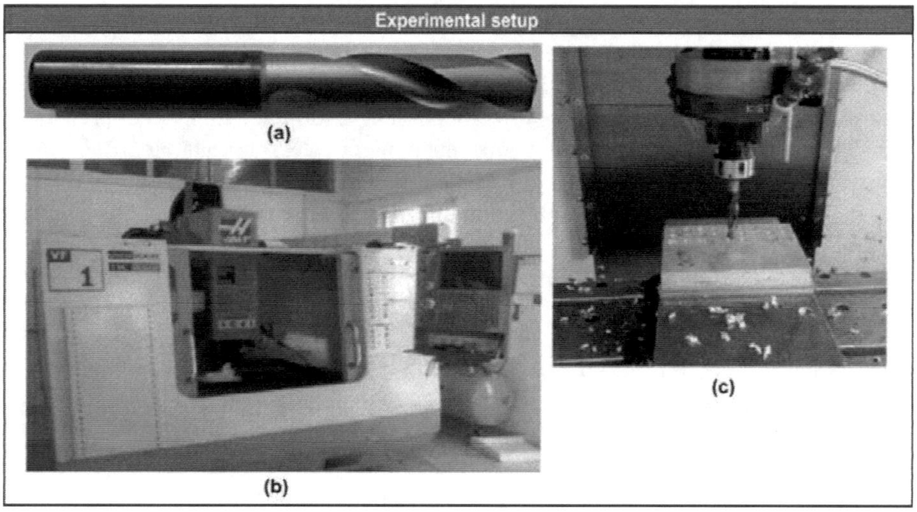

Figure 1. The setup of the experimental process.

2.2. *Overview of the simulation process*

The 3D cutting simulations of the drilling experiments were carried out with the commercially available finite element analysis software DEFORM3D™ ver. 11.3. To accelerate the process, the simulations were halted when the thrust force and thus the torque reached steady state. In order to simplify the workpiece model it was designed as a thin cylindrical piece. Due to the fact that the area of interest on the workpiece is located in the middle (based on the circular design), a spot according to the point angle of the tool was created (see Table 1). This design ensures that no time will be lost during the penetration of the drill bit to the workpiece. It was modelled as plastic and meshed with about 200000 elements. Moreover, a ratio of 10:1 was used for the localized meshing of the center spot, for increased accuracy (Figure 2b). On the other hand, the tool was modelled to be fully rigid and approximately 25000 triangles were used for the mesh. Figure 2a depicts the CAD model of the used tool. A 4:1 ratio was applied to the tip (Figure 2b) because most of the phenomena that occur during drilling are located in this area. Finally, Figure 2c illustrates the tool - workpiece model setup; to further simplify the problem only the tip of the drill was used.

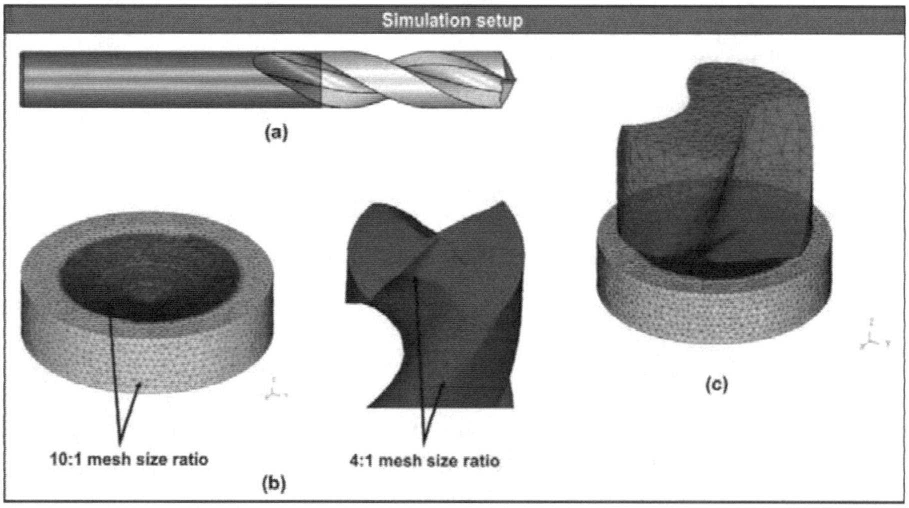

Figure 2. The setup of the simulation process.

The Johnson-Cook flow stress model is widely adopted when the problem involves high strain, strain rate and temperatures, therefore it was used in the present numerical study. In addition, the material constants have been determined for a large number of materials that exist in current literature. Equation 1 describes the aforementioned model [9], whereas Table 2 includes most of the important material properties of the Al7075-T6 alloy, as well as the constitutive model parameters.

$$\sigma = \left(A + B\varepsilon^n\right)\left(1 + C\ln\frac{\dot{\varepsilon}}{\dot{\varepsilon}_0}\right)\left[1 - \left(\frac{T - T_0}{T_m - T_0}\right)^m\right] \tag{1}$$

Table 2. Basic thermo-mechanical properties [10] and model constants of Al7075-T6 [11].

Al7075-T6 properties						
Young's modulus [GPa]		Density [kg/m³]	Poisson's ratio	Thermal expansion [μm/m°C]	Thermal conductivity at 20°C [W/mK]	
71.7		2810	0.33	$2.2\text{x}10^{-5}$	130	
Constitutive model						
A [MPa]	B [MPa]	C	n	m	T_0 [°C]	T_m [°C]
546	678	0.024	0.71	1.56	20	635

In equation 1 A is the initial yield stress, B is the strain hardening modulus, C is the strain rate dependence coefficient, ε denotes the plastic strain, n is the strain hardening exponent, m represents the thermal softening coefficient, $\dot{\varepsilon}$ is the plastic strain rate, $\dot{\varepsilon_0}$ is the reference plastic strain rate, T is the reference temperature, T_0 is the bulk temperature of workpiece material and finally T_m stands for the melting temperature of workpiece material. In the present numerical study, a reference strain rate of 1/s was used.

In order to approximate the physics of material separation that occur in the deformation zones, the damage model of Cockcroft and Latham was selected [12]. This model depends on the maximum principal stress. The contact pressure at the tool-workpiece interface generates frictional forces that can be determined with the aid of a hybrid model [13]; the constant shear friction coefficient is $m = 0.7$ for the sticking zone and the constant Coulomb friction coefficient is $\mu = 0.6$ for the sliding zone.

The boundary conditions and the movement controls were set in such a manner that the workpiece remained fixed and the tool could rotate around Z axis based on the given cutting speed. In addition, linear motion along –Z axis was applied to the tool so that it could travel towards the workpiece with the given feed. Furthermore, the heat exchange between the workpiece and the environment was defined, as well as the heat exchange between the tool tip and the workpiece (contact region). The default coefficient values for heat transfer were applied to the model for both convection and conduction.

3. Results and discussion

The cutting conditions that were applied to both the experiments and the simulations are: cutting speed of 100m/min combined with three different feeds (0.15mm/rev, 0.20mm/rev and 0.25mm/rev respectively). The produced numerical results of thrust force and torque and the equivalent diagrams versus time were exported for further study. To avoid any unrealistic values for thrust force and torque due to remeshing, the first order exponential smoothing was applied to the simulated results. Figure 3a depicts a sample thrust force diagram, whereas Figure 3b illustrates a sample torque diagram. It is pointed out that both the thrust force and the torque for the given conditions increase rapidly until the point where steady state occurs (at approximately 0.01 sec). Moreover, this trend was noticed in all three simulations.

The obtained numerical results were compared to the experimental ones for validation purposes. Comparison study shows a good agreement between the numerical results and the experimental ones for both the thrust force and the torque. The relative error for F_z was calculated 0.81% for the first combination of cutting conditions (V_c = 100m/min and f = 0.15mm/rev), whereas for the other two (V_c = 100m/min and f = 0.20mm/rev, f = 0.25mm/rev) was found to be –3.4% and –4.7% respectively. Similarly, the relative error for M_z was determined 2.1%, 1.2% and –3.3% accordingly for each of the three aforementioned cases. Furthermore, the value of standard deviation of F_z for the three simulation tests was calculated 47.9N, 28.4N and 64.6N accordingly. On the other hand, standard deviation for M_z was determined 132.9Nmm, 146.8Nmm and 164.3Nmm for each of the cutting conditions combination. Finally, it is noticed that the maximum value of simulated thrust force (F_z = 751.6N) and torque (M_z = 3269Nmm) was found to be in the test with the highest value of feed (f = 0.25mm/rev), similarly to the experiments. This fact indicates that the increase of feed from 0.15mm/rev to 0.25mm/rev rises F_z by 31% and M_z by 46.7%.

Figure 3c presents the comparison between the numerical and the experimental results for thrust force *Fz*, in the same way, the chart of Figure 3d illustrates how the simulated values of torque M_z correlate to the experimentally acquired values.

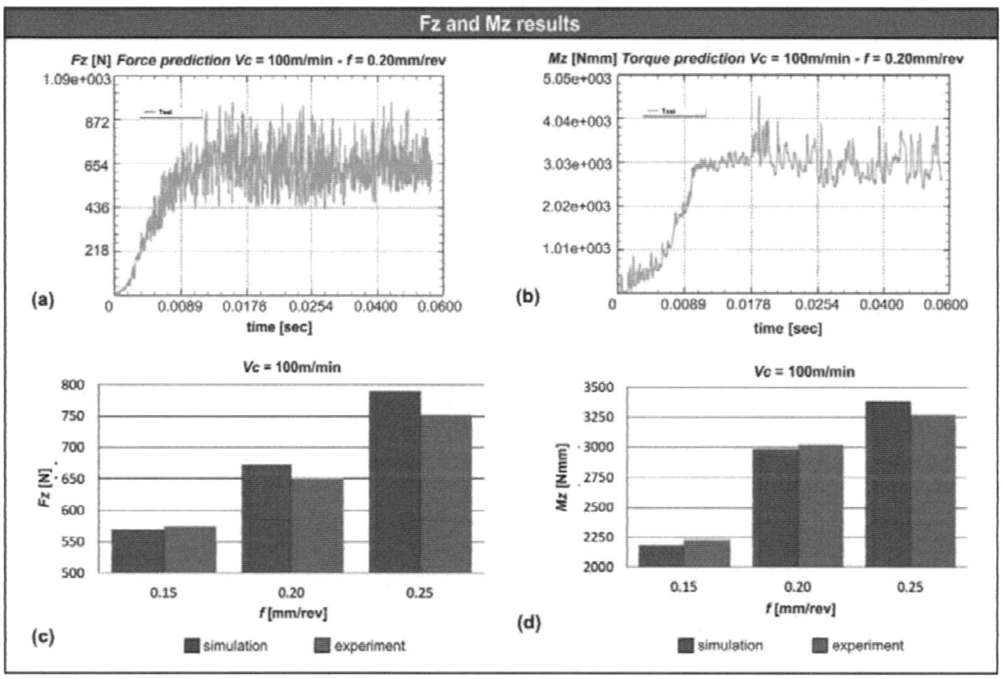

Figure 3. Comparison between the simulated and the experimental results.

4. Conclusions

In the present study, three drilling simulation tests were performed with the implementation of 3D FE modelling. The selected workpiece material is Al7075-T6 alloy and the used tool a solid carbide drill of 10mm diameter. Moreover, the chosen cutting conditions are a combination of cutting speed of 100m/min and feed of 0.15mm/rev, 0.20mm/rev and 0.25mm/rev respectively. The generated numerical results for thrust force and torque were verified via experimental testing, the comparison study showed that the simulated results are in good accordance with the experimental ones. Eventually, the findings of the numerical study lead to the following conclusions:

- Feed strongly affects the produced thrust force and torque. Specifically, an increase in feed from 0.15mm/rev to 0.25mm/rev and for the same cutting speed, increases thrust force by 31% and torque by 46.7% correspondingly.

- The relative error for both thrust force and torque was found to be relatively low which proves the accuracy of the 3D FE model.
- Moreover, the standard deviation values for F_z were determined between 28.4N to 64.6N and for M_z between 132.9Nmm to 164.3Nmm which further proves the validity of the 3D FE model.

References

[1] Maranhão C and Davim JP 2010 *Simul. Model. Pract. Theory* **18** 2
[2] Maurel-Pantel A, Fontaine M, Thibaud S and Gelin JC 2012 *Simul. Model. Pract. Theory* **22**
[3] Attanasio A, Faini F and Outeiro JC 2017 *Procedia CIRP* **58**
[4] Dou T, Fu H, Li Z, Ji X and Bi SS 2019 *Int. J. Adv. Manuf. Technol.* **103** 1-4
[5] Nan X, Xie L and Zhao W 2016 *Int. J. Adv. Manuf. Technol.* **84** 9
[6] Xiang J, Pang S, Xie L, Gao F, Hu X, Yi J and Hu F 2018 *Materials (Basel)* **11** 1
[7] Tzotzis A, García-Hernández C, Huertas-Talón JL and Kyratsis P 2020 *Micromachines* **11** 9
[8] Tzotzis A, Markopoulos A, Karkalos N and Kyratsis P 2020 *MATEC Web Conf.* **318**
[9] Tzotzis A, García-Hernández C, Huertas-Talón JL and Kyratsis P 2020 *Strojniški Vestn. - J. Mech. Eng.* **66** 7-8
[10] Kyratsis P, García-Hernández C, Vakondios D and Antoniadis A 2016 *Design of Experiments in Production Engineering* (Cham: Springer)
[11] Brar NS, Joshi VS and Harris BW 2009 *AIP Conf. Proc.* **1195** 1
[12] Cockcroft MG and Latham DJ 1968 *J. Institue Met.* **96**
[13] Tzotzis A, García-Hernández C, Huertas-Talón JL and Kyratsis P 2020 *Mech. Ind.* **21** 4

IOP Conf. Series: Materials Science and Engineering 1037 (2021) 012010 doi:10.1088/1757-899X/1037/1/012010

Improving the performance characteristics of human hip-joint implants by increasing the quality of processing and geometric accuracy of their spherical surfaces

Raul Turmanidze[1*], Giorgi Popkhadze[1], Ketevan Inasharidze[2]

[1]Georgian Technical University (GTU), Tbilisi, Georgia
[2]Batumi Shota Rustaveli State University (BSU), Batumi, Georgia

[*]Corresponding author's e-mail: inform@gtu.ge

Abstract. In view of the fact that the endo-prosthesis heads of human hip-joint are operated in extreme conditions, in respect of load, the selection of corresponding material and also increase of precision and quality of machining of spherical surfaces is rather topical task. In the submitted work are reviewed the problems connected with definition of the influence degree of orientation of the sapphire crystal on its workability during diamond grinding with a butt of the ring and elaboration of the perspective, original scheme of formation of the incomplete spherical surface, particularly, of the sapphire head of endo-prosthesis of the human hip-joint.

1. Introduction

In the last thirty years, one of the current trends in the field of material processing by cutting is the production of implants from different materials for hip-joints with high performance properties, which will ensure their use for more than 30-40 years. This is due to the fact that the heads of the said endoprosthesis, in terms of the nature and magnitude of the load, are operated under extreme conditions. Therefore, in each specific case, the selection of the necessary material with the appropriate physical and mechanical characteristics, as well as improving the accuracy and quality of processing of the most significant part of the endoprosthesis - spherical surfaces, is a very important task, the severity of which has been rapidly increasing in recent years. This is due to the fact that if earlier the need for such operations was mainly caused by the age factor of a person or traumatic fractures, then in the last twenty years the number of patients at a young age has sharply increased, even at 30-40 years old, both men and women, without any injuries or fractures [1, 2, 3, 4].

According to doctors, the main reasons for this disaster are the inactive lifestyle of young people, the composition of modern artificial foods and metabolic disorders, as well as the intensive growth of various fractures as a result of transport and other accidents. All of the above reasons determine the number of endoprostheses used - several tens of millions of pieces per year and statistics show that, unfortunately, this number is increasing annually [2, 5, 6].

2. Analysis of literature data and problem statement

Medical practice proves that the repeated prosthetics of the human hip joint is associated with great problems. In many cases, the implementation of such transactions becomes almost impossible. Therefore, the service life of a hip joint endoprosthesis until the end of life for a patient, especially after operations at a young age, is extremely important [1, 3, 4, 5, 7, 8, 9, 10].

This circumstance explains the fact that in such leading countries of the world as the USA, Germany, England, Japan, France, etc., large-scale research work is being carried out to develop optimal shaping schemes, characteristics of abrasive tools for finishing operations and the technological process as a whole, for the manufacture of spherical endoprosthesis heads from different materials with a minimum form error and with high surface quality indicators.

In addition, the wear resistance of the aforementioned materials in most cases does not meet the requirements for them. And our comparative experiments prove that the most biocompatible with the human body, wear-resistant and durable material for the production of the above-mentioned product is an artificial monocrystal of sapphire [3, 7, 8, 10, 11, 12].

Today, in world practice, these heads are made from various alloys, composite materials and ceramics, which are mainly isotropic materials. Therefore, the data of the above-mentioned works do not provide the necessary information on the processing of anisotropic materials, in particular, an artificial sapphire crystal.

In this regard, an international project was carried out by the Ukrainian Science and Technology Center, the executors of which were the Georgian Technical University (Tbilisi), the Institute of Super hard Materials of the National Academy of Sciences of Ukraine (Kiev) and the Institute of monocrystal of the National Academy of Sciences of Ukraine (Kharkov).

The project participants solved certain scientific problems, in particular: the Georgian Technical University investigated the effect of sapphire crystal anisotropy on the machinability and quality of the material surface during grinding; promising, theoretical schemes for the shaping of an incomplete spherical surface have been developed in order to improve the geometric accuracy of the spherical part of the endoprosthesis, which is of fundamental importance for increasing the service life of their operation by increasing the contact area between pairs of endoprostheses and, accordingly, reducing the load per unit area their contact surfaces. The Institute for Super hard Materials investigated the relationship between the friction coefficient and the crystallographic features of sapphire and annealing modes; machinability of materials with anisotropy of properties, in particular sapphire, according to traditional technology to assess the effect of anisotropy on the accuracy of shaping a spherical surface; Recommendations have been developed for the process of diamond finishing of the heads of endoprostheses of the hip joint made of artificial sapphire monocrystal. The Kharkov Institute of monocrystals investigated the optimal growth regimes that ensure the maximum purification of the sapphire material, as a result of which a sapphire with a crystallographic direction having a minimum anisotropy was obtained.

At the same time, prototypes of implants made by us within the framework of the project were used by Kharkov orthopedists in real operations and the Kharkov participants of our project have been observing them together with surgeons until today.

3. The purpose and objectives of the study

The aim of the presented work is: to determine the degree of influence of the orientation of the sapphire crystal on its processing by grinding by the NSP method created at the Department of Mechanical Engineering Technology of the Georgian Technical University and the development of a promising, original scheme the shaping of an incomplete spherical surface, in particular, a sapphire head of a human hip joint endoprosthesis. A detailed description of the low-temperature grinding process (LGP) method, the kinematic diagram of the laboratory setup and the experimental technique can be found in previous publications [13, 14, 15, 16].

Studies of the influence of the orientation of the sapphire crystal on the workability of the material were carried out on sapphire samples with the orientation (0001), (1010) and (1012). Sample sizes: 10x10x6 mm and Φ 10x6 mm.

Experimental studies were carried out on a laboratory setup equipped with a special precision head. We used diamond wheels with grain size 14/10 and 28/20 on ceramic, metal and organic bonds, form 6A2.

The output parameters of the process were: processing productivity, linear minute material removal - q, μm / min; height of surface roughness irregularities - Rz, μm; relative reference profile length at level 03 - t_{P03}, %; under-relief damaged layer - H, μm.

Process factors cutting speed - V, m / s; pressure in the cutting zone P, kPa and characteristics of the diamond tool: grain size - d_3, μm; concentration bundle - K, %.

The experiments were carried out under the following conditions: cutting speed range - V = 1 ... 12 m / s; pressure in the cutting zone - P = 100 ... 1500 kPa. Cooling liquid - filtered running water.

Such a methodological approach, in which almost all possible orientations of the processed material are investigated - sapphire crystal, diamond-abrasive tools with different performance characteristics in terms of the grain size of diamond grains and the type of bond, as well as different processing modes and repetition of each experiment at least 8-10 times provide high reliability of determining the above output parameters and, accordingly, generated conclusions.

Based on the analysis of the data of our complex, comprehensive experimental studies, we can draw the following conclusion:

The nature of the influence of process factors on the output parameters for the selected orientations of the sapphire crystal ((0001), (1010), (1012)) is constant [3, 6].

All other things being equal, the most difficult to machine is the orientation (0001). For all tested diamond wheels, the ratio of linear material removal - q is within the range

$$q_{0001} / q_{1010} = 0,25 \ldots 0,5,$$

while,

$$q_{1012} / q_{1010} = 0,75 \ldots 1.$$

In the studied ranges of cutting modes V and P, the processing productivity increases at V = ... 6 m / s, and at V> 6 m / s - it remains constant, also at P = 100 ... 1500 kPa - it grows, at P> 1500 kPa - remains almost constant.

Of the characteristics of a diamond tool, the grit size and bond of the diamond tool predominantly affect productivity. The effect of concentration is negligible. With an increase in the grain size within d_3 = 14/10 ... 28/20, the productivity increases by 1.5 ... 2.5 times.

The maximum value of productivity is achieved by a tool on a ceramic bond - SK6 - 130 ... 300 µm / min, for all selected orientations of the sapphire crystal. This increases the ratio

$$q_{0001} / q_{1010} = 0,4 \ldots 0,5.$$

The tool on this bond works in self-sharpening mode.

All other things being equal, processing on the (0001) orientation achieves a higher surface quality than on the other two. The difference is in 1 ... 1.5 grades of roughness.

The surface quality is predominantly influenced by the grit size and the material of the tool bond. With an increase in the grain, in the investigated range, the height of the irregularities Rz grows within 1 ... 1.5 class, and the depth of the disturbed layer H - 1.5 ... 2 times.

In terms of the quality of the treated surface, the best results are obtained by diamond wheels on organic bonds VS-11 and organic special ones, which have been developed by us and are in the process of patenting. Compared to diamond wheels on metal and ceramic bonds, the Rz parameter is an order of magnitude lower ~ 0.25 µm, the t_{P03} parameter is 1.5 higher than ~ 35 ... 45% and the H parameter in 3 ... 5 is less than ~ 2...5 µm.

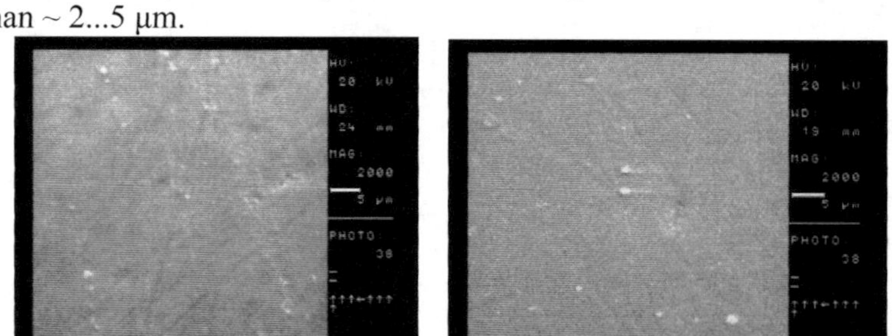

a) b)

Fig. 1. Micrographs of surfaces of experimental sapphire samples processed by the NSP method:

a- Orientation 1010, b- 1012. Diamond wheel - ACM 14/10, special organic bond, 50%. Cutting conditions: V = 1m / s, P = 750 kPa.

IOP Conf. Series: Materials Science and Engineering 1037 (2021) 012010 doi:10.1088/1757-899X/1037/1/012010

The study of the morphology of the treated surface proved the possibility of cutting sapphire material by plastic deformation of the removed layer at low cutting speeds V = 1 ... 3 m / s, with the smallest depth of the damaged under-relief layer (Fig. 1).

This result is worthy of special attention, since the processing of glassy materials, in particular, a sapphire crystal, by plastic deformation of the removed layer is a guarantee of obtaining a treated surface with practically no hereditary defects - without a disturbed under-relief layer. The value of H turned out to be the smallest in these sapphire samples.

The development of a new, or improvement of the existing technological process of mechanical processing of a sapphire head poses an urgent task of creating new, highly efficient shaping schemes. Optimization criteria for technological operations, such as productivity, surface quality indicators and, most importantly, processing accuracy, the spherical surface of the implants. As shown above, this parameter basically determines the service life of each pair of implants. Therefore, when evaluating new effective schemes of shaping in the technological process or when comparing with existing ones, the advantage is determined by this parameter [5, 17, 18, 19, 20].

The methods of abrasive processing of an incomplete spherical surface of general-purpose products used today, as separate operations of the technological process of machining an incomplete sphere, can be divided into two groups: grinding with abrasive diamond wheels, finishing with a free abrasive [21, 22, 23, 24].

Our method belongs to the first group, however, as a method for diamond abrasive processing of flat surfaces, in contrast to existing methods, it combines the positive aspects of the methods of the above two groups, namely: with high processing accuracy and surface quality.

To develop a theoretical scheme for diamond processing of sapphire spheres, taking into account the kinematic and other positive features of the NSP, the known schemes for the formation of an incomplete spherical surface were considered and analyzed.

The closest in kinematics to the NSP method is the method of grinding an incomplete sphere with the end of a grinding wheel with double rotation of the part with angular velocities ω_2 and ω_3 (Fig. 2), which is taken as the basis for the development of an original version of the theoretical scheme of diamond processing of sapphire spheres, taking into account the kinematic and other positive features of the NSP method.

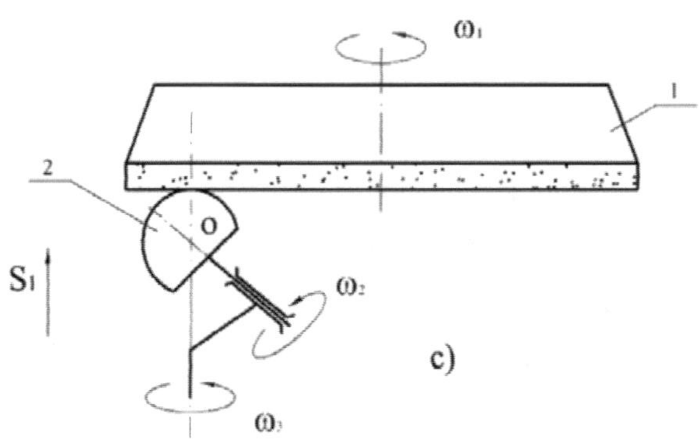

Fig. 2. Scheme of forming an incomplete spherical surface with an end diamond wheel with double rotation of the part:
1 - diamond grinding wheel;
2 - work piece - incomplete sphere.

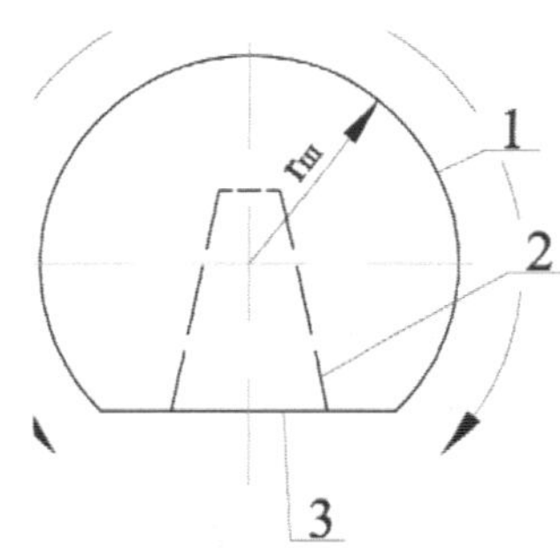

Fig. 3. Sapphire Head:
1- Spherical surface;
2- Conical holder;
$r_{Ш}$ is the radius of the sphere;
γ is the angle of the spherical surface segment.

We have developed several versions of theoretical schemes for the formation of an incomplete spherical surface. One version of the original kinematic scheme for shaping an incomplete spherical surface of a sapphire head (Fig. 3) by an end grinding wheel is shown in (Fig. 4). The cutting tool is a special, combined face grinding wheel with two concentric diamond-bearing layers, with cutting surfaces in the form of internal, at point A, and external, at point B, cut off cones.

The grinding wheel rotates with an angular velocity ω_1. The workpiece being machined - the ball - performs a double rotary motion with angular velocities ω_2 around its own axis 4 and ω_3 around axis 3, which is the axis of the part's spindle. The direction of the angular velocities ω_1, ω_2 and ω_3 is the same. The workpiece 2 to be machined is pressed against the cutting surface of the grinding wheel at two points A and B by a spring-loaded force P. In this case, the cutting surfaces of the grinding wheel in the axial section form an angle β.

Distinctive features of the theoretical scheme from the NSP such as the shape of the working surface of the tool, additional movement of the workpiece - rotation of the ball around its own axis at a speed ω_2, the form of contact of the workpiece surface with the cutting surface of the grinding wheel.

To create equal processing conditions at two points A and B, it is necessary to observe the equality of both the speeds and the cutting forces.

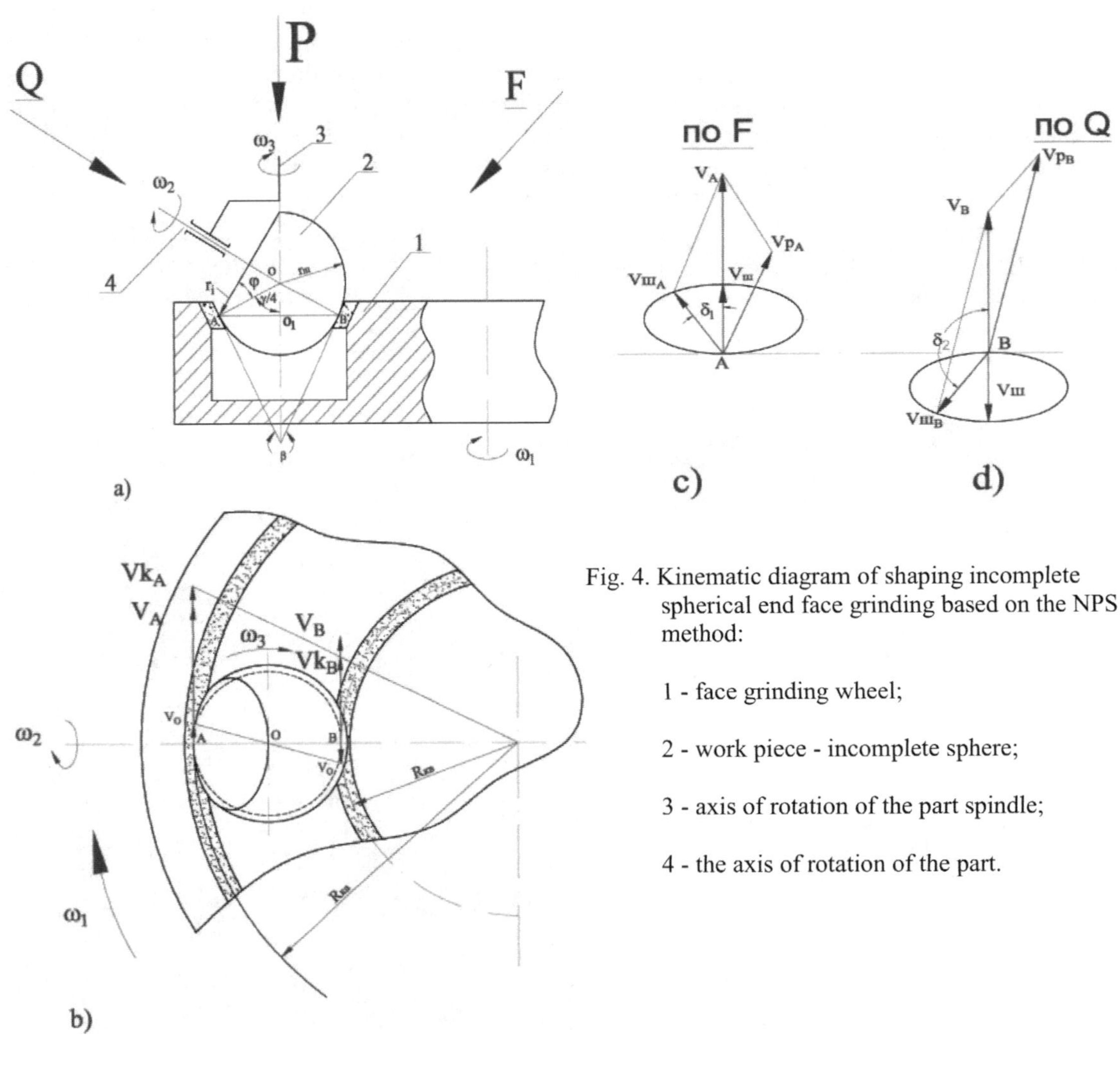

Fig. 4. Kinematic diagram of shaping incomplete spherical end face grinding based on the NPS method:

1 - face grinding wheel;

2 - work piece - incomplete sphere;

3 - axis of rotation of the part spindle;

4 - the axis of rotation of the part.

Equality of processing conditions in terms of cutting speed means equality of the maximum values of cutting speeds and the identity of their laws of change in the cycle in two cutting zones.

Cutting speeds when grinding an incomplete sphere according to the proposed scheme at points A and B are given in Fig. 4 c, c, and d.

Resulting cutting speed v_{PA}- at point A

$$\vec{v}_{PA} = \vec{v}_B - \vec{v}_{ШВ} \tag{1}$$

$$v_{P_A} = \sqrt{v_A^2 + v^2_{ш_A} - 2 v_A \cdot v_{ш_A} \cdot \cos\delta_1}\ , \tag{2}$$

where

$$v_A = v_{KA} - v_0 = R_{KA} \cdot \omega_1 - r_ш \sin\frac{\gamma}{4} \cdot \omega_3 \tag{3}$$

and

$$v_{ШA} = r_{iA} \cdot \omega_2\ . \tag{4}$$

Similar to point A, at point B the resulting cutting speed is

$$\vec{v}_{PB} = \vec{v}_B - \vec{v}_{ШВ}\ , \tag{5}$$

$$v_{P_B} = \sqrt{v_B^2 + v^2_{ш_B} - 2 v_B \cdot v_{ш_B} \cdot \cos\delta_2}\ , \tag{6}$$

where

$$v_B = v_{KB} - v_o = (R_{KA} - 2r_ш \cdot \sin\frac{\gamma}{4}) \cdot \omega_1 + r_ш \cdot \sin\frac{\gamma}{4} \cdot \omega_3 \tag{7}$$

and

$$v_{ШB} = r_{iB} \cdot \omega_2\ . \tag{8}$$

Linear speed of the part from rotary motion with angular speed ω_2

$$v_{Шi} = r_i \cdot \omega_2 = \omega_2 r_Ш \cdot \cos\varphi = \omega_2 \cdot r_ш \sqrt{\sin^2\frac{\gamma}{4} \cdot \cos^2\alpha + \frac{1}{4} \cdot \sin^2\frac{\gamma}{2}(1+\sin\alpha)^2}\ , \tag{9}$$

Where is φ the angle of inclination of the axis 4 relative to the generatrix of the working surface of the circle in the axial section, α is the angle of rotation of the axis 4 relative to the axis 3 with an angular velocity ω_3.

The current radius of the point of the surface of the part, when you rotate axis 4 around axis 3 by 180^0, changes within $0 \le r_i \le r_ш$. Thus, for a complete description of a spherical surface during one cycle, one revolution of the part around axis 4 is sufficient, when axis 4 is rotated about axis 3 by half a revolution, i.e. by 180^0. Consequently, under these conditions, the processed spherical surface of the part is described twice during one cycle. Once in each treatment area.

Based on the foregoing, between the angular velocities ω_2 and ω_3 we have the dependence

$$\omega_2 = 2\,\omega_3. \tag{10}$$

v_{P_A} reaches a maximum at $\alpha = \pi\frac{3}{2}$, and therefore, $\delta_1 = \frac{\pi}{2}$.

$$v_{P_{Amax}} = v_A = R_{KA} \cdot \omega_1 - r_ш \cdot \sin\frac{\gamma}{4}\omega_3 \tag{11}$$

v_{P_B} reaches a maximum at $\alpha = \frac{\pi}{2}$, and therefore, $\delta_2=\pi$.

$$v_{P_{Bmax}} = R_{KA} \cdot \omega_1 - 2r_ш \sin\frac{\gamma}{4}\omega_1 + r_ш \cdot \sin\frac{\gamma}{4} \cdot \omega_3 + r_ш \sin\frac{\gamma}{2}\omega_2. \tag{12}$$

On condition $v_{P_{Amax}} = v_{P_{Bmax}}$, by expressions (10), (11) and (12) the ratio of angular velocities

and ω_3 и ω_1

$$\frac{\omega_3}{\omega_1}=\frac{1}{1+2\cos\frac{\gamma}{4}} \quad . \tag{13}$$

Thus, according to the proposed scheme for shaping an incomplete spherical surface, in order to achieve equality of processing conditions in terms of cutting speed, it is necessary to observe the ratios of angular speeds ω_1, ω_2 and ω_3, according to expressions (10) and (13).

Clamping force. The equality of conditions at two points in terms of the clamping force is achieved by the location of the workpiece in relation to the grinding wheel. The spring-loaded force P passes through the center O of the sphere and coincides with the bisector of angle β and thus passes at equal distances from two A and B processing zones. In each zone, equal efforts on P / 2 are created, which in turn are composed of two components, normal PN and tangential Pτ (not shown in the figure). Wherein:

$$P_N=\frac{P}{2}\cdot\cos\frac{\gamma}{4}, \qquad P_\tau=\frac{P}{2}\cdot\sin\frac{\gamma}{4}.$$

The kinematic similarity of the new shaping scheme with the LPS method, taking into account the preservation of the physics of the cutting process with the LPS, which determined its name "Low temperature" (low compared to conventional grinding, the cutting speed is -1.6 m / s and, as a consequence, the low temperature in contact is -100^{0}C), also "Precision" (high accuracy of the shape of the working surface of the tool is achieved during straightening and is maintained in operation by the kinematics of the process, which ensures high processing accuracy - plane-parallelism on a 10x10 mm <1μm plate), allows the proposed shaping scheme. The name, as coming from the first, is called Low-temperature precision grinding of an incomplete sphere, in abbreviated form - NPSHNS, which is in the process of patenting.

It is assumed that the NPSHNS method, in comparison with the traditional methods of diamond grinding, which are used today in the technological process of mechanical processing of a sapphire head, all other things being equal, will allow: to significantly (at least twice) increase the processing productivity without deteriorating performance surface quality and shape accuracy; to significantly improve the quality of the processed surface and the accuracy of the shape of the product, without increasing the productivity of processing, thereby also significantly reducing the operating allowance, as a result, the time and cost of processing in finishing operations (transitions).

Thus, in our opinion, the NPSHNS method can be quite competitive in preliminary operations, up to finishing, in the technological process of machining a sapphire head.

The long-term program provides for the organization of a European project with the creation of the necessary equipment, tools, technological equipment and the entire technological process for manufacturing high-precision spherical surfaces from various brittle materials of increased strength, including sapphire.

Conclusion

As a result of the study of the effect of the crystallographic orientation of a sapphire monocrystal on the rate of removal of the processed material, as well as the state of the subsurface layer during surface finishing by the method of low-temperature flat grinding, it was found that:

1. The processing capacity of a sapphire monocrystal significantly depends on the crystallographic orientation. The relative values of material removal for samples with different crystallographic orientations (1010) are in the range q_{0001} / q_{1010} = 0.25… 0.5, and q_{1012} / q_{1010} = 0.75… 1. Under other equal conditions of the LPS process, the highest quality of the polished surface is achieved for the crystalographic orientation (1010). The difference with the other two orientations of the crystallographic plane is in the range of 1 ... 1.5 roughness class.
2. The study of the morphology of the processed surface proved the possibility of cutting sapphire material by plastic deformation of the removed layer without cracking at a low cutting speed.

Under such machining conditions, the smallest depth of the damaged subsurface layer was found.

3. New kinematic schemes for the processing of incomplete spherical surfaces have been proposed, which provide a higher geometric accuracy of parts, in particular, the spherical heads of the human hip joint endoprosthesis.

Acknowledgment

This work was supported by Shota Rustaveli National Science Foundation (SRNSF) [PHDF-19-2224, Improving the efficiency of mechatronic systems in order to ensure the reform of "Industry-4.0"].

References

[1] Triclot P. Metal-on-metal: History, state of the art (2010) Int. Orthop. 2011;2:201–206. doi: 10.1007/s00264-0101180-8.
[2] Topolovec M., Cör A., Milošev I. Metal-on-metal vs. metal-on-polyethylene total hip arthroplasty tribological evaluation of retrieved components and periprosthetic tissue. J. Mech. Behav. Biomed. Mater. 2014;34:243–252. doi: 10.1016/j.jmbbm.2014.02.018.
[3] Turmanidze, R.; Dašić, P. & Popkhadze, G.: Selection criteria of optimal characteristic material and technologies for precision processing of basic working surface of human hip-joint implant. MATEC Web of Conferences, Vol. 184, (2018), Article no. 01015: pp. 1–6. eISSN 2261-236X. doi: 10.1051/matecconf/201818401015.
[4] Affatato S., Spinelli M., Zavalloni M., Traina F., Carmignato S., Toni A. Ceramic-on-metal for total hip replacement: mixing and matching can lead to high wear. Artif. Organs. 2010;34:319–323. doi: 10.1111/j.15251594.2009.00854.x.
[5] R. Turmanidze, G. Popkhadze, Selection criteria of optimal characteristic material and technologies for precision processing of basic working surface of human hip-joint implant. 9TH INTERNATIONAL CONGRESS ON PRECISION MACHINING. ATHENS, GREECE-2017, ISSN 0377-6883.
[6] R. Turmanidze, D. Butskhrikidze, E. Kutelia, M. Beridze, Influence of the Sapphire Crystal Anisotropy of Medical Purpose on Workability of the Endoprosthesis Material at a Low Temperature Precision Grinding. "MicroCAD 2008" International Scintific Confer¬ence. 20-21 March, 2008. Section N: Production Engine¬ering and Manufactu-ring Systems. University of Mishkolc, Hungary. ISBN 978-963-661-812-4.
[7] Ihaddadene R., Affatato S., Zavalloni M., Bouzid S., Viceconti M. Carbon composition effects on wear behaviour and wear mechanisms of metal-on-metal hip prosthesis. Comput. Methods Biomech. Biomed. Engin. 2011;14:33–34. doi: 10.1080/10255842.2011.591623.
[8] Affatato S., Spinelli M., Squarzoni S., Traina F., Toni A. Mixing and matching in ceramic-on-metal hip arthroplasty: an in-vitro hip simulator study. J. Biomech. 2009;42:2439–2446. doi: 10.1016/j.jbiomech.2009.07.031.
[9] Ching, H. A., Choudhury, D., Nine, M. J., and Abu Osman, N. A., 2014, "Effects of Surface Coating on Reducing Friction and Wear of Orthopaedic Implants," Sci. Technol. Adv. Mater., 15(1), p. 014402. 10.10 88/1468-6996/15/1/014402
[10] Balagna, C., Faga, M. G., and Spriano, S., 2012, "Tantalum-Based Multilayer Coating on Cobalt Alloys in Total Hip and Knee Replacement," Mater. Sci. Eng. C, 32(4), pp. 887–895. 10.1016/j.msec.2012.02.007
[11] Molli R.G., Lombardi A.V., Berend K.R., Adams J.B., Sneller M.A. Metal-on-metal vs Metal-on-improved polyethylene bearings in total hip arthroplasty. J. Arthroplast. 2011;6:8–13. doi: 10.1016/j.arth.2011.04.029.

[12] Wei, X., Li, W., Liang, B., Li, B., Zhang, J., Zhang, L., and Wang, Z., 2016, "Surface Modification of CoCr-Mo Implant Alloy by Laser Interference Lithography," Tribol. Int., 97, pp. 212–217. 10.1016/j.triboint.20 16.01.039

[13] Batiashvili B.I., Butskhrikidze D.S., Mamulashvili G.L., Mgaloblishvili O. B., Turmanidze R.S., Kromp K., Mills B., Steinkellner W., Schafler E., Rösel F. G., Peterlik H.. Evaluation of surface preparation techniques, SFG: Swing Frame Grinding and LPG: Low Temperature Precision Grinding, by comparison of results on alumni and siliconcarbide model materials. FRACTOGRAPHY OF ADVANCED CERAMICS. International Conference Stará Lesná, High Tatras, May 2001.

[14] Batiashvili B.I., Butskhrikidze D.S., Mamulashvili G.A, Turmanidze R.S., Kromp K., Mills B., Mgaloblishvili O. Technological Possibilities of Low Temperature Precision Grinding Process when Machining Hard and Brittle Materials. FRACTOGRAPHY OF ADVANCED CERAMICS, International Conference, Stará Lesná, High Tatras, May 2001.

[15] Turmanidze R.S, Burtskhrikidze D.S, Kromp K., Mills B., "Low temperature precision grinding of hard and brittle materials". Problems of mechanics and physiko-chemistry of the process of abrasive machining, Kiev 2002, 490-499 pp.

[16] Turmanidze R.S., Butskhrikidze D.S., Mamulashvili G.L., Kromp K., Mills B., Morgan M., Mgaloblishvili O. Low-temperature precision grinding of hard and brittle materials and Outlook of its development and application. Proceedings of 19th NCMR Conference. Glasgow, September 2003.

[17] Choudhury, D., Vrbka, M., Mamat, A. B., Stavness, I., Roy, C. K., Mootanah, R., and Krupka, I., 2017, "The Impact of Surface and Geometry on Coefficient of Friction of Artificial Hip Joints," J. Mech. Behav. Biomed. Mater., 72, pp. 192–199. 10.1016/j.jmbbm.2017.05.011

[18] Sedlaček, M., Podgornik, B., Ramalho, A., and Česnik, D., 2017, "Influence of Geometry and the Sequence of Surface Texturing Process on Tribological Properties," Tribol. Int., 115, pp. 268–273. 10.10 16/j.triboint.2017.06.001

[19] Choudhury, D., Roy, T., Krupka, I., Hartl, M., and Mootanah, R., 2015, "Tribological Investigation of UltraHigh Molecular Weight Polyethylene Against Advanced Ceramic Surfaces in Total Hip Joint Replacement," Proc. Inst. Mech. Eng. Part J- J. Eng. Tribol., 229(4), pp. 410–419. 10.1177/13506501145 41106

[20] Borjali, A., Monson, K., and Raeymaekers, B., 2018, "Friction Between a Polyethylene Pin and a Microtextured CoCrMo Disc, and Its Correlation to Polyethylene Wear, as a Function of Sliding Velocity and Contact Pressure, in the Context of Metal-on-Polyethylene Prosthetic Hip Implants," Tribol. Int., 127, pp. 568–574. 10.1016/j.triboint.2018.07.005

[21] Gadow R., Kern F. Advanced Engineering Materials. Volume 12. John Wiley & Sons; Hoboken, NJ, USA: 2010. Novel zirconia-alumina nanocomposites combining high strength and toughness; pp. 1220–1223.

[22] Weisse B., Affolter C., Stutz A., Terrasi G.P., Köbel S., Weber W. Influence of contaminants in the stem—ball interface on the static fracture load of ceramic hip joint ball heads. Proc. Inst. Mech. Eng. Part H J. Eng. Med. 2008;222:829–835. doi: 10.1243/09544119JEIM374.

[23] Kelly J.R., Denry I. Stabilized zirconia as a structural ceramic: An overview. Dent. Mater. 2008;24:289–298. doi: 10.1016/j.dental.2007.05.005.

[24] Anguiano-Sanchez J., Martinez-Romero O., Siller H.R., Diaz-Elizondo J.A., Flores-Villalba E., Rodriguez C.A.Influence of PEEK Coating on Hip Implant Stress Shielding: A Finite Element Analysis. Comput. Math. Methods Med. 2016;2016:6183679. doi: 10.1155/2016/6183679.

Shrinkage Factor Analysis in the Dry Machining of UNS A92024 Alloy by FEM

F J Trujillo[1*], C Bermudo[1], S Martín-Béjar[1], D Svensson[2], T Andersson[2] and L Sevilla[1]

[1] Department of Manufacturing Engineering, University of Malaga C/ Dr. Ortiz Ramos s/n, E-29071 Malaga, Spain
[2] Virtual Manufacturing Processes Research Group, University of Skövde, School of Engineering Science, Högskolevägen S-541 28 Skövde, Sweden

*Corresponding author's e-mail: trujillov@uma.es

Abstract. Chip evacuation becomes a relevant factor in the dry machining of aeronautical aluminium alloys, either when they are machined as isotropic material or hybridized with other materials. Chip morphology and geometry highly depend on cutting parameters. Several works can be found in the literature focusing on the analysis of chip morphology of aluminium alloys at high cutting speeds but there is a lack of studies that apply low cutting speeds. FEM simulation may be useful to reduce the experimental time and cost of this kind of analysis. Therefore, in this work, the influence of feed-rate on several chip geometrical parameters (height of peaks, height of valleys and shrinkage factor) of dry turned UNS A92024 alloy has been analysed by FEM simulation. Three different energy fracture modes have been tested as damage evolution criterion, in order to get the mode that better fit the experimental chip morphology. In general, the mixed mode was the fracture energy mode that best suited the chip morphology behaviour, within wider feed-rate range.

1. Introduction

Aluminum alloys are widely used in the manufacturing of structural parts for aircraft, both individually and in combination with composite materials, such as CFRP (Carbon Fiber Reinforced Polymer), to form fiber metal laminates structures (FML) [1]. In the dry machining of these alloys, the chip evacuation becomes a critical factor and it may causes problems in the process continuity and deviations from the surface quality requirements. Therefore, the chip monitoring is an important factor to keep in mind to improve these processes [2,3].

The experimental analysis of the chip morphology usually involves complex metallographic and microscopy techniques, which results in a large amount of economic resources and time consuming. To overcome these difficulties, the actual trend is using simulation techniques, such as Finite Element Method (FEM). This numerical method allows predicting thermal and mechanical behaviour of both the material and the tool [4-6]. In this regard, several works can be found in the literature which analyse the chip formation process and its morphology in the dry machining of aeronautical aluminium alloys (2000 and 7000 series). However, these works are usually focused on the high speed machining of this alloys. Under these conditions, the chip morphology is segmented or saw-toothed [7]. Notwithstanding, their hybridized use with composites or other light alloys (usually titanium alloys) requires the use of low cutting speeds, which may result in higher fragmentation. In this context, the shrinkage factor is a geometrical parameter that provides useful information about the total chip strain in the cutting process [2,8].

Hence, in this work, an analysis of chip formation process at low cutting speeds in the dry turning of UNS A92024 alloy has been carried out, by using FEM analysis. The influence of the feed-rate on several chip geometrical parameters (height of peaks, height of valleys and shrinkage factor) has been studied. A damage evolution criterion based on the fracture energy has been used. Several fracture modes have been tested in order to get the mode that better fit the experimental chip morphology. The simulation results have been compared with the experimental data, obtained by metallographic and microscopy techniques.

2. Experimental Methodology

Several tests were performed in order to analyze the shrinkage factor evolution as a function of the cutting parameters in the dry turning of the UNS A92024 Al-Cu alloy. The selected cutting parameters are shown in Table 1. Previous works have revealed that the most influential parameter on the chip morphology (in the dry machining of this alloy) is the feed-rate (f) [2]. Therefore, four different values of f were tested, from finishing to roughing intervals. The cutting speed (v_c) and cutting depth (a_p) remained constant.

Table 1. Cutting parameters.

v_c (m/min)	80			
f (mm/r)	0.05	0.10	0.20	0.30
a_p (mm)	1			

The experimental methodology was divided in two steps: on one hand, the software Abaqus/Explicit was used to simulate the chip generation process by the Finite Element Method (FEM). On the other hand, several dry machining tests were carried out to validate the simulation results.

2.1. FEM modelling

Lagrangian formulation was used to simulate the chip formation process in a 2D orthogonal cutting model. The influence of temperature on stress (and vice versa) was considered using a fully coupled thermo-mechanical algorithm. A dynamic explicit scheme was used as numerical time integration procedure.

The workpiece material was modeled as an elastoplastic material with damage, whereas the tool was considered as a rigid body. The physical properties of both materials, the workpiece (UNS A92024) and tool (WC-Co) are shown in Table 2 [8].

Table 2. Physical properties of the tested alloy (UNS A92024) and the tool material (WC-Co).

Property	UNS A92024	WC-Co
Density, ϱ (kg/m^3)	2780	11900
Elastic modulus, E (GPa)	73.1	612
Poisson's coefficient, μ	0.33	0.22
Thermal conductivity, K (W/m·K)	121	86
Specific heat, C (J/kg·K)	875	337
Thermal expansion, α (m/m·K)	$2.47\cdot10^{-5}$	$4.9\cdot10^{-6}$

Johnson-Cook model was used to simulate the visco-thermo-plastic behaviour of the workpiece (Equation 1) [8-10], as well as the damage initiation (Equation 2) [11]. The parameters for these models are shown in Table 3.

$$\sigma = (\mathrm{A} + \mathrm{B}\varepsilon^{\mathrm{n}})\left[1 + \mathrm{C}\ln\left(\frac{\dot{\varepsilon}}{\dot{\varepsilon}_0}\right)\right]\left[1 - \left(\frac{T - \mathrm{T}_0}{\mathrm{T_m} - \mathrm{T}_0}\right)^{\mathrm{m}}\right] \quad (1)$$

where σ denotes the equivalent flow stress, ε the equivalent plastic strain, $\dot{\varepsilon}$ the plastic strain rate and T the temperature. The model parameters are: $\dot{\varepsilon}_0$ (reference plastic strain rate), Tm (melting point), T_0

(room temperature), A (yield strength), B (hardening modulus), n (strain hardening coefficient), C (strain rate sensitivity) and m (thermal softening).

$$\varepsilon_f = \left[\mathrm{D}_1 + \mathrm{D}_2\exp\left(-\mathrm{D}_3\frac{\sigma_p}{\sigma}\right)\right]\left[1 + \mathrm{D}_4\ln\left(\frac{\dot{\varepsilon}}{\dot{\varepsilon}_0}\right)\right]\left[1 + \mathrm{D}_5\left(\frac{T - \mathrm{T}_0}{\mathrm{T_m} - \mathrm{T}_0}\right)\right] \tag{2}$$

where ε_f denotes the equivalent plastic strain at failure, σ_p the pressure stress, σ the von Mises stress and $\dot{\varepsilon}$ the plastic strain rate. The model parameters are: $\dot{\varepsilon}_0$ (reference plastic strain rate), T_m (melting point), T_0 (room temperature) and D_1 to D_5 (damage constants).

Table 3. Johnson-Cook's plasticity and failure parameters for the UNS A2024 alloy.

Plasticity model	**A (MPa)**	**B (MPa)**	**n**	**C**	**m**	**$\dot{\varepsilon}_0$ (s^{-1})**	**T_m (K)**	**T_0 (K)**
	352	440	0.42	0.0083	1	1	793	298
Failure model	**D_1**	**D_2**	**D_3**	**D_4**	**D_5**	**$\dot{\varepsilon}_0$ (s^{-1})**	**T_m (K)**	**T_0 (K)**
	0.13	0.13	-1.5	0.011	0	1	793	298

A damage evolution criterion based on fracture energy (G_f) with exponential softening was applied, Equation 3.

$$d = 1 - exp\left(\int_0^{u_p} \frac{\bar{\sigma}}{\mathrm{G_f}} du_p\right) \tag{3}$$

where d denotes the damage evolution, u_p the plastic equivalent displacement and $\bar{\sigma}$ the average von Mises stress. Three different failure modes (I, II and mixed) have been tested, in order to get the mode that best fits the chip geometry experimentally obtained, Table 4 [9].

Table 4. Fracture energy (mJ/mm^2).

G_f^I	G_f^{II}	G_f^{mixed}
16.7	12.4	14.0

A penalty friction formulation based on Zorev's model (Equation 4) was used to model the tangential contact between the chip and the tool [12]. The heat transfer at the tool-workpiece interface was modelled using a high thermal conductance value ($h = 5{\cdot}10^4$ $kW/m^2{\cdot}K$), in order to reach the steady state quickly.

$$\tau = \begin{cases} \mu N_f, & \tau < \tau_{\text{limit}}\ (sliding) \\ \tau_{\text{limit}}, & \tau \geq \tau_{\text{limit}}\ (sticking) \end{cases} \tag{4}$$

where τ is the frictional stress, μ is the apparent friction coefficient at the tool/chip interface, τ_{limit} is the material shear stress and N_f is the friction normal stress along the contact length. For this work, μ and τ_{limit} remained constant ($\mu = 0.3$ and $\tau_{limit} = 283$ MPa).

Regarding meshing, the element type selected for most workpiece and tool areas was CPE4RT (4-node plane strain thermally coupled quadrilateral, bilinear displacement and temperature, reduced integration, hourglass control). In addition, the element type CPE3T (3-node plane strain thermally coupled triangle, linear displacement and temperature) was used to make easier the transition between areas with different elements size and to better fit complex geometry areas such as the tool edge.

The workpiece was divided in three different areas (Figure 1) in order to optimize the element size and the time computing: 1, the chip; 2, the interface area; 3, zone far away from the cutting area. A starting distortion of 45° was used for the elements in area 1, to facilitate the chip segmentation. In a similar way, the tool was divided in two areas: 4, the tool edge; 5, zone far away from the cutting area. The element size and shape, as well as the meshing control for each area, is shown in Table 5.

Table 5. Meshing features.

	Area	Element size (mm)	Element shape	Meshing control
	1	0.02	Quad	Structured
Part	2	0.02-0.10	Quad-dominated	Free
	3	0.10-0.50	Quad-dominated	Free
Tool	4	0.06	Quad-dominated	Free
	5	0.06-0.50	Quad-dominated	Free

The mechanical boundary conditions are shown in Figure 1. Displacement was restricted to zero on the top and right tool sides, in both *x* and *y* directions. Displacement was restricted to zero on the left and bottom workpiece sides, in the *y* direction and a velocity equal to the cutting speed was applied to both sides, in the *x* direction. With regard to the thermal boundary conditions, a starting temperature of 298 K was applied to the workpiece and the tool in the initial step. The temperature in both elements was computed in the following step.

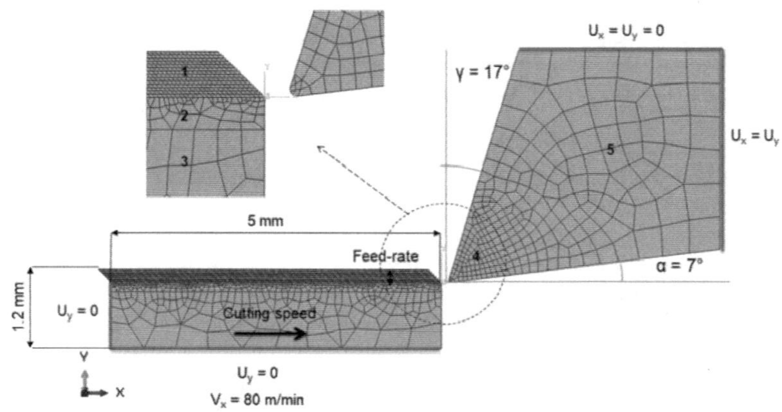

Figure 1. Model geometry, meshing and boundary conditions.

2.2. Machining tests

Several dry machining tests were carried out with bars of UNS A92024 (Al-Cu) alloy (100 mm long and diameters between 40 and 60 mm). The tested alloy composition, obtained by arc atomic emission spectroscopy (AES), is shown in Table 6. The cutting parameters applied in the tests are shown in Table1. In order to guarantee the repeatability, each test was repeated 4 times.

Table 6. Tested alloy composition (% weight).

Zn	Mg	Cu	Cr	Si	Mn	Al
0.12	1.25	4.85	0.01	0.15	0.54	Rest

A new WC–Co uncoated turning insert, with ISO DCMT 11T308F2 code, was used for each test. The tool geometry and holder provided a close to orthogonal cutting configuration. Metallographic techniques (resin drawing, polishing and chemical etching) were used to prepare the chip samples for observation by Stereoscopic Optical Microscopy (SOM). Longitudinal chip section images were captured by an optical capture system (up to 400X). Digital image processing software was used to get the geometrical chip parameters. To evaluate the shrinkage factor (ξ, Equation 5), the height of peaks (h_c) and the height of valleys (h_v) were measured, both in the real and the simulated chip, Figure 2. In Equation 5, κ_r is the main cutting edge angle (62.5°).

$$\zeta = \frac{f \, sin \, \mathrm{Kr}}{\frac{h_c + h_v}{2}} \tag{5}$$

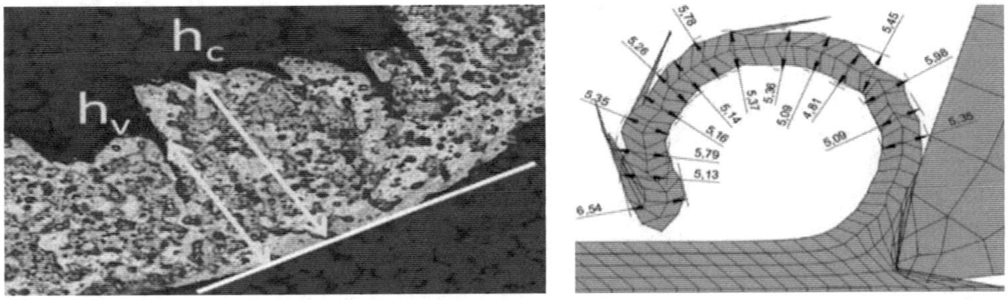

(a) (b)

Figure 2. Chip geometrical parameters measurement: (a) real chip; (b) FEM simulated chip.

3. Results

Figure 3 shows the chip morphology evolution as a function of the feed-rate (*f*), both for experimental tests and FEM simulation (for fracture energy G_f^I, G_f^{II} and G_f^{mixed}). In the experimental tests, long washer-type helical morphology was obtained within the lowest *f* range (0.05-0.10 mm/r), whereas tubular chips were obtained for the highest *f* range (0.20-0.30 mm/r), with higher fragmentation at *f* = 0.30 mm/r. Therefore, at low *f*, the cutting forces were not high enough to break the chip, which showed a general trend to form chip nests. For higher *f* values, the shear stress limit was easier to reach (higher cutting forces) and the chip segmentation and fragmentation were higher. These observations are in good agreement with previous works performed under similar cutting conditions [2]. With regard to the FEM simulation test, long continuous chip was obtained within the lower *f* range (0.05-0.10 mm/r), for all fracture energy modes. However, G_f^{II} and G_f^{mixed} simulated the chip curvature better than G_f^I. For *f* = 0.20 mm/r, the three fracture energy modes showed an adequate degree of segmentation. Notwithstanding, the G_f^{mixed} mode exhibited the best curvature. For *f* = 0.30 mm/r, G_f^I and G_f^{II} resulted in excessive segmentation from the first machining instants. However, the G_f^{mixed} mode showed good segmentation degree and chip curvature, with a little longer chip before breaking. Therefore, in general, the G_f^{mixed} mode was the fracture energy mode that best suited the chip morphology within wider feed-rate range.

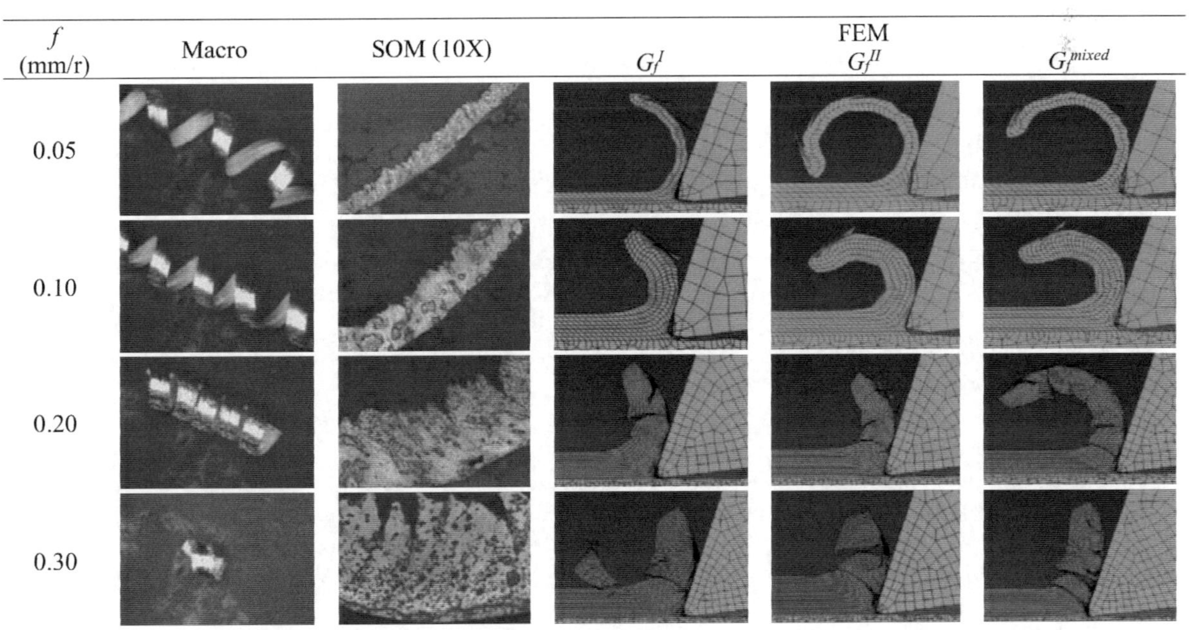

Figure 3. Chip morphology (experimental and FEM).

Figure 4 shows the evolution of the height of peaks (h_c), height of valleys (h_v) and shrinkage factor (ξ) as a function of the feed-rate (f), for the fracture energy mixed mode (G_f^{mixed}). Regarding h_c, higher differences between the experimental and the simulated values can be observed within the lowest range of f (0.05-0.10 mm/r), being the highest difference for f = 0.05 mm/r (around 50%). Nevertheless, these differences were much lower for f = 0.20 and 0.30 mm/r, without ever exceeding 5% in the average values. With regard to h_v, the differences between the experimental and the simulated data were practically negligible (lower than 5%) in the whole range of f.

These differences in h_v, for f = 0.05 mm/r, resulted in higher ξ deviations (Equation 5) for this f (around 50%). Nevertheless, for f = 0.10 mm/r, experimental ξ value started to converge with the simulated values (deviation around 15%) and, finally, both experimental and simulated data were practically identical for f = 0.20 and 0.30 mm/r. This higher deviation for f = 0.05 mm/r may be due to lower chip segmentation, which makes more difficult to identify the peaks and valleys, both in the experimental and FEM methodology. For higher f, the chip segmentation was more evident and the peaks and valleys were easier to measure. In addition, the element size remained constant for all tests and, thus, the number of elements included in the deformed chip were lower for f = 0.05 mm/r. Therefore, FEM simulation with G_f^{mixed} fracture energy mode has shown a good fit to the experimental data from f = 0.10 to 0.30 mm/r, under the tested conditions.

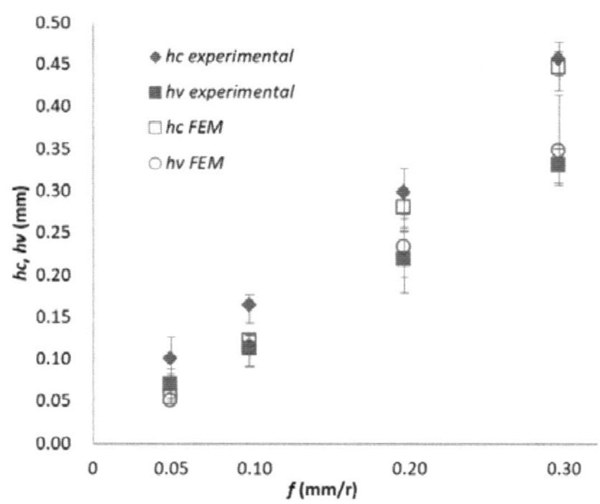

(a)

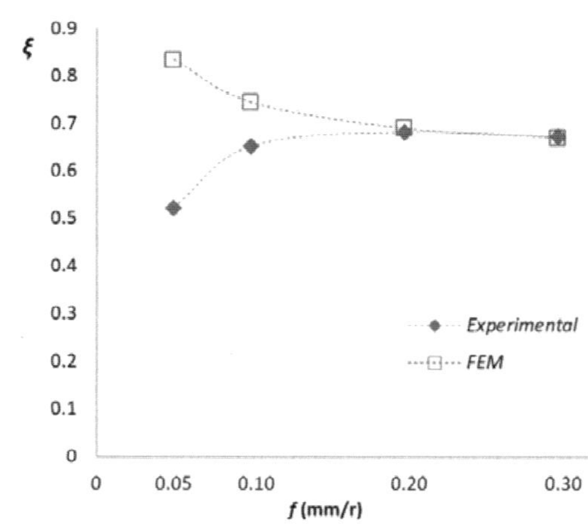

(b)

Figure 4. (a) Height of peaks (h_c), height of valleys (h_v) and (b) shrinkage factor (ξ) as a function of the feed-rate (f) (experimental and FEM simulation for G_f^{mixed} fracture energy mode).

4. Conclusions

In this work, the influence of feed-rate on several chip geometrical parameters (height of peak, height of valleys and shrinkage factor) of dry turned UNS A92024 alloy has been analyzed by FEM simulations, for low cutting speed (80 m/min). Three different energy fracture modes have been tested as damage evolution criterion (G_f^{I}, G_f^{II} and G_f^{mixed}), in order to get the mode that better fit the experimental chip morphology.

The chip morphology exhibited long washer-type helical morphology at low feed-rate range (0.05-0.10 mm/r), whereas tubular chips were obtained for higher values (0.20-0.30 mm/r). The highest fragmentation was obtained for 0.30 mm/r, with shorter chip. In general, the G_f^{mixed} was the fracture energy mode that best suited this chip morphology behaviour, within wider feed-rate range. For this fracture energy mode, the heights of peaks and valleys, as well as the shrinkage factor, exhibited a good fit to the experimental results between 0.10 and 0.30 mm/r. For 0.05 mm/r, the dispersion was higher, due to lower chip segmentation, which made the geometrical parameters more difficult to measure. Additionally, the number of elements included within the deformed chip was lower for this value (f = 0.05 mm/r).

It is necessary to point out that this work is only the first stage of the chip morphology analysis by FEM (under the specified cutting conditions) for this alloy. Other geometrical parameters, such as the chip segmentation ratio, shear angle, chip cross section and width, will be addressed in further works. In addition, for low feed values, the element size should be reduced, in order to analyze the results convergence. Finally, the generality of these results should be tested in a wider range of cutting speed and cutting depth.

5. References

[1] Santos M, Machado A, Sales W, Barrozo M and Ezugwu E 2016 *Int. J. Adv. Manuf. Technol.* **86** 3067
[2] Trujillo F J, Sevilla L, Martín F and Bermudo C 2017 *Appl. Sci.* **7** 132
[3] Sánchez Y, Trujillo F J, Bermudo C and Sevilla L 2018 *Materials* **11** 1260
[4] Markopoulos A P 2012 Finite Element Method in Machining Processes *Springer Briefs in Applied Sciences and Technology (Manufacturing and Surface)* ed J P Davim (London: Springer)
[5] Arrazola P J, Özel T, Umbrello D, Davies M, Jawahir I S 2013 *CIRP Annals Manufacturing Technology* **62** 695
[6] Haddag B, Atlati S, Nouari M and Moufki A 2016 *Metals* **6** 197
[7] Jomaa W, Mechri O, Lévesque J, Songmene V, Bocher P and Gakwaya A 2017 *J. Manuf. Process.* **26** 446
[8] Seshadri R, Naveen I, Srinivasan S, Viswasubrahmanyam M, VijaySekar K S and Pradeep Kumar M 2013 *Procedia Eng.* **64** 1454
[9] Ijaz H, Zain-ul-abdein M, Saleem W, Asad M and Mabrouki T 2017 *J Mech.* **6** 777
[10] Mahnama M, Movahhedy M R 2012 *Journal of Manufacturing Processes* **14** 188
[11] Nasr M, Ammar M 2017 Procedia CIRP **58** 134
[12] Melkote S, Grzesik W, Outeiro J, Rech J, Schulze V et al 2017 *CIRP Annals Manufacturing Technology* **66** 731

Acknowledgments
The authors thank the University of Malaga-Andalucía Tech Campus of International Excellence for its economic contribution on this paper.

Cutting parameters influence analysis on fatigue behaviour of dry turned UNS A97075 alloy after corrosion

S Martín-Béjar[1], F J Trujillo[1], C Bermudo[1], M Herrera[1] and L Sevilla[1]

[1] Department of Manufacturing Engineering. University of Malaga C/ Dr. Ortiz Ramos s/n, E-29071 Malaga, Spain

*Corresponding author's e-mail: smartinb@uma.es

Abstract. Aluminum alloys, particularly series 2000 (Al-Cu) and 7000 (Al-Zn), are widely used for structural elements in aircraft. Machining processes are frequently applied in manufacturing these parts. The actual trend in the machining of these components is to reduce or eliminate the use of lubricants, due to environmental reasons. Among the different techniques, dry machining is commonly used for aluminum alloys. However, dry machining generates higher severe cutting conditions, which may negatively affect the surface integrity and the mechanical properties of the machined parts. In this regard, fatigue life is one of the most important mechanical properties to take into account. Micro-cracks generation and nucleation strongly depend on the surface conditions of the machined parts. Additionally, the addition of Zn to pure aluminum (7000 series) reduces the corrosion resistance. Despite its importance, there is a lack of research devoted to analyze the cutting parameters influence on fatigue behaviour, before and after corrosion, in dry machining of these alloys. Therefore, in this work, the cutting speed and feed rate influence on fatigue behaviour of the UNS A97075 (Al-Zn) alloy has been analyzed. Several rotating bar bending fatigue tests have been carried out, studying and comparing the fatigue life curves obtained as a function of the cutting parameters, before and after an immersion corrosion process. The experimental results have revealed that an increment in the cutting speed and feed rate gives rise to fatigue life reduction.

1. Introduction

Actually, the aeronautical industry takes special interest due to the extensive use of aircrafts. Aluminium alloys (2000 and 7000 series) are widely employed for the manufacture of the airplanes due to their excellent mechanical properties and density relationship. Particularly, UNS A97075 (Al-Zn) alloy is used in the manufacturing of the wings, spar caps and the upper skins [1].

Machining operations are commonly employed in the structural aircraft manufacturing (milling and drilling), where turning operations are used to obtain revolution pieces designated to fit. Furthermore, the actual trend is to employ environmental friendly techniques in machining operations, reducing or eliminating lubricants (dry machining, cryogenic, minimum quantity of lubricant, etc.). Nevertheless, these cutting conditions affect negatively to the process performance (functional, economic and energetically) due to the action of high temperature and pressures [2].

Manufacturing processes (machining operations among others) generate variations in the surface conditions of the manufactured parts. In this sense, the surface integrity evaluates those variations, being an appreciate quality requirement in aeronautical industry. Three levels can be considered for the surface integrity evaluation : macro-geometrical (size, shape, geometrical deviations, etc.), micro-geometrical (roughness profile, micro-cracks, etc.) and physicochemical properties (fatigue behaviour, corrosion, etc.) [3].

Fatigue behaviour is one of the most important mechanical properties to take in to account in the evaluation of aircrafts structural parts in service due to security reasons. Fatigue fracture occurs in three phases: microcracks generation and nucleation (usual in aluminium alloys surface), crack growth and ductile fracture due to the reduction of the section that support the stress [4]. Different researches relate fatigue behaviour with other surface integrity properties such as microhardness, residual stress and surface roughness [5,6]. Therefore, the surface integrity of manufactured parts affects the initial phase of the fatigue fracture.

On the other hand, cutting parameters (cutting speed, feed rate and cutting depth) has been related with microhardness, residual stress and surface roughness of machined parts. Hence, it can be stablished that the cutting parameters affect the fatigue behaviour [7].

In general terms, aluminium alloys show a good corrosion resistance due to the alumina layer that is generated on the free surface under corrosion conditions. In spite of this, in the series 7000, the zinc addition to aluminium reduces this corrosion resistance compared with pure aluminium (series 1000). In aluminium-zinc alloys, corrosion is usually localized under the appearance of pitting corrosion, as well as intergranular corrosion, which can be considered as the generation and nucleation points of microcracks [8,9]. Thus, the corrosion process affect negatively to the surface integrity of manufacturing parts and, therefore, fatigue behaviour is negatively affected too, due to the reduction of the generation and nucleation of microcrack phase.

In previous research of aluminium alloys machined parts, the inmersion corrosion process has been affect negatively the surface conditions of machined parts. In this sense, in [10] the machining of the fatigue test notch provoke an increase of the fatigue crack length in corroded parts compared with other process.

Additionally, the influence of initial surface conditions in the corrosion activity has been considered in [11]. In this case, higher surface roughness previously the corrosion process, affect negatively to the surface conditions of corrode parts. Hence, the previous surface conditions in machined parts take especial relevance in the surface of corroded parts. Furthermore, the surface conditions affect fatigue behaviour as previously commented.

Therefore, in this work, a preliminary study of the cutting parameters influence in the fatigue behaviour of corroded parts in dry turning UNS A97075-T6 (Al-Zn) alloy has been performed. From the experimental results, the influence of the cutting parameters in the surface roughness and the fatigue behaviour previous and post immersion corrosion has been analysed.

2. Methodologhy

Several dry-turning tests were carried out in order to evaluate the cutting parameter influence on the fatigue behaviour of UNS A97075-T6 aluminium alloy corroded parts. The test alloy composition (% mass), obtained with arc atomic emission spectroscopy (AES), is shown in Table 1.

Table 1. Test alloy composition (% mass).

Zn	Mg	Cu	Cr	Si	Mn	Al
6.01	2.61	1.88	0.19	0.08	0.07	Rest

Different standards can be used to evaluate the material fatigue behaviour. Among them, ISO 1143:2010 standard [12] has been selected due to aeronautical parts are usually subjected to bending stress. The selected specimen geometry and dimensions are shown in Figure 1.

Different specimens were turned under dry conditions. A finishing step was carried out for each sample using different cutting speed (v_c) and feed-rate (f) values, remaining constant the cutting depth for every test. The selected values are shown in Table 2. Although this alloy is not usually machined with low cutting speed values, it requires low v_c values when it is hybridized with other materials.

Four specimens were machined for each cutting parameter combination to ensure a confidence level of 95% and a 50% failure probability, according to the ISO 12107:2017 standard [13]. Therefore, 96 specimens were manufactured for each tested surface state (corrode and non-corrode).

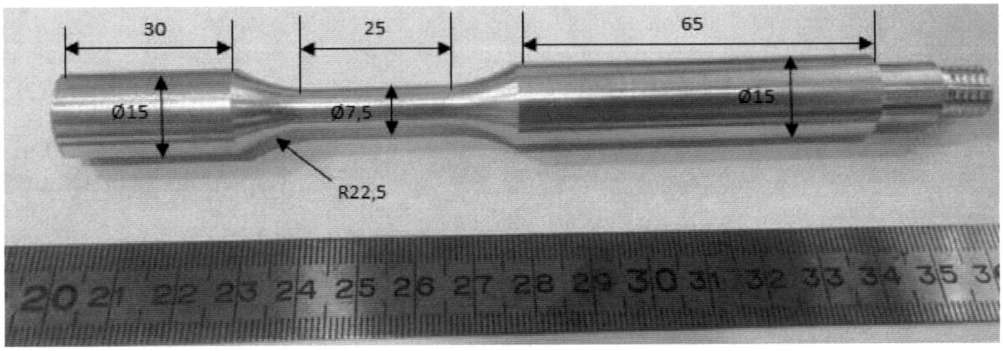

Figure 1. Rotating bar bending test geometry (mm).

Table 2. Cutting parameters.

v_c (m/min)	f (mm/r)	a_p (mm)
40 60 80	0.05 0.10 0.15 0.20	1

After the turning tests, a micro-geometrical control was carried out. The average roughness (*Ra*) and the maximum height of the profile (*Rz*) were used as parameters to evaluate the surface quality. For this purpose, a portable roughness tester (Mitutoyo SURFTEST SJ-210) was used.

Once the surface roughness was measured, 4 samples of each cutting parameter combinations were corroded. The corrosion process was carried out in a solution of deionized water and NaCl (3.5% concentration), keeping the specimens in that corrosive environment for 72 h. Hence, to ensure the corrosion process homogeneity on the surface of all samples, a pump was used inside the solution.

The rotating bar bending tests were carried out in a rotating bar bending bench, manufactured in the University of Malaga. Before the test start, once the specimen is placed in the fatigue test bench, the concentricity between the extreme of the specimen and the chuck was controlled, according to the ISO 1143:2010 standard [12]. A single load was applied in the extreme of the specimen, assuming 244.40 MPa stress in the expecting fatigue fracture section.

Finally, a fractography image of the fracture sections was performed. A stereoscopic microscope, NIKON SMZ 2T, with a Kappa Image Bases CF11 DSP camera and PSI SC module capture card was used for this purpose.

3. Results

Figure 2 shows the *Ra* results as a function of v_c and f, before and after the corrosion process. A general trend to increase *Ra* with f is observed, regardless v_c. The *Ra* highest values have been obtained for 0.20 mm/r, regardless v_c.

For samples before corrosion, from f = 0.05 mm/ to 0.10 mm/r, a slightly *Ra* increase can be observed. Notwithstanding, from f = 0.10 mm/r to 0.20 mm/r, *Ra* increase with f is more noticeable. The cutting speed influence in *Ra* is less noticeable. Only for f = 0.15 mm/r, a difference in the *Ra* values obtained can be appreciated for different v_c. In this case, for v_c = 40 m/min, the *Ra* mean value is higher than for v_c = 60 and 80 m/min. This effect can be explained due to the appearance of some vibrations during the turning process. These results are in good agreement with previous research [14,15].

It can be appreciated how the corrosion process has affected negatively the surface roughness, increasing the *Ra* values. For samples after corrosion, *Ra* increase can be observed clearly along all f values rang, regardless v_c. In this case, v_c influence is less noticeable too. The corrosion process has increased the *Ra* results dispersions, being more relevant at higher f values (0.15-0.20 mm/r).

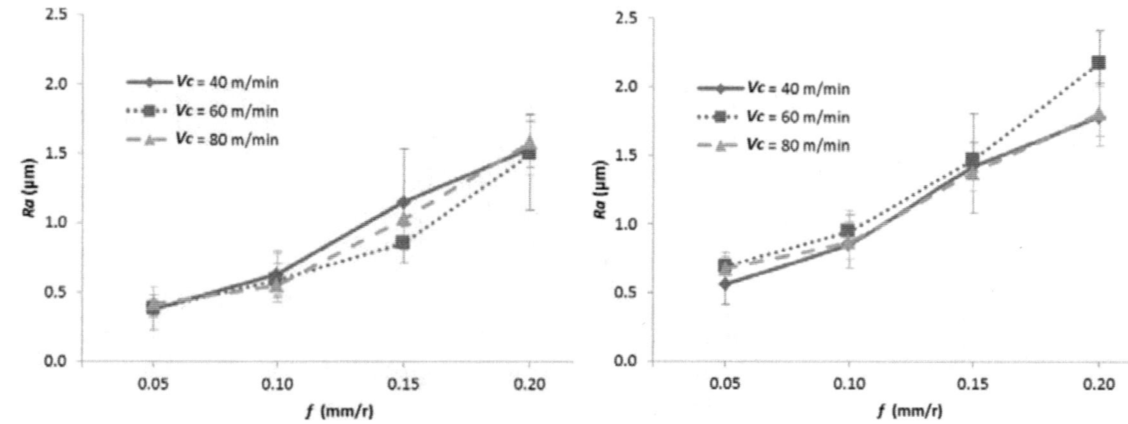

(a) (b)

Figure 2.1 Average Roughness (*Ra*) as a function of feed (*f*) and cutting speed (v_c): (a) previous corrosion process, (b) post corrosion process.

Figure 3 shows the *Rz* results as a function of v_c and *f*. The influence of the cutting parameters (v_c, *f*) in *Rz* is similar to *Ra*. Nevertheless, the corrosion process has provoked a higher *Rz* increase than observed for *Ra*. This effect can be explained taking into account the corrosion process in aluminium alloy. The appearance of pitting corrosion may have affected the values of the microgeometrical deviations at one point (localized corrosion). This effect was softened in the calculation of the average values to obtain *Ra*.

It is important to highlight that the *Ra* and *Rz* measurements, represented previously, were carried out in the specimens that were corroded, being the specimen *Ra* and *Rz* values in good agreement with the fatigue test that were performed without the corrosion process.

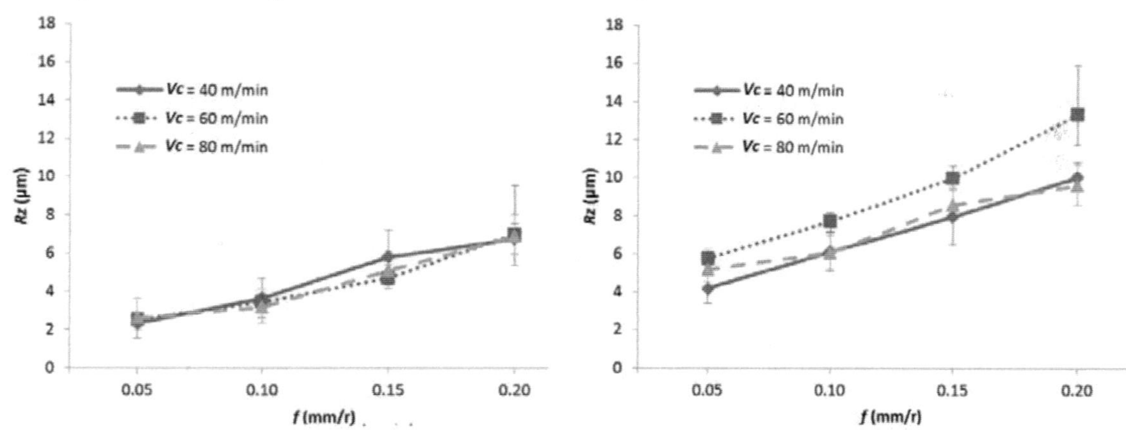

(a) (b)

Figure 3.2 Mean roughness depth (*Rz*) as a function of feed (*f*) and cutting speed (v_c): (a) previous corrosion process, (b) post corrosion process.

Figure 4 shows several SOM images from the fatigue fracture section of a non-corroded test (Figure 2a) and of a corroded test (Figure 2b). The initial cracks can be observed in the surface. This effect occurs in corroded and non-corroded specimens. In spite of this, in the corroded fracture section (Figure 4b) two initiation crack point can be observed. This is due to the pitting corrosion, that facilitates microcracks generation and nucleation.

The corrosion affects a maximum depth of 0.40 mm [6]. Therefore, cracks growth and ductile fracture surface show a similar performance, in corroded and non-corroded specimens, taking into account that they are under the same stress in the fracture section (S = 244.40 MPa).

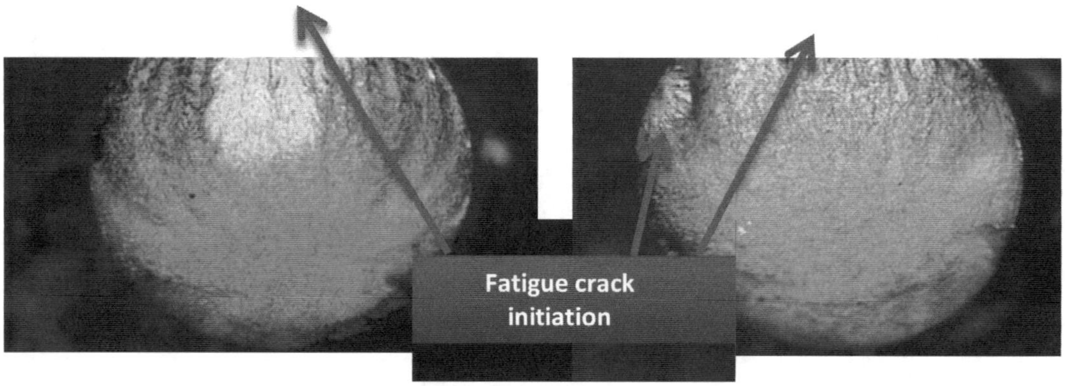

Figure 4. Fracture section SOM capture (x30): (a) previous corrosion process, (b) post corrosion process for v_c = 40 m/min and f = 0.10 mm/r.

The number of cycles (*N*) obtained for the rotating bar bending non-corroded tests are represented in Figure 5a, and results of the corroded specimen are shown in Figure 5b, as a function of v_c and f. Figure 5a shows that, in general, an increase of f tends to reduce *N*. This trend can be observed for v_c = 40 and 80 m/min. For v_c = 60 m/min, an increase of *N* can be observed in the range of 0.05 to 0.10 mm/r. The lower *N* values were obtained for higher f values (0.15-0.20 mm/r) and v_c (80 m/min). Taking into account these results, the fatigue behaviour (*N*) of non-corroded specimens is influenced by f, associated with a *Ra* and *Rz* increase and v_c related with the appearance of residual stresses and microstructural changes [7].

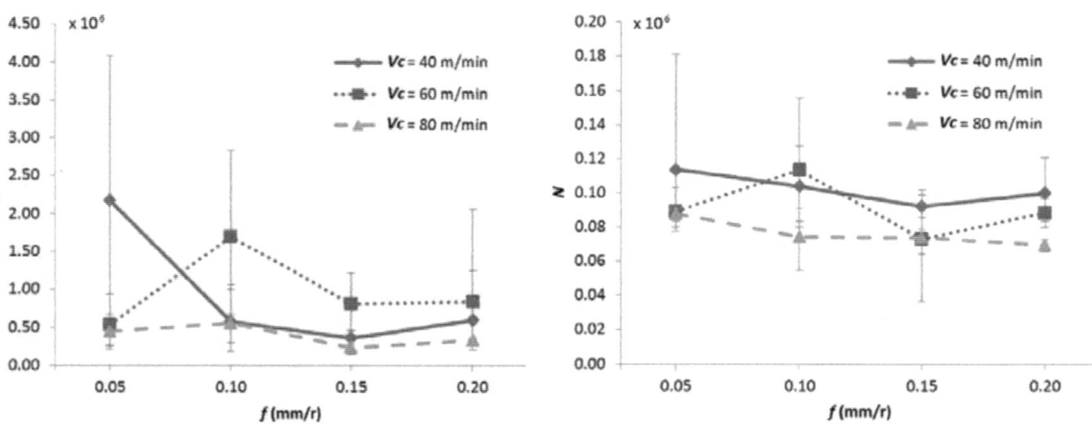

Figure 5.3 Number of cycles (*N*) as a function of feed (f) and cutting speed (v_c): (a) previous corrosion process, (b) post corrosion process.

It can be observed that the corrosion process affects negatively the fatigue behaviour (*N*) due to the changes in the specimen surface, being influenced by v_c and f. In Figure 5b, a general trend to reduce *N* with f can be observed, regardless v_c. The increase of surface irregularities, associated with a f increase, provokes that the microcracks initiation and nucleation phase requires less number of cycles. In addition, the corrosion process affects the surface residual stresses and microstructure, that encourage the acceleration of the initial fatigue fracture step.

4. Conclusions

In this work, an analysis of the influence of the cutting speed and feed-rate on fatigue behaviour of UNS A97075 dry turning corroded specimens has been carried out. The surface roughness (*Ra* and *Rz*) has been evaluated on the fatigue test samples, before and after the corrosion process.

Regarding the surface roughness, the feed-rate has been shown as the most influential variable in both parameters studied (*Ra* and *Rz*). However, the cutting speed influence has been less noticeable. The corrosion process has affected negatively *Ra* and *Rz*, increasing their values in all the *f* values rang. *Rz* has experimented a higher increase than *Ra*, due to the pitting corrosion, that is originated in the corroded specimens surface. These observations are in good agreement with previous works performed under similar cutting conditions.

Fracture section fractography has shown that the microcracks initiation and nucleation occur in the surface for both corroded and non-corroded specimens. Furthermore, in the corroded samples fractography several initial microcrack points can be observed. Crack grown and ductile fracture surface are independent of the surface conditions, depending on the load applied.

The fatigue behaviour experimental data showed that increments in v_c and *f* results in a fatigue life (*N*) reduction, according with a surface roughness increase. The cutting parameters influence in fatigue life (*N*) of corroded parts is less noticeable. This effect may be explained taking into account the surface alterations, reducing the number of cycles during microcrack initiation and nucleation phase.

5. References

[1] Goncharenko A V 2018 *Int. J. Aerosp. Eng.* **2018** pp 75-92
[2] Batista Ponce M Del Sol Illana I. Fernandez-Vidal S and Salguero Gomez J 2018 *Materials (Basel)* **9** 1598.
[3] Davim J P 2010 Surface integrity in machining 53-69
[4] Schijve J 2008 Fatigue of Structures and Materials 13–58
[5] Novovic D Dewes R C Aspinwall D K Voice W and Bowen P 2004 I*nt. J. Mach. Tools Manuf.* **44** 125–134
[6] Rotella G 2019 *J. Manuf. Process.* **41** 83–91
[7] Surya Sundara Rao K and Viswanath Allamraju K 2017 *Materials Today: Proceedings.* **4** 975–981
[8] Zupanc U and Grum J 2010 J. Mater. Process. Technol. vol 210 n 9 pp 1197–1202
[9] Shen Y 2019 *Bioelectrochemistry* 107408.
[10] Arunachalam S R Galyon Dorman S E Buckley R T Conrad N A and Fawaz S A 2018 Int. J. Fatigue **111** 44–53
[11] Bienvenido R Díaz Vázquez J E Botana F J Cano M J and Marcos Bárcena M 2010 *Adv. Mater. Res.* **107** 117–121
[12] International Standaring Organization ISO 1143:2010. Metallic materials-Rotating bar bending fatigue testing 2010.
[13] International Standaring Organization ISO 12107:2017 Metallic materials - Fatigue testing-Statistical planning and analysis of data 2017.
[14] Bermudo C Trujillo F J Herrera M and Sevilla L 2017 *Procedia Manuf.* 13 81–88
[15] Martín S Trujillo F J Bermudo C and Sevilla L 2019 *Metals (Basel)* **9** 11.

Acknowledgments
The authors thank the University of Malaga-Andalucía Tech Campus of International Excellence for its economic contribution on this paper.

IOP Conf. Series: Materials Science and Engineering 1037 (2021) 012013 doi:10.1088/1757-899X/1037/1/012013

Friction - wear modeling in drilling process of H-13 tool steel

G S Panagopoulos[1,2*], M S Panagopoulou[2] and P G Nikolakopoulos[1,2]
[1]Machine Design Laboratory, Department of Mechanical Engineering and Aeronautics, University of Patras, GR-26504 Patras, Greece
[2]BPharm (Hons), PhD candidate School of Chemistry, College of Science and Engineering, the University of Edinburgh, Edinburgh, Scotland, UK

[*]Corresponding author's e-mail: up1070396@upatras.gr

Abstract. Simulation of metal cutting is complex both from numerical and physical perspective. The scope of this work is to elaborate on the process and physics of metal cutting, and more specifically drilling. The description of H-13 steel dry drilling with the use of a single-layer TiN coated carbide twist drill is presented. H-13 is a high strength alloy steel used in demanding applications and elevated temperatures. A simulation model was developed with use of Deform-3D software, and an experiment was conducted in order to evaluate the acquired simulation results, meaning the wear mechanisms and thermal loads implemented during the process of drilling. The Usui wear model was used for the calculation of the wear rate of the drill. Regarding the experiment, a thermal camera was used in order to record the temperature of the drill after the process. The comparative analysis of the simulation models and the experiment showed that adhesion is the dominant wear mechanism. Higher wear values were observed in the rake faces of the drill, where the coating starts to fade, due to built-up edges (BUE). The tool and workpiece temperatures reach near steady state conditions during process. Elevated temperatures are applied on the removed material. Good chip evacuation and chip length indicate that coated drills with optimized geometry are suitable for drilling tough materials.

1. Introduction

Drilling or hole production is a major material removal process and is one of the most common machining processes in the manufacturing industry for the production of holes in components. In [1], Horvat *et al.* investigated the influence the drill point geometry has on the life of the tool. The research included experiments of coated and uncoated twist drills on drilling Hardox 500 specimens. The results showed that under the same conditions the TiN coated drills outperformed the other types. Most of the wear was observed on the flank faces and the margins of the drills.

Sharif *et al.* got involved with drilling of Ti6A14V with TiAlN-PVD coated and uncoated carbide drills [2]. Ti6A14V is a titanium alloy with poor thermal properties which determines the plastic deformation at the cutting point. Also, titanium alloys are known to be quite reactive with most of the tool materials, and their elasticity module reduces at moderate temperatures leading to bending, considering that the temperatures in the tool-workpiece contact area are very high during machining. The tools used were all made from similar substrate, meaning WC-CO and their grades were K30/40. The results showed that coated drills on low cutting velocities (25-35 m/min) indicated gradual flank wear, while on higher speeds (45-55 m/min) the wear was increased rapidly. Later, the same authors tested the effect of cutting geometry of the drills on the same material in [3]. For their research they used two different kinds of split point uncoated WC-CO solid drills. The main wear mechanism that occurred was adhesion wear on the tools' cutting edges.

The beneficial effects of lubrication during drilling have been already described by Grzesik *et al.*, however it is also acknowledged that lubricants bring oxygen in the tool-chip interface, leading to oxidation of the material [4],[5].

Prior knowledge suggests that TiN coating is generally considered to be prone to oxidation, as the gas oxygen molecules have the tendency to replace nitrogen on the coating, leading to the formation of a layer of oxidized titanium, in a process which is thermodynamically favoured at temperatures higher than 500°C. The oxidized titanium has reduced hardness and can be removed during drilling exposing a fresh layer of TiN available for oxidation, leading to repeated cycles of corrosion of the tool according to Chauhan and his colleagues in [6] and Knotek *et al.* in [7] . On the other hand, it has been shown that although TiN coating offers relatively poor performance when pure water is used for lubrication, it can exhibit better tribological behaviour under sea water, as it has corrosion-resistant properties, suggesting its suitability for marine applications as seen in the work of Matei and colleagues in [8] and in [9] by Wang *et al.*

On the above studies, the main wear mechanism is adhesion. Also, what is more interesting is the fact that despite changes in the cutting lips' geometry of the drill, flank faces tend to be prone to wear, determining the drills' life. In addition, flaking is very typical in most of the drills' periphery due to the high rake angle they are grinded with.

1.1. Wear mechanisms

Wear is the gradual removal of material at contacting surfaces in relative motion. While friction results in energy losses, wear correlates to increased maintenance costs as well as energy loss. Wear phenomena are linked closely to frictional process. Friction forces are generally the result of shearing and ploughing. Moreover, wear is heavily influenced by the fact that most surfaces are rough at some point, no matter the preceded treatments they have.

There are many types of wear, classified according their influence on the cutting tool. The performance of a cutting tool relies on its ability to endure the loads applied during metal cutting. The main categories of wear mechanisms that occur are thermal and mechanical wear, chemical (oxidation and diffusion), abrasive and adhesive wear. In the present work the adhesive wear mechanism will be examined thoroughly. In addition, the wear mechanisms are mostly dependent on temperature generated during cutting as seen on the following illustration (Figure 1). On lower temperatures, abrasive and adhesive wear mostly affect the tool, while on higher ones, diffusion has a major role, and adhesive wear is reduced significantly.

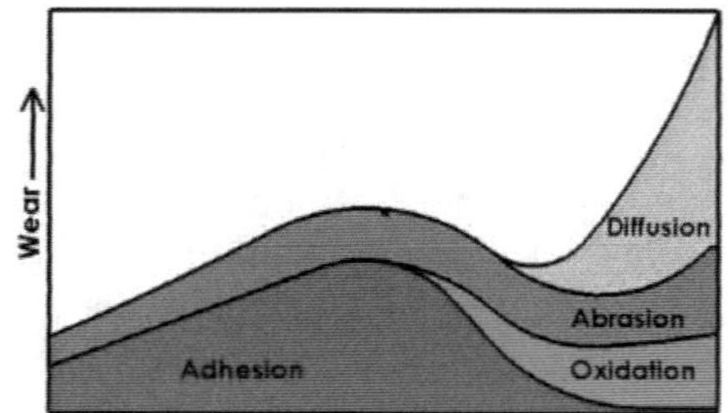

Figure 1. Wear mechanisms in metal cutting [5].

Adhesive wear mechanism is the most common wear type that is presented in machining. It is characterized by two surfaces under plastic contact that have sufficient bonding strength in between them preventing them from sliding. This results in deformation due to compression and shearing, and crack propagation usually initiates. When the crack reaches their interface then a wear particle is created and is ripped off. In machining of ductile materials, adhesive wear is found either on the tool workpiece interface, or in the tool chip interface. Another phenomenon of adhesion in machining is the built-up edge (BUE), where the parts of the material to be processed are again adhered on the cutting edges of the tool. The severity of adhesion wear depends on temperature as show in Figure 1, where after a specific value it starts to decline. Also, below this specific temperature there is a lower tendency for material to adhere on the tool's surfaces.

Many models have been designed to estimate adhesive wear. In 1953 Holm and Archard presented an equation for the calculation of adhesive wear. Based on Archard's law, a number of general models around the hardness of the adhered material have been created. In machining, the workpiece hardness is essential to be known, and this is the one that adheres on the tool. Moreover, the sliding velocity has to be known, as well, which is the length of the cut, and in drilling is the hole depth. Finally, the normal contact pressure at plastic deformation has to be of a known value. The worn volume in its simplest form can be expressed as:

$$V = \frac{1}{3}\frac{WL}{H} \quad (1)$$

where V is the volume that is worn , W is the normal load applied, L is the sliding distance and H is the hardness of the material that is worn. In this form of equation, the worn volume is increased when normal load and sliding distance are also increased and inversely proportional to the hardness of the material. Also, the tougher the drill's coating the less wear will occur [10].

With the use of Archard's law as a fundamental equation of wear calculation, Usui and colleagues resulted in the following equation in [11]:

$$\frac{\mathrm{d}W}{\mathrm{d}t} = C_1 \sigma_t V_c \exp(-\frac{C_2}{\theta}) \quad (2)$$

This equation refers to the wear rate, meaning the fraction $\frac{\mathrm{d}W}{\mathrm{d}t}$, and is known as Usui's Wear model. In the present study it will be used for the calculation of adhesive wear in carbide drilling tools [11]. The parameters C_1 and C_2 are experimentally calculated, σ_t is the normal stress between the friction couple, θ is the temperature of the chip surface and V_c is the cutting speed.

1.2. Drilling wear

In general, there are many different wear types in machining. Most of them are easily explained for turning based on their appearance, but can also be applied in drilling, with slight modifications. The most common one is flank wear.

Flank wear is the type of wear that occurs on the flank face of the tool due to shear stress from normal pressure while the tool slides over the machined surface. It firstly develops close to the cutting edge, and then grows away from it evenly distributed on the remaining flank face. It is the most typical type of wear that appears during machining. Variations can be found on this which depend on the material that is processed, cutting data and tool shape. For example, smaller lip relief angle on a drill bit is more prone to flank wear due to the higher contact with the machined surface of the material.

Flank wear is usually measured according to ISO 3685 standard which is also used by Grzesik [4], where a VB_x or VB_{max} value is produced, which is the distance measured from the cutting-edge perpendicular to lowest worn area of the tool. Tool life limits often range at about 0.3 mm for flank wear, and average value for the VB parameter.

2. Design of experiment

In order to study the adhesive wear mechanism and wear characteristics during the process of drilling, a friction couple model is determined. This means that a specific type of metal with known geometry and mechanical properties is used as the workpiece material and also a drill with known geometry and material characteristics is used as the tool. The friction couple is designed in the 3D CAD system and it is simulated with use of the Deform-3D software. Moreover, in order to validate the data collected after the finite element analysis, experiments will be conducted under the same cutting conditions.

Regarding the material that is used as a specimen, this is H-13 steel, also known as Orvar [23]. This material is used as a tool in quite demanding applications where high thermal loads are applied such as extrusion, plastic molds or die castings. Also, the yield strength R_p of Orvar is $880\,\mathrm{MPa}$ and tensile strength R_m is $550\,\mathrm{MPa}$ in soft annealed condition.

The tool that is used plays a major role on the system that is studied. The most common style of drills are the broadly known twist drills, and this is the drill which is implemented in this work. The drill is made out of tungsten carbide (WC) with a single layer coating. A commercial drill which meets the above requirements was found. It is TiN coated via the PVD process with a layer thickness of $1.5 - 4\,\mathrm{mm}$ at $500^{\circ}\,\mathrm{C}$. Also, the coating's hardness reaches $2400\,\mathrm{HV}$ and more importantly, its friction coefficient μ is measured at 0.5 according to the manufacturer.

The friction couple generated is the specimen to be processed and the drill. The specimen is H-13 steel which was secured on the CNC machine's table in an appropriate way in order to provide the best possible rigidity. The cutting conditions that were used during the process were chosen according to the manufacturer's recommendations for the specified drill series. Also, the cutting process was dry, which means that no cutting oil or emulsion was used. Table 1 shows the cutting feeds and speeds that were used both for the simulation and the experiment. The holes produced were of 20 mm depth.

Table 1. Cutting speeds and feeds for the simulations.

Parameters	Parameters' variables	Values
Cutting speed (m/min)	Vc	40
Spindle speed (rpm)	N	1270
Feed per revolution (mm/rev)	f_{rev}	0.1
Feed per minute (mm/min)	f	127

3. Results

What is interesting about the split point geometry is the fact that the temperature plots on both the tool and the workpiece tend to get a shallow slope. This means that the system starts to reach a steady state condition in machining. However, as there is no coolant during process except air, and knowing that the tool wear is gradually increasing, it is reasonable that the temperature plots will always show elevating temperatures Figure 2.

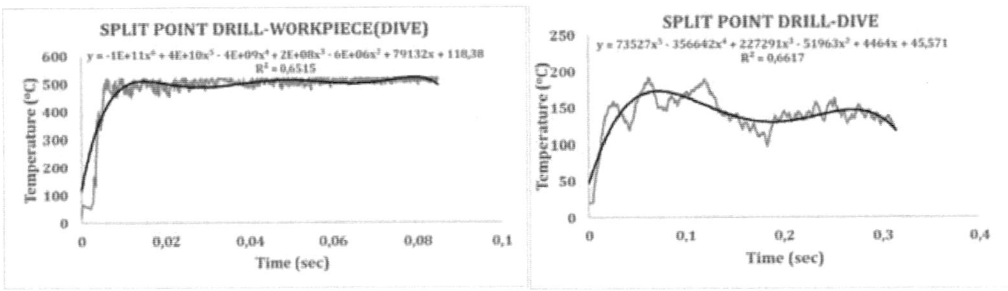

a b

Figure 2. Temperatures generated on the workpiece (a), and temperature on the tool (b) during process.

When the whole cutting edge of the drill is cutting material, the wear rate increases, and the areas that tend to be subject to high friction and stresses, and wear faster, are the ones in the periphery of the drill due to their highly positive rake angle and the higher amount of material they have to remove.

Figure 3. H-13 specimen of experiment, chips produced and simulation temperatures.

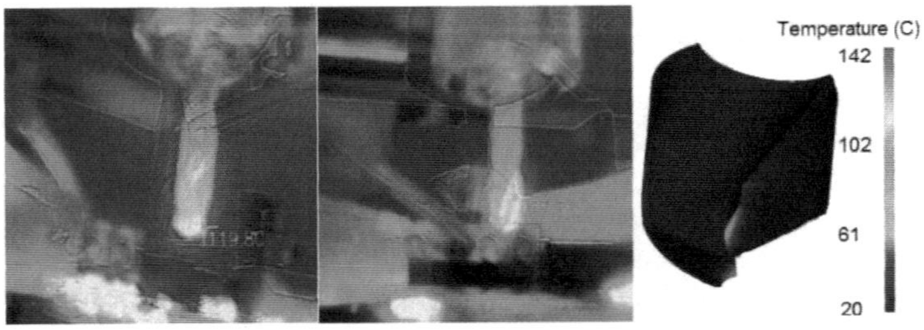

Figure 4. Temperatures of the high performance drill.

Regarding the temperatures, the drill showed results pretty close to the simulations. In Figure 4, the temperatures are illustrated. More specifically, the thermal camera used to record the temperature during machining showed the following results:

- 120 °C on the third hole and
- 135 °C on the fifth hole.

Finally, after inspecting the chips' microstructure on SEM, the abrasive and adhesive wear in the tool-workpiece interface is presented (Figure 5). In this figure, it can be seen that abrasion takes place on the workpiece, leading to the creation of chips. In addition, adhered masses can be seen, which are probably fragments of the drill's coating.

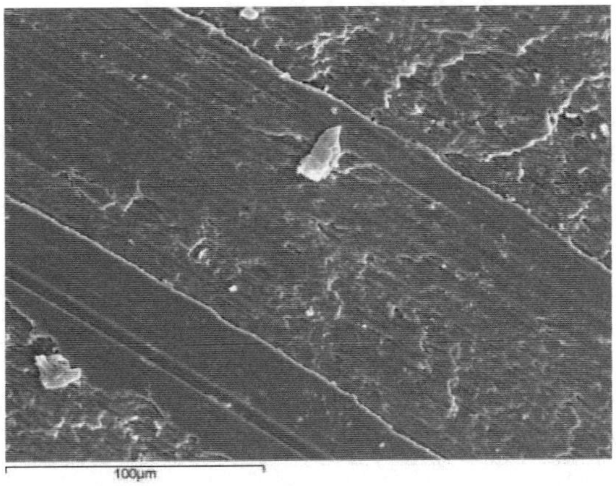

Figure 5. SEM image indicating the abrasive and the adhesive wear of the piece.

4. Conclusions and future work

The temperatures calculated from the model ranged approximately around 140 °C. The differences on the model and the thermal camera were caused by the thermal camera's calibration, and the difficulty in recording the exact tip of the tool. The optimization of geometry in twist drills is critical in order to make them more efficient and productive during drilling, especially in cases where tough materials are machined. Also, the coating used is an important factor which is responsible for low friction and wear on the tool. Nevertheless, the experiment comes to good agreement with the simulations.

Taking the results of this paper a step further, future work can focus on the chemical properties of the materials during drilling such as the oxidation and the effects of lubrication/cooling in elevated temperatures. Assessing the performance of the tool under altered tribological conditions (i.e. upon the addition of lubricant) can provide a better understanding of the suitability of each tool material for different applications. In other words, it is important to carefully select the appropriate combinations of metals and coatings for the drill and the specimen in order to achieve more efficient drilling with reduced tool wear due to high temperature and oxidation.

Based on the previous findings of other researchers, our future work will therefore elaborate on the effects of different lubricating materials in the level of oxidation and consequently, endurance of the drill.

References

[1] Horvat M and Botak Z 2018 The influence of drill point geometry on tool life *Teh. Glas.* 12 1–4
[2] Sharif S and Rahim E A 2007 Performance of coated- and uncoated-carbide tools when drilling titanium alloy-Ti-6Al4V *J. Mater. Process. Technol.* 185 72–6
[3] Wong F R, Sharif S, Kamdani K and Rahim E A 2008 The Effect of Drill Point Geometry and Drilling Technique on Tool Life when Drilling Titanium Alloy , Ti-6Al-4V 21–3
[4] Grzesik W 2008 *Tribology of Metal Cutting Adv. Mach. Process. Met. Mater.* 149–61
[5] Dahlström A 2015 Wear mechanisms in austenitic stainless steel drilling: A comprehensive wear study 69
[6] Chauhan K V and Rawal S K 2014 A Review Paper on Tribological and Mechanical Properties of Ternary Nitride based Coatings *Procedia Technol.* 14 430–7
[7] Knotek O and Leyendecker T 1987 On the structure of (Ti, Al)N-PVD coatings *J. Solid State Chem.* **70** 318–22
[8] Matei A A, Pencea I, Branzei M, Trancə D E, Ţepeş G, Sfət C E, Ciovica E, Gherghilescu A I and Stanciu G A 2015 Corrosion resistance appraisal of TiN, TiCN and TiAlN coatings

deposited by CAE-PVD method on WC-Co cutting tools exposed to artificial sea water *Appl. Surf. Sci.* 358 572–8

[9] Wang Q and Zhou F 2017 Progress in tribological properties of nano-composite hard coatings under water lubrication *Lubricants* **5** 1–21

[10] Kato K and Adachi K 2001 Wear Mechanisms 7.1 7.2

[11] Usui E, Shirakashi T and Kitagawa T 1984 Analytical prediction of cutting tool wear *Wear* 100 129–51

IOP Conf. Series: Materials Science and Engineering 1037 (2021) 012014 doi:10.1088/1757-899X/1037/1/012014

Surface modification of tool steel by using EDM green powder metallurgy electrodes

M Balanou[1], P Karmiris-Obratański[1,2,*], E L Papazoglou[1], N I Galanis[2], A P Markopoulos[2]

[1] School of Mechanical Engineering and Robotics, AGH University of Science and Technology, Cracow, Poland
[2] School of Mechanical Engineering, National Technical University of Athens, Athens, Greece

*Corresponding author's e-mail: pkarm@mail.ntua.gr

Abstract. Electro-discharge machining (EDM) is mostly used for machining difficult-to-cut materials which are difficult to machine by conventional machining processes. In recent years, EDM is also used for surface modification. This work aims to investigate the phenomenon of surface modification of tool steel Calmax (Uddeholm) by EDM process using Cu-ZrO_2 powder metallurgy green compact electrode. Results show that the higher Material Transfer Rate of 46.5 mgr/min is achieved on the combination of I_p=9A and T_{on}=25 μs. The effect of peak current and pulse duration on the material transfer rate and surface roughness (Ra and Rz) was investigated. Machined surfaces were characterized by scanning electron microscopy (SEM).

1. Introduction

Tool steels used in applications that require improved surface properties such as high hardness at high temperature, wear resistance, and corrosion resistance. Hence, the surface modification process is necessary. Conventional methods such us chromium plating, carburizing, nitriding, chemical vapor deposition (CVD)/ physical vapor deposition (PVD), plasma arc spraying and ion implantation are used for surface modification to improve the performance and life of the material [1]. Nevertheless, the high cost of equipment and the special surface preparation makes these processes complicated and time consuming. These issues with conventional methods lead to the use of an alternative surface modification process, such as the electrical discharge machining [2,3].

Electrical discharge machining is one of the most commonly used thermo-electrical non-conventional machining processes, used for machining of difficult-to-cut materials. It is based on the principle of removing material using electric discharge that occurs between an electrode and a workpiece in the presence of a dielectric fluid. In this process, there is no contact between the tool and the workpiece, so, no tool forces are applied [4]. EDM is used in a lot of industries such as aerospace, automotive, micro-electronics, biomedical, die and mould production. Additionally, EDM is used for machining hard materials, and complex shapes with high precision [5].

Two conductive electrodes, a cathode and an anode are immersed in a dielectric fluid (oil or deionised water). A voltage difference is applied between the electrode and the workpiece, separated by a dielectric fluid, causing the ionization of the dielectric fluid. When the applied voltage overcomes the dielectric strength of the fluid, a plasma channel (electric discharge phase) is formed in between the electrode and the workpiece. The electric discharge generates high temperatures (6,000– 12,000 °C) and causes melting and evaporation of the material at the point of discharge, leading to material removal from both the electrodes. As a result, a small crater is created both on the tool electrode and workpiece surfaces [5, 6]. A part of the molten and all of the evaporated material is flushed away by the dielectric fluid, while another part of the molten material from the tool and the constituents of the dielectric get re-solidified on the machined surface, generating the white layer (WL) or recast layer. The formation of recast layer is attributes of material transfer and is useful in surface modification of the base material by EDM process.

Surface modification has been observed in the EDM process for over five decades. It can be done in different ways which includes conventional electrode materials, powder metallurgy electrodes and powder-mixed dielectric fluid [7]. The P/M electrodes can be classified into two types, green compact type and sintered type. Samuel and Philip compared conventional solid electrodes with P/M electrodes. The P/M electrodes were found to be more sensitive to changes in peak current and pulse duration [8]. Compaction and sintering parameters conditions play an important role in the electrode performance and hence the EDM process performance [9]. The machined surface with P/M electrode improved the hardness, wear, fatigue, corrosion, oxidation resistance.

A lot of research works for EDM surface modification using PM electrodes have been carried out. Patowari et al. [10] studied the surface integrity of C-40 steel in EDM. WC-Cu powder metallurgy (P/M) green compact tools were used. MTR, TWR and SR were considered as the output responses. It was found that WC was transferred over the work surface in the form of a hard and uniformly deposited layer. T_{on} and I_p and settings have significant influence over the process. Gill and Kumar [2] machined hot die steel (H11) using Cu-Mn powder metallurgy electrode. Taguchi method was used to design the experiment. The formation of cementite, ferrite and manganese carbide phases were responsible for increase in MH. Gülcan et al. [11] investigated the effect of Cu–Cr and Cu-Mo powder metal tool electrodes on electrical discharge machining performance outputs. SAE 1040 steel workpiece was used as a workpiece material. They revealed that electrode material was deposited as a layer over the work surface which yields high surface hardness, corrosion resistance and strong abrasion. Gill and Kumar [3] investigated and compered the surface micro-hardness of H11 die steel using composite tool electrode (Cu–Cr–Ni) manufactured by powder metallurgy and conventional copper tool. Taguchi method was used to design the experiment. They have shown the improvement in micro hardness and good wear resistance of the machined surface due to the presence of cementite, intermetallic compound of iron, chromium and nickel and chromium carbide. Kumar et al. [12] analyzed MTR and SR on OHNS workpiece using CrB_2-Cu powder metallurgy electrode. It was established that desired deposition of hard composite layer was found on workpiece. Chundru et al. [13] studied the surface modification of Ti6Al4V alloy using TiC/Cu P/M electrode made with particle size varying from nano- to micron size. Taguchi method was used to design the experiment. They indicated that high reactive surface area of nanoparticles made greater surface alloying than the other tool electrodes. Hence, better surface roughness and improved hardness values were obtained.

The present study investigates the surface modification of tool steel Calmax (Uddeholm) by using a Cu-ZrO_2 green compact P/M tool electrode. The effect of peak current (I_p) and pulse-on duration (T_{on}) on the material transfer rate and the surface roughness was investigated. The machined surface was characterized with scanning electron microscopy (SEM) and energy-dispersive spectroscopy (EDS).

2. Experimental procedure

In the present work, experiments were carried out on tool steel Calmax (Uddeholm) using tool electrodes manufactured by Powder Metallurgy process. The chemical composition of the workpiece is shown in Table 1. The workpieces were rectangular blocks with dimensions 85×29×7mm. The electrode was produced by powder metallurgy technique by mixing Cu and ZrO_2 powder at 70:30 weight ratio. The powders were homogeneously mixed. The powder mixture was compacted under 100 MPa pressure using a hydraulic press. The electrode was made in cylindrical shape with a diameter of 20mm and height of 15mm. Kerosene is used as dielectric fluid.

Table 1. Chemical composition of workpiece material Calmax (Uddeholm).

	C	Si	Mn	Cr	Mo	V
Typical Analysis %	0.6	0.35	0.8	4.5	0.5	0.2

A die sinking electrical discharge machine (Agie Charmilles Roboform 350 Sp) was used to perform the experiments. The selected input parameters are given in Table 2. Peak current and pulse duration were used to investigate the effect of the material transfer rate (MTR) and surface roughness (SR). Fig. 1 shows the experimental set up of EDM and the workpiece after EDMachining.

Table 2. Input parameters.

Parameters	
Peak Current - I_p	5, 7 and 9 A
Pulse-on duration - T_{on}	12.8, 25 and 50 μs
Duty Factor η	0.5
Dielectric Fluid	Kerosene

The material tranfer rate was calculated by measuring the weight difference of the workpiece before and after EDM for a particular time period. MTR calculated using equation 1:

$$MTR = \frac{W_i - W_f}{t} \quad (1)$$

where W_i and W_f the weight of the workpiece before and after the machining (g) and t the machining time (min). Surface roughness of the machined surface is measured on random surface area by TOPO 01P contact profilometer. ISO 25178-2 was used to analyze the experimental results. The roughness parameters that analyzed are the sum of the maximum valley depth and maximum peak height, Rz and arithmetical mean of local peaks, Ra within a definition area. The results of MTR and surface roughness parameters are shown in Table 3. The surface morphology was further analyzed by SEM, equipped with EDS to confirm the material transfer.

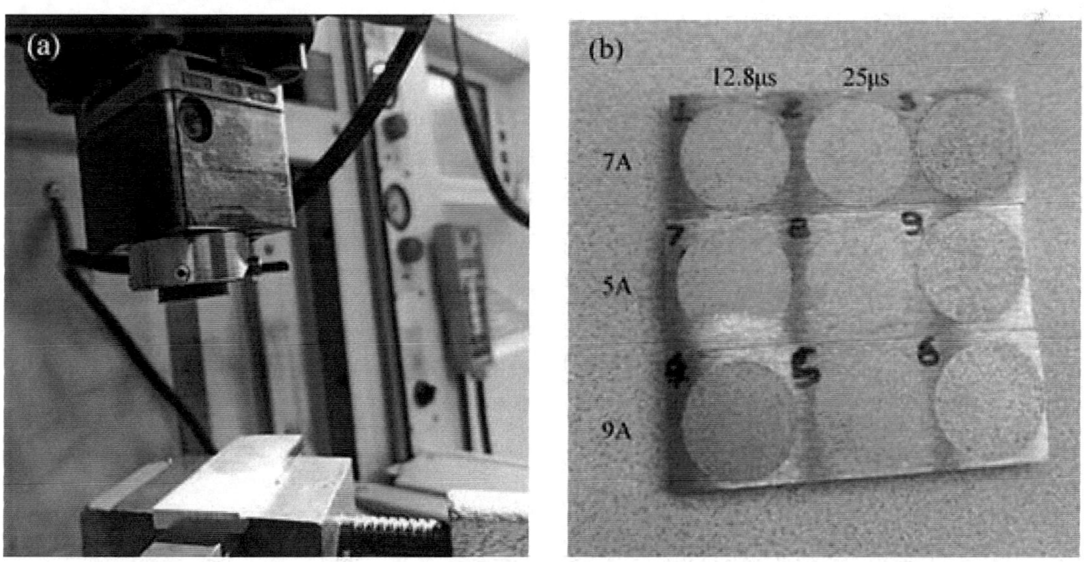

Figure 1. (a) Experimental set up and (b) Workpiece after EDM process.

Table 3. Experimental results.

I_p (A)	T_{on} (μs)	MTR (gr/min)	Ra (μm)	Rz (μm)
5	12.8	0.0228	3.72	61.08
5	25	0.0072	4.34	88.04
5	50	0.0117	6.27	101.96
7	12.8	-0.2493	5.75	99.93
7	25	0.0103	4.89	84.8
7	50	0.0082	7.12	129.14
9	12.8	0.0333	4.31	64.62
9	25	0.0465	5.07	68.21
9	50	0.0205	6.29	111.38

3. Results and discussion

3.1 Material Transfer Rate

The MTR is one of the important performance measures and is dependent on the process parameters. MTR was measured by taking the mass difference of the workpiece before and after EDM for a particular time period. The MTR for each sample are listed in Table 3. The negative value of MTR represents the net material removal and the positive values represent net material addition. The material transfer rate during EDM for different I_p and different T_{on} are illustrated in Fig. 2.

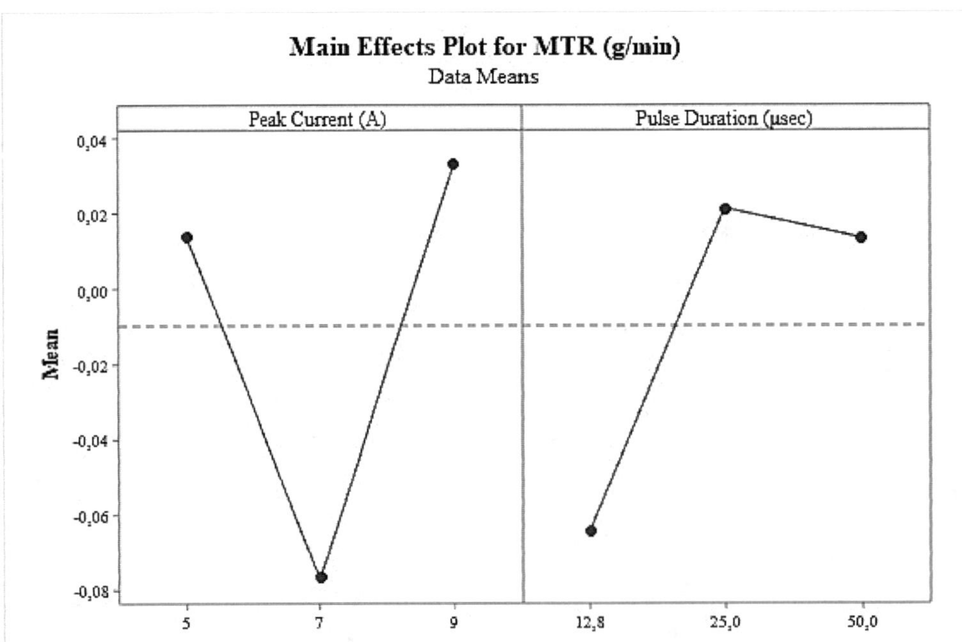

Figure 2. Main effect plot of the MTR.

From the main effect plot it can be observed that the MTR is reduced by an increase in peak current from 5 to 7A and receives a negative value for I_p= 7A. With a further increase in peak current, MTR is also increased. As the pulse on duration is increased from 12.8 to 25 μs, MTR is also increased but it appears to have a small decrease with longer pulse duration. From the interaction plot it becomes obvious that the material transfer rate is influenced by the machining parameters.

3.2 Surface roughness of the machined surface

The morphology of the machined surface is due to the amount of energy released during the EDM process. This morphology includes micro-irregularities, macro-deviations and surface waviness. The surface roughness is evaluated by the sum of the maximum valley depth and maximum peak height, Rz and an arithmetical mean of local peaks, Ra within a defined area of the micro-irregularities. The peak current and pulse duration influence the SR. Pulse duration allows the plasma column to expand beyond the point of electrical discharge, leading to larger diameter craters. On the other hand, the peak current affects the depth of the crater. The measured surface roughness parameters, Ra and Rz, for each sample are listed in Table 3.

Figure. 3 shows the main effect plots and the interaction plots Ra and Rz. From the main effect plot in Fig. 3 it can be observed that the Ra is increased with an increase in peak current. In pulse duration main effect plot it can be observed that Ra elevates slowly with an increase in pulse duration from 12.8 to 25 μs, and with further increase in pulse duration, Ra is also increased up to 42.9%. Fig. 3 (b) shows that in the peak current main effect plot, the Rz is increased as the peak current increases from 5 to 7 A, and it is getting lower for 9 A. However, Rz is increased up to 51.7% with an increase in pulse duration.

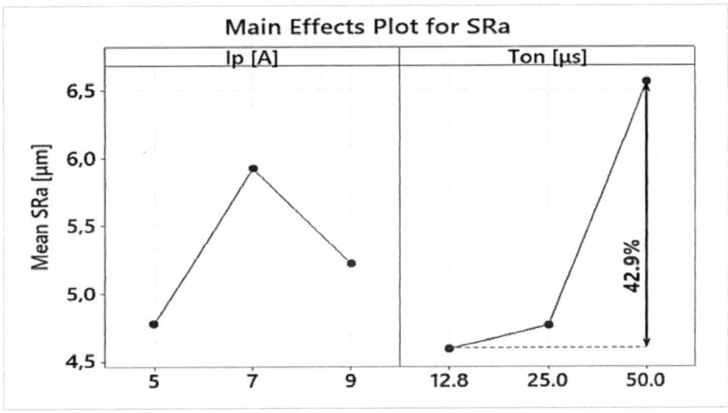

(a)

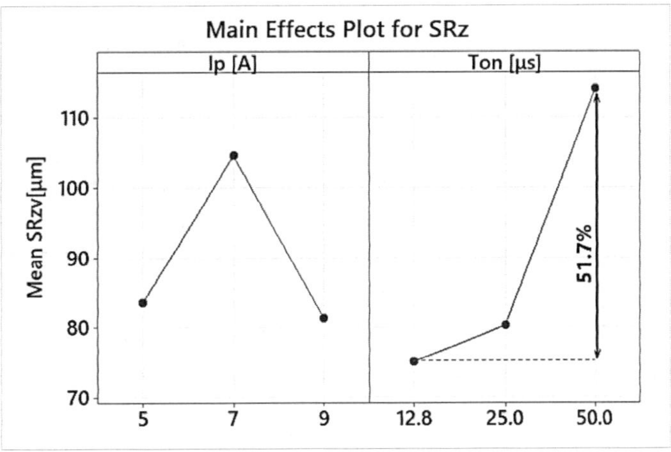

(b)

Figure 3. Main effect plot (a) for Ra, and (b) for Rz.

3.3 Surface analysis

As has been stated above, the morphology of the machined surface is due to the energy that released during EDM, which causes melting and evaporation of the material, followed by rapid cooling. Only a small amount, about 15% or less, of the molten material is flushed away by the dielectric fluid [14]. The un-expelled molten metal which had not been flushed away by the dielectric re-solidified and gets deposited in the crater and in the surroundings to form the recast or white layer.

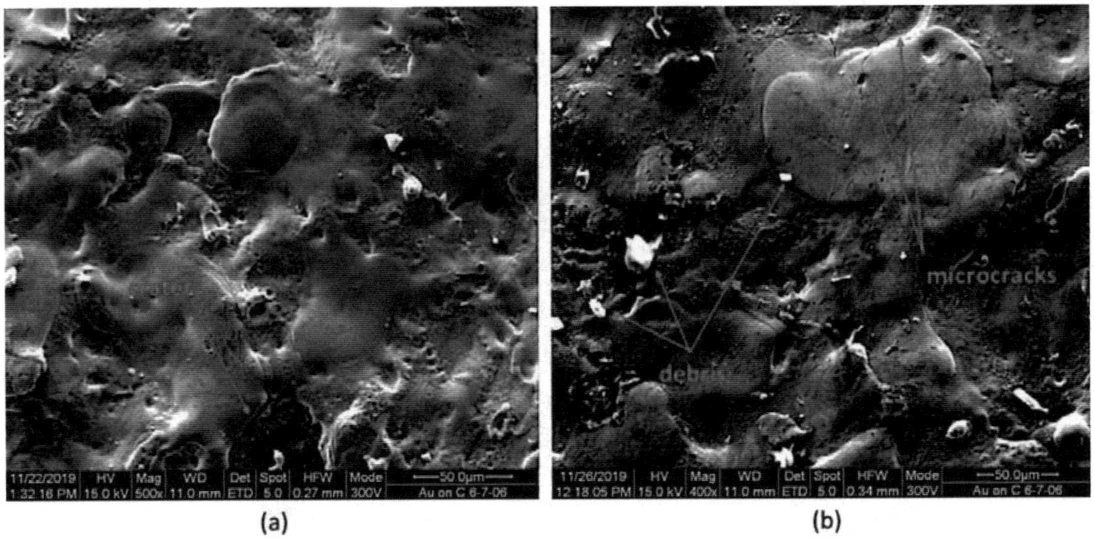

Figure 4. SEM micrographs of the machined surface
(a) I_p= 5A and T_{on}= 50 μs, (β) I_p= 9A and T_{on}= 50μs.

Examinations by SEM (Fig. 4) revealed that the machined surface covered with a distribution of craters along with irregular flow marks of molten metal. Also, micrograph shows the presence of globules of debris formed by expelled molten metal which re-solidified and gets deposited on the workpiece surface. Micro cracks and micro voids were observed on the surface of sample. The rapid heating and cooling cycles and the micro-structural transformation during machining induce residual stresses, i.e. thermal and tensile stresses, within the machined surface, that lead to micro cracks formation on the surface. Figure 4 shows that as the pulse on current and time is increased the surface has higher roughens with an intense presence of debris and the formation of microcracks.

4. Conclusions

The present study was carried out using Cu-ZrO_2 P/M green compact electrode for surface modification of tool steel Calmax (Uddeholm). Peak current (I_p) and the pulse-on time (T_{on}) are used input parameters. Material transfer rate has been calculated, surface roughness has been measured, and the machined surface has been observed with scanning electron microscopy. From the experimental results, the following conclusions are drawn.

- Surface modification of tool steel Calmax (Uddeholm) was successfully carried out by using Cu-ZrO_2 P/M green compact electrode.
- Peak current and pulse on duration parameters have significant influence over MTR and SR.
- The experimental results show that the higher depositions was achieved 46.5 mgr.min for the combination of pulse current 7A and pulse on time 25 μs.
- The Ra varies from 3.72 μm for the combination of I_p=5A and T_{on}=12.5 μs to 6.29 μm for the combination of I_p=9A and T_{on}=50μs.
- Micro cracks and micro voids are observed in the workpiece surface.

References

[1] Hamilton S., 'Overview of surface treatments currently available to the moulding industry', In: 21st Mould making Conference '95—Innovative Technology for Mould Manufacturers & Users, Solihull, UK, Institute of Materials, 1995.

[2] Amoljit Singh Gill, Sanjeev Kumar, 'Surface Roughness and Microhardness Evaluation for EDM with Cu-Mn Powder Metallurgy Tool', Materials and Manufacturing Processes, Vol. 31, Issue 4, 2015, pp. 514-521.

[3] Amoljit Singh Gill, Sanjeev Kumar, 'Micro-hardness evaluation for surface alloying of H11 die steel with Cu–Cr–Ni powder metallurgy tool in electrical discharge machining', Journal Engineering Manufacture, Vol. 232, 2016, pp.1-13.

[4] H. S. Payal, B. L. Sethi, 'Non-conventional Machining Processes as Viable Alternatives for Production with Specific Reference to Electrical Discharge Machining', Journal of Scientific & Industrial Research, Vol. 62, 2003, pp. 678-682.

[5] M. P. Jahan (ed.), 'Electrical Discharge Machining (EDM): Types, Technologies and Applications', Nova Science Publishers, New York, 2015, pp.1-17.

[6] M. Kunieda, B. Lauwers, K. P. Rajurkar, B. M. Schumacher, 'Advancing EDM through Fundamental Insight into the Process', CIRP Annals - Manufacturing Technology, Vol. 54, Issue 2, 2005, pp. 64-87.

[7] Sanjeev Kumar, Rupinder Singh, T. P. Singh, B. L. Sethi, 'Surface modification by electrical discharge machining: A review', Journal of Materials Processing Technology, Vol. 209, 2009, pp. 3675–3687.

[8] M. P. Samuel, P. K. Philip, 'Powder metallurgy tool electrodes for electrical discharge machining', International Journal of Machine Tools & Manufacture, Vol. 37, Issue 11, 1997, pp. 1625–1633.

[9] Naveen Beri, Sachin Maheshwari, Chitra Sharma, Anil Kumar, 'Technological Advancement in Electrical Discharge Machining with Powder Metallurgy Processed Electrodes: A Review', Materials and Manufacturing Processes, Vol. 25, 2010, pp. 1186–1197.

[10] P. K. Patowari, U. K. Mishra, P. Saha, P. K. Mishra, 'Surface Integrity of C-40 Steel Processed with WC-Cu Powder Metallurgy Green Compact Tools in EDM', Materials and Manufacturing Processes, Vol. 26, 2011, pp. 668–676.

[11] Orhan Gülcan, ˙Ibrahim Uslan, Yusuf Usta, Can Çogun, 'Performance and surface alloying characteristics of Cu–Cr and Cu–Mo powder metal tool electrodes in electrical Discharge machining', Machining Science and Technology, Vol. 20, 2016, p. 523-546.

[12] P. Mathan Kumar, K. Sivakumar, N. Jayakumar, 'Surface Modification on OHNS Steel Using Cu-CrB_2 Green Compact Electrode in EDM', Materials Today: Proceedings, Vol. 5, 2018, pp. 17389–17395.

[13] Venkata Rao Chundru, Ramji Koona, Srinivasa Rao Pujari, 'Surface Modification of Ti6Al4V Alloy Using EDMed Electrode Made with Nano- and Micron-Sized TiC/Cu Powder Particles', Arabian Journal for Science and Engineering, Volume 44, Issue 2, 2018, pp. 1425–1436.

[14] L. C. Lim, L. C. Lee, Y. S. Wong, H. H. Lu, 'Solidification microstructure of electro discharge machined surfaces of tool steels', Materials Science and Technology, Vol. 7, 1991, pp. 239-248.

IOP Conf. Series: Materials Science and Engineering 1037 (2021) 012015 doi:10.1088/1757-899X/1037/1/012015

Axiomatic Design as an innovation framework

Miguel Cavique[1,2*], Oana Dodun[3], Christopher Brown[4]

[1] Unidemi, Faculdade de Ciências e Tecnologia, 2829-516 Caparica, Portugal
[2] Department of Sciences and Technology, Naval Academy, Base Naval de Lisboa, Alfeite, 2810-001 Almada, Portugal
[3] Department of Machine Manufacturing Technology, Faculty of Machine Manufacturing and Industrial Management, "Gheorghe Asachi" Technical University of Iaşi, Bdul. Mangeron, 59A, 700050 Iaşi, România
[4] Department of Mechanical Engineering, Worcester Polytechnic Institute, Worcester, MA, USA

*Corresponding author's e-mail: miguel.cavique@cest.pt cavique.santos@marinha.pt

Abstract. Innovation can be defined as a set of processes for altering human activity, with economic or social relevance. Innovation has been the economic driver of several companies and countries. This article presents laws with preconditions for innovation, and it proposes a framework for innovation based on Axiomatic Design (AD) theory. There are three laws which set out necessary preconditions to initiate and maintain innovation. Research, financing, and companies that can carry out an innovation project, must be available for a project. AD can be a tool for innovation. Three main elements to AD are: Suh's axioms, to maintain functional independence and minimize information content; a structure, domains and hierarchies in which solutions develop; and a process, decomposing from abstract to detailed components and physically integrating components into a complete solution. Suh's axioms ensure that design solutions are adjustable, controllable, and robust. These are essential elements in innovation. Design matrices for functional–physical relationships select for independence, then information content is used to select the best solution. Design structures are completed by zigzagging between functional, physical, and process domains, creating the hierarchical decompositions of abstraction while applying Suh's axioms at each level.

1. Introduction

Innovation is a process of applying new solutions with economic or social profitability to satisfy the needs of society. To be innovative, solutions must be deployed, satisfying societal needs.

The most successful companies always bet on innovation. The so-called technological companies (information and related technologies) have been strongly investing in innovation. Apple Inc. grosses more than many European countries' GDP. The first ten largest technology companies in the world in 2019 include five from the USA, two from Japan, one from South Korea, one from Taiwan, and one from China. There is not a single EU company in this rank, which is a sad statement on Europe's position in a new technological world.

Creativity is coming up with new, potentially useful ideas. Developing them into complete solutions with Axiomatic Design (AD) assures technological viability and sustainability, key requirements for ethical innovation [1].

There has always been innovation, even without the aid of current methodologies and theories. A century ago, Henri Ford's phrase is paradigmatic of how to think innovation: "If I had asked my customers what they wanted, they would have said a faster horse." Ford was able to identify a need outside of the physical context of horses and innovate a new means of affordable transportation, the Ford Model-T. This new concept satisfied a need and led the company to produce 15 million of this type of vehicle. Achieving technological viability apparently took through "T" in the alphabet to label models.

Innovative solutions need a lot of work. Therefore, it is advantageous to support innovation efforts with methods or theories that guide reasoning. Innovation by a flash of genius is overrated. Innovation usually occurs after significant efforts rather than in a flash.

Regarding new solutions, there are two types of problems, those that can be solved with current knowledge and those that need new knowledge. C-K theory [2] addresses this issue of connecting innovative concepts and the knowledge required to realize them.

In many common situations, there are algorithms to achieve a solution, even though the application of these algorithms may require a lot of previous knowledge. The dimensioning of a steel structure follows standard codes; most engineering problems use codes. Similarly, in the medical field, doctors use protocols. Algorithmic problems need to be framed in a type of algorithm, asking for analysis to reach a solution. Many issues in day-to-day life use defined analyses for selecting solutions.

Synthesis reasoning is applied in ill-defined problems that need new solutions. Universities must play decisive roles by training for synthesis processes, by focusing on synthesis reasoning [3]. Synthesis reasoning includes abduction and induction. Synthesis can use techniques such as brainstorming. These methods are experiential, relying on the experiences and intellects of participants. Assertiveness of participants can dominate solution selections when there is no tool to evaluate the quality of new proposals. Therefore, scientific methods and theories for innovation, which guide and verify the quality of proposed solutions, are welcome. It is these theories, with laws, principles, and axioms, that distinguish scientific disciplines, with broad applications and powerful problem-solving abilities from experiential ones, which have more limited applicability.

When fostering innovation in an organization, a region, or a country, it is necessary to define how innovation occurs. It is essential to know when the innovation process starts and where, to what extent it is socially inclusive or determined by a department, and what is the implicit social responsibility [4]. Homogeneous innovation asks all persons of an institution to commit; conversely, in heterogeneous innovation, only specific areas of institutions work on innovative processes. Choosing one or the other type of innovation defines ways to finance innovation. Providing resources to a department or spreading them around the institution, is a crucial decision.

Research is a fundamental pillar in innovation. Several research institutions must be financed. Some that present disruptive projects must be chosen. In Europe, the tendency has been to share large sums through different users. However, bringing disruptive projects usually requires heterogeneous financing options.

Moreover, research supporting innovation can be open or closed. Entities may share innovation with their counterparts by developing innovation in an open environment [5]. Open innovation occurs in universities, where publications, ideas, and solutions spread around the world. Researchers share ideas and receive new ideas from other researchers, building an increasingly better and more competitive system. In specific situations, entities may not want to share their conclusions. Thus, innovation occurs within the entity, avoiding competition, but reducing the benefits of exchanging ideas with other institutions. In the XV century, maps, and navigation techniques were Portuguese secrets for sailing over the oceans. Similarly, NASA innovated in the sixties, hiding patents.

In conclusion, strategies for innovation must be consistent with an entity's objectives.

Some of the most well-known innovation methods are, Value Analysis, QFD, and TRIZ can be used together in systematic models of innovation [6]. TRIZ is commonly used with AD to solve couplings and create innovative frameworks [7].

This paper uses AD theory to identify frames of innovation defined by AD entities. It presents a different way of looking at AD as an innovation tool rather than a design tool.

The following section describes different types of innovation and laws that govern innovation. Emphasis is placed on the first Innovation law, which establishes necessary conditions for innovation. Then, in the next section AD is described as a framework for innovation with Suh's axioms, structure with domains and hierarchies; and processes with decomposition and physical integration. These ensure that design solutions are adjustable, controllable, and robust, which are key elements in innovation.

2. Innovation Laws

Nam P. Suh explained Axiomatic Design theory in 1990 in Principles of Design [8] and expanded it recently with more examples of applications [9]. In 2010 Suh defined three laws of innovation [10]. The first law defines conditions required for innovation to occur:

- "For innovation to occur, there can be no lack of any element or link in the innovation continuum."

This continuum of innovation requires an infrastructure for innovation with solid bases for fundamental research and technological invention, a financial community willing to provide capital and take risks, and industries with knowledge necessary to develop, manufacture and market new products.

To innovate, problems must be identified. Therefore, many new ideas need to be raised and evaluated to achieve one winning solution. Or in other words, research processes must launch many projects for each new product that succeeds in the market. Moreover, innovation stops if any part of the innovation structure fails. There is no innovation without risking capital, without companies able to take risks to produce and commercialize an innovative product. There is no innovation without a strong and productive research apparatus. In other words, universities or research companies cannot work alone. A country with an increasing number of researchers may fail to grow its GDP if there is no capital and companies capable of leading innovation. A risk-averse structure does not innovate.

Finally, it is necessary to choose the best solutions by using a method, or theory that measures solutions' viability, which AD can do.

We will now look briefly at the second and third laws of innovation. The second law says:

- "An Innovation hub will be nucleated if the initial nucleate size exceeds the critical size needed for stability, and if the activation energy barrier for nucleation can be overcome."

And the third law:

- "For innovation hubs to nucleate, the nucleation rate of innovation in a region must be greater than the rate at which innovative ideas, people, and financial capitals can move away from the region."

Both laws explain why innovation develops in exceptional locations. Moreover, once established and productive, innovation hubs attract skilled persons and resources, making it more difficult for hubs to appear in other regions. These laws explain why there is no Silicon Valley in Texas or Europe, and why wealthy areas tend to be more prosperous in a free market.

Innovation processes respond to societal or institutional problems. Therefore, raising problems is paramount. Knowing "what the problem is" and asking questions helps to define needs. Needs and problems are two sides of the same coin. The need must be identified neutrally and fundamentally. A technique to identify the needs, while avoiding faster horses, is by asking "why" successively until reaching high levels of abstraction, thereby enlarging solution spaces.

Problem-solving involves knowing the organization. The higher the approach, the more significant the impact of innovation. Starting by clarifying an organization's mission, creating its vision, and defining its values are clues for starting innovation processes. The impact will be progressively less if action is taken on policies, departments, procedures, or technologies. Creating a culture of innovation allows organizations to act on all these levels [11]. On the other hand, identifying a disconnected problem will enable conceptualization the needs of the organization if there is an in-depth knowledge of it. How to advance problems to their solution is the aim of design theories.

3. Axiomatic Design Theory

Axiomatic design theory states that the best design solutions can be selected using Suh's two axioms [8].

- First, Independence: "Maintain the independence of functional requirements";
- Second, Information: "Minimize the information content of the design."

This study focuses mainly on the first Axiom, independence. AD develops design solutions in four domains: customer, functional, physical, and process. In the customer domain, elements are customer needs (CNs). Functional requirements (FRs) belong to the functional domain. The functional domain contains answers to "What needs to be achieved?". The physical domain answers, "How can it be achieved?", with the design parameters (DPs). Finally, the process domain contains process variables (PVs), answering the question "How can it be built?".

CNs define needs, bounded by constraints, that design solutions must comply with. Figure 1 depicts the four domains and the mapping between the domains. Functional mapping occurring between CNs and FRs, can be called the conceptual project. Physical mapping is about the relationships between the FRs and the DPs, which can be called the physical design. Finally, each DP needs to be manufactured, or obtained in the market. The relationship between DPs and PVs can be called process mapping.

A complete design solution representation should contain all four of these aspects, although most only contain physical designs, drawings of physically integrated DPs. This kind of limited representation inhibits design modifications because design intent, contained in the FRs and CNs, is not included.

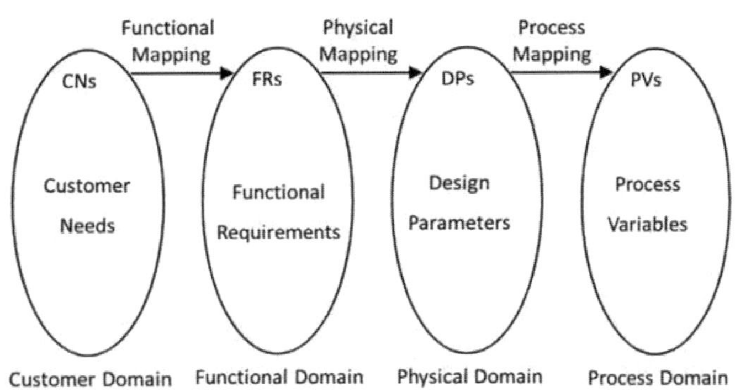

Figure 1. Design Domains and Mapping

After defining the CNs, the highest level FRs are defined. This deserves attention. Good FRs are essential to innovation. Design solutions can be no better that their FRs. Each FR has a corresponding DP, which constrains child FRs. If a DP in one branch proposes pneumatics for transferring loads, then all lower level FRs should address pneumatics, as opposed to electromechanics or hydraulics. However, at the lower levels, if it is found that a pneumatics solution violates Suh's axioms or a constraint, then an alternative DP should be selected.

This decomposition process is called zigzagging. It continues until reaching a set of DPs that are obvious and can be purchased or manufactured. At each level of a decomposition, children should be collectively exhaustive with respect to their parents and mutually exclusive with respect to each other (CEME), which complies with Suh's axioms. In this way, each level in each domain can be represented as vector with independent, i.e. orthogonal, components.

Figure 2 shows zigzagging, decomposing between functional and physical domains. The solid arrow from the physical domain to the functional domain reveals that a choice of a design parameter has implications on FRs at the next level of decomposition. Zigzagging is one of the fundamental aspects of Suh's AD theory. Complete design solutions are developed simultaneously across domains, at each level of their abstraction hierarchies, rather than one then another. A good solution has just the minimum number of FRs to guarantee satisfaction of the CNs. In turn, each FR should ideally be fulfilled by a

IOP Conf. Series: Materials Science and Engineering 1037 (2021) 012015 doi:10.1088/1757-899X/1037/1/012015

single DP, unique to that FR. This is required to comply with Suh's first axiom, independence. DPs are adjusted to control FRs and keep them in tolerance. If a single DP influences more than one FR, independence is diminished. One FR cannot be adjusted by changing that DP without changing other FRs influenced by that DP.

Design equations express relationships between FRs and DPs, $[FR]=[A]\cdot[DP]$, where [*A*] is a design matrix (DM), and [*FR*] and [*DP*] are vectors. DMs in Equation 1 shows three different types of designs, where unspecified elements of [A] are zero. From left to right, a) is diagonal and uncoupled, b) triangular and quasi or decoupled, and c) full and fully coupled.

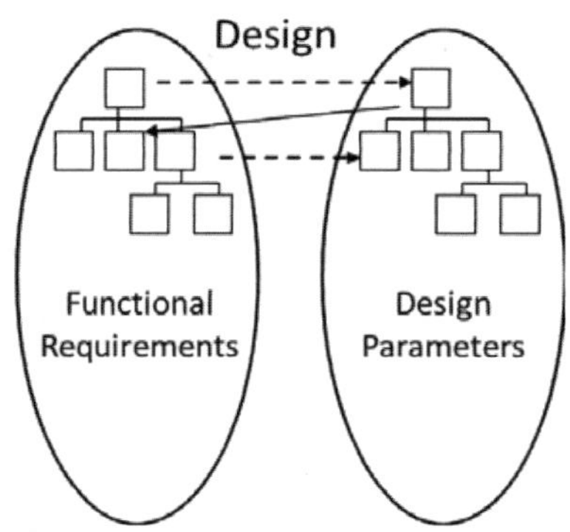

Figure 2. Zigzagging between the functional and the physical domains

Uncoupled designs (1a) allow FRs to be brought into tolerance by adjusting their corresponding DPs in any order without iterating. In these designs FRs are sensitive to variations in only one DP, because each DP influences only its unique, corresponding FR.

$$\begin{bmatrix} FR_1 \\ FR_2 \\ FR_3 \end{bmatrix} = \begin{bmatrix} a_{11} & & \\ & a_{22} & \\ & & a_{33} \end{bmatrix} . \begin{bmatrix} DP_1 \\ DP_2 \\ DP_3 \end{bmatrix} a) \quad \begin{bmatrix} FR_1 \\ FR_2 \\ FR_3 \end{bmatrix} = \begin{bmatrix} a_{11} & & \\ a_{21} & a_{22} & \\ a_{31} & & a_{33} \end{bmatrix} . \begin{bmatrix} DP_1 \\ DP_2 \\ DP_3 \end{bmatrix} b) \quad \begin{bmatrix} FR_1 \\ FR_2 \\ FR_3 \end{bmatrix} = \begin{bmatrix} a_{11} & a_{12} & a_{13} \\ a_{21} & a_{22} & \\ a_{31} & & a_{33} \end{bmatrix} . \begin{bmatrix} DP_1 \\ DP_2 \\ DP_3 \end{bmatrix} c) \quad (1)$$

Equation 1b shows a decoupled design, which are represented by a lower or upper triangular matrix. In these designs, DPs must be defined in a specific order to adjust their FRs without iterations. DP_1 is the first parameter to be adjusted to bring FR_1 into tolerance. DP_1 will influence the remaining FRs. The following DPs can adjust their FRs in order without taking any of the previous ones out of tolerance. In Equation 1b, DP_2 or DP_3 can be defined interchangeably to reach FR_2 and FR_3, which would not be the case if a_{32} were non-zero.

If FRs depend on several DPs simultaneously, and the design matrix cannot be arranged to make it triangular, then the design is coupled. It is a poor design (Equation 1c). This type of design can produce unstable systems that might not be adjustable and that could produce unintended consequences.

New concepts in an innovation process can be evaluated for technical viability using Suh's two axioms. At any level of a decomposition, the design to choose should be uncoupled, if possible, or decoupled, if not. Then the Information Axiom should be applied. The information content (I) can be calculated from the probability of success (p), Equation (2). Information content ranks remaining DP candidates. The one with the least information in the functional domain would be the best and should be selected.

$$I = \ln(\frac{1}{p}) \quad (2)$$

IOP Conf. Series: Materials Science and Engineering 1037 (2021) 012015 doi:10.1088/1757-899X/1037/1/012015

4. AD frames of innovation

AD is more than a design theory. It is a tool, a method, for innovation. By defining FRs in a physically neutral environment, and by zigzagging decomposition, the designer creates a semantical representation selecting the best solutions at each level of detail. Moreover, AD explains reverse engineering by representing what gave rise to an entity in the previous domain.

Essential elements of AD are, Suh's axioms, independence, represented with DMs (Eq.1) and information content (Eq.2); structure consisting of domains and hierarchies; hierarchical decompositions in detail and physical integrations. These elements are used for innovating and designing.

The next subsections focus on innovation by defining problems in AD domains, developing hierarchical decompositions in detail by zigzagging, examining independence (axiom one) using DMs, and evaluating the information content (axiom 2).

4.1. Innovation by identifying the domains

Problem definitions generally start in the customer domain. However, innovation problems still belong to a broader organization, entity, or country. Before defining CNs, designers might need to check to which domain problems belong, by looking at a more comprehensive system. Therefore, the first step of innovation could be to define problems in the context of a larger system considering ethics and sustainability.

If the problem regards manufacturing, it could belong to the process domain of a larger system. By checking PVs, DPs, FRs, and CNs, new ideas can arise to define new CNs, FRs, DPs, or PVs for that problem. Figure 3. shows innovation regarding a problem in a manufacturing system, relating one PV to one DP and one FR. This relation occurs most clearly in an uncoupled system.

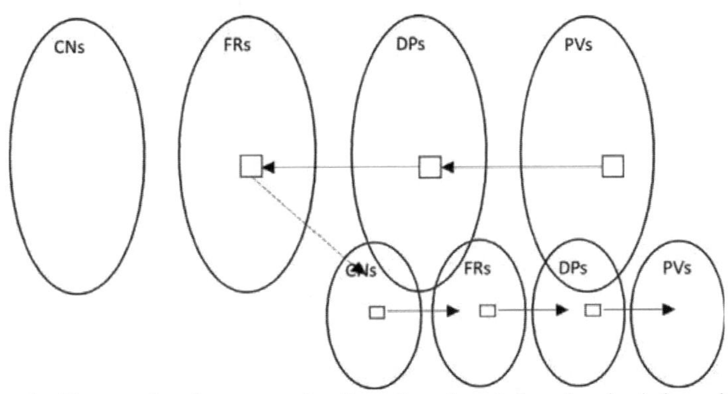

Figure 3. Zigzagging between the functional and the physical domains

In this example, a problem identified in a manufacturing system provides checks or redefines the corresponding DP and FR, in a reverse engineering context. Thus, it was possible to create a new design by defining a CN that defines the problem. However, a PV may influence many DPs, and each DP can influence many FRs, making the problem harder to define.

The broader the system, the better, because a more comprehensive DM can be achieved. However, larger systems can require greater knowledge to understand all the relations.

Moreover, by knowing the DP of a design solution, a new PV can be innovated. It is a widespread process in the industry. More seldom, an innovation department knows a company's FR and decides to implement a new DP. The impact in the organization will be higher if the FR is a child of the company mission. These approaches apply to problems other than industry. Many universal issues occur due to the lack of a global vision. The next examples show some huge mistakes that have compromised the entire society.

In the Republic [12] Plato discussed "Who must govern?". He applied this question to timocracy, oligarchy, democracy, and tyranny. "Who" belongs to the PV domain, so Plato discussed the design of regimes in the process domain! If instead of "who?", Plato had asked "how to govern?", or "what do we want?", the solution would have been different [9]. It would have avoided a lot of historical errors. Now CNs are the freedom and prosperity of people. We want to be live in peace and freedom in sustained environments. Thus, extreme conflicts are avoided by providing access to the government to different parties with different ideas.

This, 2020, has been the year of the COVID pandemic. Facing the disease, governments looked at medicine and epidemiology as the DPs. They focused the problem in the physical domain, e.g., hospitals, treatments, and vaccines. If they had started by defining CNs and FRs, they might have realized that hospitals are a piece of the problem. The problem involves more than medicine. The union between political power and medical power create unintended consequences, like economic issues, ethical issues based on eugenics or personal incomes, social repression and disruption, and phycological problems [13].

4.2. *Innovation by zigzagging*

The zigzagging decomposition process at each new level develops FRs in a solution neutral environment. Child FRs and DPs can be imagined in a vast space of solutions, as long as they comply with their parents, axioms, upper level DPs, and constraints, like sustainability [1].

Thus, once the innovation process reaches the functional domain, by direct or reverse engineering (Figure 3), many possibilities can match the requirements.

Urban electric mobility is a trend in any city for reducing pollution and global emissions. Suppose a DP is a bus with an electric motor, then a child FR will be "provide electrical energy". Electricity may be provided by overhead wires, as in the trams of old cities. Batteries can be an alternative, which has been the solution for electric cars.

On-line electric vehicles (OLEVs) include a bus developed at KAIST that receives energy by electromagnetic induction from a buried cable at high tension. The OLEV design team developed the project using AD and reached a completed new innovative solution, supply electric energy by induction. OLEV receives power from new municipal infrastructures, cables buried in roads for a significant part of the route. Moreover, vehicles have small batteries that charge at stops, allowing the bus to run for a while on ordinary roads. Trams, trucks, and other heavy transports can use a similar approach.

Therefore, the same FR "provide electrical energy" can have many physical solutions. Getting power through induction from cables in the road required generating new knowledge. If instead of discussing bus mobility, the FR is about electrical mobility for people, the number of solutions will be enormous. We might see all those solutions coming to the market soon.

4.3. *Innovation by checking the DM*

A good design must be uncoupled or decoupled. Realizing a design is coupled is the first step for changing it.

History is a useful repository of AD applications. Implementation of a new, innovative FR decided the naval battle of Hampton Roads in 1862, in the American Civil War. The CSS Virginia, a confederate ship, was advanced for that time. The steam ship had a significant firepower, was protected by metal plates, and had a configuration that deflected direct shots. In turn, the Union had the USS Monitor, a low-profile steam ship with only one cannon. This was the first ship to have a rotating turret to aim its cannon independently of the ship's direction. In this way, the FR: "aim the cannon" became independent of: "navigate in a direction". CSS Virginia was neutralized by a smaller, slower, and outgunned Monitor. The Virginia retreated up a river, where it remained under the protection of land-based cannons. The independence of satisfying CNs with a new FR fulfilled by a new, independent DP, decoupled a design solution and allowed a smaller ship with less firepower to pose a lethal threat to the CSS Virginia. This Navy example is about creating a new uncoupled solution. Sadly, many things in life are coupled, and difficult to uncoupled. An example of coupled designs are relationships between countries that, moreover, vary with time.

At the end of World War II, the US State Department comes with a proposal for creating an international structure to ensure peace and security. In 1941 US President Roosevelt and the UK Prime

Minister Churchill made the first draft of The United Nations (UN). They defined an executive branch, the International Assembly, and an intervention force called the four policemen (US, UK, USSR, and China). The relationships of nations have always been coupled. Decoupling high levels of international relationships is accomplished by creating international political organizations. That is the case of the EU and African Union, the Arab Maghreb Union, and many more international organizations. All are intellectual creations to help to establish good relationships between countries.

Reducing, or eliminating, the role of international organizations exacerbates unwanted couplings between countries, violates Suh's independence axiom and the unintended consequences that could lead to disasters.

4.4. Innovation by evaluating the information content

Design managers decide on solutions often based on intuition or on a weighted decision process. Depending on the weights, any solution could be the chosen one. Computing the innovation content in the functional domain avoids psychologism when FRs are correctly defined. In such situations, FRs have bounds making it possible to evaluate the system performance against an acceptable range. At high-level definitions of designs, information content can be computed in a fuzzy way, to choose between two or more global options.

5. Conclusions

The first law of innovation defines the necessary conditions for innovation to occur. "For innovation to occur, there can be no lack of any element or link in the innovation continuum". The elements are "fundamental research and technological invention; financial community willing to provide capital and take the risk; industry with the necessary knowledge to develop, manufacture and commercialize the products of the invention."

Once the first law of innovation is satisfied, it is possible to think about several solutions by defining "what is the problem?". The innovation process of AD uses domains, zigzagging decomposition, design matrices, and evaluations of information content.

The process of innovation starts by identifying the AD domain in which the problem arises in a more global architecture. The higher the entity is in the abstraction hierarchy in which the innovation starts, the greater the impact on the process, organization, system, or artifact.

Acknowledgements

The first author acknowledges *Fundação para a Ciência e a Tecnologia* (FCT-MCTES) for its financial support via the project UIDB/00667/2020 (UNIDEMI).

References

[1] Brown C A and Rauch E 2019 Axiomatic Design for Creativity, Sustainability, and Industry 4.0 *The 13th Int Conf on Axiomatic Design (Sydney)* vol 301 (MATEC Web of Conferences, EDP Science) p 00016
[2] Le Masson P, Weil B and Hatchuel A 2017 *Design theory* (Springer International Publishing)
[3] Mourão A, Cavique M and Gonçalves-Coelho A M 2011 Different Perspectives in Engineering Education According to the Mission – The Balance between analysis and synthesis *The 17th Int. Scientific Conf. Knowledge-Based Organization (Sibiu, România)* pp 496-501
[4] Stilgoe J, Owen R and Macnaghten P 2013 Developing a framework for responsible innovation *Research Policy* **42** 1568– 1580
[5] Eelko K.R.E. Huizingh 2011 Open innovation: State of the art and future perspectives, *Technovation* **31** 1 2-9
[6] Sheu D D and Hei-Kuang Lee H-K 2011 A proposed process for systematic innovation, *International Journal of Production Research* **49** 3 847-868.
[7] Furrer O, Sudharshan D, Tsiotsou R H and Liu B S 2016 A framework for innovative service design *The Service Industries Journal* **36** 9-10 452-471

[8] Suh N P 1990 *The Principles of Design* (New York: Oxford University Press)
[9] Suh N P, Cavique M and Foley J (Eds.) 2021 *Design Engineering and Science* (Springer International Publishing)
[10] Suh N P 2010 A Theory of Innovation and Case Study *International Journal of Innovation Management* **14** 5 893–913
[11] Gutierrez E, Olundh Sandstrom G, Janhager J and Ritzen S 2008 Innovation and decision making: Understanding selection and prioritization of development projects *The 4th IEEE Int. Conf. on Management of Innovation and Technology*, Bangkok, pp. 333-338
[12] Plato – translated by Cornforf F M 1970 *The Republic of Plato* (New York: Oxford University Press)
[13] Lévy B-H 2020 *The virus in the age of madness* (Yale University Press)

Modelling the dependence between regular reliefs ridges height and the ball burnishing regime's parameters for 2024 aluminum alloy processed by using CNC-lathe machine

S D Slavov[1] and D M Dimitrov[2]

[1] Technical University of Varna, Department of Manufacturing Technologies and Machine Tools, 9010 Varna, Bulgaria
[2] Technical University of Varna, Department of Mechanics and Machine Elements, 9010 Varna, Bulgaria

*Corresponding author's e-mail: sdslavov@tu-varna.bg

Abstract. New approach for obtaining regular reliefs onto cylindrical and tapered surfaces, by using ball burnishing process, performed on CNC-lathe machine having C-axis installed, is presented. It allows to simplify the burnishing tool and to perform the burnishing operation on the same lathe machine as previous cutting operations. The equations for the points coordinates calculation of the needed complex ball tool trajectory are shown. For their approbation, and in order to obtain a stochastic model which established the relation between the ball burnishing regime parameters (deforming force F, and feedrate f) and resulting height (Rz) of regular reliefs cells boundary ridges, an experimental investigation is carried out, using 2024-T3 aluminum alloy, and the methodology of the rotatable experimental plans. The corresponding response surface of the model is given and discussed. Conclusions about the advantages and disadvantages of the approach are given, and also for the suitable application of the obtained stochastic model.

1. Introduction

Machine parts with cylindrical or tapered shape, which are made of non-ferrous metals like bronze or aluminum alloys sometimes are combined with rubber elements or coatings and work together in different mechanisms. As example can be highlighted silicone rubber rollers for peripheral hot stamping of cylindrical shapes and roll on decorating flat or curved surfaces, high pressure rubber roller for plastic printing machines, water lubricated rubber shaft bearings for propellers, heavy duty hydro-cylinders or rollers for rubber bands from industrial conveyors, or packaging machines, etc.

Operational conditions in which such pairs work, however, in most cases are apart from the optimal. They include high contact pressures combined with high temperatures, bad lubrication conditions, contact with abrasive substances or rough contact surfaces, etc. Often these operational conditions lead to intensive wear and fatigue failures of the inner metal parts, and of the outer rubber elements (or coatings). Intensive wear of the contact surfaces of rubber and metal parts also often happens, when they slide between each other [1-5]. As a result, the operation characteristics of such parts deteriorate fast, and this leads to loss of tightness in the contact seals, unacceptably large clearances, uneven tightness of the contact, etc. The results, obtained from the conducted experimental investigations, show that one of the most important factors, which have great influence on the operation parameters of such machine parts, is the roughness characteristics of their contact surfaces [4, 5]. If any of the well-known finishing processes, such as finish turning, grinding, or superfinishing was applied, the resulting roughness (in the vertical section) has sharp peaks and valleys (albeit with a small height), combined with small steps between them, which do not ensure the low rate of wear. If deep rolling or classical ball burnishing (BB) was applied as a finishing process, the resulting surface will be very smooth (see Figure 1, a), but in this case the ability to hold-up enough lubricant between the sliding surfaces will be decreased significantly. Moreover, the expulsion of the dust and/or the wear's products out of the contact zone also will be more difficult, which additionally will increase the effects of wear. Therefore, in order to avoid the negative effects due to sharp peaks in the roughness profile after the commonly used cutting finishing processes, or very smooth surfaces, obtained after classical roll or ball burnishing, a little different finishing process is subject of interest in the present work.

The first goal of the work is to find a way for effective implementation of the BB process in order to form specific roughness onto cylindrical and tapered surfaces of the parts. The second one is to define a stochastic model and establish the relation between the RR's roughness height (by using Rz-criterion) and the basic regime parameters of BB for a workpiece, made of 2024-T3 aluminum alloy. This material is chosen because it is used in applications, requiring high strength to weight ratio, as well as good fatigue resistance.

2. Advantages of the Regular Reliefs, formed by Vibratory Ball Burnishing process

There is a modification of BB, which can significantly improve the above-mentioned drawbacks of the classical finishing processes. It is called "Vibratory Ball Burnishing" (VBB) process [5]. The VBB possess has many of the advantages of classical BB in regard to increase the surface integrity parameters. In contrast to BB, however, when VBB is applied, it is possible the so called "Regular Reliefs" (RR) to be formed onto planar, cylindrical, tapered, and even complex surfaces [5, 6, 7]. In fact, the VBB has the same basic process parameters as BB, but in this case the deforming element (which can have only a spherical shape) performs an additional oscillation movement (see Figure 1, b) with a certain amplitude and frequency, depending on the desired shape and size characteristics of the obtained RR.

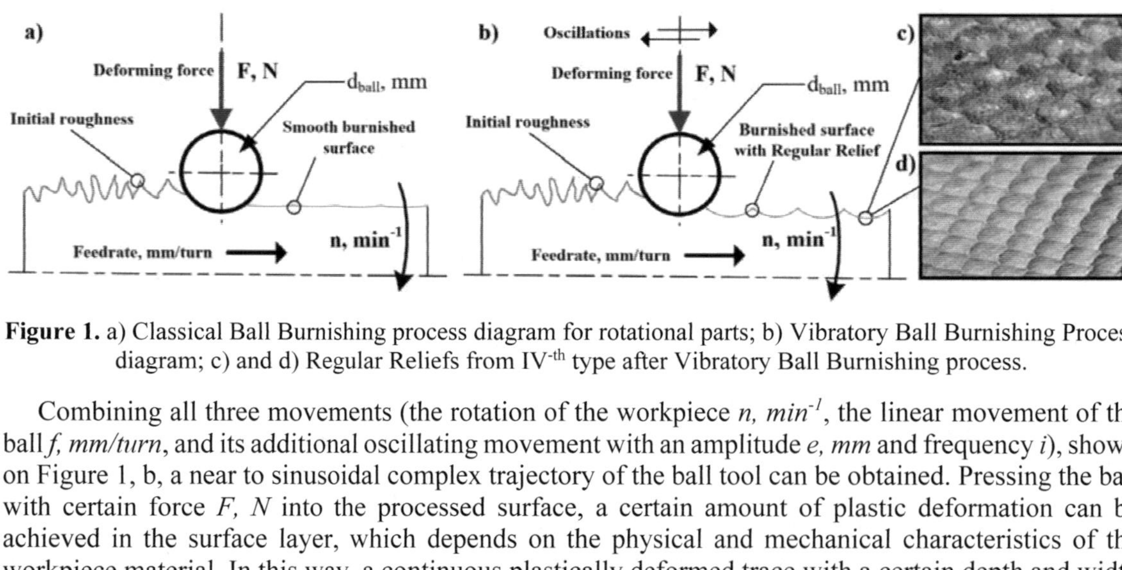

Figure 1. a) Classical Ball Burnishing process diagram for rotational parts; b) Vibratory Ball Burnishing Process diagram; c) and d) Regular Reliefs from IV^{-th} type after Vibratory Ball Burnishing process.

Combining all three movements (the rotation of the workpiece *n, min*$^{-1}$, the linear movement of the ball *f, mm/turn*, and its additional oscillating movement with an amplitude *e, mm* and frequency *i*), shown on Figure 1, b, a near to sinusoidal complex trajectory of the ball tool can be obtained. Pressing the ball with certain force *F, N* into the processed surface, a certain amount of plastic deformation can be achieved in the surface layer, which depends on the physical and mechanical characteristics of the workpiece material. In this way, a continuous plastically deformed trace with a certain depth and width is formed after the ball tool passage onto the burnished surface. Because the ball tool performs a complex relative movement, the plastically deformed trace produces specific patterns (i.e. RR), which can have different patterns. The possible combinations of RR are classified into five different types [5]. Of particular interest are the RR of the IV^{-th} type, because they fully cover the burnished surface and the formed cells with a hexagonal (see Figure 1, c) or a rectangular (see Figure 1, d) shape. The inner zone surface of the RR's cells has very small roughness heights ($Ra = 0.02 - 0.5$, *μm*), but the cells boundaries ridges (formed from self-crossing of the plastically deformed trace) have greater heights ($Rz = 15 - 110$, *μm*) [5]. The distances (or the so-called "steps") between the RR's cells ridges varies from *200* to *1500, μm*, which, in combination with the ridges heights, defines the small slopes of the peaks profile. These topology characteristics of RR provides increased ability to hold lubricant and products from wear into the resulting relief cells, and in this way facilitates the contact conditions, in comparison with the smooth burnished surfaces. Last, but not least, the VBB method is a waste-free technology, which makes it particularly suitable for use with non-ferrous metals, due to their higher cost, in comparison with ferrous metals.

3. Application of VBB to form RR onto rotational parts, using CNC-lathe machines

The VBB traditionally is performed, using manually controlled lathe machines, equipped with (specially designed) additional VBB-tool(s) that provide the oscillations, in order to assure the needed complex trajectory of the ball [5, 8] (see Figure 2, a). This approach, however, has some basic drawbacks, related to the manually controlled lathe machines kinematical limitations and the excitation of forced oscillations. They can be avoided, if the VBB is adapted for implementation with contemporary CNC-lathe machines (see Figure 2, b).

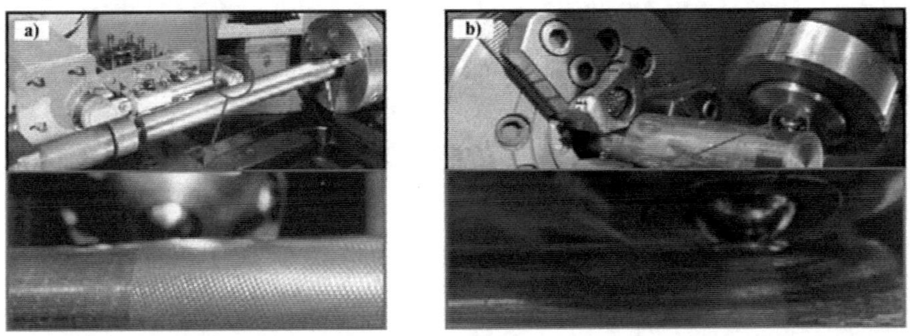

Figure 2. a) VBB process performed on manually controlled lathe; b) BB process performed on CNC-lathe.

The complex tool trajectory, needed for forming RR, can be mathematically generated in advance and converted into a numerical control (NC) program for CNC lathe. Using the integrated capabilities for interpolation of two or more linear and/or rotary axes in contemporary CNC machines avoid the need to use excited oscillations of the deforming tool. This, important in VBB regime component, can be skipped in the BB process implemented on CNC lathe, but without losing the capabilities to form RR of all types. As a consequence, the deforming tool design becomes much more simplified in comparison with VBB, and this allows utilization of standardized (commercial) ball tools for BB. Using standardized tools gives opportunity to perform the BB operation on the same lathe machine (after finishing of previous cutting operations), and thus there is no need to move the workpiece to another, specially equipped for VBB, machine. Furthermore, the contemporary CNC lathes offer much more stability, rigidity, positioning accuracy, and speed/feed rates than the manually controlled ones, which results in higher productivity and lower cost of production.

To calculate the ball tool trajectory points X,Y,Z-coordinates for cylindrical and tapered surfaces, the following equations are used:

$$Z_i = L \cdot \left(\frac{i}{p} + \frac{d}{D-d}\right) + e \cdot \sin\left(\frac{2 \cdot \pi \cdot i \cdot j}{p}\right) \quad (1)$$

$$X_i = Z_i \cdot \left(\frac{D-d}{L}\right) \cdot \cos\left(\frac{2 \cdot \pi \cdot i \cdot n}{p}\right) \quad (2)$$

$$Y_i = Z_i \cdot \left(\frac{D-d}{L}\right) \cdot \sin\left(\frac{2 \cdot \pi \cdot i \cdot n}{p}\right) \quad (3)$$

where: *D, mm* – diameter of the cylindrical surface (greater diameter of the tapered surface); *d, mm* – smaller diameter of the tapered surface; *L, mm* - length of the cylindrical /tapered surface; *2.e, mm* – amplitude of the ball tool oscillation; *n, mm*- distance (or step along to Z-axis) between ball tool positions for one full rotation of the workpiece; *p* - number of trajectory points; i ($i \in 0 \ldots p-1$) the index of a trajectory point; *j, osc./rev*– the number of ball-tool forward-backward movements, per one revolution of the workpiece.

When the processed surface is cylindrical, the condition d ≈ D must be satisfied. Otherwise, the denominator (D – d) of the formula for calculating the Z-coordinates will be zero. It is advisable the difference (D – d) = 0.0001, mm, if the movement increment of the CNC axes is 1, μm.
Using the following formulas:

$$R_i = (X_i^2 + Y_i^2)^{1/2} \quad \text{and} \quad \theta_i = tan^{-1}\left(\frac{Y_i}{X_i}\right), \tag{4}$$

calculated by (1) - (3) points coordinates can be transformed from Cartesian into Polar coordinates, because they are more suitable to place in the NC code, when the CNC-lathe has C-axis on the spindle installed. These points coordinates can be transferred into NC code, using a suitable CAM software and postprocessor. The NC code also can generate directly, if a mathematical software (such as Math Lab, Mathcad, SMath studio, etc.) is used, to concatenate the coordinates into appropriate G-code.

4. Stochastic modelling RR's topography heights depending on the BB regime parameters

An experimental investigation is conducted to obtain a stochastic model that describes the relations between main regime parameters: deforming force – *F, N* and federate – *f, mm/min,* over resulting height *Rz μm* of RR's cells boundary ridges for 2024-T3 aluminum alloy. The 2024 aluminum alloys approximately include 4.3-4.5% copper, 0.5-0.6% manganese, 1.3-1.5% magnesium, and less than 1% other elements. T3 tempered rod has a tensile strength of 421 MPa, yield strength of 272 MPa, and elongation of 11%.

The present experimental investigation is conducted in accordance with the methodology of rotatable design of experiment [9]. For modelling the RR's cells ridges height, the following second order stochastic model has been introduced:

$$Y = b_0 \cdot X_0 + b_1 \cdot X_1 + b_2 \cdot X_2 + b_{12} \cdot X_{12} + b_{11} \cdot X_1^2 + b_{22} \cdot X_2^2 \tag{5}$$

The matrix of two-factorial design with interaction and varying variables (deforming force F=500 ± 212 N and federate, f=300 ± 106 mm/min) is shown in section "Experimental design" in Table 1. These two regime parameters are chosen, because it has been established that they have the major impact on plastically deformation of the surface layer at BB [6, 7]. The other parameters of BB are, as follows: the amplitude of the ball tool oscillation is *e = 1 mm*, the number of ball-tool forward-backward movements per one revolution of the workpiece *j = 2200, osc./rev*, the diameter of the deforming ball tool is *∅12 mm*, and as a lubricant Mobil DTE 25 hydraulic oil is used.

Table 1. Matrix of coded variables, measuring and calculating results of RR's cells ridges height, Rz

Experimental design									Experimental results		
Exp.	Coded values						Physical values		Rz, *μm*	Rz, *μm*	Difference
№	X_0	X_1	X_2	X_{12}	X_1^2	X_2^2	*F, N*	*f, mm/min*	Experimental	Stochastic	%
1	1	-1	-1	1	1	1	288	194	6.72	6.47	3.92
2	1	+1	-1	-1	1	1	712	194	9.23	9.56	3.48
3	1	-1	+1	-1	1	1	288	406	17.55	17.51	0.21
4	1	+1	+1	1	1	1	712	406	26.82	27.37	3.92
5	1	-1.414	0	0	2	0	200	300	14.78	15.04	1.74
6	1	1.414	0	0	2	0	800	300	24.77	24.19	2.36
7	1	0	-1.414	0	0	2	500	150	0.63	0.64	1.04
8	1	0	1.414	0	0	2	500	450	21.34	21.04	1.45
9	1	0	0	0	0	0	500	300	16.82	16.72	0.58
10	1	0	0	0	0	0	500	300	16.63	16.72	0.62
11	1	0	0	0	0	0	500	300	16.91	16.72	1.17
12	1	0	0	0	0	0	500	300	16.52	16.72	1.22
13	1	0	0	0	0	0	500	300	16.70	16.72	0,02

The experimental plan is characterized with a minimal number of experimental data. It contains a basic part 2^k (where k is number of varying variables in the process), a symmetric set of eight points (1 – 8) around the center of the design, and five points of repetition (9 – 13) in the center of design (see Figure 3, a). A cylindrical rod with *∅40×350 mm,* made of 2024-T3, is used as a workpiece. The outer diameter of the rod is previously machined by finishing, turning to *∅38 mm* and initial roughness Rz =

8.2 μm. Then the workpiece is set up on Haas ST20Y CNC-lathe who has interpolation C- axis of the spindle installed. Using formulas (1) – (4) and the specially designed algorithm in sMath Studio software, the corresponding NC code for the needed BB operations is generated. Executing the NC-code, 13 cylindrical areas (with 20 mm length) are burnished, in accordance with "Physical values" of the two BB's regime parameters, included in the experimental matrix (see Table 1). For RR's ridges height assessment, the Rz-criterion (ISO 4287) is used. The Rz-values for all 13 areas with RR are determined by carrying out measurements of the roughness profile, within sampling length L=8 mm, using Mitutoyo, SurfTest SJ-301.

The dispersion homogeneity of the cells ridges height results is examined, using Cochran's criterion for level of reliability P=0.95. The model's coefficients b_0, b_1, b_2, b_{12}, b_{11}, and b_{22} are calculated and checked for significance, on the basis of the measured experimental results (see Table 1). As a result, their coded values are obtained, as follows: $b_0 = 16.723$, $b_1 = 3.238$, $b_2 = 7.213$, $b_{12} = 1.690$, $b_{11} = 1.449$, $b_{22} = -2.945$.

The stochastic model (5) with its coded coefficients passed successfully the check for significance, using Student's F-criterion, and after decoding, its physical form is obtained:

$$Rz(F,f) = 0.000032 \cdot F^2 - 0.000262 \cdot f^2 + 0.000075 \cdot F \cdot f - 0.039 \cdot F + 0.188 \cdot f - 15.576 \quad (6)$$

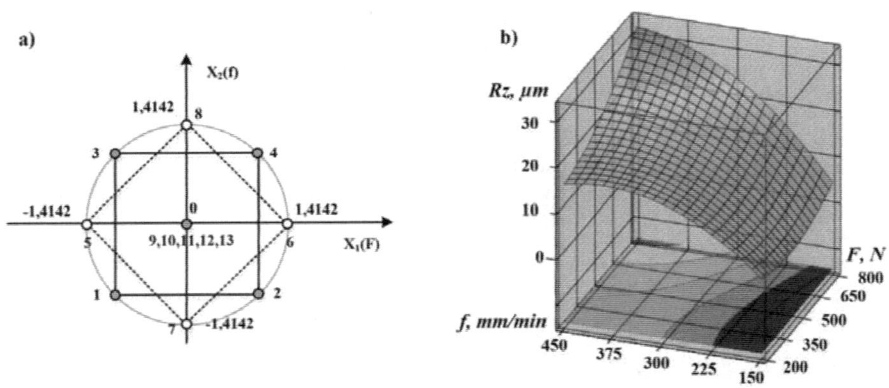

Figure 3. a) Diagram of rotatable plan for two variables $X_1(F)$ and $X_2(f)$; b) The response surface which corresponding to the stochastic model (6).

A comparison of the experimental and the stochastic results is shown in Table 1, and a graphical representation of the response surface, which corresponds to the (6) is presented in Figure 3, b.

5. Results and discussion

The results obtained from stochastic analysis and the multi-factorial experimental rotatable design carried out show that the results of Rz-criterion, established in stochastic model (6) are very close to the experimental ones. The maximum percentage differences (see Table 1) are observed only at 1st, 2nd and 4th points of the experiment, but they are not exceeding 4%, thus the derived stochastic model can be considered adequate.

The response surface, created on the basis of the stochastic model (6), is a segment of a hyperbolic paraboloid and is presented in Figure 3, b. It reveals the way of influence of the two regime parameters on the Rz-roughness criterion of RR. Analyzing it, it is seen that the minimum Rz value (1.24 μm) is obtained, when the deforming force has value of F ≈ 500 N and feedrate f ≈ 150 mm/min, and the maximum Rz = 31.9 μm is observed, when both regime parameters have maximum values (F= 800, N and f= 450 mm/min). When the deforming force and the feedrate have minimum values, however, the Rz-criterion increases its values up to Rz = 2.53 μm. This can be explained with the small rate of plastic deformation in the surface layer, when the deforming force and feedrate have low values. In this case, the initial roughness can have an impact on the measured values for Rz. If the deforming force is increased, the roughness height is decreased to a certain point, and with a further increase of the F, the

Rz starts to increase its values again. This can be explained with the comparatively greater amount of metal that pushes the deforming element in front of it, as it follows its complex trajectory. The increasing of the feedrate intensifies the degree of plastic deformation in the surface layer of the workpiece, and contributes further to increasing the height of the ridges of the RR's cells.

6. Conclusions

Based on the analysis of the BB's process requirements for creating RR and the functional capabilities of the contemporary CNC-lathe machines, the following conclusions could be made:

- Ball burnishing, as a finishing process was successfully implemented in order to form regular reliefs of the IVth type onto cylindrical outer surfaces, using CNC-lathe machine, which has C-axis;
- On the basis of equations (1) – (4), it is possible to calculate the needed ball tool trajectory points coordinates for both cylindrical and tapered surfaces. They can transform easily into a corresponding NC-code for CNC-machine, using any CAM or a suitable software for engineering calculations;
- The main advantages of the implementation of the BB on CNC-lathe machines are: simplifying the design of the tools for BB (even standardized commercial tools could be used); the BB operation can be performed on the same machine as other lathe operations, i.e. there is no need to move the workpiece from one to another equipment.
- As a disadvantage of the present approach can be listed the need of additional C-axis (which allow interpolation) installed of the CNC-lathe machine, and also in some cases the appropriate software for generating the NC-code, which can be costly;

On the basis of the conducted experimental investigation, the stochastic model (6) is derived. It describes the influence of the regime parameters of BB deforming force F, and feedrate f. The model can be used for prediction what will be the RR's cells ridge heights (expressed with Rz roughness criterion), when certain values of the regime parameters are set. It is also suitable, if the RR must have cells ridges with a certain height, what BB's regime parameters values have to be chosen, in order the needed Rz-criterion value to be achieved.

The authors consider the present work as first steps in the study of the presented issues. Our future work will focus on the study of the influence of the other regime parameters of the BB process on some other topographic characteristics of the RR, as well as on other materials.

7. References

[1] Lhymn C, & Lhymn Y 1989 *Friction and wear of rubber/epoxy composites. Journal of Materials Science, **24(4)**, 1252–1256*

[2] Natarajan N, Krishnaraj V, Davim J P 2015 Wear Analysis. In: Metal Matrix Composites. Springer, Briefs in Applied Sciences and Technology. Springer, Cham

[3] Hakami F, Pramanik A, Ridgway N, Basak A K 2017 *Developments of rubber material wear in conveyer belt system. Tribology International, **111**, 148–158*

[4] Jones D R H, Ashby M F 2019 Friction and Wear. Engineering Materials **1**, 469–484

[5] Schneider Yu G 2001 Operational properties of parts with regular microrelief, publishing IVA, St. Petersburg, ISBN 5-7577-0166-8

[6] Slavov S, Dimitrov D 2018 *A study for determining the most significant parameters of the ball-burnishing process over some roughness parameters of planar surfaces carried out on CNC milling machine. In MATEC Web of Conferences* Vol. **178**, p. 02005 EDP Sciences

[7] Slavov S, Dimitrov D, Iliev I 2020 *Variability of regular relief cells formed on complex functional surfaces by simultaneous five-axis ball burnishing, UPB Scientific Bulletin, Series D: Mechanical Engineering* **82(3)**, 195-206

[8] Krastev K 2002 A d*evice for ball burnishing, Manufacturing engineering and technology journal*, vol. **2**, ISSN 1312-0859 12 - 16

[9] Georgiev D, Slavov S 2008 *Computer based program for automatically design and data processing of experiments with two-, three-and four-factorial symmetrical experimental plans–EXPLAN. Manufacturing engineering and technology journal.* vol.**1** ISSN 1312-0859 73-78

A study on casting of structural mesh-like metal parts

D Vlagkoulis[1], A Vasileiou[2] and G-C Vosniakos[1*]

[1]National Technical University of Athens, School of Mechanical Engineering, Manufacturing Technology Section, Heroon Polytehniou 9, 15780 Athens, Greece
[2]The University of Manchester, Department of Mechanical, Aerospace and Civil Engineering, Manchester M1 3BU, United Kingdom

*Corresponding author's e-mail: vosniak@central.ntua.gr

Abstract. Mesh-like parts are employed in order to reduce weight of mechanical structures. This study focused on cylindrical shaped mesh-like spacers. First, the mesh shape was parametrically defined and mechanical strength and strain under specific combination of axial and shear loads were determined by numerical simulation. Manufacture of the best design was then investigated focusing on investment casting under both gravity and vacuum modes with a simple, yet uncommon, top feeding system. The best combination of casting mode and parameter values was determined by numerical simulation on commercially available dedicated software, considering solidification time and lack of defects as primary quality indices accompanied by an investigation of residual stress and deformation fields. To implement simulation-based decisions the part was investment-cast employing 3D printed dies for constructing the wax models.

1. Introduction

Recently, mesh-like components are increasingly used as light-weight replacements of full body parts in structural applications [1]. This trend has been particularly enhanced by Additive Manufacturing (AM) technologies having established themselves firmly due to their allowing for design freedom. In fact, a large stream of research is directed towards topological optimization of parts in view of 3D printing [2]; optimization may refer to the external shape or profile of the part, i.e. macro-scale [3], which is also the focus of this paper. In addition, optimization may also refer to part density, which is achieved by applying lattice structures at meso-scale [4], typically striving to achieve graded mechanical properties matching the expected loads [5]. Mesh-like structures at macro scale have been proposed for building machine frames [6], whereas truss-like structures, best known for a long time in connection to large structures such as bridges, have also been studied and optimized [7].

Mesh-like structures are conveniently manufactured by casting methods [8]. In order to save time and cost, before building molds to try out feeding architectures and casting parameters, it is desirable to evaluate them by simulation [9]. Due to the complexity of the casting process, involving multiphase flows, heat transfer and stress–strain fields coupled to temperatures in the presence of phase changes and microstructure formation, special software platforms are necessary as opposed to general purpose numerical simulation environments [10]. Such platforms have been reportedly exploited to assess casting quality by determining mold filling completeness, rate of solidification and solidification time, porosity and emergence of other defects, residual stresses, deformation including warpage, hot cracking sensitivity etc. [11], In addition, such simulations have been leveraged to produce data for establishing corresponding machine learning models [12].

This paper presents a case study of cylindrical structural mesh-like metal spacers. The aim is to advocate a work flow regarding improved design and manufacturing practices, starting with parametrically defined shapes, as presented in Section 2, proceeding with examining alternative casting deployment, as presented in Section 3, and ending up with prototype manufacture, as explained in Section 4. Conclusions of this study are outlined in Section 5 with a generalization flair.

2. Parametric part design study

The part that was studied was a cylindrical spacer of length 100 mm and diameter (D) of 10, 20 or 30 mm. The cylindrically shaped lattice is formed by a number of vertical rods of the same diameter (d) of 2,3,4 or 6 mm resulting in an angular spacing between rods (α) and a number of horizontal struts of circular cross section (diameter t of 1, 2 or 3 mm) forming circular patterns on horizontal equidistant planes (spaced at h), see figure 1(a). Parameters α, h and L have fixed values. The 24 alternative dimension combinations studied are shown in Table 1.

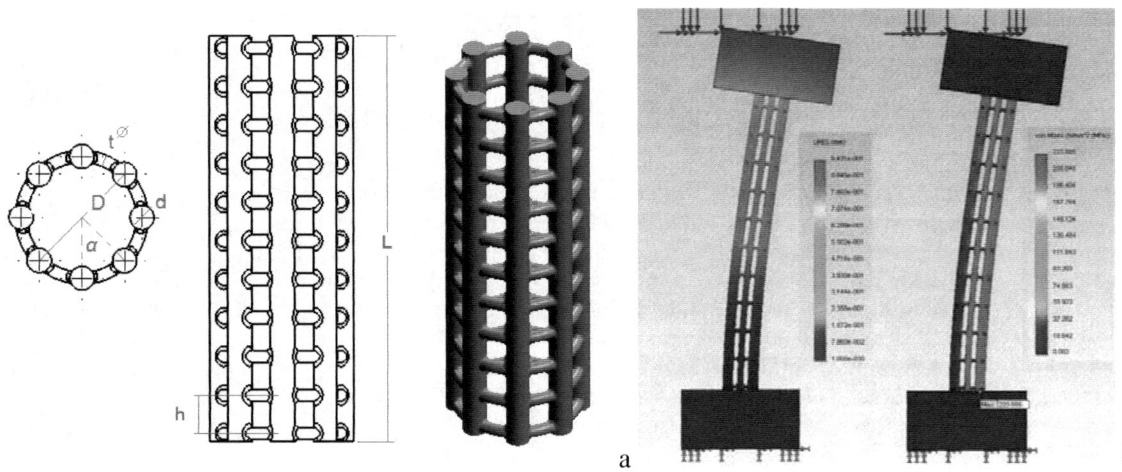

Figure 1. The spacer part (a) design involving 3 variable dimensions (D, d, t) and L=100mm, h=9mm α=45° (b) static loading simulation example (displacement)

Table 1. Specimen specifications and simulation results.

specimen ID	1	2	3	4	5	6	7	8	9	10	11	12
D (mm)	10	10	10	10	10	10	20	20	20	20	20	20
d (mm)	2	2	2	3	3	3	2	2	2	4	4	4
t (mm)	1	2	3	1	2	3	1	2	3	1	2	3
Mass (gr)	7.3	8.7	11.6	15.7	16.4	17.9	8.1	11.6	18.2	28.1	30.5	34.6
Max stress (MPa)	224	184	193	77	98	93	221	203	148	101	52	40
Max displacement (mm)	-	-	-	0.40	0.38	0.36	-	-	0.24	0.14	0.08	0.07
specimen ID	13	14	15	16	17	18	19	20	21	22	23	24
D (mm)	20	20	20	30	30	30	30	30	30	30	30	30
d (mm)	6	6	6	2	2	2	4	4	4	6	6	6
t (mm)	1	2	3	1	2	3	1	2	3	1	2	3
Mass (gr)	61.7	63.0	65.1	8.8	14.6	24.8	28.9	33.4	41.2	62.5	65.9	71.7
Max stress (MPa)	30	24	19	268	182	122	112	50	39	51	28	18
Max displacement (mm)	0.04	0.03	0.03	-	-	0.15	0.22	0.07	0.04	0.06	0.03	0.02

The loads which the part needs to withstand are an axial force of 30 N and a shear force of 1500 N, applied at one end of the part, see figure 1(b). Both ends of the part are fixed on corresponding horizontal parallel plates which are considered rigid. These conditions are imposed by the machine environment in which the spacers are to function.

The spacer's material is aluminum alloy A356 T61 with minimum tensile strength 241 MPa and minimum yield strength 179 MPa, liquidus temperature: 616°C and solidus temperature: 556°C.

Solidworks Simulation™ was used in order to calculate maximum stress and, if that was lower than yield stress, maximum displacement of the spacer under static loading, see figure 1(b). The results are shown in Table 1.

In order to determine the best case, a suitable evaluation function was defined as follows:

$$\textit{Evaluation function} = \textit{mass (gr) X maximum displacement (\%) X maximum stress (\%)} \quad (1)$$

where *maximum displacement (%)* is the maximum displacement observed in the simulation divided by the maximum allowable displacement, which in this case was taken as 0.4 mm. Similarly, *maximum stress (%)* is the maximum stress observed in the simulation divided by yield strength of the material. Seven cases with max stress exceeding yield stress were crossed out. Note that the mass of the spacer is determined from its 3D model. The evaluation function value is plotted for all 17 valid specimens in figure 2(a). It becomes minimum for specimen No 24 followed by specimen No 15. In fact, for approximately the same mass widely varying stress and displacement corresponding to different specimen may be observed in figure 2(b). This is due to the different mass distribution corresponding to different dimensions of the respective specimen, see also Table 1.

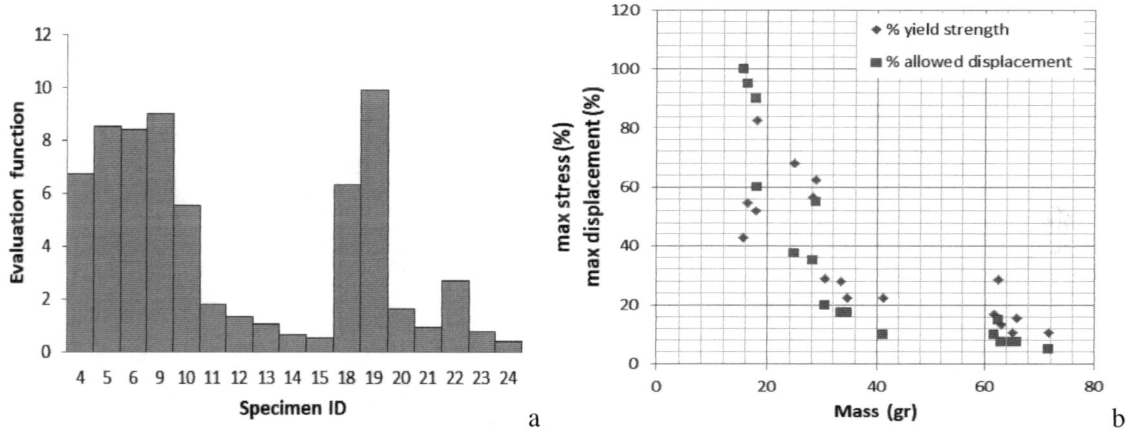

Figure 2. Part design simulation results (a) Evaluation function value per specimen (b) Normalised maximum stress and maximum displacement versus specimen mass for all 17 specimens

3. Casting method investigation

Gravity casting and vacuum assisted casting were the two methods studied by simulation in order to manufacture specimen No 24, which happens to be the largest one in the range examined. ProCast™ was used for this purpose, which supports filling as well as cooling for both thermal and mechanical domain. A direct top feeding system was designed, involving a top and bottom cylindrical plate of diameter 40 mm, at both ends of the spacer part. This is quite different from commonly applied solutions involving a sprue, runners and risers. In fact, the two plates act as risers in their own right, but feeding performance needed investigation anyway.

Mass inflow was approximately calculated using Bernoulli's equation for two different points of the conical sprue, i.e. one on the melt surface (A) and one on the inlet cross-section (B):

$$m = \dot{Q}\,\gamma = \pi \frac{d^2}{4} \sqrt{\frac{2(p_A - p_B)}{\gamma} g + 2\,g\,|z_A - z_B|\,\gamma} \tag{2}$$

where z: z-coordinate, p: pressure at the respective points, γ: fluid density, d: sprue inlet diameter. For example, mass inflow was calculated at 1.14 Kg/s for gravity casting.

The main casting process parameters, i.e. melt temperature and mold initial temperature, were also investigated using two values for each one of them (650-700°C and 250-400°C respectively).

Note that the heat transfer coefficient which is needed in casting simulation setup [13] was available in the simulation platform's library referring to a standard aluminium-plaster interface.

3.1. Filling

Irrespective of the particular casting parameter values used filling of the mold in both gravity and vacuum cases was fully completed. Characteristic snapshots of the process are shown in figure 3. Filling time was 0.5 sec and 0.09 sec for gravity and vacuum casting respectively, see figure 3(a), whilst maximum velocity of the melt does not affect structural integrity of the mold, see figure 3(b).

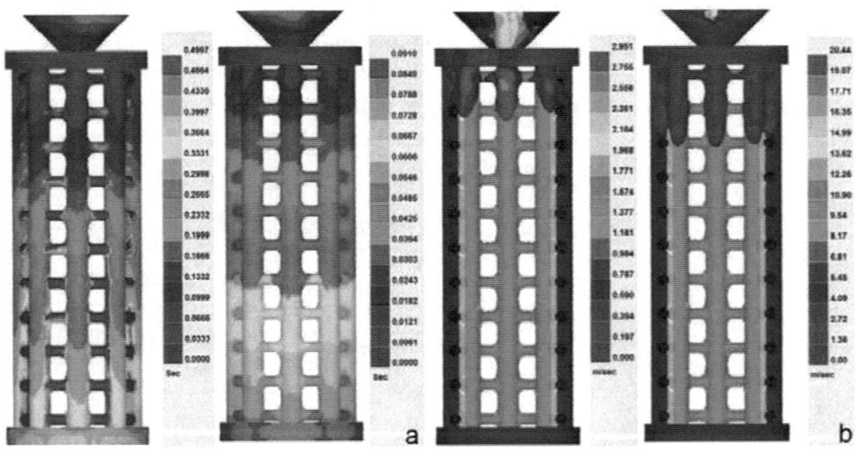

Figure 3. Mold filling (a) filling time (b) characteristic filling velocities (left: gravity, right: vacuum)

3.2. Solidification

Solidification study pertains to determining total solidification time and porosity. The fastest solidification proceeds the most advantageous it is for the resulting microstructure. Porosity is measured on a 0-1 scale and is distinguished into micro- (<0.01) and macro-porosity (>0.01). In general a value lower than 0.2 is considered acceptable locally, however, overall acceptability depends also on distribution of pores. In experimenting with mold geometry it was decided to remove the bottom plate and compare results with those corresponding to the original two plate design, see Table 2. Characteristic snapshots regarding solidification time and porosity are shown in figure 4.

Table 2. Solidification simulation results (G: gravity, V: Vacuum, N: negligible)

Case	1	2	3	4	5	6	7	8	9	10	11	12	13	14	15	16
Casting type	G	G	G	G	G	G	G	G	V	V	V	V	V	V	V	V
Number of plates	2	2	2	2	1	1	1	1	2	2	2	2	1	1	1	1
Melt temperature (°C)	650	650	700	700	650	650	700	700	650	650	700	700	650	650	700	700
Mold temperature (°C)	250	400	250	400	250	400	250	400	250	400	250	400	250	400	250	400
Solidification time (sec)	236	544	268	588	224	682	251	717	234	526	268	575	221	655	251	696
Porosity	.011	.218	.012	.52	.012	.343	.011	.52	N	N	N	N	N	N	N	N

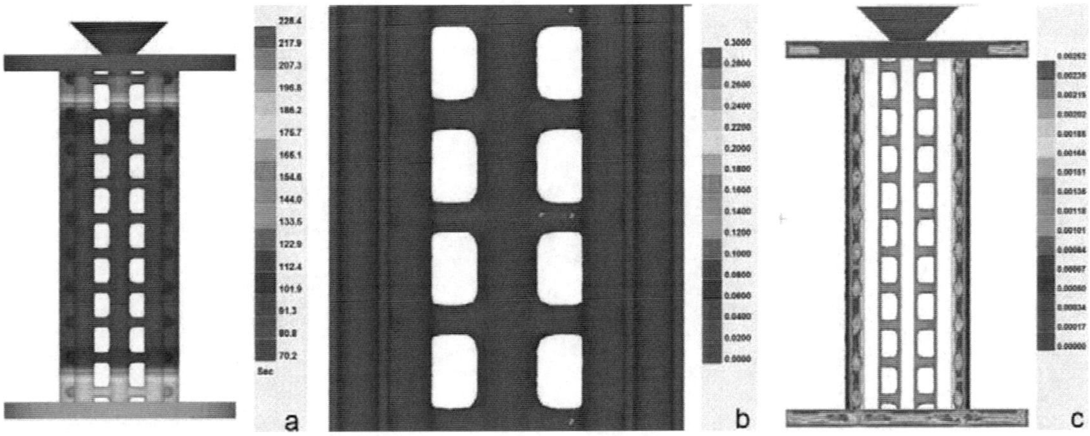

Figure 4. Calculation examples (a) solidification time (b) macro-porosity (c) micro-porosity

In summary, referring to Table 2, vacuum casting is much preferable to gravity casting, on the grounds of much reduced porosity, whereas no significant differences are observed in solidification time. Thus, Cases 1-8 in Table 2 are neglected in favour of cases 9-16. In the latter, removal of one plate is comparable to using two plates regarding solidification time for low mold temperature, compare Cases 9 and 13; thus, case 13 is preferable since it also results in material savings.

3.3. Stresses and deformation

In this part of the simulation study, the mold was defined as rigid. In the simulation program's database mechanical properties of the material (A356), including thermal expansion coefficient, are available as a function of temperature. Computation load is high, therefore only the best case of Table 2 is examined, namely case 13. Of particular interest is the distribution of residual stresses as a result of inhomogeneous cooling and contraction of the casting, hot cracking sensitivity and deformation resulting from contraction. Simulation results are shown in figure 5.

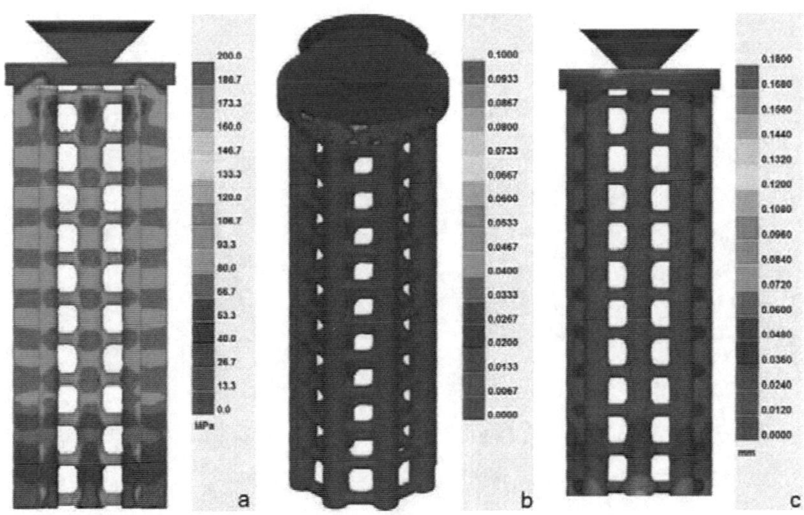

Figure 5. Stress analysis for case 13 (a) residual stresses (b) hot cracking sensitivity (c) deformation

Areas at the junction of two sections of largely different casting modulus as well as mass, notably junctions of plate and vertical rods, exhibit large differences in contraction (displacement) hence deformation is developed which is associated with the emergence of residual stresses, compare figures 5(a) and (c). This is also obvious on the finite element mesh in its initial and final state, see figure 6.

Figure 6. Initial (black, pre-contraction) and final (at room temperature, red) finite element mesh.

Hot cracking sensitivity (HCS) prediction is based on the RDG criterion [14], i.e. effectively in the pressure drop that is observed locally in solid-liquid coexistence areas due to deformation / shrinkage. This is an integral part of the casting simulation platform used. HCS index is the inverse of the maximum local deformation rate, high values corresponding to high probability of cracking. In our case the part as such is not prone to cracking, whereas some areas of the plate are mildly susceptible, see figure 5(b).

Deformation (displacement) is observed at the plate and sprue, which is unimportant, since these parts are cut off eventually. Deformation at the two ends of the spacer is deemed tolerable, see figure 5(c).

4. Investment casting prototyping

A prototype specimen was cast with the lost wax method. A wax model was initially manufactured manually by welding together wax rods of the appropriate shape and cross section, but it soon became apparent that the result was not of industrial calibre. Thus, an appropriate mold accounting of 1/16th of the original part was 3D printed using a resin and powder mix, see figure 7(c).

This mold was used in a wax casting machine at a relative pressure of 0.25 bar και temperature of 85°C in order to produce a number of identical module wax models, see figure 7(d). Sixteen of these models were then welded together by use of an electric soldering iron thereby assembling a full wax model, see figures 7(a) and 7(b).

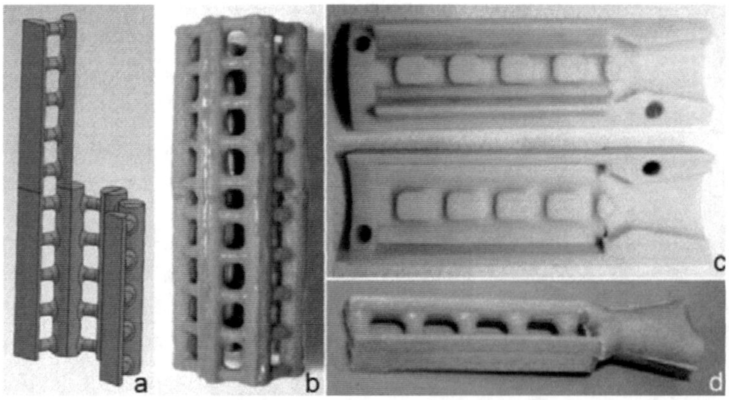

Figure 7. Wax model (a) partial assembly design (b) assembly (c) 3D printed mold (d) wax module

The alternative of using a special 3D printer to make the wax model directly, thereby avoiding wax molding, was considered expensive, because 3D printing a mold allows exploiting it indefinitely to create as many wax models as desired.

Following the investment casting process prescription, the wax model was put in a flask and plaster (Argentum Jewelry Investment by Ransom &Randolph) was poured around it. A vacuum machine was employed in order to eliminate all air bubbles trapped in the plaster while being in the flask. Upon solidification of the plaster a dewaxing oven was used in order to remove the wax model through melting it at 200°C for 90 min. Then, the plaster mold was baked in a special oven for 6 hrs to 750°C to increase its strength This plaster mold was employed in a vacuum casting machine, see figure 8(a), casting conditions corresponding to Case 13 of Table 2. The resulting casting was only partly satisfactory, discrepancies being mainly due to surface roughness and presence of surface defects, see figure 8(b). This was directly attributable to inadequate pressure difference between the vacuum and casting chambers of the machine due to wear and tear.

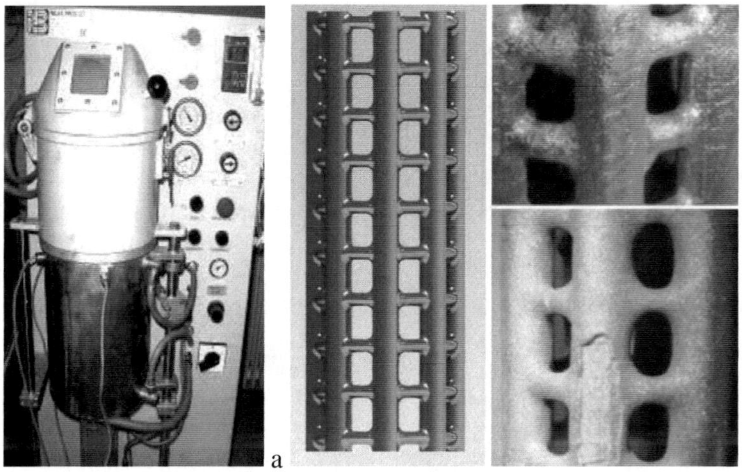
a b

Figure 8. (a) Vacuum casting machine used (b) casting with roughness and surface defects

5. Conclusions

A case study was presented regarding design and manufacture of a mesh-like cylindrical spacers intended for structural applications under a combination of axial and mostly shear loads. As a roadmap for such engineering tasks, the following conclusions are drawn.

The most favorable shape results from static elastic loading simulation for several versions of the parametrically designed part based on a composite evaluation function involving both maximum observed stress and maximum observed deformation, compared to allowable values, as well as mass of the part.

This shape is then considered for casting, a process that can achieve considerable shape complexity at low cost. Casting simulation on a specialized platform is most valuable in studying intended feeding system as well as casting type and parameters. Filling completeness, solidification time and porosity defects are the immediate criteria on which alternatives can be judged. The best alternative is then examined in terms of residual stresses, deformation and hot cracking sensitivity. Vacuum casting proved superior to gravity casting; similarly, low compared to high mold temperature favored low solidification time. Overall adequacy of an unorthodox feeding system was also ensured.

Symmetrical shapes benefit from 3D printed molds for casting wax modules that are assembled by surface heating. Lost wax investment casting involves a number of successive stages, which are well documented, but the final quality of the part depends heavily on possible vacuum casting machine's wear especially at the interface surfaces between flask, casting and vacuum chambers.

References

[1] Helou M and Kara S Design, analysis and manufacturing of lattice structures: An overview 2018 *Int. J. Comput. Integr. Manuf.* **31** 243–61
[2] Panesar A, Abdi M, Hickman D and Ashcroft I Strategies for functionally graded lattice structures derived using topology optimisation for Additive Manufacturing 2018 *Addit. Manuf.* **19** 81–94
[3] Tsavdaridis K D, Nicolaou A, Mistry A D and Efthymiou E Topology optimisation of lattice telecommunication tower and performance-based design considering wind and ice loads 2020 *Structures* **27** 2379–99
[4] Daynes S, Feih S, Lu W F and Wei J Design concepts for generating optimised lattice structures aligned with strain trajectories 2019 *Comput. Methods Appl. Mech. Eng.* **354** 689–705
[5] Daynes S, Feih S, Lu W F and Wei J Optimisation of functionally graded lattice structures using isostatic lines 2017 *Mater. Des.* **127** 215–23
[6] Maliaris G and Sarafis E Mechanical behavior of 3D printed stochastic lattice structures 2017 *Solid State Phenom.* **258 SSP** 225–8
[7] Huang Y, Xue Y, Wang X and Han F Effect of cross sectional shape of struts on the mechanical properties of aluminum based pyramidal lattice structures 2017 *Mater. Lett.* **202** 55–8
[8] Xue Y, Wang X, Wang W, Zhong X and Han F Compressive property of Al-based auxetic lattice structures fabricated by 3-D printing combined with investment casting 2018 *Mater. Sci. Eng. A* **722** 255–62
[9] Pagratis N, Karagiannis N, Vosniakos G C, Pantelis D and Benardos P A holistic approach to the exploitation of simulation in solid investment casting 2007 *Proc. Inst. Mech. Eng. Part B J. Eng. Manuf.* **221** 967–79
[10] Khan M A A and Sheikh A K Simulation tools in enhancing metal casting productivity and quality: A review 2016 *Proc. Inst. Mech. Eng. Part B J. Eng. Manuf.* **230** 1799–817
[11] Vosniakos G-C, Michael S and Vasileiou A Digital Manufacturing Process Chain for One-Off Replacement Parts: A Precision Casting Case Study 2017 *J. Manuf. Mater. Process.* **1** 17
[12] Krimpenis A, Benardos P G, Vosniakos G C and Koukouvitaki A Simulation-based selection of optimum pressure die-casting process parameters using neural nets and genetic algorithms 2006 *Int. J. Adv. Manuf. Technol.* **27** 509–17
[13] Vasileiou A N, Vosniakos G C and Pantelis D I Determination of local heat transfer coefficients in precision castings by genetic optimisation aided by numerical simulation 2015 *Proc. Inst. Mech. Eng. Part C J. Mech. Eng. Sci.* **229** 735–50
[14] Drezet J M and Rappaz M Prediction of Hot Tears in DC-Cast Aluminum Billets 2013 *Essent. Readings Light Met.* **3** 912–8

IOP Conf. Series: Materials Science and Engineering 1037 (2021) 012018 doi:10.1088/1757-899X/1037/1/012018

A case study on 3D scanning, digital reparation and rapid metal additive manufacturing of a centrifugal impeller

N Kladovasilakis[1,2], T Kontodina[1], P Charalampous[1], I Kostavelis[1], D Tzetzis[2] and D Tzovaras[1]

[1] Centre for Research and Technology Hellas – Information Technologies Institute (CERTH/ITI), Thessaloniki, Greece
[2] International Hellenic University, School of Science and Technology, Digital Manufacturing and Materials Characterization Laboratory, Thessaloniki, Greece

*Corresponding author's e-mail: nikoklad@iti.gr

Abstract. This paper demonstrates a comprehensive method for digital restoration and rapid manufacturing of a fully functional mechanical component. A damaged centrifugal impeller of an automotive engine was reverse-engineered via a 3D laser scanner, with which the 3D model of the investigated object was digitally captured. Utilizing advanced design software and surface modelling, the missing fragments were restored and the impeller's 3D digital model was reconstructed. Metal additive manufacturing technology and more specifically the Selective Laser Melting (SLM) technique was employed to construct a functional metal impeller. Stainless steel 17-4 PH powder was used as the printed material. To ensure the exploited powder was appropriate for the part to be manufactured, we applied characterization of the feedstock material was conducted through two different characterization methods, i.e. granulometry and scanning electron microscopy (SEM) combined with energy-dispersive X-ray spectroscopy (EDX). Furthermore, Surface Distance Maps (SDM) were calculated to correlate the geometrical deviations between the physical 3D printed part and the digital model in order to examine the accuracy of the applied method.

1. Introduction

Additive Manufacturing (AM) is a rapidly evolving fabrication technology that is utilized among others for the production of complex and advanced constructs [1]. Nowadays, there is a plethora of different AM methods, such as material extrusion, powder bed fusion (PBF), material jetting etc. [2]. Almost any material can be 3D printed (plastics, ceramics, metals etc.) using the proper AM method [3]. In the last three decades, AM techniques were mainly used for research as well as for rapid prototyping purposes. However, in recent years, fully functional components are produced in various scientific fields like automotive, aeronautic and medical industries. More specifically, metal AM technologies are proved to be a sustainable and cost-efficient alternative to machining and casting for the fabrication of relatively small and geometrical complex metal components [4]. The most widespread metal AM method is the Selective Laser Melting (SLM) which belongs to the PBF category. PBF processes utilize a high-powered laser in order to selectively fuse small particles of metal, plastic or ceramic. The metal powder is the feedstock material and is spread on a build plate where the laser beam melts the powder particles according to the part's cross-section geometry creating that way a thin layer of the desired geometry. In the next step, the powder bed descends for a layer thickness and the process is repeated until the entire geometry of the product is created [1]. Combining this metal AM technique with reverse engineering methods, it is possible to rapidly redesign and construct metal spare parts with particularly complex geometrical characteristics like impellers, turbines, customized parts etc. 3D scanning technologies are suitable for reverse engineering applications providing with high accuracy the digital depiction of a physical object. The most well-known 3D scanning techniques that can provide reconstruction with accuracy adequate for real-life applications are the structured-light and laser scanning [5]. 3D Laser scanning is a non-destructive method that digitally captures the external geometry of an object, utilizing a line of laser light. The acquired data from these scanning procedures could be processed in order to extract a digital 3D model that is fully editable in a CAD software. Therefore, the process of rapid manufacturing of a fully functional spare part could be divided into three steps. The first one is the 3D scanning procedure of the physical object; the second one requires the handling of the acquired data to obtain the digital 3D model and the last once comprises the metal fabrication of the desired component.

The main objective of the current paper is to prove that reconstruction and manufacturing of a functional mechanical component is feasible in order to replace an already damaged one. To this end, a metal centrifugal impeller from an automotive turbocharging system, which was heavily damaged, constitutes the use case of the present work. Centrifugal impellers for automotive applications are relatively small objects with high geometrical complexity. Hence, the majority of them are fabricated via 5 axis CNC machining processes in order to achieve the precise geometry coupled high-quality surface finish [6]. However, material removal procedures are generally time consuming and unprofitable for the fabrication of a prototype or a small batch of products. Hereupon, this study proposes a method to obtain the digital 3D model of a damaged component through 3D scanning methods and by employing rapid manufacturing techniques like metal AM technologies to replace the impaired part. Figure 1 illustrates an overview of the applied methodology.

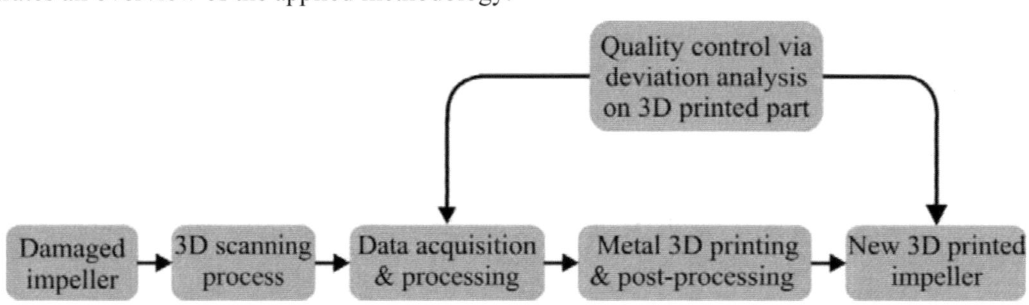

Figure 1. Flowchart of the overall procedure.

2. Methods and Materials

2.1. 3D Scanning & Metal 3D printing

According to the flowchart exhibited figure 1, the first phase of the employed procedure includes the 3D scanning of a damaged impeller. Taking into account that the impeller's smallest dimension is the thickness of each blade (≈300μm), a high-resolution 3D scanner is required. Moreover, the 3D scanner should be able to acquire data from multiple angles due to the object's high complexity. Thus, the NextEngine 3D laser scanner has been employed for this process as presented in figure 2a, where the scanning process is portrayed using the specific laser scanner. The 3D laser scanner contains two arrays of four solid state lasers with 650nm wavelength, two 5.0-megapixel cameras and two lights for capturing images. Furthermore, a 360° turn-table was used in order to fully capture the geometry of the object and to facilitate the mechanical alignment between scans during the post-processing procedure. Moreover, the scanner provides texture density up to 500dpi and dimensional accuracy up to ± 125μm with scanning speed of 50,000 points per second. Then, ScanStudio software was used for the data acquisition process as well as for various post-processing procedures. Finally, MeshMixer and SolidWorks software were applied for further processing of the scanning data extracting that way an editable CAD file. In order to reconstruct the geometry of the damaged impeller, 3D scanning and surface modelling techniques were employed to identify the missing parts of the inspected model. In particular, taking into account the symmetry of the object, it was feasible to assess the missing fragments of the damaged impeller. Hereupon, the digital 3D model of the impeller was defined including recovery of the damaged part.

Table 1. 3D printing parameters used for the metal AM process.

Printing parameters	**Values**
Layer height	25μm
Hatching distance	30μm
Scan speed	1200 mm/s
Laser power	107W
Volumetric Energy Density (VED)	118.89 J/mm^3
Spot size	50μm
Oxygen percentage	1.00% (upper limit)

In the next step, the fabrication of a new impeller utilizing a metal AM technology (SLM) took place. For the metal 3D printing process, the ORLAS Creator metal 3D printer was employed which belongs to the PBF/SLM metal 3D printing category. The specific printer possesses a continuous Yb fibre laser beam with 250Watt maximum power coupled with a wavelength of 1067 nm. Furthermore, ORLAS Creator achieves high printing accuracy of 25μm at z direction due to the respectively small layer height. A wide range of metal powder materials could be printed via this metal 3D printer. In the context of this study, stainless steel 17-4 PH powder was used as the feedstock material due to its advanced mechanical properties [7]. Table 1 documents the major printing parameters used for the construction of the new impeller. It is worth mentioning that the lowest available values were chosen for hatching distance and layer height in order to achieve the maximum dimensional accuracy. According to the literature [8], the optimum energy density was attained through the regulation of the laser power and scan speed values. Figure 2b illustrates the employed metal 3D printer and the printing process inside the building chamber.

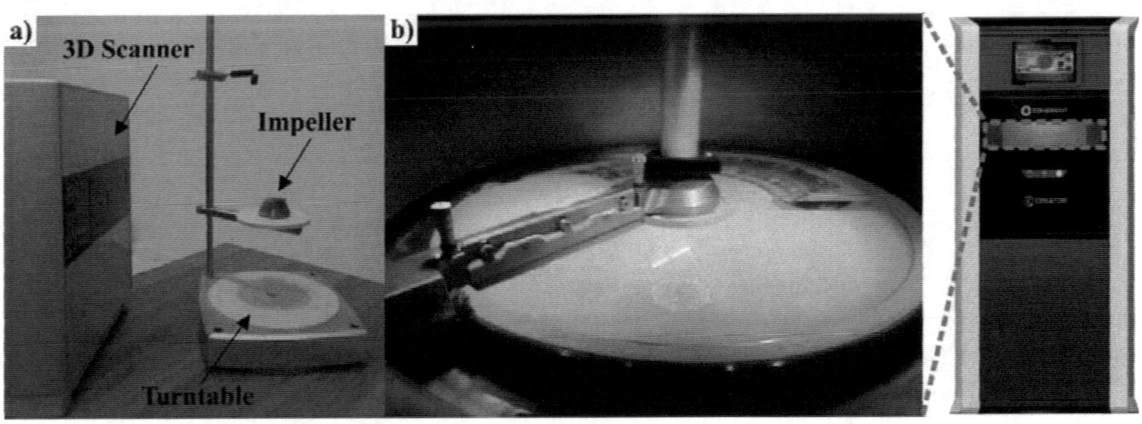

Figure 2. (a) Scanning procedure; (b) Metal 3D printing process.

2.2. *Material Characterization*

One the most crucial parameters for SLM technology that possesses a severe impact on the outcome of the metal 3D printing process is the quality of the powder itself [9]. For that reason, it is essential to conduct material characterization procedures on the feedstock metal powder. In the present work, for the characterization phase, the technique used was Scanning Electron Microscope (SEM) coupled with and Energy-Dispersive X-ray Spectroscopy (EDX). The Phenom ProX Desktop SEM was employed for the investigation of the powder, which was attached on the stub with a carbon tape. The grain morphology of SS 17-4 PH powder sample was studied and the results have shown certain irregular grain shapes deviating from the spherical pattern. Furthermore, an indicative particle size distribution (PSD) was obtained measuring various powder's samples on different scattered regions. The EDX analysis was conducted in order to identify the precise chemical composition of the metal powder and test the presence of any impurities. Figure 3a presents the electromagnetic emissions spectrum of the SS 17-4 PH powder. This analysis incorporates spectra exhibiting peaks corresponding to the elements of the investigated sample. Thus, the chemical composition of the metal alloy at weight percent was identified and listed at the right part of figure 3a. It is worth noting that the EDX analysis is in good agreement with the manufacturer datasheet (OC Oerlikon). Figure 3b illustrates an indicative PSD of a virgin SS 17-4 PH powder, which has been sieved with a 63μm mesh. PSD curve follows a typical positive skew distribution (shifted to the left). PSD outcomes are used to identify the median particle sizes of the inspected powder material in terms of size gauges (D_{10}, D_{50} and D_{90}). In the conducted experiments, D_{10}=11.5 μm indicating that the 10% of the powder sample is less than 11.5 μm, D_{50}= 16.6 μm which corresponds to the median value and D_{90}=27 μm displaying that 90% of the particles of the examined sample are smaller than this size. Figure 3c examines the morphology of the powder's grain, where it can be observed that the majority of the particles possess a spherical shape; however, there are some grains that retain an elliptical shape and few with irregular shapes.

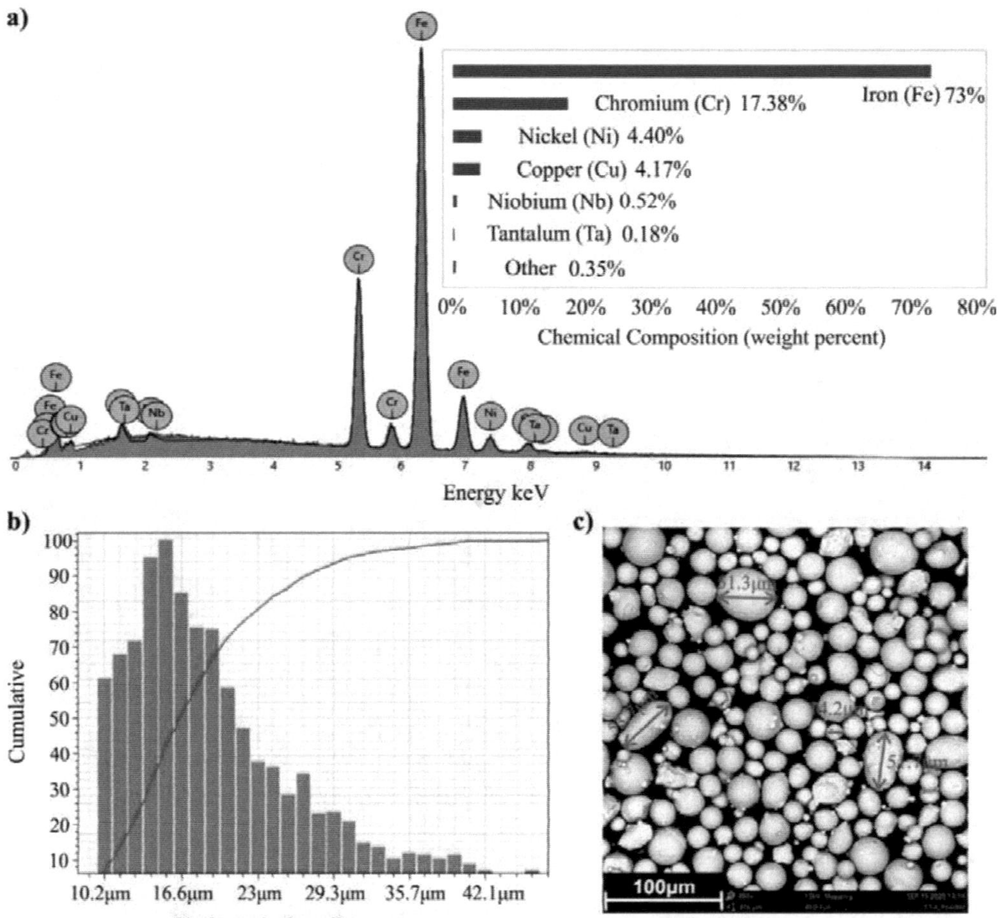

Figure 3. (a) EDX analysis and chemical composition of SS 17-4 PH powder; (b) Indicative PSD of powder's sample; (c) SEM image of SS 17-4 PH particles.

3. Results

The evaluation of the proposed methodology includes the dimensional inspection of the printed metal impeller compared to its digitally reconstructed 3D digital model. In order to maximize the precision, various parameters were adjusted on the 3D laser scanner regarding the positioning, the number of divisions, the resolution (i.e. points/mm^2), the target and the range of the scanning process. Initially, the object was coated with white powder to avoid noise and reflections. Afterwards, the object was placed on a turntable base, approximately 230mm from the scanner, on various orientations (i.e. horizontally, vertically and in front of the scanning field) ensuring that a sufficient number of points were captured for its geometrical reconstruction. Furthermore, several post-process steps were conducted utilizing the appropriate editing tools like trimming, alignment and fusion of the scans in order to optimize and refine the geometrical characteristic of the metal 3D printed impeller.

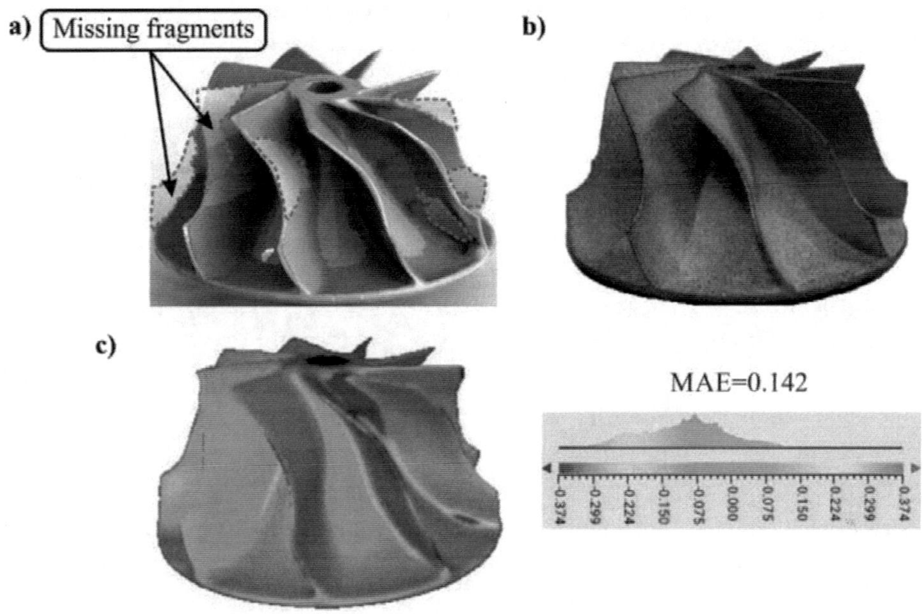

Figure 4. (a) Depiction of missing fragments on the digital 3D model before reparation; (b) 3D printed metal impeller; (c) Surface distance map for the reconstructed impeller.

Figure 4 summarizes the results of the present work. In figure 4a, the missing fragments of the damaged impeller as well as the reconstructed digital 3D model are illustrated. Figure 4b portrays the 3D printed metal impeller as it was constructed without any post-processing process (sandblast, shot-peening etc.). The Artec Studio software was employed to convert the input data (point clouds) from the scanning process to a fully contoured map. In order to quantify the accuracy of the applied methodology and to specify a numerical quantity concerning the quality of the printed impeller, Surface Distance Maps (SDM) were calculated correlating the dimensional deviations between the 3D digital model and the physical printed one as shown in figure 4c. Blue color represents negative maximum deviation and the red color is equivalent to the maximum positive deviation. In contrast, green color corresponds to the minimum dimensional deviation between the examined surfaces. Finally, the accuracy of the procedure was assessed via the calculation of the Mean Absolute Error (MAE) exhibiting that the printed object is in sufficient agreement with the digital one, as the MAE is equal to 0.142 mm. Therefore, implementing the suggested scan-to-print technique, it was feasible to accurately depict and manufacture a complex component in a short time period. Hence, this procedure provides an additional tool for reserve engineering approaches of geometrical complex parts compared to traditional methods. Finally, utilizing metal AM, the fabrication process was completed in a matter of hours compared to time-consuming CNC milling procedures.

4. Conclusions

The current paper presented the workflow of a scan-to-print methodology in order to replace damaged spare parts with new ones constructed via a metal AM 3D printer. More specifically, a damaged centrifugal impeller was scanned and by utilizing surface modelling and advanced design software, it was feasible to identify the object's missing fragments obtaining that way its digital 3D model. Prior to the metal 3D printing process, the characterization of feedstock was conducted through granulometry and SEM coupled with EDX in order to ensure good quality of the selected powder. In the next step, the optimal printing parameters were selected and the metal 3D printing procedure was initiated. The 3D printed impeller's dimensions and morphological characteristics were evaluated via a laser scanner and the geometrical deviations from its digital 3D model were computed. The results showed that the reparation of functional spare parts could be achieved providing high dimensional accuracy as the deviations of the printed impeller was relatively small. Finally, in order to elaborate the research on the functionality of customized metal AM components, it is necessary to focus in the future on the evaluation of their mechanical behavior under various operating conditions.

Acknowledgments

We acknowledge support of this work by the project "Intelligent και Automated Systems for enabling the Design, Simulation and Development of Integrated Processes and Products - ODYSSEAS" (MIS 5002462) which is implemented under the "Action for the Strategic Development on the Research and Technological Sector", funded by the Operational Programme "Competitiveness, Entrepreneurship and Innovation" (NSRF 2014-2020) and co-financed by Greece and the European Union (European Regional Development Fund).

References

[1] Gibson I, Rosen DW and Stucker B 2010 *Additive Manufacturing Technologies*, First ed., New York: Springer, https://doi.org/10.1007/978-1-4939-2113-3

[2] Charalampous P, Kostavelis I and Tzovaras D 2020 Non-destructive quality control methods in additive manufacturing: a survey, *Rapid Prototyping Journal*, **26**:4:777-790, https://doi.org/10.1108/RPJ-08-2019-0224

[3] Frazier W E 2014 Metal Additive Manufacturing: A Review, *Journal of Materials Engineering and Performance*, **23**, 1917–1928, https://doi.org/10.1007/s11665-014-0958-z

[4] Gebisa A W and Lemu H G 2017 A case study on topology optimized design for additive manufacturing, *IOP Conference Series: Materials Science and Engineering*, **276**:12-26, https://doi.org/10.1088/1757-899x/276/1/012026

[5] Karatas O H and Toy E 2014 Three-dimensional imaging techniques: A literature review *Eur J Dent*. **8(1)**:132-140, https://doi.org/10.4103/1305-7456.126269

[6] Huber M. et al. 2018 Process Setup for Manufacturing of a Pump Impeller by Selective Laser Melting (Meboldt M and Klahn C. editors) *Industrializing Additive Manufacturing - Proceedings of Additive Manufacturing in Products and Applications AMPA2017* (Springer, Cham.) https://doi.org/10.1007/978-3-319-66866-6_24

[7] Lin X, Cao Y, Wu X, Yang H, Chen J and Huang W 2012 Microstructure and mechanical properties of laser forming repaired 17-4PH stainless steel, *Materials Science and Engineering: A*, **553**, 80-88, https://doi.org/10.1016/j.msea.2012.05.095

[8] Aboutaleb A M, Bian L, Elwany A, Shamsaei N, Thompson S M and Tapia G 2017 Accelerated process optimization for laser-based additive manufacturing by leveraging similar prior studies, *IISE Transactions*, **49**:1, 31-44, https://doi.org/10.1080/0740817X.2016.1189629

[9] Maamoun A H, Xue Y F, Elbestawi M A and Veldhuis S C 2018 Effect of Selective Laser Melting Process Parameters on the Quality of Al Alloy Parts: Powder Characterization, Density, Surface Roughness and Dimensional Accuracy, *Materials*, **11**:2343, https://doi.org/10.3390/ma11122343

Direct 3D Printing of a hand splint using Reverse Engineering

J Kechagias[1], K Kitsakis[1], A Zacharias[1], K Theocharis[1], K-E Aslani[1,4], M Petousis[2], N A Fountas[3] and N M Vaxevadnidis[3*]

[1]University of Thessaly, General Department, Larissa, Gaiopolis, GR 41500, Greece.
[2]Hellenic Mediterranean University, Mechanical Engineering Department, Estavromenos, GR 71410, Heraklion, Crete, Greece.
[3]School of Pedagogical and Technological Education (ASPETE), Department of Mechanical Engineering Educators, Amarousion, GR 15122, Greece.
[4]University of West Attica, Department of Mechanical Engineering, Aigaleo, GR 122 44, Greece.

[*]Corresponding author's e-mail: vaxev@aspete.gr

Abstract. The present work is focused on the direct manufacturing of a hand splint using free-open access software and a low-cost three-dimensional printer (3DP). The hand digital model was created using panoramic photos by a common mobile phone camera. The photos were used as input to the "3DF-Zephyr" free software for creating the hand surface model. Then, the hand surface model was transferred into the "Autodesk fusion 360" free software and the surface model of the hand splint was generated and modified according to the design requirements. Sequentially, both hand and hand splint were translated to Stereolithography (STL) files and transferred to open access "MakerBot" 3D printing software in order to prepare the G-codes for 3D printing. A low cost 3D printer was used for building the models while Polylactic acid (PLA) was the material of the customized 3D physical models.

Keywords: 3D printing; reverse engineering; open software; hand splint

1. Introduction
3D printing and especially Fused Deposition Modelling (FDM) is one of the most widespread Additive Manufacturing processes for customized plastic parts directly from digital data [1]. Nowadays, FDM is known as Fused Filament Fabrication (FFF) and becomes more available to small firms or home users due to the low cost of the materials and the equipment used [2].
Reverse engineering refers to techniques that create CAD models from physical parts (damaged or broken) by data digitization method (MRI, CMM, photos etc.) [3]. These models are transferred into special CAD/CAM programs in order to create codes (similar to G-Codes) [4]. Finally, they are usually remanufactured with the use of 3D printing. Every 3D printing process comprises three stages: pre-processing of the STL-file, actual building and, post-processing of the prototype [5].

One of the most widespread 3D printing processes is Fused Filament Fabrication (FFF) [6]. This method generally uses polymers [7], while a large number of process parameters affect the final printed parts quality [4, 8].
In this project, direct manufacturing of a hand splint using free-open access software and an FFF 3D printer was attempted. A variety of studies related to 3D printing for fast production of customized three-dimensional-printed hand splints were found [9-11]. In our project, low cost 3D printing parts using free software is the main concept, as well as the evaluation of the proposed procedure as a sustainable choice. First, a hand digital model was created using panoramic photos of a common mobile phone camera. Then, the photos were used as input to the '3DF-Zephyr' free software for creating the hand surface model. After that, the hand surface model was transferred into 'Autodesk fusion 360' free software and the surface model of the hand splint was generated and modified according to the design requirements. Sequentially, both hand and hand splint, were saved as Stereolithography (STL) files and transferred to the open access 'MakerBot print' 3D printing software in order to prepare the g-codes for 3D printing. An FFF printer was used for models' building. PLA was the printing material of the customized 3D physical models.

2. Materials and methods

2.1. Photogrammetry software

The program used to scan the hand was '3DF Zephyr', which, although originally downloaded to the lite version, had to be upgraded to its Professional version. The upgrade was valid for one month. The tool used to photograph the hand was a common mobile phone. Another program that was proposed was the Autodesk Recap photo. The scanning process had to be repeated several times to achieve the desired result.

2.2. 3D design of the Splint

The scanned hand was inserted into the Autodesk fusion 360 software. Some other software tools such as Rhinoceros 5 or Meshmixer could have done the same process, too. Before the splint was designed, the triangles of the STL mesh file had to be lowered, to make the hand smoother on the 3D printer (Fig. 1).

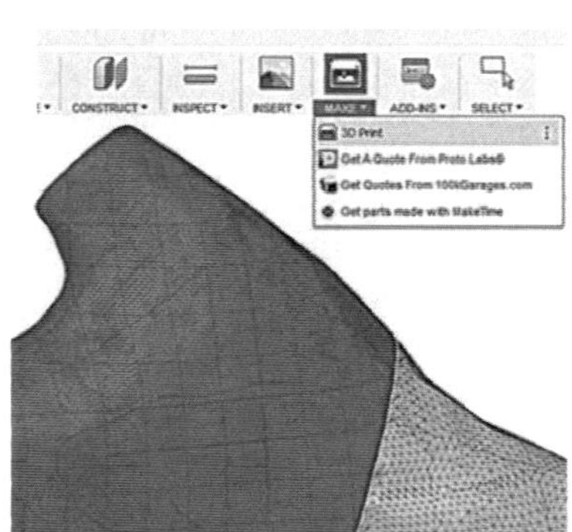

Figure 1. Hand 3D model taken using 3DF Zephyr.

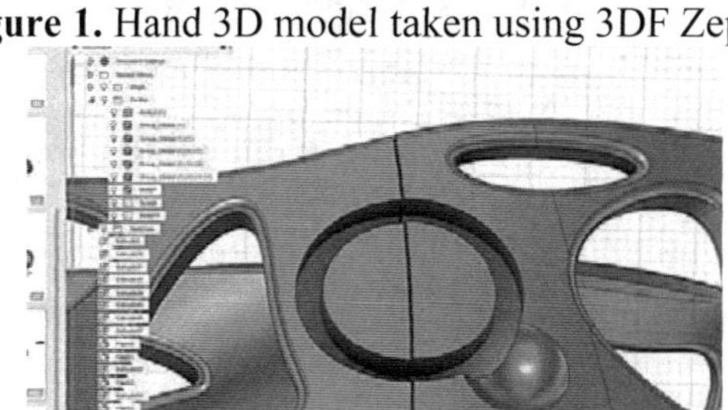

Figure 2. Hand splint 3D model.

Initially, a tube was designed which was moved around the hand and using the 'edit form' command it came close enough to the curves of the hand. The command used to create the tube is: sculpt/create/cylinder. The next command utilized was "thicken" to give the required thickness.
Then, after the splint took the shape of the hand, its buttons were designed, which are circular in the form of a groove, so that they can be closed using a washer. In order to facilitate the installation and the removal of the washer, recesses were designed that allow the placement of the finger (Fig. 2). The dimensions of the buttons are: Splint length: 13.78 cm; Splint thickness: ≈ 5mm; Button diameter: 20mm (outer) - 15mm (inner); Slot thickness: 5mm; Slot depth: 3.5mm; Diameter for finger on the clasp (Depth: 3mm, Tolerance: ± 0.3mm, the lowest point in oval shape).
The number of buttons designed is four. Their location was chosen based on the ease of splitting, the splint was into two parts, which was then performed. Next, the object was divided into two parts. To do this, a line was drawn with a sketch that passes through the centres of all four buttons, and with the use of the "extrude" command, the line was drawn, so that it could pass through the object to be split.
Finally, for the separation of the object, the "split body" command was utilized, in which the designed line was selected, and the object that will be divided into two parts was used.
Afterwards, holes were drilled to ventilate the human limb and to save material. In order to open the holes, it was necessary to decide the suitable pattern. It was decided to design holes in the formation, of the part, with triangular and oval shape. For the formation of the holes, each hole was designed separately. After the design was made, the hole was moved near the splint, placed at the appropriate height and angle and with the use of the "extrude" command the opening was made. In order for the splint to have a smoother surface at the points where the holes were opened, the fillet command was used. All fillet commands were made with a radius of one millimetre (1 mm).

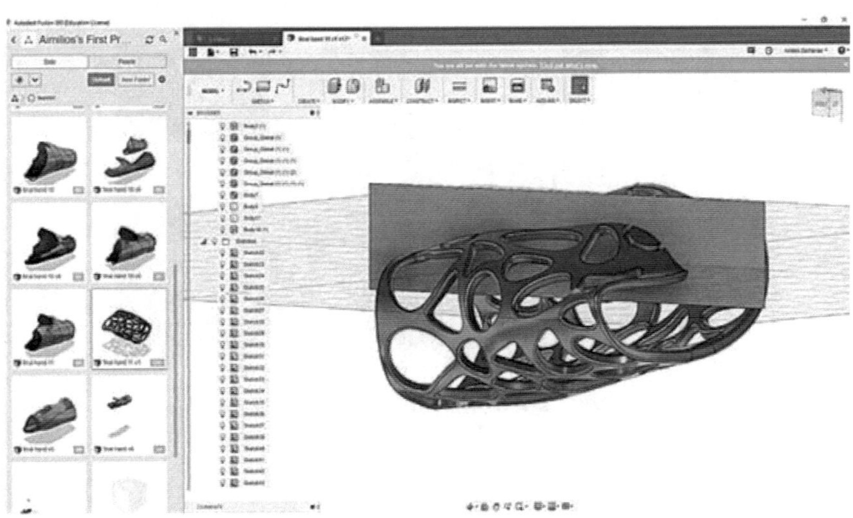

Figure 3. Hand splint 3D model.

2.3. STL files preparation
The next step was the STL files preparation.
First the hand was divided in two parts to facilitate the printing but also to fit the hand in the printer working volume. Then, part of the fingers was removed as this saved material and time, in addition they do not significantly affect the work, so with that in mind they were removed.
After that, separation follows: the first part consists of the point where the fingers were cut and about 8.1 cm to the right, and the second was the part that remained.

After cutting, each section needed to be exported in a file format suitable for the 3d printer software (STL files). The program used for this was 'meshmixer', a software that will specify the format for exporting the file. With the same way hand-splint split into two parts with the same commands used for the hand and then exported to 'meshmixer'. Prior to printing, the "sculpt" command was executed via 'meshmixer' to make the surface of the splint smoother in areas where many pixels appear.

2.4. Hand and hand splint printing

MakerBot Print was used for orientation and layer build style settings (Fig. 4). Wanhao Duplicator 4X FFF 3D Printer was used for printing the STL files (Figs. 5, 6, 7). The printer prepared with the following settings: Extruder temperature: 220° C; Platform temperature: 90° C; Travel speed: 150mm/s; Minimum layer duration: 5.0 s; Device settings: Low. Material used was PLA.

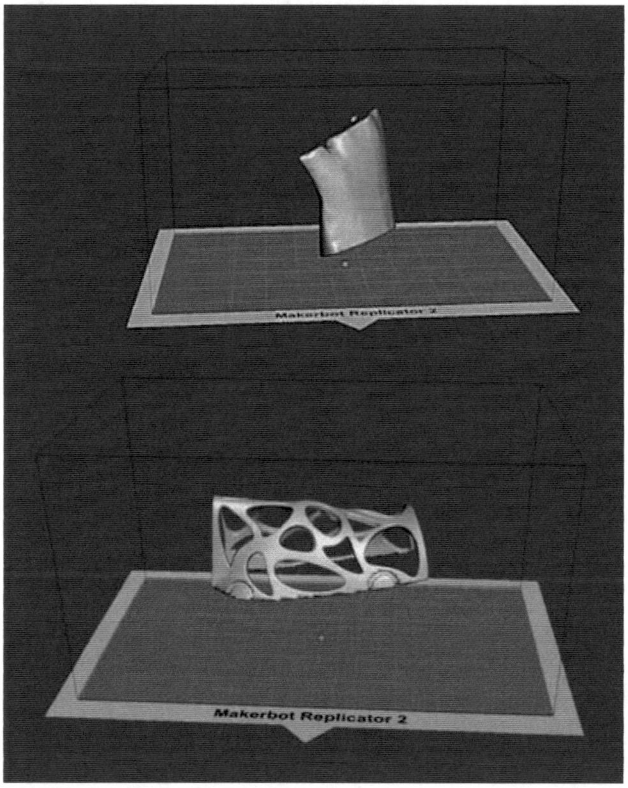

Figure 4. Hand and hand splint preparation for printing.

Figure 5. Hand physical models.

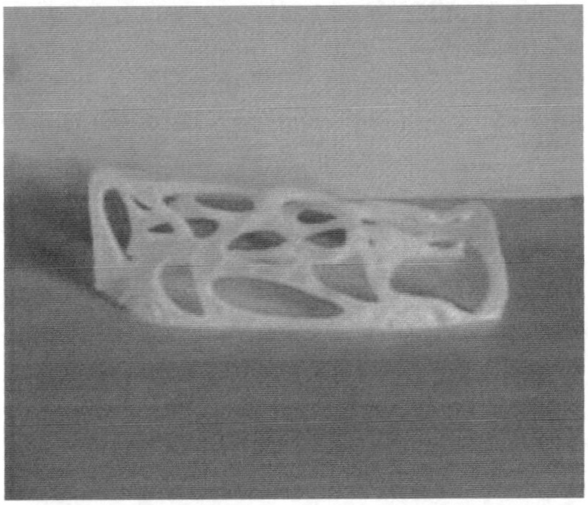

Figure 6. Hand splint physical models.

Figure 7. Assembly of hand and hand splint.

3. Results and discussion

The printing process of the hand and the hand splint was time consuming and, in some cases, required some corrections and reprinting from the beginning. The hand takes about four (4) hours for each part with the resolution setting at low. The difficulty encountered in this case was related to the extraction of the material by the extruders. During material delivery, as the extracellular material was moved, the material that was injected was carried away and detached from the printer base. This problem, in the end, was overcome by using hairspray, on the paper tapes that were already on the base of the printer as well as by increasing the base temperature so as not to entice the material with the movement of the extruder.

After the hand was 3D printed in two parts, it was then required to scrape the piece with a spatula, from the wrist to the fingers at the point of the thumb so that the piece would be smoother and the splint would fit better. After sanding the piece, the two parts were glued together with benzine glue so that the hand 3D printing process could be completed.

As for the 3D printing of the splint, the time it took was about four (4) hours for each piece. The difficulties in this case are related to the stability of the 3D print since there are holes in the splint. In order to overcome these difficulties, the parameter of the auxiliary material was selected during the processing of the 3D print settings in order to make the printing more robust. Moreover, during the 3D printing process, it was required to use the lacquer, so that the material does not come off the base. When the 3D printing of the two parts of the splint was completed, the auxiliary material was removed using the spatula, which was also used for smoothing in order to have the desired result.

4. Conclusions

In this project a hand splint and the hand prototype were 3D printed using FFF 3D printing technology and photogrammetry software. All software used in this project are open (free). Material used for 3d printings was the PLA biocompatible material. The procedure takes a lot of work to select the appropriate software for the manipulation of the panoramic photos (scanning data) and to translate them to STL 3D digital hand geometric model. Hand splint produced using appropriate software tools that produce CAD data from digital data (photogrammetry data). Finally, STL preparation for 3d printing and 3d printer settings are very important in order to have good quality physical models. Optimization of the process parameters is proposed as future work in order to have even better results.

Acknowledgements

The authors wish to thank the Special Account for Research of ASPETE for supporting this work through the funding program "*Strengthening research of ASPETE faculty members*".

References

[1] Chaunier L, Guessasma S, Belhabib S, Valle GD, Lourdin D, Leroy E 2018 *Addit. Manuf* **21** 220-233
[2] Kitsakis K, Alabey P, Kechagias J, Vaxevanidis N 2016 *IOP Conf. Ser.: Mater. Sci. Eng.* **161** 012025
[3] Bagci E 2009 *Adv. Eng. Softw* **40** 407-418
[4] Tsiolikas A, Mikrou T, Vakouftsi F, Aslani KE, Kechagias J 2019 *IOP Conf. Ser.: Mater. Sci. Eng.* **564** 012021
[5] Kechagias J, Maropoulos S, Karagiannis S 2004 *Rapid Prototyp. J.* **10** 297-304
[6] Kitsakis K, Aslani KE, Vaxevanidis N, Kechagias J 2019 *IOP Conf. Ser.: Mater. Sci. Eng.* **564** 012022
[7] Vidakis N, Vairis A, Petousis M, Savvakis K, Kechagias J 2016 *Acad. J. Manuf. Eng.* **14** 87-94
[8] Valerga A.P, Batista M, Salguero J, Girot F 2018 *Materials* **11** 1322
[9] Popescu D, Zapciu A, Tarba C, Laptoiu D 2020 *Rapid Prototyp. J.* **26** 134-144
[10] Chen PJ, Du YC, Shih JB, Xu ZF 2017 Development of a 3D printing system for hand splint by object mirroring and surface reconstruction technology *Proc. of the 2017 IEEE Int. Conf. on Inf., Comm. Eng.* (Xiamen) p 132
[11] Nam HS, Seo CH, Joo SY, Kim DH, Park DS 2018 *Ann. Rehabil. Med.* **42** 634-638

IOP Conf. Series: Materials Science and Engineering 1037 (2021) 012020 doi:10.1088/1757-899X/1037/1/012020

Reconstruction, processing and smoothing of surface geometry of a patient specific ascending aortic aneurysm

C Manopoulos[1*], A Raptis[1], W Krishan[2], C Mavratzas[2], M Drandakis[2], S Astraka[2], I Kouerinis[3], and N M Vaxevanidis[2]

[1]Biofluid Mechanics and Biomedical Engineering Laboratory, Fluids Section, School of Mechanical Engineering, National Technical University of Athens, Zografos, GR 15780, Greece.
[2]Laboratory of Manufacturing Processes and Machine Tools (LMProMaT), School of Pedagogical and Technological Education (ASPETE), Department of Mechanical Engineering Educators, Amarousion, GR 15122, Greece.
[3]Department of Cardiothoracic Surgery, "Hippocration" Hospital, Athens, GR 11527, Greece.

[*]Corresponding author's e-mail: manopoul@central.ntua.gr

Abstract. This work presents the creation of geometric surface models representing the heart chambers and the aorta for two patient specific cases. The first one concerns an acute ascending aortic aneurysm in the anterior Valsalva sinus and the second one a normal aorta for comparison reasons. The surface modelling was implemented in a medical image reconstruction software (Materialise Mimics), based on the computed tomography scans of the patients. For each case, the two atriums and ventricles, the ascending aorta, aortic arch and descending aorta, the pulmonary artery and the superior vena cava were also reconstructed and modelled. Special attention is given to the morphology of the vessels and the heart chambers that are in contact with the ascending aortic aneurysm, due to the acute aneurysmal extension of the Valsalva sinus. A comparison of the non-aneurysmal and the aneurysmal models is necessary to infer whether this disease affects the geometries and/or other vital organs that come in contact with the aortic aneurysm. Special care is given to the surfaces' smoothing process to preserve the morphology of the structures, avoiding any deviations from the actual geometries of the corresponding vessels and heart chambers.

1. Introduction

An aneurysm is an expansion or bulge of a blood vessel to more than 1.5 times its normal size. The aorta – the body's largest artery that is responsible for carrying oxygen-rich blood away from the heart – is prone to developing aneurysms. Aortic aneurysms (both abdominal and thoracic) represent the 15th leading cause of death in individuals older than 55 years, and the 19th leading cause of death in individuals of all ages. [1] Mostly, male subjects get affected by the aneurysmal disease. The risk factors include age, family history, arterial pressure, and lifestyle, among other.
If an aneurysm occurs in the ascending aorta, the proximal aortic section after blood leaving the heart through the arterial valve, it is called ascending thoracic aortic aneurysm (aTAA). aTAAs do not always cause symptoms, especially in the early stages and when small. As they increase in size, aTAAs may

begin to cause problems such as: chest pain, back pain, tenderness in the thoracic region, hoarseness, shortness of breath, cough.
Not all patients with an aTAA will experience symptoms, even when the bulge is large. A ruptured aneurysm, on the other hand, is a medical emergency. Symptoms include clamminess, difficulty breathing, difficulty swallowing, dizziness, light-headedness, loss of consciousness, low blood pressure, rapid heart rate, sudden and intense pain in the chest or back, weakness or paralysis on one side of the body.
The heart is a complex organ that pumps blood through the body with an intricate system of muscle layers, chambers, valves, and nodes. It has its own circulation system and receives electric impulses that make it contract and relax, which triggers a sequence of events forming the cardiac cycle. The heart has a middle muscular layer, the myocardium, made up of cardiac muscle cells, and an inner lining called the endocardium. The inside of the heart (heart cavity) is divided into four 4 chambers – two atria and two ventricles – separated by cardiac valves that regulate the passage of blood. The atria receive blood returning to the heart, while the ventricles receive blood from the atria – via the atrioventricular valves – and pumps it into the lungs and the rest of the body. The left atrium and left ventricle are separated from the right atrium and right ventricle by a band of tissue called the septum. The right atrium receives deoxygenated blood from the head and neck and from the rest of the body via the superior and inferior vena cava, respectively. The right ventricle then pumps blood into the lungs (through the pulmonary trunk, which divides into the right and left pulmonary arteries), where it is oxygenated. The oxygenated blood is returned to the left atrium via the pulmonary veins and passes into the left ventricle through the cardiac valves. From the left ventricle, it is delivered to the whole body through the aorta.[2]
Multi–slice computed tomography (CT) image series are a valuable source of information to extract shape and motion parameters of the heart. Indicative 3D reconstruction of the heart chambers from CT images can be found in the literature [3, 4]. Here we present the reconstruction of two cases, one normal and one aneurysmal, including all main chambers of the heart (both ventricles and atria) and the connected vessels (arteries and main vein trunks). The pathological case refers to a patient with an aTAA. A concise 3D reconstruction in a difficult case (like the aTAA) in comparison to a normal case can highlight the extend of the disease and possible effects in surrounding organs, arteries, and veins, that are not clearly visible when examining the 2D CT images. The quality of the reconstruction is crucial for increased understanding of the disease from a clinical point of view but also as a prerequisite for 3D printing applications.

2. Materials and methods

2.1. Material
The patient-based analysis relies on the computed tomography (CT) scan of a 70 years old male patient, who suffered from aTAA and underwent aortic repair surgery at "Hippocration" General Hospital of Athens in 2013. The normal case refers to a 69 years old person who had CT scan at the same hospital for irrelevant reason and was free of cardiovascular diseases. The CT data was made available for further processing, with the informed consent of the patients, by attending physician (cardiac surgeon) at the Cardiac Surgery Department of "Hippocration" General Hospital of Athens.

2.2. CT imaging protocol
The CT scan was performed on a 16-slice multidetector CT (MDCT) scanner (Activion 16, Model TSX-031A, Toshiba). A protocol of 1 mm slice thickness was used with 120 kV, 60 mA tube current, 330 mm field of view for reconstruction and a sharp reconstruction kernel. The CT was performed in the craniocaudal direction, from immediately inferior to the aortic arch to the level of the diaphragm in a single breath-hold. A contrast agent is occasionally injected in the blood flow stream of the patient, to increase the imaging resolution, clearly indicating the lumen. A contrast-enhanced CT scan, known as computed tomography angiography (CTA) was performed in the aTAA case while a typical non-contrast gated CT was performed in the normal case.

2.3. Segmentation & Reconstruction

The Mimics software (Materialise inc.) was used for the processing of the CT imaging series, as an advanced medical technology toolbox enabling the segmentation and reconstruction of medical imaging data. Researchers and engineers are using the software to develop state-of-the-art medical device designs, individual implants, and life-saving surgical instruments. The CT scans in Mimics can be examined in four views: 1) the coronal at the upper left window in Fig. 1, 2) the sagittal at the bottom left, 3) the axial at the upper right, and the 4) 3D view at the bottom right window, where the reconstructed anatomic part of interest is accessible. The four views are interrelated, adapting to the selected intersection and the crosshair pinpoint. Moreover, it is possible to adjust the threshold of gray value making the various anatomical parts more distinct, facilitating the segmentation and reconstruction process.

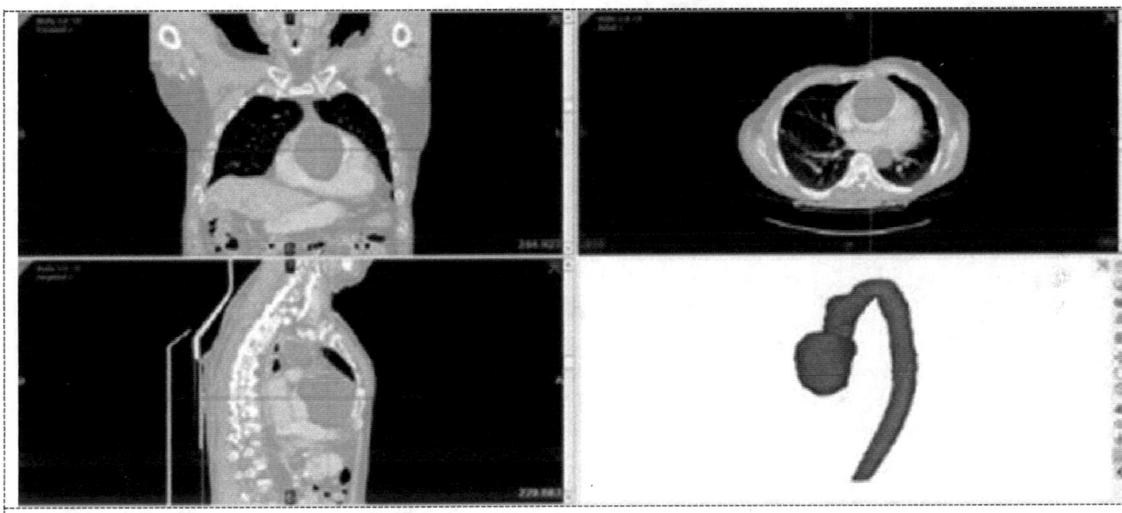

Figure 1. The Mimics user interface.

The anatomic parts of interest are the 4 chambers of the heart, the superior vena cava, the pulmonary artery, and the thoracic aorta. All anatomic parts were separately processed, and a different color-coded mask was assigned to each one.
The thoracic aorta, the pulmonary artery, and the superior vena cava were initially isolated from the surrounding bones and organs using the automated "Dynamic region growing" tool in Mimics that is based on gray value connectivity. The initial automated segmentation is rough most of the times and a manual slice-by-slice refinement is certainly required afterwards. Specifically, the anatomic parts were further restricted to the desired limits by batch-removing the out-of-bounds segmented cross-sections and any leftover segmented artifacts. Special focus was given to the aortic wall calcifications not included in the final reconstruction. The thoracic aorta included the aortic root (immediately after the left ventricle), the ascending part and the descending part until the diaphragm level, as well as its main branches, namely the innominate artery, the left common carotid artery and the left subclavian artery. Concerning the pulmonary artery and the superior vena cava, only the proximal (to the heart) parts were taken into consideration while the pulmonary artery was truncated after its bifurcation. The "Dynamic region growing" tool fails in the case of heart chambers and a fully manual segmentation process was instead carried out. All operations were executed by biomedical engineers under the supervision of the cardiac surgeon, to ensure that the reconstruction accurately corresponds to the anatomic parts of interest.

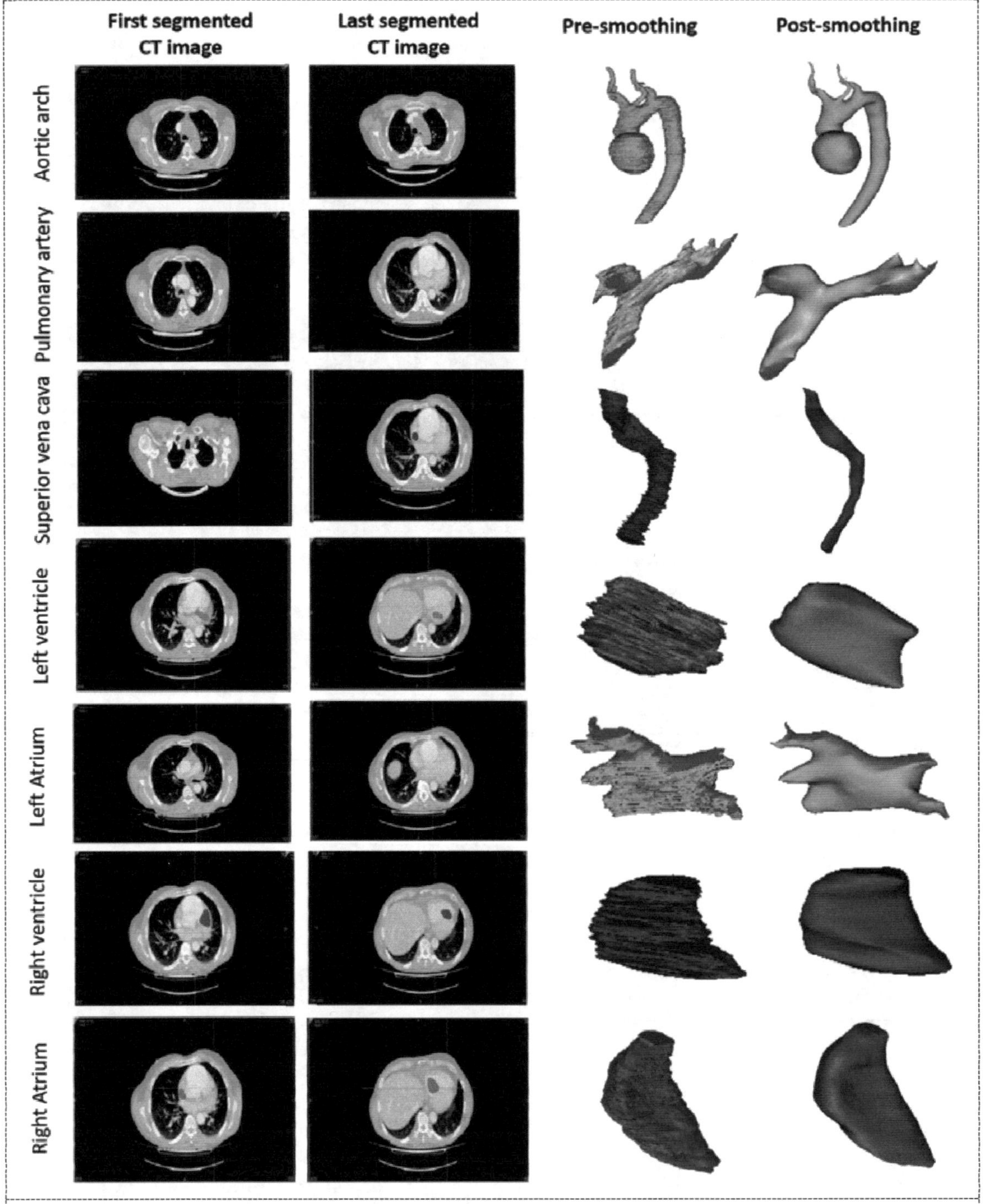

Figure 2. The reconstructed anatomical parts for the aTAA case presented separately before and after smoothing operations. On the left side, the first and last segmented image for each part in the CT series.

After completing the segmentation process, Mimics provides tools for the transformation of the segmented 2D images into 3D models. The initial reconstructed models feature rough artificial edges on their surfaces due to the resolution of the medical images and the slice thickness. The models need further refinement prior to their use either for computational simulations or 3D-printing.

2.4. Smoothing

Smoothing is an iterative process for the reduction of the surface roughness provided that the least deviation from the original data is ensured. During the smoothing process, it was observed that the cross-sectional area (and the volume of the models in extend) depended on the number of smoothing iterations and the smooth factor. More iterations were leading to a more natural result, while the smooth factor had a heavy impact on the shape of the models. Conclusively, it was decided to perform several iterations (500) and keep the smooth factor to a minimum value (0.1). It is also worth noting that during the segmentation process, we applied a small expansion of the segmentation mask to compensate for the volume shrinkage due to the smoothing, ending up with volume-preserving models with respect to the original data.

3. Results and discussion

In Fig. 2, the reconstructed models of all anatomic parts of interest for the aTAA case are presented before and after smoothing, along with the first and last segmented images in the CT series for each part. Fig. 3 illustrates the superposition of all reconstructed models providing a detailed view of the heart, its chambers, and the connected vessels from different perspectives. The Figs. 4 and 5 respectively illustrate the anatomic parts separately and in combination for the normal case.

There are characteristic differences in the position of the heart cavities between the two cases. Due to the aTAA bulge, the right ventricle of the aTAA patient has shifted to the left pushing the left ventricle posteriorly. The other two cavities have also slightly shifted. As the aTAA developed in the posterior direction, the right atrium has been compressed detaching to an extend from the right ventricle. The pulmonary artery has also been compressed due to the aneurysm. From the patient follow-up, it is known that a rupture occurred in the non-coronary Valsalva sinus almost anti-diametrically from the right-coronary sinus where the aneurysmal bulge was primarily concentrated. The aTAA case is rare as rupture did not occur at the distended part of aTAA wall but on the opposite direction. A patient-based biomechanical analysis could further enlighten the special features of the aTAA case, highlighting the stress condition and the critical areas of the aTAA wall.

IOP Conf. Series: Materials Science and Engineering 1037 (2021) 012020 doi:10.1088/1757-899X/1037/1/012020

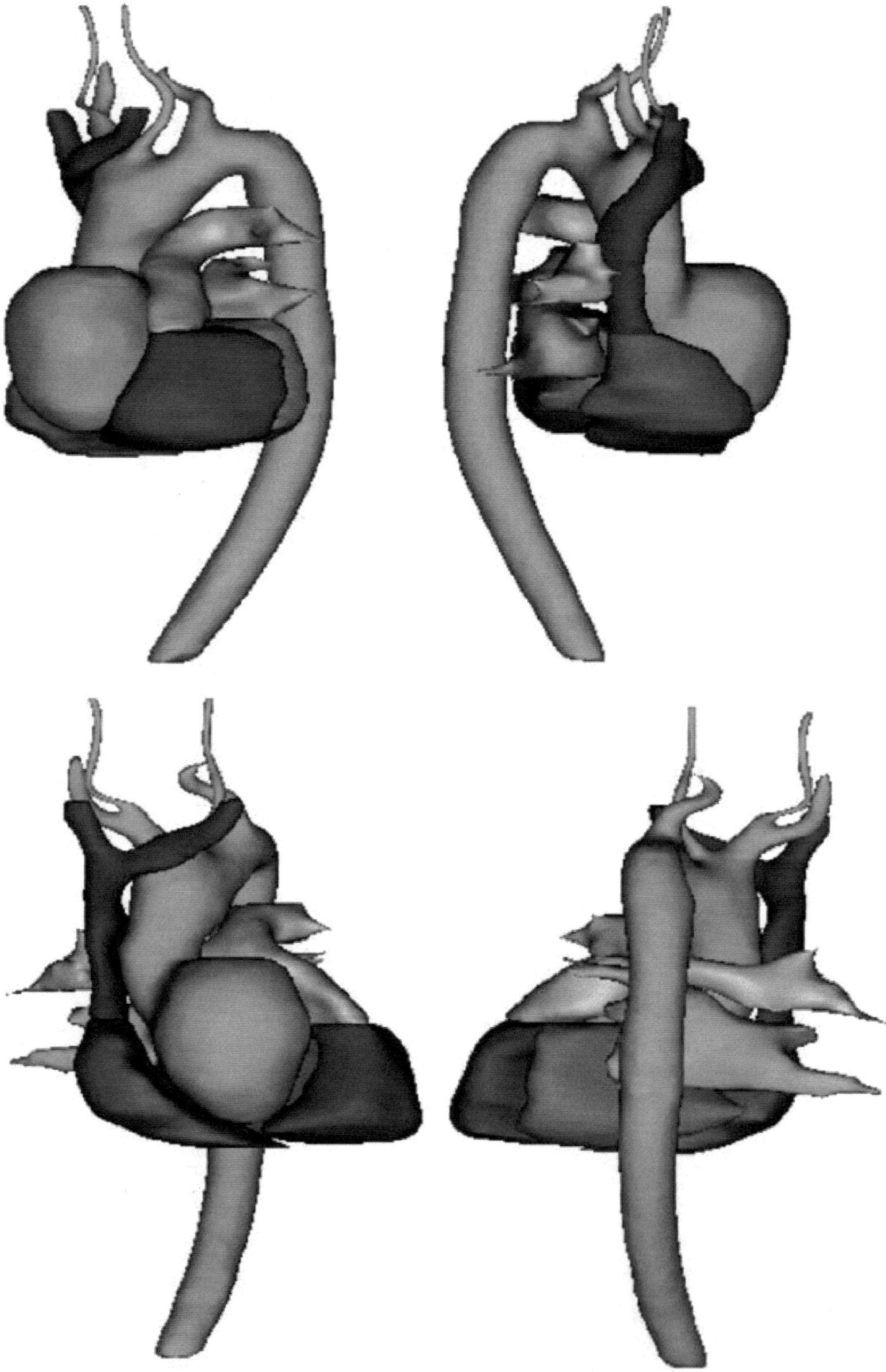

Figure 3. Superposition of the geometrical surfaces corresponding to the chambers of the heart and the connected vessels of the aTAA case.

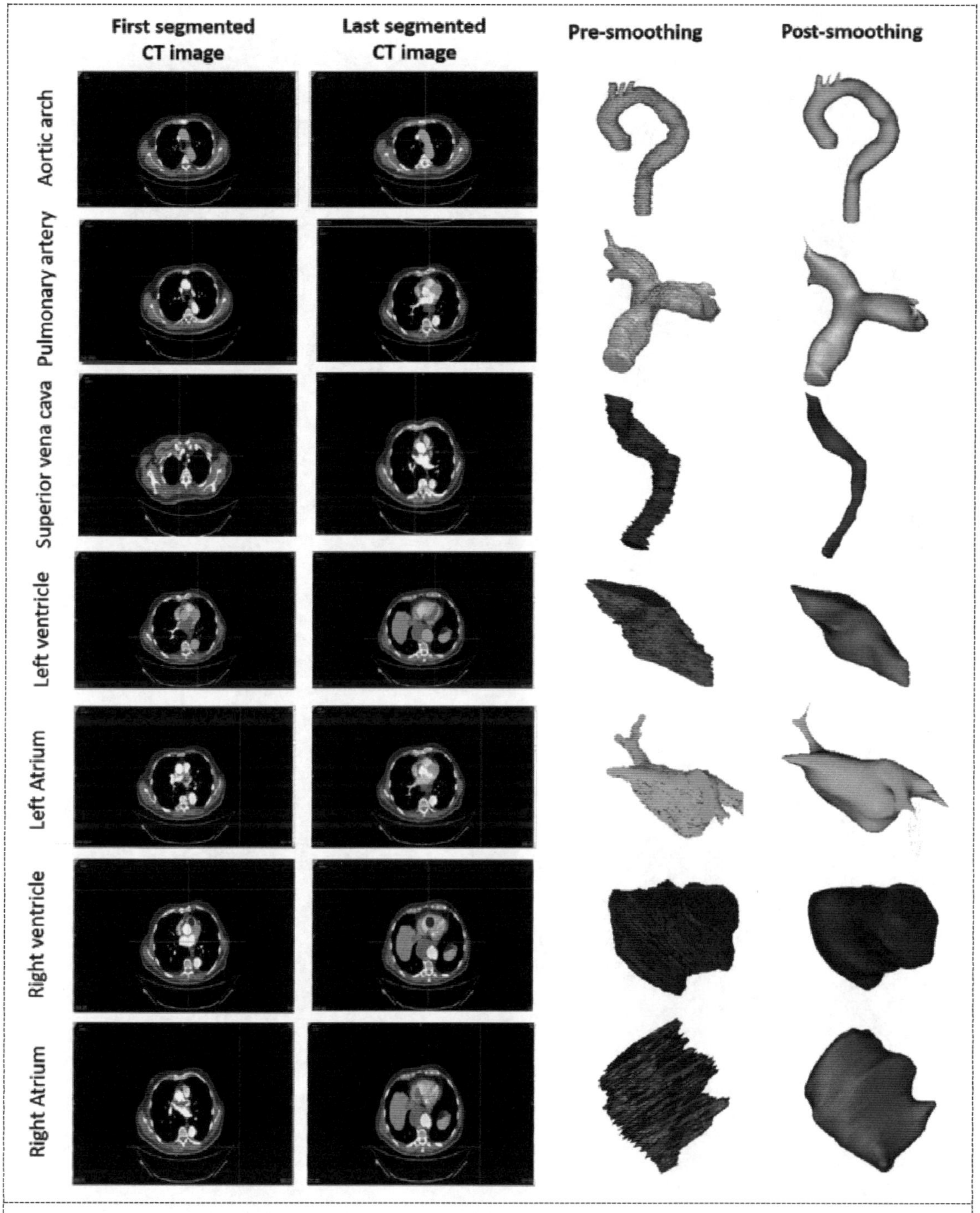

Figure 4. The reconstructed anatomical parts for the normal case, presented separately before and after smoothing operations. On the left side, the first and last segmented image for each part in the CT series.

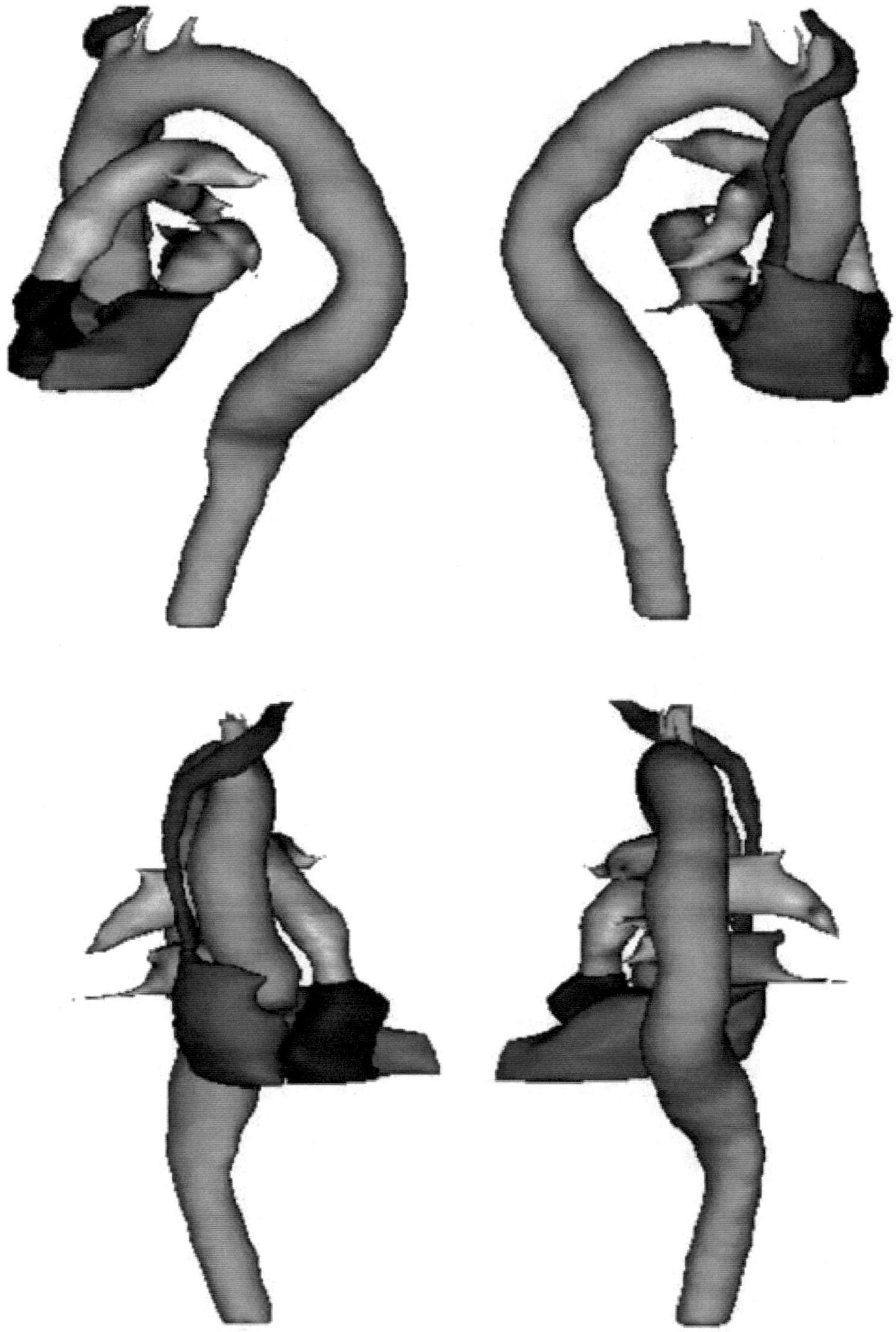

Figure 5. Superposition of the geometrical surfaces corresponding to the chambers of the heart and the connected vessels of the normal case.

Acknowledgements
The authors wish to thank the Special Account for Research of ASPETE for supporting the presentation of this work through the funding program "*Strengthening Research of ASPETE faculty members*".

References
[1] Bickerstaff LK, Pairolero PC, Hollier LH, Melton LJ, Van Peenen HJ, Cherry KJ, Joyce JW, Lie J 1982 *Surgery* **92** 1103-1108
[2] Iaizzo PA 2009 Handbook of cardiac anatomy, physiology, and devices *Springer Science & Business Media*
[3] Lorenz C, von Berg J 2006 *Med. Image Anal.* **10** 657-670
[4] Zheng Y, Barbu A, Georgescu B, Scheuering M, Comaniciu D 2008 *IEEE Trans. Med. Imaging* **27** 1668-1681

A computational wall mechanics study of an ascending thoracic aortic aneurysm under hypertensive conditions

C Manopoulos[1*], A Raptis[1], W Krishan[2], C Mavratzas[2], M Drandakis[2], S Astraka[2], I Kouerinis[3], and N M Vaxevanidis[2]

[1]Biofluid Mechanics and Biomedical Engineering Laboratory, Fluids Section, School of Mechanical Engineering, National Technical University of Athens, Zografos, GR 15780, Greece.
[2]Laboratory of Manufacturing Processes and Machine Tools (LMProMaT), School of Pedagogical and Technological Education (ASPETE), Department of Mechanical Engineering Educators, Amarousion, GR 15122, Greece.
[3]Department of Cardiothoracic Surgery, "Hippocration" Hospital, Athens, GR 11527, Greece.

***Corresponding author**: manopoul@central.ntua.gr

Abstract. A wall mechanics study is performed on two human aortic models, reconstructed from computed tomography (CT) image data using the Materialise Mimics software. The first model represents a rare ascending thoracic aortic aneurysm (aTAA) case with an excessive aortic ballooning that has displaced the cardiac cavities, and the second one a normal case free of cardiovascular diseases. Special attention was paid for the reconstruction of realistic models that do not deviate from the original data. The study presents the workflow from medical imaging data to structural simulation with the use of various software, aiming to examine the stress state of a normal aorta and an aneurysmal one (both patient specific) under a range of systolic blood pressure loads. Using the FEBio software, the effective (Lagrange) strain and the effective stress (von Mises) distributions are calculated for assessing the risk of rupture or dissection of the aorta.

Keywords: aorta wall mechanics; ascending aortic aneurysm; hypertension; aorta rupture risk

1. Introduction

Ascending thoracic aortic aneurysms (aTAAs) are defined as the pathological bulging of the proximal aortic section where blood ejects through the arterial valve from the left ventricle of the heart. The dilatation of the aortic wall is a consequence of an irreversible gradual weakening caused by a degenerative process. [1] aTAA is a silent, though fatal, disease that can lead to the rupture or dissection of the arterial wall. [2] Most patients do not experience symptoms. It is even more striking that half of the symptomatic cases, emergently presenting at a hospital, had an undiagnosed aTAA according to post-mortem studies.[3]

The estimation of the aTAA rupture or dissection risk traditionally depends on computed tomography (CT) or ultrasound (US)-based measurements of maximum diameter and growth rate. As the aortic size does not account for the regional vessel wall heterogeneity, which is evident especially in aneurysmal segments, the diameter-based complication risk prediction is possibly underinformed thus less trustworthy. There are strong advocates of wall stress and strength as more reliable risk predictors than aTAA diameter and growth rate. [4] As it is impossible to measure wall stresses in vivo, researchers have turned to the computational approach for a solution. The Finite Element Analysis (FEA) is a validated and widely used technique but, still, the calculation of arterial wall stresses requires prior knowledge of the aortic tissue mechanical behaviour. The review by Martufi et al. [1] delineates how a multidisciplinary approach involving biomechanics and medical imaging can provide accurate predictions of aTAA behaviour, based on structural and fluid mechanics studies.

The most common application of biomechanical modelling is the evaluation of wall stresses in abdominal aortic aneurysms (AAAs), while literature is limited in aTAAs. Recently, Wang et al. [5] determined the wall stress distribution in aTAAs with diameter <5.0 cm and >5.0 cm, which is the threshold for elective surgical repair according to the medical guidelines. The patient-specific aneurysmal geometries were obtained from echocardiogram-gated CT scans and the structural response was determined by the LS-DYNA FEA software (LSTC Inc, Livermore, California, USA). The peak wall stress was found greater in aTAAs > 5.0 cm but the correlation of size and stress in aTAAs < 5.0 cm was poor.

A considerable percentage of aTAAs patients (15%) also suffer from Bicuspid Arterial Valve (BAV), which is a common congenital aortic valve defect [6].The aortic size as an index ignores by definition other comorbidities, such as BAV, that might play a role in the risk of aTAA rupture or dissection. In a biomechanics study of 17 BAV patients and 19 subjects with a physiological Tricuspid Aortic Valve (TAV), the circumferential and longitudinal stresses were found greater in BAV- than TAV- aTAAs, suggesting that patient-specific aneurysm wall stress analysis is essential for accurate complication risk prediction. [7] Besides BAV, in plenty of cases, a large brachiocephalic trunk gives rise to both subclavian arteries and a bicarotid trunk, termed as the Bovine Aortic Arch (BAA) congenital variation. Comparing aTAA patients with BAV, BAA and none of the two, Martin et al. [8] concluded that the aortic size index was insufficient for delineating between patients at moderate and high risk, while no correlation existed between BAV and BAA aTAA rupture risk. The study also involved tissue mechanical testing data thus combining experiments with computational simulations. Avanzini et al. [9] provide a comprehensive analysis of the experimental methodologies for the determination of aTAAs biomechanical properties.

The current pilot study is an application of computational biomechanics in a rare aTAA case with an excessively developed aneurysmal bulge that has displaced the cardiac cavities. The modelling procedure is thoroughly described and the response of the aTAA is studied under various pressure conditions in comparison to a normal case that is free of cardiovascular diseases. A presurgical biomechanical analysis, especially in difficult cases, can enhance the clinical decision-making process.

2. Materials and methods

2.1. Material

The biomechanical analysis relies on the computed tomography (CT) scans of a 67 y.o. male patient and a 69 y.o. normal male subject who gave informed written consent for the use of their medical imaging data. The patient suffered from aTAA and underwent a successful aortic repair at "Hippocration" General Hospital of Athens in 2013. The CT data was made available for further processing by the Attending Physician (cardiac surgeon) at the Cardiac Surgery Department of "Hippocration" General Hospital of Athens.

2.2. CT imaging protocol

The CT scan was performed on a 16-slice multidetector CT (MDCT) scanner (Activion 16, Model TSX-031A, Toshiba). A protocol of 1 mm slice thickness was used with 120 kV, 60 mA tube current, 330 mm field of view for reconstruction and a sharp reconstruction kernel. The CT was performed in the craniocaudal direction, from the aortic arch until the diaphragm level in a single breath-hold. Fig. 1

shows the cross-section in the CT data featuring the maximum aTAA diameter and Fig. 2 the 3D volume rendering of the whole CT series, clearly disclosing the big aTAA bulge.

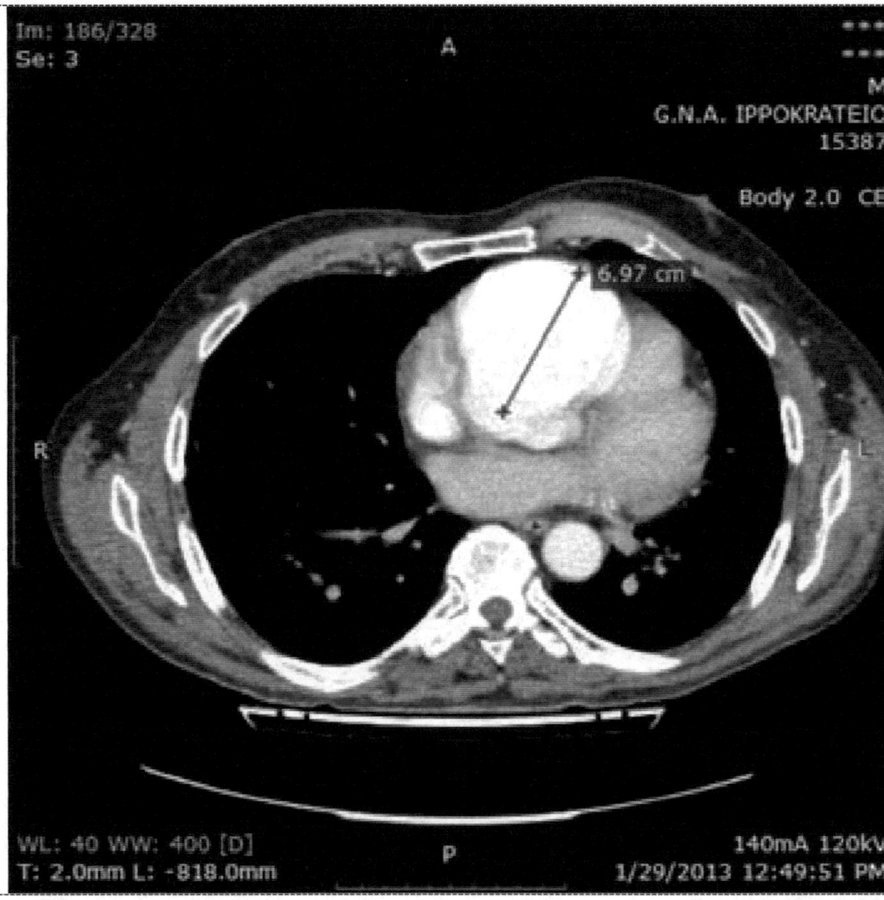

Figure 1. CT image annotated with the maximum aTAA diameter value.

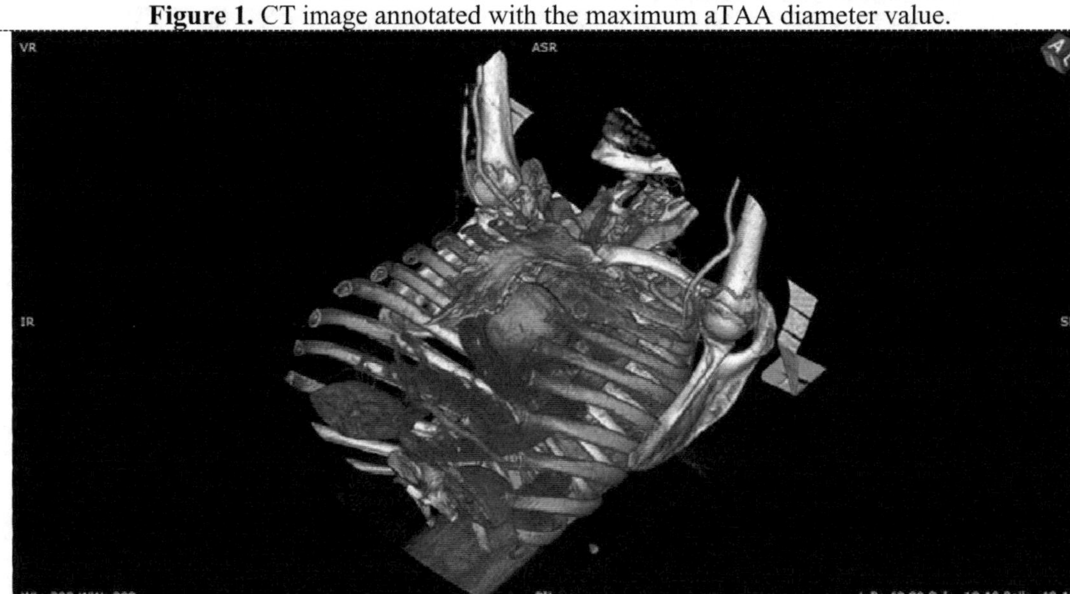

Figure 2. 3D volume rendering of CT data featuring the big aneurysmal bulge.

2.3. Segmentation & Reconstruction

The Mimics software (Materialise inc.) is an advanced software with sophisticated tools for the segmentation and reconstruction of medical imaging data. The aorta was initially isolated from the surrounding bones and organs using the "Dynamic region growing" tool that is based on the gray value connectivity. The aorta was further restricted to the desired limits (the aortic arch upwards and the descending aorta downwards) by batch-removing the out-of-bounds segmented cross-sections. The segmentation of the aorta was further optimized, with a special focus on the aortic wall calcifications (not included in the final reconstruction), via a slice-by-slice manual process for the removal of any leftover segmented artifacts. After completing the segmentation process, the 2D images were transformed into a 3D model. The initial reconstructed models feature rough artificial edges on their surfaces due to the resolution of the medical images and the slice thickness. Further refinement is required prior to the use of the models for computational simulations.

Smoothing is an iterative process for the reduction of the surface roughness, but special care should be given to minimize the deviation of the post-smoothing model from the original data. During the smoothing process, it was observed that the cross-sectional area (and the volume of the aorta in extend) depended on the number of smoothing iterations and the smooth factor. More iterations were leading to a more natural result, while the smooth factor had a heavy impact on the shape of the aorta. Conclusively, it was decided to perform several iterations (500) and keep the smooth factor to a minimum value (0.1). It is also worth noting that during the segmentation process, the segmentation mask was slightly expanded to compensate for the volume shrinkage due to the smoothing, ending up with a volume-preserving models (with respect to the original data).

To identify the aortic valve, the left ventricle of the heart was also segmented and reconstructed. Specifically, a hole, representing the aortic valve without its cusps, was opened on the aortic surface by removing the intersected area between the left ventricle and the aortic surfaces using the Visualization Toolkit (VTK) library. The post-reconstruction procedure leading eventually to the production of the simulation-ready aTAA model is illustrated in Figure 3.

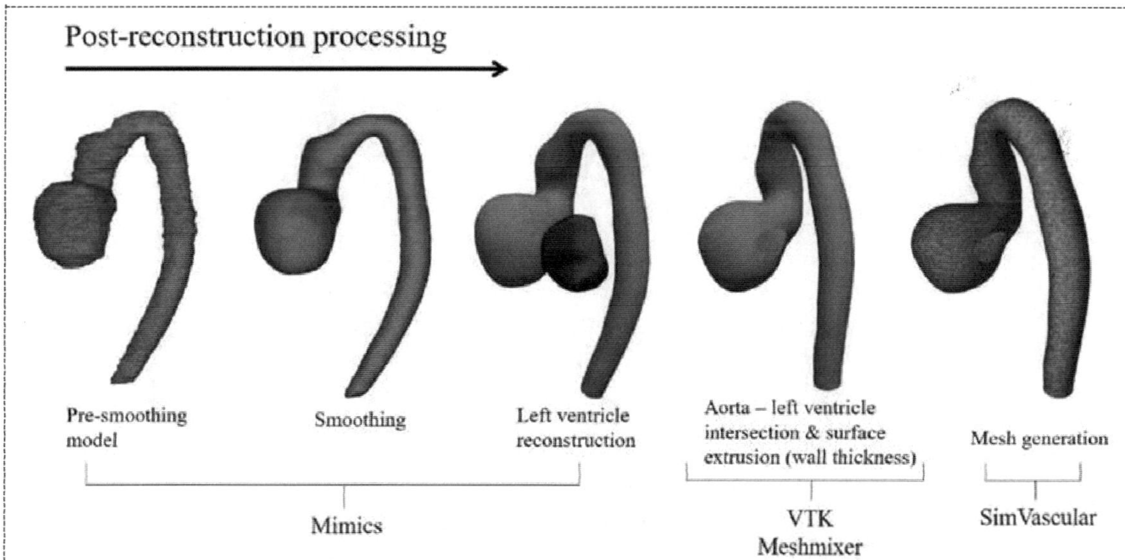

Figure 3. The procedure after the initial 3D reconstruction involving the smoothing, the left ventricle reconstruction, the opening of the aortic valve, the extrusion of the surface and the generation of the computational mesh.

2.4. Mesh generation

The aTAA model after the segmentation and reconstruction processes, is represented by an open surface without thickness, in STL format. Towards performing the structural simulations, a solid model with a given wall thickness and the corresponding computational mesh need to be generated. In this direction, a multi-step workflow involving open-source software and Python libraries, was put in place. The aTAA model (STL file) was initially imported into the free Autodesk Meshmixer (Autodesk inc.) software where the surface was extruded outwards in the normal direction with an offset of 1.526 mm corresponding to the wall thickness of a typical aTAA, as mentioned in [10]. The extruded surface, also in STL format, was subsequently imported into the SimVascular software for the generation of the computational mesh. SimVascular is a fully open-source software package that provides a complete pipeline from medical image data segmentation to patient specific blood flow simulation and analysis [11]. Upon importing the model, SimVascular identifies the parts of the model exterior where boundary conditions are applied, i.e. the inner and the outer surfaces of the wall and the rings taking shape at the inlet (aortic valve) and outlet (descending aorta) sections upon extrusion (Fig. 4). The aTAA model was finally discretized into a computational mesh of 180K elements that was converted from VTU (unstructured grid format used by VTK) to INP format which is operational in the FEBio software that was the FEA package of choice in the current study (more information in the next section).

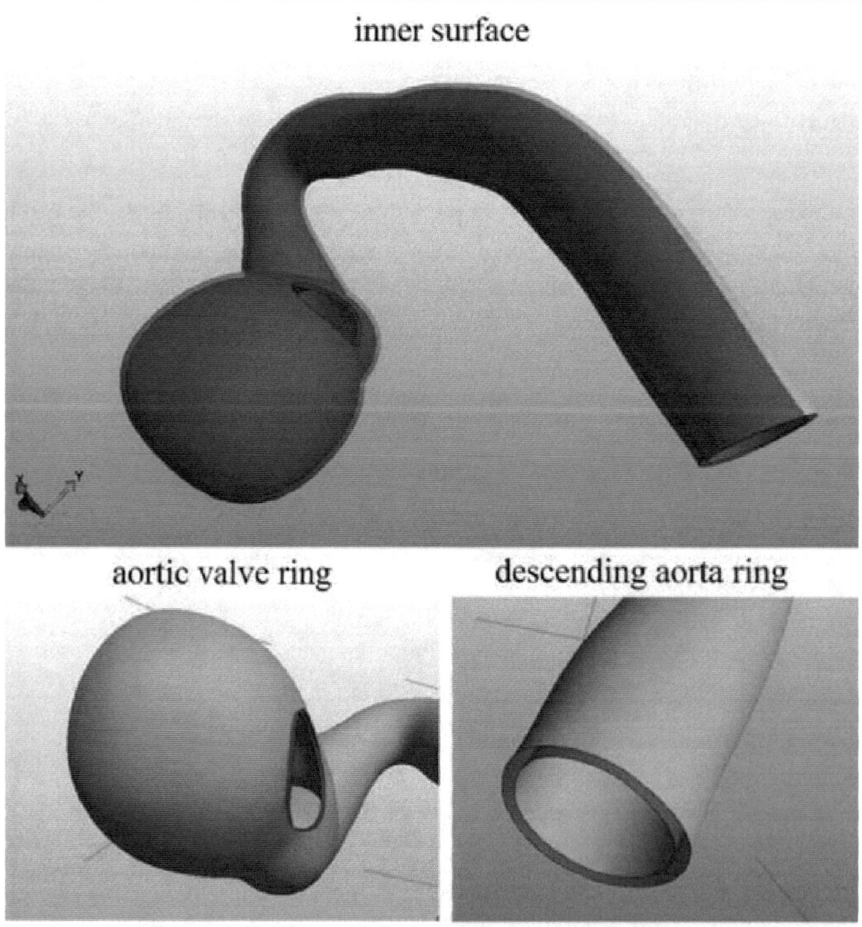

Figure 4. The parts of the aTAA exterior surface where boundary conditions are applied. Specifically, pressure is prescribed at the inner surface of the wall, while the aortic valve and the descending aorta rings are fixed in all directions.

2.5. Material model and structural simulations

The material models for the aTAA and the normal case were based on the study by Pasta et al. [12] who pursued the mechanical response of tissue specimens combining image-derived parameters and tensile testing data. Specifically, they examined the collagen fiber dispersion and alignment in experimentally dissected halves of the aTAA and normal tissue specimens, mechanically characterizing both the intimal and the adventitial arterial wall layers. The fiber-reinforced structural model introduced by Pasta et al. [12] is based on the widely known Gasser-Ogden-Holzapfel model [13], prescribed with the strain energy function,

$$W = \frac{C}{2}(I_1 - 3) + \frac{k_1}{k_2}\left\{exp\left[k_2\left(kI_1 + (1-3k)I_{4i} - 1\right)^2\right] - 1\right\} \quad (1)$$

where C, k_1 and k_2 are the material parameters, k and γ are the structural parameters, $I_1=\lambda^2_{CIRC} + \lambda^2_{LONG} + (\lambda_{CIRC}\,\lambda_{LONG})^{-2}$ is the first invariant of the right Cauchy–Green strain tensor, and $I_{41} = I_{42} = \lambda^2_{CIRC}\sin^2\gamma + \lambda^2_{LONG}\cos^2\gamma$ is a tensor invariant assuming that the two fiber families (i=1, 2) for each layer (intimal and adventitial) were mechanically equivalent. In the current study, the arterial wall was modelled by a single-layered fiber-reinforced material using the mean value of the parameters provided by Pasta et al. for the intimal and adventitial layers, as reported in Table 1 for the aTAA and the normal case. Moreover, the density was considered to be $\rho=1.07\times10^{-6}$ Kg/mm^3, [14] and the Bulk modulus was set to 12.4 MPa [15] both for the aneurysmal and the non-aneurysmal case.

Table 1. The parameters of the single-layered fiber-reinforced material model used in the structural simulations for the aTAA and normal case.

	C (kPa)	k_1 (kPa)	k_2	γ (deg)	k
aTAA	4.8	135.8	14.8	32.8	0.25
CONTROL	6.85	324.8	19.3	13.9	0.22

The FEBio Studio software was used for the performance of the structural simulations. FEBio is a nonlinear finite element solver specifically designed for biomechanical applications. It offers modelling scenarios, constitutive models and boundary conditions that are relevant to numerous research areas in biomechanics. Development, distribution and support is a joint effort between Jeff Weiss's lab at the University of Utah and Gerard Ateshian's lab at Columbia University. [16] The boundary conditions for the solution of the structural problem were defined as follows: a range of systolic pressure values (100-300mmHg with a step of 20mmHg) applied at the inner surface of the wall, and fixation in all directions of the aortic valve and descending aorta rings, as depicted in Fig. 4.

The post-processing analysis involves the calculation of the effective (Lagrange) strain and the effective stress (Von Mises). The effective strain is defined is terms of the deviatoric part of the strain tensor. If E is the (Lagrange) strain tensor, then the deviatoric strain is defined by

$$\mathbf{E}_{dev} = \mathbf{E} - \frac{1}{3}\mathrm{tr}(\mathbf{E})\mathbf{I} \quad (2)$$

and the effective strain by,

$$\mathbf{E}_{ef} = \sqrt{\frac{2}{3}\mathbf{E}_{dev}:\mathbf{E}_{dev}} \quad (3)$$

while the effective stress accordingly by,

$$\sigma_{ef} = \sqrt{\frac{3}{2}\sigma_{dev}:\sigma_{dev}} \quad (4)$$

3. Results and discussion

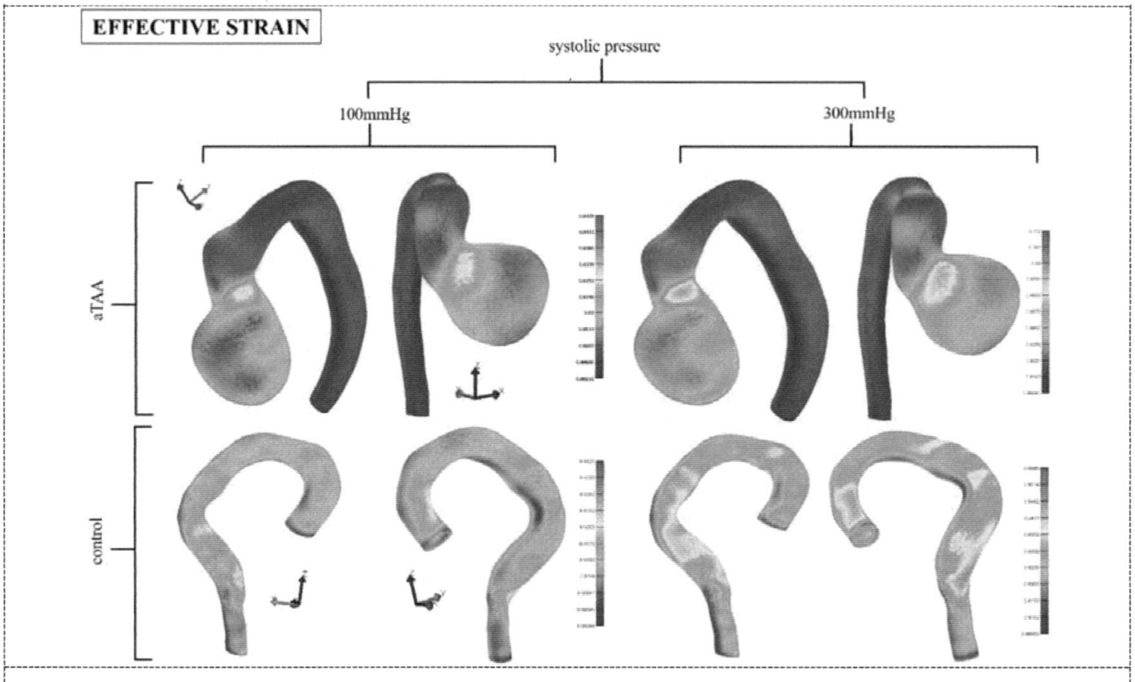

Figure 5. Contours of the effective strain distribution for the aTAA and the control case under a systolic blood pressure of 100 and 300mmHg.

The results are outlined by the 3D contour rendering of the effective strain and stress distributions (Figs. 5 and 6) and the plotting of their maximum values with respect to systolic pressure (Fig. 7) both for the aTAA and the control case. The purpose of the post-processing is to highlight the areas on the surface of the aTAA bulge where high stress and strain occurs, in comparison with a normal aorta, and to identify any differences in the peak strain and stress - pressure curves between the two cases. The effective stress is a measure of rupture risk, while the effective strain describes the increase in the size of the aTAA bulge under the effect of an elevated arterial pressure condition.

The contours in Figs. 6 and 7 highlights that the risk areas of the aTAA are concentrated around the shoulder of the aneurysm and are clearly extended when applying a high systolic pressure boundary condition (300 mmHg compared to 100 mmHg). It is therefore evident that a rupture event could take place at the area where the dilation fades and the normal aortic diameter is restored, and not at the maximum dilated part of the aTAA bulge as one could expect from the clinical perspective. Similarly, the area of maximum effective strain indicates the place where the dissection of the aorta could initiate which is indeed the expected one, as found in most aortic dissection cases (the initial part of the aortic arch). On the other hand, both under 100mmHg and 300mmHg blood pressure, the control case is on a more alleviated stress state compared to the aneurysmal one, with the normal tendency of attracting stresses at areas where surface curvature changes. The effective strain of the normal case under both pressure conditions is comparable to the effective strain of the aTAA case under 100mmHg, suggesting that an aTAA patient under controlled and low blood pressure might have lower risk of dissection and rupture.

The plots of effective strain and stress with systolic pressure clearly indicate that the peak values of both variables are increasing with a different slope between the two cases under increasing blood pressure. Specifically, both are higher in the aTAA case while the plateu in their values is reached earlier in the normal case (around at 200 mmHg, while at 240 mmHg for the aTAA case). It is therefore suggested that an aTAA patient should avoid circumstances of abrupt blood pressure escalation as the stress state experienced by the aorta gets more intense while a normal aorta seems more adaptable to the increase of pressure.

A hypertensive state is common between athletes that lift weights where maximum systolic pressure can reach as high as 360/370 mmHg. [17] Such measurements were recorded in a study of 10 athletes where blood pressure was continuously being monitored through a catheter in the radial artery. Every athlete in the study was asked to perform a series of activities and physical exercises such as double-leg press at 85% and 100% of max capacity. Of course, such an elevated blood pressure requires intense physical activity but is achievable.

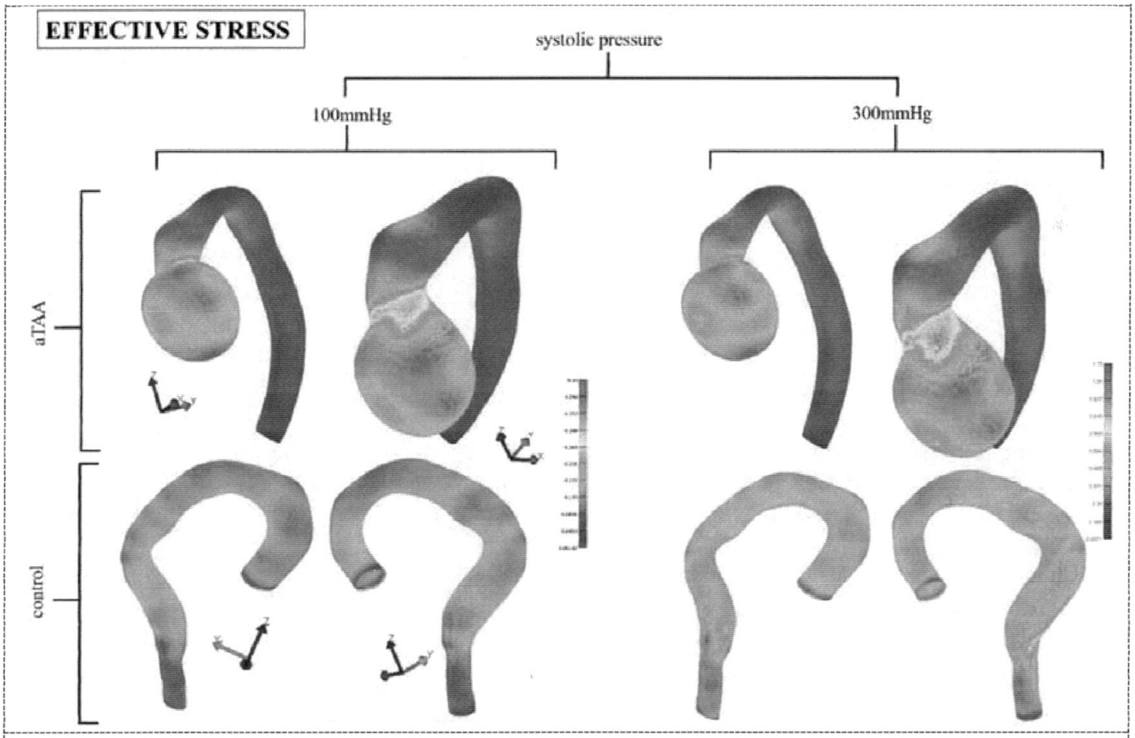

Figure 6. Contours of the effective stress distribution for the aTAA and the control case under a systolic blood pressure of 100 and 300 mmHg.

Besides athletes, there are rare cases of patients where blood pressure had elevated to 300mmHg without performing any physical activities. Moreover, there are references in medical forums about cases with systolic pressure higher than 250 mmHg. Most of the subjects under high blood pressure, experience intense medical conditions, such as excessive headaches, dizziness, even loss of consciousness. In a case of a patient at an end stage renal disease, a maximum of 290 mmHg blood pressure has been recorded during the hospitalization at the intensive care unit. [18] For the above reasons, a range of pressure from 100 mmHg to 300 mmHg has been considered in the current study, covering all possible stress states experienced by a human aorta.

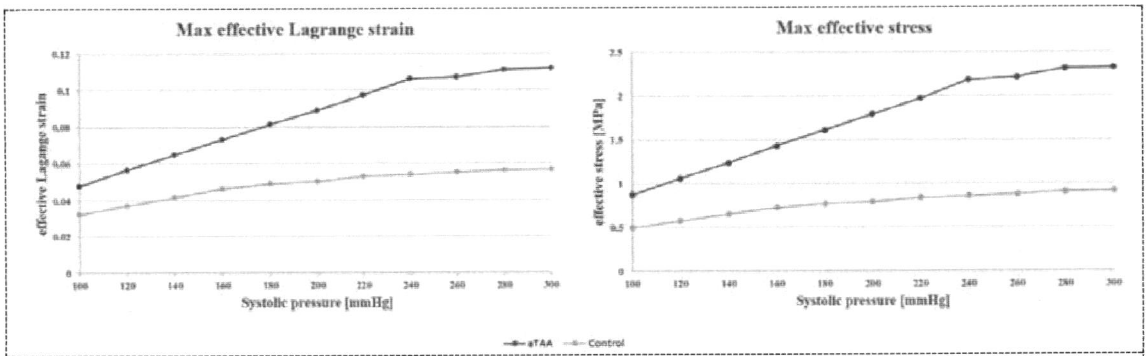

Figure 7. Plots of max effective strain and stress with respect to systolic pressure for the aTAA and control case.

Acknowledgements
The authors wish to thank the Special Account for Research of ASPETE for supporting the presentation of this work through the funding program "*Strengthening Research of ASPETE faculty members*".

References

[1] Martufi G, Forneris A, Appoo JJ, Di Martino ES 2016 *Ann. Thorac. Surg.* **101** 390-398
[2] Elefteriades JA, Sang A, Kuzmik G, Hornick M 2015 *Open Heart* **2**, e000169
[3] Elefteriades J 2007 Acute aortic disease *CRC Press*
[4] Smoljkić M, Fehervary H, Van den Bergh P, Jorge-Peñas A, Kluyskens L, Dymarkowski S, Verbrugghe P, Meuris B, Vander Sloten J, Famaey N 2017 *Biomech. Model. Mechan.* **16** 705-720
[5] Wang Z, Flores N, Lum M, Wisneski AD, Xuan Y, Inman J, Hope MD, Saloner DA, Guccione JM, Ge L, Tseng EE 2020 *J. Thorac. Cardiovasc. Surg.* doi: 10.1016/j.jtcvs.2020.02.046
[6] Davies RR, Kaple RK, Mandapati D, Gallo A, Botta Jr DM, Elefteriades JA, Coady MA 2007 *Ann. Thorac. Surg.* **83** 1338-1344
[7] Xuan Y, Wang Z, Liu R, Haraldsson H, Hope MD, Saloner DA, Guccione JM, Ge L, Tseng E 2018 *J. Thorac. Cardiovasc. Surg.* **156** 492-500
[8] Martin C, Sun W, Elefteriades J 2015 *Am. J. Physiol. Heart Circ.* **308** H1306-H1316
[9] Avanzini A, Battini D, Bagozzi L, Bisleri G 2014 *BioMed Res. Int.* **2014** 20385 (9 pages) doi: 10.1155/2014/820385
[10] Van Puyvelde J, Verbeken E, Verbrugghe P, Herijgers P, Meuris B 2015 *European Journal of Cardio-Thoracic Surgery* **49** 756-762
[11] Updegrove A, Wilson NM, Merkow J, Lan H, Marsden AL, Shadden SC 2017 *Annals of biomedical engineering* **45** 525-541
[12] Pasta S, Phillippi JA, Tsamis A, D'Amore A, Raffa GM, Pilato M, Scardulla C, Watkins SC, Wagner WR, Gleason TG 2016 *Med. Eng. Phys.* **38** 121-130
[13] Gasser TC, Ogden RW, Holzapfel GA 2006 *J. R. Soc. Interface* **3** 15-35
[14] Schiavone A, Zhao L 2015 *Mech. Adv. Mater. Mod. Process.* 1 (15 pages) doi: 10.1186/s40759-014-0002-x
[15] Selmi M, Belmabrouk H, Bajahzar A 2019 *Appl. Sci.* **9** (11 pages) doi: 10.3390/app9061216
[16] Maas SA, Ellis BJ, Ateshian GA, Weiss JA 2012 *Journal of Biomechanical Engineering* **134** 011005 doi: 10.1115/1.4005694
[17] Narloch JA, Brandstater ME 1995 *Arch. Phys. Med. Rehabil.* **76** 457-46
[18] Sikka G, Khan Z, Bachan M 2016 *Crit. Care Med.* **44** 489 doi: 10.1097/01.ccm.0000510330.58620.69

Change from one Method to Another for Processing by Cutting a Complex Rotary Surfaces

D Vasileva[1], **E Lefterov**[1]

[1]Technical University of Varna, 1 Studentska Str., Varna, 9000, Bulgaria

[*]Corresponding author's e-mail: vasilevadimka@gmail.com

Abstract. Many authors involved in the processing of materials by cutting make different definitions for complex rotary surface. They are reduced to the definition published in [1]: A complex rotating surface is called one that is composed as a combination of two or more of the following elementary surfaces - conical, frontal, cylindrical, curvilinear. The methods and devices used in machining parts that are constrained by complex rotary surfaces depend on following several factors: type of production (single, serial, mass); type of available equipment and shape accuracy and surface quality.

1. Analysis of the methods for machining complex rotary surfaces.

The considered methods are turning method by using standard turning tools and profile turning tools, milling and broaching method.

1.1. Advantages of turning method: The usedstandard turning tools are with simplified construction and cutting geometry. They do notrequire profiling or special equipment; Using of profile turning tools makes the processing possible for the realization of general purposemachines, semi-automatic and automatic lathe machines in serial and mass production.

1.2. *Disadvantages of turning method:* The machined surface of the workpiece consists of a number of screw grooves obtained from the relative motion of a portion of the points of the cutting edge. Therefore, the quality of these surfaces cannot be assessed as sufficient; Machining which requires the use of a profile cutting tool is not applicable for serial and mass production because the shape of its cutting edge is not self-recovering after sharpening; The three types of radial profile cutting tools are unsuitable when the high requirements for the accuracy shape and size on the machined surfaces are made [1]. The reasons are: The inclusion of all points from the cutting edge at the end of the cutting process adds considerable instability to the whole technological system. For this reason, they are only applicable to machining short work pieces with a large cutting depth; The inclination of the front surface at an angle greater than 0° does not allow all points of the cutting edge to be positioned in the diameter plane of the work piece. For this reason, most of the machined rotary surfaces are deviated from the desired shape and size, with a size that depends on the size of the front angle of the tool.

1.3. Advantages of milling method

Milling tools are with high durability because the cutting of stock allowance is distributed on more cutting teeth; Using of profile milling cutters, allows the original shape of the cutting edges to be restored after each sharpening.

1.4. Disadvantages of milling method:

The combination of the relative trajectory of points from the cutting edge with character of extended hypocycloid and radial beating of the tool mandrel converts the desired rotary surface of the work piece into a polyhedron with parts of hypocycloid. It becomes a real surface with low shape accuracy and bad surface quality;It is difficult to form rational values of the geometrical tool parameters, which are consistent with the nature of the relative trajectory. In some cases, it makes impossible using method into practice.

1.5. Advantages of broaching method:

It is realized with only one movement of the kinematic scheme, which is a prerequisite for using of machine tools with simplified kinematics and construction; The feed movement is achieved with the special design of the tool, as a hidden (constructive) feed movement [2].

1.6. Disadvantages of broaching method:

Moving the tool to the starting position after each work movement makes this processing method low-performing;It is necessary to perform long machine motion, which requires using machines with large dimensions and production areas; The workflow automation capabilities are limited.

Compared to the turning and milling methods of complex rotary surfaces, broaching is the most rational method that fully meets the formulated and accepted criteria. Finding the best processing option (if it is possible) after improvement of the broaching method should be done while preserving its advantages and by overcoming to some extent the identified disadvantages.

2. Change from one method to another for processing by cutting a complex rotary surfaces.

The scraping and broaching of flat surfaces are carried out by methods based on the fundamental kinematic cutting schemes (FKCS) as it is already clear [3, 4].

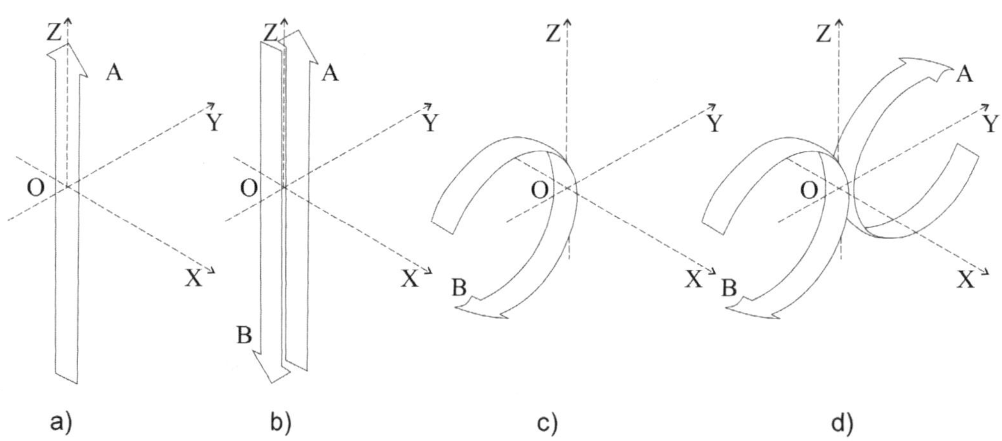

Figure 1. Change from one to another fundamental kinematic cutting schemes (FKCS).

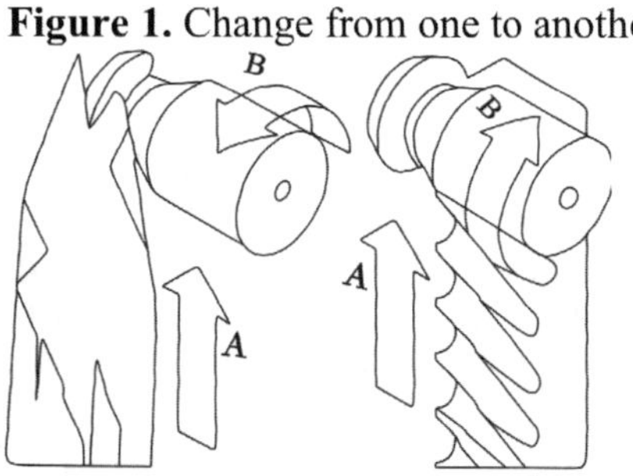

Figure 2. Rotary part processing with different tools, according to FKCS.

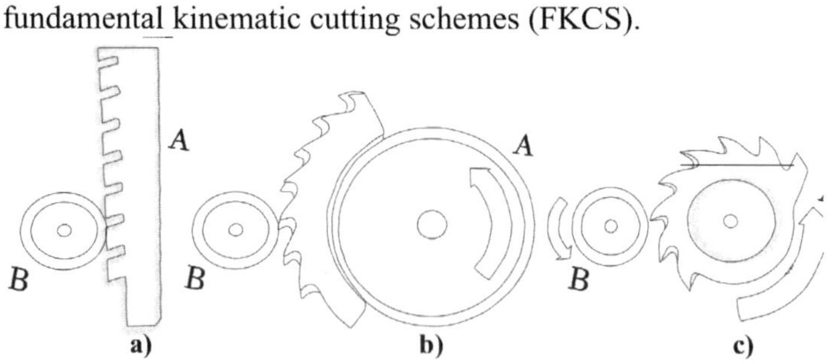

Figure 3. Change from one method to another for processing a complex rotary surfaces.

The only rectilinear motion A in the structure of this kinematic scheme is also the main cutting motion and belongs to the tool when the stock is stationary. If movement A is joined by other movement B in the same way but in the opposite direction, a peculiar transition from (Figure 1a) to FKCS (Figure 1b) is made. If the stock moves in accordance with B, this change will cause relative displacement of the stock against to the tool at increased speed and will not lead to any other changes in operating conditions. The two uniform, straight and gradual motions A and B can be regarded as uniformly rotating, with infinitely distant centers of rotation from their straight lines. If the center of rotation of movement A remains infinitely distant and the center of linear motion B approaches its straight line at some distance of finite value, a transition from FKCS (Fig. 1a and Fig. 1b) to FKCS will occur (Figure 1c). With the same belongings of the two movements (A of the tool - feeder, or ancillary and B of the stock - mainly for cutting) and in the ratio of their speed:

$$\varepsilon = \frac{V_A}{V_B} = 0{,}005 \quad (1)$$

Two methods of machining complex rotary surface are implemented - through tangential cutting tool (Figure 2a) [5] and through the external profile broach (Figure 2b) [6]. The relative trajectory is a flat spiral extending into a circle with a radius equal to the radius of the rotating surface for a given point on the cutting edge. In fact, this trajectory is a special case of an extended cycloid. The uniform nature of the relative trajectories in the final part of the circle-shaped workflow is particularly appropriate because it fully corresponds to the type of stock rotated at each point.The two methods considered have specific features that the rest, built for a similar purpose, do not possess. With regard to the machining of the rotary part, only one point from the cutting edge is theoretically included at work at any point in the work process, and in practice a limited part of it, in the vicinity of the same point. This means that in addition to the two movements A and B (Figure 2), another movement C is involved, which is not included in the structure of the kinematic scheme, but actually moves the contact point from the cutting edge and the stock in two directions: along the cutting edge from on one side and against the stock axis in the diameter and the plane on the other.There are two additional motions not figured on the FKCS (Figure 2) but actually realized when machining rotary surface with an external profile extension: in the same sense as in the tangential profile tool and D – due to the special construction of this type of tool – the location of each cutting edge is higher than the previous one with a value corresponding to the transverse feed motion of one tooth. Whereas, when considering the machining of planar surfaces, the flat broach is regarded as a set of scraper blades arranged with the so-called "Tooth lifting". Whereas, when machining rotary surfaces, the outer profile extensions can be considered as a set of tangential blades, also with the teeth lifted relative to one another. If it is assumed that the radius of the linear motion A of the FKCS of Figure 1 gradually loses its infinite value and at some point is established with a finite radius, there will be a new transition from one kinematic cutting scheme to another, this time from Figure 1c, to the FKS (Figure 1d) - with two uniform rotary motions A and B with the same direction of rotation. Changing the type of kinematic scheme causes a change in the design of the tool. In the transition from FKCS (Figure 1c to Figure 1d), the outer profile broach (from Figure 3a) with a straight line movement A is transformed into an arc (sector) from Figure 3b, with a uniform rotary motion A. If the radius of the arc extends to a radius of even smaller value, there is transformation from an arc to a helical broach (Figure 3c).In order to answer the question unequivocally, whether or not the construction of a method of machining complex rotary surfaces with helical broaches is possible or not. Determining the character of the relative path of a point on the cutting edge and to determine the extent and according to the type of the processed surface of the stock at various values of the ratio:

$$\varepsilon = \frac{V_A}{V_B} \quad (2)$$

To determine whether it possible to form the elements of the cutting mode and the geometrical parameters of the tool within rational limits, tailored to the nature of the relative trajectory and characteristics of workflow. If the affiliation of both motions is in accordance with “A” of the tool (figure 3c) and “B” of the stock with V_u and V_g are indicated respectively the peripheral speeds of the tool and the stock, the first value of the ratioε_1 will be of the value:

$$\varepsilon_1 = \frac{V_g}{V_u} < 1 \quad (3)$$

At sufficiently small values ofε_1 any point on the cutting edge of any tooth of the helical broach will make contact with the stock on the arc with the beginning of the first and last points of contact. This is possible by rotating the broach with angleω_u in a counter-clockwise direction and by rotating the part with angleω_g in the same direction. The maximum thickness of the chip a_z can fit in the range a_z=0,003-0,2 mm.

Under these restrictive conditions, the workflow is possible, but it would hardly make practical sense, since the stock has an irregular curved multi-wall outline in a cross-section that differs significantly from the original design. The working process of machining complex rotary surfaces with a helical broach fully imitates the milling process, since in both cases the character of the relative trajectory is an extended hypocycloid [7]. The trajectory and the outline of the stock can be illustrated (Figure 4) if the center of the tool with radius*r*contacts the center with the center of a circle with radius A_0 rolling without slipping to another with radius B_0 and with a center coinciding with the center of the stock of radius *r*. The indicated nature of the relative trajectory will be obtained if the conditions are met:

$$\frac{A_0}{B_0} \ll 1 \; and \; R \gg A_0 \tag{4}$$

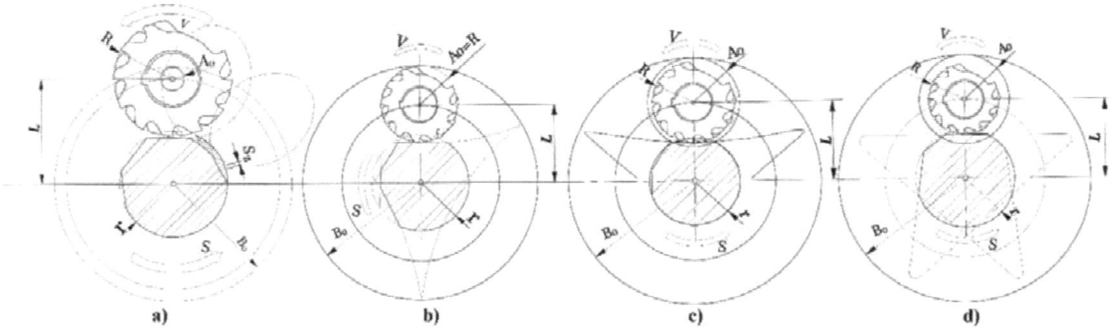

Figure 4. Relative trajectories at broaching of rotary parts with helix broach.

If the ratio (1) is increased to ε_2 but so that:

$$\varepsilon_2 = \frac{V_g}{V_u} < 1, but \; \varepsilon_2 > \varepsilon_1 \tag{5}$$

The maximum thickness of the chip is obtained with the values according to the equation:

$$a_z > 0{,}2 \; mm \tag{6}$$

It turns out that at this value of $a_z$the thickness of the layer of material from the stock taken from one tooth of the helical broach is unacceptably large. This is a circumstance that calls into question the practical implementation of the work process in general. Mathematically, an accurate hypocycloid is obtained under the following conditions:

$$\frac{A_0}{B_0} < 1 \; and \; R = A_0 \tag{7}$$

The processing under the specified restrictive conditions is also devoid of practical meaning, because the curvilinear concave multi-wall outlines of the stock in cross-section (Figure 4b) are completely inconsistent with the contours of a rotary surface in the shape of a circle in the same section. Emerging mathematical - accurate hypocycloid reduce progressively their curvature with increasing ratio A_0 / B_0. In relation to:

$$\frac{A_0}{B_0} < 0{,}5 \; and \; R = A_0 \tag{8}$$

the relative trajectory is a straight line. This particular case is a theoretical but not practical interest because processing is impossible. The stock has zero dimensions, so there is no stock at all.Further increase of the ratio A_0 / $B_0$and decrease of*R* in relation to A_0 gives the appearance of the relative trajectory, while maintaining its character of a shortened hypocycloid. In the new relation ε_3.

$$\varepsilon_3 = \frac{V_g}{V_u} < 1, but \; \varepsilon_3 > \varepsilon_2 > \varepsilon_1 \tag{9}$$

The relative trajectory has the look of a concave (Figure 4c), upright (Figure 4d), and convex (Figure 5a) truncate hypocycloid.For the reasons described above, machining carried out in accordance with the above restrictive conditions is irrelevant when stretching rotary surfaces and is therefore not subject to analysis.Another particular case occurs at the same rotational speed of the two elements of the kinematic pair:

$$\frac{A_0}{B_0} = 1 \tag{10}$$

Theoretically, the type of relative trajectory is a point, but in practice the realization of the machining is impossible because the centers of the tool and the stock coincide, so that the distance L between them is in accordance with the equation:

$$L = 0 \tag{11}$$

In order to use all possible inter-positions of the centers of the tool and the stock, the ratio between their peripheral speeds continues to increase. It already has a new value - as follows:

$$\varepsilon_4 = \frac{V_g}{V_u} > 1 \tag{12}$$

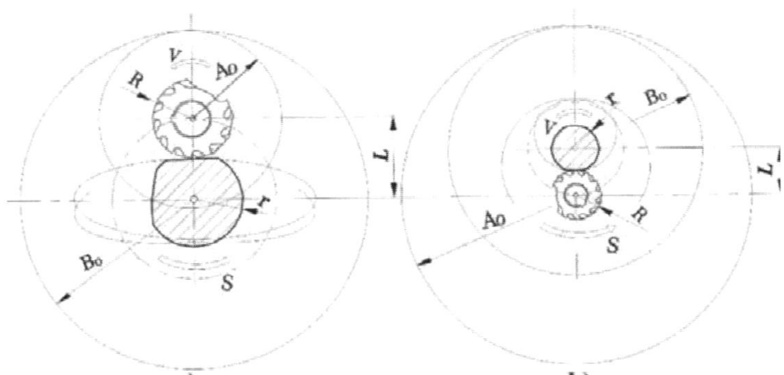

a) b)

Figure 5. Relative trajectories at broaching of rotary parts with helix broach.

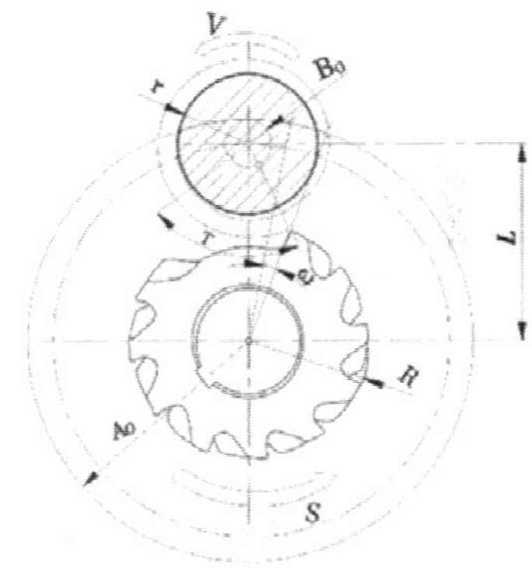

Figure 6. Optimal relative trajectory in machining rotary parts.

The relative trajectory (Figure 5b) changes from hypo to epicycloid, and the rotational movements A and B of the tool and the stock are now an auxiliary and major cutting movement. And this relative trajectory does not correspond in nature to the type of surface being treated and is not subject to further examination.If the ratio εto values is increased, according to the equation:

$$\varepsilon_5 = \frac{V_g}{V_u} > 100 \tag{13}$$

The relative trajectory (Figure 6) is a peri-cycloid with a planar spiral, cyclically extending into a circle with a radius equal to the radius of the rotary stock for a given point on the cutting edge of the helical broach. Constructing a similar trajectory by known analytical and graphical methods is a technically difficult task, almost always leading to gross errors that corrupt the end result. In this case, the complex peri-cycloid [8] curve was constructed using computer software. It turns out that the character of the obtained relative trajectory, completely corresponds to the type of rotary surface of the stock.

Its outline in the form of a circle in cross section for each point of the cutting edge defines the same outline of the machined rotary surface, also in cross section.As for the ability to make value-appropriate elements of cutting modes, it can be answered in the affirmative. The magnitude of the cutting speed frequency depends solely on the speed of the rotation stock, given the low speed rotation frequency of the tool. Submission, which is concealed (constructive), is determined by the value of the tooth lift. This is a value that depends on the type of material being machined, and the depth of cut depends on the half difference between the largest and smallest diameter of the stock.It is not a problem to form the geometrical parameters of the cutting part of the tool with rational values, in view of the wide recommended ranges for the different machinable materials. In order to fully complete the present study, the degree of accuracy of the resulting complex rotary surface resulting from the application of the treatment method under consideration should also be evaluated. Existing common features in machining rotary parts with a tangential profile blade and spiral broach, as this tool is considered to be composed of a number of tangential blades with teeth raised relative to one another. The magnitude of the deviations in this machining method does not exceed 0.002 mm for parts with a maximum diameter of the rotating surface equal to 80 mm.

3. Analysis and conclusions.

The nature of the relative trajectory obtained corresponds entirely to the type of rotary surface of the stock. The outline and the shape of a circle in cross section for each point of the cutting edge, determines the high quality of the machined surface;

Insignificant deviations from the precise geometric shape of the complex rotary surface, even for relatively large-sized parts, make this method of processing uncompromising;

The helical broach retains all the benefits of a flat broach and ranks among the instruments with a high degree of structural perfection [9];

The two movements included in the structure of the kinematic scheme are a prerequisite for the relaxed kinematics and construction of the machine tool used, and the additional two movements outside it, but actually realized due to the special design of the tool, predetermine the facilitated working conditions.

The spiral broach overcomes the three drawbacks of machining:

- The idle of the tool in the initial position and the associated loss of production time are eliminated. Upon completion of the next duty cycle, the next immediately begins;
- The metal-cutting machines used are of overall dimensions, which do not require large production areas;
- The method allows the workflow to be automated if necessary.

Extension and the associated method of machining complex rotary surfaces with helical broach are determined to be optimal when compared to other methods used like turning and milling.This conclusion is made after analyzing all the methods and associated means of treating the surfaces indicated. It is also related to the current level of branches of science and technology that are relevant to machining by cutting.

References

[1] Yoshitaka Morimoto, Souichiro Emoto, Takayuki Moriyama, Hideharu Katoa, Katsuhiro Nakagaki, Naohiko Suzuki, Yoshiyuki Kanekoand Minoru Isobe bS, 2012, Creation of Curved Surface by Lathe Turning, Procedia CIRP.

[2] S. Radzevich, "Kinematic Geometry of Surface Machining," 2013, Algorithmica, CRC Press Inc,978-1-4200-6341-7.

[3] G.I. Granovsky, Cutting Kinematics, 1948, Vishchaia Shkola, Kiev.

[4] Xingqiao Deng, Jie Wang, Shike Wang, Shisong Wang, Yucheng Liu, Ge He, An optimal process of machining complex surfaces of anti-backlash roller enveloping hourglass worms, 2020.

[5] E. A. Kudryashov, I. M. Smirnov, Yu. S. Stepanov, N. A. Khizhnyak, Improving Cutter Performance in Turning Complex Profiles, 2018.

[6] O. A. Kobelev, S. V. Albul, N. L. Kirillova, Research and development of broaching methods on mandrel of large-sized pipe forgings, 2020.

[7] Tomas Dodok, Nadežda Čuboňová, Miroslav Císar, Dorota Więcek, Influence of CNC milling strategies on complex surface machining, 2020.

[8] E. A. Kudryashov, I. M. Smirnov, D. V. Grishin, Technological features of parts with complex surfaces machining, 2018.

[9] Ali Hosseini, On the quality and integrity of broached surfaces, 2019.

Review of existing methods and tools for part processing limited to polyhedral surfaces

D Vasileva[1], **E Lefterov**[1]

[1]Technical University of Varna, 1 Studentska Str., Varna, 9000, Bulgaria

*Corresponding author's e-mail: vasilevadimka@gmail.com

Abstract. Looking retrospectively at the many years of efforts by a number of researchers to develop a method for machining by cutting of polyhedral parts, it is found that the obtained results were not particularly satisfactory. For this finding, it is not possible to substantiate justify the basis in this paper, but the reason can be briefly indicated. One of the less successful methods has been created, though unintentionally, on the basis of fundamental kinematic cutting schemes (FKCS). The uniform rotary motion B, of the stock is kinematically related to the straight-line reciprocating motion C of the simple lathe cutting tool. Uniform, straight forward motion is also performed from the stock and it is a longitudinal feed movement. This method has some disadvantages: changing the actual angles of the tool during very wide cuts between 0° to 45°; variable size chip which is a source of periodic force loading of the tool and the stock and the correct reciprocating motion is realized by a CNC machine and the deviations from the exact shape being copied relative to the shape of the part surface of the machined area.

1. Introduction

In order to overcome the shortcomings of the method under consideration, another method has been established for a similar purpose but on the basis of fundamental kinematic cutting schemes (FKCS). The uniform rotational motion B of the stock is related to the rotary reciprocating motion C of the tool and the uniform straight and forward motion A is also of the stock and is a longitudinal feeding motion [1]. The choice of rotary reciprocating motion method is dictated by the need not to alter the actual cutting angles within wide limits. However, this method has one major disadvantage: the complex kinematics of the machine tool.

Given the disadvantages of the two methods considered, neither of them can be considered as rational and reasonably recommended as the most suitable for use when needed.

2. Existing methods and tools for part processing limited to polyhedral surfaces

In order to create a more rational alternative to the methods discussed here, it is necessary to focus on the analysis of those kinematic cutting schemes which, at certain values of the relations *i* between the angular velocities of their motion, have relative trajectories with straight lines sections [2]. These are shown in Figure1(a, b).

In analyzing a fundamental kinematics scheme it is found that when:

$$i = \frac{\omega_A}{\omega_B} \tag{1}$$

within limits: $2 < i < 3$ and $R < a_0$

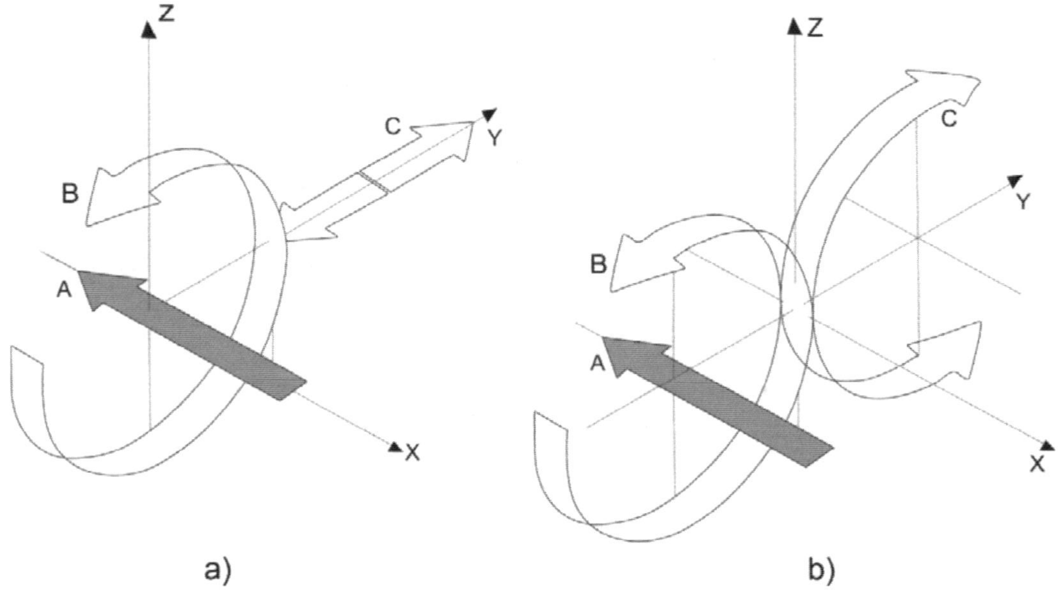

Figure 1. Kinematic cutting schemes used in polyhedral part machining.

In Figure 2, the trajectory of the relative working movement is of the nature of a shortened hypocycloid with rectilinear sections. All other relationships outside the specified range are concave or convex hypocycloids and they are unsuitable for the case at hand.

Despite their straight sections, the relative paths of Figure 2 described by points A_1, A_2, A_3, ...and B_1, B_2, B_3... of the cutting edges of the two tools of the tool head do not correspond to the machined multi-wall surface of the part for the following two reasons:

- The outline of the multi-walled part in cross-section is asymmetrical and with different lengths;
- The relative trajectory is open within a single work cycle.

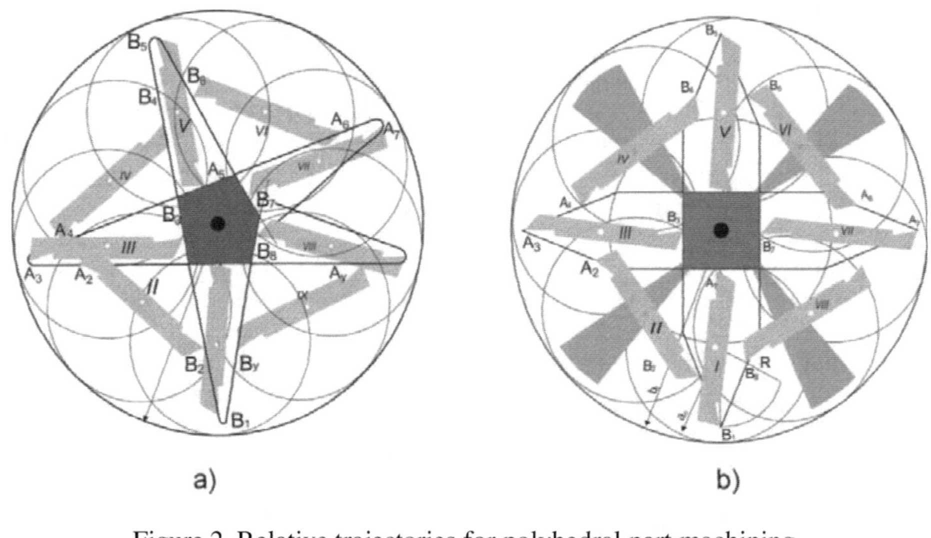

Figure 2. Relative trajectories for polyhedral part machining.

In fact, at the indicated limits of the *i* ratio, a machining method can only be established if periodically outside the cutting area, the tool head reorients itself (slows down its rotation) to the next working process. This means that the kinematic relationship between the rotation of the tool and the stock should be preserved only for the duration of the work process and not beyond. This possibility is illustrated in Figure 2.

As it is mentioned slowing down of the rotational movement of the tool should be performed in the shadow sections shown in the Figure 2. It turns out that such a possibility is only theoretical. So far, it is not known that a mechanism with the indicated kinematic capabilities is practically implemented, due to the significant technical difficulties in its design and manufacture [3].

To determine if it is possible to build a multi-wall machining process with capabilities that FKCS offer, the solution must be sought beyond the limits of the "ï" ratios taken above, despite the fact that relative trajectories would have a curvilinear type [4]. However, the requirement must be respected that the relative trajectories represent closed contours within a single duty cycle.

After FKCS (Figure 1a) analyze, it is found that such trajectories (Figure 3) are possible when:

$$i = \frac{\omega_A}{\omega_B} = \frac{b_0}{a_0} = 2 \tag{2}$$

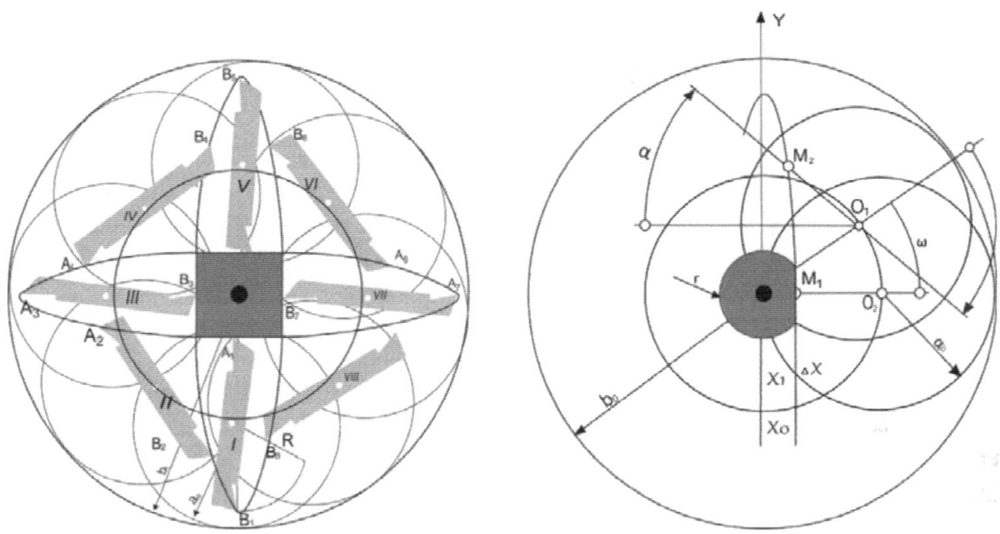

a) b)

Figure 3. Variant of relative trajectory in polyhedral part machining.

The two closed relative trajectories with the character of shortened hypocycloids and the appearance of elliptic curves with mutually perpendicular axes, described by cutting edges of the two knives of the tool head, form on two opposite sides of the multifaceted stock, in the accomplished of kinematic relationship between the rotational movement of the tool and detail. In order to determine the appropriateness of using the obtained relative trajectory, should be determined the maximum deviation *p* from the rectilinear shape of the stock, upon receipt of a curvilinear outline of each side of the multi-walled part in cross-section [5, 6]. This can be achieved by using the scheme of Figure 3b.

The stock with radius r is connected to a circle with radius b_0, and the tool with a second circle with radius a_0 which rolls without sliding along the first, which providing internal contact. When the tool is rotated and the circle associated with it around the angle α, its center from position O_1 moves to position O'_1, on angle ω from his original position.

If:

$$O_1M_1 = O_1'M_2 = e \tag{3}$$

Coordinates for position M_2 can be determinate with the following mathematical equations:

$$x = (b_0 - a_0)cos\omega - e\ cos\alpha \quad (4)$$

$$y = (b_0 - a_0)sin\omega + e\ sin\alpha \quad (5)$$

Figure 3, b shows that:

$$\alpha = \phi - \omega \quad (6)$$

and from the condition that the two circles roll against each other without slipping the following:

$$b_0\omega = a_0\varphi \quad or \quad (7)$$

$$\varphi = \left(\frac{b_0}{a_0}\right)\omega \quad (8)$$

So in that case:

$$\alpha = \left[\frac{b_0 - a_0}{a_0}\right]\omega \quad (9)$$

In substitution of (5) and (9), the coordinates of point M_2 of the relative trajectory are determined by the equations:

$$x = (b_0 - a_0)cos\omega + e\cos\left[\frac{b_0 - a_0}{a_0}\right]\omega \quad (10)$$

$$y = (b_0 - a_0)sin\omega + e\sin\left[\frac{b_0 - a_0}{a_0}\right]\omega \quad (11)$$

The maximum deviation Δx from the exact rectilinear shape of the multi-walled part is at a point of relative trajectory with the coordinate $x=x_0$, obtained at $\omega=0^0$. Then:

$$\Delta x = x_0 - x_1 \quad (12)$$

The second term $x_1=x$ on the right side of the equation (12), his coordinate $y_1=y$ determine the position of the outline point from origin line area of part, for angle $\omega_1=\omega$.
Substitution of the last three equations in (10) and (11) and after minor processing results we get:

$$x_1 = (a_0 - e)cos\omega_1 \quad (13)$$

$$y_1 = (a_0 + e)sin\omega_1 \quad (14)$$

Considering that: $b_0=2a_0$,

$$x_0 = (a_0 - e) \quad (15)$$

Substitution of (13), (14) and (15) in (12), the magnitude of the maximum deviation of the part from its theoretical rectilinear shape is determined by the equation:

$$\Delta_x = (a_0 - e)(1 - cos\omega_1) \quad (16)$$

With the number of sides of the multi-walled part n=2,4,6,8... the ordinate y_1, are determined by the equation:

$$y_1 = r.\sin\left(\frac{\pi}{n}\right) \quad (17)$$

After replacing the right side (17) in the equation (14), we get equation for determinate the value for angle ω_1:

$$\frac{\omega_1 = arc\sin\left[r.\sin\left(\frac{\pi}{n}\right)\right]}{a_0 + e} \quad (18)$$

It is know that the radius of curvature ϱ at a point of any non-linear trajectory represented by its Cartesian coordinates is determined by the equation:

$$\rho = \frac{\left[1+\left(\frac{dy}{dx}\right)\right]^{\frac{3}{2}}}{\frac{d^2y}{dx^2}} \tag{19}$$

So the last equation will be:

$$\frac{dy}{dx} = \frac{cos\omega+\frac{e}{a_0}cos\left(\frac{b_0-a_0}{a_0}\omega\right)}{-sin\omega+\frac{e}{a_0}sin\left(\frac{b_0-a_0}{a_0}\omega\right)} \tag{20}$$

In the perceived relationship between the rotational motions of the tool and the stock, the relationship between the circles connected to the two elements of the kinematic pair is represented as follows: $b_0=2a_0$. In that case the equation for dy/dx, gets his final appearance:

$$\frac{dy}{dx} = \frac{(e+a_0)cos\omega}{(e-a_0)sin\omega} \tag{21}$$

It is know that:

$$\frac{d^2y}{dx^2} = \frac{\frac{d^2y}{d\omega^2}\cdot\frac{dx}{d\omega}-\frac{d^2x}{d\omega^2}\cdot\frac{dy}{d\omega}}{\left(\frac{dx}{d\omega}\right)^3} \tag{22}$$

then:

$$\frac{d^2y}{dx^2} = \frac{-(e+a_0)}{(e-a_0)sin^3\omega} \tag{23}$$

Substitution of dy/dx and d^2y/dx^2 in equation for determined the radius *r* of the curvature of the multi-walled part is obtained:

$$r = \frac{\left[-(e+a_0)^2\right]}{e} - a_0 \tag{24}$$

3. Conclusions

Determining the deviation *Dx*, according to an algorithm developed in accordance with the described study, allows to estimate the magnitude of the deviation when machining multi-walled parts with different number of sides (Table 1.)

Table 1. Machining multi-walled parts with different number of sides.

	Number of sides	Deviation
For parts with	n=4	$\Delta_x = 0.09 - 0.17$ mm
For parts with	n=6	$\Delta_x = 0.05 - 0.11$ mm
For parts with	n=8	$\Delta_x = 0.02 - 0.007$ mm

This method of machining allows receipt multi-walled parts with 3-4 degrees of accuracy, they meet the engineering products to which medium technical requirements are imposed. The proposed method is the most rational in comparison to the other two designed for a similar purpose, one of them practically

IOP Conf. Series: Materials Science and Engineering 1037 (2021) 012023 doi:10.1088/1757-899X/1037/1/012023

unrealized. Its application has been appropriate for both technical and economic reasons where such production is required.

This method can be implemented on existing equipment for example Lathe CNC machine, such as the type of lathe 401 FKCS, in accordance with which this machine tool operates, get upgrade with another rotary motion, kinematically related to the available. The transition between 401 to 701 FKCS would ensure the appropriate nature of the relative trajectory. For this purpose it is necessary to design and construct a relatively simple device, which should be integrated into the basic structure of the lathe.

Another approach is selected by the German company HAHN& KOLB - Stuttgart. According to the proposed by Academician "Granovsky G.I." [7] kinematics, the same company produces automatic machines for the production of multi-walled parts with maximum profile sizes in the range of 3 to 40mm. The machines manufactured by this company find a market not only in the Federal Republic, but also in Italy, Austria, Australia and the countries of South and North America.

This method allows standard equipment with tools such as multi-wall heads or milling cutters to produce outer or inner multi-wall surfaces, with the minimum radius of the shaping tool being determined by internal machining.

The transition from kinemtic scheme (Figure 1a) requires using different construction tools, and in the design of the manufacturing process it is necessary to look for a combination between the speeds of elementary movements and the scheme of use of the additive.

References

[1] Kumar K., Kalita H., Zindani D., Davim J.P., Machining Process, 2019, ISBN: 978-3-030-21066-3.

[2] S. Radzevich, "Kinematic Geometry of Surface Machining," 2013, Algorithmica, CRC Press Inc, 978-1-4200-6341-7.

[3] I. Inasaki, K. Kishinami, S. Sakamoto, Y. Takeuchiand F. Tanaka, 1997, Shape Generation Theoryof Machine Tools. Yokendo Press.

[4] N. Musa, The Use of Genetic Approach to the Kinematics of Cutting, 2016, ISSN: 2160-0503, World Journal of Mechanics.

[5] Nikolay Baranov, Alexander Feofanov, Tool path optimization in complex geometry parts machining process, 2020.

[6] Haoran Zhu, On the Polyhedrality of Cutting-plane Closure, 2019.

[7] G.I. Granovsky, Cutting Kinematics, 1948, Vishchaia Shkola, Kiev.

Improving of preventive management for flat rolling products quality indices

V V Kukhar[1,*], O H Kurpe[1], A H Prysiazhnyi[1], O A Khliestova[1], V A Burko[1], E Yu Balalayeva[1], N Yu Yelistratova[1]

[1] Pryazovskyi State Technical University, Universytetska str., 7, Mariupol, 87555, Ukraine

*Corresponding author's e-mail: kvv.mariupol@gmail.com

Abstracts. Based on the combination of ISO and Lean system tools, the methodology of continuous optimization of technology and quality improvement of rolled steel has been improved. To assess the condition and determine the production level to set limits to process indices the use of colored markers was suggested. The improved flat rolling quality control methodology is based on the use of Deming cycle and includes the use of such analysis tools as Fishbone (Ishikawa) diagram, Pareto principle, distribution diagrams, 3σ rules and Shewhart control charts. For the first time, differentiated corrective solutions for plate steel which is produced with different levels of process stability and quality indices are suggested. The inspection was carried out with the data from the mechanical properties of flat steel products 14 mm thick from K60 steel, produced by thermo-mechanical controlled process on a plate mill 3600. Introduction of the methodology allowed to reduce the standard deviation of yield strength, tensile strength and percentage elongation after fracture of the studied rolled products by 44%, 31% and 46%, respectively, indicating an overall increase in technology stability and quality indices and implementation of an effective tool for their preventive control.

1. Introduction

For proper functioning and achieving development of an enterprise or a company, the product quality system must be implemented in all process areas [1–5], starting with raw materials supply and ending with the processes of working with the consumers (customers) [6–8], and can be combined with the production system [9–12]. The analysis of enterprises operation showed that their quality system operates only in the field of technology, and the use of this system in other areas that interact with technology is only formal. Therefore, further improvement of approaches to preventive product quality control and their systematic use in all process areas of production is a relevant problem.

The analysis of data [6, 9–13] allowed us to conclude that to solve this problem it is effective to combine the quality system ISO 9000, ISO 9001 with the tools of the Lean production system. This is a system of organization and control of product development, production, relationships with suppliers and consumers, when products are manufactured in strict accordance with consumer demands and with lower losses [13–16]. In particular, one of the approaches provided by the ISO 9001 quality system and can be used to solve the problem is the use of "Risk-oriented thinking" [9, 12, 13, 17–19]. Its essence is that an organization must perform a risk assessment, i.e. identify factors that may cause deviations of its processes and quality control system from planned results, and develop actions towards the risks to prevent adverse effects, based on available data on manufactured products and its quality indicators. At the same time, it is necessary to plan how to integrate and implement actions into the processes of the quality control system of an organization and evaluate their effectiveness, i.e. the Deming cycle or PDCA cycle should work – "Plan", "Do", "Check", "Act" [6, 16, 20–22].

Among the tools of the Lean production system which should be used in combination with the ISO 9001 quality system, we can note the following [13, 16, 21–24]: (a) benchmarking, which is to compare indicators, results, quality indices, etc. with similar industries, manufacturers in order to find the best experience; (b) Ishikawa diagram, as a graphical way to study and determine the most significant cause-and-effect relationships between factors and consequences in the studied problem; (c) Pareto principle – an empirical rule according to which for many events 80% of the consequences are caused by 20% of the causes. At the same time, the analysis of the experience metallurgical enterprises showed that they do not fully use the methodology, which would be based on a combination of ISO quality systems and Lean production system, which complicates the improvement of technology and product quality. Therefore, the aim of the paper is to provide the necessary quality indicators of flat rolling products (on the example of heavy plate steel) based on the combination of the ISO and Lean systems principles.

2. Research methodology

The improved rolling quality control methodology is based on the use of the Deming (PDCA) cycle and includes the establishment, determination of process indices values that affect to product quality, their control, accumulation, processing and improvement mechanisms, Fig. 1. The proposed methodology is applied to products that have passed the stage of development and are produced on an industrial scale. In order to control the production process and ensure the quality of flat rolled sheet steel, it is necessary to install and control the appropriate process parameters by the Lean tools. After establishing a list of process indices, it is necessary to determine their values, which can be established based on analysis of similar products production, statistical processing of available information, research, published sources.

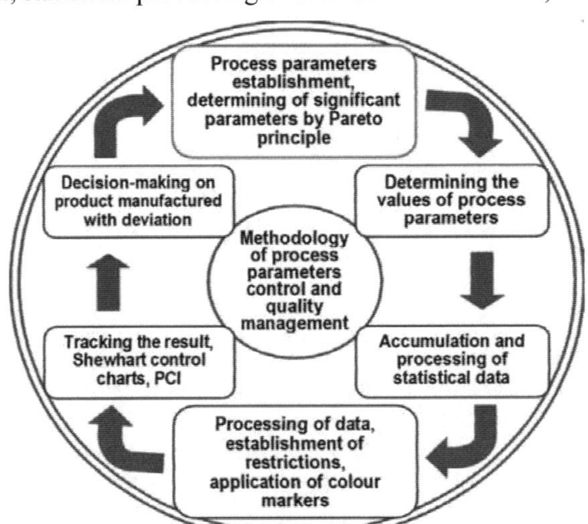

Figure 1. The methodology example of rolling quality control in the PDCA cycle.

To optimize the technology and improve the quality of rolled products, it is necessary to accumulate statistical information and process it. To obtain statistical data in the production of rolled products, it is vital to control the values of all set process indices and keep this information. These indices should be combined with product quality parameters using a tracking system, which is mandatory for modern production. Each of the rolled quality parameter should be analyzed separately or in a group of the same level indices (for example, the mechanical properties of the steel can be combined into a groups). The first step of analyzing the obtained statistical data is creating the distribution diagrams for each of the rolling quality indices, establishing the restrictions in accordance with 3σ or 6σ rules, depending on the process. Restrictions should up to existing regulations or stricter. On the second step of processing the accumulated statistical data it is necessary to build distribution charts for each of the established process indices that affect this rolling quality parameter, and set limits that ensure the required quality production. It is recommended to use colored markers to visually assess the condition and determine the levels of process when setting restrictions on process indices. In particular, the red marker can be used when the process index deviates from the limits that ensure receiving needed quality index of rolled products in accordance with the standard (this means a violation of the process). The yellow marker can be used when the process index deviates from the limits that provide increased requirements set within the quality control system, while this indicator of rolling quality is still within the standard (this means that the process needs attention). The green marker should be used when the process index meets the limits of the increased requirements for the rolling quality, and the process is stable. Using this approach within the combination of ISO and Lean systems allows not only to stabilize the process and improve product quality, but also to reduce production costs by applying restrictions on obtaining overly high quality levels. For example, consistently obtaining the average requirement level of mechanical properties indices of flat rolling (sheet metal) for the main pipelines allows saving micro-alloying elements during steelmaking.

To monitor the result and control compliance with the established restrictions on process indices there can be used direct control by an employee of the quality control department at the place of fixing the process index, visualization (control) of process indices or automatic process control system. Shewhart control charts are also widely used, which are graphs of sampling indices change, usually the average value and standard deviation, which is calculated in the process of data collection. Obtaining information on deviations allows for managing the future of such products in accordance with the requirements of ISO 9001. Thus, products produced in batches, for each unit of which all parameters are within the green marker, is considered appropriate. If the individual units of the batch have parameters that are within the yellow marker, then additional quality control is assigned to these units. Products with yellow markers are considered appropriate only after the additional control results confirmation. In case when individual units of the party have parameters that are within the red marker, the products, if not agreed with the Customer, are considered non-compliant.

3. Results and discussion

Below is an assessment of the proposed methodology effectiveness in order to establish the current level of quality and areas of improvement of such quality indices as yield strength σ_{ys}, tensile strength σ_{uts} and percentage elongation after fracture δ in the production of plate made of pipeline steel grade K60 (API 5L Standard), 14 mm thickness, that was produced with thermo-mechanical rolling with accelerated cooling on the mill 3600 at PJSC "AZOVSTAL IRON & STEEL WORKS" (Mariupol, Ukraine). The list of process indices that affect the level of these quality parameters is determined by practical data. This list includes the chemical composition with the elements that affect to the mechanical properties of the steel (C, Mn, Nb, V, Mo, Cr, Ni, Cu), as well as temperature of rolling in start in roughing stand (TRSr), the temperature of the second stage rolling start in roughing stand, which is due to recrystallization inhibition start (TRI), temperature of the rolling start and finish in finishing stand (TRS, TRF), controlled cooling start and finish temperature (TCS, TCF) [25–29].

To determine the process indices that affect the studied quality parameters, Pareto diagrams were constructed for each parameter. The basis for constructing Pareto diagrams is the value of reliability of approximation R^2 and the value of Pearson's correlation coefficient (by module), which were obtained when determining the relationship between each quality parameter and process index. Fig. 2, for example, shows Pareto diagrams constructed for the yield strength σ_{ys}. For the data set used in the calculations, the critical value of the Pearson's correlation coefficient is 0.17 for the significance level $p = 0.05$ (see Fig. 2(b)). Process indices which Pearson's correlation coefficient towards the corresponding rolling quality parameter is less than 0.17 are not significant.

Table 1 shows the process indices, which according to the results of statistical data processing are set as significant. Actions to improve the technology and the rolling quality were developed only for significant process indices, and insignificant indices were left at the level set by the technological process. In accordance with the proposed methodology for rolling quality control there were also constructed distribution diagrams for these quality parameters of rolled thick sheets and heavy plates. Fig. 3, for example, shows distribution of the rolling yield strength. The analysis of the constructed diagrams established that the distribution of the yield strength has cases of obtaining results below the required regulatory standard and level with minimum required regulatory standard. In relation to the distribution of temporary resistance, there are also cases of obtaining lower than the regulatory standard of requirements. The distribution of percentage elongation after fracture has a significant number of cases obtaining results below the regulatory standard of requirements, which indicates the instability of the process to obtain this quality parameter of sheet steel.

Table 1. Parameters those are significant in determining the appropriate rolling quality index.

Parameter	Pearson's correlation value according to the rolling quality indices		
	σ_{ys}	σ_{uts}	δ
TRI	0.316	0.379	-
V	0.236	0.252	-
TCS	– 0.224	– 0.334	-
Ni	– 0.213	– 0.219	-
Mo	– 0.196	-	-
Cu	– 0.188	– 0.202	-
C	0.178	0.246	-
TCF	-	– 0.241	0.412

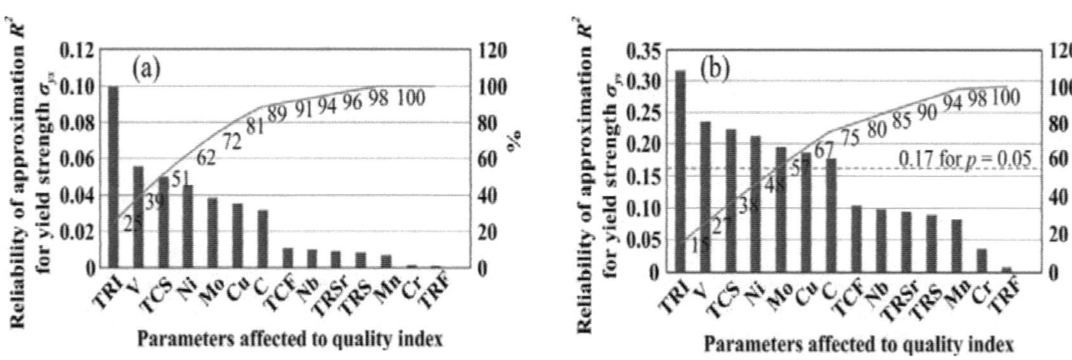

Figure 2. Distribution diagram of the reliability of approximation (a) and the Pearson's correlation coefficient (b) on the indices that affect the yield strength of rolled products made of K60 grade steel.

To stabilize the rolling technology and improve the rolling quality on studied parameters, the restrictions were set and the process levels determined, using color markers presented in Table. 2. In order to obtain the required level of technology indices according to the established limits of rolling quality parameters in studied data set, using filters only those data which corresponded to the stable level of the process (indicated by the green marker in Table 2) were kept and their limits set. Filtration was started with the most unstable quality parameter, which in this case is the specific elongation. The list of significant process indices for all three parameters of sheet quality was the same. When setting yellow and red markers and the corresponding levels of the process, the values of Pearson's correlation in terms of rolling quality parameters were taken into account (see Table 1). Determining the limits of violation of the rolling technology (marked with a red marker in Table. 2 was carried out according to the data that meet the limits of regulatory requirements for plate steel quality parameters. It should be noted that the quality of originally developed (basic) technology plays an important role in using the proposed methodology to establish the levels of process.

In the studied case the wrong basic technological process of plate steel thermo-mechanical rolling was developed, it did not allow to obtain the required mechanical properties parameters of rolled products by accelerated cooling, which can be seen in negative correlations of cooling indices. The desired level of plate steel properties was achieved mostly through the rolling process indices. After establishing levels of process indices (see Table 2) required to obtain stable quality parameters of plate steel, the diagrams of yield strength, tensile strength and specific elongation of the metal were constructed. Fig. 4 shows an example of the diagram of the yield strength distribution of rolled products 14 mm thickness, made of K60 grade steel for the green level process.

Table 2. Restrictions on process indices and rolling quality parameters.

Quality parameter/process index	Process levels				
	violation of the process	needs attention	stable	needs attention	violation of the process
Yield strength, (σ_{ys}), MPa	<505	505–510	511–570	571–580	-
Tensile strength, (σ_{uts}), MPa	<590	590–600	600–660	661–710	>710
Elongation after fracture, %	<22	22.0–22.4	22.5–25.0	25.1–26.0	-
TRI, °C	<876	876–908	909–1008	-	>1008
V, %	<0.055	0.056–0.059	0.06–0.08	-	>0.08
TCS, °C	<717	718–720	721–737	-	>737
Ni, %	-	-	0–0.29	-	>0.29
Mo, %	-	-	0–0.07	-	>0.07
Cu, %	-	-	0–0.28	-	>0.28
C, %	-	-	0.08–0.11	-	>0.11
TCF, °C	<541	542–595	596–640	-	>640

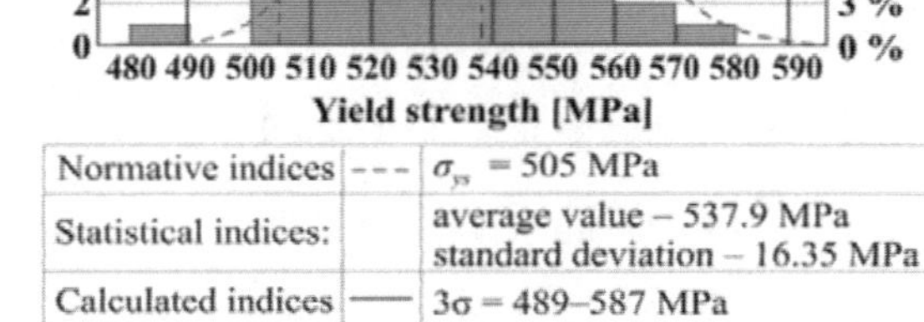

Figure 3. The initial distribution of the yield strength of 14 mm thick rolled products made of K60 grade steel.

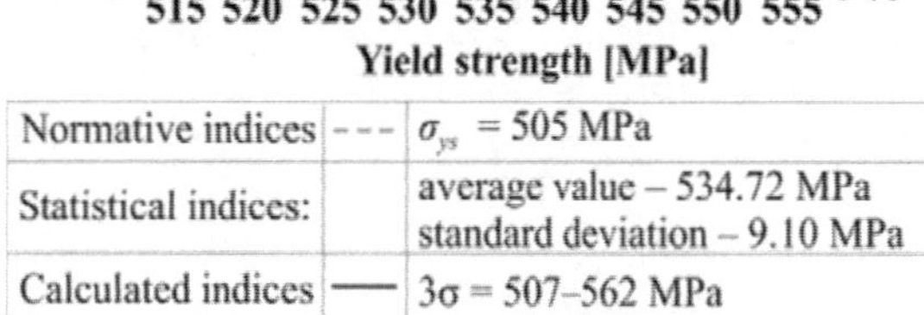

Figure 4. Distribution of the yield strength of 14 mm thick rolled products made of K60 grade steel for the green level of the process.

The analysis of obtained results showed that the green level process provides the parameters of plate steel mechanical properties that are within the regulatory requirements. At the same time, the standard deviation along the yield strength, tensile strength and specific elongation of rolled products decreased by 44%, 31% and 46%, respectively, indicating an overall increase in technology stability.

4. Conclusion

The methodology of product quality control was improved and performed the efficiency evaluation of its application according to the conditions of obtaining plate steel with thickness of 14 mm and above from steel of K60 strength grade on a plate mill "3600". The results of this assessment showed that suggested methodology allows to obtain the mechanical properties parameters of plates that are in line with the regulatory requirements, as well as to control the level of these indices to obtain economic benefit, combining the ISO and Lean systems.

References

[1] Backman J, Kyllönen V and Helaakoski H 2019 *IFAC PapersOnLine* **52(13)** 1174-1179
[2] Grudzien Ł and Hamrol A 2016 *International Journal of Information Management* **36(4)** 599-606
[3] Su H-C, Kao T-W and Linderman K 2020 *European Journal of Operational Research* **283(2)** 530-540
[4] Kukhar V, Kurpe O, Klimov E, Balalayeva E and Dragobetskii V 2018 *International Journal of Engineering & Technology (UAE)* **7(4.3)** 35-39
[5] Artiukh V, Mazur V and Butyrin A 2018 *Advances in Intelligent Systems and Computing* **692** 212-219
[6] Martínez-Costa M, Choi T Y, Martínez J A and Martínez-Lorente A R 2009 *Journal of Operations Management* **27(6)** 495–511
[7] Dragobetskii V, Naumova E, Shapoval A, Shlyk S and Moloshtan D 2019 *Proc. Int. Conf. on Modern Electrical and Energy Systems (MEES 2019)*(Sept. 23-25, 2019, Kremenchuk, Ukraine) 506-509
[8] Konstanciak E, Budzik R and Waszkielewicz W 2003 *Metalurgija* **42(2)** 123-127
[9] Antsev V Yu, Vitchuk N A and Miroshnikov V V 2017 *Procedia Engineering* **206** 950-957
[10] Markov O E, Gerasimenko O V, Shapoval A A, Abdulov O R and Zhytnikov R U 2019 *International Journal of Advanced Manufacturing Technology* **103(5-8)** 3057-3065
[11] Orlov G A and Orlov A G 2020 *Solid State Phenomena* **299** 693-698
[12] Liu C and Sun Y 2009 *Proc. 2nd Int. Conf. on Future Information Technology and Management Engineering (FITME 2009)* (Dec. 13-14, 2009, Sanya, China) 77-80
[13] Priede J 2012 *Procedia – Social and Behavioral Sciences* 58 1466-1475
[14] Kukhar V V, Grushko A V and Vishtak I V 2018 *Solid State Phenomena* **284** 408-415
[15] Kukhar V, Balalayeva E, Prysiazhnyi A, Vasylevskyi O and Marchenko I 2018 *MATEC Web of Conferences* **178** 02003
[16] Silva A J V 2015 *Proc. 2nd Int. Conf. on Computer, Intelligent and Education Technology (CICET 2015)* (Apr. 11–12, 2015, Guilin, China) 823-826
[17] Popova L, Yashina M, Babynina L, Ryzshakova A, Yefremova N and Andreev A 2019 *Quality – Access to Success* **20(170)** 58-63
[18] Hunchenko O 2019 *International Journal of Recent Technology and Engineering* **8(2)** 5787-5790
[19] Kukhar V, Yelistratova N, Burko V, Nizhelska Y and Aksionova O 2018 *International Journal of Engineering and Technology(UAE)* **7(2.23)** 216-220
[20] Emelianova D, Kliuchareva N, Kolesnichenko-Yanushev S and Yakovlev A 2020 *E3S Web of Conferences* **164** 10013
[21] Badea D M, Stanciu V M, Constantin P, Stefănescu G and Tănăsescu F T 2010 *Quality – Access to Success* **11(1-2)** 33-39
[22] Garza-Reyes J A, Torres R J, Govindan K, Cherrafi A and Ramanathan U 2018 *Journal of Cleaner Production* **180** 335-348
[23] Djapic M, Popovic P and Lukic L 2019 *IOP Conference Series: Materials Science and Engineering* **682** *012017*
[24] Kohl H and Neumann K 2020 ZWF Zeitschrift Fuer Wirtschaftlichen Fabrikbetrieb **115(1-2)** 27-30
[25] Kurpe O, Kukhar V, Klimov E, Chernenko S and Balalayeva E 2020 *Lecture Notes in Mechanical Engineering* (Advances in Design, Simulation and Manufacturing II, DSMIE-2019) 418-429
[26] Kurpe O, Kukhar V, Klimov E and Prysiazhnyi A 2018 *Solid State Phenomena* **291** 63-71
[27] Artiukh V, Kukhar V, and Balalayeva E 2018 *MATEC Web of Conference* **224** 01036
[28] Efremenko V G, Zurnadzhi V I, Chabak Y G, Tsvetkova O V and Dzherenova A V 2017 *Material Science* **53** 67-75
[29] Smirnov E N, Smirnov A N, Sklyar V A, Belevitin V A, Eron'ko S P and Pivovarov R E *2018 Steel in Translation* **48**(6) 381-387

FEM analysis of stress-strain state and material consumption of RHS with longitudinal gap and weld during the tests

V V Kukhar[1,*], Y G Sahirov[1], V M Hornostai[2], O E Markov[3], M M Nahnibeda[1]

[1] Pryazovskyi State Technical University, Universytetska str., 7, Mariupol, 87555, Ukraine
[2] National Technical University of Ukraine "Igor Sikorsky Kyiv Polytechnic Institute", Peremohy ave., 37, Kyiv, 03056, Ukraine
[3] Donbass State Engineering Academy, Akademichna str., 72, Kramatorsk, 84313, Ukraine

*Corresponding author's e-mail: kvv.mariupol@gmail.com

Abstract. The paper implements a method for analyzing the stress-strain state of rectangular hollow sections (RHS) by finite-element modeling (FEM) of tests for three-point bending and torsion. Design schemes, 3-D solid-state and deformable models have been developed using the automated analysis and CAD/CAE system software, made it possible to obtain equivalent stress distributions and displacements in models. A simulation of tests for RHS with a cross section of 40 mm × 50 mm, manufactured in two ways, was carried out: (a) by direct-forming of galvanized steel strips on roll-forming mill in a semi-closed section with a longitudinal gap of 0.5 mm between the edges formed on a 40 mm web (DF-RHS); (b) similar direct-forming to the closed section and next welding the edges to a longitudinal weld along the web middle of 50 mm (DFW-RHS). RHS with various wall thicknesses (t = 1.93 mm, 1.84 mm and 0.7 mm) was investigated, given the design features that depend on the manufacturing processes of structural sections. It was found DFW-RHS is stiffer by at least 50% compared to DF-RHS, which allows to savings the metal by reducing the RHS wall thickness by 62% while maintaining the same stiffness and ensuring high strength of structural section.

1. Introduction

Rectangular hollow sections (RHS) are great importance for much purposes, such as construction (load-bearing frames, columns, posts) and reinforced concrete structures, uPVC window systems of buildings and constructions, some areas of civil or mechanical engineering, shipbuilding, etc. [1]. Due to the prevalence and increasing demand for structural section, the requirements for the geometric parameters and strength of such products are becoming more stringent, for that the different standards have been developed in the European Union [2, 3] and most developed world countries [1, 2].

For the manufacture of structural sections with required service properties different fabrication methods are used [1, 4–8]: skew roll piercing process, pilger process, Fretz-Moon pipe process, induction welding process, spirally bent strip welding. Hot-finished process (HFP) [9, 10] and cold-formed process (CFP) [9] is the most common manufacturing for RHS. These consist in the roll-forming of circular tube from a strip, welding of edges and re-forming into rectangular tube section (round to square or rectangular section forming) with heating (HFP) or without heating (CFP), respectively. Recently, the direct-forming process (DFP) [9, 11] for RHS has been very justified and economical.

This process of directly forming to square or closed rectangular section by sequential bending on multi-roll roll-forming mills with longitudinally welding of edges at finishing have proved to be satisfactory [1, 5, 8, 9, 11].

The improvement of designs, process of production and quality control of these products is constantly given considerable attention due to their high importance and adaptation to different ranges of use. Therefore, the development of methods of modeling of the section behavior of various (including improved) structures remains extremely relevant [2–5, 12, 13, 15], because allows to identify the advantages and disadvantages of new solutions and develop recommendations for improving the processes of manufacturing structural sections. Research on stiffness of structural sections is conducted by bending and torsion tests.

The bending tests are carried out according to standardized methods by loading sections by four-point or three-point scheme [2, 12]. It is quite common to combine experimental tests with computer simulation of the test process [2, 13, 14]. The latter requires an account of the physical conditions of the test deformation, the complexity of the geometric shapes of the section and the material properties of the specimens to be as accurate as possible, which presents certain difficulties [2–4, 12–16]. Studies of the complex characteristics of strength and stiffness are performed for sections of different shapes (open and closed) [17], of different materials (including non-metal) [18, 19], non-welded and welded [13, 18, 20, 21], taking into account the complexity of the shape of the final product, location and load in real construction [17, 20, 21]. This defines the deflection values as the characteristics of sections stiffness, and magnitudes of stresses as characteristics of sections strength. Certain progress has been made in testing and modeling the section loading process by three-point bending, when the section is viewed as a beam supported on round supports (pins) and loaded along the axis of symmetry by a rounded loading pin. As a result, the behavior of the walls and webs (faces) of the section, different buckling shapes [12, 17], revealed critical values of forces and stresses have been described. The main result of the tests on the stiffness of the sections is to obtain a graphical dependence of the deflection values on the load force [12–16]. In some cases, instead of the deflection values, they determine the parameters of the beam curvature [3] at loading or analytically calculate the bending angle between the bent beam and the horizontal axis [22]. It is particularly difficult to consider the multilayer sections when using coated specimens [16, 19].

Torsion test is carried out for structural sections of various designs and materials [23–25], including those hardened by various processes [26–28]. Modern packages for finite element modeling (FEM) make it possible to evaluate the behavior of sections for both static torsion and torsion with the application of various types of loads [29–31]. The main evaluation criterion is to identify the dependence of torsion angle (or equivalent displacement) on the applied torque (or force impact) [21, 23–25, 31, 32]. For basic cases static conditions are usually investigated.

For the production and, accordingly, for testing of structural sections, galvanizing carbon steel is most often used as a material. When testing galvanized steel sections, the thickness of the zinc coating is generally neglected as an absolutely thin film [33]. Changing the design of the structural section [34–36] or introducing of consistent direct-forming with longitudinal welding of the edges of the closed section requires evaluation of the indices of stiffness and strength (stresses) of RHS to identify the advantages or disadvantages of new processes.

2. Research objectives and methodology

So, there is a certain enterprise that produces of RHS by direct-forming (DF-RHS) for reinforcing of uPVC window (Fig. 1). According to production plans, it is desirable to establish the series release of RHS by direct-forming with welding of edges (DFW-RHS) in longitudinal seam. Strength and stiffness (stresses and displacements) of the sections should be evaluated by bending and torsion tests for estimate of prospects of material consumption reducing for improved RHS, as a practical value of the research. The results of the study of the behavior of sections of the same shape, but different variants of manufacture, after the tests will allow to establish dependencies between the force & torque magnitudes and deflection & displacement, to estimate the section deflect or twist distortions at the same load, and to compare the stresses in the webs, which is a scientific originality of research.

The aim of this work is to determine, using finite-element modeling of bending and torsion tests, differences in the strength and stiffness performance characteristics of sections with 40 mm × 50 mm width of webs, manufactured by directly multi-roll forming of steel beam:

(a) 40*50 DF-RHS – as semi-closed section with a gap between the edges not more than 0.5 mm, formed along a middle of the web (face) with a size of 40 mm;

(b) 40*50 DFW-RHS – as closed sections and subsequent welding of the edges with the longitudinal weld on the middle of the web (face) with a size of 50 mm.

Comparing these types of RHS with the gap and the weld will allow to see differences in the behavior of products of similar cross-section and different variants of manufacture with differences in thickness, and, as a consequence, to evaluate the prospects of reducing the material consumption of the DFW-RHS while ensuring equivalent stiffness.

The research of the RHS stress-strain state was performed on a solid-state deformable model using the CAD/CAE systems of SolidWorks. The geometric parameters of the model are fully consistent with the full-scale specimens (Fig. 2) and are made in a scale of 1:1 (for geometric similarity). The simulation did not take into account the dead weight of the specimen. As an assumption, a RHS wall with weld was considered as solid body without discontinuities, but with insignificant lateral longitudinal cut in the middle of the external web face. The study of the stress-strain state was carried out with loading options for three-point bending (I) and torsion (II). For each option, diagrams of equivalent stresses and linear deflections in the model were determined. The calculated loads and their combinations were taken such that they correspond to the experiment on full-scale specimens.

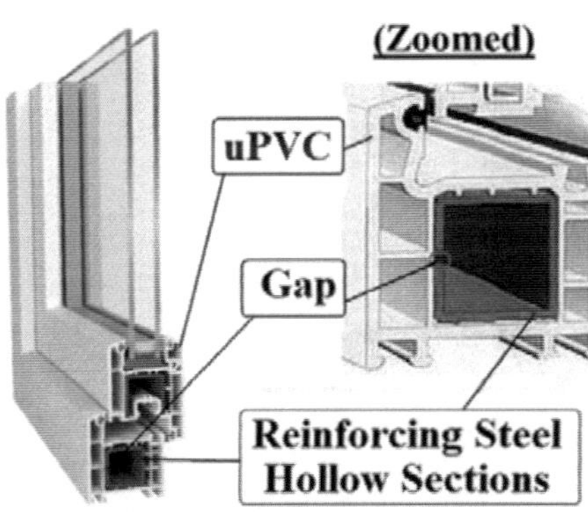

Figure 1. Reinforcing RHS in uPVC window system.

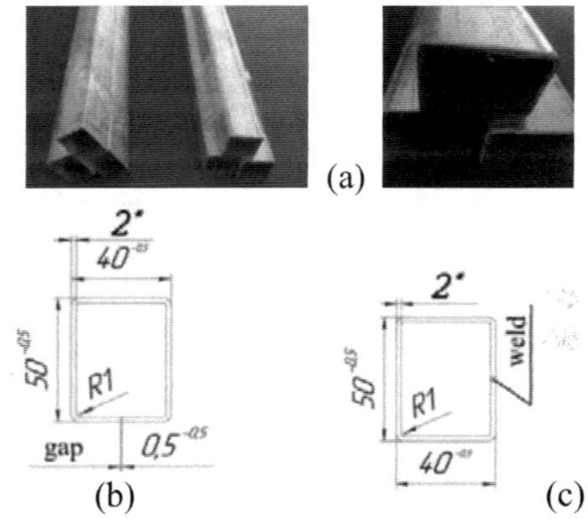

Figure 2. Delivered RHS for research (a), DF-RHS(b) and DFW-RHS (c) with webs of 50 mm × 40 mm (*t = 2 mm – nominal thickness of the section wall).

The specimen models under research: (i) DF-RHS with cross section 40 mm × 50 mm and thickness of the wall t = 1.93 mm; (ii) DFW-RHS with cross section 40 mm × 50 mm and thickness of the wall t = 1.84 mm; (iii) DFW-RHS with cross section 40 mm × 50 mm and thickness of the wall t = 0.7 mm (validation of results). Plain carbon steel was selected from the standard material library as model material. When calculating these models, the specimen's own weight was not taken into account. In order to find solutions for reducing the RHS material consumption, it is important to have a general picture of the stress-strain state. Obtaining this data is possible by creating a calculated three-dimensional model. To determine the actual stress-strain state, FEM was used, which was implemented in the CAD/CAE system environment. A finite element grid was created by "Simulation" means. Obviously, lower stresses, deflection and twist displacement correspond to the strengthliest and stiffest section.

3. Results and discussion

3.1 Three-point bending

Specimen model was placed horizontally, bottom web is simply supported on supporting pins located at the distance of 1000 mm (Fig. 3). The finite element mesh of the specimen model is shown in Fig. 3(b). Since the external load acting on the specimen is transmitted through the loading pin, such element with a diameter of 11.8 mm was taken into account in the model (Fig. 3(c)). The most unfavorable loading options for RHS were considered, leading to maximum deflections at the same force, i.e. loading was acting on a web of 50 mm, see Fig. 3(d),(e). Thus, if DF-RHS was loaded, then the gap was on lateral web of 40 mm; if DFW-RHS was loaded, then the weld was at the bottom web. The applied load P(50) = 500 N was act as static in the midpoint between the support pins, at an equal distance of 500 mm (Fig. 3).

The simulation results are given in Fig. 4 – Fig. 6.

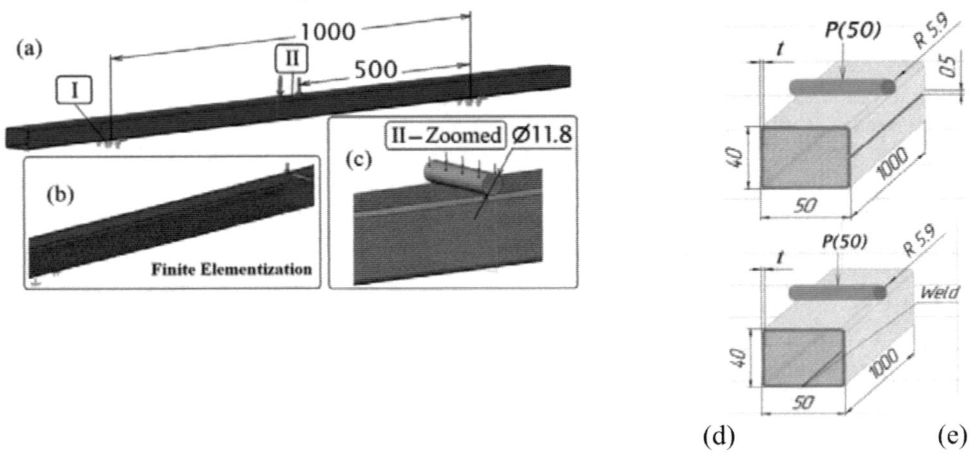

Figure 3. Solid-state model of RHS (a) with supporting (I) and loading (II) pins, illustration of left part finite elementization (b), load acting (c) on DF-RHS (d) and DFW-RHS (e) at modelling.

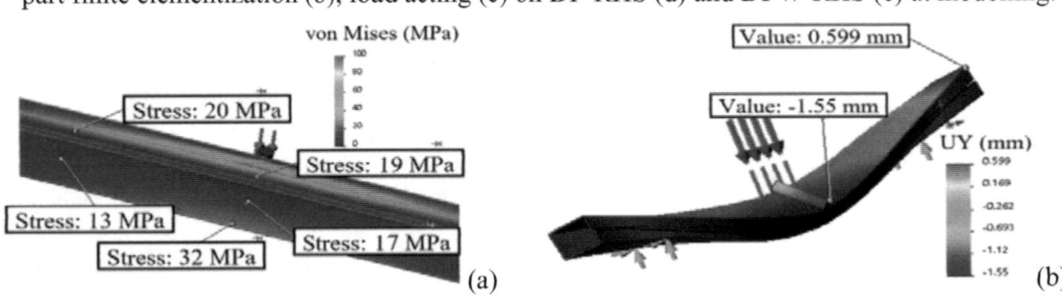

Figure 4. Stresses in the bottom web (a) and deflections (b) of 40*50 DF-RHS, t = 1.93 mm.

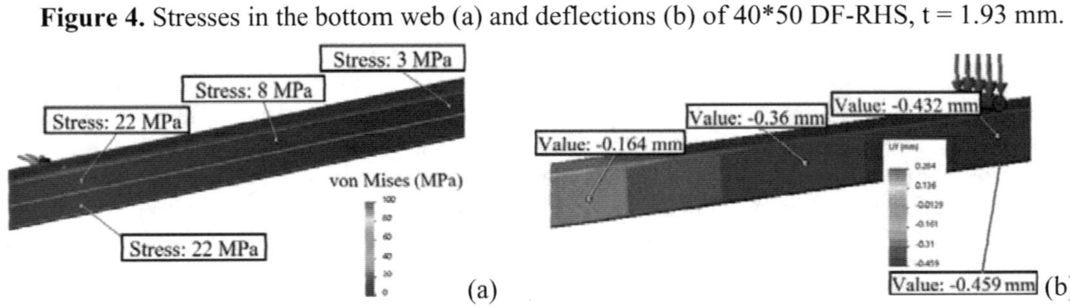

Figure 5. Stresses in the bottom web (a) and deflections (b) of 40*50 DFW-RHS, t = 1.84 mm.

The following was established, as a result of the analysis results of stress-strain state modelling for identical loading schemes and equivalent points of sections:

(i) DFW-RHS with t = 1.84 mm compared to DF-RHS with t = 1.93 mm has higher strength. Maximum values of equivalent stress in DFW-RHS with t = 1.84 mm were 22 MPa (Fig. 5(a)), and in DF-RHS with t = 1.93 mm − 32 MPa (Fig. 4(a));

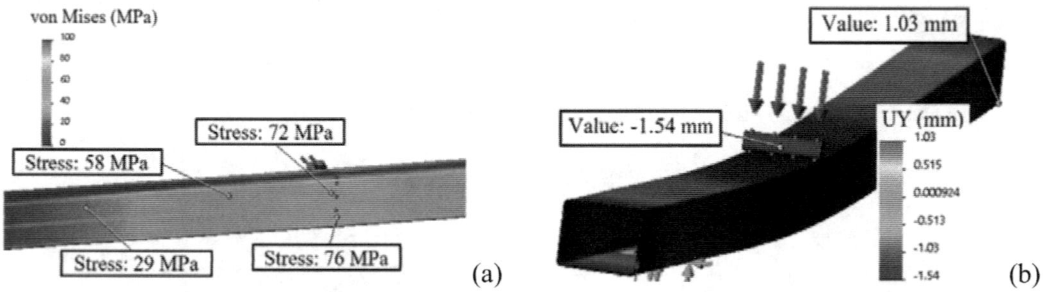

Figure 6. Stresses in the bottom web (a) and deflections (b) of 40*50 DFW-RHS, t = 0.7 mm.

(ii) DFW-RHS with t = 1.84 mm compared to DF-RHS with t = 1.93 mm has higher stiffness. Maximum deflection for DFW-RHS with t = 1.84 mm were *f* = 0.432 mm (Fig. 5(b)), and for DF-RHS with t = 1.93 mm – *f* = 1.55 mm (Fig. 4(b));

(iii) DFW-RHS with t = 0.7 mm compared to DF-RHS with t = 1.93 mm has a less strength, since maximum equivalent stress in DFW-RHS with t = 0.7 mm were 76 MPa (Fig. 6(a)), and in DF-RHS with t = 1.93 mm − 32 MPa (Fig. 4(a)). But in this case, the resulting stresses are still much smaller than the yield stress of a given carbon steel (y_s = 175–195 MPa);

(iiii) DFW-RHS with t = 0.7 mm compared to DF-RHS with t = 1.93 mm has slightly greater or equal stiffness, i.e. maximum deflection for DFW-RHS with t = 0.7 mm were *f* = 1.54 mm (Fig. 6(b)), and for DF-RHS with t = 1.93 mm – *f* = 1.55 mm (Fig. 4(b)).

3.2 Torsion

Solid-state models of specimens of considered RHS with a length of 1000 mm were used in the torsion test modeling. One end of specimen model was fixed at the simulation and torque of T = 50,000 N·mm was applied to the other end. The twist displacement value [37, 38] was estimate for each RHS design, i.e. measured which travel path at the farthest point from the axis of section rotation. Simulation results in the form of distribution of equivalent stresses and twisting displacements for different RHS design are shown in Fig. 7–9.

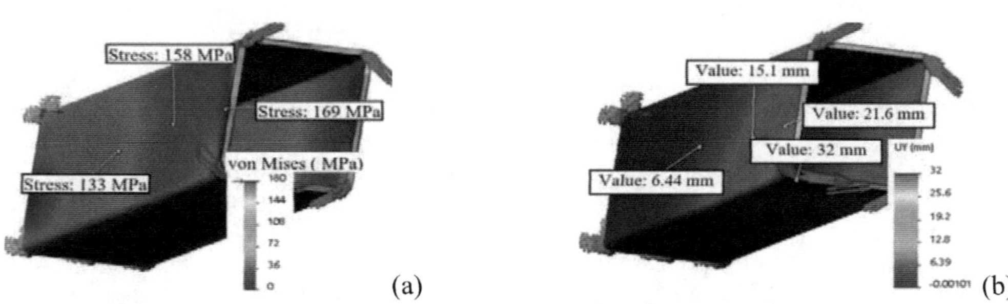

Figure 7. Stresses (a) and twist displacements (b) of 40*50 DF-RHS, t = 1.93 mm.

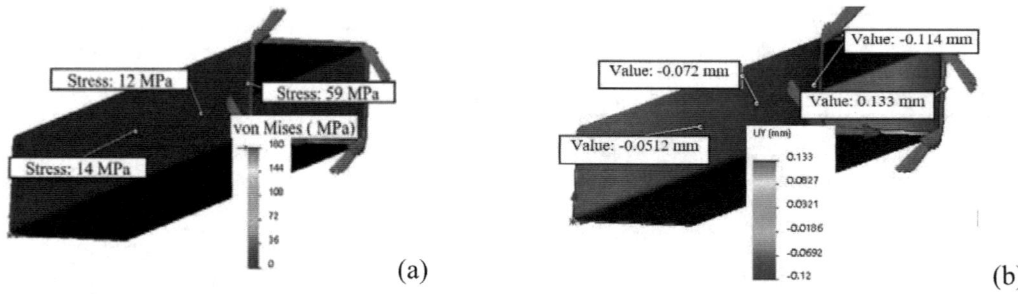

Figure 8. Stresses (a) and twist displacements (b) of 40*50 DFW-RHS, t = 1.84 mm.

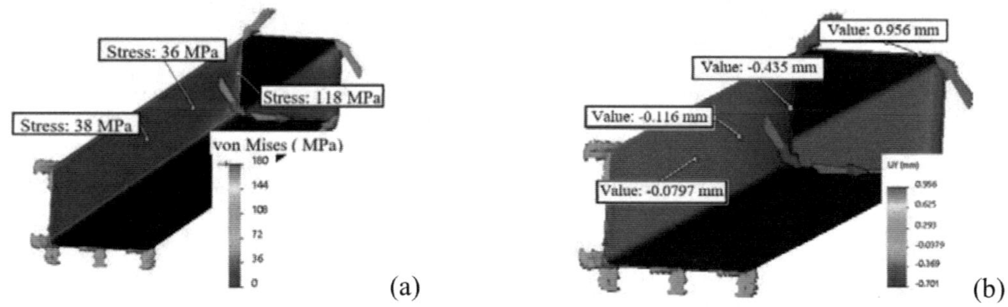

Figure 9. Stresses (a) and twisting displacements (b) of 40*50 DFW-RHS, t = 0.7 mm.

As a result of modeling the torsion test and analysis of the stress-strain state at the same torques, it should be concluded:

(i) DFW-RHS with t = 1.84 mm is more strength than with DF-RHS with t = 1.93 mm. The maximum equivalent stress in DFW-RHS with t = 1.84 mm were 59 MPa (Fig. 8(a)), and in DF-RHS with t = 1.93 mm – 169 MPa (Fig. 7(a)), respectively;

(ii) DFW-RHS with t = 1.84 mm is more stiff compared to DF-RHS with t = 1.93 mm. The maximum values of twist displacements for DFW-RHS with t = 1.84 mm were 0.133 mm (Fig. 8(b)), and for DF-RHS with t = 1.93 mm – 32 mm (Fig. 7(b)), respectively;

(iii) DFW-RHS with t = 0.7 mm is more strength than DF-RHS with t = 1.93 mm. The maximum values of equivalent stress in DFW-RHS with t = 0.7 mm were 118 MPa (Fig. 9(a)), and in DF-RHS with t = 1.93 mm – 169 MPa (Fig. 7(a)), respectively;

(iiii) DFW-RHS with t = 0.7 mm is more stiff compared to DF-RHS with t = 1.93 mm. The maximum twist displacement for DFW-RHS with t = 0.7 mm were 0.956 mm (Fig. 9(b)), and for DF-RHS with t = 1.93 mm – 32 mm (Fig. 7(b)), respectively.

4. Conclusions

Based on the FEM using, RHS with a cross section of 40 mm × 50 mm was investigated, that have a difference in manufacturing options by direct-forming (DF-RHS) with longitudinal gap and direct-forming with the next edge welding (DFW-RHS) into longitudinal weld seam. It was found that the existing initial differences affect to the results: (a) the gap at DF-RHS is located in the middle of the width of the smaller web (40 mm), and the weld seam at DFW-RHS is in the middle of the width of the larger web (50 mm); (b) the average wall thickness of DF-RHS is t = 1.93 mm, and DFW-RHS is t = 1.84 mm (i.e. the difference is 4.8%). Based on the proposed methodology for three-point bending test and torsion test, the distribution of equivalent stresses and displacements in RHS was analyzed. It is shown that the use of DFW-RHS as a reinforcing beam has advantages (especially in terms of stiffness) compared to DF-RHS. It was found that with identical loading schemes, DF-RHS with t = 1.93 mm has the same stiffness characteristics as a DFW-RHS with t = 0.7 mm, i.e. it becomes possible to reduce metal consumption by 62%. Strength of DFW-RHS with t = 0.7 mm are worse than DF-RHS with t = 1.93 mm due to a significant difference in wall thickness, however, the maximum bending stresses in DFW-RHS with t = 0.7 mm, equal to 77 MPa, i.e. do not exceed the yield strength of the section steel (y_s = 175–195 MPa). For the most unfavorable loading for DF-RHS to the wider web (50 mm), when the unwelded gap is on the lateral web, the forming a weld gives an improvement in stiffness at least 50%. That is, despite the smaller wall thickness, DFW-RHS is operationally stable under the most unfavorable load case. Thus, changing the manufacturing process from DF-RHS to DFW-RHS will improve stiffness and keep strength values at a sufficient level.

References

[1] Wardenier J, etc. 2010 *Hollow sections in structural applications* (Geneva: CIDECT).
[2] Huang Y and Young B 2013 *Thin-walled Structures* **73** 216-228
[3] Wang J, Afshan S, Gkantou M, Theofanous M, Baniotopoulos C and Gardner L 2016 *Journal of Constructional Steel Research* **121** 97-109
[4] Lubenets A N, Smirnov E N and Shcherbachev V V 2002 *Stal'* **8** 102-106
[5] Puzyr R, Kukhar V, Maslov A and Shchipkovskyi Y 2018 *International Journal of Engineering & Technology (UAE)* **7(4.3)** 30-34
[6] Kukhar V, Balalayeva E, Prysiazhnyi A, Vasylevskyi O and Marchenko I 2018 *MATEC Web of Conferences* **178** 02003
[7] Sikulskiy V, Kashcheyeva V, Romanenkov Y and Shapoval A 2017 *Eastern-European Journal of Enterprise Technologies* **4(1)(88)** 43-49
[8] Orlov G A 2015 *Metallurgist* **58(11–12)** 1001-1005
[9] Sun M and Packer J A 2014 *Journal of Constructional Steel Research* **92** 67-78
[10] Gardner L, Saari N and Wang F 2010 *Thin-Walled Structures* **48(7)** 495-507
[11] Forming Technology of Cold Rolling Square Rectangular Tube and Prospect on New Technology 4, http://www.rishbin.com/services/cold-rolling-square-rectangular-tube-technology.html
[12] Bock M, Arrayago I and Real E 2015 *Journal of constructional steel research* **109** 13-23
[13] Osadchii V A, Gorbatyuk S M, Filippov D I and Kuprienko N S 2019 *Metallurgist* **63** 658-664
[14] Podstawka A, Kovar M, Foglar M and Kristek V 2015 *Advanced Materials Research* **1106** 118-121
[15] Markov O E, Kukhar V V, Zlygoriev V N, Shapoval A A, Khvashchynskyi A S and Zhytnikov R U 2020 *FME Transactions* **48** 946-953
[16] Singh A K, Tyagi R, Ranjan V and Sathujoda P 2018 *Vibroengineering PROCEDIA* **21** 248-252
[17] Li Y-L, Li Y-Q and Shen Z-Y 2016 *Thin-Walled Structures* **107** 66-79
[18] Laksono P W, Rochman T, Setyanto H, Pujiyanto E and Diharjo K 2014 *Advanced Materials Research* **893** 504-507
[19] Chen J H and Luo W S 2017 *Micromachines* **8(7)** 206
[20] Kuchta K and Tylek I 2018 *MATEC Web of Conferences* **163** 07005
[21] Anishchenko O S, Kukhar V V, Grushko A V, Vishtak I V, Prysiazhnyi A H and Balalayeva E Yu 2019 *Materials Science Forum* **945** 531-537
[22] Larour P, Hackl B, Leomann F and Benedyk K 2012 *Proc. IDDRG 2012: Lightweight: Possibilities & Challenges* 203-215
[23] Devi S V, Singh T G and Singh K D 2019 *Journal of Constructional Steel Research* **162** 105730
[24] Tafsirojjaman T, Fawzia S, Thambiratnam D, Zhao X-L 2020 *Structures* **24** 610-626
[25] Amin A and Bentz E C 2018 *Structural Concrete* **19(3)** 684–694
[26] Efremenko V G, Zurnadzhi V I, Chabak Y G, Tsvetkova O V and Dzherenova A V 2017 *Material Science* **53** 67-75
[27] Vaganov V E, Aborkin A V, Alymov M I and Berbentsev V D 2015 *Russian Metallurgy (Metally)* **9** 732-738
[28] Tarelnyk V B, Gaponova O P, Konoplianchenko I V, Evtushenko N S and Herasymenko V A 2018 *Metallofiz. Noveishie Tekhnol.* **40(2)** 235-254
[29] Gunalan S and Mahendran M 2013 *Thin-Walled Structures* **73** 1-17
[30] Shats'kyi I P, Lyskanych O M and Kornuta V A 2016 *Materials Science* **48(3)** 469-472
[31] Sputo T and Tovar J 2005 *Thin-Walled Structures* **43(12)** 1852-1881
[32] Anishchenko A, Kukhar V, Artiukh V and Arkhipova O 2018 *MATEC Web of Conferences* **239** 06007
[33] Shatskyi I P, Makoviichuk M V and Shcherbii A B 2017 *Proc. 11th Int. Conf. on Shell Structures: Theory and Applications (SSTA 2017)* (October 11-13, 2017, Gdansk, Poland) **4** 165-168

[34] Markov O, Gerasimenko O, Aliieva L, Shapoval A and Kosilov M 2019 *Eastern-European Journal of Enterprise Technologies* **2(1)(98)** 39-46
[35] Dragobetskii V, Naumova E, Shapoval A, Shlyk S and Moloshtan D 2019 *Proc. Int. Conf. on Modern Electrical and Energy Systems (MEES 2019)* (Sept. 23-25, 2019, Kremenchuk, Ukraine) 506-509
[36] Castellucci M A, Pillinger I, Hartley P and Deeley G T 1997 *Thin-Walled Structures* **29(1-4)** 159-174
[37] Yoon K and Lee P-S 2012 *Structural Engineering and Mechanics* **43(4)** 411-437
[38] Wang A, Xue H, Bayraktar E, Yang Y, Saud S and Chen P 2019 *Metals* **10(1)** 31

A new and lean finite element model to predict the out of plane crash behaviour of aluminium honeycomb structures

Kevin Chacko[1], Mehrdad Asadi[1], Ahad Ramezanpour[1], Angelos P. Markopoulos[2]

[1] Faculty of Science and Technology, Anglia Ruskin University, UK
[2] School of Mechanical Engineering, National Technical University of Athens, Greece

[*]Corresponding author's e-mail: kevin.chacko@pgr.anglia.ac.uk

Abstract. Aluminium honeycombs are well-known anisotropic structures which are commonly used as energy absorber units. In the past decades, several researchers have focused on developing finite element models credible enough to predict the crash behaviour of honeycomb structures. A number of modelling methodologies have been developed and validated to closely represent the structural behaviour under various boundary conditions. Due to the limitations with modelling the structure accurately as well as the adhesive failure and trapped air, presents a serious challenge to predict the crushing behaviour of honeycomb structures. This paper introduces multiple simplified FE models and the results from simulations are compared with experimental data to evaluate the most effective model. The results have also been compared with our existing article which used LSDYNA for numerical analysis. The techniques proposed in this paper show good correlation with the mentioned experimental data and the LSDYNA model. Thus, the simplified FE models are regarded as validated and credible enough to predict out-of-plane crash behaviour of aluminium honeycomb. Altair's HyperMesh and HyperCrash were used to create the FE models and the dynamic explicit code RADIOSS to solve the models.

1. Introduction

Aluminium honeycomb structures are used in a wide range of engineering applications such as rail, marine, automotive and aircraft industries due to its ability to absorb high impact energy and the highly desired strength to weight ratio. To ensure good prediction of their crushing behaviour, it is necessary to validate the FE models in order to accurately predict the physical and numerical results when the structure is subjected to variety of boundary conditions. However, due to the limitations with modelling the structure accurately as well as the adhesive failure and trapped air, presents a serious challenge to predict the crushing behaviour of honeycomb structures. Hence, current focus is to develop simplified FE models for honeycomb structure and assess its credibility to represent the crush strength.

2. Literature

2.1. Review

A study into the impact behaviour of honeycomb structures with various cell specifications [1, 2] investigated the effect of internal angles and foil thickness on the crash behaviour of aluminium honeycomb structures using the explicit FEA code DYNA3D. Impact tests using a drop hammer with a velocity of $10 m/s^{-1}$ and the corresponding quasi-static tests were also performed. The experiment showed that the compressive stress increases under impact due to the air enclosed in the honeycomb cells. For the numerical analysis a simple "Y" cross-section column (double flange) was used due to geometric symmetry. Branch angles of the column were varied from 30° to 180°. The numerical work from their investigation shows the cyclic buckling mode occurs in every case and crash strength increases with smaller branch angle as well as increased foil thickness. The numerical results were reported to show good correlation with experimental data.

A simplified model was developed by researchers [3-6] who analysed the crash behaviour of a simple "Y" cross-section column (single flange) and compared the numerical results with experimental data. A total of fifteen samples of honeycomb cores produced by Cellbond were used in the experimental tests. The effect of adhesives, trapped air and curves at the cell corners were not considered in the model. In spite of the initial buckling phenomenon that resulted in a large instantaneous peak, the numerical results from this work shows close representation of the energy absorption, deformation behaviour as well as compressive strength.

A study on the yielding function of aluminium honeycomb [7] focused on developing a more accurate yielding function of aluminium honeycomb block. Experimental tests were done to evaluate the behaviour of honeycomb block under compressive load. Observations from the work show that the yielding stress of aluminium honeycomb highly depends on the direction of compression. Using the test data a new yielding function was proposed as a function of volumetric change and angle of compression. This newly developed yielding function was then introduced to MAT126 in LSDYNA. Frontal crash tests were performed with the yielding function and showed better correlation with the test results in comparison with MAT126 without the yielding function.

An investigation was carried out focusing on the mechanical behaviour of aluminium hexagonal honeycombs subjected to out-of-plane dynamic indentation and compression loads [8, 9]. The finite element models have been verified using experimental data in-terms of deformation pattern, stress-strain curve and energy absorption. The model reveals that for velocities of $5m/s^{-1}$ under indentation, plastic buckling of the honeycomb cell walls from the end that is adjacent to the indenter, while under compression the buckling of honeycomb cell walls occurs from both ends of the honeycomb.

Analytical study conducted by a group of researchers [10, 11] focused on developing a beam element based aluminium honeycomb FE model with lateral and frontal crash load. The intention behind the work was to achieve realistic deformation behaviour of the anisotropic honeycomb structure. The models generated with beam elements showed reasonably good results with that of physical testing during the validation process. Initially the model was validated with cladding sheets; the corresponding force displacement curve seemed to be representative of the curves produced during the physical testing phase. The study concluded stating that special attention must be given to local phenomena such as sheet metal failure, glue failure and modelling the air trapped in the hexagonal pockets.

To improve the accuracy of the results with a shell element-based honeycomb barrier, a group of researchers carried out an investigation [12-14] and compared the results with a solid element model. Upon solving both shell and solid element models, the results were compared to the physical test. To approximate the real crash pattern, ends of the honeycomb core were bent locally to simulate the pre-crashed surface of the samples prior to solving the model. Modelling of the adhesive with the cladding sheet was also considered in the model. With regard to the bonding between aluminium honeycomb and cladding sheet, beam elements were used in order to simulate the rupturing of the adhesive. The study reports that the shell element model is capable of predicting the physical result more accurately than solid element model however the disadvantage is the higher computational cost.

Another investigation [15-18] examined the influence of cell wall thickness, node length, cell size and loading conditions. The crush patterns of the honeycomb structure have been studied in both out-of-plane and in-plane crash modes. Aspects such as geometrical core configuration and geometric property like cell size, cell wall thickness and node length have a significant importance in administering the potential of honeycomb according to this research. The study also reports that the cladding sheet material and thickness majorly affects the energy absorption credibility of the core during impulsive impacts.

IOP Conf. Series: Materials Science and Engineering 1037 (2021) 012026 doi:10.1088/1757-899X/1037/1/012026

Wide range of analytical and experimental work has been performed by researchers in-order to characterise the anisotropic behaviour of honeycomb structures. However, further study is essential to characterise the crash behaviour of honeycomb structures more accurately whilst not compromising computational cost.

3. Experimental Test

Data and results from our previous experimental tests and existing paper have been used to validate the proposed simplified FE models. The aluminium honeycomb samples used in the previous experiment was produced by Cellbond. The samples used are regarded as 1.8-3/4-3003 with a foil thickness of 0.06mm. The sample cores used were 250mm x 250mm x150mm in length, width and height respectively with a wall to wall thickness of 19mm.

Table 1: Specification of Aluminium A3003

Mass Density (ρ)	2100kg/m^3
Young's Modulus (E)	70GPa
Poisson's Ratio (υ)	0.33
Yield Stress (σ_{ys})	190MPa
Ultimate Tensile Strength	800MPa

Figure 1: Experimental Test Set up – Honeycomb Out-of-Plane Crush Test

A drop trolley of mass 72kg was used to crush the honeycomb samples with an impact velocity of 5.24m/s^{-1}.

Aluminium honeycombs are hexagonal structures made of thin aluminium foils and due to the anisotropic behaviour, honeycomb structures are approximately ten times stronger in the principle "T" direction in comparison to "L" and "W" directions.

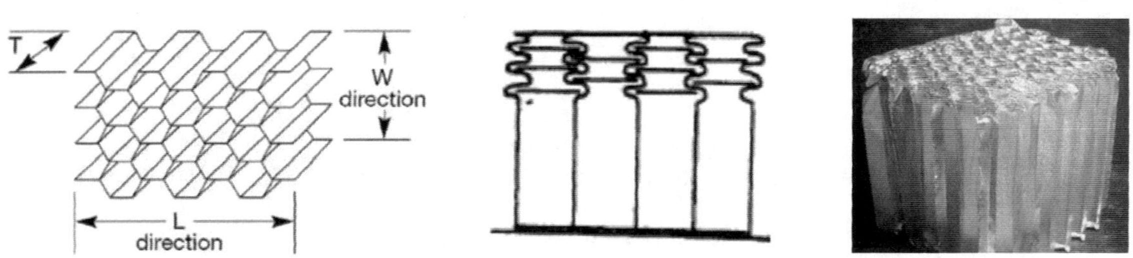

Figure 2: Deformation behaviour of Honeycomb Structures

When loaded in the T direction, the structure shows a particular buckling behaviour. Initially the structure buckles elastically and when the impact load reaches a critical point, the core buckles for the first time and is indicated by a distinct peak. Progressing past this first buckling point, the core then continuously buckles under constant load. It should be noted that the buckling behaviour of honeycomb structure is initiated from the top and progresses down the structure in "T" direction. During test, the impact force was measured and the compressive stress profile was calculated using the equation below,

$$\sigma = \frac{F}{A} \quad (1)$$

Where, σ is the compressive stress, F is the impact force and A is the out-of-plane cross-sectional area of the honeycomb samples.

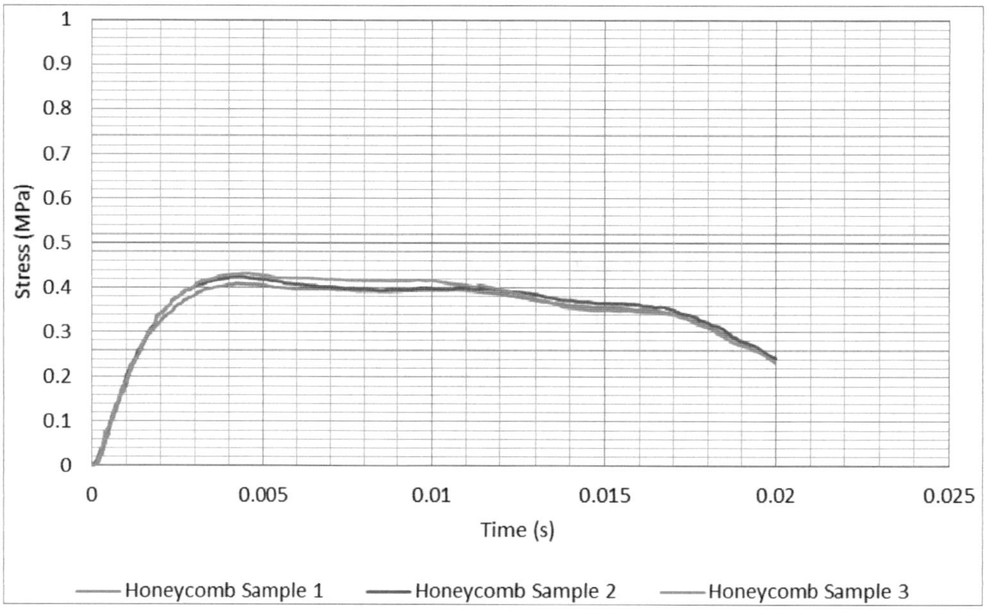

Graph 1: Compressive stress as a function of Time – Honeycomb Samples

A total of fifteen crash tests were done, however, results of three samples are presented in this paper. The corresponding stress vs time curves for the three samples are shown above. The measured curves from the tests gives an average compressive stress which is defined in σ=F/A. The full cross-sectional area (A) of the honeycomb samples are 62500mm^2. A sharp peak appeared on all graphs immediately after the initial trolley contact, thereafter a smooth curve which gives an average crash strength of approximately 0.37MPa. In all samples, the deformation process was initiated at the top side progressing down the principle "T" direction. The curves shown in Graph 1 are filtered for visual clarity. Figure 2 shows one of the crushed honeycomb sample.

4. Finite Element Model

4.1. Simplified FE Model development

As shown in Figure 3, due to the geometrical symmetry within the structure, a "Y" cross-section was selected for this research. In doing so, the effects of trapped air and the marginal curvatures in cell corners are eliminated and thus the FE models simplified significantly. The three FE models based on the "Y" cross-section columns will be referred to from here on as "Unit Cell (UC)". All three UC models are broken down into three surfaces and a localised co-ordinate system assigned to each surfaces.

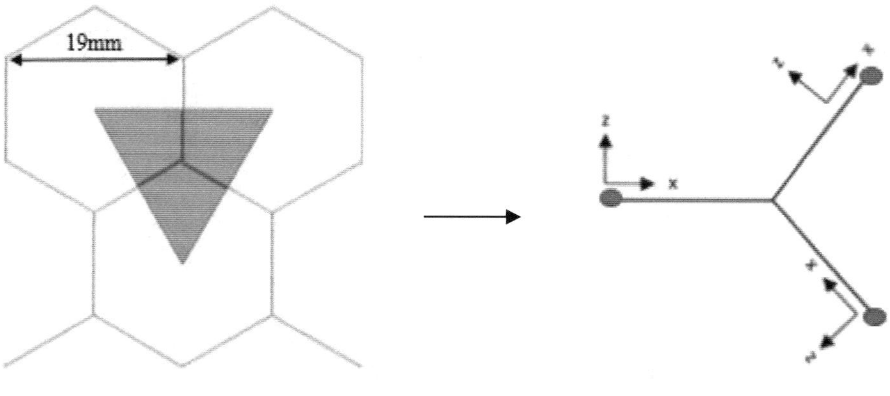

Figure 3: Geometric pattern in Honeycomb Structures

Constraints were applied to the nodes (labelled as green dots in Figure 3) of the three surfaces of the UC model. These constraints were applied to ensure that the UC behaves as if it was part of the overall honeycomb structure during an out of plane crash scenario. The applied constraints are described in Figure 10

4.2. Unit Cell FE Model

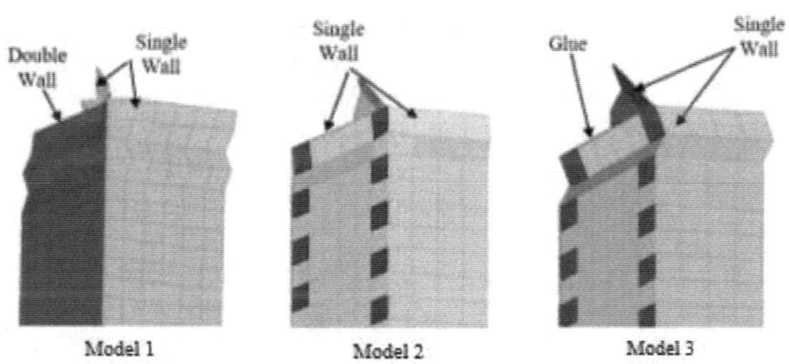

Figure 4: The proposed variants of the UC Model

The three FE models presented in this paper was developed using HyperMesh and HyperCrash and the models were solved using the explicit code RADIOSS.

4.2.1. Model 1: A "Simplified Single Flange Model" consists of a "Y" cross-sectioned column with two walls representing the single flanges and one wall is modelled as double sided to represent two foils joined by adhesive. This is achieved by controlling the thickness of the shell property.

4.2.2. Model 2: A "Simplified Double Flange Model" consists of two halves of a "Y" cross-sectioned column. All walls in this model are represented as single sided and a type 2 contact is utilised to define the adhesive connection between the two flanges. With the HyperCrash Type 2 tied interface contact, it is possible to rigidly connect a set of slave nodes to a master surface. Nodes belonging to the two neighbor walls are defined as slave and master with no priority. Sliding of the slave nodes is restricted and voids are not present in between contact either.

4.2.3. Model 3: A "Double Flange Model with Adhesive" consists of a "Y" cross-sectioned column with the walls modelled same as the Model 2. MAT LAW59 is applied between the neighbour walls to represent adhesive. MAT LAW59 is typically used to model the mechanisms of connection material in spot-weld, welding line or adhesive layers in laminated material. Normal tangential stress functions were defined to characterise the elastic and plastic behaviour in the normal and shear directions of the adhesive layer.

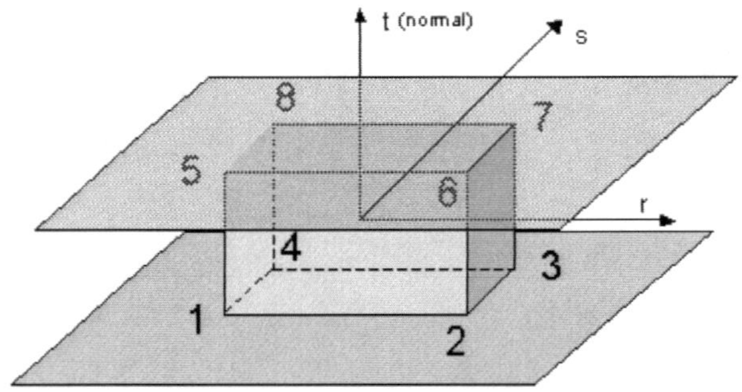

Figure 5: Illustration of adhesive modelling using Solid connection element (HyperMesh Theory Manual, 2018)

MAT LAW59 is only applicable to solid hexahedron elements and the material time-step does not depend on element height, therefore, the adhesive layer was modelled with a uniform thickness of 0.05mm solid 8-noded elements.

4.3. Description of Material Card
Walls in all three models are assigned the ELASTO-PLASTIC JOHNSON-COOK Plasticity Model LAW2. The strain hardening behaviour of materials is a major factor in material characterization and a robust definition of this parameter is imperative with large plastic strains. For a number of plasticity problems, the hardening behaviour of the material is simply defined by the stress-strain curve. Although the stress-strain curve is adequate for the linear loading phase, a more detailed definition is necessary to characterize the material's behaviour in its plastic deformation phase. Most isotropic elastic-plastic material laws in RADIOSS use Von-Mises yield criteria.

In JOHNSON-COOK LAW2, the material behaves as linear elastic when the stress is lower than the yield stress. As for higher values of stress experienced, the material behaviour becomes plastic. This law is valid for brick, shell, truss and beam elements. The relation between describing stress during plastic deformation is defined by the formulae,

$$\sigma = (\alpha + b\varepsilon_p{}^n)(1 + c\ln \frac{\dot{\varepsilon}}{\dot{\varepsilon}_0})(1 - T^m) \quad (2)$$

where,

σ = Flow Stress (Elastic + Plastic Components)
ε_p = Plastic Strain (True Strain)
α = Yield Stress
b = Hardening Modulus
n = Hardening Exponent
c = Strain Rate Coefficient
$\dot{\varepsilon}$ = Strain Rate
$\dot{\varepsilon}_0$ = Reference Strain Rate
m = Temperature Exponent
$T = \frac{T-298}{T_{melt}-298}$
T_{melt} = is the melting temperature in Kelvin degrees. The adiabatic conditions are assumed for temperature computation
$T = T_f + \frac{E_{int}}{\rho C_p\,(Volume)}$

Where:
ρC_p = the specific heat per unit of volume
T_f = initial temperature (in degrees Kelvin)

4.4. Unit Cell Crush Analysis – Observations

Proposed FE model are validated by taking to consideration the crash pattern and the structure's compressive strength/stiffness.

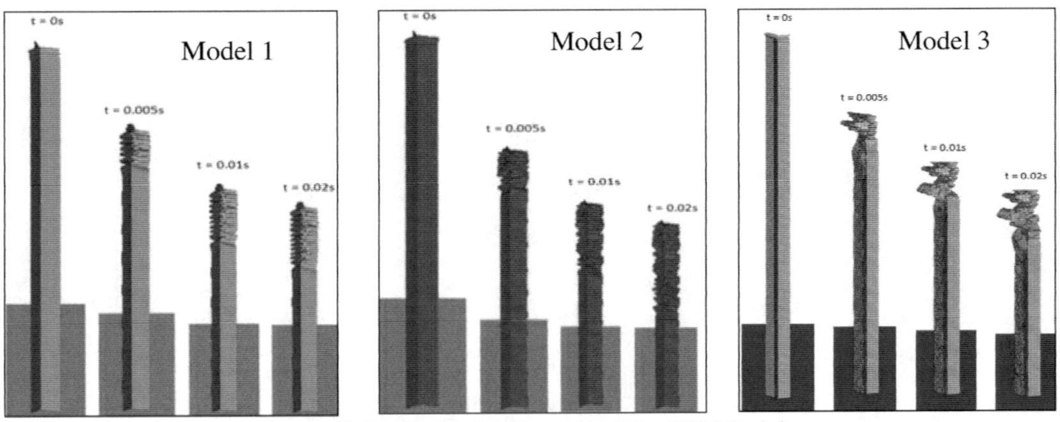

Figure 6: Deformation Pattern of the UC Models

Figure 6 shows the UC models being simulated through the experimental test scenario and the predicted deformation behaviour when crushed with the rigid mass (impacting at $5ms^{-1}$). Identical deformation pattern between experimental test and simulation was observed: when the "Unit Cells" were impacted in the out-of-plane (T) direction, deformation of the structure was initiated from the top and propagated down the axis of compression which closely represents the deformation pattern of the crushed honeycomb samples.

Identical deformation patterns are found between the three proposed FE models as well. The "Unit Cell" models of the single and double flange deformed with regular crash pattern whereas the double flange model with adhesive deformed in a more irregular manner. Nevertheless, the realistic deformation pattern of the crushed honeycomb samples are closely predicted by all three models.

Model 1 Model 2 Model 3

Figure 7 Deformation pattern of the Unit Cell Models

The average compressive strength of the three samples and FE models are also compared in Graph 2.

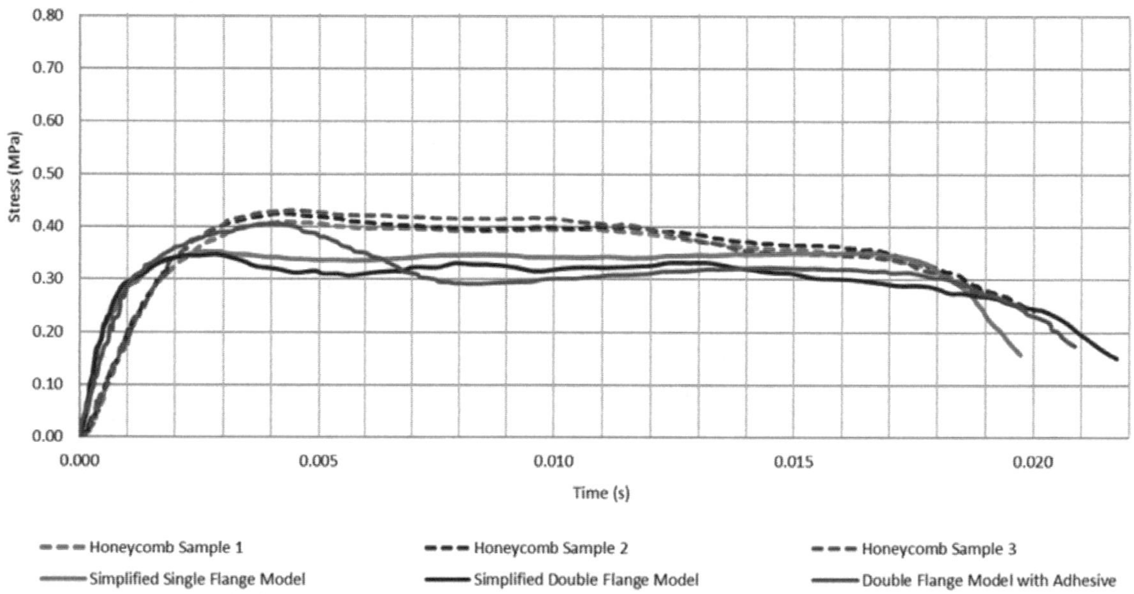

Graph 2: Compressive Strength Samples vs Unit Cell FFE Models

Generally, similar trends in stress curves were observed between the experimental data and FE simulations. The typical distinctive peak is not visible in the graph due to filtering of the data. It is suspected that modifying the coefficient of friction value between honeycomb and the baseplate could further represent the stiffness more accurately.

Comparing the average crash strength of all three FE models shows good correlation with test results. The Double Flange Model with Adhesive is more accurately predicting the real crash strength.

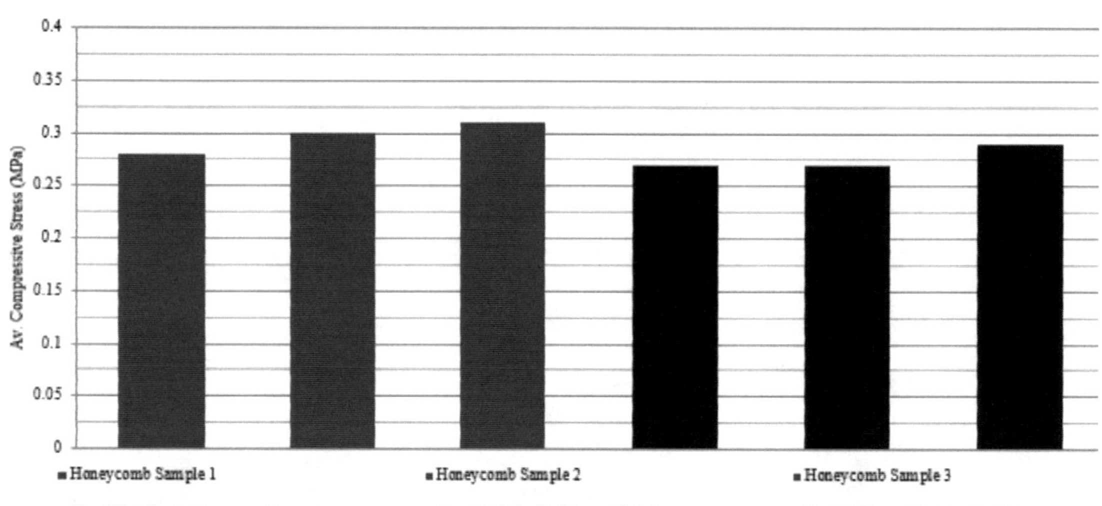

Graph 3: Average compressive stress

The Simplified Single and Double Flange models seem less stiff under compression. Nevertheless, the proposed FE models show good correlation with test results.

Comparison of computational time,

Variants of Unit Cell	Solve Time
Simplified Single Flange	10 minutes
Simplified Double Flange	19 minutes
Simplified Double Flange with Adhesive	21 minutes

Table 2: Comparison of solve time

Table 2 shows a breakdown of the solve time required by the three simplified UC models. With a clear indication in the reduced solve time, the computational cost is proved to be significantly lower for the "Simplified Single Flange" model.

4.5. Generalised Valid Discrete Model – Development of a New Modelling Approach for Aluminium Honeycomb Structures

In the current (conventional) FE models of honeycomb, the structure is modelled as a continuum shown in Figure *8* of layers of foils bonded together. After simulating the conventional model, it was observed that the compressive stress/stiffness is representative of the test data, however the deformation pattern was not predicted accurately. Deformation occurred at the bottom of the structure and an irregular deformation pattern was observed at the top of the structure.

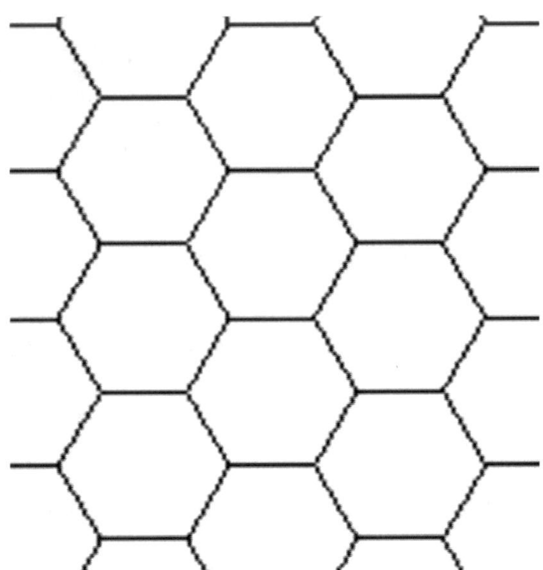

Figure 8: Conventional FE Modelling Approach

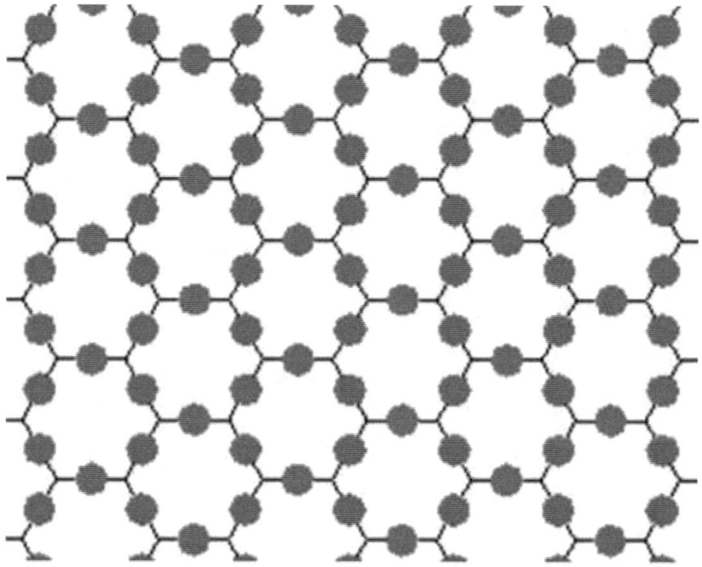

Figure 9: The New Modelling Approach with localised co-ordinate system and constraints

Based on the approach taken and findings from the Unit Cell models, a new modelling approach was identified. This method differs by two ways in comparison to the conventional method. The new approach is such that the validated "Unit Cell" model (with the local co-ordinate systems and the localised nodal constraints) is patterned in the "in-plane" directions to form the FE honeycomb model as shown in Figure 9.

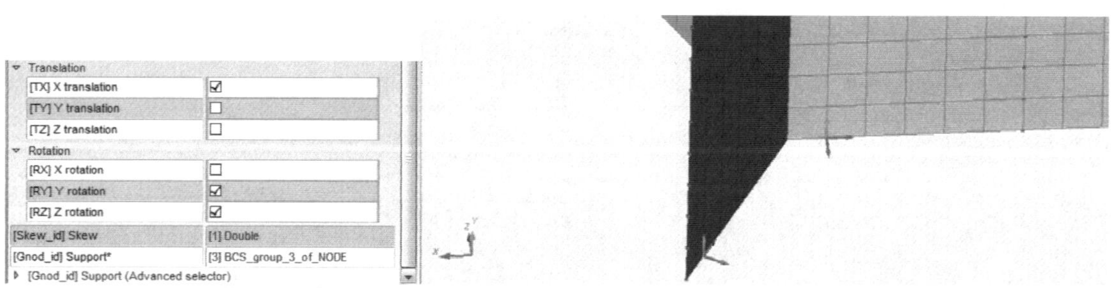

Figure 10: Generalised co-ordinate system and Nodal constraints

The full scale honeycomb structure (representative of the samples used in testing) was generated by generalising the "Unit Cell" model in both the in-plane directions. The model was then modified into three series of parallel elements and each series were assigned a local co-ordinate system. Constraints were applied to the edge nodes as shown in Figure 10.

The model was then simulated in-order to compare the deformation pattern and crush strength/stiffness to test data.

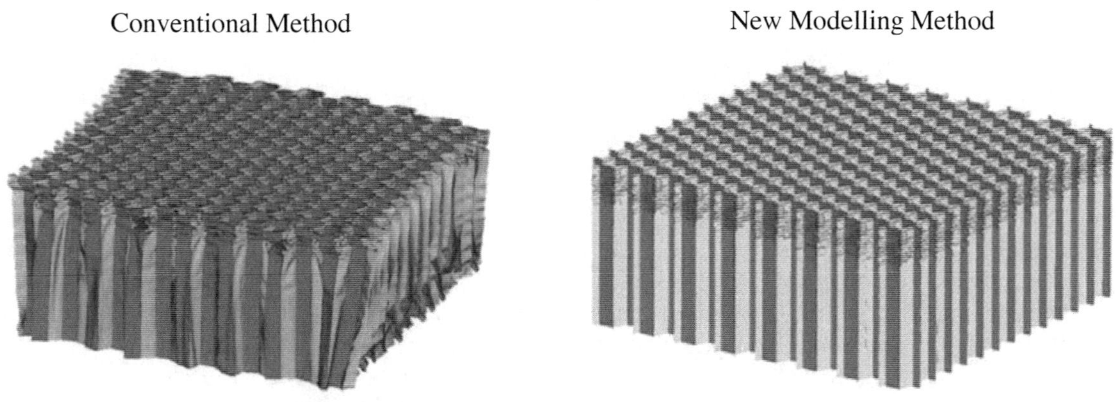

Figure 11: Comparison of the deformation behaviour

Figure 11 above shows a direct comparison of the deformation pattern between the current (conventional shell method) and the new FE method. From observations, it can be seen that the new method captures the realistic deformation pattern much more accurately and is more representative of the typical honeycomb deformation behaviour when subjected to impact loading. The new method shows a consistent regular deformation pattern and the deformation process is initiated from the top of the structure.

Graph 4 shows the compressive strength of the FE models as a function of time during the crash scenario and compares against test data. The distinctive peak anticipated at the initial stage is not visible in the graphs below due to filtering of the data.

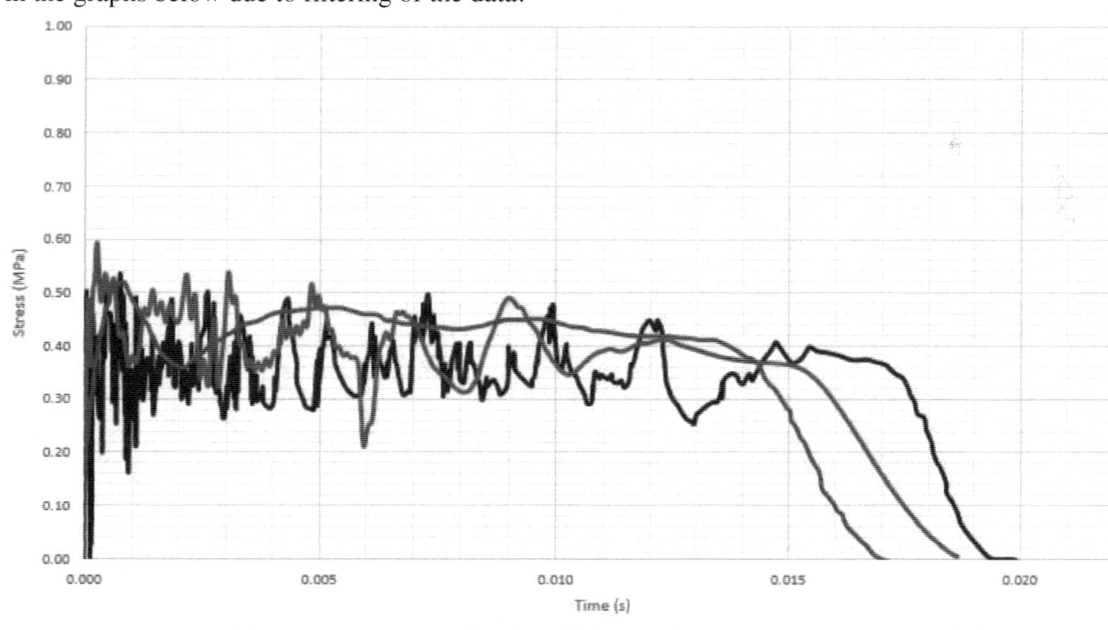

Graph 4: Compressive Stress curve

Although the current FE model is capable of representing the stiffness characteristics reasonably well, the model still failed to capture the distinctive peak in stiffness right at the beginning of the crushing process.

However, the new method distinctively captured the sharp rise in stiffness at the very beginning of the crushing process and represents the stiffness characteristics quite accurately. It is anticipated that, by modifying the coefficient of friction values between the base plate and the structure could potentially improve the accuracy of the stiffness curves. Nevertheless the average crush strength of both FE models are around 0.37MPa and is comparable with test data.

5. Conclusion

Multiple simplified finite element models have been developed using the pre-processor HyperCrash. The intention was to evaluate the credibility of simplified models to represent the mechanical deformation behaviour of honeycomb structures under out-of-plane impact. The proposed FE models have been validated using previous experimental results in terms of crush pattern and compressive strength. Good correlation have been established between the FEA and experimental results.

The research showed that the simplified models are reasonably credible in predicting the crush behaviour of honeycomb structures. The “Simplified Single Flange” model was the most favourable due to the reduced analysis time, deformation procedure and pattern. Taking to consideration the approach taken with the “Unit Cell” model, the full honeycomb block was generalised to simulate the crash test scenario. Existing method and the new method has been compared against test data and the new model demonstrates the improved accuracy in predicting the deformation pattern and good representation of the stiffness characteristics.

References

[1] Yamashita, M.; Gotoh, M.; Impact Behaviour of honeycomb structures with various cell specifications – numerical simulation and experiment. Int. J. Impact Engng, Avaliable online 25 November 2004

[2] Zhao, H and Gray, G, 1998. Crusing behaviour of aluminium honeycomb under impact. International journal of impact engineering, pages: 827-836

[3] Aaron, V, Nagarajan, H, Adibi-sideh, A, 2003. Determination of strength characteristics of aluminium honeycomb material subjected to out of plane compression using finite element analysis, Advances in structures, pages: 427-432

[4] Alavi Nia, A; Sadeghi, M.Z. An experimental investigation on the effect of strain rate on the behaviour of bare and foam filled aluminium honeycombs. Mater. Des. 2013, 52, 748-756

[5] Asadi, M.; Shirvani, H.; A simplified model to simulate crash behaviour of honeycomb, Available online 15 January 2019

[6] Asadi, M, 2009. An Investigation to compare the application of shell and solid element honeycomb in ODB. Conference paper, 7th European LS Dyna conference paper, pages: 3 – 7

[7] Kojima, S.; Yasuki, T.; Mikutsu, S.; Takatsudu, T. A study on yielding function of Aluminium honeycomb. Available online 20 January 2019

[8] Asadi, M, Walker, B, Shirvani, H, 2006. Advanced Finite Element Model for AE-MDB side Impact barrier. Conference Paper, 6th European LS Dyna user’s conference paper, pages: 2-9

[9] Ayman, A.S.M, A.; Ruan, D.; Lu, G; Bhuiyan, A. Finite Element Analysis of Aluminium Honeycombs Subjected to Dynamic Indentation and Compression Loads. Mater. Des 2015, 74, 138-149.

[10] Jost, T, 2008. A new method to model Aluminium honeycomb based crash barriers in lateral and frontal crash load cases, LS Dyna Anwenderforum, DYNAmore GmbH 2008, Whole Text

[11] Jost, T.; Heubrandtner, T.; Ruff, C.; Fellner, B. A new method to model Aluminium honeycomb based crash barriers in lateral and frontal crash load cases. Available Online 13 March 2019.

[12] Kojima, S.; Yasuki, T.; Oono, K. Application of shell honeycomb model to IIHS MDB model, 6th European LS-DYNA users conference, Gothenburg, 2007, Session 1.3.1, p. 1.71-
1.80

[13] Lee, Y, Chun, M, 1996. An Experimental study on the behaviour of aluminium honeycomb sandwich panels, Transactions of the society of naval architects of Korea, pages:106-123

[14] Paik, J, Thayamballi, A and Kim, G, 1999. The strength characteristics of Aluminium honeycomb sandwich panels , Thin walled structures, pages: 205-231
[15] Takahira, Y, Kojima, S, 2007. Application of Shell Honeycomb Model to IIHS MDB Model, 6th European LS Dyna User's Conference, pages: 2-8
[16] Thomas, T.; Tiwari, Gourav. Crushing behaviour of honeycomb structure: a review. Available online 20 March 2019
[17] Tryland, T, 2005. Hydro Automotive Structures. Alternative model of the Offset Deformable Barrier, 1st, pages: 2-6
[18] Yang, H 2010. Finite Element Modelling and validation of the honeycombs for automobile crash MDB and ODB", 9th International LS-Dyna users conference, Whole Text

Validating dynamic crush response of unidirectional carbon fibre tube via finite element analysis method using LS-DYNA

R Samuel[1], M Asadi[1], A Tarda[2] and G Simbotin[2], A P Markopoulos[3]

[1] Faculty of Science and Technology, Anglia Ruskin University, UK
[2] Tecosim Ltd., UK
[3] School of Mechanical Engineering, National Technical University of Athens, Greece

*Corresponding author's e-mail: rs939@pgr.aru.ac.uk

Abstract. Crashworthiness of composite structures is of significant interest to manufacturers and operators as composites become more commonly used in automobiles and aircraft. Experimental crush testing remains an important tool in the design of subfloor fuselage structures for crashworthiness, as numerical simulations are still largely unable to accurately predict the crushing response of complex composite structures. In this paper, dynamic crush testing of carbon-fibre/epoxy specimens experimental result is validated using the modelling and simulation software LS-DYNA to understand impact behaviours under dynamic loading for finite element simulations. The tubular finite element model was modelled using MAT54 and was impacted by a 65KG rigid wall impactor with an impact velocity of 55 km/h. Material card parameters were evaluated to understand crush behaviour of the FE model. Upon analysis the experimental results and the FEA results showed a good correlation. Modelling criteria used for this model can be further used to simulate crush response of unidirectional carbon tubes under dynamic impact.

1. Introduction

A well-designed carbon fibre part has better energy absorption capability compared to metallic energy absorbing structures [1, 2]. Hence, in various sectors of the automotive industry Carbon – Fibre composites are being used as crush energy absorbers for improvised impact energy absorption during collision [3-6].

Under impact, progressive crushing of axially loaded metallic tube structures occurs via plastic buckling which is referred to as progressive folding [7]. Whereas in case of carbon fibre structures it occurs via brittle fracture and is referred to as progressive crushing if, the crush is guided [8]. Guided crushing of carbon fibre structures does not only depend on material structure but can also be accomplished by changing component geometry as geometrical changes can affect the failure mechanism thus causing change in energy absorption levels [9]. Since the energy absorption mechanism varies based on the part's geometry and features, its design methodology is also distinctively different. Cross sectional features and stacking sequence of the carbon fibre highly affects the energy absorption capability and fracture behaviour of the carbon fibre part [10]. Unlike simulation of metal structures, simulation of composite structures is deemed to be more complex as the model's parameters needs to be examined and studied to assure that the model replicates real life test behaviour with the right parameters. Nevertheless, the part's crush and mutual interaction behaviour can be predicted through numerical finite element simulation software's such as LSDYNA, ABAQUS Explicit, PAM-CRASH and RADIOSS which uses explicit integration formulation for generating results. However, ample simulations are required to find the apt criteria for modelling.

Many studies have investigated on these parameters for modelling and validating real life test behaviour of carbon fibre composites. However, only limited research data is available on validation of dynamic guided progressive crushing of carbon fibre composite using a double sided plug.

In this paper, explicit finite element code LS-DYNA has been utilised for comparing and evaluating performance of guided crushing of a carbon fibre tube with double sided plug. It aims to examine and validate the ability of the LS-DYNA software to depict progressive crush absorption response of the carbon fibre tube. The LS-DYNA software is widely used to perform explicit dynamic post-failure simulations [11]. Several studies of carbon fibre composite impact modelling has been performed using the pre-existing material models such as MAT54 in LS-DYNA as it closely depicted the characteristics of carbon fibre [12] Hence, the material card used for validation the model for this research is the *MAT_ENHANCED_COMPOSITE_DAMAGE (MAT054). The study investigates on how material card parameters affects the crush performance behaviour or carbon fibre. Depending on analysis results, the study proposes suitable simulation criteria that could be used to model unidirectional carbon fibre tubes under dynamic loading. The model can be used as a base model to select apt parameters for achieving optimum dynamic crush response behaviour. The main focus of the study is achievement of progressive crush behaviour through unidirectional fiber delamination. For the study, parameters such as material geometry, impactor speed and impactor mass are defined however, the models design parameters involving material card and boundary conditions are evaluated for apt impact response.

The final simulation results are compared with the dynamic crush test experimental results obtained from Ueda et al. [13] to validate if the simulation model depicts same behaviour as the experimental model.

Design dimensions and test results for the model constructed in this paper is extracted from an existing journal article published [13] where they used a double-sided plug for controlled and improvised energy absorption. The two-piece double-sided plug constrains the outer and inner walls of the carbon fibre tube which under crush conditions allows the damaged fibre to eject through the built cavity.

2. Experiments and results

Experimental test results for validation were obtained from an existing published [13]. The test specimen was a circular CFRP tube fabricated from a PYROFIL TR30S/#380, Mitsubishi Rayon prepreg sheet. The prepeg sheet was wrapped onto a 50 mm inner diameter aluminium mandrel which was then covered using a heat resistive film. Twelve prepeg sheets were stacked onto the aluminium mandrel and the stacking sequence was unidirectional with about 60% fibre volume fraction. Under vacuum conditions, the prepeg sheet was cured for 80 minutes at 130°C. Once cured the aluminium mandrel was removed and the consolidated carbon fibre tube which had a thickness of 2.9 mm and was cut into 80 mm length sections.

To increase energy absorption of the carbon fibre part, an outward-splaying and inward-folding double sided plug was used. The double sided plug which was attached to the base of the CFRP tube ejected the fractured part through its gaps preventing blooming failure and moderating peak force which is generally observed at the beginning of fracture.

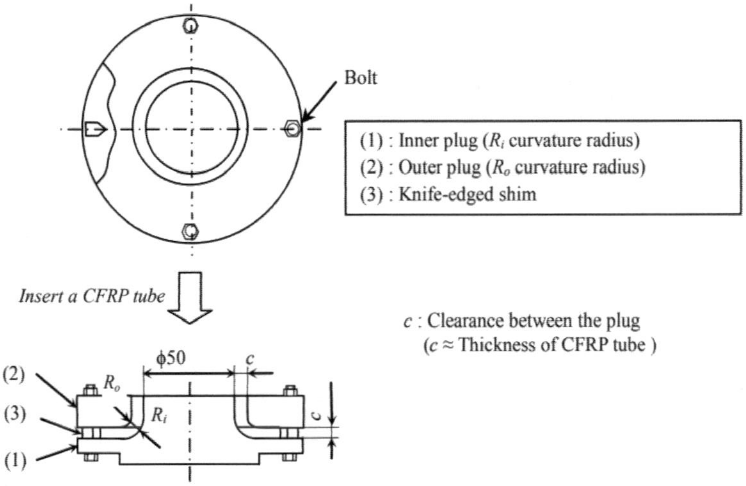

Figure 1. Assembly drawing of the double-sided plug for a circular CFRP tube [13]

The experimental test data considered for this report is the dynamic test results at 55 km/h. The double-sided plug used for this test has an inner radius curvature of 8 mm and outer radius curvature of 5 mm. Experimental test result data is shown in figure 2. From the experimental data it was observed that the average crush forces was about 10kN.

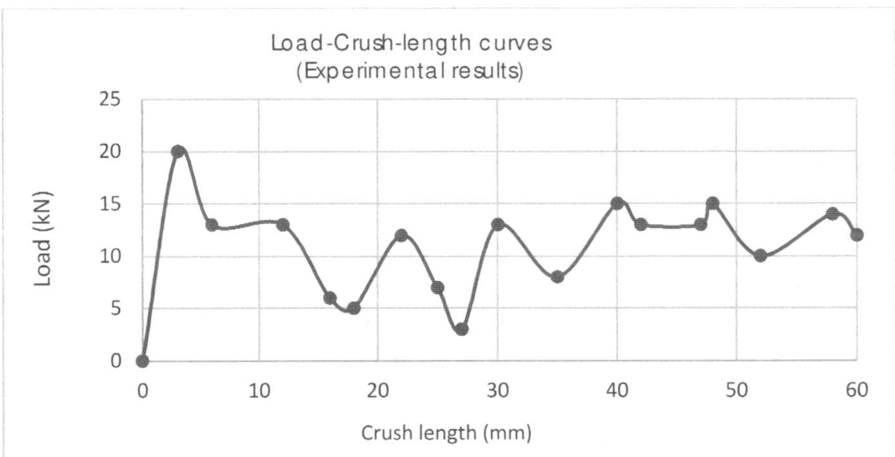

Figure 2. Dynamic test results with plug of (Ro,Ri) = (5,8) [13]

3. FEA analysis of the CFRP plug

3.1. 3D modelling

The model and the plugs were initially designed using Autodesk Inventor which is a 3D modelling platform. The part was designed as a 'sheet metal' which behaves as a thin flat piece of metal with uniform thickness throughout the entire part. The advantage of the sheet metal design is that when the model is exported to LS-PREPOST for meshing it can easily be defined as a 'shell element'. Since Autodesk Inventor has a much approachable user interface for 3D designing it was preferred over designing using LS-PREPOST.

Upon completion of the 3D model it was exported as a STEP file and then imported into LS-DYNA. The simulation software used to depict failure of the modelled CFRP tube was LS-DYNA. However,

IOP Conf. Series: Materials Science and Engineering 1037 (2021) 012027 doi:10.1088/1757-899X/1037/1/012027

initial drawings were done using Autodesk Inventor which was then exported to LS-PREPOST. The inventor 'sheet metal' STEP files were imported to LS-PREPOST and a thickness of 2.9 mm was set for the CFRP tube. Meshing of the model was done using LS-PREPOST and each parts (CFRP tube and plugs) were individually meshed. The CFRP tube was given a fine mesh of 0.5 mm whereas the double-sided plugs were given a slightly coarse mesh of 1.5 mm.

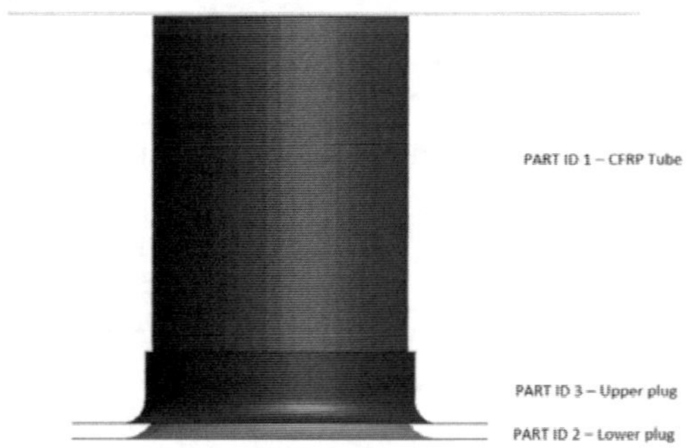

Figure 3. 3D model components

3.2. Material Card

According to [12, 14], MAT54 can successfully be used to simulate axial crushing behaviour of a sinusoidal composite specimen. The model can accurately capture the peak and average crush force, overall load-displacement response and most importantly it exhibits stable and progressive element failure and deletion. Hence, the material card used for the CFRP tube was MAT 54 (MAT054_ENHANCED_COMPOSITE_DAMAGE), which is widely used for dynamic test performance of CFRP tubes. The part had 12 layers of CF sheet layered around it and this was simulated using the option PART_COMPOSITE and a specific thickness of 0.24166 mm was given for each sheet. The complied thickness of 12 such sheets would add up to the original thickness of 2.9 mm.

3.3. Boundary conditions

The impactor for the model was designed to be rigid wall to reduce computational time. The wall was assigned a mass of 65 KG (6.5E-2 ton) and velocity of 55 km/h (15300 km/mm). A small distance was kept between the CFRP tube and impactor to replicate the initial hit and a coefficient of friction of 0.3 was given.

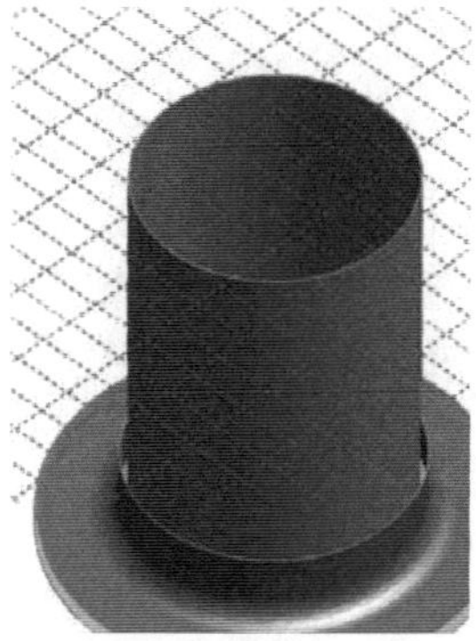

Figure 4. Rigid wall impactor set up

The control cards used for the model were mainly for hourglass, termination and timestep. The hourglass was set as LSDYNA standard itself of 0.1. The termination ENDTIM was set to be 4.92E-3 and ENDMAS was set to be 1.0E8. For the model to be reliable the percentage increase should be less than 5% for mass scaling. This data can be found in the d3hsp file when opened in a WordPad. In the added mass calculation, the model should show less than 5% percentage increase [15]. In this section we have achieved a percentage increase of 2.1% which is exceptionally good.

Table 1. Added mass calculations

Calculation with mass scaling for minimum dt	
Added mass	1.3759E-06 t
Physical mass	6.5145E-02 t
Ratio	2.1120E-05
Added mass	1.3759E-06 t
% Increase	2.1121E-03 %

The timestep used for the model is -1.0687E-7 with a TSSFAC value which is scale factor for computed time step value of 0.9. The contact set between the plugs, CF part and the impactor was 'automatic surface to surface contact' (CONTACT_AUTOMATIC_SURFACE_TO_SURFACE).

In crash analysis, predetermination of where and how contact will take place may be difficult or impossible as the deformations can be exceptionally large at times. Due to this, LSDYNA recommends the automatic contact options as these contacts are non-oriented, meaning they can detect penetration coming from either side of a shell element. Hence, the contact card used for contact between the CFRP tubes and the plugs and the impactor and the CFRP tube is CONTACT_AUTOMATIC_SURFACE_TO_SURFACE. The CFRP tube was set as the master side as it had more nodes than the support plugs.

Figure 5. Cross-sectional view of the part

4. Results

The simulated crush pattern represented the original shattering behaviour of the CFRP tube. The crush initiated at the bottom of the plug when it hit the curved surface of the lower plug and propagated upwards like a shock wave. Close observation on the initial crush behaviour also portrayed delamination properties of CFRP tubes. However, due to element deletion it did not visually portray the damaged strands. An illustration of crush behaviour between the experimental model and simulation model is shown in figure 7.

Figure 6. Crushing of the CFRP tube in LSDYNA

Figure 7. Comparison between original model and simulation model

The model represented the working mechanism of the real-life test model as it sheared in the strong X ply direction.

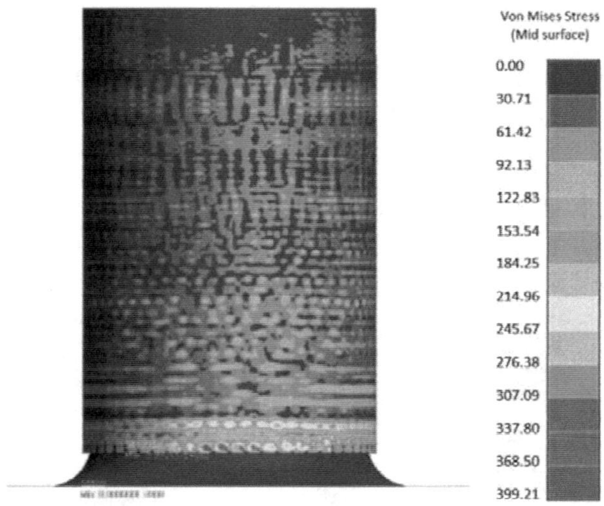

Figure 8. Crack initiation at 0.000100 sec

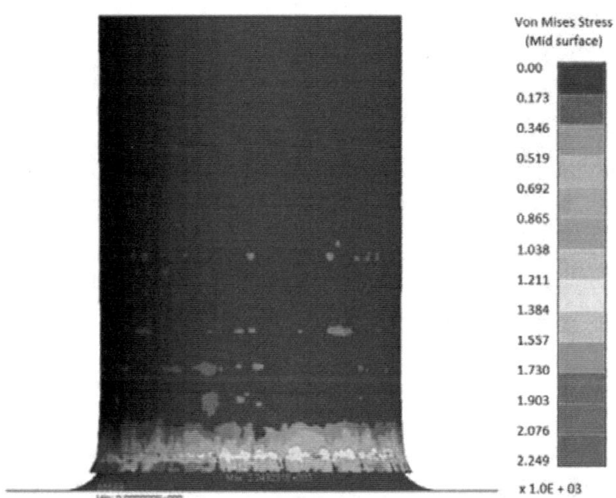

Figure 9. Crack initiation at 0.000220 sec

Figure 10. Crack propagation at 0.004300 sec

Once the visual simulation results correlated with the real life model a load vs. crush-length graph was generated to compared experimental and simulation results.

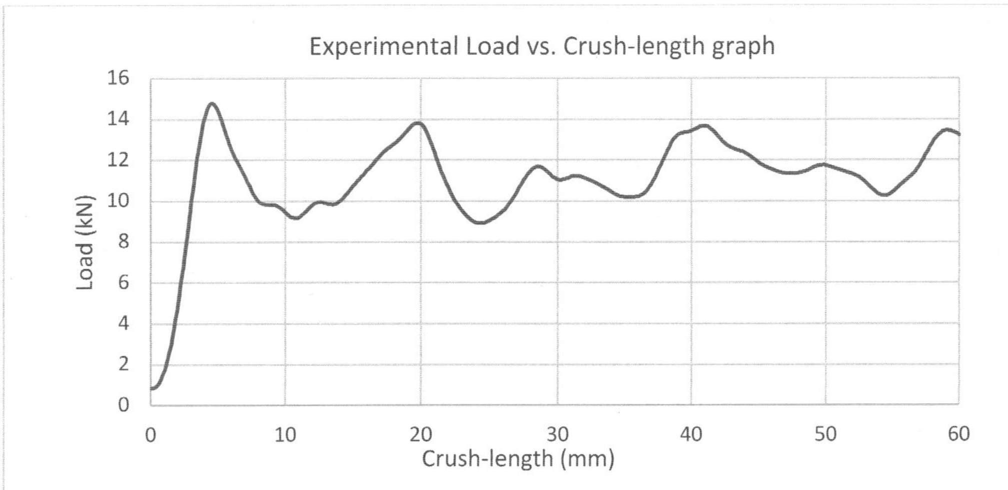

Figure 11. Dynamic test result of simulated model

From the above graph it was observed that the average force of the graph in the simulated model is about 11 kN and in the real life model average force was approximately 10 kN. A good correlation has been established between the real life model and the simulated model.

It is also noteworthy that the simulation model's delaminated fiber strands and the real life models delaminated strands were not visually identical. However, the results showed a good relationship because when carbon fibre physically disintegrates it loses structural stiffness and that's why delamination in the FE model does not affect simulation results. Although the visual data was not fully as expected the numerical data correlated very well. A comparison of both experimental and simulation results is shown in figure 12.

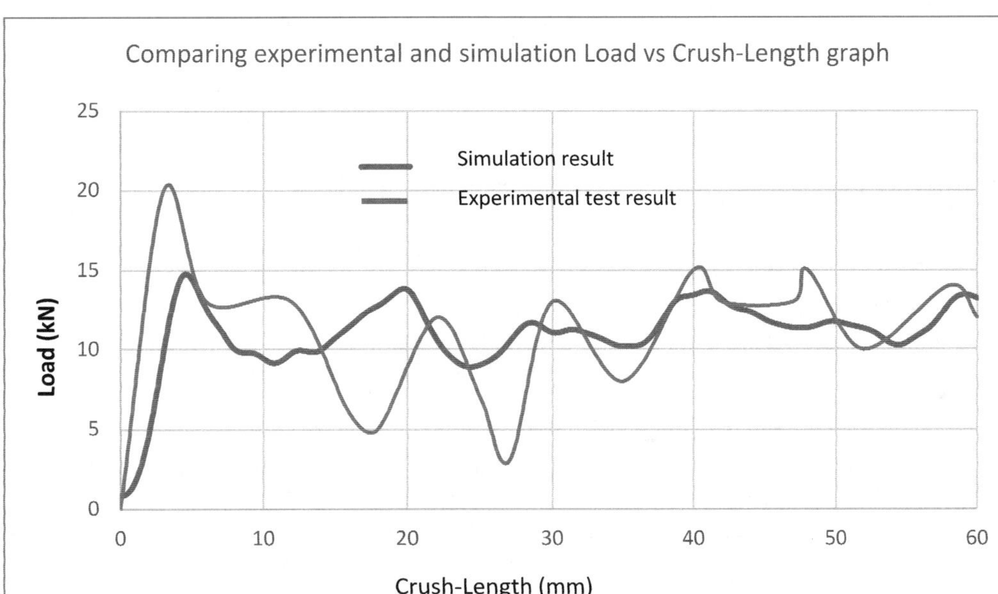

Figure 12. Comparing experimental and simulation results

5. Conclusion

The main aim of the study was to validate dynamic crush response behaviour of carbon fibre via finite element simulation. In order to attain this aim, three objectives had to be achieved.

1. Selecting the right material card and obtaining apt material card parameters.
2. Defining boundary conditions to replicate real-life experimental setup.
3. Validating simulation data with experimental data.

From the results a good visual and graphical correlation was obtained between the experimental and simulation model. Hence, it can be concluded that MAT54 portrays real-life dynamic crush behaviour of unidirectional carbon fiber tubes and contact forces can be simulated using AUTOMATIC_SURFACE_TO_SURFACE_CONTACT. For future research, researchers could use these simulation parameters to analyse dynamic crush response of carbon fibre tubes.

6. References

[1] Beardmore, P., and C.F. Johnson. "The Potential for Composites in Structural Automotive Applications." Composites Science and Technology, vol. 26, no. 4, Jan. 1986, pp. 251–281, www.sciencedirect.com/science/article/abs/pii/0266353886900023, 10.1016/0266-3538(86)90002-3. Accessed 13 Oct. 2020.
[2] Lyu, Min-Young, and Tae Gyun Choi. "Research Trends in Polymer Materials for Use in Lightweight Vehicles." International Journal of Precision Engineering and Manufacturing, vol. 16, no. 1, Jan. 2015, pp. 213–220, link.springer.com/article/10.1007/s12541-015-0029-x, 10.1007/s12541-015-0029-x. Accessed 20 May 2019.
[3] Hou, T., et al. "Pressurised Composite Tubes as Variable Load Energy Absorbers." Composite Structures, vol. 120, Feb. 2015, pp. 346–357, 10.1016/j.compstruct.2014.09.060. Accessed 13 Oct. 2020.
[4] Wu, Suzhen, et al. "On Design of Multi-Cell Thin-Wall Structures for Crashworthiness." International Journal of Impact Engineering, vol. 88, Feb. 2016, pp. 102–117, 10.1016/j.ijimpeng.2015.09.003. Accessed 13 Oct. 2020.
[5] Zhou, Hongyu, et al. "Crashworthiness Characteristics of a Carbon Fiber Reinforced Dual-Phase Epoxy–Polyurea Hybrid Matrix Composite." Composites Part B: Engineering, vol. 71, Mar. 2015, pp. 17–27, 10.1016/j.compositesb.2014.10.053. Accessed 13 Oct. 2020.
[6] Zhu, Guohua, et al. "Energy Absorption of Metal, Composite and Metal/Composite Hybrid Structures under Oblique Crushing Loading." International Journal of Mechanical Sciences, vol. 135, Jan. 2018, pp. 458–483, www.sciencedirect.com/science/article/pii/S0020740317325821?via%3Dihub, 10.1016/j.ijmecsci.2017.11.017. Accessed 13 Oct. 2020.
[7] Pantazopoulos, George A. "Failure Mechanisms in Alloys." Metals, vol. 10, no. 1, 13 Jan. 2020, p. 117, www.researchgate.net/publication/338556602_Failure_Mechanisms_in_Alloys, 10.3390/met10010117. Accessed 11 Aug. 2020.
[8] Liu, P.F., and J.Y. Zheng. "Progressive Failure Analysis of Carbon Fiber/Epoxy Composite Laminates Using Continuum Damage Mechanics." Materials Science and Engineering: A, vol. 485, no. 1–2, June 2008, pp. 711–717, www.researchgate.net/publication/222860192_Progressive_failure_analysis_of_carbon_fiberepoxy_composite_laminates_using_continuum_damage_mechanics, 10.1016/j.msea.2008.02.023. Accessed 30 Aug. 2020.
[9] Beaumont, Peter W. R., and Costas Soutis. "Structural Integrity of Engineering Composite Materials: A Cracking Good Yarn." Philosophical Transactions of the Royal Society A: Mathematical,

Physical and Engineering Sciences, vol. 374, no. 2071, 13 July 2016, p. 20160057, www.ncbi.nlm.nih.gov/pmc/articles/PMC4901257/, 10.1098/rsta.2016.0057. Accessed 13 Oct. 2020.

[10] Quaresimin, Marino, et al. "Energy Absorption in Composite Laminates under Impact Loading." Composites Part B: Engineering, vol. 44, no. 1, Jan. 2013, pp. 133–140, 10.1016/j.compositesb.2012.06.020. Accessed 13 Oct. 2020.

[11] ANSYS. "LS-DYNA: Explicit Dynamics Materials Deformation Simulation | Ansys." Www.Ansys.Com, 2020, www.ansys.com/products/structures/ansys-ls-dyna.

[12] Feraboli, Paolo, et al. "LS-DYNA MAT54 Modeling of the Axial Crushing of a Composite Tape Sinusoidal Specimen." Composites Part A: Applied Science and Manufacturing, vol. 42, no. 11, Nov. 2011, pp. 1809–1825, 10.1016/j.compositesa.2011.08.004. Accessed 13 Oct. 2020.

[13] Ueda, Masahito, et al. "Controlling the Energy Absorption Capability of a Unidirectional Carbon Fiber Reinforced Plastic Tube Using a Double-Sided Plug." Open Journal of Composite Materials, vol. 05, no. 01, 2015, pp. 30–40, pdfs.semanticscholar.org/d93b/6681af149824e943286ea453050211c06b2a.pdf, 10.4236/ojcm.2015.51006. Accessed 13 Oct. 2020.

[14] LS-DYNA. "Contact Modeling in LS-DYNA." Welcome to the LS-DYNA Support Site, 2020, www.dynasupport.com/tutorial/ls-dyna-users-guide/contact-modeling-in-ls-dyna. Accessed 13 Oct. 2020.

[15] Waimer, M., et al. "Experimental Study of CFRP Components Subjected to Dynamic Crash Loads." Composite Structures, vol. 105, Nov. 2013, pp. 288–299, 10.1016/j.compstruct.2013.05.030. Accessed 13 Oct. 2020.

Prediction of roughness of planar surfaces processed with toroidal milling through an artificial neural network

A R Osan[1,2], M Banica[1] and V Nasui[1]

[1] Technical University of Cluj-Napoca North University Center of Baia Mare, Dr. Victor Babeş Street No. 62A, Romania

*Corresponding author's e-mail: osan.andrei@yahoo.com

Abstract. This paper is intended to create an artificial neural network capable of generating new values for the roughness on the basis of experimentally obtained data bases. Experimentally you will measure the roughness of the flat surfaces processed with the toroidal milling, the process factors being the input neurons of the neural network, following the roughness values being the output neurons. It aims to modify the input neurons from the same neural network and generate new roughness values.

1. Introduction

In recent years, artificial neural networks (RNA) are offered in front of the largest model offerings that can be kept and are now successful in certain engineering fields for modeling complex relationships difficult to describe with physical models.

Artificial neural networks have been widely applied in modeling many cutting operations, such as turning, drilling and milling [1]. Several researchers have used artificial neural networks to predict the influence of cutting parameters on production rate, production cost [2] or to predict the influence of cutting parameters on surface roughness [3], [4], tool wear [5]], [6] or the cutting force [1], [7].

Examples of optimization attempts can be found in the work of Mohana et al. [8] aimed at modeling surface roughness using neural networks. Genetic algorithms have been used in their research to optimize the weighting factors of the network. Ortiz-Rodrigues et al. [9] proposed the use of Taguchi methods (DOE technique) for robust RNA-driven design by the back propagation algorithm and develop a systematic and experimental strategy that emphasizes the simultaneous optimization of artificial neural network parameters under different conditions.

Several examples can be found in the works of Assarzadeh and Ghoreishi [11], meant to optimize the surface roughness using neural networks. The authors stated the effectiveness of using RNA to predict the removal rate of materials and Ra. In Hossain et al. [11], an RNA model was developed to investigate and predict the relationship between milling parameters and surface roughness during high-speed milling of Inconel 718 alloy. A very good predictive performance of the neural network was observed. Other approaches include the work of Panda and Mahapatra [12] in which the main components were used to model the drill wear. The main components of the drilling parameters were calculated and the networks were trained to predict them. The networks were able to classify low wear and high wear with an accuracy of 90% and to predict the wear of the main edge with an error of ± 6.5%.

MatLab® (MATrix LABoratory) is a high-performance, interactive software package for mathematical, scientific and engineering calculus. MatLab integrates calculation, programming and visualization, in a friendly work environment, solving problems involving the use of classical mathematical notations.

With tools and functions for managing large data sets, MatLab offers different types of specialized tools with machine learning, neural networks, deep learning, computer vision and machine learning.

Improving the generalization of the network helps prevent overload, a common problem in neural network design. Overload occurs when a network has memorized the training set, but has not learned to generalize to new entries. Overload produces a relatively small error on the training set, but a much larger error when new data is obtained in the neural network.

2. Approximation of functions with the artificial neural network

RNA approximation can be solved in MatLab:
From the command line, using specific functions Neural Network Toolbox;
Using the Neural Network Fitting Tool (nftool) graphical interfaces.

Before starting to create an artificial neural network, it is necessary to create input and output data. This is where computer resources based on previous research come in [13], [14], [15].

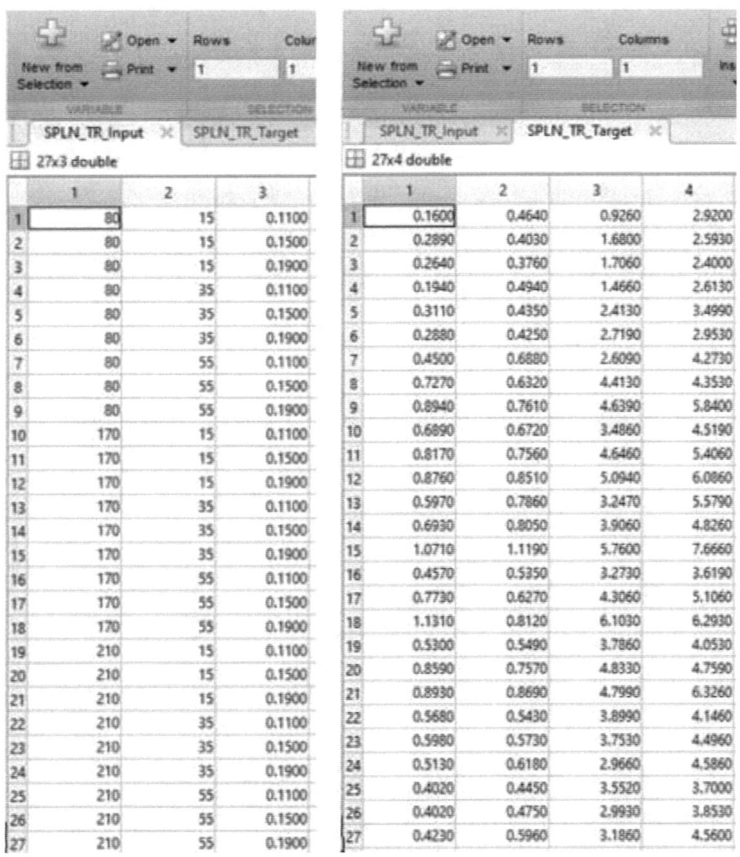

SPLN_TR_Input — 27x3 double

	1	2	3
1	80	15	0.1100
2	80	15	0.1500
3	80	15	0.1900
4	80	35	0.1100
5	80	35	0.1500
6	80	35	0.1900
7	80	55	0.1100
8	80	55	0.1500
9	80	55	0.1900
10	170	15	0.1100
11	170	15	0.1500
12	170	15	0.1900
13	170	35	0.1100
14	170	35	0.1500
15	170	35	0.1900
16	170	55	0.1100
17	170	55	0.1500
18	170	55	0.1900
19	210	15	0.1100
20	210	15	0.1500
21	210	15	0.1900
22	210	35	0.1100
23	210	35	0.1500
24	210	35	0.1900
25	210	55	0.1100
26	210	55	0.1500
27	210	55	0.1900

SPLN_TR_Target — 27x4 double

	1	2	3	4
1	0.1600	0.4640	0.9260	2.9200
2	0.2890	0.4030	1.6800	2.5930
3	0.2640	0.3760	1.7060	2.4000
4	0.1940	0.4940	1.4660	2.6130
5	0.3110	0.4350	2.4130	3.4990
6	0.2880	0.4250	2.7190	2.9530
7	0.4500	0.6880	2.6090	4.2730
8	0.7270	0.6320	4.4130	4.3530
9	0.8940	0.7610	4.6390	5.8400
10	0.6890	0.6720	3.4860	4.5190
11	0.8170	0.7560	4.6460	5.4060
12	0.8760	0.8510	5.0940	6.0860
13	0.5970	0.7860	3.2470	5.5790
14	0.6930	0.8050	3.9060	4.8260
15	1.0710	1.1190	5.7600	7.6660
16	0.4570	0.5350	3.2730	3.6190
17	0.7730	0.6270	4.3060	5.1060
18	1.1310	0.8120	6.1030	6.2930
19	0.5300	0.5490	3.7860	4.0530
20	0.8590	0.7570	4.8330	4.7590
21	0.8930	0.8690	4.7990	6.3260
22	0.5680	0.5430	3.8990	4.1460
23	0.5980	0.5730	3.7530	4.4960
24	0.5130	0.6180	2.9660	4.5860
25	0.4020	0.4450	3.5520	3.7000
26	0.4020	0.4750	2.9930	3.8530
27	0.4230	0.5960	3.1860	4.5600

Figure 1. Images with input data and target data of the neural network.

To create the input data and the output data, we chose to debate the quality of the flat surfaces obtained with the toroidal milling cutter. Thus, the input data are the three process variables. These are the cutting speed, the feed on the tooth and the angle of inclination. As for the output data, the target information, they are represented by the average values of Ra, measured parallel and perpendicular as well as by the average values of Rt, measured parallel and perpendicular. Therefore, the input data and the output data are shown in Figure 1.

To approximate the functions with artificial neural networks, using the graphical interfaces, the Neural Network Fitting Tool interface will be opened with the command "nftool".

This type of neural network is able to associate an input data set with a target data set for estimating certain values. This application has the function of helping to select data, to create and train a neural network, as well as to evaluate performance.

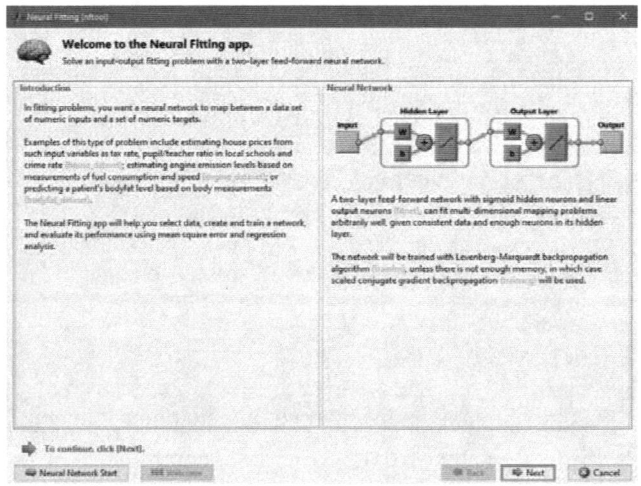

Figure 2. Neural Fitting Tool.

It has a power supply network with two layers of hidden neurons being trained with a back propagation algorithm as shown in Figure 2.

The next step in creating a neural network is to load the input and target data, as well as select the type of matrix, as shown in Figure 3.

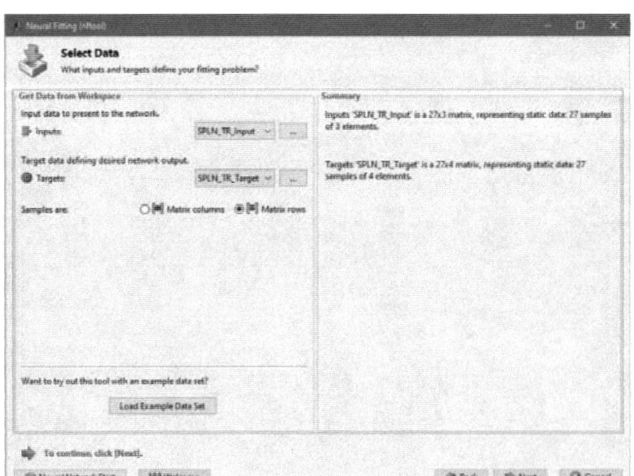

Figure 3. Selection of input and target data for RNA creation.

The next step is illustrated in Figure 4 and represents the establishment of the data assigned to the neural network training, but also the establishment of the number of values assigned to the network validation and testing function.

In this case, we chose to use 25 values for the training function, leaving only one for validation and testing.

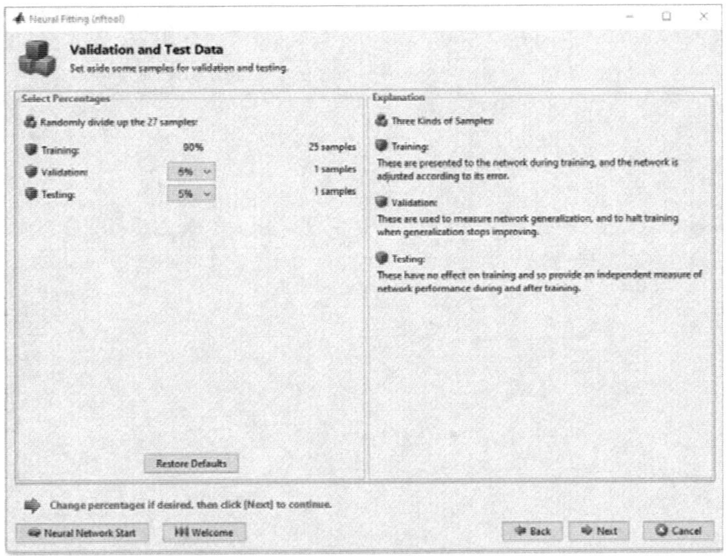

Figure 4. Selection of percentages assigned to neural network training, validation and testing.

Regarding the network architecture, here are defined the number of hidden neurons of the neural network, in our case we decided to use 20 neurons for the network to function in optimal conditions, as shown in Figure 5.

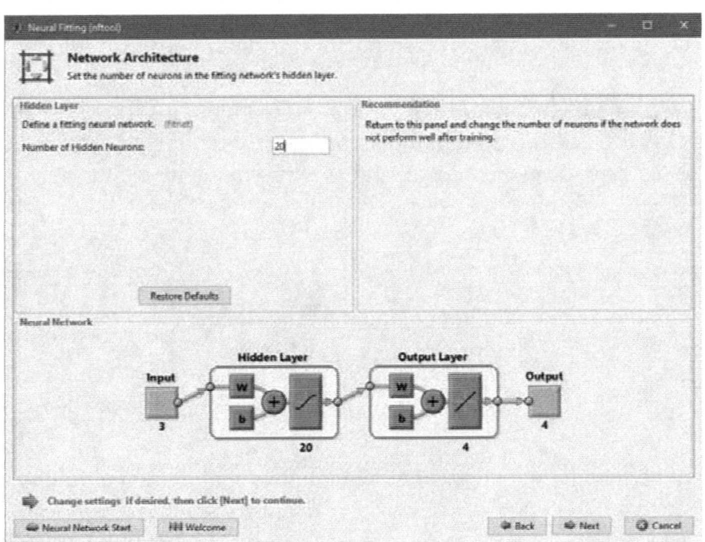

Figure 5. Neural network architecture.

Once the input data are established, the percentage of training, validation, testing, as well as the number of hidden neurons can only be transmitted to the network to learn the working algorithm, as in Figure 6. Following the training of the neural network, it transmits a series of indices, these are shown in Figure 7.

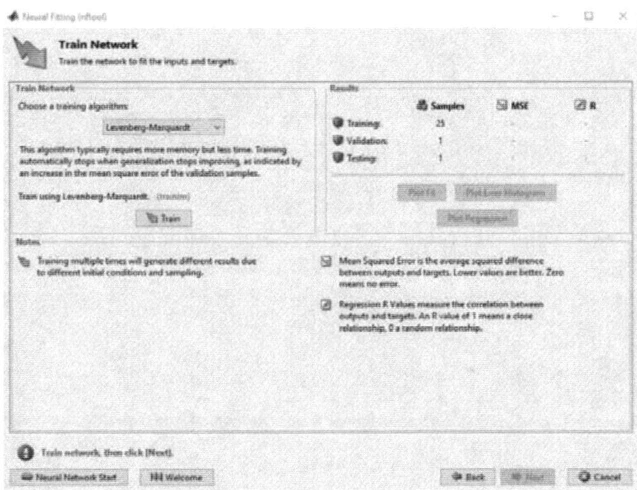

Figure 6. Neural network training model.

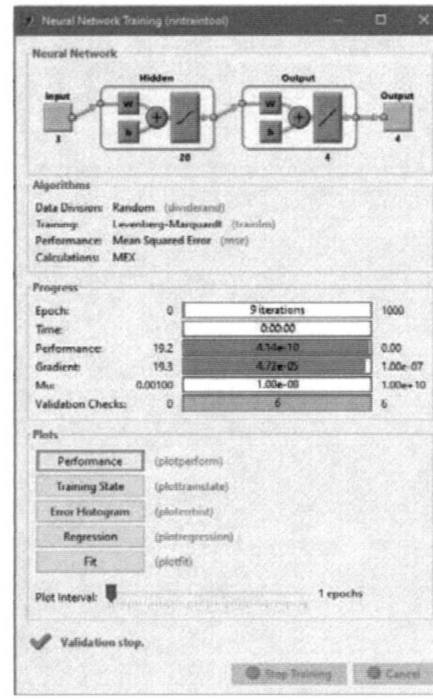

Figure 7. Neural network training.

Following the training of the network, it provides some information about the network, one of the information is supported by the number of interactions within the network, in this case it is about 9 interactions or 9 periods as shown in Figure 8.

In terms of network performance, the best performance was recorded during interaction number 3.

The training conditions of the neural network are dependent on the value of the gradient, which is a vector field whose vectors are directed in the direction of the highest growth rate of the scalar field. Thus, the mode is the highest rate of change, the maximum value being performed in the case of interaction number 9. The validation checks were 6, also at interaction 9, being presented in figure 9.

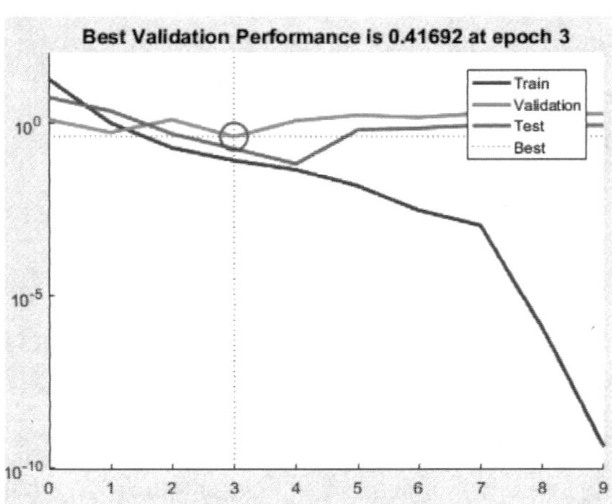

Figure 8. Neural network performance.

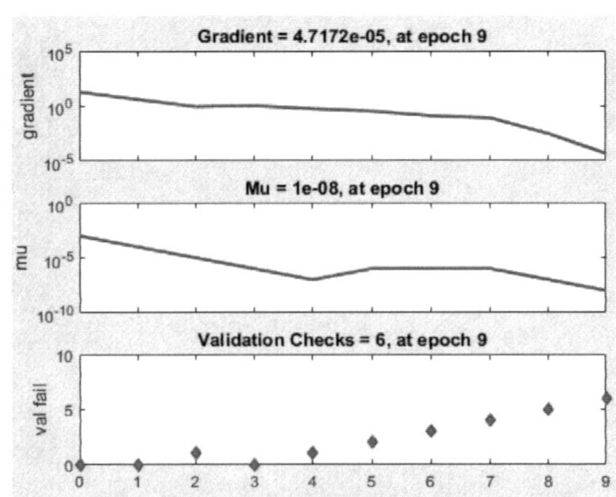

Figure 9. Training conditions.

Any system also has a series of errors, in the case of the neural network created, it has the most common type of error of 0.01469, as shown in Figure 10.

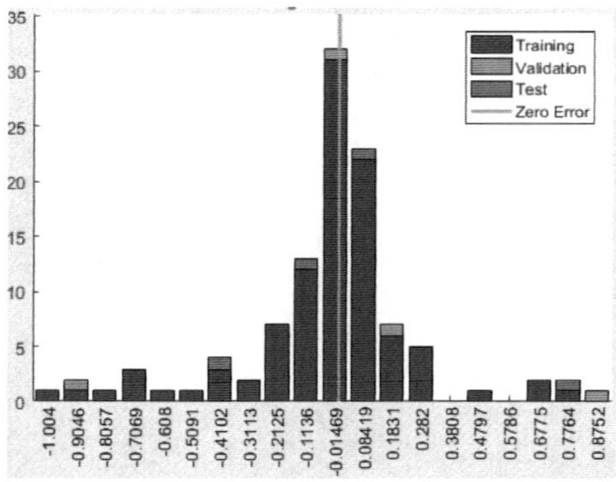

Figure10. Histogram of errors.

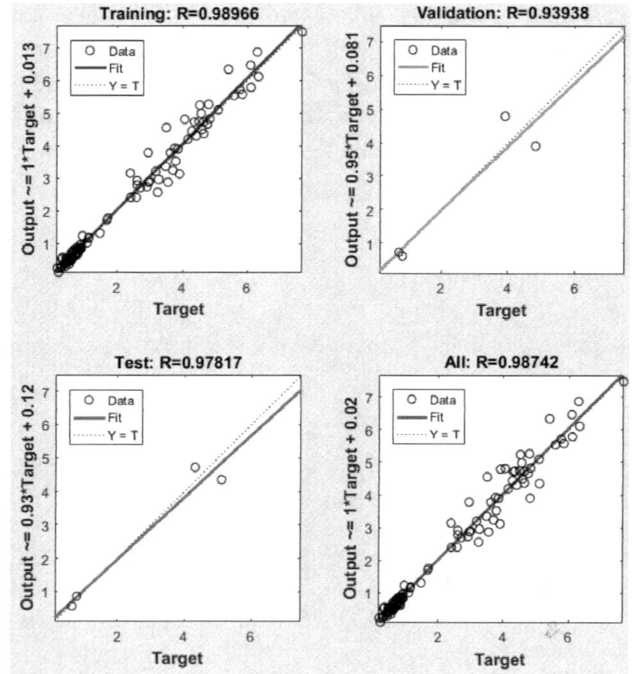

Figure 11. Linear regression of the neural network.

Regarding the linear regression, in the case of the artificial neural network studied and created on the surface quality, this is illustrated in figure 11, both in terms of training and validation and testing of the network.

For the implementation of the solutions in the next subchapter is presented the MatLab function as matrix support. The neural network diagram consists of 3 input neurons, 20 hidden neurons and 4 output neurons, as in Figure 12.

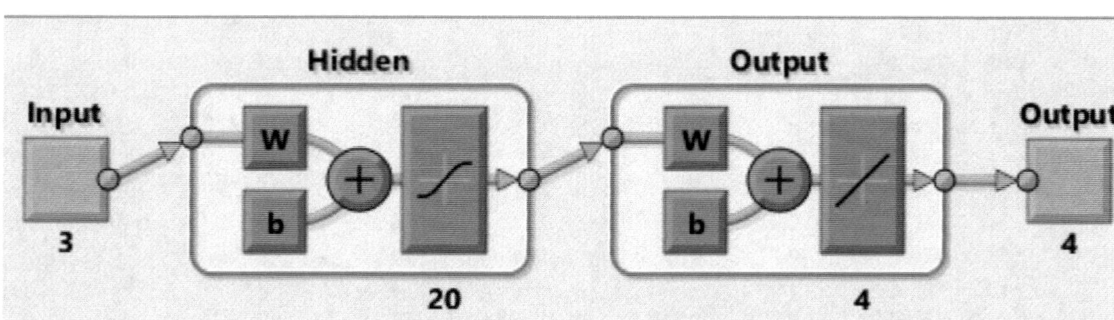

Figure 12. Diagram of the neural network.

Once the whole process of creating and training the neural network is completed, all that remains is to put the target data face to face with the outputs that the network offers.

In the next chapter are highlighted the measured data of roughness, having as input the three variables, and in comparison are the approximate values obtained after training the neural network, with a column with network errors attached.

IOP Conf. Series: Materials Science and Engineering 1037 (2021) 012028 doi:10.1088/1757-899X/1037/1/012028

3. Comparison of approximations

According to the previous chapter, where a neural network was created and trained using MatLab software, this network is able to approximate the output values based on the input values. As established, the network was created on the flat surface processed with toroidal milling cutter to approximate the roughness values. As input data, the input neurons are the three variables, the cutting speed, the feedrate on the tooth and the angle of inclination of the tool axis. The target values, output neurons are represented by the measured values both parallel and perpendicular to the roughness Ra and Rt, so there are 4 output neurons.

Table 1. Approximate values of RNA for the Ra quality of the flat surface processed with the toroidal milling cutter.

Surface type	Roughness R_a[μm]				Parallel Ra approximation error	Perpendicular approximation error Ra
	Real values		Approximate RNA values			
	The direction of measurement in relation to the direction of advance					
	Parallel	Perpendicular	Parallel	Perpendicular		
SPLN -TR-1	0.160	0.464	0.241	0.426	-0.081	0.038
SPLN -TR-2	0.289	0.403	0.270	0.397	0.019	0.006
SPLN -TR-3	0.264	0.376	0.279	0.372	-0.015	0.004
SPLN -TR-4	0.194	0.494	0.116	0.505	0.078	-0.011
SPLN -TR-5	0.311	0.435	0.566	0.517	-0.255	-0.082
SPLN -TR-6	0.288	0.425	0.525	0.552	-0.237	-0.127
SPLN -TR-7	0.450	0.688	0.462	0.684	-0.012	0.004
SPLN -TR-8	0.727	0.632	0.857	0.578	-0.130	0.054
SPLN -TR-9	0.894	0.761	0.975	0.651	-0.081	0.110
SPLN -TR-10	0.689	0.672	0.610	0.802	0.079	-0.130
SPLN -TR-11	0.817	0.756	0.822	0.811	-0.005	-0.055
SPLN -TR-12	0.876	0.851	0.884	0.947	-0.008	-0.096
SPLN -TR-13	0.597	0.786	0.499	0.691	0.098	0.095
SPLN -TR-14	0.693	0.805	0.737	0.622	-0.044	0.183
SPLN -TR-15	1.071	1.119	1.016	1.157	0.055	-0.038
SPLN -TR-16	0.457	0.535	0.555	0.463	-0.098	0.072
SPLN -TR-17	0.773	0.627	0.859	0.569	-0.086	0.058
SPLN-TR-18	1.131	0.812	1.182	0.848	-0.051	-0.036
SPLN -TR-19	0.530	0.549	0.540	0.579	-0.010	-0.030
SPLN -TR-20	0.859	0.757	0.813	0.731	0.046	0.026
SPLN -TR-21	0.893	0.869	0.840	0.874	0.053	-0.005
SPLN -TR-22	0.568	0.543	0.464	0.548	0.104	-0.005
SPLN -TR-23	0.598	0.573	0.517	0.698	0.081	-0.125
SPLN -TR-24	0.513	0.618	0.362	0.745	0.151	-0.127
SPLN -TR-25	0.402	0.445	0.423	0.416	-0.021	0.029
SPLN -TR-26	0.402	0.475	0.298	0.452	0.104	0.023
SPLN -TR-27	0.423	0.596	0.445	0.640	-0.022	-0.044

Table 1 shows the roughness values Ra measured parallel and perpendicular to the direction of advance, as well as the approximate values using the neural network, also on the two measuring directions.

From a percentage point of view, the nearest value is in the case of the flat surface with number 11, processed with the toroidal milling cutter (SPLN-TR-11) with an approximation error of 0.61% with a difference of 0.005 μm for the value measured in parallel , and for the value measured perpendicularly, the nearest value is in the case of the flat surface number 7, processed with the toroidal milling cutter (SPLN-TR-7) with an error of 0.58%, with a difference of 0.004 μm. The graphical representation is shown in Figure 13.

The largest error identified is in the case of the SPLN-TR-5 surface, measured in parallel with an approximation error of 81.9% and a difference of 0.255 μm, and in the case of perpendicular measurement, the largest error obtained is 22.7 %, for the surface of SPLN-TR-14 with a difference of 0.183 μm.

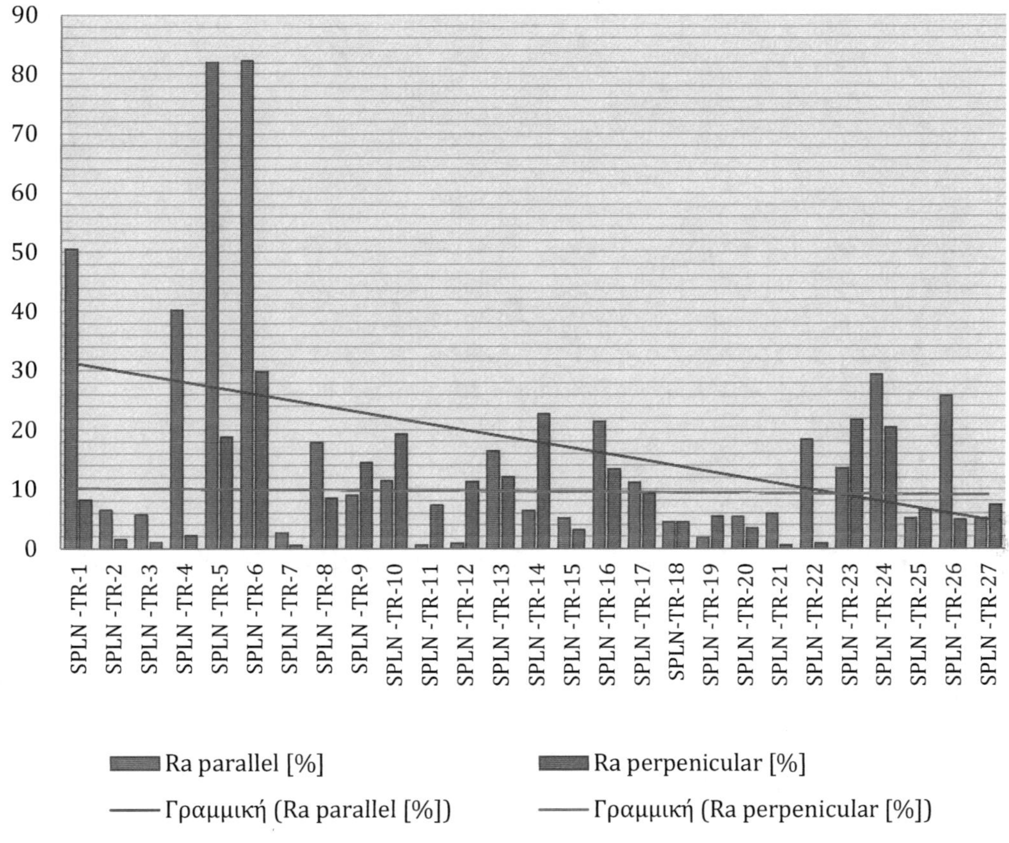

Figure 13. Graphical representation of the percentage error diagram for the approximate values Ra.

Table 2 represents the values of the roughness Rt measured on the flat surface processed with the toroidal milling cutter, and in comparison the approximate values are passed with the help of the neural network, following in the columns on the right to represent the approximation errors of the neural network.

In terms of percentage, the lowest value was made on the flat surface with number 12, processed with toroidal milling cutter (SPLN-TR-12) with a percentage of 0.27% and a minimum difference of 0.003 μm, determined value on the measuring direction parallel to the feed direction.

The lowest percentage value recorded in the direction of perpendicular measurement shall be 0.86% over the flat surface number 13 (SPLN-TR-13) with a difference of 0.048 μm.

Table 2. Approximate values of RNA for the Rt quality of the flat surface processed with the toroidal milling cutter.

Surface type	Roughness R_t[μm]				Parallel Rt approximation error	Perpendicular Rt approximation error
	Real values		Approximate RNA values			
	The direction of measurement in relation to the direction of advance					
	Parallel	Perpendicular	Parallel	Perpendicular		
SPLN -TR-1	0.926	2.920	1.235	2.733	-0.309	0.187
SPLN -TR-2	1.680	2.593	1.715	2.410	-0.035	0.183
SPLN -TR-3	1.706	2.400	1.776	3.147	-0.070	-0.747
SPLN -TR-4	1.466	2.613	1.311	2.920	0.155	-0.307
SPLN -TR-5	2.413	3.499	2.406	4.552	0.007	-1.053
SPLN -TR-6	2.719	2.953	2.704	3.786	0.015	-0.833
SPLN -TR-7	2.609	4.273	2.786	4.445	-0.177	-0.172
SPLN -TR-8	4.413	4.353	4.303	4.733	0.110	-0.380
SPLN -TR-9	4.639	5.840	4.723	5.571	-0.084	0.269
SPLN -TR-10	3.486	4.519	3.361	5.236	0.125	-0.717
SPLN -TR-11	4.646	5.406	4.372	6.337	0.274	-0.931
SPLN -TR-12	5.094	6.086	5.097	6.465	-0.003	-0.379
SPLN -TR-13	3.247	5.579	2.563	5.531	0.684	0.048
SPLN -TR-14	3.906	4.826	4.788	3.901	-0.882	0.925
SPLN -TR-15	5.760	7.666	5.722	7.477	0.038	0.189
SPLN -TR-16	3.273	3.619	2.961	3.782	0.312	-0.163
SPLN -TR-17	4.306	5.106	4.716	4.352	-0.410	0.754
SPLN-TR-18	6.103	6.293	5.785	6.865	0.318	-0.572
SPLN -TR-19	3.786	4.053	3.523	4.806	0.263	-0.753
SPLN -TR-20	4.833	4.759	4.820	4.660	0.013	0.099
SPLN -TR-21	4.799	6.326	5.262	6.114	-0.463	0.212
SPLN -TR-22	3.899	4.146	3.127	4.193	0.772	-0.047
SPLN -TR-23	3.753	4.496	3.920	4.741	-0.167	-0.245
SPLN -TR-24	2.966	4.586	2.880	4.492	0.086	0.094
SPLN -TR-25	3.552	3.700	2.871	3.239	0.681	0.461
SPLN -TR-26	2.993	3.853	2.892	3.900	0.101	-0.047
SPLN -TR-27	3.186	4.560	3.206	4.985	-0.020	-0.425

The maximum approximate percentage value was obtained on the palmar surface number 1 (SPLN-TR-1) with an error rate of 33.37% and a difference of 0.309 μm on the parallel measuring direction and on the perpendicular measuring direction, the maximum percentage value given is 30.09% with a difference of 1.053 μm.

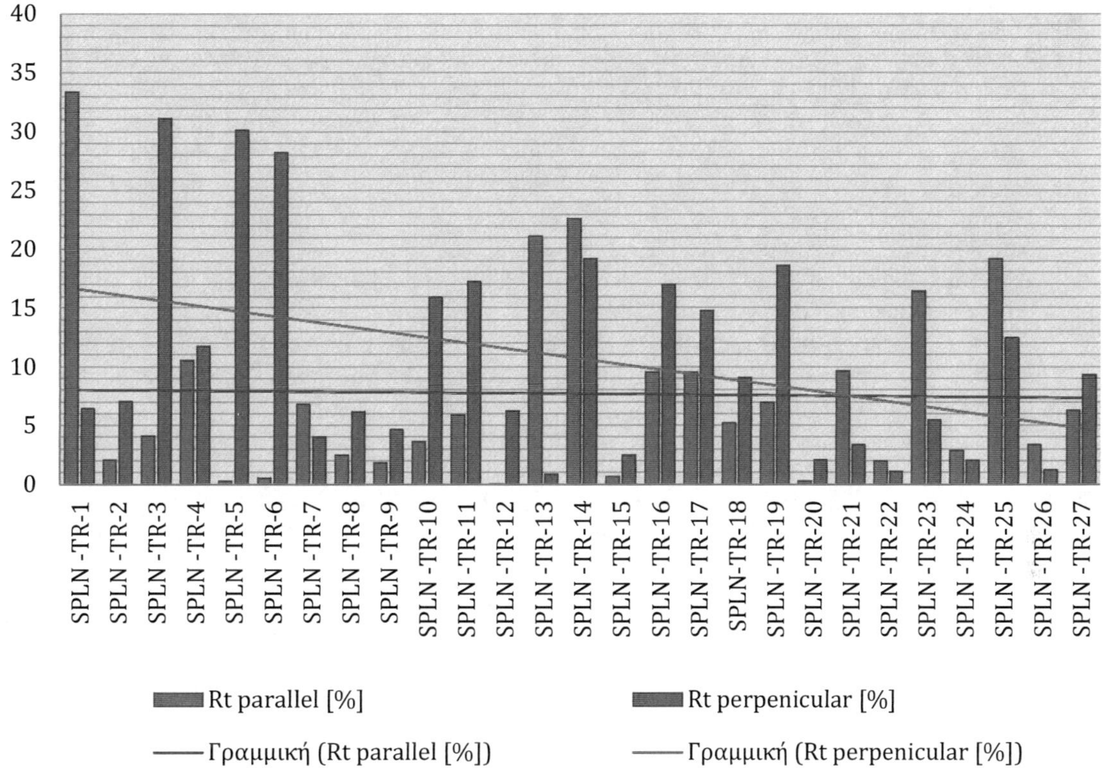

Figure 14. Graphical representation of the percentage error diagram for the approximate values Rt.

Following the graphical illustration of the error percentages in figure 14, a constant can be observed in the approximation of values Rt measured perpendicularly, as well as in the approximate values Rt measured in parallel.

4. Conclusions

The research and data modeling plan starts from the measurements, 3 measurements for each surface performed parallel to the feed direction and 3 measurements on each surface performed perpendicular to the feed direction. These measurements deserve as output neurons of the artificial neural network.

Data processing using MatLab software made it possible to create an artificial neural network, capable of approximating roughness values, having as input neurons the process variables during processing, and as output parameters the measured roughness.

With the creation and training of the artificial neural network it is able to learn the connections between neurons so as to generate new values of roughness.

These new values generated by the neural network for both the arithmetic roughness Ra and the total roughness Rt were compared with the real values. The comparisons resulted in a series of errors, the smallest error being 0.27% with a difference of 0.003 μm. This result certifies the adaptability of the neural network in generating new values of roughness.

5. References

[1] Golcu S, Semiz M, Ergur S and Aykut H S 2007 Modeling of cutting forces as function of cutting parameters for face milling of satellite 6 using an artificial neural network, Journal of Materials Processing Technology, vol. 190, pp. 199–203.
[2] Zuperl F and Cus U 2006 Approach to optimization of cutting conditions by using artificial neural networks Journal of Materials Processing Technology, vol. 173, pp. 281–290.
[3] Khani R A, Fakrabadi N, and Mahdavinejad S 2012 Optimization of milling parameters using artificial neural network and artificial immune system, Journal of Mechanical Science and Technology, vol. 26, pp. 4097-4104.
[4] Meenu A, 2013 Optimized Prediction and Modeling Under End Milling Machining By ANOVA And Artificial Neural Network , IJERT, vol. 2.
[5] Teti D and D'addona R, 2013 Image data processing via neural networks for tool wear prediction, Procedia CIRP, vol. 12, pp. 252 – 257.
[6] Celent D, Jozic L and Bajic S, 2012 Modeling of the Influence of Cutting Parameters on the Surface Roughness, Tool Wear and Cutting Force in Face Milling in Off-Line Process Control, Journal of Mechanical Engineering, vol. 58, pp. 673-682.
[7] Cus T, Paulic F, Balic M, and Irgolic J, 2014 Prediction of Cutting Forces with Neural Network by Milling Functionally Graded Material," Procedia Engineering, vol. 69, pp. 804 – 813.
[8] Rangajanardhaa GK, Hanumantha G, Rao G, Sreenvasa D, Rao M 2009 Development of hybrid model and optimization of surface roughness in electric discharge machining using artificial neural networks and genetic algorithm, J Mater Process Technol, vol. 209, pp. 1512–1520.
[9] Martinez-Blanco J, Vega-Carrillo E, Ortiz-Rodriguez H 2006 Robust design of artificial neural networks applying the Taguchi methodology and DOE, Proceedings of the Electronics, Robotics and Automotive Mechanics Conference, vol. 2, pp. 131–136.
[10] Ghoreishi S and Aaaarzadeh M 2008 Neural-network-based modeling and optimization of the electro-discharge machining process, Int J Adv Manuf Technol, vol. 39, pp. 488–500.
[11] Amin M, Patwari A and Hossain A 2008 Development of an artificial neural network algorithm for predicting the surface roughness in end milling of Inconel 718 Alloy, Roceedings of the International Conference on Computer and Communication Engineering, pp. 1321–1324.
[12] Mahapatra S and Panda S 2009 PCA fused NN approach fordrill wear prediction in drilling mild steel specimen, Proceedings of the 2nd IEEE International Conference on Computer Science and Information Technology, pp. 85–89.
[13] Oșan. A, 2019 Experimental Research on the Processing of Convex Spherical Surfaces with Toroidal Mills Versus Spherical Mills, Magazine of Hydraulics, Pneumatics, Tribology, Ecology, Sensorics, Mechatronics. Hidraulica, vol. 4.
[14] Oșan A, Bănică M and Năsui V 2019 The influence of inclination of the axis of the toroidal on a flat surface roughness. The 23nd edition of Innovative Manufacturing Engineering&Energy International Conference – ImanE&E 2019 Pitești vol 564.
[15] Oșan. A., Bănică. M. and Năsui. V. 2020 Processing of plane surfaces with toroidal milling versus ball nose end mill, Acta Technica Napocensis - Series: Applied Mathematics, Mechanics and Engineering, vol. 64.

Investigation of the suitability of fiber reinforced polymer matrix composites for facilities operating in marine environment

D Spasova[1], Y Argirov[1], T Mechkarova[1], N Atanasov[1]

[1] Technical University of Varna, Material Science and Technology Department, Studentska Street 1, Varna, Bulgaria

*Corresponding author's e-mail: danielats@tu-varna.bg

Abstract. This paper is related to the production and investigation of the structure, the mechanical and physicochemical properties of two kinds of fibre reinforced polymer matrix composites in order to assess their suitability as materials used in equipment operating in marine environment. The one of the investigated composites has a polyester resin matrix and a reinforcement phase of six-layer fibreglass, and the other one has a vinylester resin matrix and a reinforcement phase which is a combination of three-layer fibreglass and three-layer biaxial fibreglass. A comparative analysis has been made of the relative weight, tensile strength and bending strength of the two types of composite materials.

1. Introduction

The rapid growth in the manufacturing industries necessitates the need for materials that have better properties than many of the conventional materials like metals in various applications [1]. The most commonly used composites, as modern engineering materials with the largest application in marine industry, are the composite with resins matrices. Most of these composites included glass-reinforced polyesters, vinylesters, phenolics, and an epoxy (epoxy-polyamide) [2,3]. The glass-fiber-reinforced resin matrices have excellent ratios between performance and price and they better properties in terms of strength, stiffness, density, weathering properties, with improved sustainability and can be easily molded into complex shapes, than metals [1, 4]. These composites are commonly fabricated to near-net shapes of the reinforcement phase which increases their mechanical properties. [5]. But the exposure to seawater caused the greatest deterioration of the bending and tensile strengths of the composites [3].

The objective of the present work is to determine the strength of the composite materials studied in order to assess their suitability for manufacturing facilities and equipment operating in marine environment. The investigated materials are net shaped glass-fiber reinforced Polymer Matrix Composites. One of the studied composites features a polyester resin matrix and a reinforcement phase of six layers of fiberglass, and the second has a vinylester resin matrix and a reinforcement phase of a combination of three layers of fiberglass and three layers of biaxial fiberglass 450 g / sm2. Biaxial fiberglass fabric consists of glass fibers intertwined on top of each other at an angle of +/- 45 °. The + 45° and - 45° layers are of equal weight. Fabric is Silane coated for easier wetting out, and is compatible with both Polyester and Epoxy Resin systems. Glass biaxial fabric is generally used on components where the force is applied constantly in the direction of the grain.

2. Experimental procedure
The composites studied are produced in the form of 6 mm thickness panels, through layer-by-layer application of resin and reinforcement phase. Upon applying each layer of resin, mechanical impact, i.e. rolling, is employed in order to remove air bubbles. The panels thus produced are used for cutting samples for conducting tensile and bending tests and calculating the density of the composite materials. Upon test completion, a macro-fractographic analysis of the fracture surface of the tensile test samples was carried out by means of a EUROMEX stereomicroscope [6].

2.1. Tensile testing
The tensile test was carried out in compliance with ISO 527-4 Tensile Test of Plastic Composites & Thin Film Sheet standard at T = 20 °C [7, 8]. For the purposes of tensile testing, three identical-size samples were cut for each of the two types of composite, as shown in figure 1.

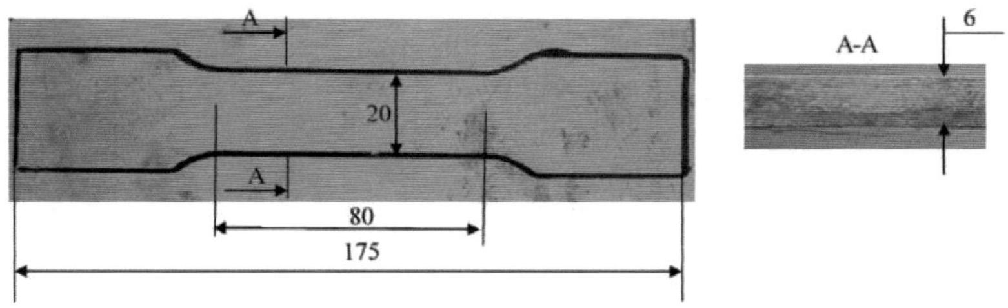

Figure 1. The samples sizes for tensile testing

The tensile strength (*Rm,* MPa) of the materials tested is determined by the following formula:

$$Rm = \frac{Fm}{So}\ [MPa] \tag{1}$$

Where: *Fm* - maximum applied load (force), N; S_0 – the initial cross-sectional area of the sample, mm^2

2.2. Flexural testing
The bending test was conducted according to ASTM D790 Testing for Flexural Properties of Plastics and Insulating Materials standard at T = 20 °C (figure 2). The determination of the bending strength was carried out using three identical samples with dimensions 180 mm / 30 mm cut for each type of composites.

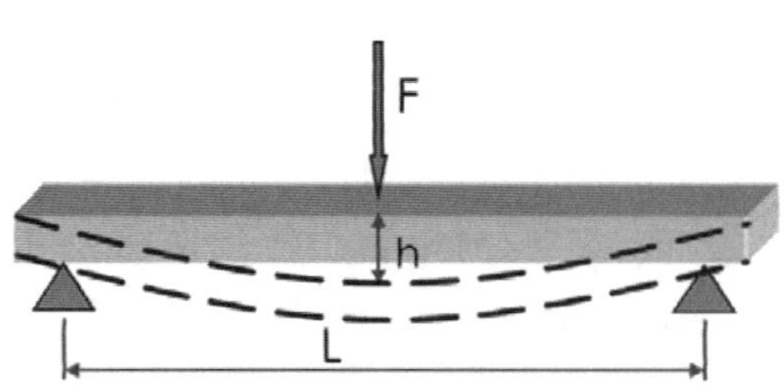

Figure 2. Scheme of the bending test

The bending strength (*Rmb,* MPa) of the materials tested is determined by the following formula:

$$R_{mb} = \frac{3FL}{2bd^2}\ [MPa] \tag{2}$$

Where: F- the load (force) at the fracture point, N; L- the length of the support span, mm (L = 107 mm); b- the width of the sample, mm (b = 30mm); d - thickness of the sample, mm (d = 6 mm);

Furthermore, the test results were used for determination of the vertical displacement (h, mm) at maximum bending load, which is an indication of flexural deformation

2.3. Determination of the density of the composites

The calculations of the density of the composites studied were performed using the conventional hydrostatic weighing method based on Archimedes' law. The conventional hydrostatic weighing method uses water as a primary standard for measuring the density. The size of the test samples used (three for each type of composite) is 30 mm / 30 mm [8, 9]. The density (ρ, g/cm^3) of the materials tested is determined by the following formula:

$$\rho = \frac{m}{V}\ [g/cm^3] \tag{3}$$

Where: m – the sample mass, g; and V – the sample volume, cm^3;

3. Results and analysing

Table 1 shows the tensile test results obtained. The tensile strength obtained for biaxial reinforced vinylester matrix composites (BRVMCs) is slightly higher than that of fiberglass reinforced polyester matrix composites (FRPMCs), which can be seen in the diagram in figure 3. The greater values of BRVMCs are most probably due to the fact that the fibers of the biaxial texture are intertwined on top of each other at an angle +/- 45 °. This is also confirmed by the macro-fractographic analysis (figure 4), where indicates that in BRVMCs the crack propagation is tangential, unlike FRPMCs, where the crack propagation is linear, perpendicular to the tensile load. Tangential crack propagation requires more work of fracture than linear crack propagation.

Table 1. The obtained tensile test results

Sample	Fm (N)	S_0 (mm^2)	Rm (MPa)
$BRVMC_1$	16440	120	137
$BRVMC_2$	16420	120	136.8
$BRVMC_3$	16500	120	137.5
$FRPMC_1$	13850	120	115.4
$FRPMC_2$	13690	120	114.1
$FRPMC_3$	13820	120	115.2

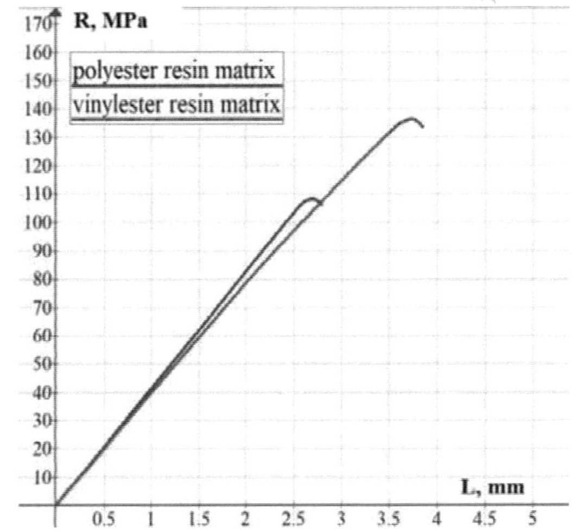

Figure 3. The tensile test diagram of the investigated composites

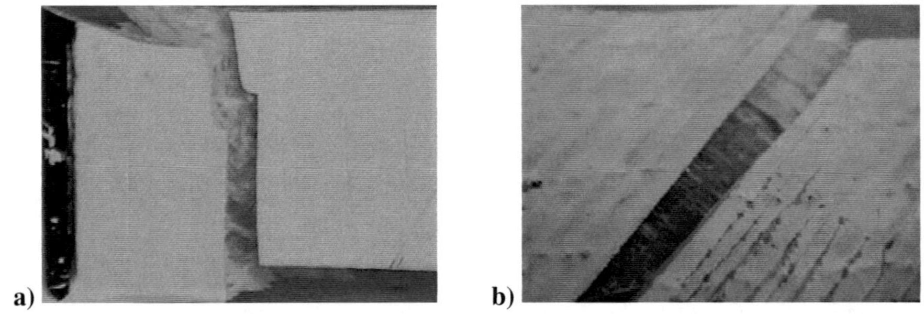

Figure 4. Macro-fractographic analyse: a) crack propagation of the FRPMCs; b) crack propagation of the BRVMCs

The results of the bending test are given in table 2. The bending strength values obtained for BRVMCs are higher than those for FRPMCs, but in this case, the difference is significant - twofold (figure 5).

Table 2. The obtained bending test results

Sample	*F* (N)	*L* (mm)	*Rmb* (MPa)
$BRVMC_1$	2080	107	289.8
$BRVMC_2$	2070	107	288.4
$BRVMC_3$	2085	107	290.5
$FRPMC_1$	1000	107	148.6
$FRPMC_2$	1020	107	151.6
$FRPMC_3$	1015	107	150.8

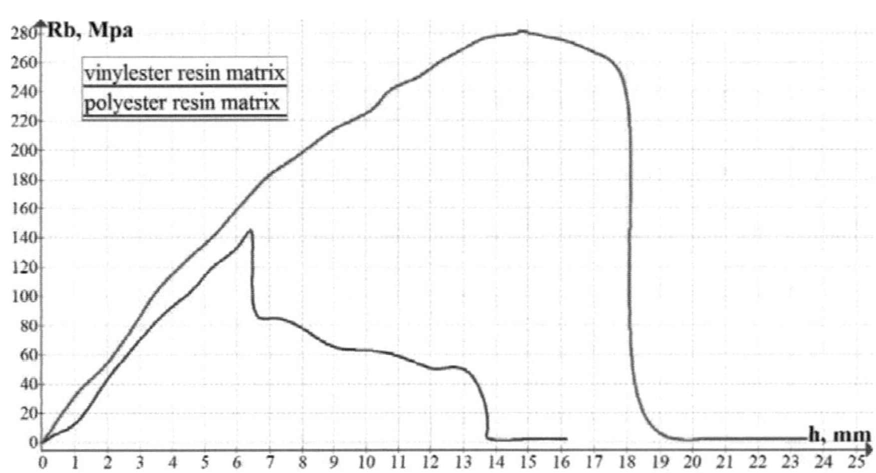

Figure 5. The bending test diagram of the investigated composites

In addition, the vertical displacement of the BRVMCs samples at maximum load is twice as great as that of FRPMCs samples, which indicate that they haves a higher degree of deformation. This is due to both the reinforcement phase of biaxial texture and the vinylester resin matrix material. From the fractographic analysis (figure 6) of the fracture surface of the FRPMCs samples, complete fracture of the layers of the reinforcement phase was observed upon removing the load, while in the BRVMCs samples the integrity of the inner layer under compression was preserved.

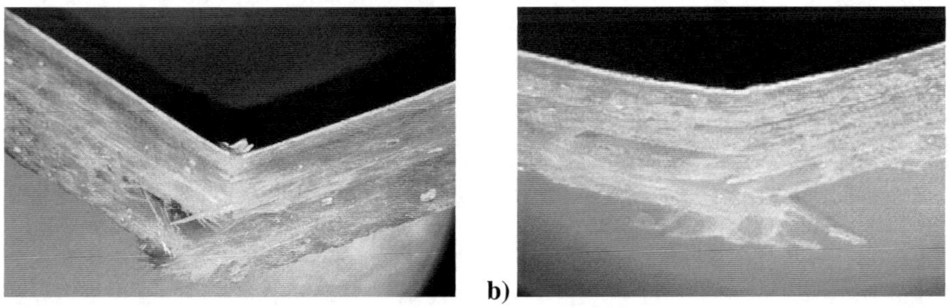

a) b)

Figure 6. Macro-fractographic analyse of the fracture surface: a) of the FRPMCs; b) of the BRVMCs

Table 3 indicates the average values of the mechanical and physical characteristics of the composites tested, which can be used for a complete comparative analysis. It can be seen that the density of the two materials is almost the same, but BRVMCs have better mechanical properties than FRPMCs, especially in terms of flexural strength. This fact is particularly important taking into account that the facilities operating in marine environment are subjected mainly to bending stresses.

Table 3. the average values of the mechanical and physical characteristics of the investigated composites

Sample	Density (g/sm^3)	Tensile strength *Rm* (MPa)	Bending strength *Rmb* (MPa)
BRVMC	1,46	137	290
FRPMC	1,49	115	150

4. Conclusion

The investigation of the suitability of Fiber Reinforced Polymer Matrix Composites for the manufacture of facilities operating in marine environment has shown that:

1. The density of both materials is approximately the same, lower than that of the traditional metallic materials used for manufacturing equipment operating in marine environment.
2. BRVMCs have higher tensile strength and significantly higher bending strength compared to FRPMCs, which is mainly due to the fact that its reinforcement phase is composed not only of fiberglass but also of biaxial fabric.
3. Both composite materials are suitable for the construction of equipment and facilities operating in marine environment, but taking into consideration that the equipment

operating in the marine environment is subjected mainly to bending stresses, BRVMCs would be the more suitable material as compared to FRPMCs.

References

[1] Gowda T, Sanjay M, Bhat K, Madhu P, Senthamaraikannan P, and Yogesha B 2018 Polymer matrix-natural fiber composites: An overview *Cog. Eng.* **5(1)** 1446667
[2] Yusriah L, Mariatti M and Abu Bakar A. 2010 The properties of vinyl ester composites reinforced with different types of woven fabric and hollow phenolic microspheres *J. of Reinfor. Plast. and Comp.* **29.20** 3066-73
[3] Caceres A, Jamond R, Hoffard T, and Malvar L 2000 Accelerated testing of fiber reinforced polymer matrix composites *NFESC Spe. Public* 4-14
[4] Liu Q, Xue Z, Liu Z, Gao H, Zhang R, Li W and Yang D 2018 Industrial polymer matrix composites and fiber-glass-reinforced plastics *Comp. Mater. Eng.* **2** p 165-04
[5] Azmi A 2012. *Machinability study of fibre-reinforced polymer matrix composites* Doctoral dissertation Research Space@ Auckland pp 4-6
[6] Spasova D 2019 Investigation of the structure and the properties of 18XГТ (1.2162) Steel Gear-Wheel Surface Hardened by Vacuum Carburizing. *Int. J. "NDT Days"* **II** 569- 75
[7] Argirov Y, Mechkarova T, Stoyanova A, Atanasov N and Konsulova-Bakalova M, 2020 Study on structure and mechanical properties of 20X13 steel welding joints. *IOP Publ.* **843** 012012.
[8] Stoyanova A, Mechkarova T, Argirov Y, Konsulova-Bakalova M, and Atanasov N 2020 Study of structure and physico-mechanical properties of welding joints on vessel tank of austenite steel SS316 *IOP Publ.* **843** 012013
[9] Sukhon R, Tulasombut V, Laopornpichayanuwat W, Pangviwat T, and Mitaree M 2011 Measurement of the volume and density of weight by hydrostatic weighing method *SPC* 1-4

Computer modelling-based selection of accelerated cooling parameters for advanced high-strength structural steel

V G Efremenko [1], D S Zotov[2], V I Zurnadzhy[1], R A Kussa[2], V I Savenko[3], R I Sagirov[2], O A Bocharova[1] and A V Efremenko[1]

[1] Pryazovskyi State Technical University, Universtiteskaya Str. 7, Mariupol, 87555, Ukraine
[2] PJSC "AZOVSTAL IRON & STEEL WORKS", Leporskyi Str. 1, Mariupol, 87555, Ukraine
[3] Kyiv National University of Construction and Architecture, Povitroflotskyi Av. 31, Kyiv, 03037, Ukraine

Corresponding author's email: vgefremenko@gmail.com

Abstract. The article describes the working principles and configuration of the computer system applied for selecting parameters of accelerating water cooling of rolled sheets made of low-carbon advanced strength steel. The system is based on mathematical models using empirical regression equations that describe the interconnections between working parameters of cooling equipment and the technological parameters of thermal processing. The equations are outlined and discussed. The results of computer modelling verification in industrial trials are presented. The case-study involved computer system concerning steel properties and microstructure is described.

1. Introduction

Advanced high-strength steels (AHSSs) are constructional steels having ultimate strength higher than 550 MPa [1]. AHSSs' sheets are intended for use in automotive and structural applications to reduce construction weight due to higher steel strength [2]. According to standard approach, AHSSs are subjected to heat treatment in order to acquire the necessary mechanical properties [3, 4]. Some of the AHSSs grades are manufactured via the process involved hot rolling (thermo-mechanical controlled processing - TMCP) with consequent accelerated cooling (AC) by water [5]. For this purpose accelerated cooling unit (ACU) is used being installed into the production lines after the hot-rolling mill [6]. ACU combines a rolling conveyor (for sheet transportation) and a cooling water supplying system. «AZOVSTAL IRON & STEEL WORKS» (further «AZOVSTAL» in short) is the only Ukrainian steel producer equipped with ACU. AC technology is energy-saving and environment friendly since the heat treatment is fulfilled without additional heating in fuel furnace. However, AC needs using precise processing parameters which are dependant on steel chemical composition, sheet size/temperature, water flow/temperature, etc. The selection of AC processing parameters should be controlled automatically to avoid the human factor. A computer-based controlling system is one of the most important parts of ACU. As usual, such systems with appropriate software are developed by ACU manufacturer to be installed together with ACU equipment. This software should use the mathematical models that are adequate for specific production conditions.

IOP Conf. Series: Materials Science and Engineering 1037 (2021) 012030 doi:10.1088/1757-899X/1037/1/012030

2. The approaches to control accelerated cooling parameters

Steel mechanical properties are greatly affected by microstructure status resulted from technological processing. The microstructure results from phase transformation occurring in steel mostly under the cooling from the austenite temperature range. The path of phase transformation depends on steel chemical composition as well as on initial metal temperature before cooling, cooling rate, and the temperature of cooling interruption [7, 8]. As reported by Tang et al. [9] accelerated water cooling of TMCP-processed low-carbon low-alloy constructional steels leads to a fine-grained structure comprising quasi-ferrite, granular bainite and martensite/austenite "islands" ensuring advanced tensile properties in combination with high ductility and high low-temperature impact toughness. In TRIP-assisted AHSSs multi-phase structure forms, consisting of polygonal ferrite, carbide-free bainite and retained austenite which is able to excellent strain hardening under the load [10-12]. The selection of AC regime parameters should be specified regarding time/temperature peculiarities of phase transformation in certain steel grade. The temperature of AC start (T_{start}) defines the starting structural condition which prerequisites the final structure [13]. The proper selection of the temperature of AC finish (T_{finish}) is also of great importance since T_{finish} defines the temperature range of overcooled austenite transformation [9, 13, 14]. Cooling rate (CR) should ensure avoiding austenite transformation in a higher temperature range which could lead to a coarse structure with poor mechanical properties [9, 15]. Thus T_{start}, T_{finish}, and CR are the key processing indicators of steel sheets accelerated cooling. They should be precisely ensured by using appropriate ACU working parameters that presumes involving an automatic controlling system.

Each accelerated cooling unit is mandatorily equipped with a controlling system for automatic regulation of steel processing. The development of such a system and appropriate software should be based on the latest approaches of modelling, simulation, and automation [16-18]. Thus Knapiński et al [19, 20] applied physical simulations and numerical modelling of the plates accelerated cooling to govern the temperature distribution on the cross-section of the cooled plates of X100 grade steel plate. Chen et al [21] developed a basic automation subsystem and process automation subsystem intended for using in device for ultra-fast cooling installed in the steel plate production line. There the cooling schedule is set by process control system using the finite element temperature field model and the temperature homogeneity model. These models are based on a multi-dimensional self-learning strategy [22] which ensures precise calculation of steel strip temperature. The automation system for ACU includes "Programmable Logic Control" (PCL) devices to operate water flow/temperature [21, 23]. Since any metal processing operates in uncertainty conditions of parameter variation [24] a "closed-loop control" concept can applicable to steel sheets accelerating cooling in order to reduce inconsistencies in product properties [25].

Routinely steel sheets are manufactured being collected in separate batches. The batch presumes the sheets of certain size made of steel of specific grade with specific mechanical properties. The on-going manufacturing process is characterized by frequent alteration of sheet batches subjected to AC; thus AC processing indicators are changed as well. Accordingly, ACU working parameters have to be quickly changed to follow the current manufacturing situation. The change could be done manually by the operator or automatically by the controlling system.

Operation of ACU in «AZOVSTAL» revealed the problem which often happens when after the automatic regulation of ACU working parameters the current AC indicators do not meet the required values. In this situation, ACU working parameters are being adjusted further manually by the operator. The adjustment is performed over and over again until AC indicators reach the target level (usually it takes 3-5 «adjusting» sheets for each batch). That means that most «adjusting» sheets may gain inappropriate microstructure and mechanical properties followed by additional mechanical testing or even additional heat treatment. This situation increases manufacturing costs therefore it should be avoided. The mentioned problem appears when ACU controlling system uses inappropriate mathematical models. This is quite a usual situation since standard controlling software can not match all steel grades which are produced all over the world. Therefore some adjustments of the controlling system should be fulfilled focusing on the peculiarities of specific steel production.

The aim of the paper is to develop the mathematical models for automatic selecting ACU working parameters in order to ensure AC processing parameters focusing on hot-rolled sheets made of low-carbon advanced high strength steels produced by «AZOVSTAL».

3. Mathematic models for ACU functioning

Mathematical models for ACU working parameters selection were developed using the database provided by «AZOVSTAL». This database covered three latest years; it consisted of approximately 5.5 thousand records manually registered by ACU operators during the accelerated cooling of steel sheets. The records included the sets of ACU working parameters which were selected by the operators in "manual" regime. All records were divided into the next categories such as:

(a) The parameters of steel sheet batch (steel grade; sheet size (thickness, length));

(b) AC processing indicators (T_{start}, T_{finish}, CR);

(c) ACU working parameters (cooling water temperature, water flow from upper and bottom sides, number of water suppliers, rolling conveyor speed (RCS)).

These records were digitalized to be further processed according to the regression procedure using MS Excel Tables. The regressions were supposed to link ACU working parameters with steel sheet parameters and AC processing indicators. The rolling conveyor speed was chosen as the target working parameter which should be defined under the postulating other parameters/indicators values.

Taking into account the effect of carbon and alloying elements combination on the kinetics of phase transformation and steel mechanical behaviour the regression equations were developed separately for different steel grade groups. For this purpose, all steel grades produced in «AZOVSTAL» and subjected to accelerated cooling were divided into three groups namely: Cr-Si-Ni-Cu (group I), Mn-Si-V-Nb (group II), and Mn-Cr-Ni-Mo-Cu-V-Nb (group III).

The functional scheme of ACU parameters calculation is presented in Fig. 1. On the first stage the total water flow (TWF) is calculated using the data of steel grade, sheet thickness (δ, mm), water temperature (TW, °C), number of water suppliers (NS), and the targeted AC indicators such as T_{start} (°C), T_{finish} (°C) and CR (K/s). Then the total water flow is divided into water flow from upper and water flow from the bottom sides of ACU. On the second stage, the speed of rolling conveyor is calculated based on all above-mentioned parameters and indicators.

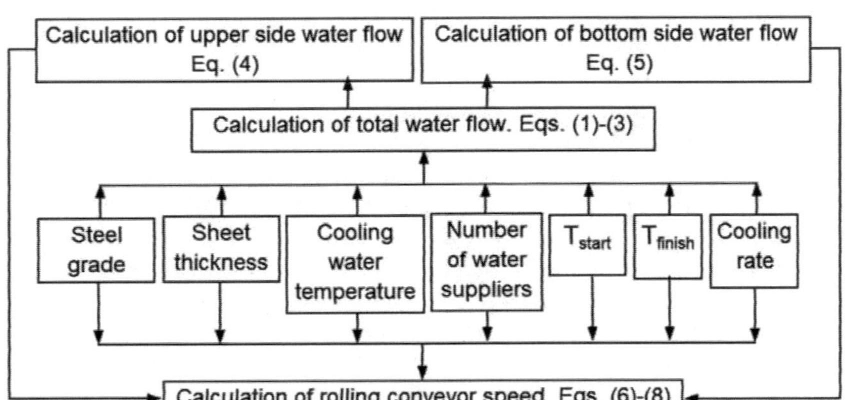

Fig. 1. Functional scheme of ACU parameters calculation

After processing the raw data the following mathematical models were obtained as regression equations (1)-(8) presented below.

The models for the total water flow (liters/min):

(a) for Cr-Si-Ni-Cu steels:

$$TWF_I = 56.7 \cdot T_W - 92.8 \cdot N_S + 34.5 \cdot \delta + 0.4 \cdot (T_{start} - T_{finish}) + 25.5 \cdot CR \ (R^2 = 0.99) \quad (1)$$

(b) for Mn-Si-V-Nb steels:

$$TWF_{II} = 9.2 \cdot T_W - 15.8 \cdot N_S + 30.6 \cdot \delta + 8.3 \cdot (T_{start} - T_{finish}) + 47.2 \cdot CR + 0.5 \cdot \delta^2 - 0.03 \cdot (T_{start} - T_{finish})^2 - 2.2 \cdot CR^2 + 0.01 \cdot N_c \cdot (T_{start} - T_{finish}) \cdot CR - 4.4 \cdot 10^{-9} \cdot \delta^2 \cdot (T_{start} - T_{finish})^2 \cdot CR^2 \quad (R^2=0.97) \qquad (2)$$

(c) for Mn-Cr-Ni-Mo-Cu-V-Nb steels:

$$TWF_{III} = 37.6 \cdot T_W + 60.6 \cdot N_S + 36.3 \cdot \delta + 0.02 \cdot (T_{start} - T_{finish}) + 23.1 \cdot CR \quad (R^2=0.90) \qquad (3)$$

Total water flow is divided into water flow from upper side (WFU) and water flow from bottom side (WFB) as:

$$WFU = TWC/(1+x) \qquad (4)$$

$$WFB = x \cdot WFU \qquad (5)$$

where "x" depends on steel grades to be in the range of 2.0 – 3.0.

The models for the rolling conveyor speed (RCS, m/s):
(a) for Cr-Si-Ni-Cu steels:

$$RCS_I = 0.006 \cdot T_W + 0.1 \cdot N_C - 0.2 \cdot \delta - 0.003 \cdot (T_{start} - T_{finish}) + 0.06 \cdot CR + + 3.9 \cdot 10^{-4} \cdot WFU + 3.8 \cdot 10^{-5} \cdot WFB \quad (R^2=0.99) \qquad (6)$$

(b) for Mn-Si-V-Nb steels:

$$RCS_{II} = 0.003 \cdot T_W + 0.09 \cdot N_C - 0.2 \cdot \delta - 0.003 \cdot (T_{start} - T_{finish}) + 0.06 \cdot CR + + 9.5 \cdot 10^{-4} \cdot WFU + 9.0 \cdot 10^{-5} \cdot WFB \quad (R^2=0.99) \qquad (7)$$

(c) for Mn-Cr-Ni-Mo-Cu-V-Nb steels:

$$RCS_{III} = -0.01 \cdot T_W + 0.2 \cdot N_C - 0.1 \cdot \delta - 0.008 \cdot (T_{start} - T_{finish}) + 0.07 \cdot CR - - 3.9 \cdot 10^{-4} \cdot WFU + 7.2 \cdot 10^{-4} \cdot WFB \quad (R^2=0.99) \qquad (8)$$

As seen, the determination coefficient for all equations is rather high (not less than 0.90) indicating the adequacy of regression models.

4. Models verification and software industrial trial approbation

Models reliability has been verified by comparison of calculated RCS values with the values taken from the registration journal for the same set of working parameters/indicators values. The comparison is shown in Fig. 2 for different steel grades. As seen, the variation of calculated RCS values well follows the variation of actual values of rolling conveyor speed. It was found that the difference between predicted and actual RCS values (shown as red line) in 90-92 % of cases does not exceed 0.25 m/s, which is acceptable for AC processing. Moreover, the cases exceeding 0.5 m/s account only 1% of the total number of measurements. These results allowed to conclude that above mentioned mathematical models are adequate for the production conditions of «AZOVSTAL».

The next stage of research was a trial approbation of developed models in real manufacturing situations. For this purpose, the software «Controlled Cooling Calculator» was developed basing on the models (1)-(8). It was written in the language C# with the adoption of WPF system. The software consists of the «Calculation unit» and «Correction unit». The first one contains the input fields for the selection of steel grade and for inserting assigned working parameters and process indicators. In «Calculation unite» the calculated values of water flow (upper and bottom) and rolling conveyor speed appear. These three values are input by the operator into ACU before the start of AC processing of the coming up sheet batch. The «Correction unite» is intended for quick correction of working parameters if some of the processing indicator (CR or T_{finish}) is out of the required range after the initial calculation.

The cooling rate is corrected by RCS alteration with corresponding water flow correction. The temperature of AC finish is corrected by water flow tuning while the initial RCS remains unchanged.

Model-based software «Controlled Cooling Calculator» was approbated upon industrial trial held in the sheet-rolling shop of «AZOVSTAL». The software was used for computer selection of ACU working parameters under manufacturing of 31 batches of steel sheet of 17.5-50 mm thick made of different steel grades. The calculated ACU working parameters (WCU, WCB, RTS) were input into ACU before the start of processing each sheet batch. The reached values of AC indicators (T_{finish}, CR) were registered accordingly.

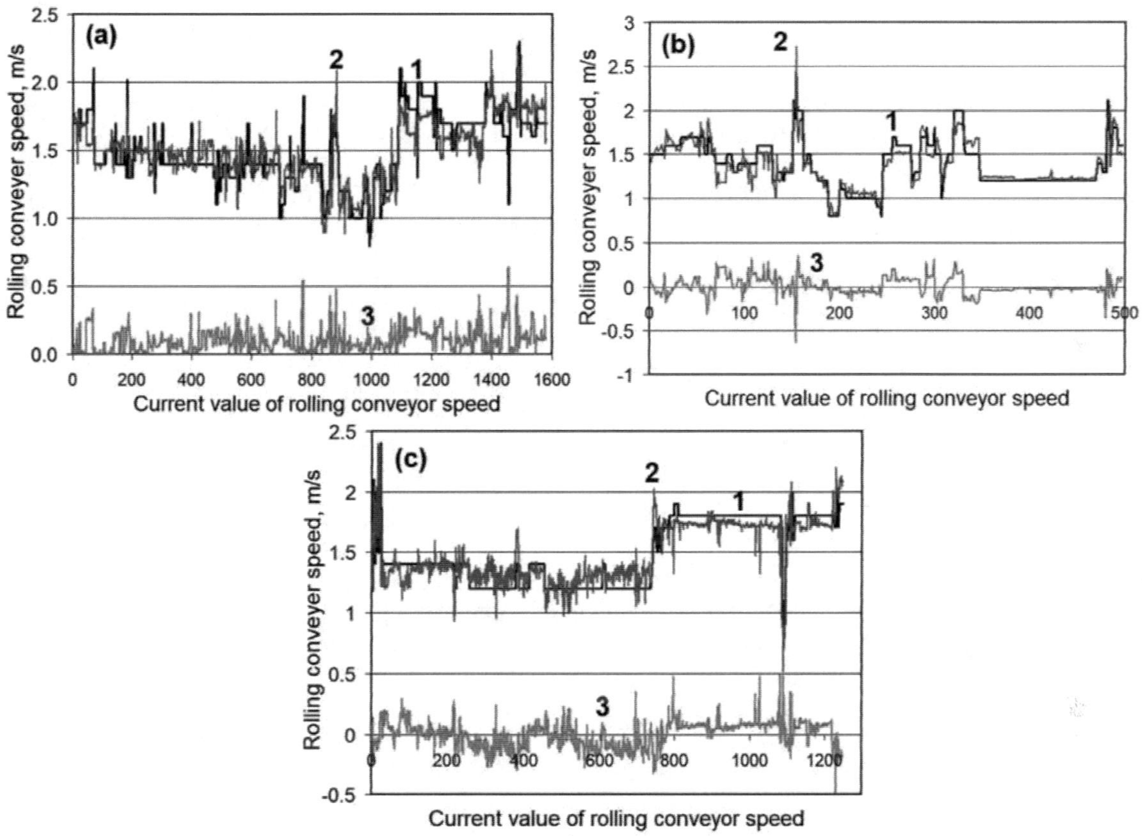

Fig. 2. The comparison of calculated RCS values and actual RCS values for steel grades (a) Cr-Si-Ni-Cu, (b) Mn-Si-V-Nb and (c) Mn-Cr-Ni-Mo-Cu-V-Nb (1 – actual values (taken from the database), 2 – calculated values, 3 – the difference).

The results showed that under «Controlled Cooling Calculator» guiding the required AC processing indicators (with acceptable deviation in the range of ±25 °C for T_{finish} and ±2 K/s for CR) were obtained in 96.7 % of batches on the first-second sheet of the batch. Moreover, the target indicators values were obtained mostly on the first sheet of the batch: in 73.3 % cases for T_{finish}, and in 76.7 % for CR. Only in one batch (3.3 %) the required CR value was gained on the third sheet. These results revealed the acceptable adequacy of the models (attaining target indicators on the first sheet) as well as the workability of correction unite.

The appropriateness of the computer-based modelling were illustrated for one batch of 30 mm-thick sheets made of steel X70 grade which belongs to Mn-Cr-Ni-Mo-Cu-V-Nb steels. This steel grade is intended for large-diameter pipe lines for gas transportation under the high-pressure thus it belongs to high-importance products. Therefore it implies the improved combination of the mechanical properties. The target AC indicators for this batch were appointed as T_{start} is 765 °C, T_{finish} is 590 °C, CR is 16 K/s.

Using software «Controlled Cooling Calculator» the ACU working parameters were calculated as: (a) water flow upper is 401 liters/min, (b) water flow bottom is 962 liters/min, and (c) rolling conveyor speed is 1.84 m/s. These parameters were generated under the water temperature is 20 °C, and the number of water suppliers is 14.

These parameters ensured actual AC indicators being in required limits, specifically: T_{start} is 766.6±2.4 °C, T_{finish} is 593.4±3.7 °C, CR is 15.8±0.4 K/s. Accordingly, steel gained advanced mechanical properties such as: yield strength (YS) of 555±8 MPa, ultimate tensile strength (UTS) of 682.5±8 MPa, elongation of 23.5±1 %. Notably, an improved low-temperature (–20 °C) impact toughness was reached as well: fracture energy (KV) of 206.5±2 J, drop weight test (DWTT) of 84.7±2 %. Mentioned values fully met the standard requirements for X70 grade which are: YS ≥ 505 MPa, UTS ≥ 595 %, elongation ≥ 22 %, KV_{-20} ≥ 125 j, $DWTT_{-20}$≥ 85 %.

These properties have resulted from optimal microstructure obtained by computer-based selection of ACU working parameters. The microstructure consists of fine-grained quasi-polygonal ferrite of 10-11 number, granular bainite and a minor amount of pearlite bands oriented along rolling direction (Fig. 3). Such structural status is beneficial for an improving complex of strength, ductility and impact toughness of the steel, intended for exploitation (as a pipe-line) at subzero temperature [7, 26].

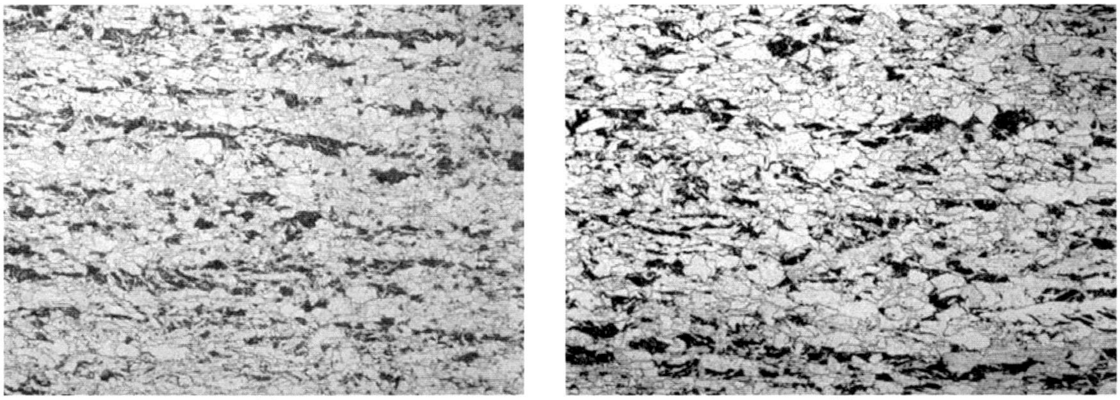

(a) (b)

Fig. 3. The microstructure of 30 mm steel sheet of X70 grade heat treated using «Controlled Cooling Calculator» software: (a) at ¼ of sheet thickness, (b) in sheet axis zone. Magnification is x500.

Summing up, the use of «Controlled Cooling Calculator» software allows shortening the number of «adjusting» sheets at least 3 times thus reducing additional production expenses in order of 25-40 $ per every sheet batch. The developed models and software prove to be perspective for full implementation into the rolled sheets manufacturing process in «AZOVSTAL IRON & STEEL WORKS».

5. Conclusion

Mathematical models were developed to calculate the working parameters of water cooling equipment in order to ensure an appropriate steel sheet heat treatment (accelerated cooling after thermo-mechanical controlled processing). The models are regression equations including sheet size, the required technological indicator (T_{start}, T_{finish}, cooling rate) and ACU working parameters (number of water suppliers, water flow upper, water flow bottom, rolling conveyor speed). The models were verified to be adequate for the manufacturing of advanced high strength steels at «AZOVSTAL IRON & STEEL WORKS».

The models were used as a basement for developing software «Controlled Cooling Calculator» (C++, WPF) for the selection of ACU working parameters focusing on target indicators of accelerated cooling for steel sheets of different thickness and steel grades (advanced high strength steels). The software was successfully tested under manufacturing of 31 batches of steel sheet of 17.5-50 mm thickness made of X70 grade. It allowed producing high-quality steel sheets reducing the additional manufacturing costs caused by adjusting the working parameters to the target values. The cost reduction may amount to 25-40 $ per every sheet batch.

6. References

[1] Da Silva E A, Malerba F, L F V., De Jesus Silva, J W., Ribeiro R B., Dos Santos Pereira M 2016 *IOSR Journal of Mechanical and Civil Engineering.* **13(5)** 21-27
[2] Bleck W, Guo X and M Y 2019 Steel Research International **88(10)** 1700218
[3] Bublíková D, Jeníček Š, Vorel I and Mašek B 2017 *IOP Conf. Series: Materials Science and Engineering.* **179** 012009
[4] Zurnadzhy V I, Efremenko,V G., Wu K M, Azarkhov A Yu, Chabak Yu G, Greshta V L, Isayev O B and Pomazkov M V 2019 *Mater. Sci. Eng., A.* **745** 307-318
[5] Nakata N and Endo S 2014 *JFE GIHO.* **33** 1–6
[6] Xiangwei K and Liangyun L 2014 *Procedia Eng.* **81** 114-119
[7] Hu J, Du L-X, Wang J-J, Gao C-R, Yang T-Z, Wang A-Y and Misra R D K 2013 *Metall. Mater. Trans. A.* **44(11)** 4937-4947
[8] Kim S, Lee S, Lee B S 2003 *Mater. Sci. Eng., A.* **359** 198–209
[9] Tang S, Liu Z Y, Wang G D and Misra R D K 2013 *Mater. Sci. Eng., A.* **580** 257-265
[10] Efremenko V, Shimizu K, Pastukhova T, Chabak Yu, Brykov M, Kusumoto K and Efremenko A 2018 *Int. J. Mater. Res.* **109(2)** 147-156.
[11] Rykavets ZM, Bouquerel J, Vogt J-B, Duriagina Z A, Kulyk V V, Tepla T L, Bohun LI. and Kovbasyuk T M 2019 *Uspehi Fiziki Metallov.* **20(4)** 620–633
[12] Efremenko V G, Zurnadzhi V I, Chabak Y G, Tsvetkova O V and Dzherenova A V 2017 *Materials Science.* **53(1)** 67-75
[13] Matrosov Y I, Kholodnyi A A, Matrosov M Y, Popov E S, Konovalov G N and Sosin S V 2015 *Metallurgist.* **59** 60–68
[14] Hu Z, Xu Z, Xu G, Yang H, Zhang C and Yu R 2014 *J. Mater. Eng. Perform.* **23(12)** 4216-4222
[15] Chabak Yu G, Fedun V I, Pastukhova T V, Zurnadzhy V I, Berezhnyy S P and Efremenko V G 2017 *Probl. At. Sci. Technol., Ser.: Plasma Phys.* **110(4)** 97-102
[16] Hong L, Han C., Chang C and Hongchao Y 2019 *Int.J. Heat Mass Transfer.* **143** 118342
[17] Qian X, Zhenyi H, Qingyu H, Long Z and Jiajia C 2019 *ISIJ International.* **59(1)** 113-121
[18] Markov O E, Gerasimenko O V, Kukhar V V, Abdulov O R and Ragulina N. V 2019 *J. Braz. Soc. Mech. Sci. Eng.* **41** 310
[19] Knapiński M, Dyja H, Kawałek A, Kwapisz M and Koczurkiewicz B 2013 *Solid State Phenom.* **199** 484–489
[20] Laber K B, Knapiński M, Kawałek A and Kwapisz M 2013 *Solid State Phenom.* **199** 478–483
[21] Chen X L, Wang B X, Tian Y, Yuan G, Wang Z D, Wang G D 2012 *Advanced Materials Research.* **605–607** 1836–1840
[22] Chen D, Li Z-L, Li Y-J and Guo Y 2019 *ISIJ International.* **60(1)** 136-142
[23] Hamzah M A, Rasha F, Nadhim K and Raheem A 2018 *IIER International Conference*, 99-104
[24] Kukhar V V, Grushko A V and Vishtak I V 2018 *Solid State Phenom.* **284** 408–415
[25] Allwood J M, Duncan S R, Cao J, Groche P, Hirt G, Kinsey B, Kuboki T, Liewald M, Sterzing A and Tekkaya A E 2016 *CIRP Annals.* **65(2)** 573-596
[26] Koval' A D, Efremenko V G, Brykov M N, Andrushchenko M I, Kulikovskii R A and Efremenko A V 2012 *Journal of Friction and Wear.* **33(2)** 153–159

IOP Conf. Series: Materials Science and Engineering 1037 (2021) 012031 doi:10.1088/1757-899X/1037/1/012031

Analyze of shape memory wires behavior under external solicitation using finite elements analysis (FEA)

L Burlacu[1], C Dănilă[2], T I Birnoveanu[1], N Cimpoeşu[1*] and LG Bujoreanu[1]

[1]Materials Science and Engineering Faculty, Technical University "Gheorghe Asachi" from Iasi, Romania
[2]Mechanical Faculty, Technical University "Gheorghe Asachi" from Iasi, Romania

*nicanor.cimpoesu@tuiasi.ro

Abstract. Nitinol shape memory alloys present many applications in the industrial areas based on exceptional memory effect properties. The behavior of a shape memory alloy (superelastic effect) underwire form (four diameters: 1, 1.5, 2 and 2.5 mm and three lengths: 75, 150 and 375 mm) at drawing was analyzed using finite element analysis (FEA). The wire shape was C letter. The experimental temperature test, in a first step, was 23 °C (room temperature) and we analyze the appliance of four solicitation forces in order to determine the specific deformations and tensions. In second simulation step we apply a higher temperature, respectivelly 80 °C, in order to analyze the behavior of the element through transformation phase from martensite to austenite. The simulation provides information on the total deformation, equivalent stress, and equivalent elastic deformation of the intelligent elements under external solicitations pointing the appearance of maximum and minimum solicitation areas.

1.Introduction

The role of shape memory alloys in the field of fire protection is clear: a shape memory element can detect an increase in temperature and, at the same time, react to the increase, performing a work [1-3]. The various applications of shape memory alloys in fire protection can be divided into three groups according to the action to be performed by the intelligent element: fire detection (detection of rising temperature), detection and operation by realiying a work and simply actuation of the element. In the last group, the shape memory element is used only to directly convert electrical energy into mechanical energy. In the case of simple actuation, all the advantages that electronics currently brings, can be used [4, 5].

Detection of a change in temperature can be achieved by sensing the modifications in resistivity associated with the martensite-austenite transformation or by the shape memory process itself. Both aspects can be exploited for practical applications. Like conventional materials, shape memory alloys show a slow increase of resistivity as temperature increases, but when heating is continued in the transformation temperature range, there will be a sudden and marked reduction of the resistivity of about 10 to 20%.

Another possibility for making a linear detector is the use of the shape memory effect itself. Figure 3 shows the contraction of a Cu-Zn-Al wire according to the temperature. This reduction (dimensional change) is sufficient to activate an alarm system using a micro-switch or an equivalent device. This change is quite complex in Ni-Ti alloys but very distinct in Copper-based alloys [6].

When a measuring circuit records a certain rate of decrease in the strength of the material, it indicates the temperature has risen and the temperature of transformation in the solid state has been reached. On the other hand, a sudden increase in the strength of the memory shape material indicates a decrease in temperature during the transformation process. Thus, a system can be secured against overheating, e.g. a fire alarm or electrical installation that can be secured against overload and undercover cooling. These types of detectors become the most interesting when large or long-length installations need to be protected, e.g. electrical cables. This can be easily achieved by using an alloy wire with shape memory [7].

If two elements with shape memory and different transformation temperatures are used and placed in series, a two-stage indicator can be made. Such detectors can activate an alarm before activating a fire extinguishing system.

Major accidents resulting in fires occur due to negligence, gas leaks, short circuits and overheating of electrical appliances, etc. During a fire-like accident, the fire should be stopped as soon as possible to avoid loss of life, properties and material goods. Currently, Halon fire extinguishing systems are being replaced by water spray systems with very fine droplets to suppress the fire even in areas where electrical appliances are involved [1, 2].

The literature investigated the effects of several important factors, such as the water spray modality, the size of the water drop and the water spray flow rate on the fire extinguishing mechanism, and proved that water spraying with solid cones and smaller size water drops become very effective in fires extinguishing [1-3].

This article analyzes, in two stages, the behavior of NiTi alloy wires with different dimensions (diameter and length) on an external stress in the first stage and in the second stage when it is heated over the transformation temperature range. The alloy and the shapes of the smart element can be used for fire detection and prevention applications.

2.Preparing the experiment with Ansys program

Using the Ansys program, four solid elements were designed, figure 1 a), made of nitinol shape memory effect alloy with different diameters (1, 1.5, 2 and 2.5 mm - figure 1 b)) three different lengths, figure 1 c) in the form of the letter C [8].

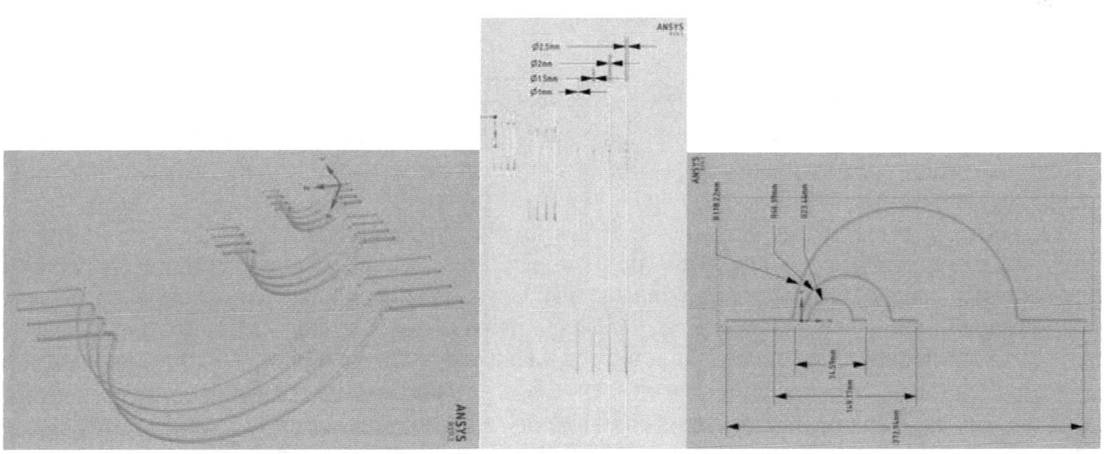

a) b) c)

Figure 1. Schematic representation of the analyzed elements and their geometric shape (letter C shape) a) geometric dimensions of the analyzed NiTi elements b) solid element diameters c) connection length and radius for obtaining the C-shaped element

For practical applications, the dimensions of the active elements are very important, and their behavioral changes are relatively large and can decisively require the use of one material or another. In this regard, replacing an element with a diameter of 2 mm and 10 wires ,with one with a diameter of 0.2 mm can solve several operating problems, such as the speed of response and cooling of thermally sensitive elements, if they can successfully replace the single-stranded material. The elements of different lengths were connected to achieve the C shape, obtaining three sets of elements with four samples of distinct diameters. The behavior of the same alloy in the form of the letter C with twelve different dimensional parameters at external stress is sought, considering the opposite end, embedded.
Volumetric elements of SOLID185 type were used meshing the model [8, 9]. Having to deal with bodies with regular geometry, these elements were created using the sweep method. For the validation of the chosen meshing, was used the condition that by doubling the density of the network (mesh), the modification of the results does not exceed a percentage of 5%. The SOLID185 element is used for 3-D modeling of solid structures, figure 2. It is defined by eight nodes having three degrees of freedom at each node: translations in the nodal directions x, y and z.

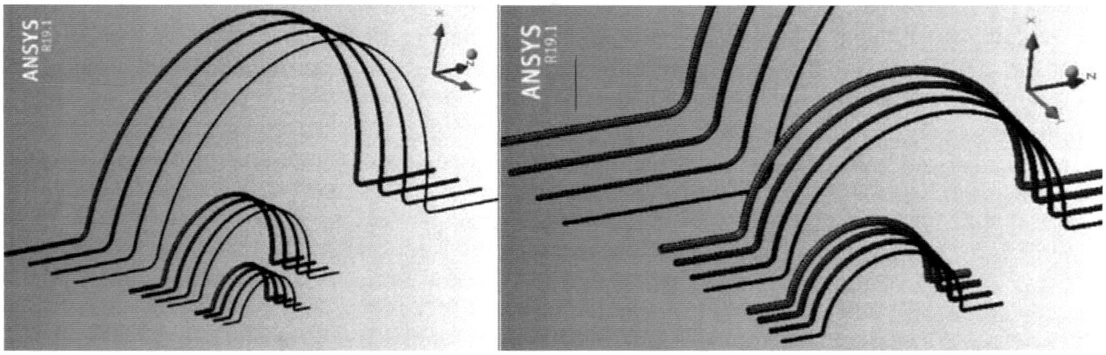

a) b)

Figure 2. Solid elements after meshing a) sets of four elements depending on diameters b) detail of the network mesh.

The element can be characterized by the following specific properties: Plasticity, hyper-elasticity, stiffness on load, creep, high deviation and high compression capacities. It also has a mixed formulation capability to simulate deformations of almost incompressible elasto-plastic and fully incompressible hyper-elastic materials.
In this experiment, two steps were applied to determine the temperature at which it is stressed, figure 3 a), for the first stage it was considered the room temperature of 22 ºC, stage in which the material is in martensitic state below M_f, for the mechanical stress that leads to the change, for example, of 5N and a second stage of heating the NiTi element with shape memory form, to 80 ºC, figure 3 b), for passing over the transformation temperature range, the second step with the cessation of the mechanical stress and the application of the temperature over A_f, to recover the initial shape[10].
After meshing, each NiTi wire from the shape memory alloy was divided into 39,984 elements using 214867 nodes. The wires, each separately, were stressed with the following forces: 0.03, 1.5, 3.5 and 5N, figure 2 a). The influence of the diameter and length of the wires on their behavior at various external demands was followed. This force produces either a phase transformation of austenite into martensite "detwinned" – the "stress-induced martensite scenario" or, in this case, leads to a self-induced reorientation resulting in "detwinned" martensite through the "reorientation scenario" ,depending on the relative position of the loading temperature and of the transformation temperature of the required NiTi alloys [11-15].

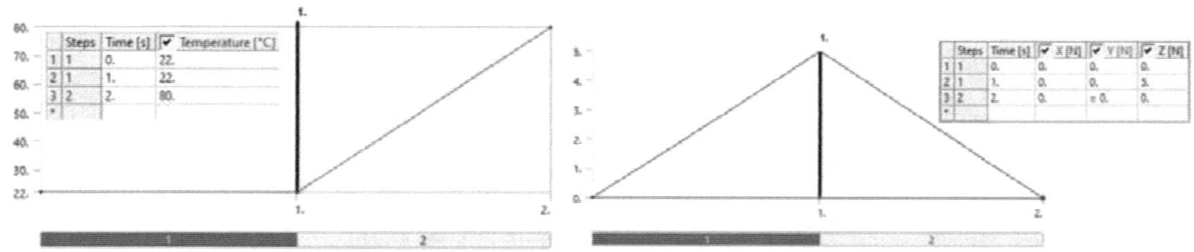

a) b)

Figure 3. Solicitation parameters a) temperature and b) stress

Increasing the hardness of the NiTi shape memory alloy and the corresponding introduction of defects in the microstructure also causes changes in the transformation temperatures and can lead to an augment in the generated stresses compared to the NiTi alloy hardened with larger grains and a lower density of dislocations. [6-8].

Properties used to make the AMF (NiTi) material model: Modulus of elasticity of austenite E^A: 70 GPa, Modulus of elasticity of martensite E^M: 30 Gpa, Poisson's ratio υ: 0.33, Thermal expansion coefficient of austenite α^A: 22x$10^{-6}K^{-1}$, Thermal expansion coefficient of martensite $\alpha^M 22x10^{-6} K^{-1}$, martensitic start transformation temperature M^{0S} 23 °C, martensitic end transformation temperature M^{0F}: -2 °C, austenitic start transformation temperature A^{0S}: 21 °C, austenitic end transformation temperature A^{0F}: 50 °C.

3. Analysis of experimental results

In the case of the initial shape recovery mechanism, the arching of a shape memory element produces a variable force, while the weight produces a force at a constant level [16], this type of actuation is desirable to keep the stress at a constant level and to prevent functional problems during the interval, such as changing transformation temperatures. Therefore, a constant level of demand is further analyzed. Under the direct action of tension, a high force, relatively small dislocations can be reached, even if under the torsion or bending load larger dislocations are achieved [17].

Therefore, the design of the "C-shape" geometry is analyzed, in which the bending loads dominate in order to reach the capacities to achieve high demands. This geometric shape can also be used for cyclic applications where it is necessary to make net displacements, for example as in the case of valves, figure 4 a).

Figure 4 shows the results obtained in the Ansys R19.1 program for b) the reaction force, c) the equivalent stress and d) the equivalent elastic deformation.

The equivalent stress is used when there is a multiaxial stress state with multiple stress components acting at the same time in the same structure. In such a case, we can use a selected criterion to transform the entire tension tensor into a single equivalent component, which can be treated as tensile stress and therefore used compared to the strength of the material [18-20]. Various criteria can be used, but among them, there is one with an incomparably higher popularity than the others - the von Mises efficiency or otherwise called the distortion energy criterion. It is commonly used in engineering, for example, finite element analysis programs use it as a default measure of stress. In the case of NiTi alloy the average stress values are between 0.225 Pa and 193.65 MPa, figure 4 b). Obviously the most requested areas are found on large and medium length wires in their middle area and in the case of long length wires at the fastening ends.

In terms of diameters, there are clear differences between thinner threads (d = 1mm) and thicker threads (d = 2.5mm). In the first case, there is a state of stress evenly distributed over the entire thread with lower values than for thinner threads, this result is confirmed for all types of lengths. Figure 4 a) shows the recesses and the direction of the applied forces as well as their values.

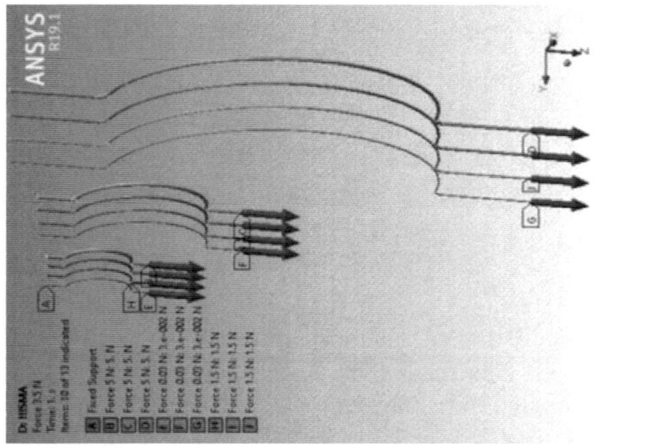

a)

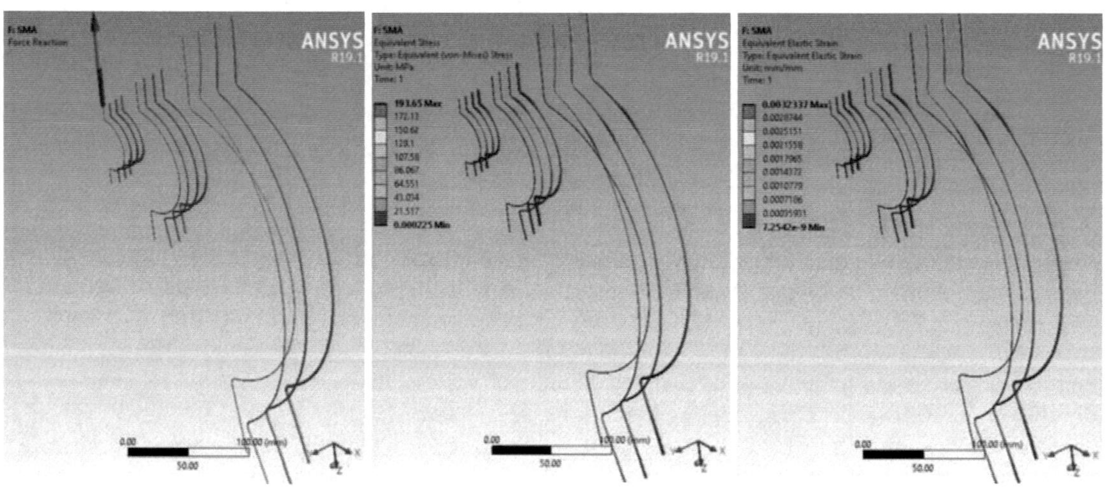

b) c) d)

Figure 4. Ansys results for a) reaction force, b) equivalent stress and c) equivalent elastic deformation

The equivalent elastic deformation is seen as an external control parameter of the deformation process at ambient temperatures. Von Mises deformation satisfies the theoretical properties of both groups, supporting its use for measuring equivalent deformation [21]. On the other hand, Hencky deformation does not satisfy the simple properties of the shear group, which assumes that it is not suitable for measuring the equivalent deformation in simple shear [22-23]. The equivalent elastic deformation has values between 7.2542 x 10^{-9} and 0.0032337 mm / mm which represents deformations from the level of picometers to that of micrometers.
All these local deformations applied to the 39984 elements, in which the Nitinol threads meshed, lead to displacements of the order of micrometers and even millimeters on a macroscopic scale.
The areas with higher deformations overlap with those of material stress, figure 4 b), in particular for the gripping parts of the elements and the middle area of the element C towards the end on which the stress is applied.

Figure 5 presents the results obtained from the Ansys analysis for Nitinol elements with various lengths and diameters a) total deformation, b) equivalent stress and c) equivalent elastic deformation. All the results below are scaled to 0.5, in reality they are deformed (elongated) 2 times more.
Figure 5 a) shows the evolutions of the total deformations of the experimental elements in Ansys after establishing the stress conditions. All elements regardless of the length or diameter of the wire are deformed from 0 to a maximum of 76.14 mm for the sample with a diameter of 1.5 mm.
The most deformed areas and marked in red are in the case of the longest wires at the end where the pulling force was applied. Further, the deformation zones decrease in intensity up to the embedded part, confirming the deformation of the material, especially in the areas with high stress, figure 5 b).
The intervals of variation of the considered elements represent a sum of the values obtained for all three wire lengths taken into account by coloring differently the actuation zones according to the values and with the indication of the maximum zones for the equivalent stress and for the equivalent elastic deformation.

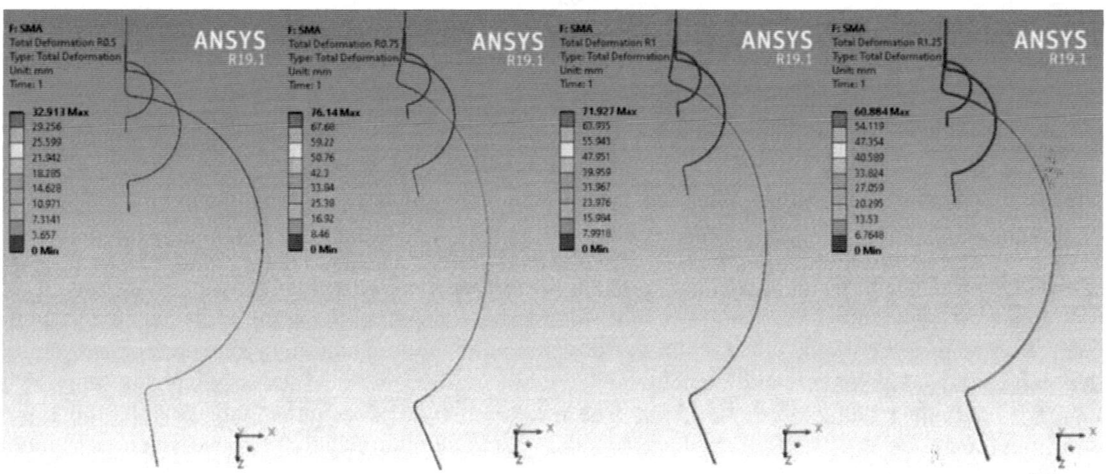

a)

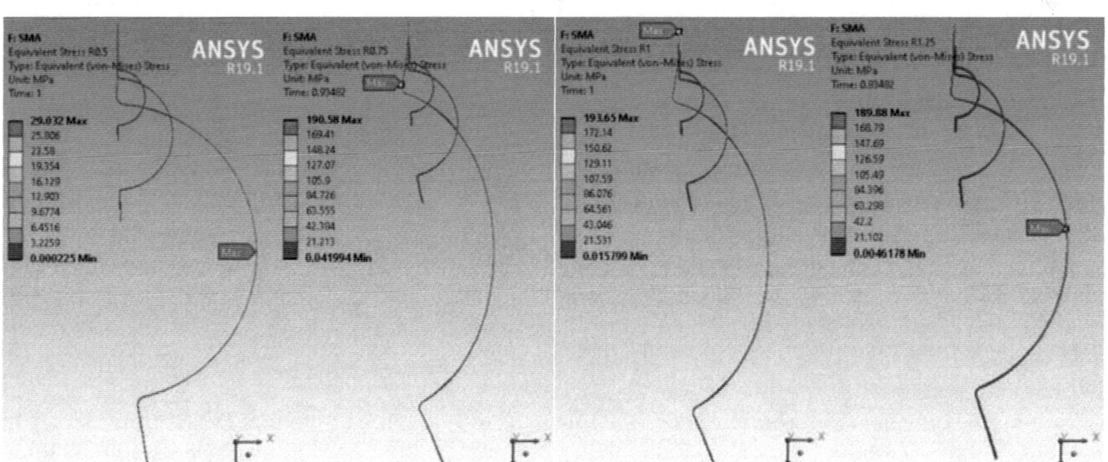

b)

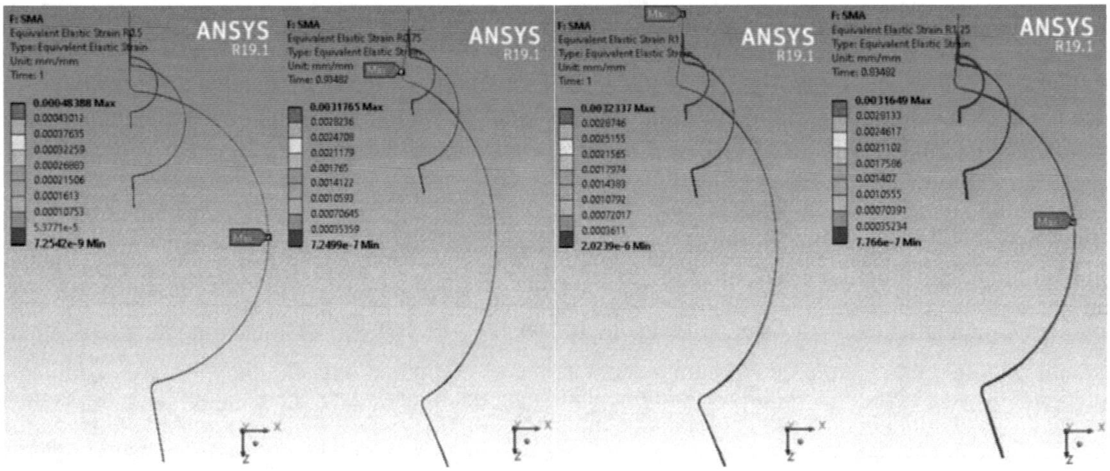

c)

Figure 5. Results from Ansys analysis for Nitinol elements with various lengths and diameters a) total deformation, b) equivalent stress and c) equivalent elastic deformation

The equivalent elastic deformation (mm / mm) shows different areas of the maxima that appear, figure 5 c), for cases R0.5 and R1.25 respectively the thinnest and thickest, taken in the analysis, the maximum areas are located in the middle of the wires (the wires with the longest lengths are considered) in the other two cases R0.75 and R1 they are located in the sampling areas. These confirm the areas with maximum equivalent demands, figure 5 b). The minimum equivalent deformations have the lowest values, obviously, for the thinner samples (R0.5) but between the other 3 thicknesses there are no big differences even the sample R1 presenting a higher minimum (the largest of them) compared to R1.25. In the case of stressing the material in a totally martensitic state and in which the reorientation stage takes place, which happens only in the first step, there is no phase transformation, this means that no heat is added to the system by plastic deformation.

4.Conclusions

Shape memory alloys, especially nitinol, have many applications in various industrial and medical fields. The properties of shape memory alloys can also be used in the field of fire protection and prevention as temperature sensitive elements.

The equivalent elastic deformation (mm / mm) shows different areas of the maxima that appear for cases R0.5 and R1.25 respectively the thinnest and thickest, taken into analysis, the maximum areas are located in the middle of the yarns (yarns with lengths are considered the largest) in the other two cases R0.75 and R1 they are located in the embedding areas of the samples. These confirm the areas with maximum equivalent demands.

The minimum equivalent deformations have the lowest values, obviously, for the thinner samples (R0.5) but between the other 3 thicknesses there are no big differences even the sample R1 presenting a higher minimum (the largest of them) compared to the sample with R1. 25.

In the case of stressing the material in a totally martensitic state and in which the reorientation stage takes place, which happens only in the first step, there is no phase transformation, this means that no heat is added to the system by plastic deformation.

References

[1] Mathew Y and Babu B G 2015 *Transactions of Famena* **39** 9
[2] Johnson A, Gilbertson R, Martynov V and San Leandro F 2010 *Shape Memory Alloy Fire Sprinkler Valve Actuator* (United States Patent Application 20100025050, CA)
[3] Ghafoori E, Neuenschwander M, Shahverdi M, Czaderski C and M Fontana 2019 *Constr. Build. Mater.* **211** 437
[4] Burlacu L, Cimpoeşu N, Bujoreanu LG and Lohan NM 2018 *IOP Conference Series: Materials Science and Engineering* **227(1)** 01
[5] Burlacu L, Cimpoeşu N, Lohan NM and Bujoreanu L G 2017 *Mater. Sci. Forum* **907** 8.
[6] Vitel G, Paraschiv AL, Suru MG, Cimpoesu N and Bujoreanu L-G 2011 *Optoelectron. Adv. Mat.* **5(8)** 858.
[7] Micu CA, Ciubotariu-Ana P, Lohan MN, Dumitru M, Bejinariu C 2019 *IOP Conference Series: Materials Science and Engineering* **572(1)** 012072.
[8] ANSYS Inc. 2009 *Ansys 12.1 Help*. Canonsburg, Pennsylvania.
[9] Lagoudas D C 1996 *Mech. Composite Mater. Struct.* **3** 153.
[10] Liang C 1992 *J. Eng. Math.* **26** 429
[11] Fuentes JMG, Gumpel P and Strittmatter J 2002 *Phase Change Behavior of Nitinol Shape Memory Alloys* (WILEY VCH, Germany) p. 437.
[12] Mirzaeifar R, DesRoches R and Yavari A 2011 *Continuum Mech. Therm.* **23 (4)** 363
[13] Auricchio F and Sacco E 2001 *Int. J. Solids Struct.* **38** 6123
[14] Cimpoesu N, Axinte M, Hanu RC, Achitei DC and Stanciu S 2010 *J. Optoelectron. Adv. M.* **12(8)** 1772
[15] Markopoulos A P and Manolakos D 2015 *Rev. Adv. Mater. Sci.* **42** 28
[16] Alipour A, Kadkhodaei M and Ghaei A 2015 *J. Intell. Mater. Syst. Struct.* **26** 554
[17] Auricchio F, Fugazza D and Desroches R 2007 J. Intell. Mater. Syst. Struct. 19 (1) () 47–61.
[18] Yu H and Young ML 2018 *Smart Mater. Struct.* **27** 015031.
[19] Arghavani J, Auricchio F and Naghdabadi R 2011 *Int. J. Plast.* **27** 940
[20] Alazzawi S 2019 *Designing a Smart Greenhouse Ventilation Window Based on NiTi SMA Actuator (*Doctorate Dissertation, SIUC, USA)
[21] Nemat-Nasser S and Guo W-G 2006 *Mech. Mater.* **38** 463
[22] Morin C, Moumni Z and Zaki W 2011 *Thermomechanical Coupling in Shape Memory Alloys under Cyclic Loadings: Experimental Analysis and Constitutive Modeling* (Elsevier Science B.V., Amsterdam) 1959
[23] Mohd Jani J, Leary M, Subica A and Gibson M A 2014 *Mater. Des.* **56** 1078

Paint removal from thermoplastic materials and its influence on the physical-mechanical properties for the recycling of the polymer

A Sover[1*], M. Zink[1], M. Michalak[1]

[1]Ansbach University of Applied Sciences, Faculty of Engineering Sciences, Residenzstr. 8, Ansbach, Germany

[*]Corresponding author's e-mail: a.sover@hs-ansbach.de

Abstract.
Laser technology offers many advantages when removing paint (coating) layers from different substrates such as metals, plastics or ceramics. Sensitive carriers with a low melting point like thermoplastic materials can easily be thermally damaged or destroyed using this technology. This paper analyses the quality of the thermoplastic material after removing the thermoset layers from its surface by laser and aims to identify the changes of the most important properties of the polymeric material. The physical-mechanical properties of thermoplastic materials which were coated were investigated before as well as after the removal of the coating layers and compared with parts which were not coated. The focus was on melt flow rate, impact strength, and tensile stress as well as strain. The results show that even a small amount of the paint (coating) can change the behaviour of the thermoplastic material and thus has an impact on further processing. Moreover, it was found that removing the thermoset coating from the thermoplastic material leads to a considerable improvement of its physical-mechanical properties and consequently also extends the possibilities of recycling (reuse) of this polymeric material.

1. Introduction

To protect resources and the environment, it is inevitable to improve the recycling paths for plastic materials. Painted plastic components, which are an integral part of numerous consumer goods and are particularly widespread in the construction and auto-motive industries [1, 2], constitute a great proportion of such materials whereas the painting/coating of such thermoplastic parts massively impedes recycling. Consequently, new paint stripping technologies are necessary [2]. During the last years, the adhesion between the thermoplastic part and the thin thermosets layer was substantially improved, and the technologies allow to have a high paint quality on plastic parts [4]. There are two different material types of such hybrid systems with different characteristics and proprieties, not only for their fabrication and use, but also for recycling [1]. While the thermoplastics can be heated and formed repeatedly, thermosets undergo a chemical change when they are heated, in which a three-dimensional network is created which impedes a re-heating and re-forming the parts. The combination of such different types of plastics is an obstacle to recycling. For example, the recycling of painted plastic parts such as car bumpers, fenders, grilles, etc. is still a challenge for the industry [1, 2].
The recycling of painted plastic parts poses numerous technical problems especially due to the thermoset layers as paint or protective varnish, which cannot be melted. Such a plastic part is shown in figure 1.

Figure 1 Painted car bumper (left) with the painted layer (right)

To recycling, a comparably small amount of coat proves to be so detrimental, that a separation of the thermoset layer from the thermoplastic carrier is necessary before starting the recycling of the material. To separate the paint from the thermoplastic carrier different treatments are available: chemical, mechanical, and physical.[1]

A new technology in paint removal is the use of lasers. Compared with the traditional surface cleaning technologies, laser cleaning has several advantages: a high cleaning rate, needlessness of toxic substances like cleaning agents, and low maintenance costs [1, 3]. Consequently, laser cleaning is widely used in the fields of paint removal which is presented in different articles [1, 2, 3, 4, 5]. A pulsed laser beam is directed onto the surface at a specific energy level. It breaks down the paint and primer coatings and produces a minimal waste stream. Powerful, very short laser pulses produce shockwaves and thermal load resulting in the sublimation and ejection of the target material [1, 4, 5, 6]. Various research works have contributed to improve this method for different substrate materials such as aluminium, steel or thermosets [4, 5, 6, 7].
In recently published articles [1, 2] it was demonstrated that this technology can be used also for paint removal of thermoplastic materials. Even on thermoplastic materials that have a lower melting and glass transition point, this technique can be used to clean the thermoplastic parts from the thermosets coating without damaging, i.e. carbonisation or melting, the plastic carrier. Damaging the surface can have an adverse effect on the quality of the recyclate materials. The above-mentioned articles illustrate this paint removal technology on polypropylene (PP) and acrylonitrile-butadiene-styrene copolymer (ABS).
However, this research paper investigates the quality of thermoplastic material after the application of this technology by testing the striped plastic parts and comparing them with the painted parts.

2. Preparation of the thermoplastic parts

For this research paper plastic parts of ABS Novodur were used as shown in figure 2. For the investigation and characterisation of the quality of the thermoplastic material the following parts for sample preparation were used: original ABS granulate, injected moulded part (not painted), paint striped part (the paint was removed by laser) and painted part (no paint was removed).

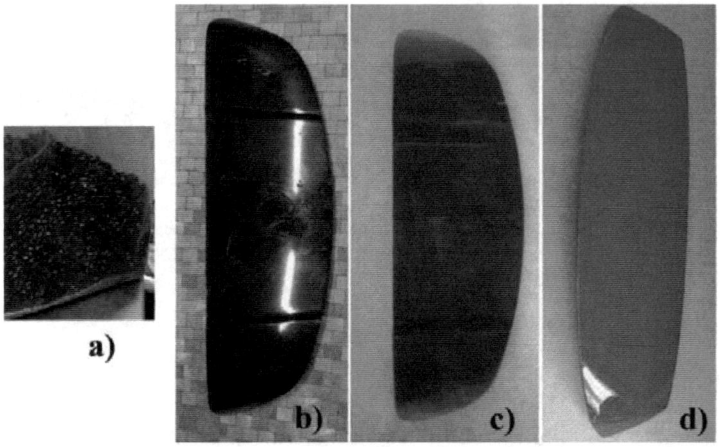

Figure 2. ABS parts used for sample preparation: a) granulate raw ABS; b) unpainted part; c) part with removed paint by laser; d) painted part

The plastic parts b), c), d) figure 2 were shredded by a cutting mill. All the plastic parts (a, b, c, d) were then processed by extrusion, using a single screw extruder, and thus granulated. Afterwards the granulate of each processed part was prepared by injection moulding to form the samples for tensile testing and Charpy impact test or for measuring the melt mass flow rate (MFR), figure 3. The samples were prepared and tested according to the standards ISO 1133 (MFR), DIN EN ISO 179 (Charpy impact properties) and ISO 527 (Determination of tensile properties). These processing steps are typical in the recycling industry for thermoplastic materials, to provide a further usage for the thermoplastic material.

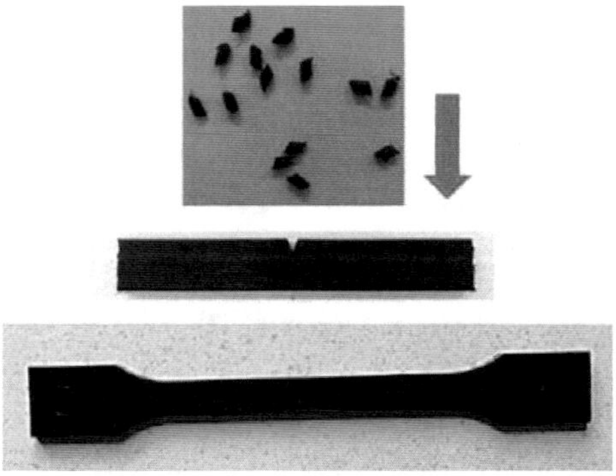

Figure 3. ABS granulate after extrusion and samples prepared by injection moulding

The measurements of the melt flow rate and mechanical properties were conducted in the laboratory of Ansbach University of Applied Sciences. Samples of different plastic parts, see figure 2, were prepared. Microscopy was also used to observe different characteristics of the samples. The main goal of the investigation is to identify the influence of the laser paint removal on the characteristics of the thermoplastic materials and the influence of the paint layer on the properties of the recycled plastic material.

3. Results

The Melt mass-Flow Rate (MFR) indicates the ease of flow of the melt of a thermoplastic polymer. It is defined as the mass of polymer, in grams, flowing in ten minutes through a capillary of a specific diameter and length by a pressure applied via gravimetric weights for different, prescribed temperatures [8]. It is one of the most important properties for processing a thermoplastic material especially by injection moulding. The next figure presents the MFR-result of each plastic material after the extrusion process according to figure 2.

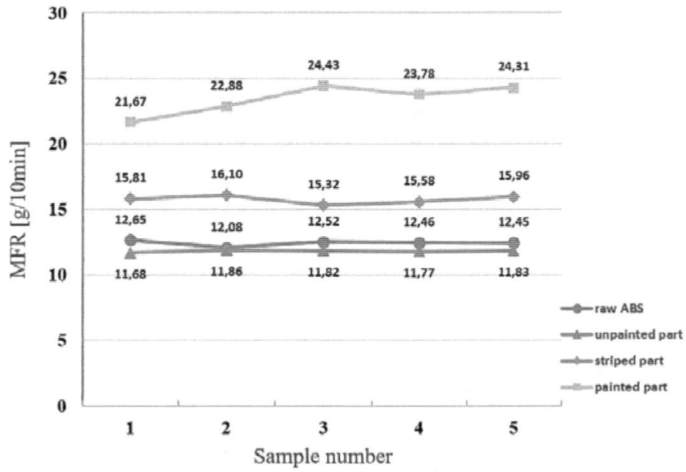

Figure 4. Results of MFR measurements from five samples of each prepared material

The average value of the MFR measurements are 12.41 g/10min for raw ABS material; 11.791 g/10min for the unpainted plastic part; 15,75 g/10min for the striped part and 23,41 for the painted part. All the tested samples were processed with the same parameter on the single screw extruder before the measurement.
One can observe that the results for raw ABS material is very close to the unpainted material. After drying the ABS raw material and again testing it, the same results as for unpainted part were achieved (11.9 g/10min for raw ABS and 11.83 g/10min for unpainted part). The MFR value of the painted part is approximately two times as high as the one of the raw ABS or the unpainted plastic part, which means that the material flows more easily. The next observation is that the after striping the paint by laser the MFR improves and reaches a smaller value close to the unpainted ABS parts. This change shows that after cleaning the ABS plastic parts of the thermosets paint, the Melt mass-Flow Rate comes close to the unpainted parts.

Tensile testing is an essential test for materials science and engineering in which a sample is subjected to a controlled uniaxial tension until failure. The properties that are directly measured via a tensile test are ultimate tensile strength, breaking strength, maximum elongation, and reduction in area. From these measurements different mechanical characteristics can be determined as strain, maximal force, deformation, Young's modulus, Poisson's ratio, etc. The next figure shows the two important properties of the tested samples, deformation at break and maximal force occurred during the uniaxial tension.

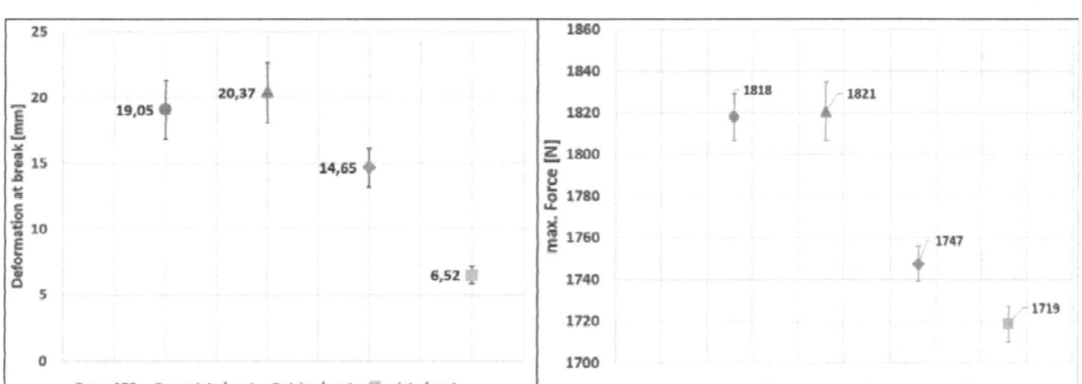

Figure 5. Deformation at break (left) and max. Force (right) of all tested samples

One can observe that the deformation at break values and the max. force are clearly different between painted and unpainted ABS parts. The samples of painted parts break at 6.52 mm comparing with the samples of unpainted parts 20.37 mm. The painted parts show a brittle behaviour compared to the unpainted parts and raw material. Due to the paint particles which are in the plastic material the mechanical behaviour of the painted samples is inferior to the unpainted samples. The mechanical characteristics of laser stripped parts are between these values which shows that the removal of the paint is an improvement regarding the mechanical properties. The measurements of deformation at break rise by 124% from 6,52 mm to 14,65 mm. The maximal force increases only by a small value from 1719 N to 1747 N.

A further mechanical characterisation was done by the Charpy impact test. This testing method is carried out at high strain-rate and determines the amount of energy absorbed by a material during fracture. Absorbed energy is a measure of the material's notch toughness. This test is widely used in the plastic product industry, as it is easy to prepare and conduct and results can be obtained quickly and cheaply. The next figure shows the testing results for 5 samples of each prepared material according to figure 2.

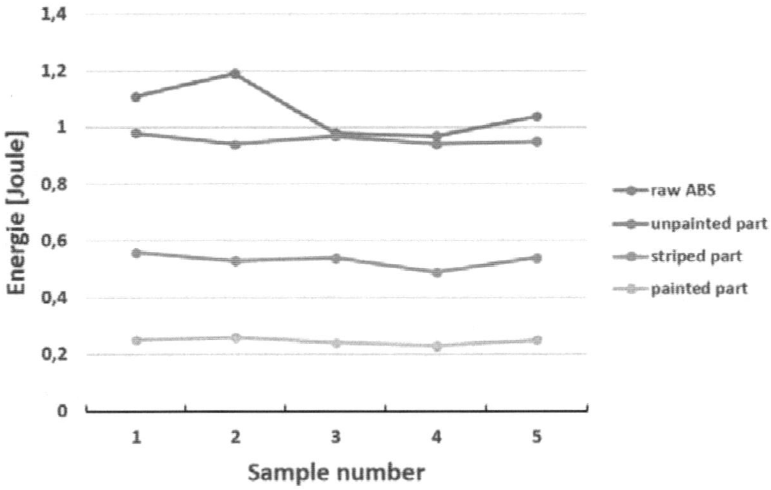

Figure 6. Results of notched impact test for different samples

The results show a significant difference between the raw material respectively unpainted ABS parts to the painted parts, see Table 1. The samples prepared with the painted material can take 288% less energy before they break. When the material is striped the values increase with 116% compared to the painted parts. The quantitative result of the impact tests is the energy needed to fracture a material and can be used to measure the toughness of the material. The paint particles in the ABS material have a significant influence on the material's notch toughness and deteriorate the mechanical behaviour under impact load. Removing the paint of the ABS material improves the impact behaviour of the material more than two times, which allows the use of the recycled stripped material in different applications. The next table presents the values of Charpy impact testing for all four materials.

Table 1. Test results of the Charpy impact testing.

material	Impact Energy [J]	notched impact strength [kJ/m²]
raw ABS	1.058	33.062
unpainted part	0.956	29.87
striped part	0.532	16.62
painted part	0.246	7.687

With the help of optical microscopy one can see the fractured surface of the samples after impact testing. Here different patterns of defects are seen but also impurities or contaminations of the material e.g. with paint particles. The next figure shows the microscopy pictures of the unpainted part, striped part, and painted part.

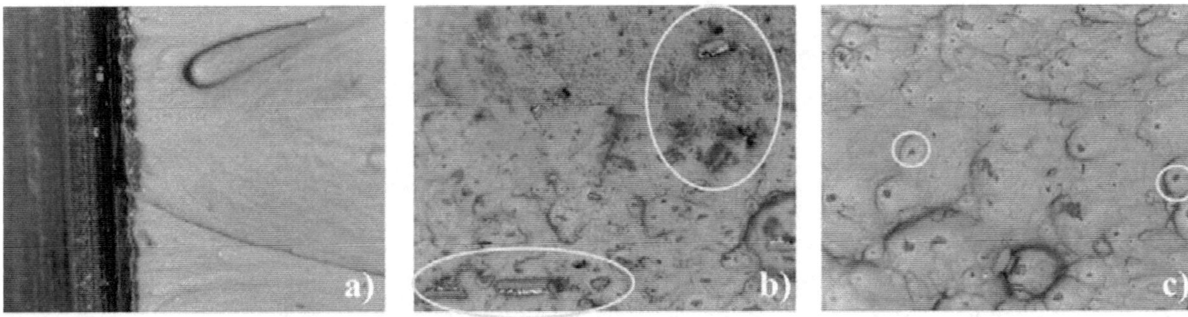

Figure 7. Microscopy of the samples of different materials after Charpy testing; a) unpainted part, b) painted part and c) striped part

The first picture in the figure shows the unpainted part at the fracture surface after Charpy test. No contamination of the material can be observed. In the second picture different impurities of the thermoset paint particles can be observed. These impurities have different dimensions and they are spread inhomogeneous overall cross section of the sample. Some of them are sharp and large. The third picture presents the laser striped surface. Here only very small remains of the paint particles are visible. They are very small and good integrated in the polymer matrix. This last information from the picture shows that some small particles of paint are still in the material after laser paint striping, which indicates that the parts should be cleaned better by laser processing and after. Further investigations are planned to remove also the last small impurities from the plastic material.

4. Conclusions

To recycle painted thermoplastic parts without a loss of quality of the recyclate, it is essential to separate the thermoset coat from the thermoplastic carrier material. The laser clean technique shows very good results in removing the paint layer from sensible thermoplastic materials. The result of the rheology and mechanical characterisation shows that properties after laser stripping the painted thermoplastic parts are superior to the painted parts. The Melt mass-Flow Rate (MFR) was improved by approx. 30% compared to the painted and striped plastic parts. The deformation at break was increased by 124% from 6.52 mm to 14.65 mm for striped parts. The notched impact strength was also improved by 116 % from 7.67 [kJ/m²] to 16.6 [kJ/m²]. All these results demonstrate that by using the laser paint striping technology we can improve the properties of the recycled material (e.g. ABS), which would enable the industry to reuse the plastic material for different applications. With this method of cleaning surfaces thermoplastic parts are turned from non-recyclable materials to recyclable ones and thus the circular economy can be supported and the waste of plastic reduced.

Acknowledgments

A special gratitude I give to the Company Höglmeier Polymer-Tech GmbH & Co. KG, which supported the project by providing different plastic parts. Special thanks for the financial support of the research goes to the German Federal Environmental Foundation (DBU). A special gratitude I give to technical staff of the Ansbach University of Applied Sciences. Special thanks for her support go to Ms Antje Sover for reviewing the article.

5. References

[1] A Sover 2019 IOP *Conf. Ser.: Mater. Sci. Eng.* 564 012028
[2] Zink, M., Sover, A., *Study on the Process of Paint Removal from Thermoplastic Materials by Laser Technology,* Proceedings of the 2nd International Conference of the University of

Applied Sciences Ansbach 3rd - 4th of July 2020 (currently in print)

[3] Xiaodi Z, Wenbin Qin, Jing L, Zhiyong W, *Research on mechanism and process of paint removal with pulsed fiber laser* MATEC Web Conf. 189 01008 (2018)

[4] Schweizer G, Werner L, *Industrial 2-kW TEA CO2 laser for paint stripping of aircraftGas Flow and Chemical Lasers:* Tenth International Symp. International Society for Optics and Photonics 57-62 (1995)

[5] Zhang H, Liu W W, and Dong Y Z, *Experimental and mechanism research on paint removal with low frenquency YAG pulsed laser* (Laser & Optoelectronics Progress vol 50) 114-20 (2013)

[6] David V, d. W., *Laser Ablation for Advanced Surface Treatments, Clean automated process improves production and increases throughput*, Lasertechnikjournal, Volume14, Issue3, June 2017, Pages 34-37

[7] Kuang Z, Guo W, et al, *Nanosecond fibre laser paint stripping with suppression of flames and sparks, Journal of Materials Processing Technology*, 266, 2019, p 474-483

[8] M. Guessoum, N. Medjdoub, S. Nekkaa and N. Haddaoui, *Rheological and mechanical properties of recycled poly(ethyleneterephtalate)/high density polyethylene blends*, MATEC Web of Conferences 5, 04026 (2013)

Structural aspects and chemical analyses on cutting process of metallic-ceramic materials

M Lutcanu[1], B Istrate[1], M Coteata[1], D L Chicet[1], I Ionita[1], C Paraschiv[2], I Stirbu[1], G Badarau[1] and N Cimpoesu[1*]

[1]"Gheorghe Asachi" Technical University of Iassy - Romania, Faculty of Mechanical Engineering, Blvd. Mangeron, No. 59A, Iasi, 700050, Romania
[2]"Grigore T. Popa" University of Medicine and Pharmacy of Iassy, 16 Univ. Street, Iasi, 700115, Romania

[*]Corresponding author's e-mail: nicanor.cimpoesu@tuiasi.ro

Abstract. Preliminary results on the behaviour of metallo-ceramic system during the cutting process were presented. The ceramic layer was obtained, after sandblasting of the substrate, through air plasma spraying with thicknesses of 30 μm on a steel substrate. Layers are chemically homogeneous, without cracks, pores or crevices. The thin layer (30 μm) present discontinuities with uncovered surfaces were the substrate is near in contact with the environment. Cutting process was realized on marking fiber laser equipment (Boron), of 30 watts' maximum power and wave length of 1064 nm. There were obtained two different grooves on the test pieces with the following parameters: laser double pass at speed of 500 mm/sec, pulsed laser of 20KHz frequency and the beam power was set at 50% (around 15W) for the first groove) and at 80% (around 24W for the second groove). Structural, morphological and chemical evaluation of the cutting kerf was realized using scanning electron microscopy (SEM, Vega Tescan LMHII, SE detector, 30 kV, 15.5 mm WD) and energy dispersive spectroscopy (EDS, Bruker X-flash) using automatic/element list mode, Point, Mapping and Line features. This analysis highlighted the type of defects along the cut, respectively the phenomena occurring at the ceramic - metal interface.

1. Introduction

Ceramics are materials with a high number of engineering applications in many industry fields [1,2]. Main characteristics are based on their great properties at upper temperatures, good corrosion resistance, oxidation stability and superior hardness [3-5]. In the same time the brittleness properties present few difficulties final dimensions for industrial applications (respecting dimensional values). The main industrial ceramics are: oxides (alumina-Al_2O_3, magnesia-MgO, Berylia – BeO, Aluminium titanate – Al_2O_3-TiO_2 and zirconia – ZrO_2); carbides (boron carbide – B_4C, tungsten carbide – WC, silicon carbide – SiC, titanium carbide – TiC); nitrides (silicon nitride Si_3N_4, aluminium nitride – AlN, titanium nitride TiN and Zirconium nitride ZrN); borides (titanium diboride Ti_2B and zirconium diboride ZrB_2); silicates (Sialon – Si-Al-O-N, magnesium silicate – $MgOSiO_2$ and mullite – $3Al_2O_3.2SiO_2$) and glass ceramics (soda lime glass – $70SiO_2.10CaO.15Na_2O$ and borosilicate glass – $80SiO_2.15B_2O_3.5Na_2O$) [3,6-9]. Cutting of ceramic materials is not a new procedure being necessary in different domains for many practical applications. These procedures are: conventional methods (abrasive wheel cutting, diamond saw cutting, mechanical mounting and wire saw cutting); non-conventional methods (wire electrical discharge machining WEDM, laser beam machining LBM, abrasive water jet machining AWJM) and hybrid machining (hybrid laser water jet machining, electrochemical discharge assisted wire machining) [10-12].

In this article we perform a laser cut line on a medium thick ceramic layer (alumina) deposited on a steel substrate and analyze (structural and chemical) the effects of the pulses impact on the material.

2. Experimental details

Ceramic layers with 30 μm thickness were realized through plasma spraying method using an isolated room with robotized equipment type SPRAYWIZARD 9MCE. The ceramic layer was deposited on a Fe-C alloy as substrate (steel P265G for petroleum transportation pipe). Thin layer was obtained from powders of alumina (Al_2O_3/Metco-105SFP) used for chemical resistance enhancement [8]. Metallic surface of the substrate was prepared before deposition process applying a sandblast operation (Rosler equipment/dry powder sandblasting/Al_2O_3 Fepa 40 (430 μm) at 3 bar pressure [9]). The cutting line and metallic and ceramic surroundings were investigated using scanning electron microscope (VegaTescan LMH II/SE-detector, 30kV/VegaTC software for 2D and 3D characterization). Chemical composition insights were taken using energy dispersive spectroscopy (EDS detector –Bruker, PB-ZAF analyze mode with Automatic/Element list, Point, Line and Mapping features). Ceramic materials present a low machinability [13-15] therefore for mechanical processing high energy processes can be used like laser cutting or EDM - electrical discharge machining [16, 17].

The aim of the laser cutting tests was to analyse the impact area between the laser pulses and the ceramic/metallic material. For cutting process was used a 30 watts max. power/1064 nm wave length marking fiber laser equipment Boron type. The cutting parameters were as follow: laser double pass at speed of 500 mm/sec, pulsed laser of 20KHz frequency and the beam power was set at 50% (around 15W) [18, 19].

3. Experimental results

Experimental sample was structurally and chemically analysed. Insights of the cutting line made with pulsation laser were taken from the ceramic layer and also from the metallic substrate. In figure 1 are presented SEM micrographs as (a) general view, (b) cutting line through ceramic + metallic system, (c) cutting line over two areas with ceramic, on the top part and without ceramic layer (d) details of the cutting line and (e) 3D image of the cutting line. From the structural images, figure 1 (a) - (e) we observe three areas near the cut affected by this process.

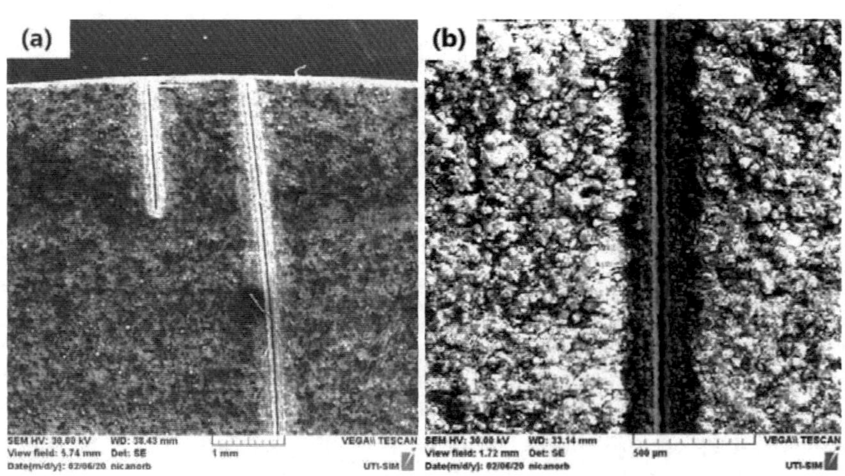

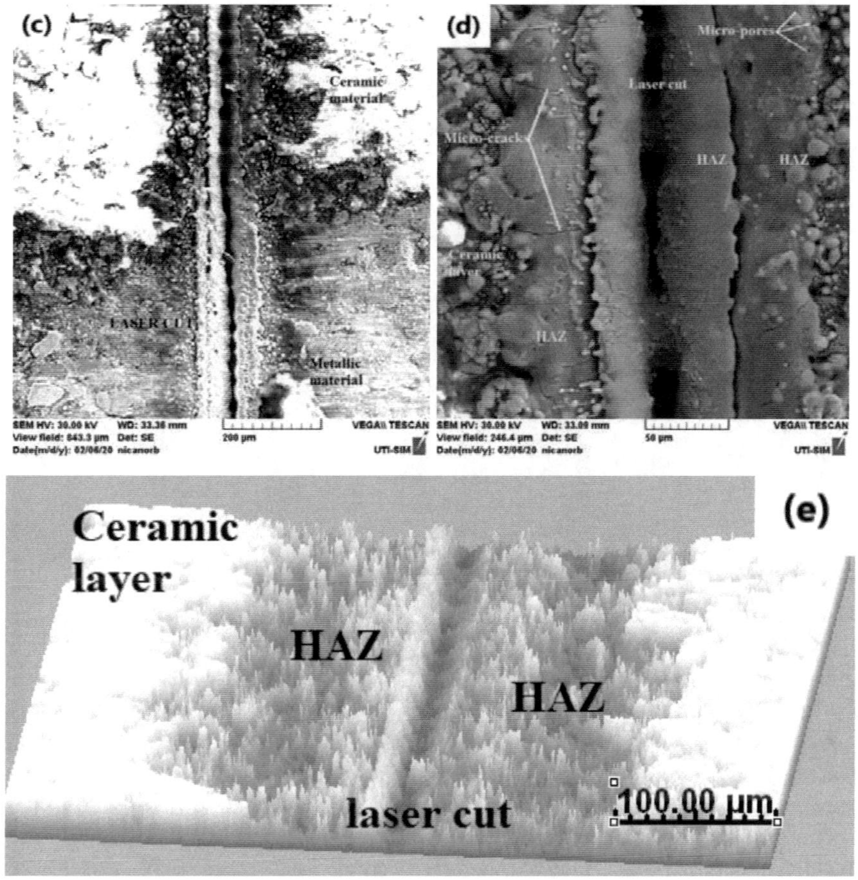

Figure 1. SEM micrographs: (a) general view, (b) cutting line through ceramic + metallic system, (c) cutting line over two areas with ceramic, on the top part and without ceramic layer, (d) details of the cutting line and (e) 3D image of the cutting line.

The linear cut is made by joining points; figure 1 (d), made with pulsation laser at a proper cutting speed. By macroscopic point of view, the cut can be considered straight and without interruptions [20, 21]. Both figure 3 (d) and (e) confirm the nature of the cut as being based on many holes connected by the laser pulse energy impact with the basic material. During the pulsation laser process an exfoliation of the ceramic layer can be observed on the entire heat affected area (HAZ) with widths around 50 μm [22]. At microscopic level few metallic material interruptions can be observed between the laser spots, figure 1 d), fact that confirm the necessity of a lower forward rate. The aspects of the cut line are similar with and without the ceramic layer, figure 1 (c) in the complex material case we observed the removal of the ceramic layer.

The area near the cut observed in figure 1 (d) and (e) present many fissures and longitudinal cracks with pores and micro-pores appeared based on the local heated area from the laser pulse (heat affected zone HAZ). The heat transfer from the laser pulse to the metallic-ceramic material is the main reason why the ceramic layer is removed from the substrate. At macro-scale the removal of the ceramic material seems uniformly done but at microscopic scale, figures (c) and (e) we observed a fluctuation of the ceramic layer based on the center of the laser pulse. HAZ appear on both sides of the cut in equal parts and as two waves of metallic material (one of 25 μm width near the cut and second larger of 50 μm) [23]. Chemical composition analyze results are presented in figures 2-4.

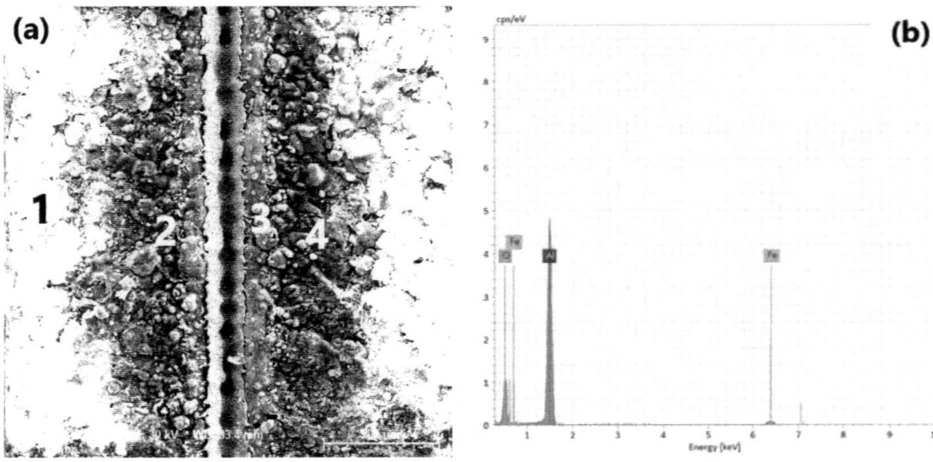

Figure 2. Chemical composition analyze in (a) areas proposed for analyze (points 1-4) and energy spectrum in (b) of identified elements O, Al and Fe.

Qualitative identification of the elements was done through EDS technique and the main elements were identified in energy spectrum from figure 2 (b). Quantitative results of chemical compositions realized in points 1-4 from figure 2 (a) are given in table 1 as weight and atomic percentages (also EDS detector error are given for each element). We follow the evolution of the main elements Al and O characteristic for the ceramic layer deposited on the steel substrate and Fe for substrate (carbon element was ignored because of the small percentage). In point 1 from figure 2 (a) we observe that the ceramic layer is structurally and chemically intact and only a small percentage of iron is observed from the substrate based on the porosity of the layer or the variation of the layer thickness. In the point number 2, figure 2 (a), near the HAZ we observe a decrease of the thin layer elemental components (Al and O) and an increase of iron content [24]. In point 3 a mixture of ceramic layer and substrate exist, based on chemical composition results from table 1, representing a crossing area between the delaminated material near the cut and the ceramic layer with a full integrity.

Table 1. Chemical composition of the experimental sample along the laser cut (points 1-4 from figure 2 (a))

	Al		**O**		**Fe**	
	wt%	at%	wt%	at%	wt%	at%
Point 1	45.03	34.03	50.46	64.32	4.51	1.65
Point 2	28.31	29.87	26.47	47.08	45.22	23.05
Point 3	33.24	28.09	43.92	62.58	22.85	9.32
Point 4	44.58	40.59	31.98	49.10	23.44	10.31
EDS error	1.01		1.8		0.6	

Chemical composition from point 4 present a reduced affected ceramic layer by the laser cut with partial removal of the ceramic layer and the presence of the iron in a bigger amount since in the left part of the cut but, as can be seen from figure 2 (a), in an affected by cut area. Figure 3 present a Line distribution of Al, O and Fe elements for (a) the selected line and (b) component elements distributions.

The Line distribution, figure 3 (a), include all interesting areas from the ceramic layer to HAZ and laser cut. Iron chemical element present a high signal on the both sides of the laser cut were no aluminum signal can be observed. The HAZ area present iron signal and also a percentage of oxygen, smaller comparative with the ceramic layer but existing and confirming an oxidized HAZ area. The inhomogeneity of the ceramic layer is confirmed by the presence of the iron signal in an area where we supposed to have only the ceramic layer, but the decreased signal comparative with the clean substrate, shows only a variation of the ceramic layer thickness and not an entire lack of the non-metallic layer [25].

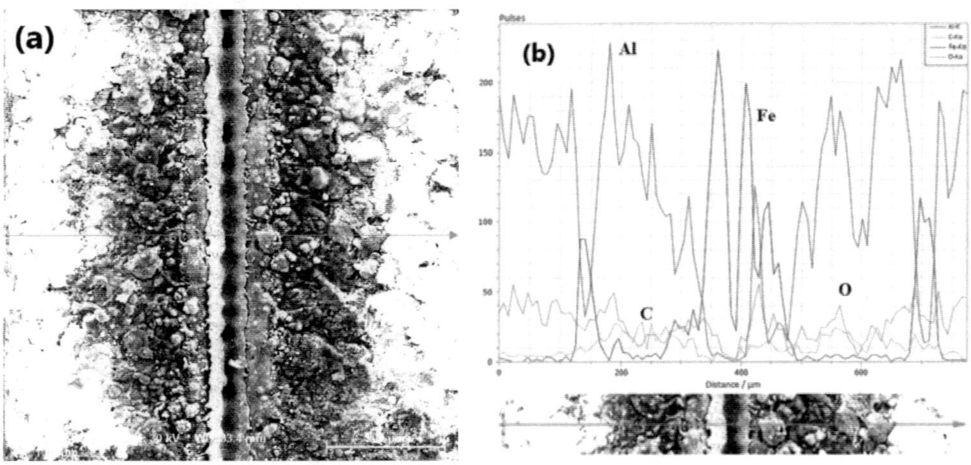

Figure 3. Line distribution of Al, O and Fe elements (a) selected line and (b) elements distributions.

In figure 4 using Mapping distributions of Al, O and Fe, C for different areas of the cut line we can observe, through chemical analyze, the areas affected by the laser cut on different positions during the cutting line. In figure 4 (a) is presented the distribution of elements on two distinct areas, one metallo-ceramic in the top part and the other only metallic both crossed by the cutting line. The laser cut line has a similar aspect in both areas. From figure 3 (b) can be observed that the ceramic layer is more affected on the right side of the cut fact that confirm the chemical analyze from points 1 and 4, table 1, explaining the difference between the iron percentages (4.51 against 23.44 wt%). Along the cutting line we can also observed an oxidized area especially on the right side.

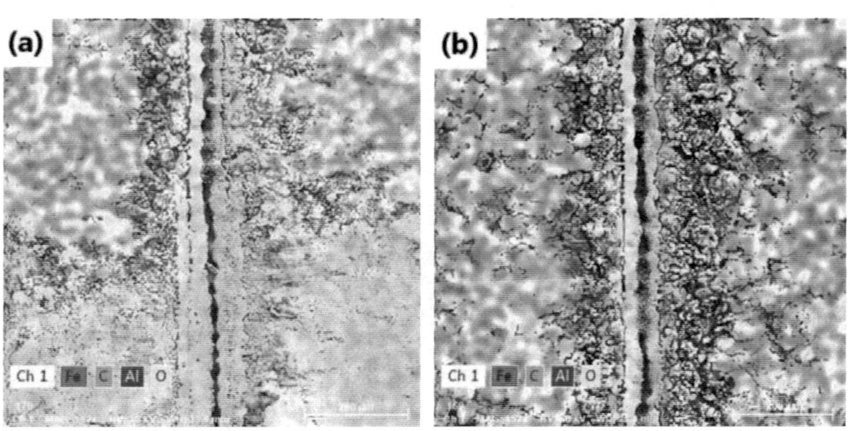

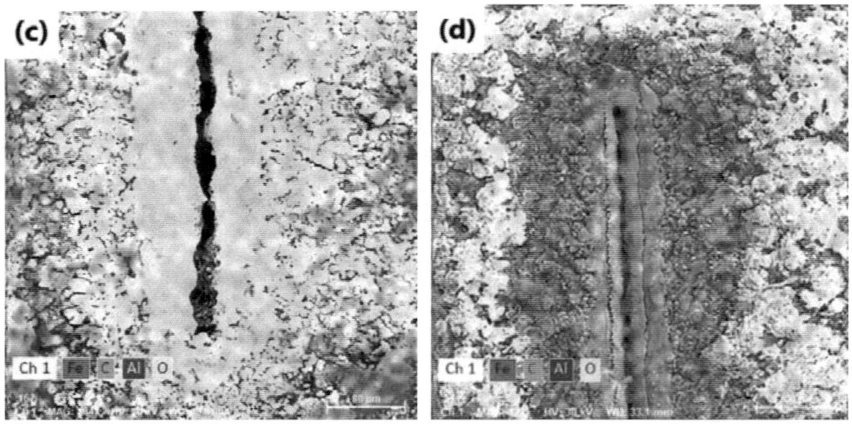

Figure 4. Mapping distributions of Al, O and Fe, C for different areas of the cut line.

In figure 4 (c) and (d) are presented end-points of the laser cut with the elements Fe, C for substrate and Al, O for the covering ceramic layer. On the left side of the cut, near the line we observed stains of the ceramic layer that roughly correspond to the margins of the laser pulse impact areas. By macroscopic point of view, we can consider the cutting line straight but under 100 µm the details, figure 4 (c), present a various line based on the stain let by the impact of the laser pulse with the ceramic/metallic material. The HAZ is big in this case with variations between 50 and 150 µm based on the fastening system of the experimental sample or technological details of the laser cutting equipment.

4. Conclusions

In specialty literature are present different methods for cutting of ceramic materials (conventional, non-conventional and hybrid). In this article few preliminary results (structural and chemical) about the cutting of a ceramic layer deposited on top of a steel substrate are presented and analysed. Aim of the article was to follow these procedure aspects: dimensional neatness, sharpness, substrate quality, surface roughness and heat affected zone (HAZ). By this method obvious flaws are present from the begging of the cutting process on the surface. A totally elimination of the micro-cracks must be considered for further studies. Fissures and micro-cracks appearance is based on the differences between the thermal or mechanical stresses induced during the cutting process and material strength. A proper solution to eliminate fissures from brittle materials is to apply a selection of operational ranges of parameters for the technological process.

As an alternative a proper solution can be hybrid machining in order to gain efficiency in cutting ceramics (specially to improve the surface characteristics near cut like roughness, cracks, pores, exfoliations, HAZ etc.

5. References

[1] Munz D and Fett T 2013 Ceramics: mechanical properties, failure behaviour, materials selection *Springer Science & Business Media*
[2] Liang Y and Dutta S P 2001 *Technovation* **21** 61–5
[3] Tuersley I P, Jawaid A and Pashby I R 1994 *J. Mater. Process. Technol.* **42** 377–90
[4] Rakshit R and Das A 2019 *Precision Engineering* **59** 90–109
[5] Bejinariu C, Munteanu C, Florea C D, Istrate B, Cimpoesu N, Alexandru A and Sandu A V 2018 *Rev. de Chim.* **69** 3586-3589

[6] Samant A N and Dahotre N B 2009 *J. Eur. Ceram. Soc.* **29** 969–93
[7] Mizuno M, Iyama T and Zhang B 2016 Analysis of the sawing process with abrasive circular saw *J. Manuf. Sci. Eng.* **130(1)** 1–15
[8] Florea C, Bejinariu C, Munteanu C and Cimpoesu N 2017 *Advanced Materials Engineering and Technology V Book Series: AIP Conference Proceeding* **1835 UNSP 020053** 10.1063-1.4983793
[9] Florea C D, Munteanu C, Cimpoesu N, Sandu I G, Baciu C and Bejinariu C 2017 *Rev. de Chim.* **68** 11
[10] Enescu F, Irimiciuc S A, Cimpoesu N, Bedelean H, Bulai G, Gurlui S and Agop M 2019 *Symmetry* **11(11) 1391** 2-14
[11] Quintero F, Pou J, Lusquin F, Boutinguiza M, Soto R and Perez-Amor M 2004 *Optic Laser Eng.* **42** 327–40
[12] Mahu G, Munteanu C, Istrate B, Benchea M and Lupescu S 2018 *IOP Mat. Science and Eng.* **444** 032010
[13] Radtke J, Abeln T, Weikert M and Dausinger F 2000 **4088** 252–5
[14] Wang X C, Zheng H Y, Chu P L, Tan J L, Teh K M, Liu T, Ang B C Y and Tay G H 2010 *Optic Laser Eng.* **48** 657–63
[15] Makarov Y, Medvedev O, Khait O, Kurin S, Usikov A, Papchenko B and Helava H *Journal of Physics* **643** 012129
[16] Vijay Praveen D, Ranga Raju D and Jagannadha Raju M V 2019 *Materials Today: Proceedings*
[17] Tuersley I P, Jawaid A and Pashby I R 1994 *J. Mater. Process. Technol.* **42** 377–90
[18] Samant A N and Dahotre N B 2009 *J. Eur. Ceram. Soc.* **29** 969–93
[19] Yamasaka M, Fujisawa M, Oku T and Masuda M 1990 *CIRP Ann – Manuf. Technol.* **39(1)** 333–6
[20] Yan Y, Li L, See T L, Ji L and Jiang Y 2013 *J. of the Europ. Ceramic Soc.* **33(10)** 1893–1905
[21] Cîrlan Paleu C, Istrate B, Paleu V and Munteanu C 2020 **724(1)** 012063-157023
[22] Paleu V, Cîrlan Paleu C, Istrate B, Bhaumik S and Munteanu C 2020 *IOP Conference Series: Materials Science and Engineering* **724(1)** 012064
[23] Panturu M, Chicet D, Lupescu S, Istrate B and Munteanu C 2018 *Acta Tech. Napoc. Series-Appl. Math. Mech. and Eng.* **61(1)** 37-142
[24] Cirlan Paleu C, Paleu V, Istrate B, Cimpoesu N and Munteanu C 2019 *IOP Conference Series: Materials Science and Engineering* **591(1)** 012007
[25] Chen X, Ji L, Bao Y and Jiang Y 2012 *J. Eur. Ceram. Soc.* **32** 2203–11

Chemical and structural analyze of experimental biodegradable ZnMgY alloy

C Panaghie[1], R Cimpoesu[1], A Alexandru[1], M Bernevig[1], V Manole[1], A M Roman[1], B A Prisacariu[2], P Paraschiv[1] and N Cimpoesu[1]

[1]"Gheorghe Asachi" Technical University of Iassy-Romania, Faculty of Mechanical Engineering, Blvd. Mangeron, No. 59A, Iasi, 700050, Romania
[2]"Grigore T. Popa" University of Medicine and Pharmacy of Iassy, 16 Univ. Street, Iasi, 700115, Romania

*Corresponding author's e-mail: nicanor.cimpoesu@tuiasi.ro

Abstract. Beside biocompatible classic metallic materials, biodegradable metals (BMs) like alloys of zinc (Zn-based) present a high potential as an alternative solution for permanent implants elements generally being applied for fractures restorations or other similar medical conditions. An experimental alloy, ZnMgY, was obtained using an induction furnace from high purity materials (Zn: 99,995 and master alloy MgY: 65-35 wt%), in Argon atmosphere. Microstructure of the alloy (after mechanical grinding and polish plus chemical etching) and chemical insights (before chemical etching) were taken using optical microscope (Zeiss+Motic digital camera for image acquisition) scanning electron microscope (SEM VegaTescan LMH II, SE, 30 kV, 16 mm WD) and dispersive energy spectroscopy (EDS Bruker, PB ZAF, Automatic mode of analyze, Point and Mapping features). The experimental alloy was five times re-melted in the induction furnace using a ceramic crucible. The experimental alloys present a good chemical homogenization without porosity, metallic inclusions or segregation.

1. Introduction

Metals have a long history in using them as a material for implants used in the medical field [1–4]. The desire to use metals is due to their unique combination of properties, including good mechanical strength, ductility, hardness, wear resistance and formability. In terms of biocompatibility, the first generation of metal materials used for applications as an implant had the first requirement to be inert in the physiological environment [5–7]. Recently, biodegradable metals such as magnesium (Mg), iron (Fe) and zinc (Zn) have been increasingly accepted as implant material [8–15,16,17]. Of the three candidate metals, zinc is the most recently introduced [9].

Biodegradable metals are best suited for implants with temporary functions in the body. Two promising medical applications of bio-resorbable metals are the manufacture of orthopaedic stents and fasteners. A biodegradable material requires specific properties before becoming viable. Research suggests that some of the problems related to magnesium and iron can be addressed using zinc [18,19]; hence the considerable increase in scientific interest in this metal.

For biodegradable Zn metals, the mechanical properties of the alloy may be influenced by (i) type, (ii) number and (iii) the quantity or concentration of alloying elements. A wide range of compositions has already been studied; i.e. (i) pure, (i) binary, (iii) ternary, and (iv) quaternary combinations of Zn with other elements [20]. More than ten types of zinc alloy systems have been developed for biomedical applications including the Zn-Mg system; Zn-Fe; Zn-Ca; Zn-Li; Zn-Zr; Zn-Al; Zn-Sr; Zn-Mn; Zn-Ge.

Zinc-based biodegradable alloys exhibit a corrosion rate between Mg- and Fe- based alloys being proper for many medical applications. These alloys possess different processing and economic advantages, a very good ability to be poured, fluency, and other metallurgical and mechanical properties which are very competitive with other ferrous and non-ferrous materials. Zinc material represents an essential trace element in biological environment and human body. In case of an adult biological system which has approx. 2–3g of Zn, is many biological functions from enzymatic catalysis to cellular neuronal systems, Zn plays as factor in all 6 groups of enzymes, also in several classes of proteins for regulatory processes. In the same time zinc element have contributions in blood pressure regulation in the arteries.

In this article the authors present some preliminary results, microstructural and compositional insights, about the obtaining of a new alloy based on zinc with addition of magnesium and yttrium for medical applications as biodegradable alloy.

2. Experimental details

A new alloy was obtained from high purity zinc (99,995%) and MgY (65-35 wt%) master alloy using an induction furnace under Ar atmosphere. A 100 grams' ingot was realized and, after wire cutting of a 10/5 mm disk, was mechanically prepared for chemical and structural analyses. The chemical composition proposed was Zn3Mg1Y (wt%) based on metallic charge calculation. For bigger weight percentages of Y extremely fragile alloys were obtained.

Microstructure of the alloy (after mechanical grinding and polish chemical etching with $FeCl_3$ was used) and chemical insights (before chemical etching) were taken using optical microscope (Zeiss+Motic digital camera for image acquisition) scanning electron microscope (SEM VegaTescan LMH II, SE, 30 kV, 16 mm WD) and dispersive energy spectroscopy (EDS Bruker, PB-ZAF, Automatic/Element list mode of analyse, Point (90 μm diameter), Line and Mapping features).

3. Experimental results

Optical micrographs, figure 1, present a good homogeneity of the material with two different components, a matrix and few compounds of white colour with two different geometrical shapes: rectangular and round.

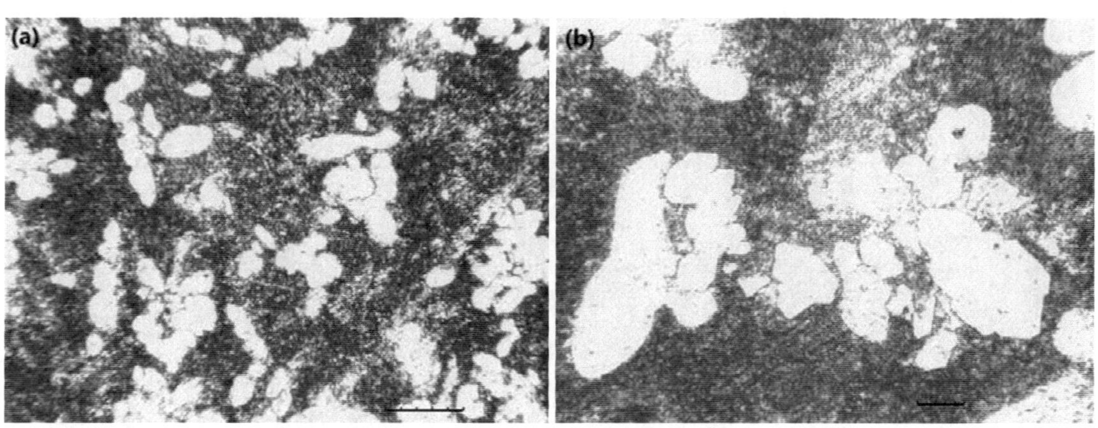

Figure 1. Optical micrographs (a) 200x and (b) 600x.

With a magnesium content around 3 wt. %, zinc grains are not recognized in optical micrographs. It is a eutectic structure containing Zn and intermetallic compounds according to Zn–Mg, Zn-Y and Mg-Y phase diagrams [21, 22].

Figure 2 shows SEM micrographs of experimental alloy after grinding and chemical etching for two different scales: (a) 500x and (b) 2500x. As a general aspect, figure 2 (a), present a homogeneous microstructure without pores, cracks or fissures. Also an evenly distribution of compounds is observed at macro-scale that usually is considered a proper fact in material degradation. The induction melting process through his turbulent currents mix the alloy components increasing the chemical and structural homogeneity of the alloy. At higher amplification powers of the structure, figure 2 (b), three different phases can be observed noted with 1 (rectangular compounds), 2 (round compounds) and 3 (matrix).

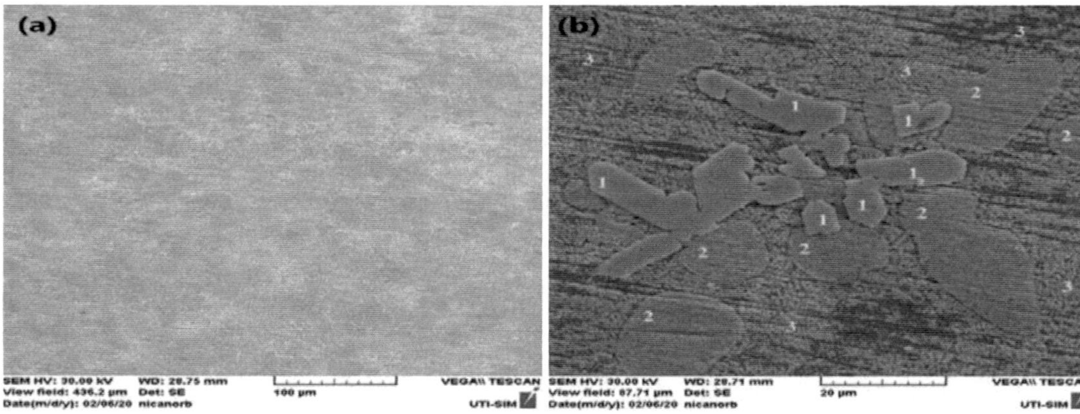

Figure 2. SEM micrographs of experimental alloy after grinding and chemical etch (a) 500x and (b) 2500x.

Chemical composition determination was realized before and after chemical etching of the surface. Energy dispersive spectroscopy (EDS) was used to identify the main elements of the experimental alloy. The energies characteristic for Zn, Y and Mg (keV) elements are presented in figure 3 (a) spectrum. Chemical composition was determined in three different areas of the sample and the quantified results are given in table 1 first 3 position and average values. The signal was taken from a one millimeterr square area and given as weight and atomic percentages.

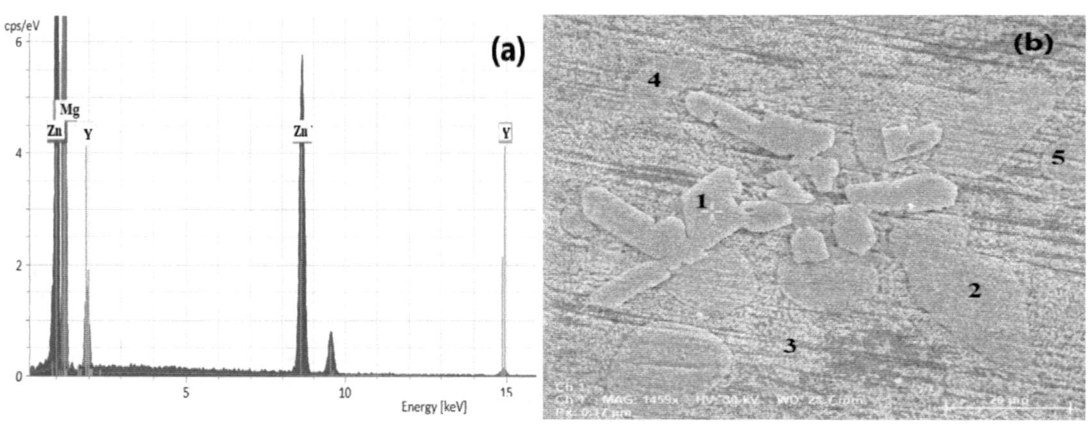

Figure 3. Chemical composition analyse (a) elemental identification and (b) points selection for zonal chemical composition identification.

In figure 3 (b) are presented five points (each 90 μm diameter) for chemical analyze of the main components identified from the microstructure (figure 2 (b)) that appear under different aspects (matrix, rectangular or round shape). The chemical composition results are presented in table 3 as compared to the average composition of the entire sample.

The selected points for analyse comprises the main formations of the alloy. In the first point of analyse (rectangular formation) (chemical composition given in table 1, point 1, figure 3 (b)) the results show the presence of a ZnY (Zn8Y) compound with approximately 11 wt% of Y which is bigger than average percentage of Y in the alloy. In case of this compound no Mg was identified so the new phase ZnY was formed during the melting and solidification of the alloy from high purity Zn and MgY master alloy. The second analysed point is on an elongated formation, figure 3 (b), that present a high content of zinc with less Mg and Y content as average and a different structural shape than rectangular formations or matrix (point 3 of analyse). This compound contains 1.4 at% Mg and 0.2 at% Y (Zn1.4Mg0.2Y).

Table 1. Experimental results after chemical composition analyse.

Elements	Zn		Mg		Y	
Analysed area	wt%	at%	wt%	at%	wt%	at%
Area 1	95.76	90.9	3.3	8.4	0.93	0.65
Area 2	96.02	91.34	3.16	8.08	0.82	0.58
Area 3	95.9	91.18	3.2	8.19	0.89	0.63
Average	**95.89**	**91.14**	**3.22**	**8.22**	**0.88**	**0.62**
Point 1 (figure 3 (b))	89.12	91.76	-	-	10.88	8.24
Point 2 (figure 3 (b))	99.2	98.4	0.52	1.4	0.28	0.2
Point 3 (figure 3 (b))	96.28	92.16	2.79	7.19	0.93	0.65
Point 4 (figure 3 (b))	99.12	99.2	0.8	0.59	0.08	0.22
Point 5 (figure 3 (b))	99.43	98.48	0.57	1.5	-	-
EDAX error %	2.25		0.4		0.4	

The third point, figure 3 (b) (chemical composition shown in table 1) is selected on the alloy matrix (more than 50% of the structure). Chemical composition of the matrix is near the average composition determined on the alloy. Point 4 (figure 3 (b)) shows an area mainly formed by zinc with reduced percentages of dissolved Mg and Y. The fifth area analyzed, point 5, figure 3 (b), shows areas with no Y dissolved in zinc and the formation of ZnMg phase. XRD analyze will explain the formation of different compounds between Zb, Mg and Y in this experimental alloy. A high percentage of magnesium is presented dissolved in the matrix and less in other compounds. The alloy shows phases with a high percentage of yttrium (in a reduced contribution at general scale) and as dissolved element in zinc matrix and complex three elements compounds.

In figure 4 elements Zn, Mg and Y chemical composition variation on a line selected are presented. The elemental variations confirm the chemical compositions presented in table 1 and the formation of compounds with more or less yttrium.

All element variations during the selected line are homogeneous and the presence of compounds with a higher Y percentage at similar distances, figure 4 (b), can help at generalization of corrosion and degradation rate. In figure 5 elements distribution is present on a selected area, figure 5 (a), with the highlight of compounds with more yttrium, figure 5 (e), and a homogeneous distribution of zinc and magnesium.

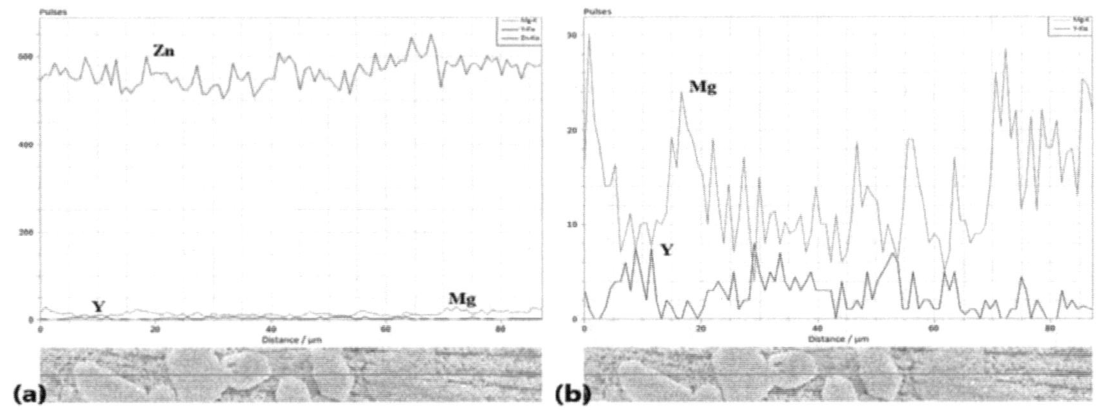

Figure 4. Elements Zn, Mg and Y chemical composition variation on a line selected on the experimental alloy (a) Zn, Mg and Y variations and (b) detail of Mg and Y signal variation.

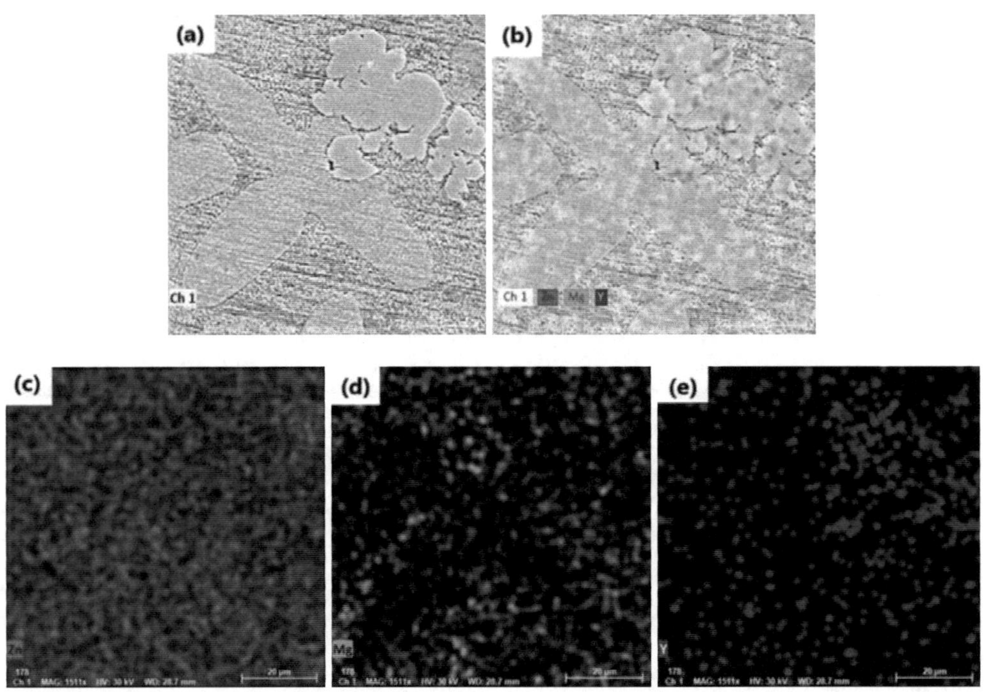

Figure 5. Elements distribution (a) selected area of distribution, (b) all elements distribution and (c) Zn, (d) Mg, (e) Y distributions.

Distributions of both Mg and Y elements confirm the results from chemical composition analyze highlighting the areas with more Mg or Y also those without these elements. No clear interfaces are observed between the matrix and the other components (phases, compounds etc.). In order to evaluate the mechanical properties, the material will be the subject of tensile/compression test, micro-hardness and scratch test and also dynamic mechanical analyze with results proposed for a further paper.

4. Conclusions

A new chemically and structurally homogenous alloy based on zinc was obtained using induction melting under Ar atmosphere. In order to avoid the fragility of the alloy we maintain the alloying percentages of Mg and Y under 3.5 wt% respectively 1.5 wt% based on previous studies of the same authors. The presence of different constituents was observed through optical and electronic microscopy (OM, SEM). The chemical composition of the matrix and other compounds was determined with respect for the main elements: Zn, Mg and Y. Compounds like ZnY, ZnMgY and ZnMg were identified.

A further investigation of the experimental alloy must be realized using XRD, XPS and EBSD equipment in order to establish all the components, chemical and structural properties. For medical applications of this alloy as biodegradable material the corrosion behaviour of all the chemical and structural components will be analysed as different materials and also as a complex alloy. Different behaviour at corrosion resistance between the matrix and the other compounds is expected.

References

[1] Fraker A C and Ruff A W 1977 *Metallurgical Surgical Implants: State of the Art* **29** 2–28
[2] Gotman I 1997 *J Endourol* **11** pp 383–389
[3] Triclot P 2010 *Int. Orthop. 2010* **35** 201–206
[4] Hermawan H, Ramdan D and Djuansjah J R P 2011 Metals for Biomedical Applications *INTECH Open Access Publisher*
[5] Saini M, Singh Y, Arora P, Arora V and Jain K 2015 Implant biomaterials: a comprehensive review *World J. Clin. Cases: WJCC 3 2015* pp 52–57
[6] Manam N S, Harun W S W, Shri D N A, Ghani S A C, Kurniawan T, Ismail M H and Ibrahim M H I 2017 *J. Alloys Compd.* **701** 698–715
[7] Sykaras N, Iacopino A M, Marker V A, Triplett R G, Woody R D 2000 *Int. J. Oral Maxillofac. Implants* **15** 675–690
[8] Witte F 2010 *Acta Biomater* **6** 1680–1692
[9] Zheng Y F, Gu X N and Witte F 2014 *Mater. Sci. Eng. R Rep.* **77** 1–34
[10] Hermawan H., Biodegradable Metals from Concept to Applications, Springer-Verlag Berlin Heidelberg, 2012, ISBN 978-3-642-31170-3.
[11] Seitz J M, Durisin M, Goldman J and Drelich J W 2015 *Adv. Healthc. Mater.* **4** 1915–1936
[12] Heiden M, Walker E and Stanciu L 2015 *J. Biotechnol. Biomater* **5** 178
[13] Li H, Zheng Y and Qin L 2014 *Proc. Nat. Sci-Mater. Int.* **24** 414–422
[14] Im S H, Jung Y and Kim S H 2017 *Acta Biomater* **60** 3-22
[15] Hermawan H 2018 *Prog. Biomater* **7** 93–110
[16] Liu Y, Zheng Y and Hayes B 2017 *Sci China Mater* **60** 377–391
[17] Crubzy E, Murail P, Girard L and Bernadou J P 1998 *Nature* **391** 29
[18] Bowen P K, Drelich J and Goldman J 2013 *Adv. Mater.* **25** 2577–2582
[19] Vojtech D, Kubasek J, Serak J, Novak P 2011 *Acta Biomater* **7** 3515–3522
[20] Mostaed E, Sikora-Jasinska M, Mostaed A, Loffredo S, Demir A G, Previtali B, Mantovani D, Beanland R and Vedani M 2016 Novel *J. Mech. Behav. Biomed. Mater.* **60** 581–602
[21] Yao C, Wanga Z, Tay L S, Zhu T, Gao W *Journal of Alloys and Compounds* **602** 101–107.
[22] Egami M, Ohnuma I., Enoki M., Ohtani H., Abe E. 2020 *Materials and Design* **188** 108452.

IOP Conf. Series: Materials Science and Engineering 1037 (2021) 012035 doi:10.1088/1757-899X/1037/1/012035

Structures for Metamaterial Superlenses Used in Medical Imaging

Daniela Ionescu[1], Gabriela Apreotesei[2]

[1,2]"Gh. Asachi" Technical University of Iasi; [1] Department of Telecommunications and Informational Technologies, Carol I Blvd. 11, 700506 Iasi, Romania;
[2] Department of Physics, Dimitrie Mangeron Blvd. 67, 700050, Iasi, Romania

*Corresponding author's e-mail: danaity@yahoo.com

Abstract The metamaterial superlenses for imaging proteins, viruses, and DNA have to present a high resolution, which cannot be ensured by classical lenses. Such a material for lenses exceeds the Abbe-Rayleigh diffraction limit, leading to a nanoscale level of resolution, a several times better than the classical diffraction limits. We have illustrated here a hotspot size of ca. 0.25 – 0.28 λ, corresponding to a numerical aperture of about 1.4. The used metamaterial structures are isotropic negative index metamaterials (NIMs), with negligible losses. Material combinations of metallic nanoparticles (with dimensions of tens of nm) inserted in a dielectric slab have been considered for study. Microcomponents periodicity in the layer is of a few hundred of naometers. Material properties evolve in function of the constituent's nature and dimensions. Refraction index in function of wavelength was determined and represented on graphs in order to illustrate the domain of negative values and the manner in which it can be controlled at structure level. Analysis was performed in visible and IR range by simulation methods, using the HFSS program and a proper algorithm based on physical considerations.

1. Introduction

A high resolution is necessary for imaging proteins, viruses, and even DNA molecules, which cannot be ensured by classical lenses. The metamaterial structures used for lenses exceeds the Abbe-Rayleigh diffraction limit and nanoscale level of resolution can be achieved, a several times better than the classical diffraction limits [1], [2].

The metamaterial slab compensates the wave decay and images can be reconstructed in the near field. The negative index metamaterial gives us the possibility to recover evanescent waves from the object in the image plane. The material amplifies the evanescent waves that carry information at very small scales [1], [3].

Superlens with a negative refractive index are obtained, working at frequencies in visible range [7]. Sub-wavelength metastructures has to be included inside the metamaterial in order to obtain subwavelength foci and resolution close to the size of the focus [9]. We have illustrated here a hotspot size of ca. 0.25 – 0.28 λ, corresponding to a numerical aperture of about 1.4.

2. Structure Description

We have analyzed a negative index metamaterial (NIM) for lenses, with a particular structure.
The material with $\varepsilon > 0$ and $\mu < 0$ supports evanescent waves and represents an MNG (μ negative material). If the object and the lens are placed along the z-axis, the rays from the object are traveling in

the $+z$ direction. Due to the negative refraction index, transport of energy in the $+z$ direction requires the z component of the wave vector to have opposite sign [2], [4], [5]:

$$= -\sqrt{\frac{\omega^2}{c^2} - (k_x{}^2 + k_y{}^2)} \tag{1}$$

Consequently, the evanescent waves are transported and amplified.

The unit cell of the considered metamaterial is a type of split ring resonator (SRR). This ensures the negative permeability in a limited frequency domain, which can be modified by dimensions tuning. A regular distribution of the unit cells (SRRs) inside a nonmagnetic dielectric has been considered.

The material is isotropic, with negligible losses. The SRR segments are metallic, with dimensions of tens of nm inserted in the dielectric slab (Fig. 1). Microcomponents periodicity in the layer is of a few hundred of nanometers.

The SRR were structured from cylindrical rods of Ag with length (L) of 100 nm to 120 nm and diameter (d) of 20 to 30 nm. Number of segments of each SRR can be 3, 5 or 7- odd number was chosen in order to ensure the negative permeability of the sample. The considered dielectric was GaP. Analysis has been performed for similar sets of materials given in literature. E.g. of material combinations: Au, Ag, Cr on Si; Ag on Al_2O_3, Si_3N_4, SiO_2, TiO_2, GaP, GaAs, InP [10] - [13].
The spatial structure of the lenses metamaterial is given in Fig. 1.

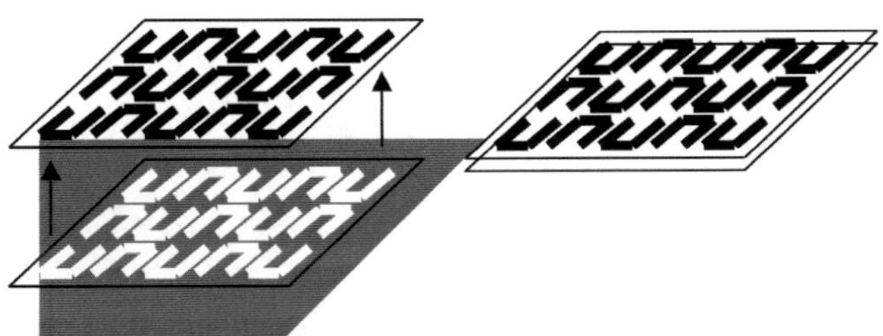

Figure 1. Structure of the metamaterial. The two successive layers represent the negative of each other regarding the metal / dielectric constituents.

One remarks the opposite orientation of the successive units (individual SRR), in the same layer, either in successive ones. This type of unit orientation was chosen in order to obtain the maximum inductive effect of the SRRs when they interact with the applied field. The geometry was tested until a convenient domain of negative permeability has been obtained.

Material properties evolve in function of the constituents nature and dimensions (size and shape of the metallic nanostructures). Subwavelength confinement and strong field enhancement have been achieved.

The wave vector of the surface plasmon polaritons is given by the relation [12], [14]:

$$_p(\omega) = \frac{\omega}{c}\sqrt{\frac{\varepsilon_d \varepsilon_m}{\varepsilon_d + \varepsilon_m}} \tag{2}$$

where ε_m and ε_d are the frequency dependent permittivity of the metal, respectively dielectric, ω is the pulsation of the incident field, and c is the light speed. In the same time,

$$_p = k_0 n = \frac{2\pi}{\lambda} \tag{3}$$

where k_0 is the free-space wavenumber, n_{eff} is the effective refraction index of the propagation medium and λ is the wavelength. The imaginary part of the wavenumber expresses attenuation per unit distance and is useful in the study of evanescent fields with exponential decaying.

3. Analyzing Method

The test configuration is represented in Figure 2. Exposure of the metamaterial slab has occurred in a multi-mode ridge channel waveguide, with dimensions of $w = 2.8$ μm, $h = 1.6$ μm. With help of the HFSS program, the samples analysis was performed in the domain of 600 … 840 nm for the incident field wavelength.

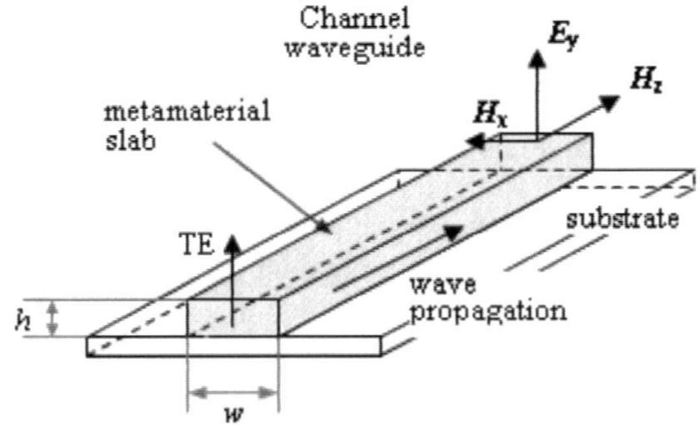

Figure 2. The ridge channel waveguide with the metamaterial sample.

The normalized frequency (V number, which determines the number of modes) of this rectangular waveguide can be written as:

$$V = \frac{2\pi}{\lambda} w \sqrt{n_{meta}^{2} - n_{cladding}^{2}} \tag{4}$$

where n_{meta} and $n_{cladding}$ are the refraction index of the metamaterial slab, respectively substrate and w is the metamaterial slab width (Fig. 2).

For the TE mode, E_z=0, $E_x >> E_y$, and $H_z \neq 0$. The considered channel waveguide is a nonplanar one and has strong optical confinement because it is surrounded on three sides by a low-index material - air.

The S-parameters have been determined by simulation with help of the HFSS program. Then the reflection coefficient Γ and the transmission coefficient T have been calculated using the simulation data, for the TE_{10} mode.

The following expressions have been used [12], [15]:

$$S_{11} = \frac{\Gamma(1 - T^2)}{1 - \Gamma^2 T^2} \quad ; \; S_{21} = \frac{T(1 - T^2)}{1 - \Gamma^2 T^2} \tag{5}$$

$$\Gamma \approx \frac{n_{meta} - n_{air}}{n_{meta} + n_{air}}; \; T \approx \frac{2n_{meta}}{n_{meta} + n_{air}} \tag{6}$$

The effective refraction index of the metamaterial sample has been calculated at different frequencies in the considered domain.

4. Results for the Negative Parameters

For the superlemses material, the refraction index in function of wavelength has been determined and represented on graphs in order to illustrate the domain of its negative values. We are interested in the manner to which the refraction index can be controlled at structure level. Analysis was performed in visible and IR range (600 … 840 nm) by simulation methods, using the finite element method and a proper algorithm based on physical considerations.

Results for n(λ) were illustrated in Fig. 3, 4 and 5 below. Different structural parameters have been varied in order to observe the n dependence on the geometry of the metamaterial constituents.

In Fig. 3 the shape of the rings of the SRRs has been modified by varying their segments number (odd, 3, 5, 7) in the unit cell.

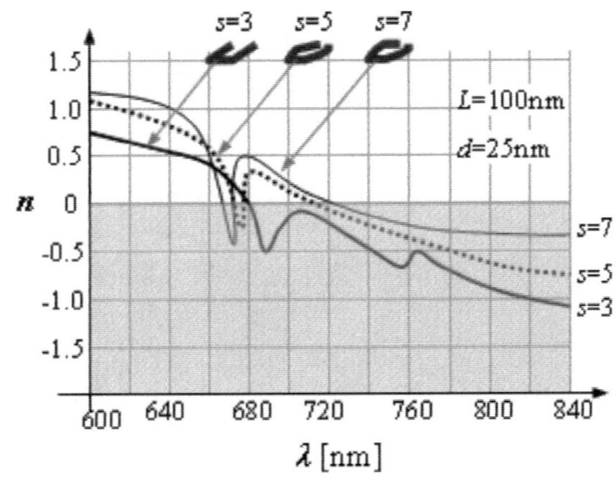

Figure 3. Refraction index evolution for $\lambda = 600 \ldots 840$ nm, for the metamaterials having metallic SRRs with different number of segments, s, included in the dielectric volume.

One observes that the segments number do not influence significantly the results, the recommendation is for the segments loop to be rather almost closed than open, for leading better the field lines.

The refraction index evolution in frequency range is characterized by a negative peak (a valley) and in a narrow frequency range is possible to have negative values. This fact depends on the structure, the nature and geometry of the metamaterial constituents. Excepting this valley, negative values for n are achieved when λ increases and is close to IR region. These results are in agreement with the data presented in literature [1], [6], [8], etc. The decreasing trend of the refraction index is limited in practice, n presenting positive values over the most of the electromagnetic spectrum. The simulation model has to be perfected in order to reproduce the real behavior of the material in a more wide frequency range that the domain considered for analysis.

The negative peak is caused by a resonant phenomenon inside the metamaterial sample and a correlation can be pointed out between the SRR dimensions, respective periodicity and the wavelength corresponding to this peak central frequency. Simulations are indicating very easy modification of the position of this negative peak when SRR perimeter, respectively the distance between the metamaterial layers are changed.

Fig. 4 indicates variation of the refraction index when the SRR segments length vary.

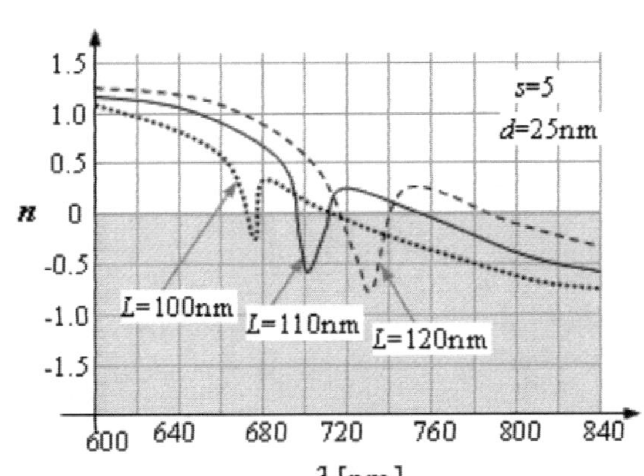

Figure 4. Refraction index in function of the wavelength, for different values of the SRR segments length, L.

IOP Conf. Series: Materials Science and Engineering 1037 (2021) 012035 doi:10.1088/1757-899X/1037/1/012035

A maximum for negative values more deep and wide have been obtained when the perimeter of the SRR loop increases. But in the same time, the domain of the negative values for n, which succeeded this maximum, is smaller. This effect is valuable for an increasing of L up about 140 – 150 nm, then the maximum is rising to positive values, not convenient for a metamaterial. The interval of values for L in which the negative peak can be obtained is correlated to the wavelength and is linked to the resonant phenomena inside the material sample.

In Fig. 5 are represented the curves for *n* when the metal of the SRR rings modify its thickness. The interval of possible modifications it is not too large, limited by a rising of the curve to positive values when the rings are thicker (see the example of the curve for a SRR segments diameter d = 30 nm). Variation of the metal thickness does not modify the position of the resonant peak, but it's magnitude. Finer the SRR ring, deeper is the negative peak, but also very narrow. The ring diameter is limited in practice by the technology.

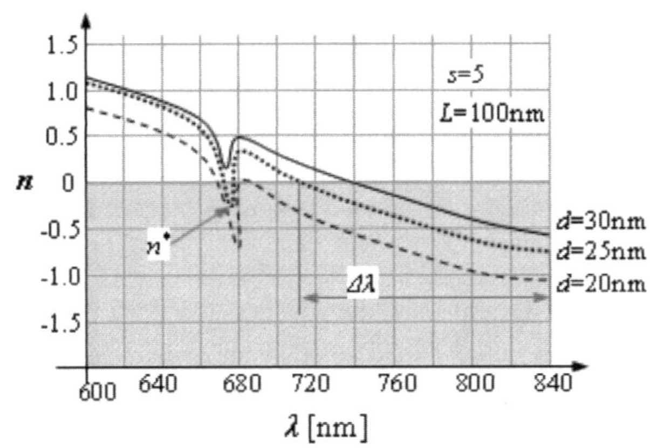

Figure 5. Refraction index in function of the wavelength, for different values of the SRR segments diameter, *d*.

If we denote the maximum negative value of the refraction index on graphs with n^* and the domain of the negative values, which succeeded this maximum, with $\Delta\lambda$, the results obtained for different internal structure of the considered metamaterial for lenses are given in Table I. Results were obtained for the five segments SRR structures (s = 5). Arrows indicate the increasing of the parameter in the table column. The lenses resolution using slabs of metamaterial with characteristics given in the table lines has been determined.

Table 1. Resolution of the Superlenses

No.	*d* [nm]	*L* [nm]	n^* negative peak	$\Delta\lambda$ [nm] for *n* negative	**Resolution**
1	20	100 ↓	-0.66	156 ↑	0.23 λ (min 138nm) ↑
2	20	110	-0.82	133	0.24 λ
3	20	120	-0.74	87	0.26 λ
4	25	100 ↓	-0.26	124 ↑	0.25 λ ↑
5	25	110	-0.52	81	0.28 λ
6	25	120	-0.73	45	0.32 λ
7	30	100 ↓	0.18 positive	98 ↑	0.27 λ ↑
8	30	110	0.05 positive	82	0.30 λ
9	30	120	-0.24	76	0.33 λ (max 277nm)

One observes that the obtained resolution values exceed the classical diffraction limit and can be adjusted by modifying the metamaterial geometrical configuration. For obtaining finer spots, finer SRR rings have to be used (e.g. d = 20 nm), with a perimeter correlated with the wavelength, in agreement with the resonant phenomena which accompany the propagation. Yellow lines in the table are indicating possible preferable solution for the metamaterial synthesis.

5. Conclusions

A model of structure for metamaterial used for superlenses have been proposed here, in order to improve the resolution and control the lens properties by modifying the geometrical parameters of the samples.

Superlens resolution of fraction of wavelength can be achieved, using metamaterial structures conceived with periodic metal SRRs included in a proper dielectric background, alternating in volume, on layers, with similar components but inversing the metal / dielectric materials.

Resonant phenomena occur in the metamaterial. A maximum recovering of the evanescent wave energy can be achieved when a periodical dimension of the metamaterial structure is a multiple or submultiple of a half wavelength of the propagating field. This determines the maxim focalization effect of the superlens.

The adopted configuration of the metamaterial which represents a mirror positive / negative structure (Fig. 1) increases significantly the amplification of the evanescent wave, in comparison with the structure with only metal SRRs inside the dielectric.

Concerning the material nature of the components, it can be pointed out that the metal nature influences the foci, while the dielectric nature influences absorption and losses inside the superlens material.

References

[1] Gangwar K, Paras and Gangwar R P S 2014 *Advance in Electronic and Electric Engineering* **4**(1) 97
[2] Venkatesan R and Singaravelu R 2013 *Proc. Conference: Progress in Electromagnetics Research Symposium* (Taipei)
[3] Rosaline S I and Raghavan S 2013 *Proc. IEEE International Conference on Computational Intelligence and Computing Research* (Madurai) pp 1-4
[4] Lee Y, Kim S-J, Park H and Lee B 2017 *Sensors* **17** (8) 1726
[5] Wong Z J, Wang Y, O'Brien K, Rho J, Yin X, Zhang S, Fang N, Yen T-J and Zhang X 2017 *J. Opt.* **19** 084007
[6] Bing W F, Zengbo Y and Wu W L 2016 *Sci. Adv.* **2** e1600901
[7] Hamid R H and Hamed H 2018 *Ortho Res Online J.* **2**(5) OPROJ.000549
[8] Lipworth G, Ensworth J, Seetharam K, Huang Lee D J S, Schmalenberg P, Nomura T, Reynolds M S, Smith D R and Urzhumov Y 2014 *Sci. Rep.* **4** 3642
[9] Yuan G, Rogers K, Rogers E T F and Zheludev N I 2018 *Proc. Conference on Lasers and Electro-Optics* (San Jose) FM3J.2
[10[Kante B, O'Brien K, Niv A, Yin X and Zhang X 2012 *Physical Review B* **85** 041103(R)
[11] Fan W, Yan B, Wang Z and Wu L 2016 *Sci. Adv.* **2** e1600901
[12] Yao K and Liu Y 2013, arXiv:1312.4614 [physics.optics]
[13] Haxha S, AbdelMalek F, Ouerghi F, Charlton M D B, Aggoun A and Fang A X 2018 *Scientific Reports* **8** 16119
[14] Smolyaninov I I, Hung Y-J and Davis C C 2007 *Science* **315** 1699
[15] Zhang J; Luo Y and Mortensen A 2010 *Optics Express* **18**(4) 3864

Study of the durability of ferritic - austenitic steel samples after cyclical fatigue impact

Georgi Georgiev[1], Yaroslav Argirov[1], Tatyana Mechkarova[1], Daniela Spasova[1], Aneliya Stoyanova [1]

[1]Technical University of Varna, Material Science and Technology Department, Studentska Street 1, Varna, Bulgaria

[*]Corresponding author's e-mail: t.mechkarova@tu-varna.bg

Abstract. The purpose of this article is to investigate the fatigue strength of corrosion resistant ferritic austenitic steels used in marine equipment. Two groups of sheet steel samples were prepared: the first group were reference parts and the second group of samples were welded. The fatigue tests were performed with specially developed vibrating equipment. The destroyed surfaces were examined by optical microscopy, and the phase composition by X-ray diffraction analysis. Fatigue strength was evaluated by the number of cycles needed for a crack at a certain alternating load. Fatigue tests show how multi-cyclical fatigue affects the formation and development of a fatigue crack, and metallographic and X-ray diffraction analysis shows the phase changes occurring in the structure of SAF 2507 steel.

1. Introduction

Very often in practice, especially in drilling in marine conditions, the use of corrosion-resistant steels is required. They are deployed at great depths and are subjected to cyclical loading [2,3,4] due to the dynamic sea processes and their connection with the drilling ship. Ferritic - austenitic steels, such as duplex steel SAF 2507 (1.4410) are very sensitive to minimal changes in the chemical composition of the technological process of their production, as well as to the modes selected for their welding and cooling, affecting their service life [5,6,7,8,9]. In this example, the causes of cracks in SAF 2507 duplex steel parts, which form drilling equipment subjected to fatigue impact, are analyzed (Fig. 1).

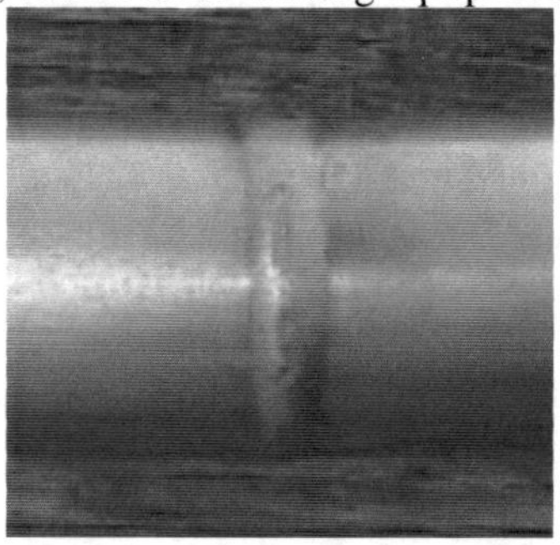

a b

Figure 1. Machine parts of duplex steel operating in sea conditions at a depth of 70m: a) pipeline with welding; b) flange connection

IOP Conf. Series: Materials Science and Engineering 1037 (2021) 012036 doi:10.1088/1757-899X/1037/1/012036

2. Experimental study

2.1. Methodology for obtaining of samples

For the purposes of this study, steel SAF2507 (ASTM S32750) with the following chemical composition was used (Table 1).

Table 1. Chemical composition of steel SAF2507 [1]

C	Si	Mn	P	S	Cr	Ni	Mo
≤0.030	≤0.8	≤1.2	≤0.025	≤0.015	25	7	4

In order to reduce the temperature effect influencing the structural changes, a laser is used to cut the test samples. Fig. 2 shows laser-cut unwelded and welded samples.

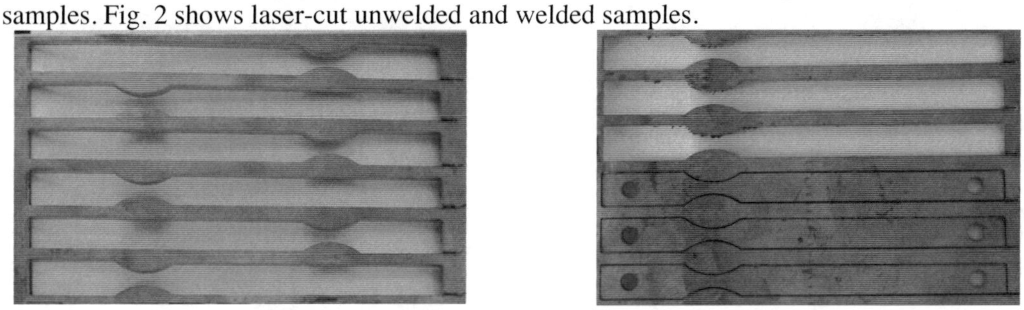

a b

Figure 2. Laser cut samples
a) unwelded; b) welded

The unwelded samples (fig. 2-a) are laser cut from a plate with dimensions 350mm by 200mm and thickness 3mm. Welded samples (fig.2-b) are made of two plates measuring 150mm by 350mm with a thickness of 3mm. The two plates are welded unilaterally to each other by manual Shielded Metal Arc Welding (SMAW). ESAB OK 68.82 electrodes were used. The plates are pressed at the four corners to avoid distortion. They are roughly ground so that they can be laser cut. The welded plate is rapidly cooled with water.

2.2. Methodology for determining the geometric dimensions of the test samples and selection of the cyclical loading mode

The determination of the sample frequency and the approximate number of cycles is performed with the software product "SolidWorks" and more precisely with the "frequency" and "fatigue" tool, through which it is possible to select the approximate operating modes of the cyclical load installation.

In order to determine the geometric dimensions of the test samples, the frequency at which the experimental installation will operate is selected. The most stable operation, without fluctuations, is achieved then the installation performs at a frequency of 50 - 80 Hz (determined experimentally). The geometric dimensions are chosen so that the resonant frequency of the samples is in the range of 50 - 60 Hz. With the parameters thus set, the dimensions should be as follows: sample length 172.5 mm, width in the gripping section in the vibrating table 13 mm, width of the loose section 10 mm.

Figure 3 shows all other geometric dimensions of the tested samples.

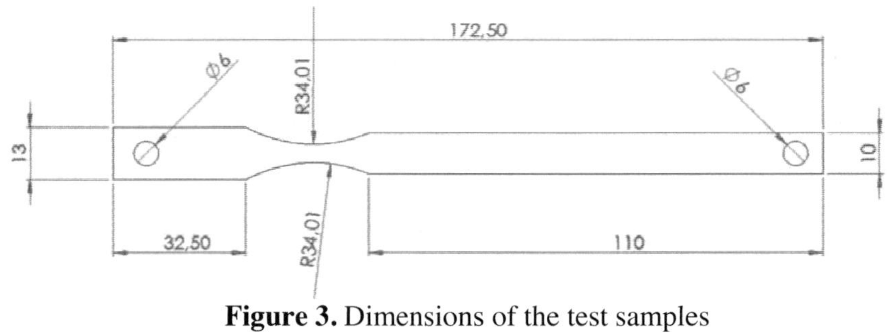

Figure 3. Dimensions of the test samples

Using the "sketch" and "features" tools of SolidWorks, a 3D model of the sample is created and its material is selected.

The first step of the simulation analysis is to select "New study" from the "Study Advisor" menu. Of the types of analyses, "frequency" is the one chose on this occasion, which searches the natural frequencies, as well as visually showing their shape and displacement in case of vibrations from the resulting resonance (fig. 4).

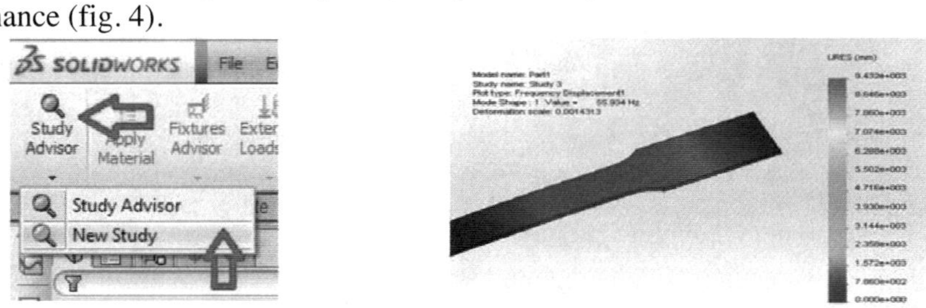

Figure 4. Determination of natural frequency in a 3D model of the test sample

The study obtained a result of 55.9Hz, and with the increase of the 3D model length, the operating frequency decreases.

The second step in the methodology is the study of the test sample for the maximum stresses and displacements that will be created. It is necessary to determine the displacement of the vibrating table at a maximum operating mode of 80N by means of "linear dynamic - harmonic". The results of the analysis show that the stresses in the gripping zone occur in the range of 800 MPa to 830 MPa, which exceeds the stiffness of the material many times and its destruction in this range is inevitable.

To determine the approximate number of cycles, we use the "fatigue" menu with the "constant amplitude" submenu, which is added to the "loading" tool, which is part of the fatigue study. After performing the simulation analysis, the required number of cycles for failure in the range of maximum stresses was established at 32000, and at the same time it was found that the destruction process will last about 10 minutes at an operating frequency of 55 Hz.

2.3. Fatigue test installation

The samples are firmly fixed in a special device made for the specific vibrating table. Accelerometers are used to measure acceleration. The software (National Instruments) enables determination of the moment of the crack formation, which in most cases is visible only at high magnification (fig.5).

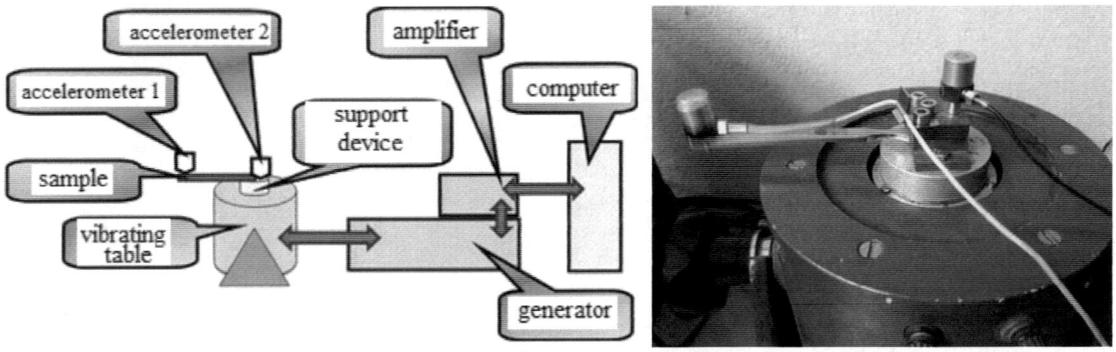

Figure 5. Diagram of the installation for determining the fatigue strength of the material

3. Results of the experimental study

3.1. Fatigue test results

The results of the cyclical tests of the samples that have not been welded are shown in Table 2. The limit observed below which no cracks form is 480 MPa (in the range up to $1x10^7$ cycles). Based on the data obtained (Table 2) the fatigue curve is drawn in fig. 6.

The samples with positions 13, 14, 15, 16, 17and 18 in table 2 were not destroyed within the study range, they are shown by an arrow in fig.6-a. The sample position 11 show values that do not fall within the normal deviation of 5%.

The results of the study of the fatigue strength of the welded samples in the area of the weld metal are shown in table 3. The cyclical tests were performed in the range from 280 MPa to 470 MPa. The fatigue limit in the range from 10^4 to 10^7 cycles was studied. The limit observed below which no cracks from is 340 MPa (in the range up to $1x10^7$ cycles). Based on the data obtained (table 3), the fatigue curve fig. 6-b was compiled. The sample position 30 in table 3 were not destroyed within the studied range, they are shown by an arrow in fig.6-b.

Table 2 The experimental results of the cyclical tests of the samples that have not been welded

Sample №	1	2	3	4	5	6	7	8	9
Load (MPa)	715	640	572	562	560	557	555	550	520
Cycle ($1x10^6$)	0.11	0.25	0.2	0.25	0.32	0.3	0.3	0.33	1

Table 2- continued

Sample №	10	11	12	13	14	15	16	17	18
Load (MPa)	495	370	485	480	470	390	320	290	240
Cycle ($1x10^6$)	0.5	1.3	2	10	10	10	10	10	10

Table 3 The experimental results of the cyclical tests of the welded samples

Sample №	21	22	23	24	25	26	27	28	29	30
Load (MPa)	470	425	415	405	395	385	380	375	360	340
Cycle ($1x10^6$)	0.07	0.21	0.5	0.92	0.94	1.2	2.1	3.3	5.56	10

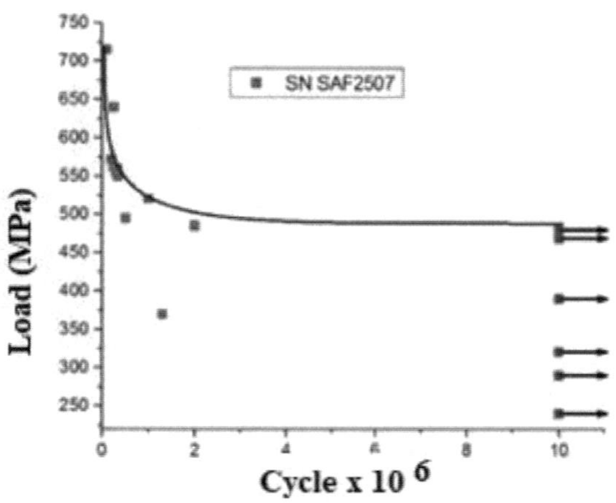

a)

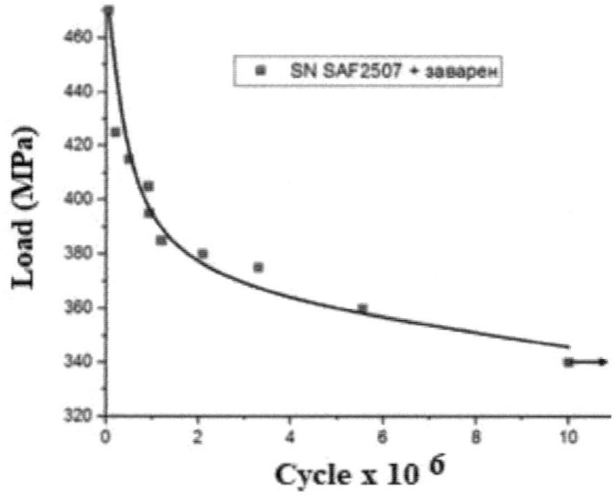

б)

Figure 6. S-N curve of SAMF2507 steel samples: a) unwelded; b) welded

3.2. Results of metallographic tests

Before the samples are placed in fatigue conditions, they must be perfectly sanded, going from sandpaper 100 to 1000. This is followed by polishing to a perfectly smooth surface and the absence of any scratches. When the samples are destroyed, only the fracture of the respective sample is visible, but it does not give a clear idea and assessment of the parameters and features of the crack. Using a specially made device (fig. 7) the parts of the already destroyed sample are pressed against each other along the crack. It was decided to place two samples simultaneously in order to polish them evenly.

This is followed by grouting with polyester resin, which helps to clearly delineate the edges and prevents the incorrect reflection of light in the subsequent microstructural analysis. Using a solution of HCl, HNO_3 and NaCl (MACC60) followed by subsequent polishing and drying, we corrode the structure around the crack. The process is repeated until the set result is achieved, and in this case it is a clear delineation of the grain boundaries. Through the developed methodology we get clear images of the sliding lines, and the reason for the crack to stop or deviate. The examination of the microstructure is performed by a metallographic microscope.

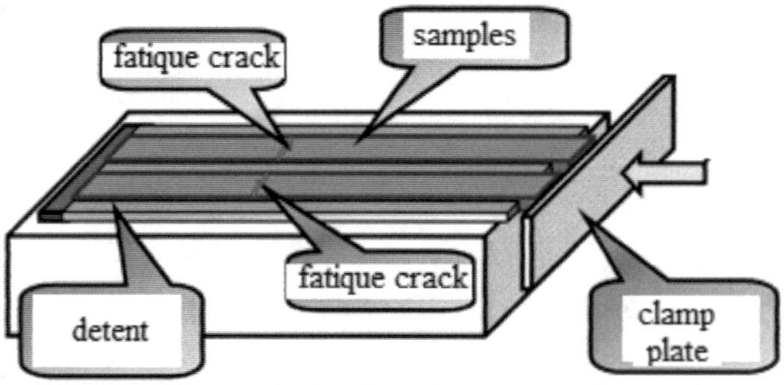

Figure 7. Schematic of a device for microstructural analysis of cracks

Figure 8 shows the crack that passed through the weld metal on the forward facing surface. The crack is offset from the perpendicular side section by a small angle. In addition, a zigzag pattern is observed, which is due to the dendritic structure in the weld metal, in contrast to the crack in the base metal, which has a linear character.

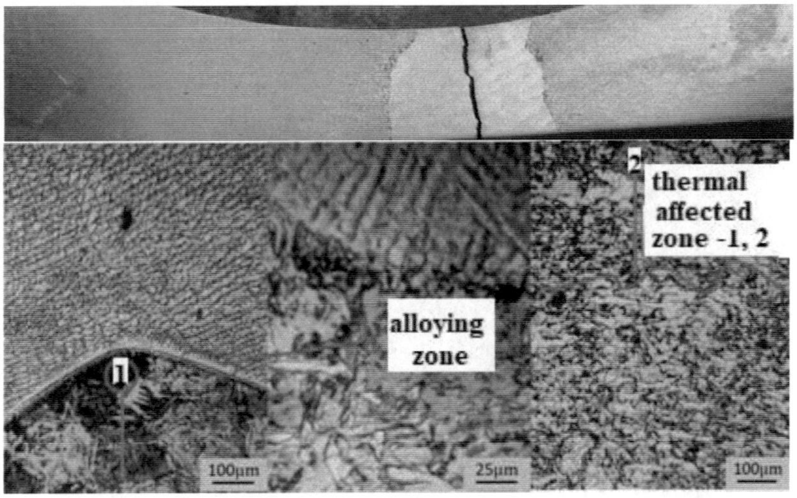

a b c

Figure 8. Microstructure of a welded sample, a) alloying zone, X100; b) alloying zone, X400; c) thermal affected zone X100.

Fig. 9 shows the microstructure of the weld metal in the side section. The development of the crack in the lateral section passes through the weld metal of the dendrite crossing the transcrystalline dendritic grains. The structures on fig. 9-a show the beginning of a crack development, fig. 9-b the lower end of the crack, and fig. 9-c shows the pattern through the middle part of the sample.

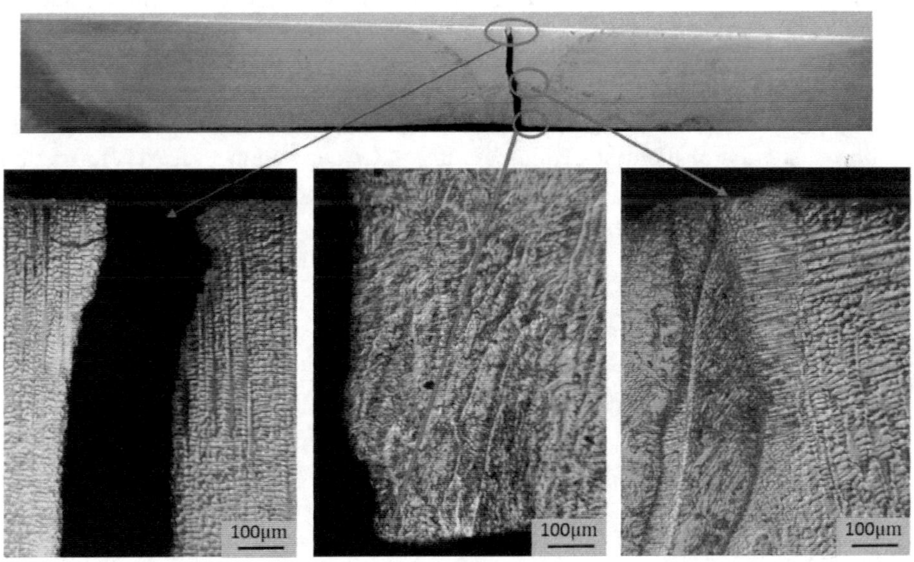

a b c

Figure 9. Microstructure of a sample in side section: a) focus of the crack X100; b) area near the crack in the welded metal X100; c) the area between the two welds X100

3.3. *X - ray diffraction analysis of ferritic - austenitic steel samples*

The studies were performed using an X-ray diffractometer URS-50IM, in the emission of a chromium anode, using a vanadium filter. The integral area of the two phases is determined using the Graph software [10].

- *in delivery condition of SAF 2507 steel*

Fig. 10 shows the radiograph of the unwelded samples in the area of the base material. The quantitative ratio between the two phases was established by the research conducted as γ - Fe (111) = 34%, and α - Fe (110) = 66%. The X-ray structural analysis performed shows the presence of γ - Fe (111) and α - Fe (110), characteristic of ferritic austenitic steels.

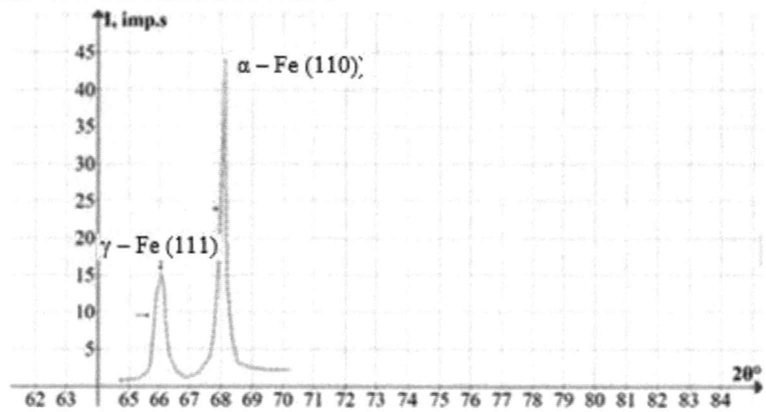

Figure 10. Radiograph of unwelded samples in the area where there is no destruction

Fig. 11 shows a radiograph of the examined samples in the fracture area. The phases and their ratio are determined in this area as γ - Fe (111) = 46%, and α - Fe (110) = 54%. The measured integral areas in the fracture area, respectively, are for γ - Fe (111) = 10.7 deg * imp / sec, and the same phase of the initial sample from Fig.12 has an integral area γ - Fe (111) = 8.5 deg * imp / sec. The broadening of the line observed in the fracture area relative to the initial sample is due to the increased dislocation density and the increased internal stresses (which deform the crystallographic planes).

The case is similar to the ferrite phase, with plane 110 considered. The integral area in the initial state is 13.7 deg * imp / sec, and in the destruction area it is 18.7 deg * imp / sec. Therefore, even in this phase (α - Fe (110)) there is a significant strengthening due to the past deformation processes, during the creation and development of the crack.

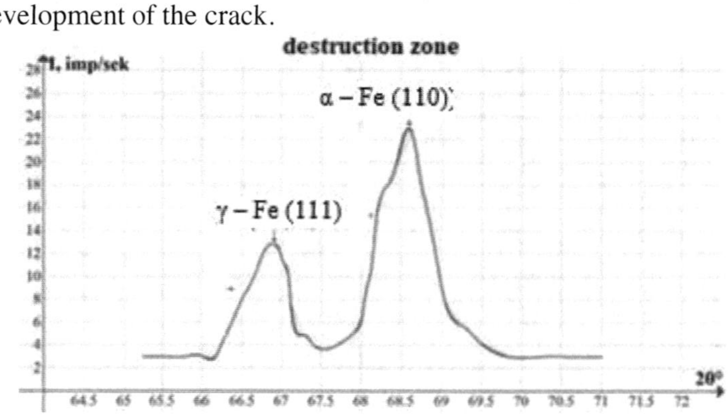

Figure 11. Radiograph in the zone of destruction

- *in the zone of welded metal*

The purpose of the conducted phased qualitative analysis is to establish whether the presence of ferritic - austenitic structure will be preserved at the conducted cooling rate after welding. After the study, the presence of both phases γ - Fe (111) and α - Fe (110) was established (Fig. 12). It was found that the ratio between ferrite and austenite was preserved, as γ - Fe (111) = 60% in the area of the weld metal, and α - Fe (110) = 40%.

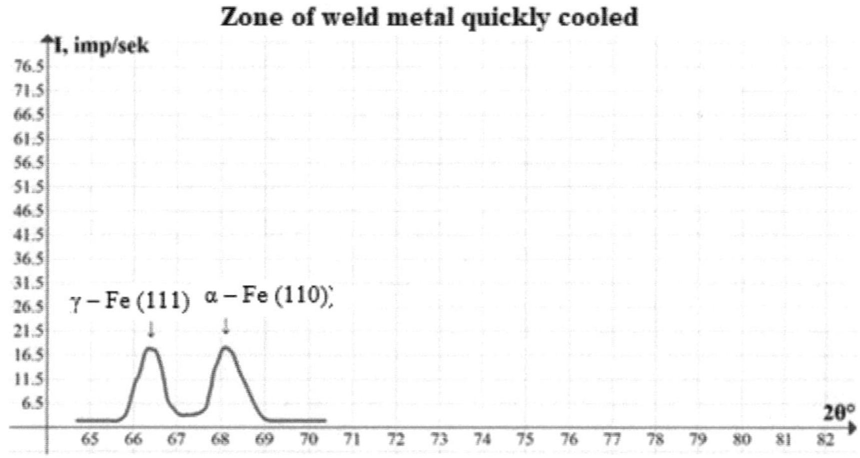

Figure 12. X-ray diffraction analysis in weld metal (γ – Fe (111) и α – Fe (110))

4. Conclusion

1. An increase in the dislocation density and surface internal stresses of γ-Fe (111) and α-Fe (110) was found from the widening of the X-ray line in the fracture area of the unwelded samples.
2. In the weld metal, the X-ray line is identical to that of the unwelded samples while maintaining the ratio between ferrite and austenite as γ - Fe (111) = 60% and α - Fe (110) = 40%.
3. The microstructural analysis found that the crack has a zigzag character in the weld metal, and it is rectilinear in the base material, and in both cases its character is transcrystalline.

References

[1] Sandvik SAF 2507 (http://www.matweb.com/)
[2] Konovalov S, Atroshkina A, Ivanov Y and Gromov V 2010 Evolution of dislocation substructures in fatigue loaded and failed stainless steel with the intermediate electropulsing treatment *MSE A* **12** 3040-3043.
[3] Johannson J and Oden M 2000 Load sharing between austenite and ferrite in a duplex stainless steel during cyclic loading *Met Mater Trans* **A31** 1557
[4] Goh Ch and Yip T 2002 Lowcycle fatigue behavior on duplex stainless steels *Mat. Trans.* **A33** 3433
[5] Duber O, Kunkler B, Krupp U, Christ H and Fritzen C 2006 Experimental characterization and two-dimensional simulation of short crack propagation in an austenitic-feritic suplex steel *Intl J Fatigue* **28** 983
[6] Chai G 2006 Microstructural aspects of duplex steel during high cycle and very high cycle fatigue *Intl J Fatigue* **28** 1533
[7] Callister W 2007 *Materials Science and Engineering: An Introduction* JW& SI USA p15
[8] Alvarez A 2010 Lowcycle fatigue behavior on duplex stainless steels *TIIM* **63** 159-65
[9] Левина А, Мальцева А, Архангельская А, Мальцева Т and Озерец Н 2014 *Разработка аустенитно-ферритной стали нового поколения для упругих элементов в точном машиностроении* Екатеринбург pp 12- 15
[10] Миркин Л 1961 *Справочник по рентгеноструктурному анализу поликристаллов* Физматгиз p 39

IOP Conf. Series: Materials Science and Engineering 1037 (2021) 012037 doi:10.1088/1757-899X/1037/1/012037

Investigation of the occurrence and development of fatigue cracks after corrosion of welded samples of ferritic - austenitic steel

Georgi Georgiev[1], Yaroslav Argirov[1], Tatyana Mechkarova[1], Daniela Spasova[1], Aneliya Stoyanova [1]

[1]Technical University of Varna, Material Science and Technology Department, Studentska Street 1, Varna, Bulgaria

[*]Corresponding author's e-mail: t.mechkarova@tu-varna.bg

Abstract. The article treats the occurrence and development of fatigue cracks in welded samples of ferritic - austenitic steel, which were previously subjected to corrosion. For the purposes of the study, two groups of sheet steel samples were prepared: the first group were reference parts, and the second group of samples were welded and then exposed to a corrosive environment. Corrosion tests were performed by immersing the samples in an acidic solution of (HCl, HNO3 and NaCl). Both groups of samples were tested with specially developed vibrating equipment for high cycle fatigue (10^6 - 10^7 cycles). The durability of SAF 2507 steel is estimated by the number of cycles until a crack is obtained at a certain alternating load. The damaged surfaces were examined by microstructural analysis to trace the shape and development of the crack in the structure of ferro-austenitic steel SAF 2507 after its welding and corrosion.

1. Introduction

Duplex and super duplex stainless steels are widely used in the oil and gas industries due to their unique structure and mechanical properties, making them indispensable in terms of their service life. In practice, they are used almost everywhere, from the surface and surf zone to the underwater parts of offshore structures and oil and gas production facilities, in desalination plants, equipment for the pulp and paper industry, tanks, vessels used in the chemical industry, vessels operating under pressure valves, shut-off valves, tankers for the transport of chemical products, protective barriers against fire and explosion of offshore oil rigs, etc. It can be seen from the literature [2,3,4,5,6,7,8], that there are not enough data for these steels in the field related to welded joints operating in a corrosive environment. These processes are inextricably linked to the structure and phase composition of the steel in the areas of welding, as well as to the properties of the resulting surface layer. There is insufficient information about the development of cracks in the areas where intergranular corrosion has taken place – such as in the area of the weld metal. In order to find optimal operating parameters to increase service life, test fixtures are made of duplex steel SAF 2507 (ASTM S32750), which is pre-welded. Some of the test samples are subjected to corrosion, and all samples are subjected to multi-cycle fatigue (10^6 - 10^7 cycles) [9,10,11].

2. Experimental study

2.1. Preparation of experiment samples

Duplex steel SAF2507 (ASTM S32750) was used in the experiments. In fig. 1 shows laser cut pre-welded samples of 3mm thick plates. The two plates are welded unilaterally to each other by manual Shielded Metal Arc Welding (SMAW). Electrodes ESAB OK 68.82 were used. In order to eliminate internal stresses after welding, the welded plate is rapidly cooled with water.

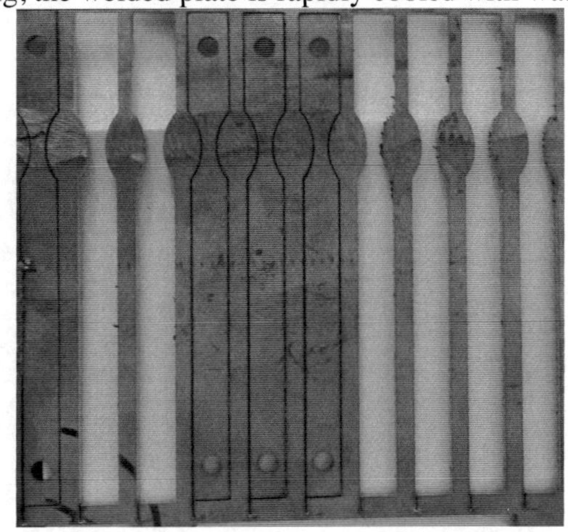

Figure 1. Laser cut pre-welded samples

2.2. Corrosion tests

For the purposes of this study, the samples were placed in a corrosive medium composed of nitric acid, hydrochloric acid and sea salt. The proportion used was 15 ‰ sea salt, 1% nitric acid and 3% hydrochloric acid in 1L of water at a temperature of 60°C (MACC60). The samples remained in the solution for a total of 368 hours, being withdrawn every 48 hours and the difference in weight recorded. The samples were placed in a glass vessel, which was pre-filled with the acid solution of (HCl, HNO3 and NaCl). The samples are suspended above the bottom of the vessel so all their surfaces are in contact with the solution. The vessel is placed in a dry chamber, which is part of a thermostatic device. The chamber is made of pure copper and its outer surface is in contact with the heated water up to the set temperature (Fig. 2). The mass of the samples is determined using precision scales.

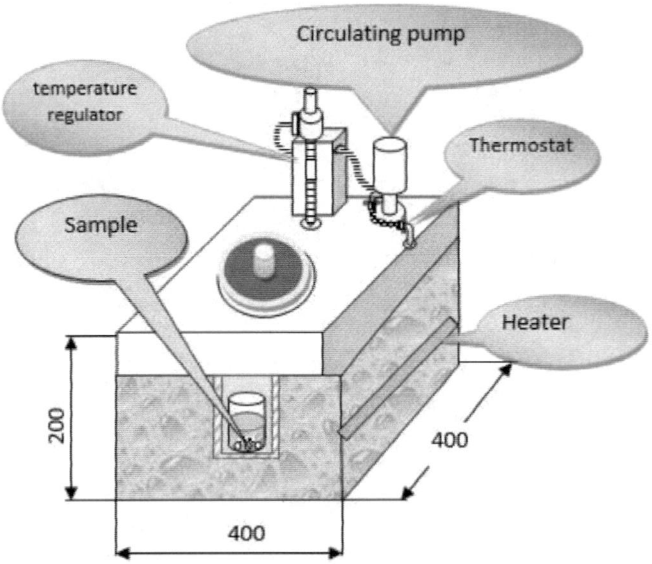

Figure 2. Scheme of thermostatic device for determining the corrosion resistance of samples

2.3. Fatigue test installation

For the purposes of this study, 14 samples were subjected to fatigue tests, 10 of which had previously been in a corrosive environment. The samples subjected to corrosion are numbered in sequence from 1 to 10 (table 1). The cyclical tests were performed in the range from 280 MPa to 470 MPa. The fatigue limit in the range from 10^4 to 10^7 cycles was studied. The experiment installation consists of a vibrating table (Fig.3-3) on which the samples are fixed in a special device (Fig. 3-2). Piezoelectric accelerometers, type K (Fig. 3-1) are attached to the test samples. Through a generator (fig. 7-4), sinusoidal signals with a certain resonant frequency are sent to the vibrating table.

Figure 3. Experimental stand for fatigue tests of samples:
1-accelerator; 2-support device; 3-vibrating table; 4- generator of sinusoidal signals

3. Results of the experimental study

3.1. Results of fatigue tests of welded samples that had not been subjected to corrosion

The results of the cyclical tests of the samples which had not been previously exposed to corrosion are given in Table 1. Their numbering is from 1 to 10. The limit below which no crack formation is observed is established at 340 MPa (in the range up to $1x10^7$ cycles). Based on the data obtained (Table 1), the fatigue curve is compiled fig.4. The samples with positions 10 in Table 1 did not break within the studied range, and they are shown by an arrow in Fig.4.

Table 1 The experimental results of the cyclical tests of the samples that had not been subjected to corrosion

№ Sample	Cycle ($x10^6$)	Load (MPa)
1	0.07	470
2	0.21	425
3	0.5	415
4	0.92	405
5	0.94	395
6	1.2	385
7	2.1	380
8	3.3	375
9	5.56	360
10	10	340

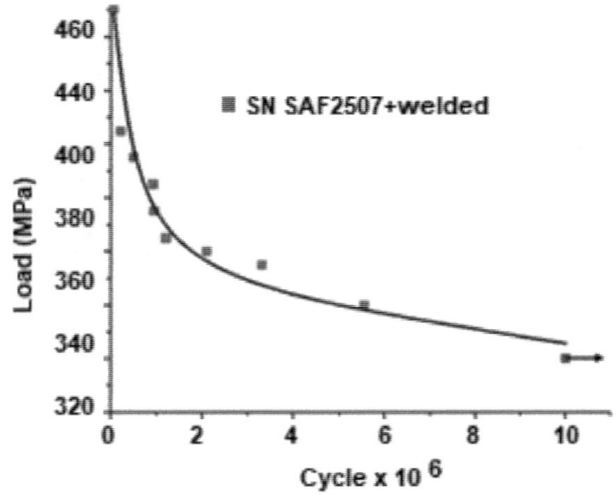

Figure 4. S-N curve of welded samples which had not been previously exposed to corrosion

3.2 Results from fatigue tests of welded samples that had previously been subjected to corrosion
For the purposes of this study, 4 samples were placed in a corrosion medium MASK60. The results of the measurements are summarized in Table 2. It contains data on the initial and final weight, and the percentage weight loss of each of samples.
Figure 5 shows a graphical representation of the results in table 2

Table 2 The experimental results of the percentage weight loss of each of samples

№ Sample	Initial weight	Weight after 368 h	Weight loss in %
1	34.116	34.011	0.31
2	34.387	34.274	0.32
3	34.234	34.122	0.33
4	34.572	34.467	0.30

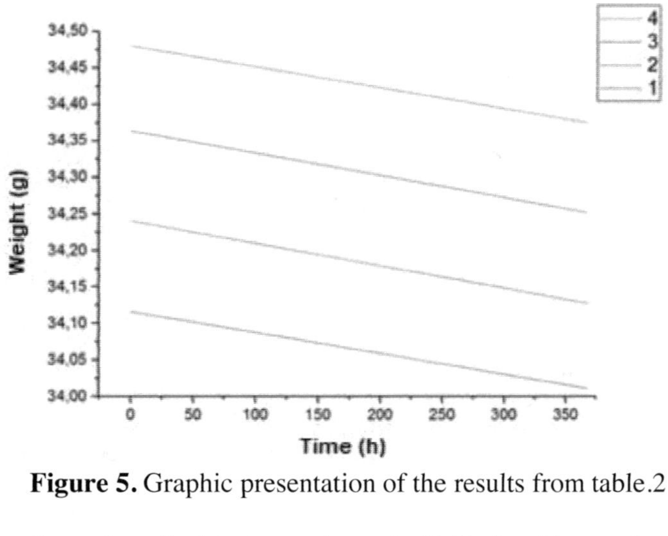

Figure 5. Graphic presentation of the results from table.2

The results of the performed cyclical tests are shown in Table 3, with samples ordered according to their applied load. (The load and the cycles given in the table correspond to the destruction of the sample). Based on the tests performed, a limit was established below which no crack formation was observed and a fatigue curve was drawn, shown in fig.6. In the samples which measured up to 0.33% weight loss from the corrosive medium, no crack formation up to 320 MPa was observed in the range up to 10^7 cycles. The sample with positions 4 in table 3 has not collapsed within the range under study, as shown by an arrow. The remaining (base) samples were destroyed in the studied range, and the results are presented in table 3.

Table 3 The experimental results of the cyclical tests of the samples that had previously been subjected to corrosion

№ Sample	Cycle ($x10^6$)	Load (MPa)
1	0.2	420
2	1	390
3	5	360
4	10	320

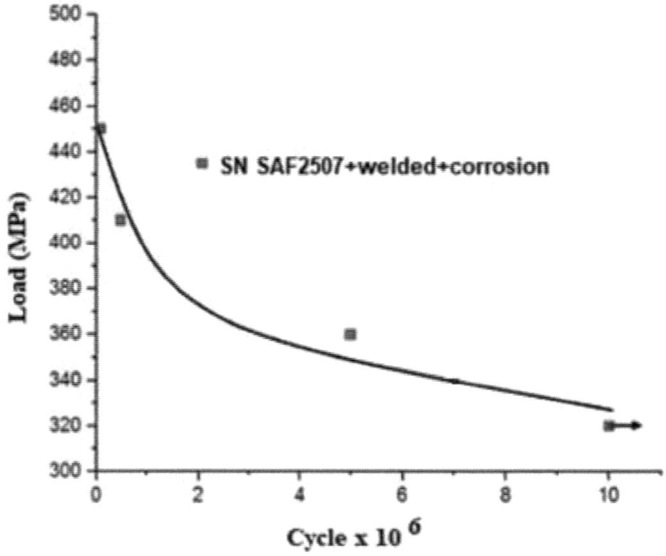

Figure 6. S-N curve of welded samples which had been previously exposed to corrosion

3.3. Results of the microstructure study

It is characteristic of corrosion-resistant steels that due to their high content of chromium, a passivating surface layer is obtained which protects the surface from further impact. In areas where, as a result of a long-term corrosion process, the surface layer has depassivated, corrosion takes place, which for these steel characteristically is intergranular corrosion. This is because corrosion occurs at the grain boundary in areas poor in chromium. After the corrosion, the sample is subjected to cyclical loading as a result of which a focus is formed, and subsequently a crack – its microstructure is shown in fig. 7, the beginning of a crack development and the crack end are shown in fig. 7a. and 7.b. The crack is considered from a lateral section, and at this magnification (X100) it is seen that its development is almost perpendicular to the front surface. In the area of the cracks in fig. 7b, a crushed structure is observed due to the cyclical loading.

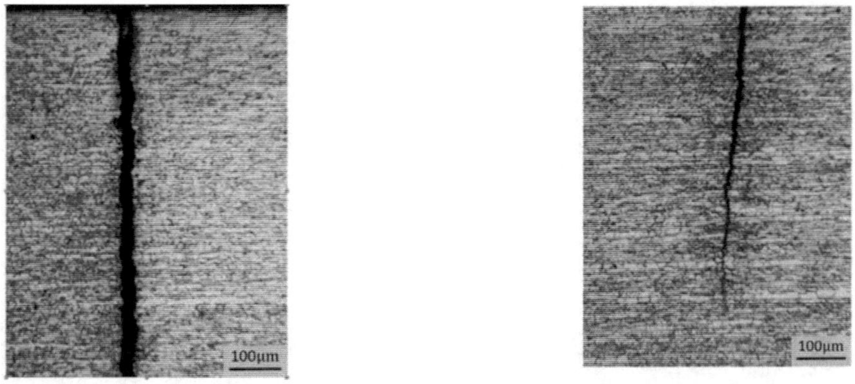

a b

Figure 7. Microstructural studies of a sample subjected to corrosion with magnification X100, a) focus of the crack; b) crack tip and hardened areas

Relatively larger magnifications are used for the sample under consideration, subjected to corrosion, for more precise analysis of the individual areas and zones (fig.8). It is clear that the focus of the developed crack is close to the corroded area, which is also a concentrator. Passing through the crystal structure in a cross section shown in fig. 8b, it can be observed how the crack passes through the grains and only in rare cases can it pass along the grain boundaries. This is observed only in the areas of crushed grains, elongated due to their fibrous structure, and in other cases the crack passes transcrystalline. In fig. 8.c different deviations of the crack along the grain boundary are observed, but due to the elongated (fibrous) grain, the crack passes along a plane with less resistance (transcrital).

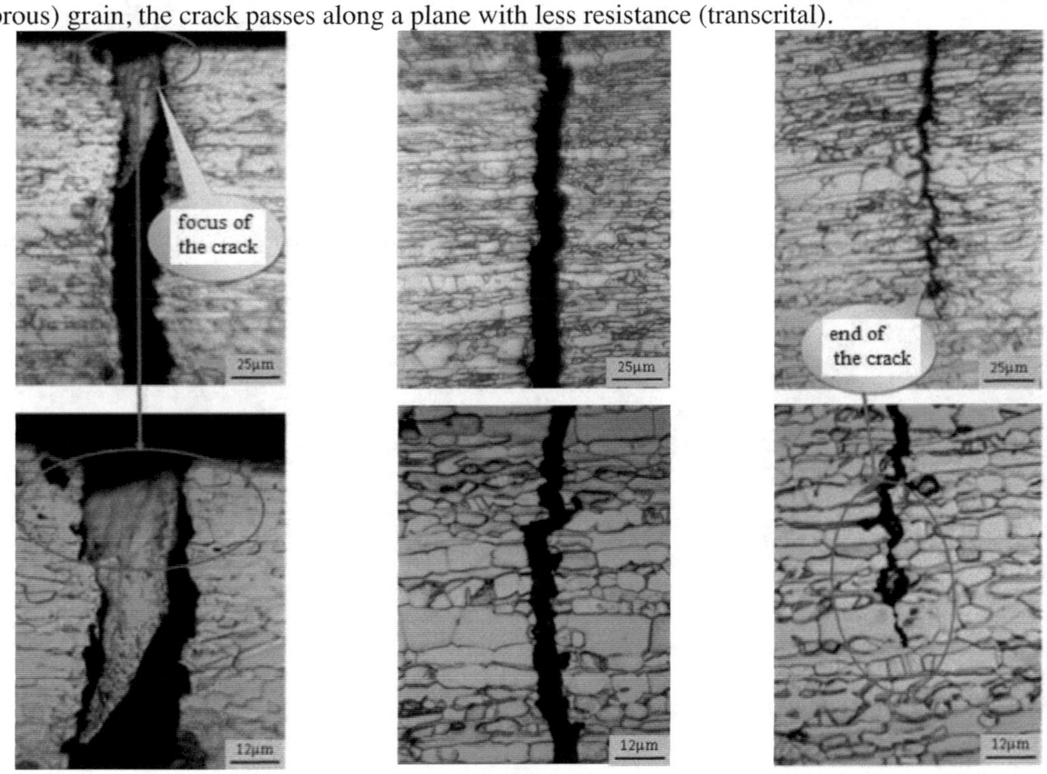

a b c

Figure 8. Microstructural studies of a samples subjected to corrosion, a) focus of the crack with magnification X400 and X800; b) crack with magnification X400 and X800; c) crack tip with magnification X400 and X800

In the case of samples that are welded and then corroded, the microstructural examination is performed in both directions, with the focus being the same. The focus is on the weld metal. Fig. 9.a shows the front surface with a part of the crack and its beginning in the weld metal as an oxidized zone is observed. The end of the crack is shown in Fig.9.b. We can conclude on the basis of all the observed structures that the crack is not linear but meanders and some deviations from its movement are observed (9.b). Figure 10 shows the movement of the crack in a side section through the weld metal, crossing the alloying zone and passing into the base metal (fig10.a-c). Characteristically, passing through the area of weld metal, the crack has a meandering character, and passing through the area of alloying and entering the base material, it has a linear character.

a b

Figure 9. Welded samples microstructure: a) crack start X400; b) end of crack X400

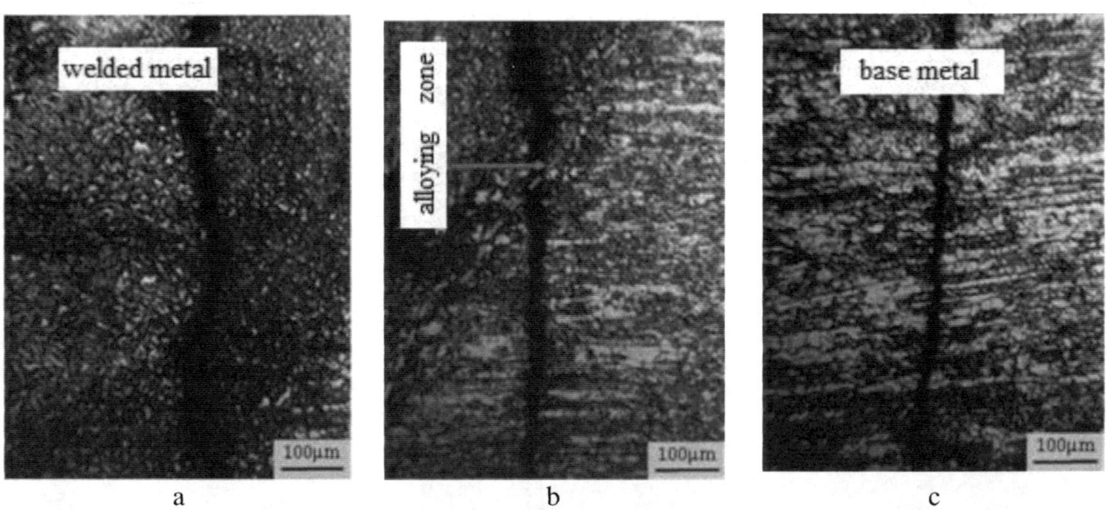

a b c

Figure 10. Movement of a crack: a) through the weld metal X100; b) through the alloying zone X100; c) through the base material X100

4. Conclusion

1. The corrosion wear of ferritic - austenitic steels in MACC60 environment was determined, as the samples subjected to corrosion were shown to have reduced durability compared to samples that were not subjected to corrosion;
2. The cracks have a pronounced transcrystalline character, which is explained by the fibrous structure of the samples studied;
3. Welded samples subjected to corrosion have a reduced durability compared to welded samples that are free of corrosion;
4. The crack has a zigzag character in the weld metal, and in the base material it is rectilinear, and in both cases its character is transcrystalline;
5. It was found that in welded samples after corrosion, a crack passes near the ZS on the front surface, crossing it several times.

References

[1] Sandvik SAF 2507 (http://www.matweb.com/)
[2] Konovalov S, Atroshkina A, Ivanov Y and Gromov V 2010 Evolution of dislocation substructures in fatigue loaded and failed stainless steel with the intermediate electropulsing treatment *MSE A* **12** 3040-3043.
[3] Johannson J and Oden M 2000 Load sharing between austenite and ferrite in a duplex stainless steel during cyclic loading *Met Mater Trans* **A31** 1557
[4] Goh Ch and Yip T 2002 Lowcycle fatigue behavior on duplex stainless steels *Mat. Trans.* **A33** 3433
[5] Duber O, Kunkler B, Krupp U, Christ H and Fritzen C 2006 Experimental characterization and two-dimensional simulation of short crack propagation in an austenitic-feritic suplex steel *Intl J Fatigue* **28** 983
[6] Chai G 2006 Microstructural aspects of duplex steel during high cycle and very high cycle fatigue *Intl J Fatigue* **28** 1533
[7] Callister W 2007 *Materials Science and Engineering: An Introduction* JW& SI USA p15
[8] Alvarez A 2010 Lowcycle fatigue behavior on duplex stainless steels *TIIM* **63** 159-65
[9] Левина А, Мальцева А, Архангельская А, Мальцева Т and Озерец Н 2014 *Разработка аустенитно-ферритной стали нового поколения для упругих элементов в точном машиностроении* Екатеринбург pp 12- 15
[10] Миркин Л 1961 *Справочник по рентгеноструктурному анализу поликристаллов* Физматгиз p 39

Characterization of metallic materials used in food packaging

D A Necsulescu[1], M Ciurdas[1], C Milea[1]

[1] University POLITEHNICA of Bucharest, 313 Spl. Independentei, Bucharest, Romania

*Corresponding author's e-mail: aliananecsulescu@yahoo.com

Abstract. The work represents a comparative study of three type of metallic food package used in Romanian food industry. In this paper, three food packages made of low-alloy steel were analyzed. To characterize their properties, spectrometric analyzes were performed to determine the chemical composition, microhardness analyzes, optical microscopy and Fourier Transform Infrared Spectroscopy. The investigations on the three samples were achieved through X-ray Fluorescence Spectrometry (XRF) analysis performed with a Spectro XEPOS analyzer, optical microscopy analyzes performed on a Reichert UnivaR optical microscope, and microhardness tests by the Vickers method using the CV - 400 DTS tester / 2 (CV Instruments) and Fourier Transform Infrared Spectroscopy (FTIR) with JASCO 6200 analyzer, equipped with Golden Gate type attenuated total reflection (ATR) device. Following the investigations, the presence of epoxyphenolic resin was confirmed in the case of the first two samples and a layer of polyester in the case of sample three. Finally, conclusions are presented about the qualities of the selected packaging and their properties to keep safe the packaged products for a period of several months to several years.

1. Introduction

Packaging technologies, packaging materials and packaging have dynamically evolved in the past decades, thus ensuring that packaging has a complex functional and expressive role.

The selection of the material for the packaging and the packaging method implies a deep knowledge of them and the processes that take place in the preserved products, during the storage period [3].

They may undergo various chemical reactions with substances from the package, or may be impregnated with various toxic or odors substances.

The properties to be considered, related to food distribution, are many and may include gas or water vapor permeability, mechanical properties, sealing capacity, thermoforming properties, (resistance to water, grease, acid, UV radiation, light, etc.), machinability (on the packaging line), transparency, anti - condensation capacity, printing capacity (printability), availability and of course cost.

Conventional food packaging materials currently used vary between metals, paper, glass, and plastics [4]. Metal packaging is made from steel or aluminum. Steel is used in the production of containers for packaging a wide range of products, such as food, paints, etc [7]. Aluminum packaging is used to make containers for food and beverages, foils and laminates.

Metallic and alloys packaging is mainly used in the food industry in the packaging of canned meat, fish, fruits and vegetables, alcoholic and non-alcoholic beverages. Consumer opinion is less favorable to metals because they can influence the taste of packaged products.

The metallic packaging is made from tinplate [1], aluminum and combined materials (plastics, cardboard and metal). Recently, the weight of aluminum packaging and aluminum alloys has increased due to the advantages that these materials offer [5][6].

The following type of packaging is made from metallic materials: metal boxes, aluminum cans, metal drums and barrels, aerosol (spray) type packaging, deformable tubes. Metal boxes are used in the food industry for food packaging and preservation. The packaging system, the material used and the type of packaging must ensure the stability of the food, prevent the transfer of toxic substances from the packaging material or during the packaging process to the food, avoid exchange of moisture with the environment, impurity with unwanted substances in the environment, contamination with microorganisms and favoring photochemical reactions or caused by microbiological contamination.

The advantage of metal boxes is rigidity, avoiding thus the risk of breakage during transport.

Smart packaging (or Intelligent packaging) has been gaining publicity lately, with an expected market of $26.7 billion by 2024 [2]. Smart packaging represents packaging systems carrying sensors that help prolong the shelf life of foods, reveal information on freshness and quality, and enhance the safety of the product and consumer.

Another technology in packaging systems is active packaging, which is the integration of additives with the packages to improve food shelf-life and quality [8].

2. Materials and methods

2.1. Materials used in the study

For the study, three types of metallic food boxes were selected from various manufacturers (figure 1), and were cut from the bottom of the boxes, with a diameter of 40 mm, which were then subjected to spectral chemical analysis, optical microscopy, microhardness analysis and Infrared Spectrometry with Fourier Transform.

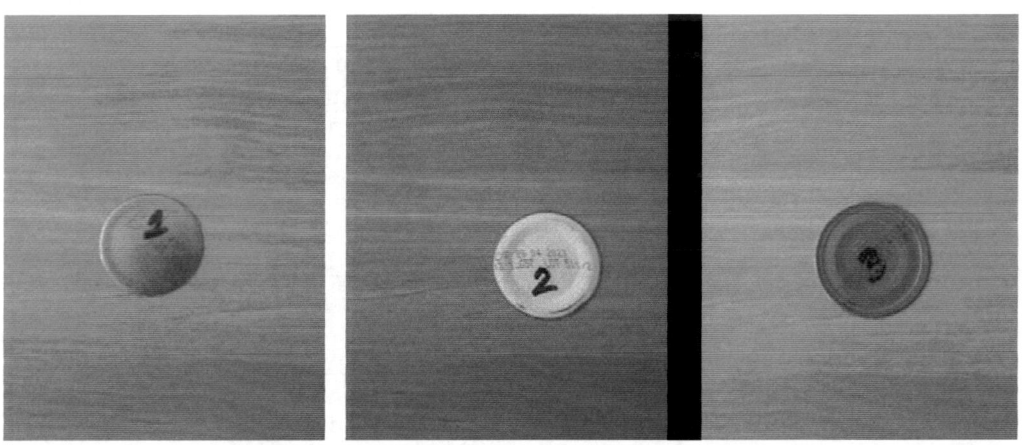

Figure 1. The three metallic food packaging samples

2.2. X-ray Fluorescence Spectrometry (XRF)

The XRF Analyzer spectrometer used for the elementary analysis of the samples of interest for this research it was Spectro XEPOS type. The analysis was performed according to the requirements [15] of the SR EN ISO 17025: 2018 standards, using the TurboQuant Alloys analytical program. Method parameters: supply voltage: 120V / 230V a.c. ± 10%, 50/60 Hz, for quantitative analysis methods: Fundamental Parameters and TurboQuant, analysis in helium atmosphere, detects elements from Na to U from the periodic table, SDD - Silicon Drift Detector with cooling system, stable peak position and high detection rate (120 kcps); X-ray tube with Pd anode, maximum power of 50 W, high tube voltage 50 kV [14].

2.3. Optical microscopy

The optical investigations, made on three polished samples, were performed on a Reichert UnivaR microscope, viewing can be done in bright field (BF), dark field (DF) and in simply reflected polarized light. It was used a compact Hitachi KPF1 digital camcorder, lightweight CCD camera , color, integrated type, using the latest 2/3 inch CCD image size type, with CCIR signal type; The actual 782x582 pixel count in combination with a high-density montage. The Analysis Program is Omnimet Express, which includes a powerful database and report generator that allows you to store images and their associated data to generate a report.

2.4. Microhardness

The Vickers microhardness tests were performed using the CV - 400 DTS / 2 testers (CV Instruments), with the following characteristics: loading force 200 gf and the action time was 30 s.[13]

2.5. Fourier Transform Infrared Spectroscopy

Fourier transform infrared spectroscopy (FT-IR) was used to determine the chemical structures of some inorganic substances, which is one of the methods we can perform a qualitative and quantitative analysis of organic compounds. The analysis was performed with a JASCO 6200 analyzer, equipped with a Golden Gate type attenuated total reflection (ATR) device.

3. Results and discussions

3.1. Spectrometry results

Following the XRF analyzes (Tables 1, 2 and 3), the mass concentrations of the 4 main elements were determined. [15] Variations in Fe concentrations can be observed in the case of the three samples between 79-84%.

Table 1. Chemical composition analysis (XRF) of sample 1

Symbol	Concentration, %	Standard deviation
Al	10.98	0.01
Fe	84.86	0.05
Co	1.960	0.048
Sn	0.7817	0.0020
C	0.06	0.0015
Others	1.3583	-

Table 2. Chemical composition analysis (XRF) of sample 2

Symbol	Concentration, %	Standard deviation
Al	14.11	0.01
Fe	81.70	0.05
Co	1.828	0.043
Sn	0.8499	0.0020
C	0.06	0.0013
Others	1.4571	-

Table 3. Chemical composition analysis (XRF) of sample 3

Symbol	Concentration, %	Standard deviation
Al	17.14	0.02
Fe	79.01	0.05
Co	1.726	0.041
Sn	1.156	0.003
C	0.06	0.0012
Others	0.908	-

Following these determinations, the brand classification is as follows: STEEL C4D1, low carbon steel according EN 10016-3 (1994); ISO / FDIS 16120-2 (2001), with the following properties: density 7.8-7.9 g/cm3, elastic modulus 200-215 GPa, melting point 1480-1526 ºC.

3.2. Optical microscopy results

Qualitative microscopy studies were performed on three cans, of different producers, but from the same type of material, respectively ferric steel covered with a protective layer. Thus, the samples taken from the three cans were cutted in cross section and they were prepared for quantitative metallographic analysis.

For the layer thickness measurements were performed eight fields with magnification of 80 x.

A comparable layer thickness can be observed, at the first test the degree of non-uniformity of the layer thickness being higher (figure 2), and the lowest degree of non-uniformity at sample three (figure 4)

Following some metallographic analyzes the thickness of the upper layer was determined. Metallographic analyzes were performed using an optical microscope Reichert Univar on cross-cut samples.

Investigations on the metallographic microscope reveals that the third sample has a uniform layer deposited on the base material and the other samples show non-uniformities in the deposited layer.

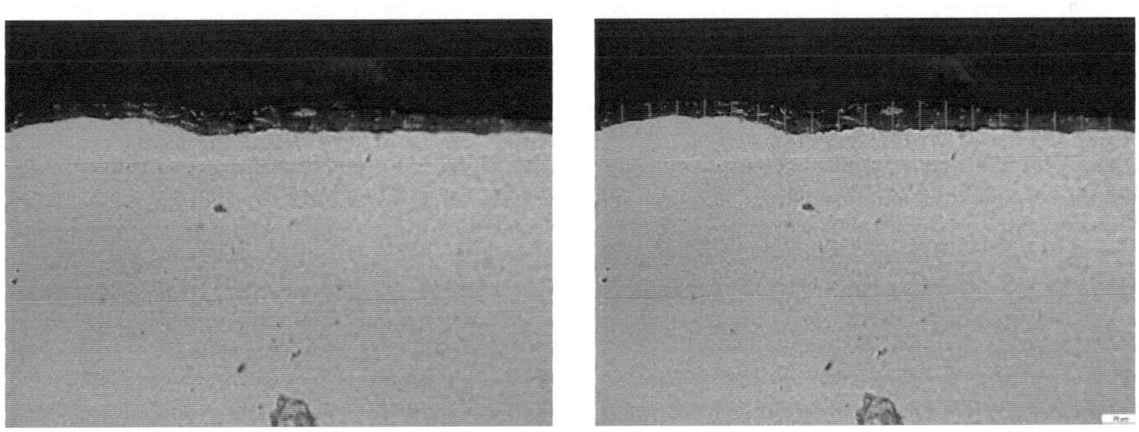

a b

Figure 2. Optical microscopy image (X80) of sample 1 a- captured; b- processed

IOP Conf. Series: Materials Science and Engineering 1037 (2021) 012038 doi:10.1088/1757-899X/1037/1/012038

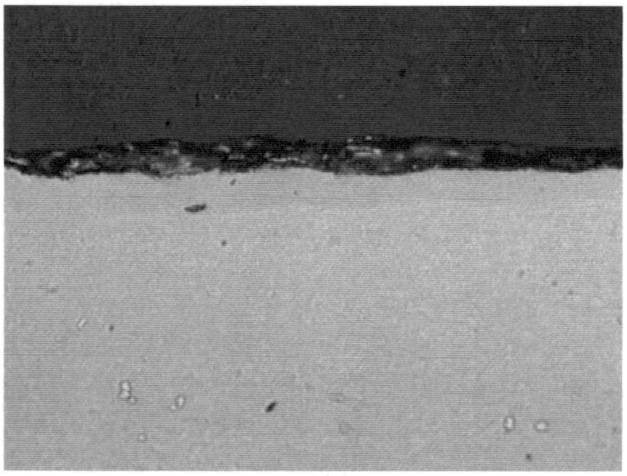

a

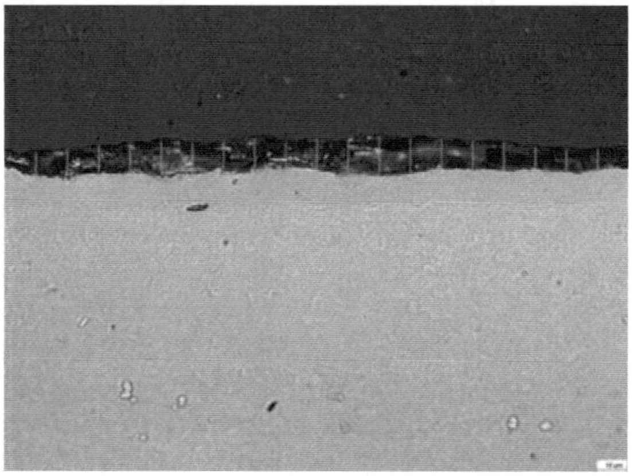

b

Figure 3. Optical microscopy image (X80) of sample 2 a- captured; b- processed

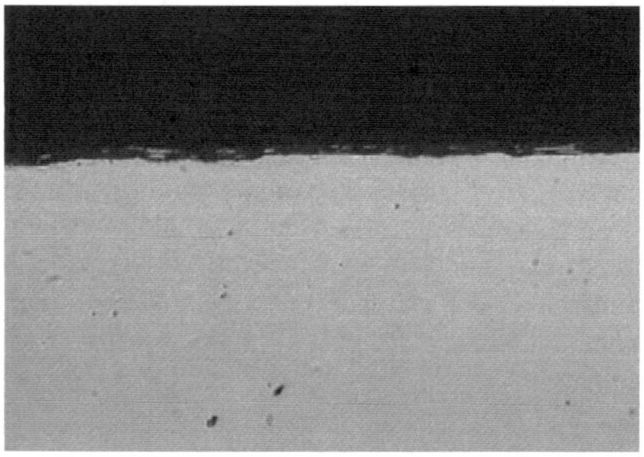

a

b

Figure 4. Optical microscopy image (X80) of sample 3 a- captured; b- processed

In the three cases, even if the samples are un-attacked, the limits of polyhedral, fine and uniform grains are observed, characteristic of a cross section through a steel sheet. The confirmation that the sheet is made of steel came from the test with the magnet, which is strongly attracted by the magnet.

From the three graphs resulting from the measurements it results that sample 1 presents a more uneven layer thickness, and sample 3 the most uniform layer thickness (figure 5-7).

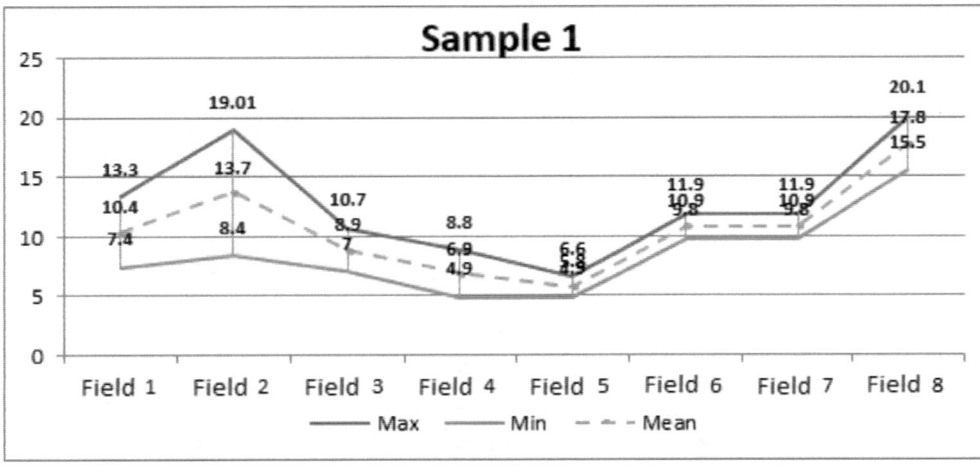

Figure 5. Analysis of the evolution of the layer thickness for the sample 1

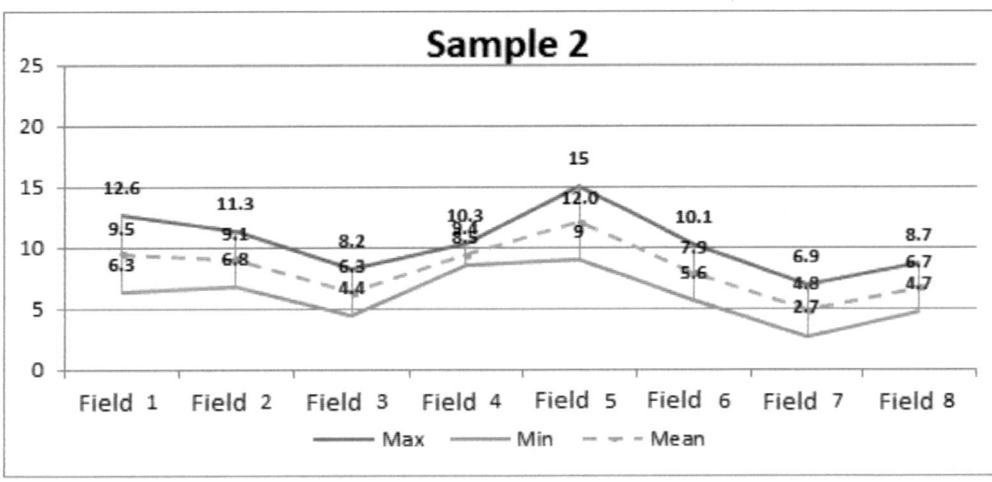

Figure 6. Analysis of the evolution of the layer thickness for the sample 2

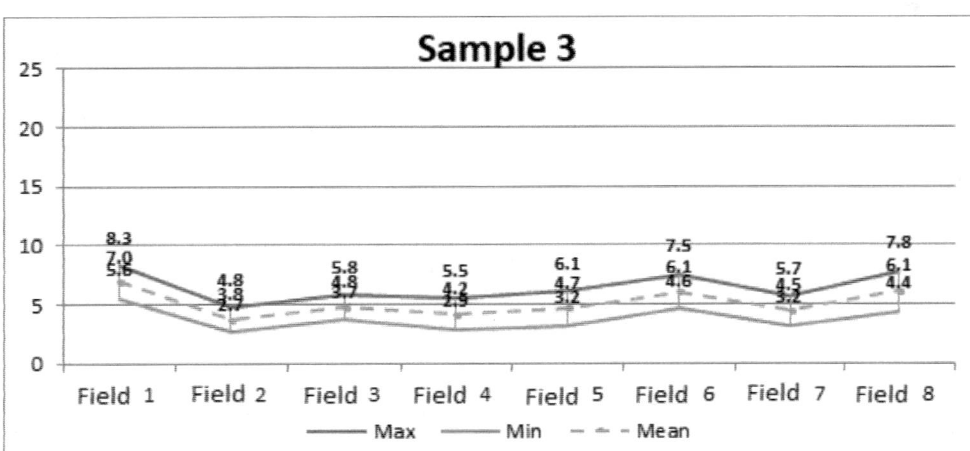

Figure 7. Analysis of the evolution of the layer thickness for the sample 3

IOP Conf. Series: Materials Science and Engineering 1037 (2021) 012038 doi:10.1088/1757-899X/1037/1/012038

3.3. Hardness results

After chemical analysis and characterization by optical microscopy, the samples were subjected to microhardness tests by the Vickers method, the resulting average values being shown in the figure 8. From the obtained data it is observed that the maximum value of the micro- hardness is found in sample 3, and the minimum value has sample 1.

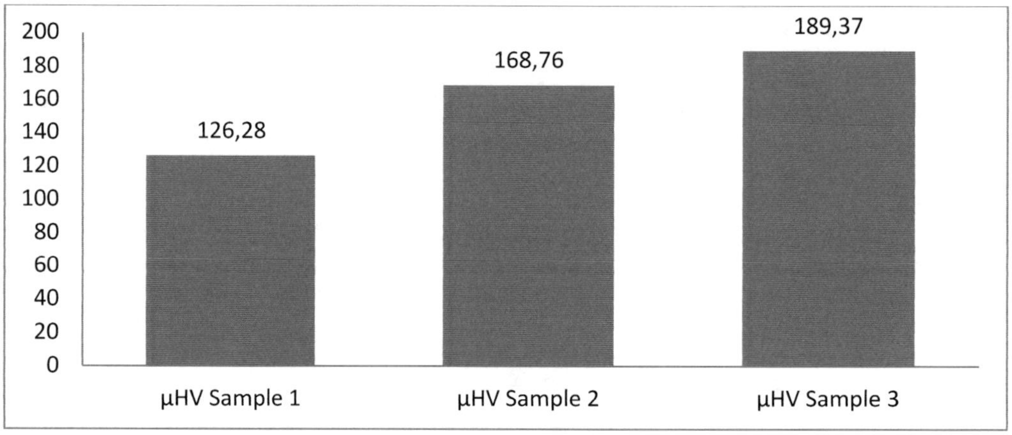

Figure 8. Vickers microhardness values

3.4. Fourier Transform Infrared Spectroscopy results

The Fourier Transform Infrared Spectroscopy (FTIR) spectra for the investigated sample are shown in Figures 9, 10 and 11.

In the spectra recorded for sample 1 and sample 2, between 3200 and 3500 cm^{-1}, a wide absorption band of the phenolic H-O bond is observed. Between 2860 and 2970 cm^{-1} it is reveals bonds due to the stretching vibrations of the C-H bonds from the methyl, methylene and methoxy groups (CH3, CH2, O-CH3), at 1605.45 cm^{-1} and 1505.17 cm^{-1} it is reveals bonds due to the stretching vibrations of the C = C and C-H bonds from the aromatic ring, at approximately 940 cm^{-1} it is reveals bonds due to the stretching vibration of the epoxy group, at approximately 1230 and 1178 cm^{-1} it is reveals bonds due to the vibrations of the C-O bond from the vicinity of the aromatic nucleus , and between 1100 and 650 cm^{-1} bands appear due to the vibrations of deformation of the C-H bonds from the methyl, methylene and methoxy groups.

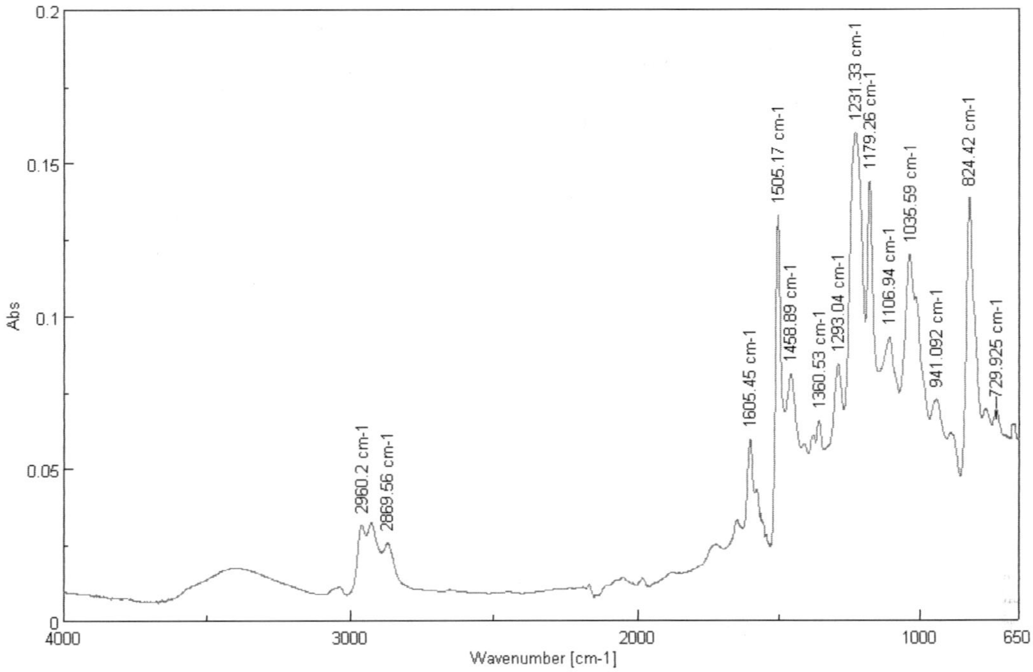

Figure 9. FTIR spectra for sample 1

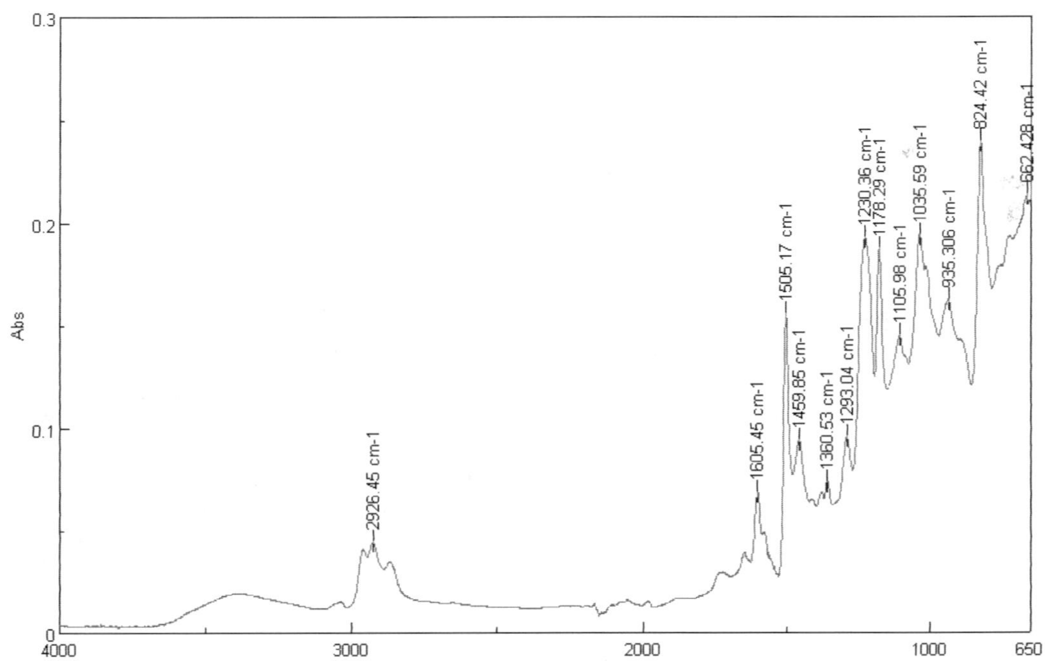

Figure 10. FTIR spectra for sample 2

The obtained spectra on sample 3 were similar to typical FTIR spectra of polyesters, being revealed characteristic peaks at 1712 cm-1 (due to the asymmetric stretching vibration of C = O bond from the carbonyl group), at 1406 cm^{-1} (due to the stretching vibration of the C-C bond in the benzene ring), at 1242 cm^{-1} (due to the bending vibration of the C = O bond in the ester group coupled with the stretching vibration of the C-C bonds from the benzene ring), at 1093 cm^{-1} (assigned to the asymmetric stretching vibration of C–O bond of esteric group), at 1015 cm^{-1} (due to the deformation vibration of the C-H bond (methine group) from the benzene ring coupled with the stretching vibration of the C-C bond from the benzene ring),at 968 cm^{-1} (assigned to the stretching vibration of C=C bond), at 721 cm^{-1} (due to the deformation vibration of the of carbonyl group C = O coupled with the stretching vibration of the C-H bond from the benzene ring).

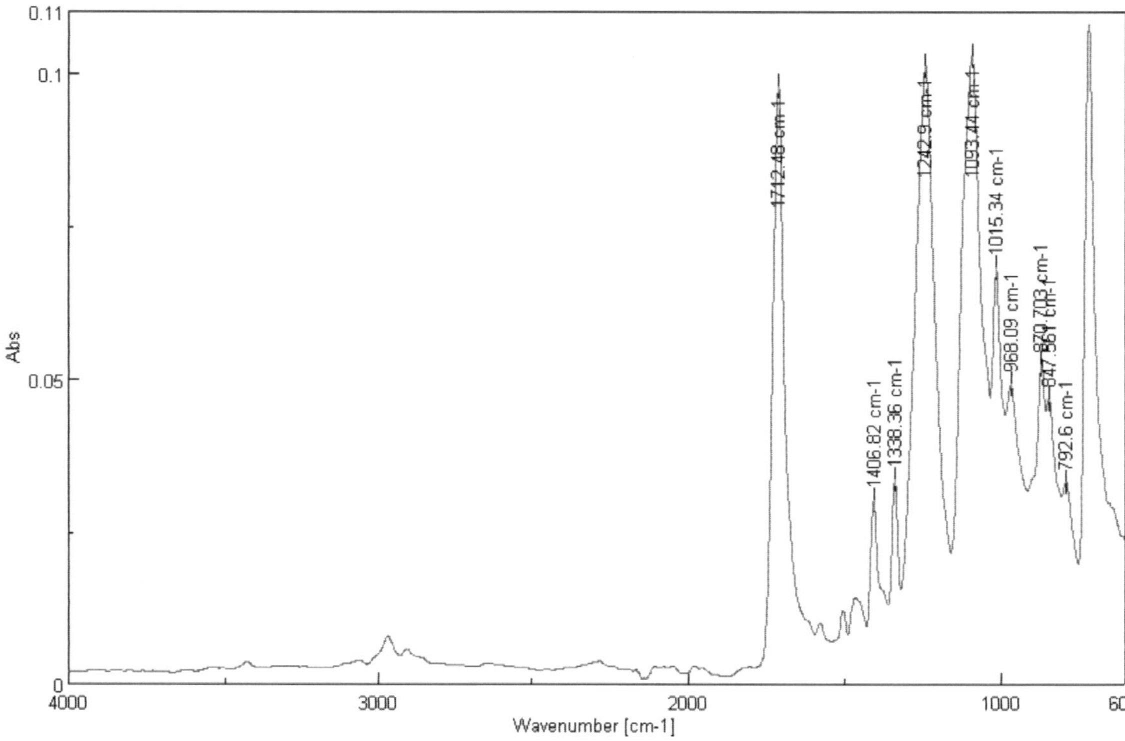

Figure 11. FTIR spectra for sample 3

After the previous analyzes, the sample were investigated by Infrared Spectrometry with Fourier Transform showed that on the investigated samples made from metallic materials is deposited a layer of epoxyphenolic resin (sample 1 and sample 2) and a layer of polyester (sample 3) .

Epoxyphenolic resins are acid-sulfur-resistant; it evenly covers the surface, dries quickly, lasts up to 200 °C and does not give the product a foreign taste or smell.

Polyesters have high tensile strength, very good chemical resistance and are stable in a wide temperature range (-60 °C ... + 220 °C). They are impermeable to liquids and have a slight impermeability to gases.

4. Conclusion and perspectives

Following the XRF analysis determinations, the metallic material for the food packaging was classified as STEEL C4D1, which is frequently used for the manufacture of these types of cans.

Based on the optical microscopy analysis, in sample 3 can be observed a good adhesion between the substrate and the covered area, with relatively small layer thickness values, and for samples 1 and 2 there it is found high degrees of unevenness of different layer thickness. In all three cases, even if the samples are unattacked, we can observe polyhedral, fine and uniform grain boundary, which is characteristic of a cross section through a ferric steel sheet.

The Vickers microhardness results show that the maximum value: 189.38 is found in sample 3, and the minimum value has sample 1: 126.28.

Fourier transform infrared spectrometry analyzes showed that a layer of epoxyphenolic resin (sample 1 and sample 2) and a layer of polyester (sample 3) were deposited on the investigated samples made from metallic materials. In case of sample 3 the FTIR spectra reveals a highlighted bands characteristic for polyesters. The FTIR spectra highlighted the presence of epoxyphenolic resins, compounds that are acid-sulfur-resistant that uniformly cover the surface of cans, dries quickly, withstands up to 200 ºC and do not give the product a foreign taste or smell. Also, polyester class compounds have been identified that have high tensile strength, very good chemical resistance and are stable in a wide temperature range (-60 ºC ... + 220 ºC). They are impermeable to liquids and have a slight impermeability to gases.

Acknowledgment
The work is financed by the University Politehnica of Bucharest, through the project "Engineer in Europe" online edition, registered at Romanian Ministry of Education and Research under no. 457 / GP / 06.08.2020. Funds are granted for special situations that cannot be integrated in the form of financing state for higher education institutions

5.References

[1] Marsh, K. & Bugusu, B. 2007 *Food Packaging – Role, Materials, and Environmental Issues* Institute of Food Technologists **72** Nr.3. 39-55.

[2] Schaefer, D. & Cheung, W. M. 2018 *Smart Packaging: Opportunities and Challenges* Procedia CIRP **72** 1022-1027.

[3] Siracusa, V. & Rosa, M. D. 2018 *Sustainable Packaging. In Sustainable Food Systems from Agriculture to Industry* 275-307.

[4] Geueke, B. et al. 2018 Food packaging in the circular economy: Overview of chemical safety aspects for commonly used materials. *Journal of Cleaner Production.* **193**. 491-505

[5] (n.d.). *Food packaging with Aluminium foil* Retrieved from: https://alfipa.com/applications/aluminum-foil-laminates-food-packaging/

[6] (n.d.). *Aluminium Foil* CNBM International Corporation. Retrieved from: https://www.aluminiumchina.com/__novadocuments/427439?v=636512583854900000

[7] (n.d.). *Tinplate and Tin Free Steel* JFE Steel Corporation. Retrieved from: http://202.229.24.177/en/products/sheets/catalog/b1e-006.pdf

[8] Majid, I. et al. 2016 *Novel food packaging technologies: Innovations and future prospective* Journal of the Saudi Society of Agricultural Sciences **17** 454–462.

[9] Atudosiei Nicole, P. Ştefănescu 2003 *Principii biologice aplicate in conservarea produselor alimentare* Revista Calita pt. ind. alimentara **4**

[10] Nastasia Belc, Mihaela Ghiduruş, Amalia Miteluţ, Mona Popa, Petru Nicoliţă, Mira Turtoi 2006 *Ambalarea modernă a produselor alimentare* Editura Agir

[11] Athanasie Trutia, Iulian Ioniţă, Floriana Iova, Gabriel Stănescu 2003 *Spectroscopie optica, atomica si moleculara* Editura Universitatii Bucuresti

[12] Sopa S. 1995 *Metode de conservare a produselor alimentare – Curs pentru studenţi* Editura Tipo Agronomica Cluj-Napoca

[13] Ion Ciucă, Octavian Trante- *Note de curs- Încercări mecanice*

[14] Pencea, I. 2011 *Metode si tehnici instrumentale de analiza elementala a materialelor", capitolul 8, "Tratat de Ştiinţa şi Ingineria Materialelor* Editura Agir **5** 1057-1152

IOP Conf. Series: Materials Science and Engineering 1037 (2021) 012039 doi:10.1088/1757-899X/1037/1/012039

Researches on the recovery of useful metals from spent Li-Ion batteries

I Bratosin[1], V G Ghica[1,*], M I Petrescu[1], M Buzatu[1], G Iacob[1], A D Necşulescu[1]
[1] University POLITEHNICA of Bucharest, 313 Spl. Independenţei, Bucharest, Romania

[*]Corresponding author's e-mail: gvghica@yahoo.com

Abstract. The present and future importance of Li-Ion batteries is immeasurable; this is evidenced by the awarding of the Nobel Prize in chemistry in 2019, to the team of John Goodenough - director, M Stanley Whittingham and Akita Yashino, who developed the technology to make this type of rechargeable battery. Used at the beginning (1990-2000) in applications such as portable electronic equipments (mobile phones, laptops, electric tools, energy storage systems, etc.), these batteries are considered the most suitable battery for powering electric vehicles. Today the number of electric cars in the world has exceeded 4 million units and the future estimates are among the most optimistic. Recycling of Li-Ion batteries is mainly done in China and South Korea, countries that have a large share in their production, but things can change. Intense research for the recovery and reuse of useful metals contained in these batteries is carried out in many research laboratories. The paper presents the current state of the research undertaken for the recovery of the cathodic paste with high content of Cobalt of waste LIBs, by ultrasonography in lactic acid solution. The hydrometallurgical method uses a non-polluting organic (lactic) acid. The working technique, the results obtained and the investigations carried out on the recovered materials are presented (analysis of optical and electron microscopy, EDX, X-ray diffraction). The paper also includes data regarding the optimization of the separation process of the active cathode paste from the aluminium foil using the orthogonal central-compositional programming of the second order.

1. Introduction

For the development of lithium-ion batteries a class of materials of interest is lithium nickel manganese cobalt oxides - NMC (abbreviated Li-NMC, LNMC, NMC or NCM) which are mixed oxides of Li, Ni, Mn and Co ($LiNi_xMn_yCo_zO_2$). The most important representatives have a composition with x + y + z = 1 and are closely related to lithium cobalt (III) oxide ($LiCoO_2$) and have a layered structure like these [1].

The most used combination is 1-1-1 (1/3Ni-1 / 3Mn-1/3Co) but there are also 5-3-2 (5 parts Ni - 3 parts Co - 2 parts Mn) and others.

This type of battery was born from the attempt to use as little cobalt (expensive and deficient metal) as possible at the cathode of Li-Ion batteries. Nickel cathodes have a lower price, higher energy density, longer life, even if they have a lower voltage.

The success of this type of battery is due to the nickel / manganese combination: manganese has a high specific energy but por stability while manganese forms a spinel type structure (low internal resistance) with low specific energy.

NMC batteries are used as Energy Cells or Power Cells (Battery storing electricity) for: electric cars (Audi e-tron, Renault ZOE, BMW i3, VW e-Golf), e-bikes, power tools, mobile electronics such as mobile phones / smartphones, laptops etc. The cell voltage of lithium-ion batteries with NMC is 3.6 - 3.7 V [2].

In Figure 1 is presented Lithium-ion batteries available for recycling by chemistry (tonnes) [3].

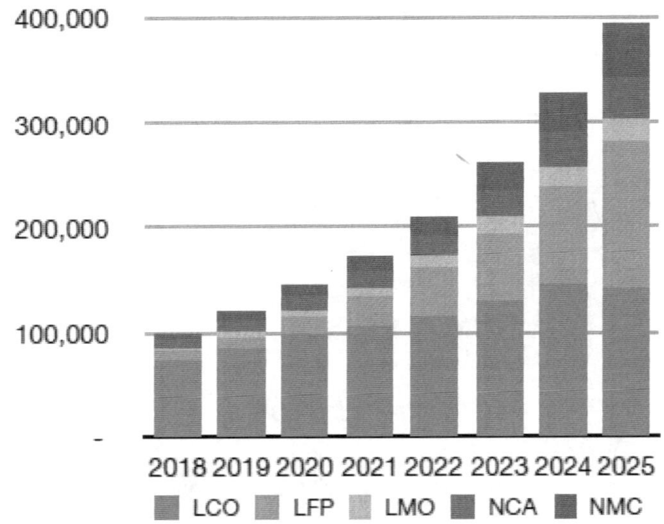

Figure 1. Lithium-ion batteries available for recycling by chemistry (tonnes) [3].

It can be seen that the market share for these batteries is increasing.

2. Materials and Methods

The experiments were performed at the Hydrometallurgy Laboratory from the Department of Engineering and Management of Metallic Materials Elaboration, Faculty of Materials Science and Engineering - University "Politechnica" of Bucharest.

In the experiments we used spent NMC Lithium-ion batteries from mobile phones which were dismantled manually in order to reach out the cathode material. The spent NMC Lithium-ion batteries were completely discharged in a salt solution for 1h. The cathode materials were manually dismantled and removed from the aluminum foil (immersed in acidic solution - lactic acid and subjected for ultrasonic cleaning) [4].

We used the ultrasonic cleaning machine (Emmi12-HC) with the following technical specifications: housing – stainless steel, cleaning frequency = 45 kHz; cleaning time = 1 - 60 min; volume = 1.2 l; heating temperature = 20 - 80°C; bath dimension 200x100x65 mm; ultrasonic power = 50/75/100W.

The samples were analyzed by SEM (Scanning Electron Microscopy) using a Quanta Inspect F50, with a Field Emission Gun (FEG) with resolution of 1.2 nm, and an EDX (Energy Dispersive X-ray) analyzer having a resolution of 133 eV at MnK. Also, the powder sample was characterized by XRD (X-ray Diffraction) using the PANalytical X'Pert PRO MRD diffractometer (wave length of l_{Cu} = 1.544).

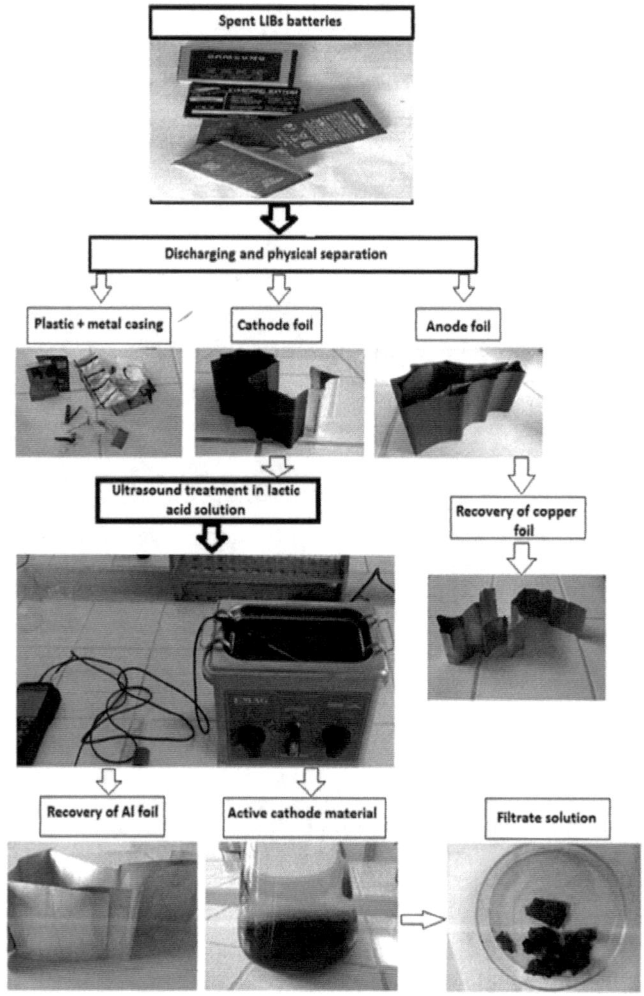

Figure 2. The separation process of the cathode material [15].

3. Results and discussions

The optimal parameters for the ultrasonic process and obtaining a high separation efficiency of the active paste were [5, 6]:

- Concentration of lactic acid solution: 1.7 M;
- Temperature of the ultrasonic bath: 50^0C;
- The power of ultrasonic bath: 80 W;
- Time range between 1.5 to 2.5 minutes.

After the optimal parameters have been established, we tested the process using an entire cathode foil. The result was a maximum separation efficiency of $\eta = 88.08$ %.

The SEM-EDX analyses, performed on the samples are captured in Figures 3 and 4.

IOP Conf. Series: Materials Science and Engineering 1037 (2021) 012039 doi:10.1088/1757-899X/1037/1/012039

Analysis	Elem	Wt %	At %	K-Ratio	Z	A	F
EDX	CK	9.90	23.81	0.0209	1.1060	0.1907	1.0005
	OK	24.18	43.64	0.0803	1.0902	0.3039	1.0018
	MnK	6.21	3.27	0.0724	0.9293	1.0043	1.2478
	CoK	44.41	21.76	0.4113	0.9313	0.9943	1.0000
	NiK	15.29	7.52	0.1480	0.9670	1.0009	1.0000
	Total	100.00	100.00				

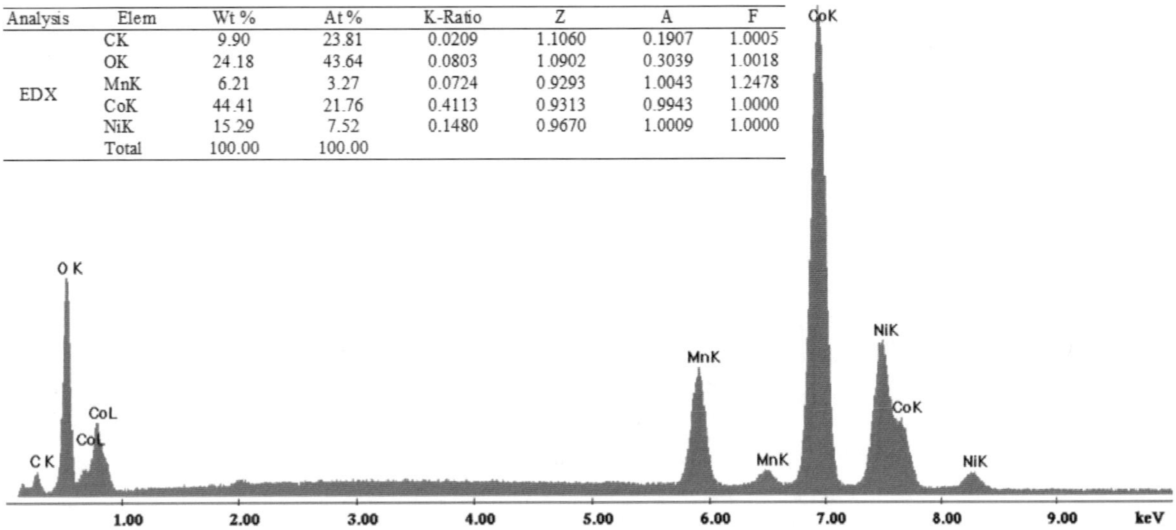

Figure 3. EDX sample investigation - $LiNiMnCoO_2$.

Figure 4. SEM image - LiNiMnCoO.

The XRD analysis of the sample is presented in Figure 5.

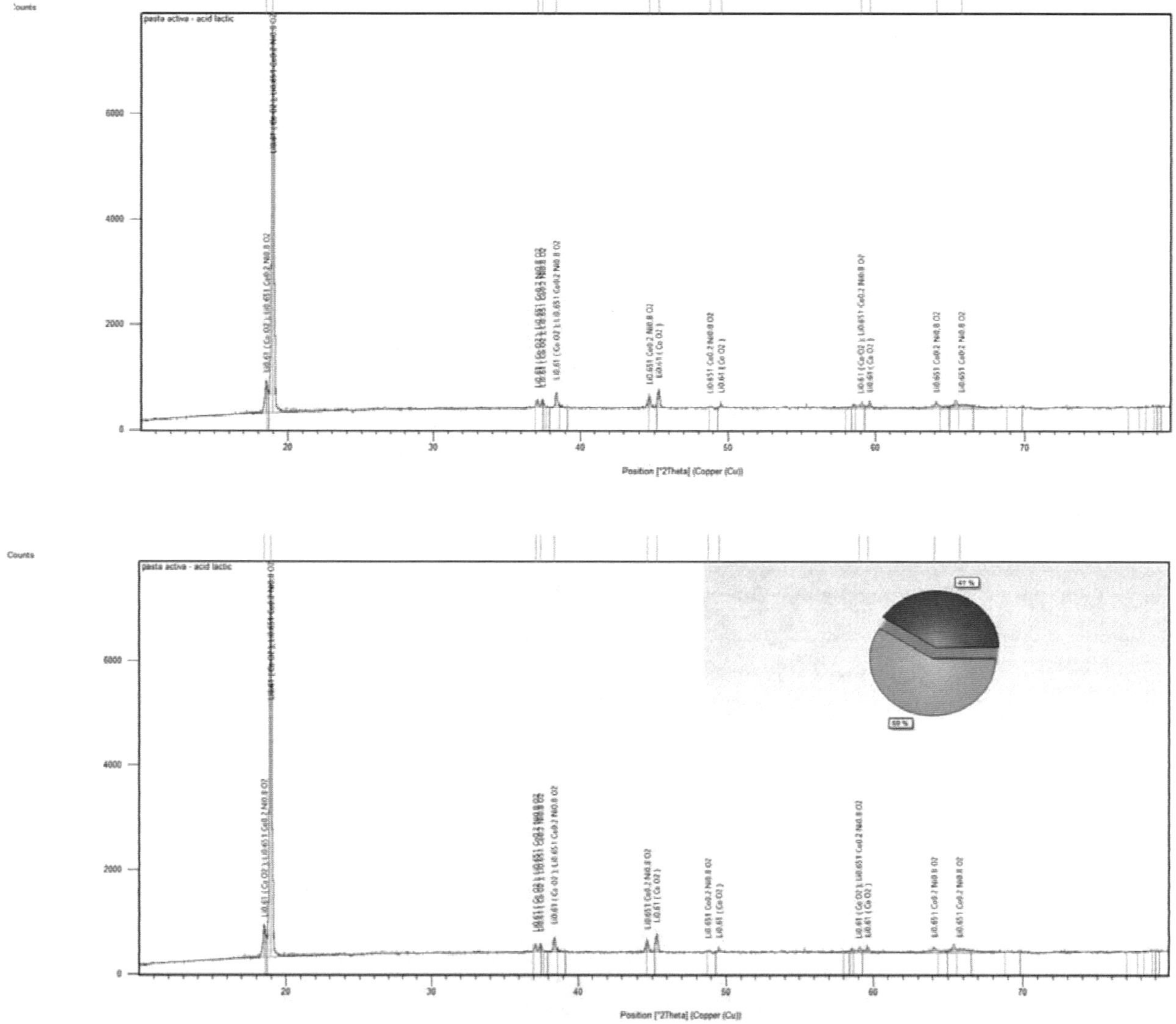

No.	Visi...	Ref. Code	Compound Na...	Chemical Form...	Score	Scale ...	Display Co...	Quality	Subfil...	Crystal System
1	☑	01-080-4976	Lithium Cobalt...	Li0.61 (Co O2)	41	0.049	Blue	S;NAT	ICSD ...	Rhombohedral
2	☑	04-017-3766	Lithium Cobalt...	Li0.651 Co0.2 ...	43	0.061	Lime	B	Inorg...	Rhombohedral

Figure 5. XRD sample investigation.

After the analysis, it was highlighted the main compounds – $LiCoO_2$ or $LiNiO_2$ (nickel dioxide is formed during the charging process), forming the solid solution LiCoxNi1-x¬O2 in which cobalt ions are substituted for nickel ions.

The crystallographic parameters are presented in Table 1 and in Table 2 the comparative indexing sheets of the compounds determined by XRD.

Table 1. Crystallographic parameters of $Li_{0.651}Co_{0.2}Ni_{0.8}O_2$ compared to $Li_{0.61}CoO_2$.

Crystallographic parameters	$Li_{0.651}Co_{0.2}Ni_{0.8}O_2$	$Li_{0.61}CoO_2$
Crystal system	Rombohedral	
Space group	R-3m	
Space group number	166	
a (Å)	2.8453	2.8076
b (Å)	2.8453	2.8076
c (Å)	14.3050	14.2129
Alpha (^{0})	90.0000	
Beta (^{0})	90.0000	
Gamma (^{0})	120.0000	
Volume of cell ($10^6 pm^3$)	100.29	97.03
Z	3.00	
RIR	4.50	5.29

Table 2. The X-ray diffraction indexation sheet in Figure 5.

Peak list

No.	h	k	l	d [A]	2Theta (deg)	I (%)
1	0	0	3	4.76830	18.593	100.0
2	0	0	3	4.73760	18.715	100.0
3	1	0	1	2.42830	36.990	30.7
4	1	0	1	2.39660	37.497	29.0
5	0	0	6	2.38420	37.699	3.8
6	0	0	6	2.36880	37.954	3.7
7	0	1	2	2.32970	38.616	8.7
8	0	1	2	2.30050	39.126	7.8
9	1	0	4	2.02910	44.621	58.7
10	1	0	4	2.00660	45.149	48.2
11	0	1	5	1.86710	48.732	7.9
12	0	1	5	1.84770	49.278	7.5
13	0	0	9	1.58940	57.979	0.9
14	0	0	9	1.57920	58.389	0.9
15	1	0	7	1.57300	58.642	10.0
16	1	0	7	1.55850	59.242	10.1
17	0	1	8	1.44720	64.318	11.9
18	0	1	8	1.43450	64.957	9.7
19	1	1	0	1.42260	65.568	12.1
20	1	1	0	1.40380	66.559	10.0
21	1	1	3	1.36330	68.808	7.7
22	1	1	3	1.34600	69.820	7.6

4. Conclusions

Cobalt is a rare and expensive metal, so any way to replace it in the cathode composition of Li-ion batteries is welcome. The recovery of cobalt and other metals present in used Li-ion batteries (Ni, Mn, and Li) is an economic necessity and an ecological responsibility.

An environmentally route for the recovery of active cathode material from spent Li-ion batteries was investigated. The ultrasound technology has the advantages to be easy to use, to work at different temperatures without the need of additional heat and to have low working times.

The solution used, i.e. lactic acid - C3H6O3, was used to completely detach the cathode material from Al-foil without degrading the material. The results obtained have shown that the technique gave good results at the established parameters.

Analyzing by SEM-EDX the samples of recovered paste, its structure and morphology were highlighted. X-ray diffraction analyzes confirmed the structural proximity of the active paste components.

5. References

[1] Rowden B and Garcia-Araez N 2020 *Energy Reports* **6**(5) 10
[2] Miller P 2015 *Johnson Matthey Technology Review* **59**(1) 4
[3] Melin H E 2018 *The lithium-ion battery end-of-life market – A baseline study* Circular Energy Storage (Global Battery Alliance) p 8
[4] Bratosin I, Toma C M, Vasile E, Ghica V G, Buzatu M, Petrescu M, Iacob G and Kovács T A 2019 *IOP Conf. Series: Materials Science and Engineering* **572** 012053
[5] Ghica V G, Toma C M, Buzatu M, Petrescu M, Iacob G, Antoniac I V, Vasile E and Veglio F 2017 *U.P.B. Sci. Bull., Series B* **79**(3) 75
[6] Turcu A M, Ghica V G, Veglio F, Buzatu M, Petrescu M I, Iacob G and Bîrloagă I 2018 *U.P.B. Sci. Bull., Series B* **80**(3) 183

Optimization Algorithms for Energy Building Applications

S Tsala[1] , I P Koronaki[1]

[1] Laboratory of Applied Thermodynamics, School of Mechanical Engineering, Thermal Engineering Section National Technical University of Athens, Heroon Polytechniou 9, Zografou Campus 15780, Athens, Greece

*Corresponding author's e-mail: koronaki@central.ntua.gr

Abstract. The creation of the structured environment affects the environment in all three phases of the construction and the demolition of buildings. The appropriate design of buildings requires full identification of the interaction between the environment and the structured environment. Energy efficiency is a mandatory requirement and integral part of green and sustainable buildings. Energy efficient design optimization is both a design philosophy and a practical technique which has been proposed and used by architects and other professionals for several decades, especially in recent years. Bioclimatic construction has been an important tool for improving the construction of buildings in the past few decades. Energy-saving options in buildings include the use of appropriate materials, exposures and the use of alternative power generation systems. In the present work, the methods and tools used for building design optimization are displayed in an effort to explore the reasoning behind their selection, to present their abilities and performance issues, and to identify the key characteristics of their future versions. In this essay, several algorithms are provided for optimizing problems which are proposed for building construction. Depending on whether limitations are given and on the general form of the optimization problem, one could choose the optimization algorithm which suits each case.

1. Introduction

Optimization is a process which seeks to identify the best solution for a given problem. Any optimization problem can be considered as a problem of mapping from decision-making to objective [1,2].

Solutions to an optimization problem are feasible when they meet all the requirements defined by the problem, while the optimal solution can minimize the cost of the process or maximize the effect of the solution to the problem. Typically, optimization problems are used to solve various algorithms.

The algorithm is a well-defined computational process which has input data of one or more variables and produces a set of values as output values. Therefore, an algorithm is a computational process which converts input data into output results [3].

The field of building construction is considered to be an interdisciplinary area as multidimensional designing skills are necessary. One pillar of this ecosystem of expertise is computational engineering which can contribute to the amelioration of designing methods for the energy consumption of buildings. Since 2000, when Hien, Poh, &Feriadi(2000)[4] postulated that real life architects cannot use dedicated algorithms and software for building optimization due to time and economic pressure, things have dramatically changed. The relevant literature has been enriched and a significant part of research is in progress aiming at the formulation of algorithms which would optimize the energy performance of buildings.

For example, a crucial part for energy consumption is lighting. Glaser et al. (2004)[5] attempted to develop an algorithm for building lighting optimization and presented its implementation to "LightSketch", a dedicated software to lighting design. On the other hand, Castro-Lacouture, Sefair, Flórez & Medaglia(2009)[6] realized that green buildings need to be not only energy efficient while in full function but also built with the proper technologies and materials as inappropriate material selection can cause energy problems. Furthermore, Baglivo, Congedo, Fazio & Laforgia(2014)[7] focused on the amelioration of buildings located in the Mediterranean. Their study was based on the implementation of a multi-objective approach in order to optimize the efficiency of external walls. The algorithm implementation included multiple factors such as thermal transmittance, periodic thermal transmittance, decrement factor, time shift, areal heat capacity, thermal admittance, surface mass, thickness.

Yu, Li, Jia, Zhang & Wang (2015)[8] introduced a multi-objective genetic algorithm based on the application of an improved multi-objective genetic algorithm (NSGA-II), a simulation-based improved Back propagation network, and the Pareto solution to obtain a building design multi-objective optimization model. According to the researchers, the application of the specific algorithm would help building designers detect the cross-section between thermal comfort and energy consumption. Touloupaki and Theodosiou(2017)[9] also focused on Multi-Objective Optimization Algorithms integrating evolutionary algorithms and energy simulation through Grasshopper for Rhinoceros 3d. On the other hand, it is well understood that the implementation of such algorithms presents a high risk of uncertainty. Bamdad, Cholette, Guan & Bell (2018)[10] introduced a modified version of the Ant Colony Optimization algorithm for Mixed Variables (ACOMV-M) in order to deal with problems arising from such uncertainty.

The issue of the high level of uncertainty presented by each algorithm has become the focus of research. Numerous studies attempt to compare and spot disadvantages among proposed solutions. Si, Tian, Jin, Zhou & Shi, Shi (2019)[11] tried to present the problematic issues related to optimization algorithms and analyze the factors affecting their effectiveness. Their study focuses on the comparison of Armijo gradient, Hooke-Jeeves, particle swarm optimization with constriction coefficient (PSOCC) and particle swarm optimization with inertia weight (PSOIW) and aims at determining weaknesses and assisting designers with their work. Accordingly, Kheiri(2018)[12] attempted to shed light on existing tools and algorithms.

The most important optimization algorithms will be analyzed and compared in this paper.

2. Local Optimization Algorithms

A local algorithm is a distributed algorithm which runs on a fixed number of synchronized circles, regardless of the number of nodes in the network. In other words, the output of a node in a local algorithm is a function of the input that is available in a fixed radius neighborhood of the node. Therefore, the output of a node in a local algorithm is a function of the available input in a neighboring fixed-radius region of the node [13].

In general, local optimization methods are based on the iterative process of finding a solution in which a simple design point alternates in the different steps of the iterative process until the optimal solution is determined. Consequently, local optimization algorithms require the expression of the problem through an objective function.

2.1. Golden Section algorithm

The Golden section method is used to find the maximum or minimum of a monotonous function, that is, a function that contains only the minimum or maximum space [α, b].

This algorithm consists of the following steps [14]:

Step 1: Two points a and b are defined between which the maximum (or minimum) of the function $f(x)$ is known to exist. Then, two intermediate points x_1 and x_2 are defined such that $x_1 = a + d$ and $x_2 = b - d$ where:

$$\boldsymbol{d} = \frac{\sqrt{5}-1}{2} \times (\boldsymbol{b} - \boldsymbol{a}) \tag{1}$$

Step 2: Values for the function $f(x_1)$ and $f(x_2)$ are estimated. If $f(x_1) > f(x_2)$ then new a, x_1, x_2 and b are defined so that:

IOP Conf. Series: Materials Science and Engineering 1037 (2021) 012040 doi:10.1088/1757-899X/1037/1/012040

$a = x_2$ (2)
$x_2 = x_1$ (3)
$b = b$ (4)
$\boldsymbol{x_1 = a + \frac{\sqrt{5}-1}{2} \times (b-a)}$ (5)
If $f(x_1) < f(x_2)$ then new a,x_1 ,x_2 and b are defined so that:
$a = a$ (6)
$b = x_1$ (7)
$x_1 = x_2$ (8)
$\boldsymbol{x_2 = b - \frac{\sqrt{5}-1}{2} \times (b-a)}$ (9)

Step 3: If $b - a < \varepsilon$ where ε is a very small number, then the maximum is determined at $\frac{a+b}{2}$ and stops the process otherwise step 2 is repeated.

The above algorithm is satisfactory when it is possible to determine a and b. Otherwise the algorithm Golden Search changes in a stricter way, in order to determine the required primary guesses for the lower and upper limits required and to find only one intermediate point (instead of two). In this case, the algorithm becomes[13]:

Step 1: For a small incremental δ size in space, e.g. $\delta = 0{,}05$, j is chosen as the smallest integer such as to apply:

$$g\left(\sum_{v=0}^{j} \delta(1{,}618)^{v}\right) g\left(\sum_{v=0}^{j-1} \delta(1{,}618)^{v}\right) \quad (10)$$

The upper and lower limit of a^i is $a_U = \sum_{V-0}^{j} \delta\ (1.618)^V$ and $a_L = \sum_{V-0}^{j-2} \delta\ (1.618)^V$

Step 2: Calculate $g(a_b)$where $a_a = a_L + 0.382(a_U - a_L)$ and $a_b = a_L + 0.618(a_U - a_L)$. Given that$a_a = \sum_{V-0}^{j-1} \delta\ (1.618)^V$the function$g(a_a)$ is known.

Step 3: Compare $g(a_a)$ and$g(a_b)$ follow steps 4, 5 and 6.

Step 4: If $g(a_a) < g(a_b)$then $a_L \leq a^i \leq a_b$ from the choice of a_a and a_b the new points $\overline{a_L} = a_L and \overline{a_U} = a_b$have $\overline{a_b} = a_a$. In this case it is calculated $g(\overline{a_a})$ where $\overline{a_a} = \overline{a_L} + 0.382(\overline{a_U} + \overline{a_L})$and proceeds to step 7.

Step 5: If $g(a_a) > g(a_b)$ then $a_a \leq a^i \leq a_u$ as in step 4 is set$\overline{a_L} = a_L and \overline{a_U} = a_b$. Calculated $g(\overline{a_b})$ where $\overline{a_b} = \overline{a_L} + 0.618(\overline{a_U} + \overline{a_L})$ and followed by step 7.

Step 6: If $g(a_a) = g(a_b)$ set $a_L = a_a$ and $a_U = a_b$ and return to step 2.

Step 7: If $\overline{a_U} - \overline{a_L}$ is too small set $a^i = \frac{1}{2}(\overline{a_U} + \overline{a_L})$ and stop the process. Otherwise, the process returns to Step 3 after the bar is switched off $\overline{a_L}, \overline{a_a}, \overline{a_b} and \overline{a_U}$.

In recent work [15] the Golden Search algorithm was used to estimate the performance of moving photovoltaics using battery, current and voltage characteristics and to examine the effect of shading on battery performance. The specific algorithm is preferred because of the rapid convergence and the noise resistance.

In general, this algorithm has been widely studied for photovoltaic systems in buildings, assessing the parameters which may affect their performance. Gayarthi & Ezhilarasi [16] also used this algorithm with a system of inverter in photovoltaic systems to determine their performance and proved that the voltage ratio is not linear and that the maximum power exists for specific current and voltage values. The maximum power varies according to the atmospheric conditions and the radiation that the system receives. In addition, by using the algorithm it was proved that the inverter performance reaches 90%. The exact form of the algorithm used is shown in the figure below:

IOP Conf. Series: Materials Science and Engineering 1037 (2021) 012040 doi:10.1088/1757-899X/1037/1/012040

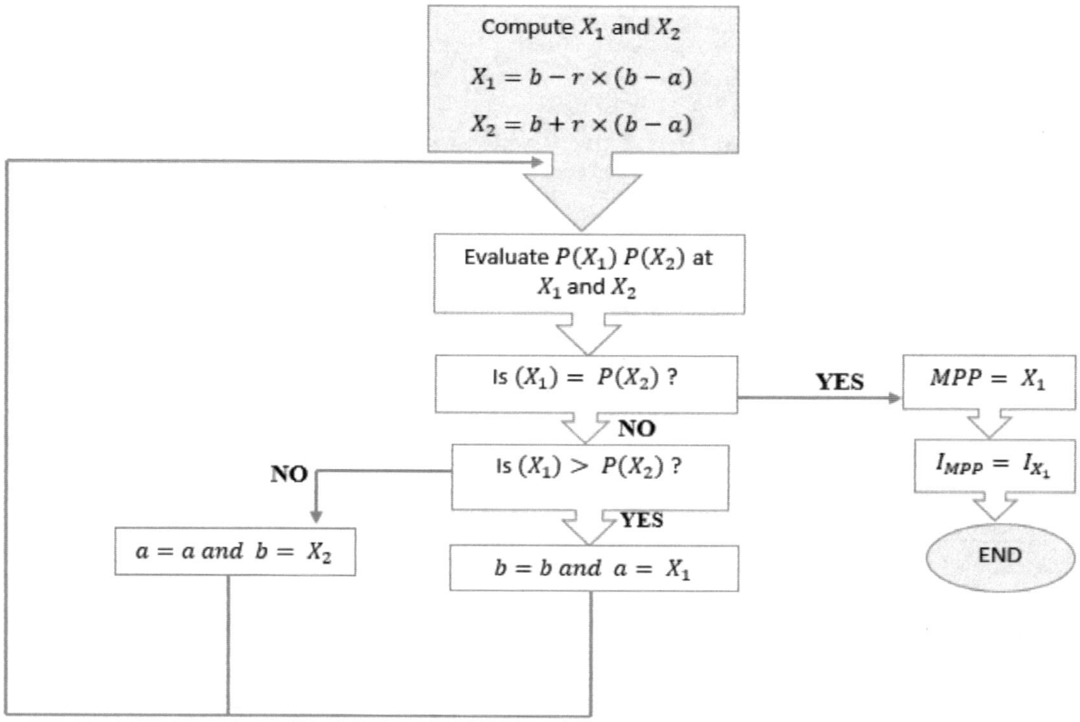

Figure 1_The Golden Method Algorithm used by Gayarthi and Ezhilaras (2017)

We will give an example of the Golden Section method. Let us minimize the function $f(x) = x^4 - 14x^3 + 60x^2 - 70x$, in the space [0,2], with 0,3 accuracy.

First of all, we define a=0, b=2 and $r = (\sqrt{5} - 1)/2 = 0{,}618$

Step_1: It is $x_1 = a + r(b - a) = 0{,}764$ and $x_2 = b - r(b - a) = 1{,}236$. The f-values of x_1 and x_2 are $f(x_1) = -24{,}36$and $f(x_2) = -18{,}96$. Since $f(x_1) < f(x_2)$, the new search space will be the [a, x_2].

Step_2:We choose $x_3 = a + r(x_2 - a) = 0{,}472$and x_4=x_1, hence $f(x_3) = -21{,}10$and $f(x_4) = -24{,}36$. Since $f(x_3) > f(x_4)$, the new search space shall be [x_3, x_2].

Step_3:We choose $x_5 = x_4$and $x_6 = x_3 + (1 - r)(x_4 - x_3) = 0{,}944$, hence $f(x_5) = -24{,}36$ and $f(x_4) = -23{,}59$. Since $f(x_5) < f(x_6)$, the new search space shall be [x_3, x^6].

Step_4: For $x_7 = x_3 + r(x_7 - x_3) = 0{,}653$ and $x_8 = x_5$, we have $f(x_3) = -23{,}84$ and $f(x_8) = -24{,}36$. Since $f(x_7) > f(x_8)$, the new search space shall be [x_7,x_2].

The algorithm terminates here, because x_2-x_7=0,292≤0,3. So the answer is that the minimum x* is at the space [0.653, 0.944]

IOP Conf. Series: Materials Science and Engineering 1037 (2021) 012040 doi:10.1088/1757-899X/1037/1/012040

2.2. Fibbonacci Search

In computer science, Fibbonacci Search is used to search into a classified array by using a divide-and-conquer algorithm that reduces the number of possible locations using Fibbonacci numbers.
The specific method examines only those locations which have the smallest dispersion, reducing the overall runtime of the algorithm [17].
We consider k as an element in F, which is the Fibbonacci numbers' array, and $n = F_m$ is the size of the array. If n is not a Fibbonacci number, F_m is considered to be the smallest number in F that is greater than n.
The Fibbonacci numbers are defined by the relationship $F_{k+2} = F_{k+1} + F_k$ when k≥ 0, $F_0 = 0$and $F_1 = 1$. Given a table of entries $R_1, R_{2,} \dots\dots, R_n$ whose keys increase in that order: $K_1 < K_2 < \cdots < K_n$, the Fibbonacci algorithm searches for a given statement K. It is assumed that $n + 1 = F_{k+1}$. The Fibbonacci search algorithm is described in the following steps:
Step 1: Set: $F_k \rightarrow i$, $F_{k-1} \rightarrow p$ and $q \rightarrow F_{k-2}$. In the algorithm, p and q are sequential numbers of Fibbonacci.
Step 2: A comparison is made. If$< k_i$, then step 3 follows, if$k > k_i$, step 4 follows and if$k = k_i$ the algorithm terminates successfully.
Step 3: Reduce$i$. If $q = 0$, the algorithm terminates without completing. Otherwise, $(p, q, p - q) \rightarrow (i, p, q)$ is set. In this way, p and $q$move a position backwards in the Fibbonacci series and the algorithm returns to step 2.
Step 4: Increase i. If $p = 1$, the algorithm terminates without completing. In the opposite case, set $(i + q, p - q, 2q - p) \rightarrow (i, p, q)$ namely p and q are moved two positions backwards in the Fibbonacci series and the algorithm returns to step 2.
Lin and Ma[18] have optimized phase shifting of building enhancing materials with integrated photovoltaic systems and concrete photovoltaic panels that are built into the building shell to increase the thermal mass of the building while the heated air generated by the photovoltaic system was used to heat space. The aim of the simulation was to maximize the signal / noise ratio of the thermal efficiency boosting factor of buildings. The optimized factors were the air flow rate, the type of material phase transformation, the thickness of the material and the insulation thickness of the wall. The combination of the Taguchi method and the Fibonacci method was used for optimization.

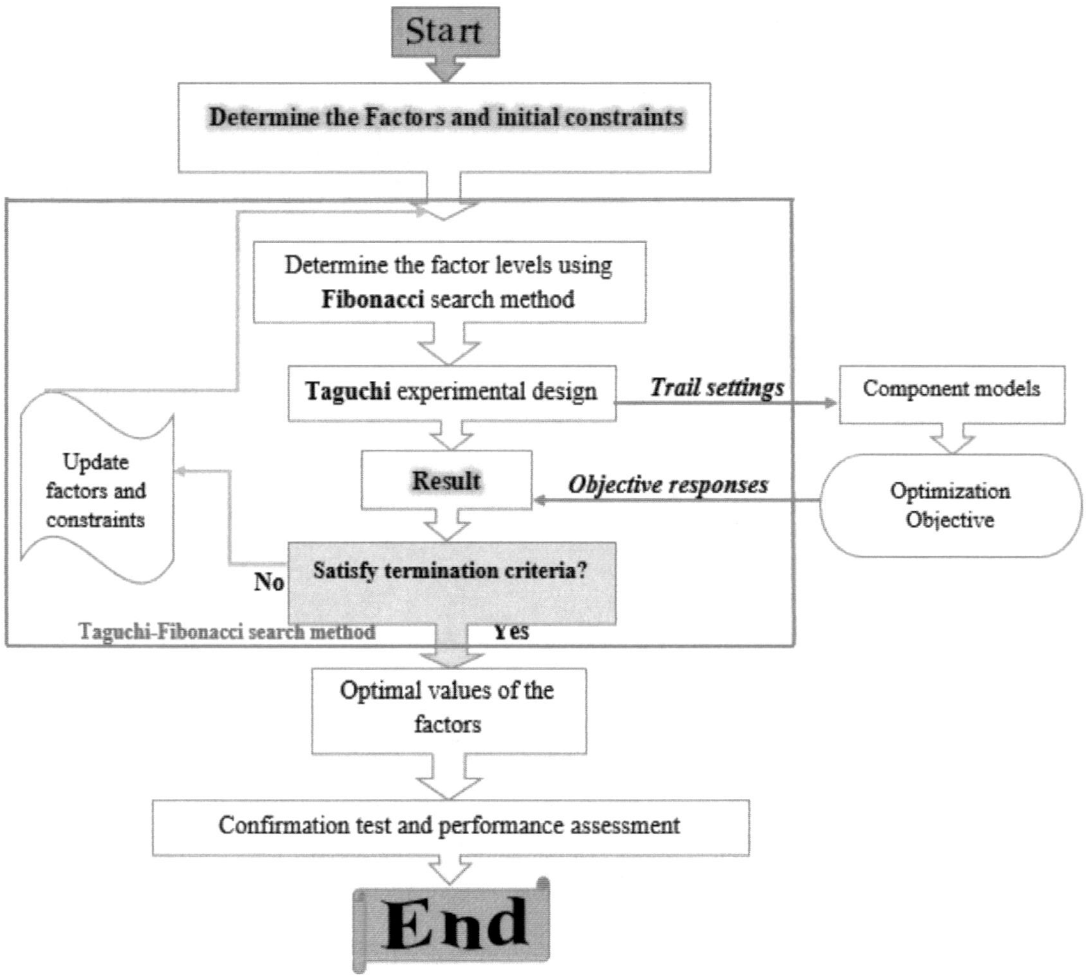

Figure 2_Outline of the optimization methodology

The following symbols are associated with the following chart:
$\boldsymbol{I^k}$ → factor interval at current iteration
$\mathbf{x_i}$ → $\mathrm{i^{tk}}$ continuous factor
$\mathbf{z_j}$ → j^{tk} discrete factor
% $\mathbf{C}$ → percentage contribution
$\mathbf{a_L^k}$ → Lower limit of continuous factor search range at current iteration
$\mathbf{b_U^k}$ → Upper limit of continuous factor search range at current iteration
$\mathbf{a_l^k}$, $\mathbf{b_u^k}$ → factor interval candidates generated by Fibonacci search at current iteration

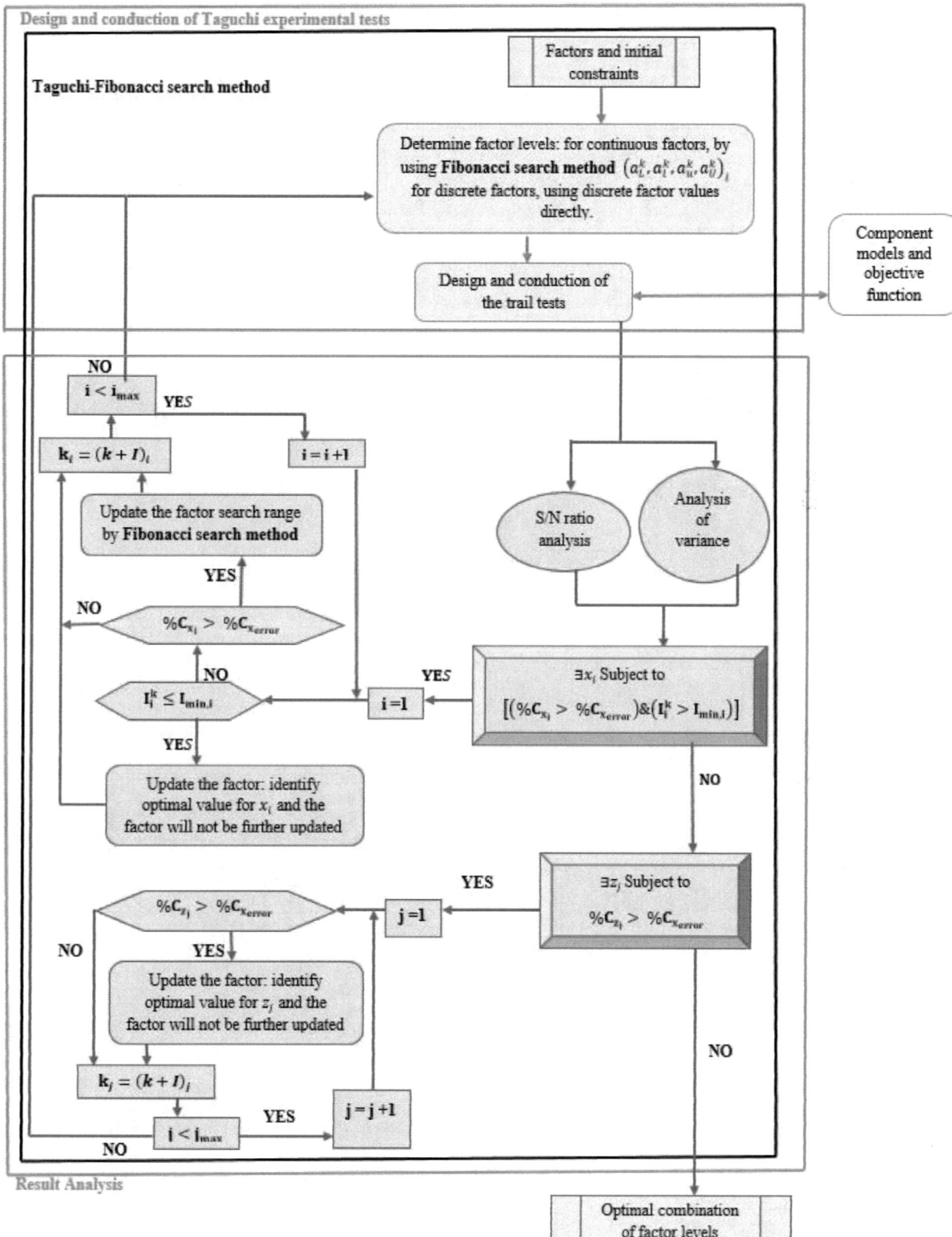

Figure 3_The Fibonacci Taguchi optimization method used by Li and Ma (2016)

The optimization results were based on a typical Australian home and specified airflow values, insulation thickness, and phase change factors of the materials.

2.3. The Fletcher-Reeves method

The Fletcher-Reeves method is a non-linear conjugate gradient method mainly used for nonlinear optimization problems [19]. The Fletcher-Reeves method is used to solve the unconstrained optimization problem: minimize f (x), x ε R * using a series of searches of the form: $x^{k+1} = x^k + a^k d^k$ after first estimatingx^k. The Fletcher-Reeves variable in the conjugate tilt algorithm generates the d_{k+1} of the relations:

$$\boldsymbol{d_{k+1}} = -\boldsymbol{\nabla f(x_{k+1})} + \boldsymbol{\beta_k d_k} \tag{11}$$

$$\boldsymbol{\beta_k} = \left(\frac{\|\nabla f(x_{k+1})\|_2}{\|\nabla f(x_k)\|_2}\right)^2 \tag{12}$$

The Fletcher-Reeves method can be used to estimate the boundary heat flows in a building and to solve the reverse convection problem without having any information about the previous heat flow [20].

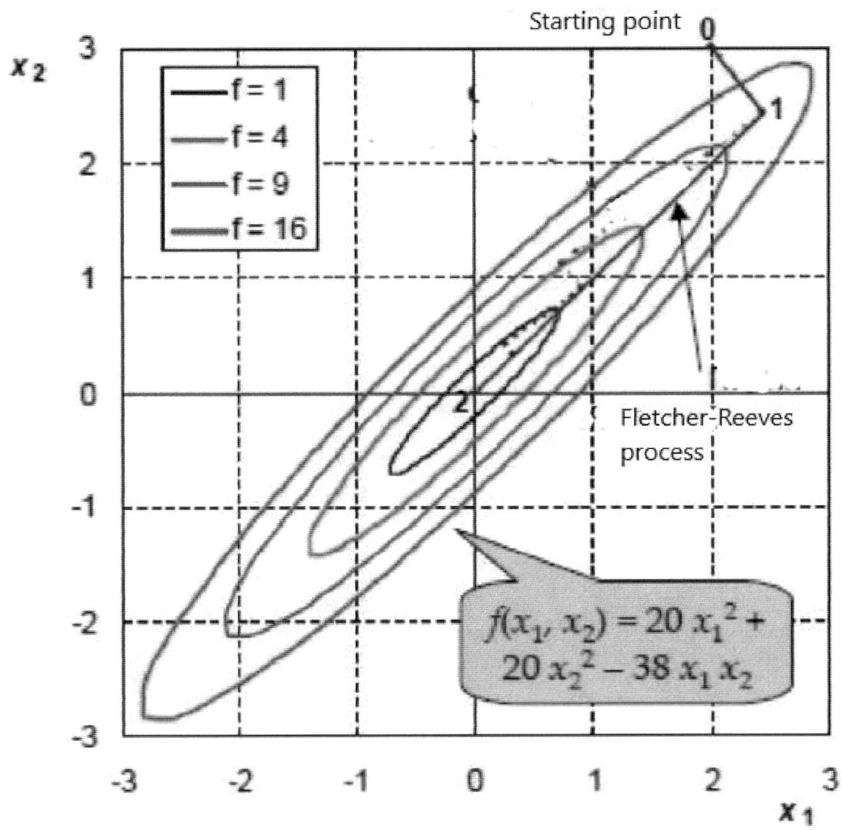

Figure 4_Example for Fletcher-Reeves process

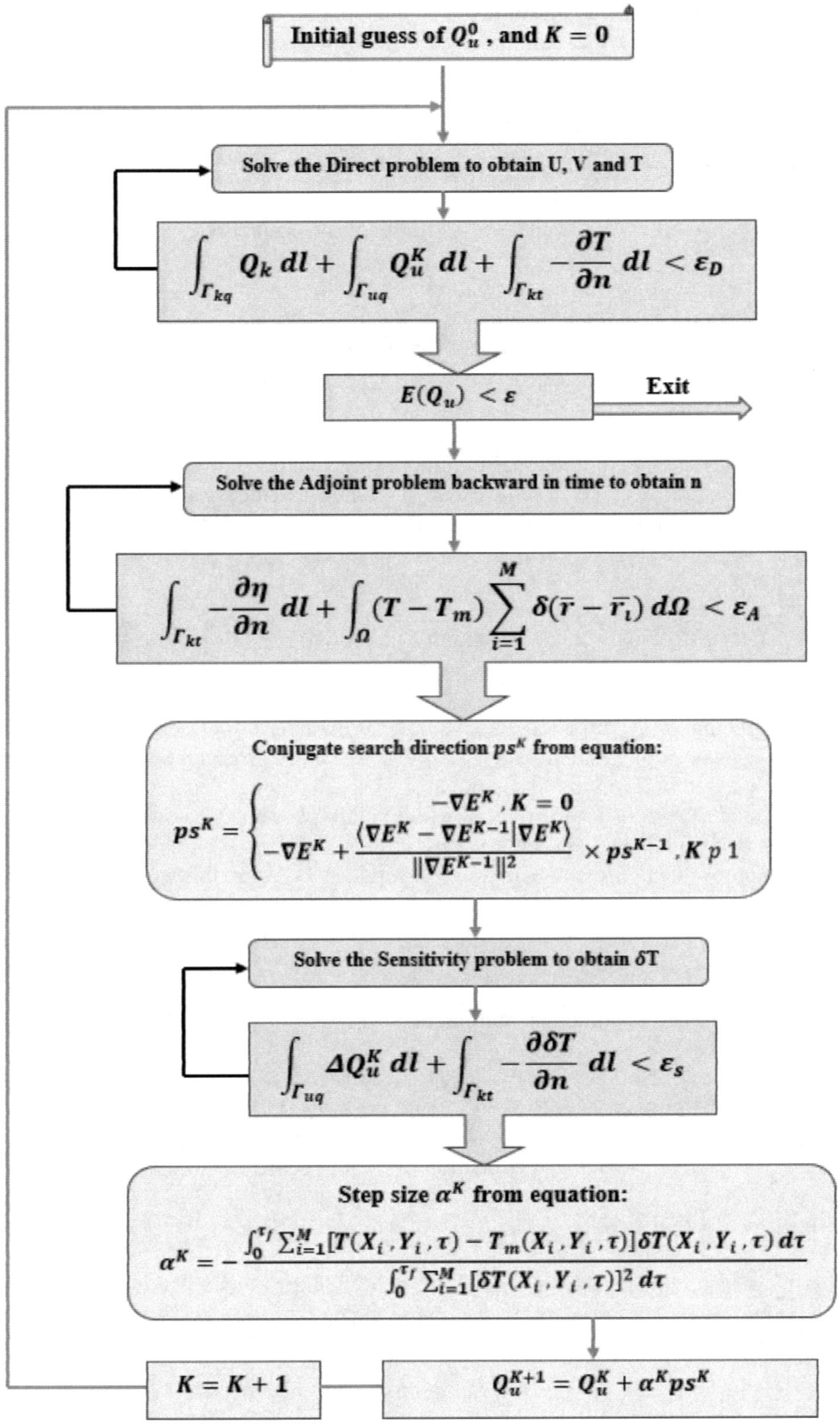

Figure 5_Solving the Fletcher Reeves Algorithm to solve the inverse conundrum problem in buildings. (Zhao et al 2009)

2.4. Successive repeat algorithm without limitations

Suppose that one wants to minimize the real function $f: X \to R$ that is likely to be subjected to constraints of the independent variable through the sequential repeat algorithm without restrictions. The main problem in convex programming is to minimize a convex function $f: R^j \to R$ for which $h_i(x) \leq 0$ for i = 1,2, ... , I and h_i are convex. If the problem is consistent, that is, there is $x$that satisfies the condition $h_i(x) \leq 0$, then the necessary and sufficient condition for solving the problem is that there exists $\lambda_i * \geq 0$ for which $\lambda_i * h_i * (x^*)$ for each i and x^* minimizes the function $f(x)+\sum_1^i \lambda_i^* h_i(x^*)$

In the k step of the sequential repetitive process without limitations, the function $G_k(x) = f(x) + g_k(x)$ is minimized to find x^kwhile $g_k(x)$ is a function selected by the user. The goal is to find a series $\{x^k\}$that converges to solution x of the original problem. Otherwise, the values of the function $f(x^k)$ subjected to fewest restrictions are searched for.

2.5. Powell Algorithm

Powell discovered an algorithm for finding a local minimum of a function. Having a function with N parameters, an initial parameter guess and N search vectors should be provided. In this way, it is possible to write the new parameter as a linear combination of all search vectors. After that, a new search direction is placed there where there is the search direction that gives the higher impact. The algorithm is repeated until convergence [21].

Step 1: Set the original p_o position
Step 2: For $i = 1,2, \dots, N$ the p_{i-1} moves to the minimum u_i direction and this point is defined as p_i
Step 3: For $i = 1,2, \dots, N-1$ set $u_{i+1} \to u_i$
Step 4: Put $p_N - p_O \to u_N$
Step 5: p_N moves to a minimum of u_N direction and this point is defined as p_O.

Powell showed that, for a square format, the iterations of the above basic procedure produce a set of u_i directions whose last k members are mutually conjugated. Therefore, the N iterations of the basic procedure, which correspond to $N(N+1)$ line minimization, minimize exactly a square form.
The disadvantage of this algorithm is that at each stage u_N in relation to $p_N - p_O$ tends to produce sets of directions that "fold over the others", resulting in linear dependencies. When this happens, the process finds the minimum of $f$only on a subset of the total space, ie it leads to incorrect results.
The example given is based on the following Powell method algorithm. It better illustrates the process followed in the example.

Step 1: Define $P_0=X_i$
Step 2: For i=1, 2, …, N , find the value of a_i that minimizes $f(P_{i-1} + a_i U_i)$and set $P_i = (P_{i-1} + a_i U_i)$.
Step 3: Set j=j+1
Step 4: Set $U_k=U_{k+1}$ for k=1, 2, …,N-1. Set $U_N=P_n - P_0$
Step 5: Find the value of a, that minimizes $f(P_0 + aU_N)$ and set $X_i = (P_0 + aU_N)$
Step 6: Repeat Steps 1 through 5.

We will see an example of Powell's Method at the function $f(x,y) = \cos(x) + \sin(y)$, using the initial point X_0 = (5.5, 2).
Firstly, $U = \begin{bmatrix} 1 & 0 \\ 0 & 1 \end{bmatrix}$and $P_0 = X_0$ = (5.5, 2). For i=1, we have in the function

$$f(P_0 + a_1 U_1) = f((5.5,2) + a_1(1,0)) = f(5.5 + a_1, 2) = \cos(5.5 + a_1) + \sin(2)$$

minimum at a_1=-2.3584042. So P_1=(3.1415958, 2). For i=2, the function

$$f(P_0 + a_1 U_1) = f((5.5,2) + a_1(1,0)) = f(5.5 + a_1, 2) = \cos(5.5 + a_1) + \sin(2)$$

has a minimum at a_2=2.7123803. So, P_2=(3.1415958,4.7123803). Set U_2'=(P_2-P_0)' and

$$U = \begin{bmatrix} 0 & -2.3584042 \\ 1 & 2.7123803 \end{bmatrix}.$$

The function
$f(P_0+a_1U_1)=f\left((3.1848261,4.6626615)+a_1(0,1)\right)=f(3.1848261,46626615+a_1)=\cos(3.1848261)+\sin(4.6626615+a_1)$
Has a minimum at a_1=0.0497117. So P_1=(3.18848261,4.7123732). For i=2, the function
$f(P_1+a_2U_2)=f\left((3.1848261,4.7123732)+a_2(-2.3584042,2.7123809)\right)=f(3.1848261-2.3584042a_2,4.7123732+2.7123809a_2)=$
$=\cos(3.1848261-2.3584042a_2)+\sin(4.7123732+2.7123809a_2)$
Has a minimum at a_2=0.0078820. So, P_2=(3.1662373,4.7337521). Set U_2'=(P_2-P_0)' and

$$U=\begin{bmatrix}-2.3584042 & -0.0185889\\ 2.7123803 & 0.0710906\end{bmatrix}$$

The function
$f(P_0+aU_2)=f\left((3.1848261,4.6626615)+a(-0.00185889,0.0710906)\right)=f(3.1848261-0.00185889a,4.6626615+0.0710906a)=$
$=\cos(3.1848261-0.00185889a)+\sin(4.6626615+0.0710906a)$
has a minimum at a=0.8035684. So X_2=(3.1698887, 4.7197876).
In the end, the function $f(x,y)=\cos(x)+\sin(y)$,, has a relative minimum at $P=(\pi,\frac{3\pi}{2})$.

2.6. Nelder-Mead algorithm

The Nelder-Mead algorithm or the downhill simplex algorithm, originally published in 1965 (Nelder and Mead, 1965), is one of the best-known algorithms for multidimensional, unrestricted optimization without the use of derivatives.
Although the method is relatively simple, it can be accomplished in various ways. In addition to some small computational details in the basic algorithm, the main difference between the different forms of implementation is found in the construction of the original Simplex (an optimization method, perhaps the most basic) and in the selection of the convergence or termination tests used in the algorithm. The generic algorithm consists of the following steps [22]:

- Construction of the original simplex S.
- Repeat the following steps to satisfy the shutdown test: The shutdown information is calculated and if it is not satisfied, then the original simplex is converted.
- Return to the best top of the current simplex and correlate the value of the function.

Home Simplex

The original Simplex S is usually constructed by creating $n+1$peaks $x_0,\dots,x_n$, around a given entry point $x_{in}\epsilon R_n$. In practice, the most common option is $x_0=x_{in}$ to allow the algorithm to restart correctly. The remaining vertices n are produced to obtain one of the two patterns of simplex.

- S is rectangular on the x_o based on the coordinate axes or:

$x_j=x_o+h_je_j, j=1,2,\dots,n$ where h_j is the step in the direction of the unit vector $e_j$at R_n.

- S is a normal simplex, where all edges have the same specified length.

Simplex Transformation algorithm

A repeat of the Nelder-Mead method consists of the following three steps.

- Layout: Determine the $h,s,1$ of the worst, second worst, and optimal peaks, respectively, in the current simplex operation.

$$\boldsymbol{F_h=max_jf_j\,,f_B=max_{j\neq h}f_j\,,f_1=min_{j\neq h}f_j}$$

In some embodiments, the vertices of S are arranged relative to the operating values to satisfy: $f_o\le f_i\le\dots\le f_{n-1}\le f_n$. Then$1=0\,,s=n-1\ and\ h=n$.

- Centroid: Calculate the centroid c of the best side that is considered the opposite of the worst peak x_h

$$\boldsymbol{c=\frac{1}{n}\sum_{j\neq h}x_j}$$

- Transformation: The new Simplex is calculated. Initially, only the worst x_h top is replaced with a better spot using reflection, swelling or shyness in relation to the best side. All test points are in the line defined by x_h and c and at most two of them are calculated in one iteration. If this succeeds, the acceptable point becomes the new peak of simplex. If this fails, the simplex is shrunk to the best topx_1. In this case, new peaks n are calculated.

The transformations of simplex in the Nelder-Mead method are determined by four parameters: a for reflection, β for contraction, γ for swelling and δ for contraction which must meet the conditions: $\alpha > 0, 0 < \beta < 1, \gamma > 1, 0 < \delta < 1$.
In most applications the values used are $= 1, \beta = \frac{1}{2}, \gamma = 2\ and \delta = \frac{1}{2}$.

- ***Reflection:*** Reflex points are calculated: $x_r = c + a(c - x_h)$ and

$f_r = f(x_r)$. If $f_1 < f_r < f_s$ is accepted the x_r and terminates the iterative process.

- ***Swelling:*** If $f_r < f_1$ calculates the bulking point $x_e = c + \gamma(x_r - c)$ and

$f_e = f(x_r)$. If $f_e < f_r$ is accepted x_e and the iterative process stops. Otherwise, x_r is accepted and the process stops.

- ***Contraction***: If $f_r \geq f_s$ computes the point inserted x_n using the best point between the x_h andx_r.

External: If $f_s \leq f_r$ <fs computes$x_c = c + \beta(x_r - c)$ and $f_c = f(x_c)$.
If $f_c \leq f_r$ the x_c is acceptable and the iterative process is terminated. Otherwise, a shrink transformation takes place.
Internally: If $f_c \geq f_h$, $x_c = c + \beta(x_h - c)$ and $f_c = f(x_c)$ are computed
If $f_c \leq f_h$, x_c are accepted, the iterative process is terminated. Otherwise, a shrink transformation takes place.

- ***Shrinking:*** New peaks $x_n = x_i + \delta(x_j - x_i)$ and $f_j = f(x_j)$ for $j = 0,1,2, \ldots, n$ with $j \neq i$ are calculated.

Termination tests
The implementation of the Nelder-Mead method must include a test that ensures the termination process in a finite period of time. The termination test refers to the convergence or termination test performed when the simplex is relatively small, the $f$function values convergence test is triggered when the f values are relatively close and the failure test is triggered when the values of the functions exceed a specific value.
The simplex method was used by Buclaghem and Letherman [23] to automatically optimize the passive performance of buildings. The optimization criterion was the objective function of minimizing comfort that combines the desired comfort temperature and the internal temperature of the building.
Al Hamoud (1997) optimized energy consumption using the simplex method and non-linear programming. The variables used are associated with the construction properties as well as the shape and orientation.
The simple algorithm of Nelder and Mead is a derivative optimization algorithm and constructs a simplex n dimension in the space of the independent variables. The cost function is estimated for each of the $(n + 1)$ variables in the simplex. At each iteration step the simplex vector with the highest cost is replaced by a new vector which is achieved either by reflecting the largest cost function vector in the centroid of simplex or by expanding the simplex [24].
At the following example, we will use the Nelder-Mead algorithm, to find the minimum of the function$f(x, y) = x^2 - 4x + y^2 - y - xy$, starting with the following vertices:
V_1=(0,0), V_2=(1.2,0.0), V_3=(0.0, 0.8)
Given the vertices above, we have the following values for the function f
f(V_1)=0, f(V_2)=-3.36, f(V_3)=-0.16
We set
B=V_2 (gives the Best f-value)
G=V_3 (gives a Good f-value)
W=V_1 (gives the Worst f-value)

We will replace the vertex W=(0,0), and we set the points $M = (B + G)/2 = (0.6,0.4)$and $R = 2M - W = (1.2, 0.8)$

Since $f(R) = -4.48 < f(G) = -0.16$, and $f(R) \leq f(B)$, we should construct the E vertex, which is $E = 2R - M = 2(1.2, 0.8) - (0.6, 0.4) = (1.8, 1.2)$.

Since $f(E) = -5,88 < f(B) = -3.36$, the new triangle has vertices

$V_1 = (1.8, 1.2)$, $V_2 = (1.2, 0.0)$, $V_3 = (0.0, 0.8)$

A sequence of triangles which converges at the solution point (3,2) is generated from the continuation of the process, and gives the value $f(3,2) = -7$.

3. General optimization Algorithms

General Optimization is intended to find the general best solution for a problem when multiple local best solutions are known. Generic algorithms are considered to provide a better solution than local algorithms. There are several kinds of generic optimization algorithms that were developed depending on the application in which they were applied. The current work presents the DIRECT algorithm which is a combination of general and local optimization.

3.1. The DIRECT algorithm

The DIRECT algorithm is used in engineering applications for which there are no convincing assumptions and there are multiple minima and black box designs.

If an objective function f and the design space are considered $D = D_0$the steps of the DIRECT algorithm are [25]:

Step 1: Design space D is normalized to be a unit super cube. Determines the center of c_i and estimates $f(c_i)$. The algorithm starts from $f_{min} = f(c_i)$ sets the estimation measure $m = 1$ and the repeat counter$t = 0$.

Step 2: Set the S group of the potential best boxes.

Step 3: Select each box $j \in S$

Step 4: Box j is divided in the following way:

1. Set Group I of dimensions with maximum length.Set δ equal to 1/3 of the maximum side.
2. Select the function in points $c \pm \delta e_i$ for all $i \in I$ where c is the center of the box and e_i is the i unit vector.
3. Divide the j box containing c into thirds along the dimensions in I starting with the dimension with the smallest value $w_i = min\{f(c + \delta e_i), f(c - \delta e_i)\}$ and replaced in the dimension with bigger w_i. Then, f_{min} and m are revised.

Step 5: Set $S = S - \{j\}$ .If $S \neq 0$ then repeat the process from step 3.

Step 6: Putting $t = t + 1$ if the repeat threshold or the limit of estimation has been reached, the algorithm stops. Otherwise the process goes to step 2.

The choice of boxes j is potentially optimal if:

$$\boldsymbol{F(c_j) - Kd_j \leq f(c_j) - Kd_j}\text{and}$$

$$\boldsymbol{F(c_j) - Kd_j \leq f_{min} - \varepsilon|f_{min}|}\text{for any } K > 0 \text{ and } i = 1,2,\dots,m$$

Diakaki et al**[26]** using the DIRECT algorithm minimized the cost of life cycle and environmental impact using wall and window insulation as variables and the simulation was performed with a steady-state treatment model. In this work, the use of target programming and the global criterion method was found to be ideal for simple problems but scaling is a problem for this method.

Below, we give an example of DIRECT that has been used on the GP (Goldstein-Price) test function:

$$f(x) = \left[1 + (x_1 + x_2 + 1)^2 (19 - 14x_1 + 3x_1^2 - 14x_2 + 6x_1x_2 + 3x_2^2\right]$$
$$\left[30 + (2x_1 - 3x_2)^2 (18 - 32x_1 + 12x_1^2 + 48x_2 - 36x_1x_2 + 27x_2^2\right]$$

where $x_i \in [-2, 2]$,for i=1,2. The global minimum value of GP function is f_{min}=3. The figure below shows the plot of the function :

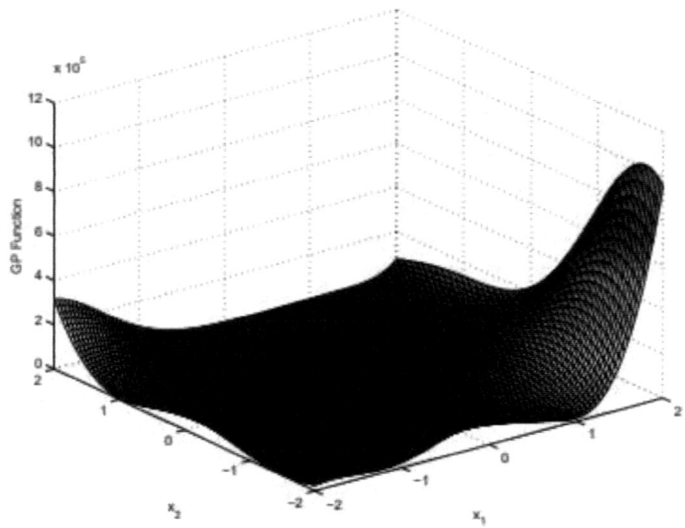

Figure_6_The plot of function

3.2. Genetic algorithms

Genetic algorithms have proven their performance in various energy optimization problems such as optimizing building heating design [27]. These algorithms are a class of algorithms which begin by selecting a randomly produced population of solutions for the problem under consideration. They move from one generation of solutions to another, evolving new solutions using objective entities of assessment, selection, intersection and mutation. Generally, genetic algorithms function with solutions represented by a code instead of a set of initial variables. Typically, it is a solution represented by a series of bits. Each bit position is called a gene and the values acquired by each gene are called alleles.
A basic genetic algorithm has three main operators performed at each iteration. The vectors are reproduction, intersection and mutation. At the breeding stage, the chromosomes or solutions of the existing breeding are copied to the next one with the same probability based on the value achieved for the objective function. At the intersection stage, randomly selected chromosome pairs are to be introduced into the next generation while the mutation stage includes an occasional random alternation of the allelic genes [28].
The selection factor for replication is useful for creating a new generation, improved in comparison with the previous one, while the intersection stage separates the population which handles the generations of the generated chromosomes and the mutation stage introduces the necessary data for the investigation area. Znuda et al 2007 developed a genetic algorithm aiming at a method for achieving a group of optimal architectural configurations based on the Goldberg algorithm. That is, they chose a coding strategy by starting the algorithm by random reproduction of the first population, and then each iteration produces a population better than the previous one based on the objective function being calculated. The algorithm terminates when the best solution is repeated in several iterations.
In another paper [29], they used the following mathematical procedure to describe the optimization problem:

$$\textbf{Minimization } F(x_1, x_2, \dots, x_n)$$
$$G(x_1, x_2, \dots, x_n) \geq 0$$
$$x_i \leq S_i$$

There is a large number of objective F functions which should be minimized by convention and there are a number of G constraints which are required to be greater than or equal to zero. Moreover, each variable x_i is structured with specific S_i variables defined either as discrete variables or from boundary values. Objective functions and constraints must be alternating depending on how the problem is formed.

Genetic algorithms have proven to be much more efficient when optimization parameters are more than 10. The literature mentions [30] the minimization of energy by a genetic algorithm from the change of manufacturing options for the air conditioning of a building in tropical climate using the method of acceptance.
The minimization of costs and energy has been studied with genetic algorithms [31] using heat recovery and constructions as variables, while comparison of multiple and single objective functions has shown the advantage of objective functions despite the fact that this method has computational cost.
Genetic algorithms and multivariate analysis as well as artificial neural networks can be combined to evaluate performance and optimize composite production units based on structural indicators [32]. This combination initially determines the inputs and outputs for oriented neural networks and the genetic algorithm is based on multivariate analysis variables and collects the input data describing the input-output relationships for the modules. The linearity of the input data is then tested to assess the performance of the modules, normalize the input data based on the relationship determined, and divide into two subgroups in order to develop the neural networks. Different neural networks are then used to estimate the relationships between the input and output data, and a sensitivity analysis is performed by altering the input subgroup elements to achieve the minimum value while testing the genetic algorithms according to the selected parameters, and finally calculating the error between the actual output and output of neural networks and genetic algorithms. This combination of algorithms can be used in all building blocks for which suppliers set different economic, technical and environmental criteria. In addition, this combination method allows the continuous monitoring and improvement of composite units based on the overall productivity and reliability of the machine.

4. Conclusions

There are many different methods of using computers to optimize engineering designs. Below are brief details of some common optimization algorithms used for sustainable building optimization. All are heuristic methods, which means that they present no guarantee as to arriving at the true optimum solution but offer an efficient method with a high probability of approaching the optimum. "Direct search" covers methods that compare test solutions to the best ones calculated so far, with a strategy based on the results presented so far in order to determine the next test. Examples used in building optimization include:

- Search on a template, e.g. Hooke and Jeeves: Every dimension in turn is being tested. When no further improvement is possible, step size is reduced by half.
- Non-linear programming: a series of extensions allowing non-linear targets and constraints.
- Linear programming, including the simplex / Nelder and Mead method [33]: if objective operation and all constraints are linear, the optimal must fall to an extreme point.

The commonly used algorithms in building energy efficient design optimization can be grouped into three categories, namely evolutionary algorithms, derivative-free search algorithms, and hybrid algorithms. Note that hybrid algorithms are not new algorithms. They are combinations of different algorithms, often evolutionary algorithms with derivative-free algorithms. A statistical analysis of the core literature shows that evolutionary algorithms are the most commonly used category of optimization algorithms, accounting for approximately 60%. The rest are derivative-free search algorithms such as Hooke-Jeeves direct search algorithm [34] and hybrid algorithms. In the category of evolutionary algorithms, genetic algorithm (GA)[35] or its variations such as non-dominant sorting genetic algorithm (NSGA)[36] are dominant. A general trend to shift from normal GA to NSGA is noted, probably because NSGA is more suitable for solving multi-objective optimization problems, which are common in architectural design. Particle swarm algorithm is another frequently used evolutionary algorithm [37].
Although most researchers in the literature apply algorithms to building energy efficient design optimization without carefully studying the algorithm itself, several studies shed light on how effective and efficient different algorithms are in finding the optimal design solution. Wetter and Wright compared the Hooke-Jeeves direct search algorithm with a genetic algorithm and concluded that the latter performs better than the previous one and that the Hooke-Jeeves direct search algorithm can be trapped locally at optimum[38].Wetter and Wright compared the performance of eight algorithms: algorithm search algorithm, Hooke-Jeeves algorithm, PSO (Particle Swarm Optimization) algorithm,

PSO grid search, PSO-HJ hybrid algorithm, simple GA, Simplex algorithm Nelder and Mead and Discrete Armijo gradient algorithm, for their performance in minimizing cost functions with different smoothness[39]. They found that the hybrid algorithm achieved the most significant cost reduction with a higher number of simulations, and that the simple GA consistently approached the best minimum. However, the performance of other algorithms was not stable. A recommendation has also been made with regard to avoiding the Simplex algorithm and the Discrete Armijo grading algorithm if EnergyPlus is used to evaluate the cost function.

Last but not least, it is worth highlighting that the energy simulation engine is a key component in the building energy efficient design optimization workflow. Its function is to calculate the energy consumption or other energy related parameters of the building being designed and optimized. Among the studies included in the basic literature, EnergyPlus [40] is by far the most common energy simulation engine, and is the representative of the category of detailed and dynamic energy simulation programs. Other power assimilation engines found in the basic literature in this category are: DOE-2[41], TRNSYS[42,43] , ESP-r[44] , and IDA ICE[45].

In conclusion, energy-efficient design optimization is a philosophy and technique which assists architects and other professionals in designing buildings with high energy efficiency and better overall performance. This section draws upon the evidence presented to conjecture about future trends and challenges in computational optimization of sustainable buildings. There is a clear increase in the popularity of optimization for sustainable building design, especially for multi-objective optimization.

Acknowledgments

This research is co-financed by Greece and the European Union (European Social Fund - ESF) through the Operational Programme «Human Resources Development, Education and Lifelong Learning» in the context of the project "Strengthening Human Resources Research Potential via Doctorate Research – 2nd Cycle" (MIS-5000432), implemented by the State Scholarships Foundation (IKY).

Ευρωπαϊκή Ένωση
European Social Fund

Operational Programme
Human Resources Development,
Education and Lifelong Learning

Co-financed by Greece and the European Union

ΕΣΠΑ 2014-2020

References

1. Al-Homoud MS.,*Optimum thermal design of office buildings.* International Journal of Energy Research 21 (1997): 941-957
2. Shi Y.,*Developmental swarm intelligence: developmental learning perspective of swarm intelligence algorithms*. International Journal of Swarm Intelligence Research 5 (2014): 36–54
3. Souomela, J.,*Survey of local algorithms.* ACM Computing Surveys 45 (2013): 24-40
4. Hien, W. N., Poh, L. K., &Feriadi, H.,*The use of performance-based simulation tools for building design and evaluation-a Singapore perspective.* Building and Environment, 35(8) (2000): 709–736.
5. Glaser, D. C., Feng, O., Voung, J., & Xiao, L. ,*Towards an algebra for lighting simulation.* Building and Environment, 39(8) (2004): 895–903.
6. Castro-Lacouture, D., Sefair, J. A., Flórez, L., &Medaglia, A. L. ,*Optimization model for the selection of materials using a LEED-based green building rating system in Colombia.* Building and Environment, 44(6) (2009): 1162–1170.
7. Baglivo, C., Congedo, P. M., Fazio, A., &Laforgia, D. , *Multi-objective optimization analysis for high efficiency external walls of zero energy buildings (ZEB) in the Mediterranean climate.* Energy and Buildings, 84 (2014): 483–492.
8. Yu, W., Li, B., Jia, H., Zhang, M., & Wang, D.,*Application of multi-objective genetic algorithm to optimize energy efficiency and thermal comfort in building design.* Energy and Buildings, 88 (2015): 135–143.

9. Touloupaki, E., &Theodosiou, T.,*Optimization of building form to minimize energy consumption through parametric modelling.* Procedia Environmental Sciences, 38 (2017): 509–514.

10. Bamdad, K., Cholette, M. E., Guan, L., & Bell, J.,*Building energy optimisation under uncertainty using ACOMV algorithm.* Energy and Buildings, 167 (2018): 322–333.

11. Si, B., Tian, Z., Jin, X., Zhou, X., & Shi, X. ,*Ineffectiveness of optimization algorithms in building energy optimization and possible causes.* Renewable Energy, 134(2019): 1295–1306.

12. Kheiri, F.,*A review on optimization methods applied in energy-efficient building geometry and envelope design.* Renewable and Sustainable Energy Reviews, 92 (2018): 897–920.

13. Thomas H. Cormen,Charles E. Leiserson,Ronald L. Rivest, Clifford Stein, Introduction to algorithms. Third Edition MIT Press (2009)

14. Ali Yalcin, Autar Kaw, Chapter 09.01 Golden Section Search Method. Holistic Numerical_Methods(2012),Website:http://mathforcollege.com/nm/mws/gen/09opt/mws_gen_opt_txt_goldensearch.pdf.

15. Patel A., Kumar, V., Kumar, Y., *Golden Section Search Optimization Technique for Maximum Power Point Tracking.* International Journal of Engineering Research and Applications 3 (2013): 1204-1209

16. R. Gayathri, G.A. Ezhilarasi,*Golden section search based maximum power point tracking strategy for a dual output DC-DC converter.,* Ain Shams Engineering Journal (2017)

17. *Fibonacci Search Technique.* Website:
https://ipfs.io/ipfs/QmXoypizjW3WknFiJnKLwHCnL72vedxjQkDDP1mXWo6uco/wiki/Fibonacci_search_technique.html.

18. Wenye Lin, Zhenjun Ma, *Using Taguchi-Fibonacci search method to optimize phase change materials enhanced buildings with integrated solar photovoltaic thermal collectors.* Energy 106 (2016): 23-37

19. Deyan Pang, Shouqiang Du, Jingjie JuPang, *The smoothing Fletcher-Reeves conjugate gradient method for solving finite minimax problems.* Science Asia 42 (2016): 40-45

20. Informatic 2013. *Optimization without Gradients: Powell's method,* Website:
http://informatik.unibas.ch/fileadmin/Lectures/HS2013/CS253/PowellAndDP1.pdf.

21. Al Baali, M., Descent *properties and global convergence of the Fletcher-Reeves method with Inexact Line Search*, IMA Journal of Numerical Analysis 5 (1985): 121–124.

22. Saša Singer and John Nelder, *Nelder-Mead algorithm.* Scholarpedia (2009), Website: http://www.scholarpedia.org/article/Nelder-Mead_algorithm

23. Buglaghem N.M., Letherman, K.M., *Numerical Optimization Applied to theThermal Design of Buildings.* Building and Environment 25 (1990): 117-124.

24. Wetter, M., Wright, J., *A comparison of deterministic and probabilistic optimization algorithms for non-smooth simulation-based optimization.* Building and environment 39 (2004): 989-999.

25. Layne T. Watson, Jian He,.*The Deterministic Global Optimization Algorithm DIRECT.* HPCS/BIS Tutorial (2008)

26. C. Diakaki, E. Grigoroudis, D. Kolokotsa,.*Towards a multi-objective optimization approach for improving energy efficiency buildings.* Energy and Buildings 409 (2008):1747–54.

27. Wright J.A., Loosemore H.A., Farmani R., *Optimization of building thermal design and control by multi-criterion genetic algorithm.*, Energy and Buildings 34 (2002): 959–972.

28. Znuda E., Ghrab Morcos, N., Hadj Alouane, A., *Optimization of Mediterranean building design using genetic algorithms.* Energy and Building 39 (2007):148-153.

29. Wang W., Rivard H., Zmeureanu R.,*An object-oriented framework for simulation-based green building design optimization with genetic algorithms.* Advanced Engineering Informatics 19 (2005): 5-23.

30. Sahu M., Bhattacharjee B., Kaushik S.,*Thermal design of air-conditioned building for tropical climate using admittance method and genetic algorithm.* Energy and Buildings 53 (2012):1-6.

31. Palonen M., Hasan A., Siren K.,*A genetic algorithm for optimization of building envelope and HVAC system parameters.* Eleventh International IBPSA Conference, In: Proceedings of the building simulation (2009).

32. Azadeh A., Mianaei, H.S., Asadzadeh S.M., Saberi, M., Sheikhalishashi, M., *A flexible ANN-GA-multivariate algorithm for assessment and optimization of machinery productivity in complex production units*, Journal of Manufacturing Systems 35 (2015): 46-75.
33. Nelder JA,. Mead R., *A simplex method for function minimization.* Computer Journal 7 4 (1965): 308–13.
34. Holst J., *Using whole building simulation models and optimizing procedures to optimize building envelope design with respect energy construction and indoor environment.* Eighth International IBPSA Conference, In: Proceedings of the building simulation (2003).
35. Wright J., Alajmi A., *The robustness of genetic algorithms in solving unconstrained building optimization problems.* Ninth International IBPSA Conference, In: Proceedings of the building simulation (2005).
36. Brownlee I., Wright A., Mourshed MM., *A multi-objective window optimization problem.* In: Proceedings of the 13th annual conference companion on Genetic and evolutionary computation (2011).
37. Rapone G., Saro O., *Optimization of curtain wall facades for office buildings by means of PSO algorithm.* Energy Buildings 45 (2012): 189–96.
38. Wetter M., Wright J.; *A comparison of deterministic and probabilistic optimization algorithms for non-smooth simulation-based optimization.* Building and Environment 39 (2004): 989 – 999.
39. Wetter M., Wright J., *Comparison of a generalized pattern search and a genetic algorithm optimization method.* Eighth International IBPSA Conference, In: Proceedings of the building simulation (2003).
40. Zhou G., Ihm P., Krarti M., Liu S., Henze G., *Integration of an internal optimization module within EnergyPlus.*Eighth International IBPSA Conference, In: Proceedings of the building simulation (2003).
41. Bichiou Y., Krarti M., *Optimization of envelope and HVAC systems selection for residential buildings.* Energy and Buildings 43 (2011): 3373–3382.
42. Carreras J., Boer D., Guillén-Gosálbez G., Cabeza L., Medrano M., Jiménez L., *Multi-objective optimization of thermal modelled cubicles considering the total cost and life cycle environmental impact.* Energy and Buildings 88 (2015): 335–346.
43. Xu J., Kim J., Hong H., Koo J., *A systematic approach for energy efficient building design factors optimization.* Energy and Buildings 89 (2015): 87–96.
44. Marco Manzan, Alberto Clarich, *Energy and daylighting optimization for an office with fixed and moveable shading devices.* Building and Environment 113 (2017): 175-184.
45. Hamdy M, Palonen M, Hasan A., *Implementation of pareto-archive NSGA-II algorithms to a nearly-zero-energy building optimization problem.* In: Proceedings of the building simulation and optimization conference (2012).

IOP Conf. Series: Materials Science and Engineering 1037 (2021) 012041 doi:10.1088/1757-899X/1037/1/012041

Energy savings strategies in Campus Buildings of Athens - Greece

S Tsala[1], I P Koronaki[1]

[1] Laboratory of Applied Thermodynamics, School of Mechanical Engineering, Thermal Engineering Section National Technical University of Athens, Heroon Polytechniou 9, Zografou Campus 15780, Athens, Greece

*Corresponding author's e-mail: koronaki@central.ntua.gr

Abstract Bioclimatic construction has been an important tool for improving the construction of buildings in recent decades. Energy-saving options in buildings include the use of appropriate materials, exposures and the use of alternative power generation systems. In the present work the basic features of the bioclimatic design and study of the energy saving in the buildings of the polytechnic area located in Zografou is presented. In particular, the RETscreen software is used to test energy savings in the buildings of library, mechanical engineering and management building using photovoltaics. It turns out that an important parameter that affects energy saving is the use of the building and the type of photovoltaic used. This is evidenced by the results of RETscreen that in the case of the Library and the Mechanical Engineering building stable photovoltaic have greater energy savings while in the case of the Management building, biaxial photovoltaics produce greater savings.

1. Introduction

The constant increase in consumer goods, the excessive growth of the population and the improvement in the standard of living of modern reality have, over the last twenty years, led to a rapid increase in energy consumption. Oil production has increased sixfold over the past decade and electricity demand has increased tenfold every ten years. The minimal use of non-renewable energy sources has contributed to an increase in emissions, which deplete the ozone layer and have seriously degraded the environment by gradually destroying ecosystems. The main culprits of this catastrophe are the industries, transport, power plants, but as the foreword states, the built environment [1].

In fact, 45% of the energy produced worldwide is spent on the building sector, with 2/3 of that being owned by private homes. Although these rates are responsible for 50% of CO2 emissions, only 15% of the needs are covered by the US, with solar as the protagonist. Globally, the contribution of the building industry to the emissions of carbon compounds (C) is close to 6 billion tons, with 4,5 billion of these being owned by industrialized countries and 2,25 billion tons of 4,5 being directly or indirectly due to building construction [2,3]. For Greek data, the building sector consumes 30% of the energy produced, with a 40% contribution to CO2. Most of the energy is consumed for cooling and heating, since thermal losses in Greek buildings are enormous and the use of energy reduction techniques is mandatory for all buildings only from 2007 [3].

A structured environment is defined as the area developed by human construction and includes all buildings covering housing, employment and leisure needs. Based on the size of buildings and

IOP Conf. Series: Materials Science and Engineering 1037 (2021) 012041 doi:10.1088/1757-899X/1037/1/012041

neighbouring infrastructure, the built environment is characterized as urban, semi-urban and rural [3]. Dealing with the urban environment of large cities is particularly important for the Greek reality, taking into account the small blocks of construction, the increased height of buildings, but also the large size of the city, which complicates the operation not only of its climate, but also of the climate in the wider region. This is linked to the thermal and aerodynamic behaviour of the largest cities in it, according to the thermal island phenomenon, which refers to the temperature fluctuations between two neighbouring towns. The first oil crisis in 1973 alarmed the scientists involved in the design and construction of buildings and led them to study and research new forms of energy, such as the creation of a more economical, more practical and more ecological building. The building is now treated as a living organism and not as consumed and vanity [3,4]. The ecological equilibrium factor in a building is now the top priority in any project, leading scholars to the solution of the immediate application of the principles of bioclimatic building and ecological construction [4].

The bioclimatic structure shall attempt to design, construct and operate a building or on a larger scale of a housing stock in harmony and cooperation with local climatic conditions and using GDP, as far as possible, so as to meet the specific specifications of the building, without placing a particular burden on its environment and to try to maintain its energy identity as ecological as possible [2,5]. Construction of more energy-efficient houses could reduce carbon emissions by 60%, equivalent to 1.35 billion tons. It also achieves savings in conventional energy resources, leading in the long term to a reduction in dependence on imported oil. In addition, the use of local construction materials, as long as permitted, can significantly help to reduce the cost of construction, as transport costs are reduced [2].

The concept of ecological footprint appeared in 1994 and is defined as the trace a building leaves in the world. It is essentially the "metabolism" of the building, with the logic of the living organism, and it is an indicator of sustainability measurement. The meaning of the trace can be extended to a settlement or even to a human, with the logic of "geomerides" [6]. It is generally understood that the limits of the land are finite and therefore also the limits of productivity. According to global economic indicators, 20% of the world's population consumes 83% of the available land resources. The geometry corresponds to the average value of the ecologically produced land corresponding to each inhabitant of the land. This does not include the land that should correspond to biodiversity, that is to say, the whole of the rest of the animal and plant kingdom, and it presupposes that all people have an equal right to the land's share. Therefore, an ecological footprint is the total soil area required to support the needs of a population, but also to absorb all its waste [7]. The ecological footprint is defined as [6]:

$$I = P \times A \times T$$

Where: I the effect or the trace, P the population, A the abundance and T the technology.

From 1961 to 2005 worldwide, the ecological footprint has tripled and according to data from W.W.F. it is estimated that in 2005 every person consumed resources corresponding to 22 acres of ground while the planet could deliver a maximum of 18. In a ranking of 147 countries in terms of per capita ecological footprint, Greece is the 17th most unfavourable, with the country's per capita ecological footprint increasing by 101% between 1975 and 2003. The ecological footprint for every Greek citizen corresponds to 50 acres per inhabitant. This unfavourable position is mainly due to energy consumption and thus to CO2 emissions into the atmosphere [7]. According to the latest data from the European Environment Agency (2006), total energy consumption in Greece is on average an annual increase of 2.7% (1990-2003), one of the highest in the EU. With regard to the two major cities, Athens and Thessaloniki, it is understood that they can be very easily unsustainable, with a particularly high ecological footprint, with more than 65% of the Greek population. As is evident, buildings affect the environment in many ways during their construction, operation and demolition. But the environment also has a big impact on buildings. In order to be able to design buildings properly, full knowledge of this interaction must be available [7].

The buildings of the major urban centers unfortunately cause a number of problems in the environment, such as the change in the balance of the main components of the atmosphere, in soil and subsoil water, mainly because of the chemical emissions from urban waste and waste. This phenomenon is particularly pronounced in most Greek cities. Also, the depletion of natural resources results from the intensity of energy use for the structure [2,3]. The disadvantaged use of non-renewable energy sources has led to a gradual increase in air pollution and the greenhouse effect. In addition, the disruption in the

geobiological cycles of water, oxygen and carbon dioxide results in unstable climate changes throughout the earth, and the anarchic construction has degraded both the urban and the rural environment, causing the disappearance of local flora and fauna. Finally, the use of radioactive and non-ecological building materials has the direct effect of causing problems for the health of the occupants and of degrading the quality of life, their quality and their environment [2,3,8].
All of this has made scientists concerned about finding a new way of building houses, which is healthier and more environmentally friendly. The result was a shift towards Bioclimatic Building using US. and other energy saving methods. Good planning can lead to a gradual reduction of the environmental crisis and to an upgrading of the urban and thus the terrestrial environment [9,10]. Bioclimatic design developed in the 1980s as a new trend in urban planning with references to the local microclimate. The term bioclimatic design lies in the architectural and urban planning of buildings and agglomerations aimed at adapting to the local climate and the natural environment, while protecting sensitive areas with rare ecosystems. The microclimate, the mid-climate and the macroclimate determine the lighting, ventilation, design and energy behaviour of buildings. In particular, the macroclimate is shaped by the average weather conditions that prevail throughout the year, the mid-climate is characterized by the impact of topography, vegetation and the nature of the region, and finally, the microclimate is the result of human intervention, which directly changes the built environment [9,10]. Bioclimatic design aims to exploit the positive environmental parameters in order to reduce the energy needs of the building throughout the year and to save conventional energy. In addition, it incorporates elements linked to the local physiology, local culture and to the traditional - and thus empirically proven - techniques for more favourable building of the area concerned [9,10,11].
Bioclimatic Structuring, in addition to local traditional building techniques, is the result of an integrated and complex synthesis, which is linked to a wide range of parameters such as orientation, appropriate choice of openings, design of the building shell, and correct selection of materials [12]. This does not mean, however, that intervention in existing buildings is limited. With low cost and user-friendly technologies, heat losses can be reduced, buildings can be protected from overheating, lighting conditions can be improved and noise reduced. The implementation of bioclimatic structuring may lead to energy independence for non-US. up to 60%. It also contributes to the increasing reduction of $CO2$ emissions and other gases, the existence of which aggravates the rational use of water and local infrastructure [12,13]. It has been observed that the traditional ecological materials of the pre-industrial period are reliable, have a long life span, are not harmful to human health and the environment, and still allow US savings. These benefits are linked to bioclimatic design and contribute to the construction of structures that meet the needs of modern lifestyle, without posing a threat to future generations [14].
In view of the above-mentioned energy savings, this study aims to investigate bioclimatic building and to analyse the ways of saving energy in residential blocks, i.e. the Library building, the Mechanical Engineering building and the Administration building.

2. Method

The aim is to study the energy consumption of the various buildings in Campus (100 ha total area) and to compare them with each other using the RETscreen tool. The RETscreen Clean Energy Project Analysis Software is the innovative decision support software for clean energy. It is provided entirely free of charge by the Canadian Government and is a tool for strengthening clean energy projects worldwide. This software significantly reduces the costs (both economically and chronically) associated with identifying and evaluating potential energy projects. RES and allows decision makers and professionals to determine whether or not a proposed renewable energy, energy efficiency, or cogeneration project is economically appropriate. The procedure followed in the RETscreen software for all buildings is summarized in the following steps:
Step 1: Select a location and automatically import climate data from the software.
Step 2: Enter installation information. The user imports:
- The extent of the building and the reference point, defined as the average value of energy consumption (as derived from experimental measurements).

- Level 2 defines the type of reference point for each case as electricity.
- The Action provides options for fuels, space usage planning, equipment, end-use of energy, and power optimization.
- The Schedule specifies the time of use of the building.
- Space Heating Demand is defined as the type used by each building.
- End-use defines the electricity applications for each building.
- Energy optimization defines each photovoltaic scenario with its characteristics.

For each power saving scenario, the software delivers:

- Energy and economic data report
- Energy efficiency
- Economic sustainability

In this case, the buildings of the Library (Services), Administration (Administrative Services) and Mechanical Engineers (Educational Building) will be studied and the installation of photovoltaics on their energy-saving roofs. In the RETscreen, the location data of the buildings will be entered and, using the databases it contains for specific types of buildings, an economic data and energy saving report will be generated. To calculate the fluctuation of the consumption trend in Polytechnic City, measurements made on one of the buildings and on an hourly basis for each day of each month from March 2015 until December 2016 will be used. For the estimation of the monthly electricity load, the average values of the measurements for each month were considered.

3. Results

3.1. Energy measurement and consumption compared by building

The Library has an area of about 8928 m^2 and operates 16 hours a day on a daily basis except on weekends. The electric load measurements show that this building is mainly used during the months of December, January, June, July and September.

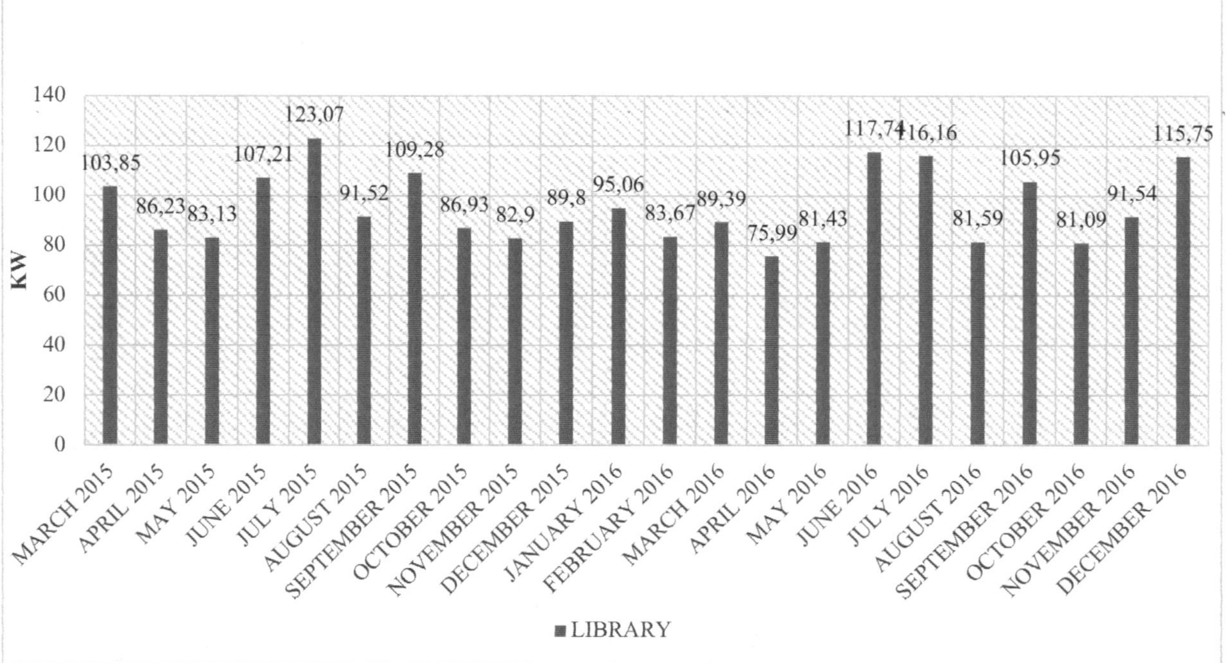

Table 1:Average electric charge for the Library building (2015-2016).

Table 1 shows that in the months of June, November and December 2016 the average electric load increased compared to the 2015 average, while in the remaining months the reverse occurs.

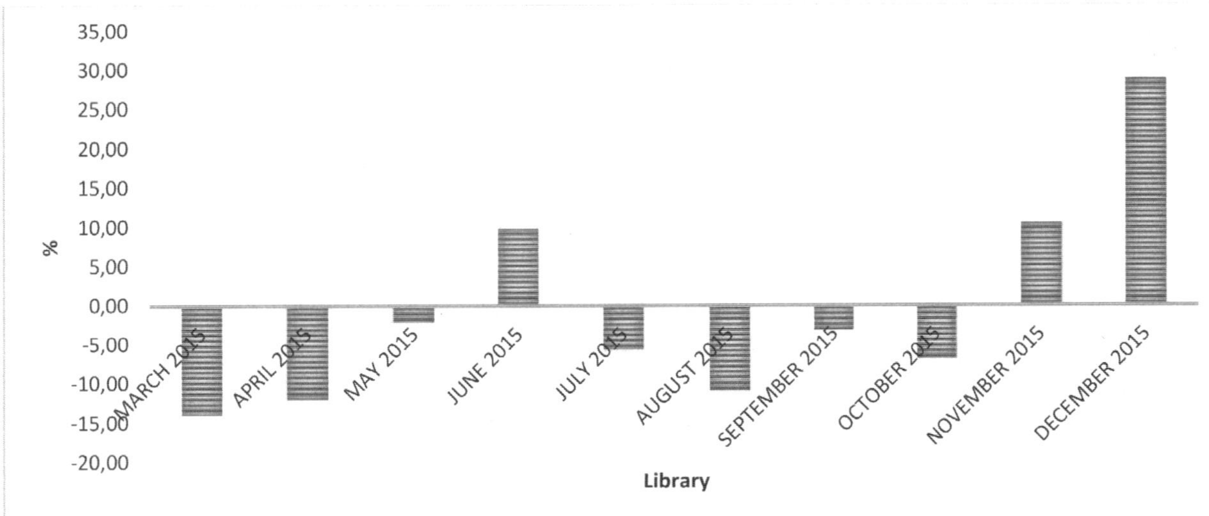

Table 2: Electrical load change rates for the Library building.

Measurements of the MN Mechanical Engineering Building show that the maximum load of this building was 101 kW in June 2016 and the minimum was 54 kW in October 2016.

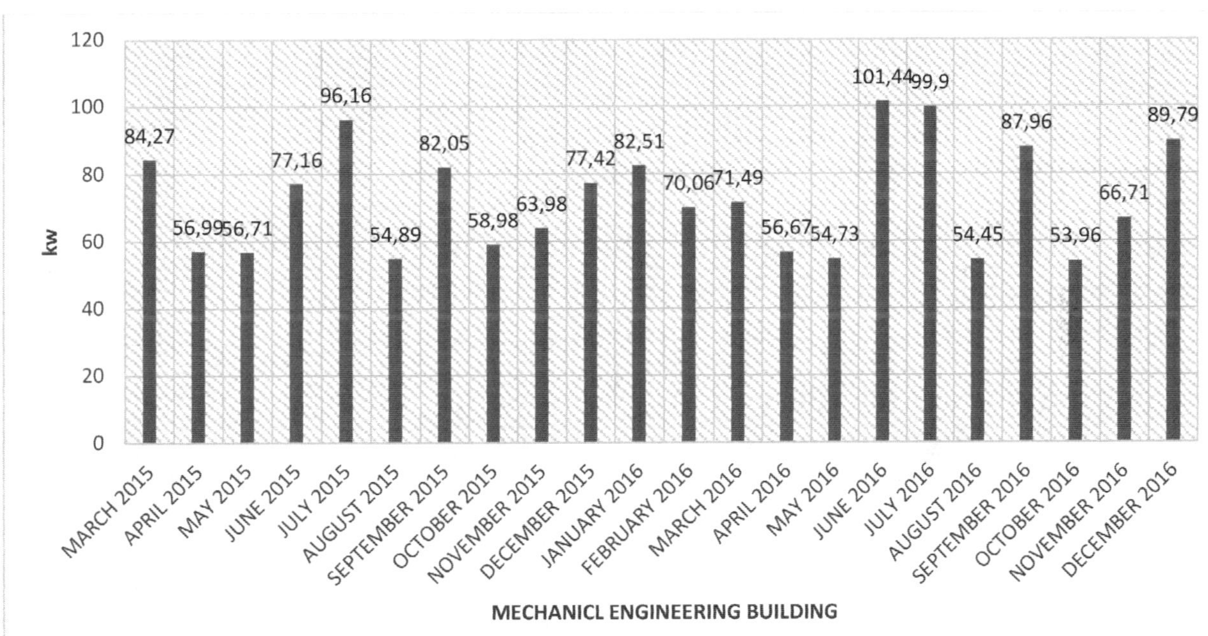

Table 3: Average electric charge for the Mechanical Engineering building.

Table 4 shows that in June, July, September, November and December 2016 the average electric load increased compared to the corresponding load in 2015, while in the remaining months the reverse is observed. A maximum increase is observed in June and December, when a higher temperature change is observed.

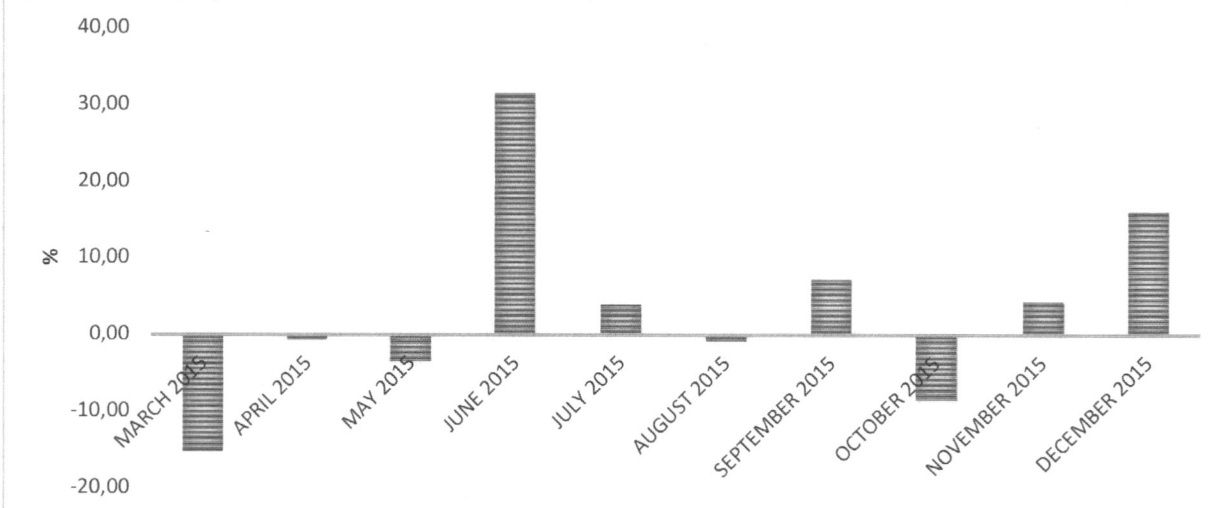

Table 4: Electrical load change rates for the Mechanical Engineering building.

The Administration building showed the largest electric charges between the buildings being compared. In particular, the maximum electric load of the building is observed for the month of July 2016 and is equal to 83,1 kW and the minimum load is observed in May 2015 and is equal to 43,8 kW.
Table 5 shows an increase in the average electric load compared to the corresponding 2015 load over all months except March (-9,73%) and October (-0,24%), where the average electric load was reduced.

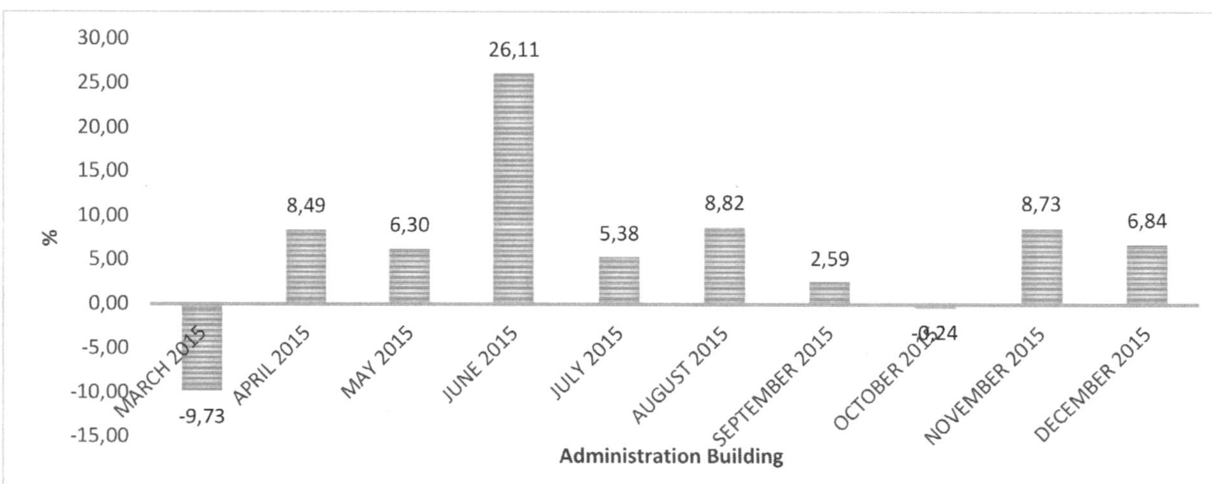

Table 5: Electrical load change rates for the Administration building.

The comparison of the previous electrical charges shows that the average electric charge is a function of the number of people using each building and the use of the building.

3.2. Energy Saving Scenarios

3.2.1. Power saving scenarios for the Library building

For the Library building, the area is 8928 m^2 and the reference point is the average energy consumption value as determined by the experimental data. The reference point shall be the average value for two years of 858.8 MWh and for its area of 0.1 MWh/m^2. The schedule is 16 hours/day for daily. Space Heating Demand defined the gas boiler system connected to ventilation, engines and fans. The electricity data shall be taken as the average monthly values of the measurements taken.

3.2.1.1. Scenario 1: Energy optimization with Twinax 5kW Photovoltaics

The annual financial flow for Scenario 1 is shown below:

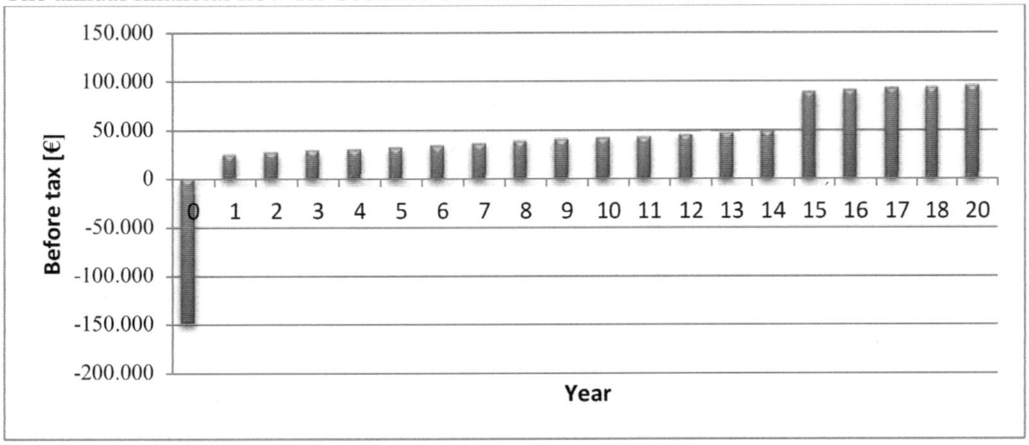

Figure 1: Annual financial flow for Scenario 1.

The energy savings and fuel consumption for Scenario 1 are shown below:

	Heating	Cooling	Electricity	Set
Fuel consumption	kWh	kWh	kWh	kWh
Basic case	1,672	9,330	104,416	115,418
Proposed case	1,672	6,015	64,134	71,821
Fuel saving	0	3,315	40,282	43,597
Fuel savings-rate	0%	35.5%	38.6 %	37.8%

Figure 2: Energy and fuel savings for the library using a 5kW two-axle photovoltaic.

The financial viability for Scenario 1 is reflected in the following Figure:

Economic parameters		
Generally		
Rolling fuel cost tax	%	2%
Inflation value	%	2%
Discount rate	%	9%
Reinvestment rate	%	9%
Project lifetime	Year	20
Financing		
Interest-in-charge	%	70%
Debt	€	12,99
Share	€	6
Borrowing rate	%	5,570
Debt period	Year	7%
Debt payments	€/	15
	Year	1,427

Cost | Saves | Revenue			
Initial costs			
Increase initial costs	100%	€	18,566
Total initial costs	**100%**	€	18,566
Annual costs and debt payments Operation & Maintenance Costs (saving)		€	-3,196
Fuel cost-proposed case		€	7,182
Debt payments-15 years		€	1,427
Total annual costs		€	5,411
Annual savings and revenue			
Fuel cost-base case		€	11,542
Total annual savings and income		€	11,542

Figure 3: Economical viability for the library using a 5kW biaxial photovoltaic.

3.2.1.2. Scenario 2: Energy optimization with 5kW Constant Photovoltaic

For this scenario only the type of voltaic used is changed, which will be stable with a 45th gradient.
The annual funding for Scenario 2 is as follows:

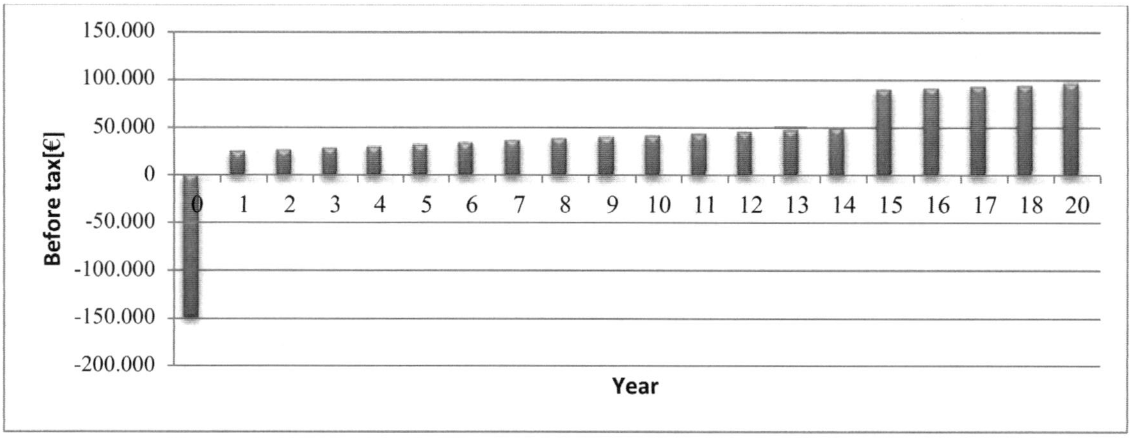

Figure 4: Annual funding for stable photovoltaic for the Library building.

The energy saving and fuel consumption for Scenario 2 is shown below:

	Heating	Cooling	Electricity	Set
Fuel consumption	kWh	kWh	kWh	kWh
Basic case	1,672	7,991	104,416	115,520
Proposed case	1,672	7,991	62,993	72,656
Fuel saving	0	1,441	41,423	42,564
Fuel savings-rate	0%	15.3%	39.7 %	37.1%

Figure 5: Energy saving for stable photovoltaic in the Library building

3.2.1.3. Compare scenarios for the Library Building

The energy savings with a twinax photovoltaic are 38.6% for electricity and 37.8% for total. Annual cash flows are positive over a 20-year horizon and annual costs are $5411 while annual savings and revenues are €11542. For the library building and the fixed photovoltaic, the total energy savings are 37.1% and for the electrical is 39.7%. Annual costs and annual revenues remain stable. It follows, therefore, that in the case of fixed photovoltaics in the Library building the energy savings are lower than in the case of biaxial photovoltaics.

3.2.2. Energy-Saving Scenarios for Engineering Building

The Engineering Engineers Building uses the Commercial/Institutes-Training standard. Site data is the same as the Library. The extent of the building is 4000 m^2 and the reference point is the average energy consumption value as determined by the experimental data, in this case the reference point is 718.46 MWh. The minimum electrical energy is 149.79MWh i.e. 0,18MWh/m^2 and the maximum of 1254,42MWh i.e. 0,31MWh/m^2. The fuel type and tele energy systems for hot and cold water shall be taken into account. The schedule is set for the highest winter temperature (January-9.3 ^{o}C) and summer temperature (July-27 ^{o}C). Daily, it is assumed that the building operates 12 hours a day and at weekend 4. For regional heating the initial cost is obtained for boiler and for regional cooling is considered to be an air heat pump. The electricity data shall be taken as the average monthly values of the measurements taken.

3.2.2.1. Scenario 1: Power optimization with 68kW Twinax Photovoltaic.

The annual and cumulative funding for Scenario 1 of the Engineering Building is shown in the following Figure:

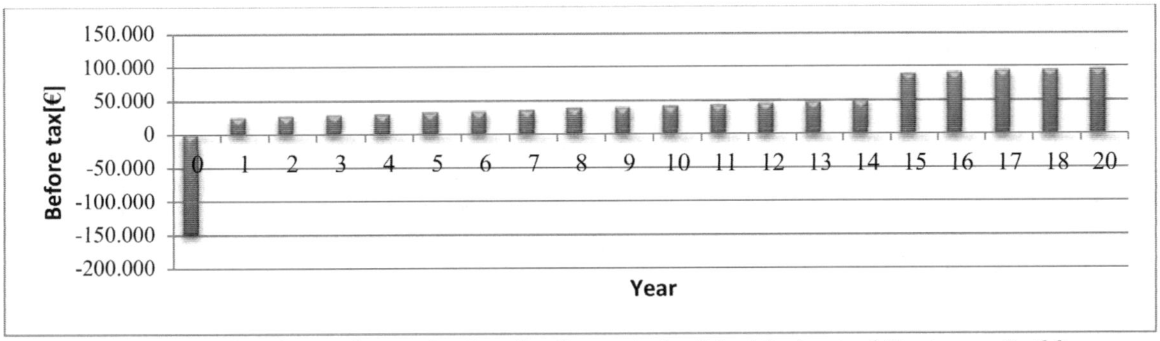

Figure 6: Annual and cumulative funding for Scenario 1 of the Mechanical Engineers Building.

The energy saving and fuel consumption for Scenario 1 of the Mechanical Engineers building is as follows:

	Heating	Cooling	Electricity	Set
Fuel consumption	kWh	kWh	kWh	kWh
Basic case	19,062	806,292	803,347	1.628,701
Proposed case	19,062	655,078	620,119	1.294,259
Fuel saving	0	151,214	183,228	334, 442
Fuel savings-rate	0%	18.8%	22.8 %	20.5%

Figure 7: Energy saving for Scenario 1 of the Mechanical Engineers building.

The financial viability for Scenario 1 of the Mechanical Engineers building is illustrated below:

Economic parameters		
Generally		
Rolling fuel cost tax	%	2%
Inflation value	%	2%
Discount rate	%	9%
Reinvestment rate	%	9%
Project lifetime	Year	20
Financing		
Interest-in-charge	%	70%
Debt	€	268,283
Share	€	115,235
Borrowing rate	%	7%
Debt period	Year	15
Debt payments	€/ Year	29,522

Cost | Saves | Revenue			
Initial costs			
Increase initial costs	100%	€	384,118
Total initial costs	**100%**	€	384,118
Annual costs and debt payments		€	-2,085
Operation & Maintenance Costs (saving)		€	128,473
Fuel cost-proposed case		€	29,522
Debt payments-15 years			
Total annual costs		€	156,910
Annual savings and revenue		€	161,917
Fuel cost-base case			
Total annual savings and income		€	161,917

Figure 8: Financial sustainability data for the engineering building.

3.2.2.2. Scenario 2: Energy optimization with Constant Photovoltaic 68kW

The annual funding for Scenario 2 of the Engineering Building is shown as follows:

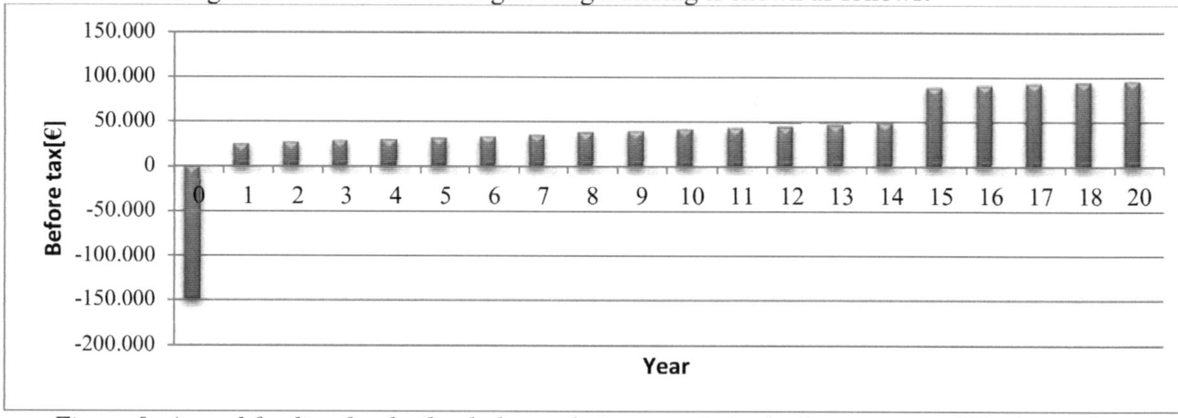

Figure 9: Annual funding for the fixed photovoltaic scenario in the Department of Mechanical Engineering.

The energy and fuel savings for the building of mechanical engineers with stable photovoltaic is shown below:

	Heating	Cooling	Electricity	Set
Fuel consumption	kWh	kWh	kWh	kWh
Basic case	19,062	806,292	803,347	1.628,701
Proposed case	19,062	655,078	610,637	1.284,777
Fuel saving	0	151,214	192,709	343,924
Fuel savings-rate	0%	18.8%	24 %	21.1%

Figure 10: Energy and fuel saving with stable photovoltaic in the engineering building.

The financial viability for Scenario 2 of the Mechanical Building is given in the following figure:

Economic parameters		
Generally		
Rolling fuel cost tax	%	2%
Inflation value	%	2%
Discount rate	%	9%
Reinvestment rate	%	9%
Project lifetime	Year	20
Financing		
Interest-in-charge	%	70%
Debt	€	268,283
Share	€	115,235
Borrowing rate	%	7%
Debt period	Year	15
Debt payments	€/ Year	29,522

Cost | Saves | Revenue			
Initial costs			
Increase initial costs	100%	€	384,118
Total initial costs	**100%**	€	384,118
Annual costs and debt payments Operation & Maintenance Costs (saving)		€	-2,085
Fuel cost-proposed case		€	127,525
Debt payments-15 years		€	29,522
Total annual costs		€	165,961
Annual savings and revenue			
Fuel cost-base case		€	161,917
Total annual savings and income		€	161,917

Figure 11: Sustainability results for stable photovoltaic in the Mechanical Engineers building.

3.2.2.3. Comparison of scenarios for the Engineering building

It is found that cash flow does not change significantly. Also, from the results of energy saving it seems that in the case of the building of engineers the stable photovoltaic presents greater energy and fuel savings than the two-axis. Economic viability shows that the total costs are lower in stable photovoltaic than the biaxial. In particular, the total energy savings are for the two-axle photovoltaic equal to 20.5% while the savings of electricity are 22.8%. The total starting cost is 348,118€, the total annual costs are 155910€ and the total savings on an annual basis are 161917€. In the case of stable photovoltaics, the savings of electricity are estimated at 24% and the total savings are 21.1%. Annual savings and initial costs are estimated at the same levels as the two-axle while annual costs are reduced to 154961€.

3.2.3. Energy Saving Scenarios for the Administration Building

For the administration building the location information is as in both previous cases. In the installation the type of building is set service and is described as a large office with an area of 2500m^2. The average value of energy consumption as determined by the experimental data is set as a reference point. The energy tab sets out the fuel elements that are natural gas and electricity. The operating schedule is now set for 10 hours as it is a service and the weekend is thought to be zero. The electricity data shall be taken as the average monthly values of the measurements received.

3.2.3.1. Scenario 1: Energy optimization with Biaxial Photovoltaic 59kW

The annual funding for Scenario 1 of the Administration Building is shown in the following Figure:

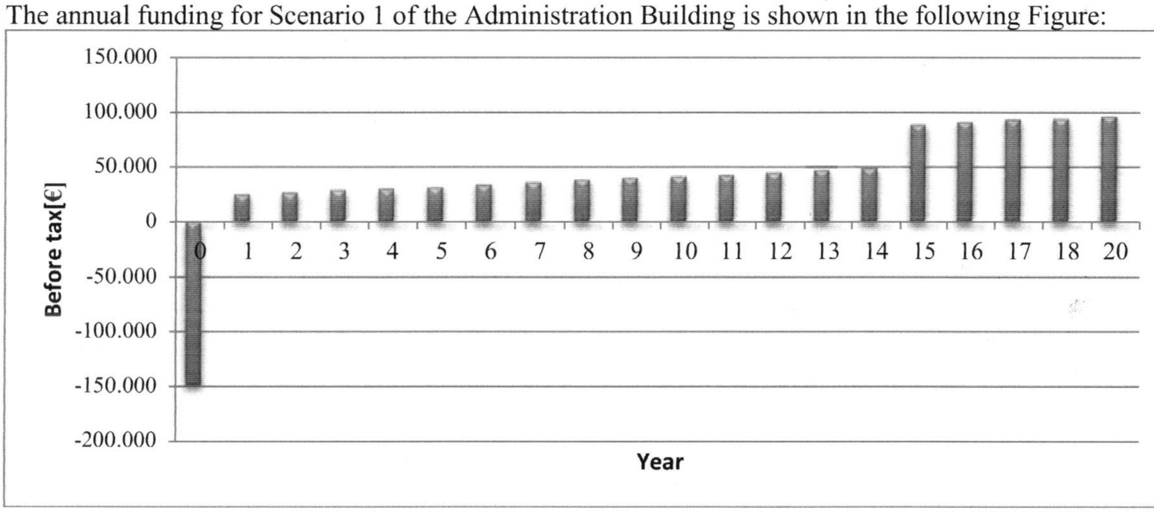

Figure 12: The annual funding for the administration building with two-axis photovoltaic.

The energy saving and fuel consumption for Scenario 1 of the Administration building is as follows:

	Heating	Cooling	Electricity	Set
Fuel consumption	kWh	kWh	kWh	kWh
Basic case	67,510	158,509	1.116,612	1.342,632
Proposed case	29,190	103,643	567,196	690,029
Fuel saving	38,320	54,866	529,960	652,603
Fuel savings-rate	56.8%	34.6%	50.1%	48.6%

Figure 13: Energy saving and fuel consumption for two-axle photovoltaic in the Administration building.

The financial parameters and storage costs and revenues from this scenario are shown in the following figure:

Economic parameters		
Generally		
Rolling fuel cost tax	%	2%
Inflation value	%	2%
Discount rate	%	9%
Reinvestment rate	%	9%
Project lifetime	Year	20
Financing		
Interest-in-charge	%	70%
Debt	€	347,656
Share	€	148,995
Borrowing rate	%	7%
Debt period	Year	15
Debt payments	€/ Year	38,171

Cost | Saves | Revenue			
Initial costs			
Increase initial costs	100%	€	496,652
Total initial costs	**100%**	€	496,652
Annual costs and debt payments Operation & Maintenance Costs (saving)		€	-4,574
Fuel cost-proposed case		€	66,925
Debt payments-15 years		€	38,171
Total annual costs		€	100,522
Annual savings and revenue Fuel cost-base case		€	129,457
Total annual savings and income		€	129,457

Figure 14: Economic sustainability analysis for Scenario 1 of the Administration Building.

3.2.3.2. Scenario 2: Energy optimization with Constant Photovoltaic 60kW

The annual and cumulative funding for Scenario 2 of the Administration Building is shown as follows:

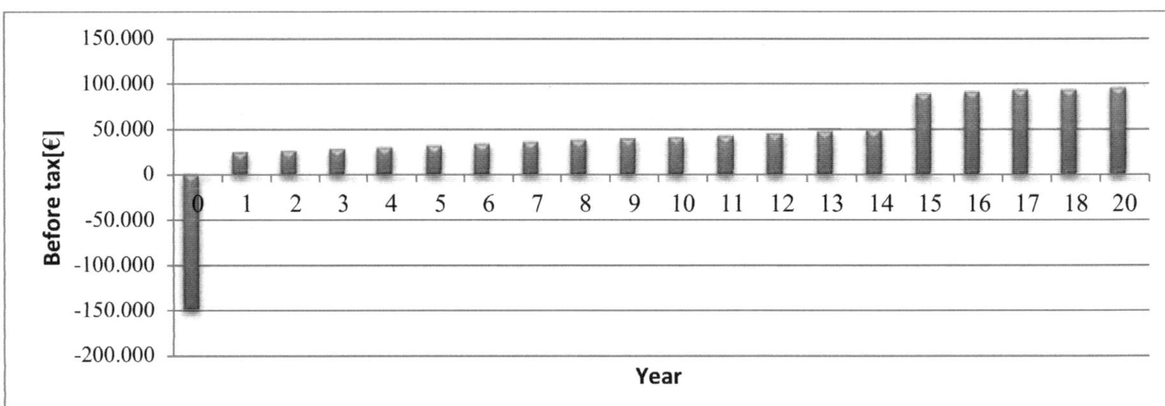

Figure 15: Cash flow for stabilized photovoltaic in the administration building.

The energy saving and fuel consumption for Scenario 2 of the Administration Building is as follows:

	Heating	**Cooling**	**Electricity**	**Set**
Fuel consumption	kWh	kWh	kWh	kWh
Basic case	67,510	158,509	1.116,612	1.342,63 2
Proposed case	29,190	103,643	586,652	719,485
Fuel saving	38,320	54,866	529,960	623,146
Fuel savings-rate	56.8%	34.6%	47.5%	46.4%

Figure 16: Energy saving for stabilized photovoltaic in the Administration building.

The financial parameters and storage costs and revenues from this scenario are shown in the following figure:

Economic parameters		
Generally		
Rolling fuel cost tax	%	2%
Inflation value	%	2%
Discount rate	%	9%
Reinvestment rate	%	9%
Project lifetime	Year	20
Financing		
Interest-in-charge	%	70%
Debt	€	347,656
Share	€	148,995
Borrowing rate	%	7%
Debt period	Year	15
Debt payments	€/ Year	38,171

Cost | Saves | Revenue			
Initial costs			
Increase initial costs	100%	€	496,652
Total initial costs	**100%**	€	496,652
Annual costs and debt payments			
Operation & Maintenance Costs (saving)		€	-4,574
Fuel cost-proposed case		€	69,870
Debt payments-15 years		€	38,171
Total annual costs		€	103,468
Annual savings and revenue			
Fuel cost-base case		€	129,457
Total annual savings and income		€	129,457

Figure 17: Economic sustainability analysis for the administration building.

3.2.3.3. Comparison of scenarios for the Administration Building

The savings of electricity from two-axle photovoltaic are 50.1% and the total energy savings are 48.6%. The total initial costs are 496652€ and the total annual costs of 100522€ and the total savings are 129457€. The use of stable photovoltaic leads to savings of 46.4% total energy while in electricity to 48.6%. From an economic point of view there is an increase in overall annual costs equal to 3%.

4. Conclusions

In this work, a number of energy-saving scenarios were studied in three buildings of the Polytechnic City, in the area of Zografou. In particular, they were used by 2 energy optimization scenarios (use of two-axial photovoltaic and fixed photovoltaic) for the Library building, the Engineering building and the Administration building. The study was conducted using the RETscreen software. In order to implement the scenarios, the consumption measurements for the above mentioned buildings had to be calculated. The measurements were made from March 2015 in December 2016 and showed that energy consumption increased linearly for all buildings. In particular, the Library building (15%) was the largest increase in the library building (15%) and smaller presented the administration building (5%).

The results of the software used for energy-saving scenarios showed that the use of fixed photovoltaics (i.e. Scenario 2) showed the highest percentage of total energy savings for the Library building (39.7% vs. 37.8%). Similarly, for the Engineering building, the use of fixed photovoltaics (i.e. Scenario 2) gave higher percentages of total savings (21.1% vs. 20.5%). Annual savings and initial costs remained at the same levels while annual costs in Scenario 2 fell to €154961. For the Administration building, the use of two-axial photovoltaics (i.e. Scenario 1) gave higher percentages of total energy savings (48.6% vs. 46.4%). From an economic point of view, however, Scenario 2 increased overall annual costs to 3%. It was found that the power of installed photovoltaics, the number of panels and the coverage of demand are the parameters that determine the cost and energy savings.

IOP Conf. Series: Materials Science and Engineering 1037 (2021) 012041 doi:10.1088/1757-899X/1037/1/012041

References

1. Energy Efficiency Policies and Measures in Greece- Monitoring of Energy Efficiency in EU 27,Norway and Croatia (ODYSSEE-MURE), CRES, Athens, September 2009.
2. KAPE, Renewable Energy Sources in Residential Totals
3. Environmental impact and energy saving for heating in Greek apartment buildings, Droutsa K., Balaras K.A., Energy Saving Group Institute for Environmental research and Sustainable Development, National of Athens.
4. IPE – Ecological Construction 2000.
5. RENEWABLE ENERGY SOURCES, ZERVOS A., Athens 2005.
6. http://www.energymining.sa.gov.au/__data/assets/pdf_file/0017/315404/Energy-Efficient-Strategies-REES-review-report-2017.pdf
7. https://www.wwf.gr/images/pdfs/LPR_2016_full%20report_low-res_embargo.pdf
8. ECOLOGICAL ARCHITECTURE, Bioclimatic architecture, ecological construction, geobiology, internal architecture, Kostas and Themis Steph.Tsithes, Kedros Publications, 2005.
9. ENERGY BIOCLIMATIC DESIGN OF BUILDINGS AND HOUSES. Kontoroupis G., Athens 2002.
10. ENVIRONMENTAL DESIGN, Randall Thomas, Max Fordham & Partners, E & FN SPON, London 1996.
11. IIP- Ministry 2000.
12. Tsipiras Th. & Kostas. St., Ecological Architecture, Kedros, Athens 2005.
13. Tsipiras Th. & Kostas. St., The Eco-House, Livani Publications, Athens 1996.
14. The Energy Research Group-School of Architecture-University College Dublin, Energy in Architecture-The European Passive Solar Handbook, Brussels 1996.
15. http://publications.gc.ca/collections/collection_2008/nrcan/M39-115-2005E.pdf

Acknowledgments

This research is co-financed by Greece and the European Union (European Social Fund - ESF) through the Operational Programme «Human Resources Development, Education and Lifelong Learning» in the context of the project "Strengthening Human Resources Research Potential via Doctorate Research – 2nd Cycle" (MIS-5000432), implemented by the State Scholarships Foundation (IKY).

Ευρωπαϊκή Ένωση
European Social Fund

Operational Programme
Human Resources Development,
Education and Lifelong Learning

Co-financed by Greece and the European Union

ΕΣΠΑ
2014-2020
ανάπτυξη - εργασία - αλληλεγγύη

IOP Conf. Series: Materials Science and Engineering 1037 (2021) 012042 doi:10.1088/1757-899X/1037/1/012042

Potential for the regional energy supply system optimization through recycling waste into energy

A A Tulegenova[1], T I Andreenko[1] and Yu Yu Rafikova[1]

[1] Lomonosov Moscow State University, Faculty of Geography, Renewable Energy Sources Laboratory. Leninskie Gori 1, Moscow, Russian Federation 119991

*Corresponding author's e-mail: ju.rafikova@gmail.com

Abstract. The use of renewable energy sources is becoming a general trend for optimizing energy supply and achieving low-carbon development goals in various regions of the world. In Russia, in connection with the introduction and implementation of government support measures for solar, wind, bioenergy, starting in 2013, a large-scale introduction of energy-generating facilities began. This factor, as well as the need to solve the problem of recycling waste from the municipal sector, agriculture and forestry, makes it necessary to use local renewable energy resources, including waste. This approach helps to optimize the regional energy system, even if they are grid connected. The paper discusses the methods and potential for optimizing energy supply in the regions of southern Russia, which has a high population density, a significant deficit in energy balance, and problems in the field of solid waste processing. The basis for this work is the use of municipal sector waste as an energy resource, with an assessment of the economic and environmental efficiency of this process. The use of geoinformation technologies provides a spatial analysis of the resource potential, a comparative analysis of the distribution of energy production and consumption within the region, as well as helps to assess the availability of territories for the optimal placement of solid municipal waste processing facilities.

1 Introduction

The renewable energy sector is actively developing in many countries. Especially high growth rates are observed in solar and wind energy. Bioenergy lags behind, although biomass and biogas plants produce a predictable amount of energy (unlike solar and wind power plants), since their operation does not depend on climatic conditions. Especially the great potential of bioenergy in solving climatic, environmental and energy problems is highlighted by a number of EU Horizon 2020 program projects (uP_Running [1], AgroBioHeat [2] projects), which work on research and implementation of energy generating enterprises based on agrobiomass. These projects prove that agrobiomass is a large, under-exploited local resource that can be used for energy purposes to help solve energy and climate problems while contributing to rural development.

In most cases, the raw material for bioenergy are wastes from other industries [3]. A stable supply of raw materials is required for the operation of bioenergy plants. Ensuring such supplies, choosing the location of the processing factory, developing efficient logistics schemes for short and reliable supplies of biomass are logistical tasks, on the effective solution of which the success of the enterprise depends.

Logistics solutions include harvesting, transporting and storing agricultural waste. The energy use of agrobiomass is expected to have a positive social impact on the regions, where it creates additional jobs on the ground and prevent the outflow of youth from rural areas. Finally, the projects are designed to promote the dissemination of affordable modern technical solutions for heating systems that use agrobiomass with high efficiency instead of traditional fuel, thereby reducing the level of harmful emissions and improving the ecological situation in the regions [4].

In the Republic of Kazakhstan, agriculture is one of the basic industries, which makes it possible to practically completely ensure food security, as well as to export significant volumes of agricultural products abroad. At the same time, the processing of organic waste is practically not developed. At the same time, due to the sufficient amount of animal and crop waste, the country has a great potential for producing biogas, heat and electricity using biogas plants and thermal power plants using waste biomass.

2 Methodology

The research methodology is based on the approaches that were developed in the work of the European Institute for Energy Research at the KIT [5]. The methodology allows to assess the potential for using bioenergy resources, taking into account the spatial distribution of raw materials, as well as to select the most optimal sites for placing biogas power plants (BGPP) within a given area. Spatial analysis carried out using GIS tools, in addition to economic feasibility, takes into account environmental constraints and infrastructural factors (proximity to gas distribution and electric networks, roads, etc.). To achieve these goals, the mathematical functions of ArcGIS [6] are used, which, based on the initial data sets, allow a feasibility study. Fig. 1 shows the algorithm of the proposed methodology.

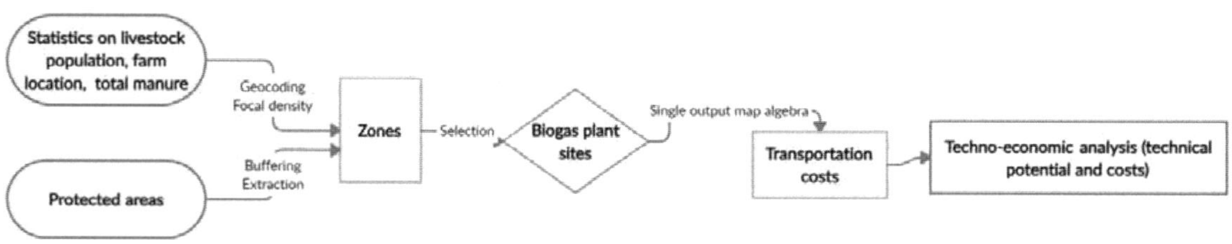

Figure 1. Algorithm of the methodology for the selection of sites for the installation of high potential BGP using ArcGIS [5]

One of the key stages in the analysis of the territory in order to select and justify the site that is most optimal for the construction of a station for processing livestock waste is to determine the transport accessibility and availability of sources of raw materials (animal and crop waste).

The sequence is as follows:

1) pre-selection of the zones suitable for biogas development with respect to the exclusive and selective criteria;
2) geocoding of the animal farms data and assessing its spatial density to identify optimal sites within those pre-selected zones;
3) mapping the arable land within certain distances from biogas plant sites to assess the share of required land for biogas dedicated crop planting in the total arable land (similar spatial analyses is carried out for selected crops);
4) assessing the costs of transporting waste to potential sites for biogas plants, the amount of waste processed and other characteristics, a technical and economic analysis, as a result of which the technical potential of biogas production from the processing of agricultural waste and its cost are determined.

The assessment of technical and economic potential in this work was carried out for energy production with combined heat and power technologies as well as the bio-methane feeding into the natural gas grid.

Thus, in the process of research, such ArcGIS tools as *Geocoding, FocalStatistic-sum, MapAlgebra, Buffering, Extract, Overlay, Proximity, Selection* were used. *Corine Land Cover* and *Open Street Map* were used as sources of digital data for mapping factors.

3 Results and discussion

The Akmola region of Kazakhstan, which is included in the group of agricultural regions with significant industrial potential, was chosen as a region for testing the methodology. The main direction of agricultural production in this region is the cultivation of cereals, vegetables, as well as meat and dairy farming (Fig. 2). In connection with the large amount of waste generation and environmental problems that may grow up as a result of waste storage (especially storage of animal waste), the problem shows necessity of processing them into energy.

In accordance with the methods presented above, an assessment of animal breeding waste energy potential was carried out, taking into account the type of livestock and poultry, its age, and housing conditions. In this case, only waste generated in large farms where it is possible to organize collection, transportation and disposal of waste (available gross waste potential) was considered (Fig. 3). It turned out that poultry farming (large poultry farms) provides the main contribution to the formation of animal breeding waste energy potential in the Akmola region.

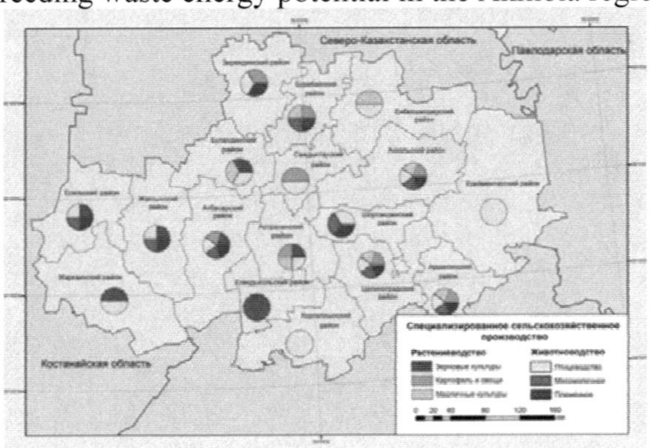

Figure 2. Specialization of agriculture in the districts of Akmola region of the Republic of Kazakhstan

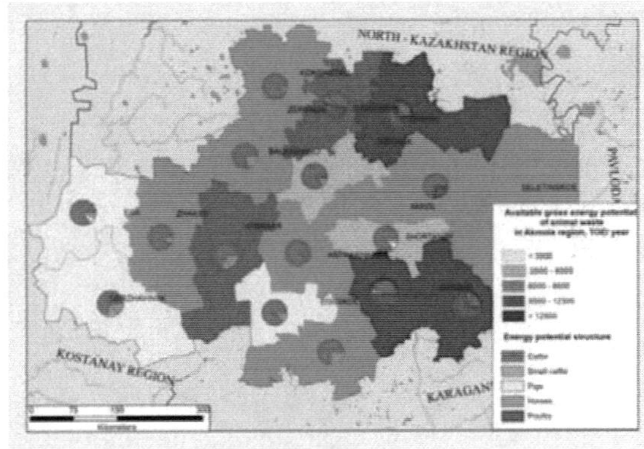

Figure 3. Available gross energy potential of livestock waste in Akmola region

Assuming that animal breeding waste can be processed at biogas stations and further processed into heat and electric energy, the task was to find the optimal sites for the placement of these stations, taking into account a number of natural, infrastructural and socio-economic factors of the territory and using mapping and geoinformatics tools (ArcGIS). Baseline information is the location of large livestock farms, the capacity of waste generation and their energy potential, as well as cartographic information, including hydrographic and road transport networks; boundaries of protected areas; settlements, forests, power lines and electrical substations. The result of combining these datasets is cartographic layers that reflect both favorable and limiting to the construction and operation of agricultural waste processing facilities factors (Tab. 1).

Table 1. Factors that influence on the possibility of waste processing facilities locating

Favorable factors	Limiting factors
Road transport network	Nature protected areas
Power lines	Forests
Low voltage power substations	Swamps
	Rivers and water protection zones
	Territories of cities and villages

To take into account the spatial distribution of waste biomass sources, the "density" of waste generation was calculated within a circle with a radius not exceeding the distance of cost-effective transportation of animal waste (40 km). Density "is considered here as a conventional value equal to the ratio of the waste source capacity (tons of waste per year) to the area of a circle of the corresponding radius. Then closely located livestock farms - as a result of addition - provided a higher spatial "density" of waste (Fig. 4).

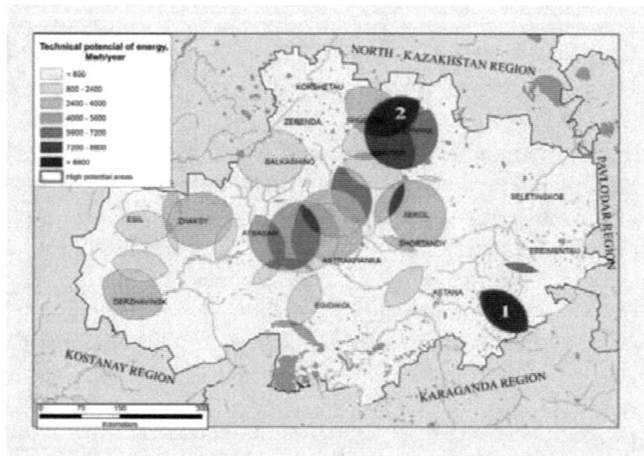

Figure 4. Cumulative waste density within 40 km zones around animal breeding farms

In addition to the size of the resource, the main factor in choosing a territory for the location of biogas stations (BGPP) is transport accessibility (logistics). Using ArcGIS tools, we created a map showing the maximum distance for waste transportation from each livestock enterprise (40 km) (Fig. 5). Then, the most suitable, taking into account resource availability and transport accessibility, for the placement of BGS were determined as polygons built at the extreme points of the road network no longer than 40 km. The map (Fig. 5) shows these landfills and highlights the most productive sites in terms of resources (animal waste).

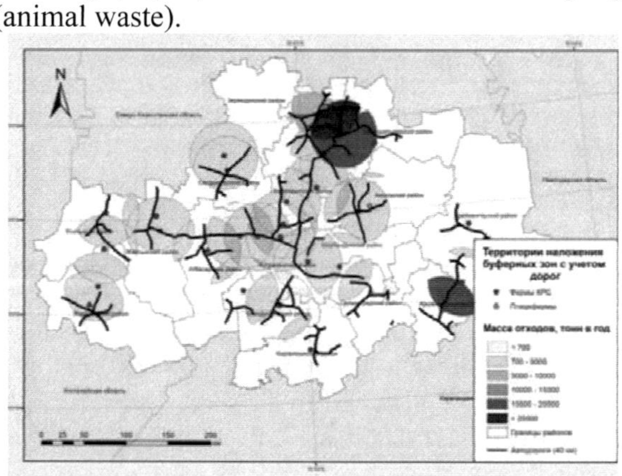

Figure 5. Road network within zones with high animal waste energy potential

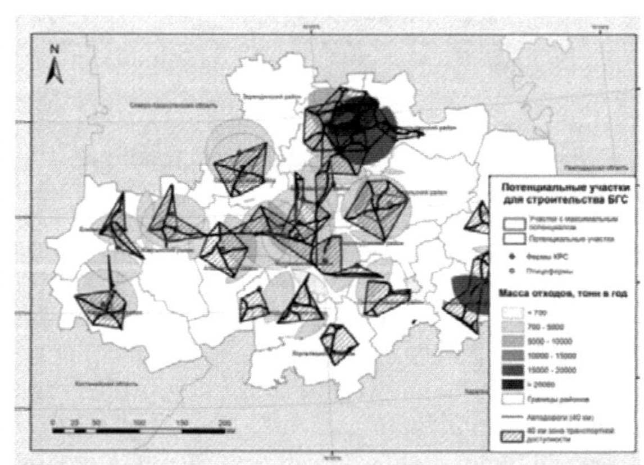

Figure 6. Potential sites for the construction of BGPP

To increase the yield of biogas, it is necessary to add plant raw materials to animal waste in order to achieve a ratio of carbon to nitrogen concentration of the order of C/N=10-20. On the territory of the Akmola region, wheat accounts for 83% of the total harvest of livestock products, therefore, the potential for using wheat straw collected within no more than 20/40 km from large livestock farms was assessed to achieve the specified ratio. The calculation results, assuming a uniform distribution of arable land over the territory, show that the need for straw can be fully satisfied with crop waste collected within 20 km from livestock and poultry enterprises.

At the final stage of the study, the amount of biogas and its cost were calculated using the example of three biogas stations (BGPP) for processing animal and poultry waste in the Akmola region. Moreover, each of the biogas plants processes waste from several livestock farms. It was assumed that the produced methane can act as an analogue of natural gas for the consumer. At present, after purification and upgrading, biogas produced at biogas stations and installations abroad is supplied to the consumer (if gas distribution systems are available) as an analogue of natural gas. Therefore, estimates

were made for the cost of the produced methane, taking into account the investment costs for the construction of biogas stations, the costs of transporting waste to the biogas plant, the current costs of operating the stations and repairs, cleaning (desulfurization of methane). To bring biomethane to the quality of natural gas supplied to the consumer, i.e. removal of hydrogen, sulphides due to their high corrosiveness and odors, water wash technology and pressure swing adsorption technology are currently used in European countries. At the same time, according to [3], these methods are used only in a few countries (Sweden, Germany), and it is difficult to estimate the average investment costs for the introduction of these technologies. As a result, when calculating the cost of biogas, the costs of the following operations were not taken into account:
- biomethane refining procedures;
- the cost of connecting BGPP to gas distribution networks;
- the cost of animal breeding waste (it was assumed that this waste is free and only the cost of transportation was estimated);
- the cost of crop waste (it was assumed that crop waste (straw), which goes to the bedding of livestock and is thus included in organic waste, increasing the biogas yield, is free and the cost of their delivery to livestock farms is zero).

When calculating the payback of biogas plants, it was assumed that a simple payback period was calculated (excluding the discount rate, capital is not borrowed).

An analysis of the typical results obtained for a plant operating on waste from several livestock farms showed that in the case when only methane is sold as a waste product, the simple payback period is 11.5 years. With the additional sale of biofertilizers (sludge), the payback period is reduced to 3.40 years. Since part of the energy produced by the biogas plant (in the form of methane in biogas) must be used for the needs of the plant itself, calculations of a simple payback period were also carried out on the assumption that only 30% of biomethane is sold to external consumers. In this case, the payback period rises slightly - up to 4.29 years, which is also a completely satisfactory result. Analysis of the cost structure shows that the maximum contribution is made by investment (capital) costs; transportation costs are only 6.2%. When constructing waste processing stations at each of the livestock enterprises, the simple payback period for BGPP projects without the sale of fertilizers is from 12.05 to 17.71 years. With the sale of 30% of biomethane and the total amount of fertilizers received to the consumer - from 3.56 to 5.24 years, which is higher than the results of calculating the payback period for the combined enterprise. This is due to a reduction in investment costs in the construction of a large enterprise in comparison with several small biogas plants and a relatively low share of waste transportation costs in the total cost structure. The results obtained can only be considered as "bottom-line estimates", since a simple payback period is calculated and capital and operating costs for a number of technological processes are not taken into account.

Evaluations of the efficiency of waste utilization of all livestock enterprises in the region showed that the total technical potential of electricity production is 54 243 MWh/year, and heat - 395,714 GJ/year. Savings of hydrocarbons (coal) in the processing of animal breeding waste by means of methane digestion followed by the use of methane in the Mini CHPP for the Akmola region of Kazakhstan amounted to: 10 848 t/year in the case of electricity generation at the CHP and 16 848 t/year in the case of heat energy production When utilizing plant waste with energy production, it is assumed that net CO_2 emissions into the environment (which were absorbed by plants during growth) does not occur. In this regard, the prevention of CO_2 emissions in the event of biogas production from waste and its further use as fuel at a thermal power plant in the region will amount to 548 tons of CO_2 per year.

4 Conclusions

In the Republic of Kazakhstan, the development of the biogas industry is currently at an early stage, the government has introduced green tariffs to stimulate bioenergy, BGPP projects are being implemented within the framework of the Concept for Sustainable Development of the Republic of Kazakhstan until 2050. Spatial analysis tools allow us to search for optimal areas for the location of waste processing facilities. The main selection criteria are: the presence of permanent sources of waste (poultry and cattle farms; crop farms), limiting (protected areas, water bodies, settlements) and favourable factors (power lines, roads, electrical substations) to the development of biogas production. Estimates of the simple payback period of biogas plants in the Akmola region showed that the most optimal is the centralized collection and processing of waste from several farms, taking into account the costs of transportation. At the same time, the payback period is reduced by 0,5-6 years compared to BGPP at each farm. The total technical potential of electricity production from waste of large livestock enterprises in Akmola region is about 54 000 MWh/year, heat energy – 395 000 GJ/year. This can provide fuel (coal) savings of more than 10 000 tons/year in the production of electrical energy and 16 000 tons/year - of thermal energy, as well as obtaining more than 35 000 thousand tons/year of valuable organic fertilizers. When utilizing crop waste, CO_2 emissions can be reduced by about 550 t/year.

Acknowledgments

Authors wish to acknowledge assistance of Sophia Kiseleva during prepairing this paper and financial support from Faculty of Geography, Lomonosov Moscow State University within the research framework «Geography and rational use of the renewable energy sources».

References

[1] uP_running. Policy Guidelines and Roadmap. 8/10/2018. Available online on: https://www.uprunning.eu/
[2] AgroBio Heat project. Available online on: https://agrobioheat.eu/
[3] Waste to Energy 2018/2019: Technologies, plants, projects, players and backgrounds of the global thermal waste treatment business» / 11th edition Ecoprog GmbH, 2018
[4] Khan I, Zobaidul K 2020 *Renewable Energy* **150** 320
[5] Sliz-Szkliniarzab B, Vogt J 2012 *Renew. Sust. Energ. Rev.* **16** **1** 752
[6] Esri, ArcGIS Pro, Environmental Systems Research Institute, Redlands, California, 2020.

IOP Conf. Series: Materials Science and Engineering 1037 (2021) 012043 doi:10.1088/1757-899X/1037/1/012043

Waste energy recovery simplify assessment

L Knapčíková[1]*, R Hricová[1], I Pandová[2] and J Piteľ[1]

[1]Technical University of Košice, Faculty of Manufacturing Technologies with a seat in Prešov, Department of Industrial Engineering and Informatics, Bayerova 1 , Prešov, Slovakia
[2]Technical University of Košice, Faculty of Manufacturing Technologies with a seat in Prešov, Department of Process Engineering, Štúrova 31 , Prešov, Slovakia

*Corresponding author's e-mail: lucia knapcikova@tuke.sk

Abstract. The paper deals with research on the energy recovery of municipal solid waste and the related impact of the recovery on the environment. Through research steps, analyses of the current production of waste, its energy recovery and the impact on the environment were gradually performed in the Prešov self-governing region in Slovakia. A simplified comparison was made between the impacts of the current landfill method and the proposed incineration method. In terms of forecasts for the development of municipal solid waste produced in the coming period, scenarios of solid municipal waste treatment possible alternatives for the future period are outlined by the help of the electronic monitoring platform.

1. Introduction

Research activities described in this contribution were done in the frame of the project "Energy Recovery from Municipal Solid Waste by Thermal Conversion Technologies in Cross-border Region" (EnyMSW). Municipal solid waste (MSW) can be defined – in very strong simplification - is a time- and region- varying heterogeneous accumulation of more kinds of solid trash produced by citizens household activities. Municipal solid waste production direct depends on socio-economic indicators can be observed in general. [1] European Waste Framework Directive 2008/98 / EC introduced a five-step waste hierarchy as a priority regulation with the prevention of peak waste and subsequent preparation for re-use, recycling, another recovery, including energy recovery and waste disposal as a last resort. [2], [3] The Directive allows classifying a municipal waste incineration plants as recovery operations, provided that they contribute to high energy efficiency production, to encourage the use of waste for energy production in energy-efficient municipal waste incineration plants and to promote innovation in waste incineration. [4] In this context, it is important to note that 'recovery' is an operation whose results waste applying for useful purposes by replacing other materials that would otherwise be used to perform the function or waste prepared for that function in the manufacturing plant.

2. A brief characterization of Slovak waste management

In August 2019, the Slovak government passed a new law on waste, the ambition is to significantly increase the sorting and recycling. Slovakia produces relatively less waste than other EU countries, but recycling is significantly lower. Two-thirds of municipal waste and more than half of all waste is deposited in landfills, which is significantly higher than in the EU. The trend of declining waste landfill and increasing recycling is very weak and will not change without stricter measures. There is also a need for consistent separation and recovery of biodegradable municipal waste. The economy thus loses significant volumes of materials that could be used secondarily.

At present, the Slovak citizen is obliged to sort MSW by law and the municipalities are obliged to create conditions for separation by a suitable system in Slovakia. A tighter waste management policy brings the risk of illegally deposited waste (illegal dumps), the removal of which is often costly. There are thousands of locations with illegally deposited waste in Slovakia, which degrade a given area, threatens the health of population and ecosystems and poses further future risks. Most waste in such landfills is mixed municipal and construction waste. [5]

While the Slovak Republic does not generate more waste per inhabitant than similar economies, it is facing significant challenges in improving the way this waste is managed. Almost 70% of municipal waste is currently still landfilled and almost 80% of landfilled municipal waste is mixed municipal waste of non-defined content. [6] Slovakia plan is to increase to 60% of the municipal waste recycling rate, including its preparation for re-use by 2030, and to reduce the land-filling rate to less than 25% by 2035. [5] A general overview of MSW production by Slovak regions is presented in Figure 1.

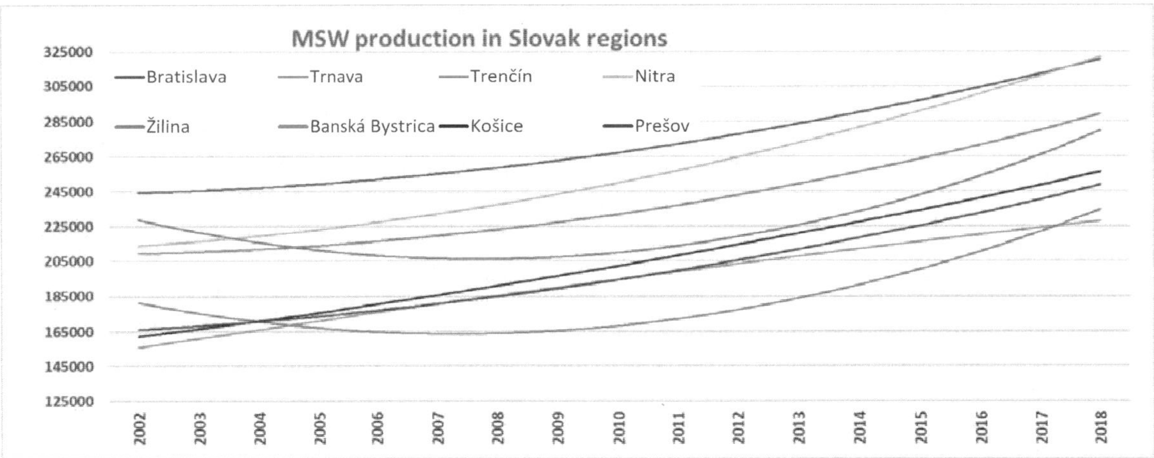

Figure 1. Production of MSW in Slovak regions [7]

The official statistic showed the first position in MSW amount generation for the Bratislava region - the most developed Slovak region. Prešov Self-Governed Region (SGR) - the region with the highest number of inhabitants in Slovakia - is the lowest MSW amount producer from the relative amount of MSW per capita point of view (Figure 2). [7]

3. Waste management in the examined region

Prešov (SGR) fills the north-eastern part of Slovakia and it stretches from the west to the east. It borders on Poland and Ukraine in the north and east respectively.

The region is with over eight hundred thousand population the most populous of all the regions in the country and is known for its hilly landscape. The population density in the region is over 90 inhabitants per km^2. The region metropolis is the town Prešov, the third biggest town in Slovakia with a population of about one hundred thousand. The Prešov Self-Governed Region consists of 13 districts. There are 665 municipalities, of which 23 are towns, where about half of the region's population live. Strong aspects of the region are well-preserved environment, high forestation of the territory, rich sources of geothermal and mineral springs, a good position at the eastern European Union border, favourable demographic progress, transport connection with the entire Slovak transport system, and qualified workforce. The weak characteristics of the region are high unemployment ratio, the lowest wage rates in Slovakia and need to solve MSW negative impacts. [8]

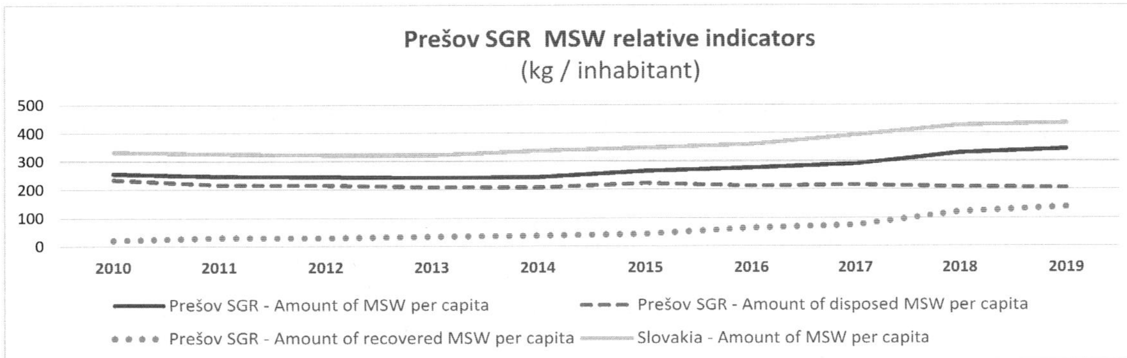

Figure 2. Relative indicators of selected Slovak regions MSW production [7]

These analyses raise the question of how to exploit produced MSW to obtain a smaller landfilled quantity of it. One of the possibilities is to use MSW for energy recovery because this MSW treatment is on a relatively low level of utilization in the region.

4. Regional analyses of MSW

For the study of MSW energy recovery capacity in the region was designed methodology:

1. Regional MSW samples collection;
2. The collected MSW Analyse - Weight fraction of collections - identification of typical categories of waste / Exclusion of non-combustible and hazardous components;
3. Experiments sample preparation - by weight fraction identification after exclusion of non-combustible and hazardous components /grounding of the samples for homogenizing of contents;
4. Samples Analyses - Humidity / Calorific analyses - Gross calorific value and Net calorific value calculation;
5. Regional MSW production analyses /landfills analyses - distribution in the region & annual and total capacities;
6. Region energy recovery potential determination;
7. Environmental impact determination.

4.1. Regional MSW samples collection

Municipal solid waste was collected in the time of transhipment operation – freshly MSW just after unloading from low capacity collection vehicles.

Five randomly selected Collections were stored in marked plastic bags in an approximate weight of ten kilograms for every one of them. [9]

4.2. Regional MSW samples analyses

Collected waste was analysed in the laboratory from a composition point of view.

Slovak Waste Catalogue [10] defines in code 20 "Municipal wastes (household wastes and similar wastes from trade, industry and institutions) including their components from the separate collection" 51 detailed components of MSW. The research team prepared a synthesis of the legally stated waste components to simplify structure by grouping similar items to 13 components. In these thirteen components are two of them created by the research team to enrich items defined by the Catalogue:

- 7 Food waste, and
- 9 Hygienic supplies

because of an important proportion in the collected samples– see "Waste structure" column in Table 1.

Table 1. Analyses of Prešov SGR landfilled MSW

	Waste structure	Average content: Full composition	Average content: Combustible composition
1	Paper	11.57%	13.74%
2	Plastics	12.25%	14.54%
3	Glass	8.04%	X
4	Metals	2.57%	X
5	Layered packaging materials (Tetrapack, etc.)	1.47%	1.75%
6	Textiles	3.38%	4.01%
6	Food waste	7.33%	8.70%
8	Biodegradable w.	25.61%	30.40%
9	Hygienic supplies (diapers, etc.)	8.04%	9.55%
10	Electrical waste	0.70%	X
11	Construction mineral waste	3.49%	X
12	Unsorted waste (Dirty plastics, Combined materials, etc.)	14.58%	17.31%
13	Dangerous waste (Bateries, ...)	0.97%	X
		100.00%	100.00%

4.3. Experiments sample preparation

MSW collection structure analyses were followed by preparing samples for the next research steps.

In this step were prepared five pieces of the one-kilogram sample following weight fraction identification after exclusion of non-combustible and hazardous components (Table 1).

Prepared Samples were grounded for homogenizing of contents.

Samples humidity evaluation was done in the standard process:

- Weighing of samples before drying - weight W_0;
- Drying of the samples in the thermal chamber with maximum temperature up to 90° C;
- Weighing of samples after drying - the result is weight W_D;
- Calculation of the collected MSW Samples humidity H.

Calculation of the collected MSW samples humidity was done by the formula:

$$H = \left(\frac{W_O - W_D}{W_D}\right).100\% \tag{1}$$

Results of the humidity analyse of the samples are presented in Table 2.

Table 2. Results of the samples humidity analyse

Description of analysed features	Sample 1	Sample 2	Sample 3	Sample 4	Sample 5	Average
W_0 weight of samples before drying	0.9862	1.0694	0.9831	1.0024	1.0687	1.021960
W_D weight of sample after drying	0.8018	0.7875	0.7136	0.8292	0.8878	0.803980
Humidity (Moisture)	**18.70 %**	**26.36 %**	**27.41 %**	**17.28 %**	**16.93 %**	**21.34%**

The wide dispersion of moisture values is due to the composition of the waste and the MSW storage conditions before collecting.

4.4. Samples Analyses

Calorific analyses serve for obtaining of MSW important characteristic for calculation of the energy recovery potential of MSW incineration. Evaluation of the potential was processed in steps:

- Calorimetric examination of the samples for determination of results of possible combustion heating characteristics (Q_s gross calorific value);
- Standardization of the calorimetric results by calculating the influence of real municipal solid waste properties – e.g. vaporization of the water, etc. (Q_n net calorific value).

Determination of calorific theoretical energy potential value was done in laboratory of Technical University of Košice built in the frame of the funded project. They were prepared samples following the identification of typical MSW categories of the collection of combustible components (Table 1).

The basic formula considering the waste sample chemical composition and moisture is [11]:

$$Q_n = [(Q_s - 212.2 * HC - 0.8 * (OC + NC)] * \left(\frac{100-H}{100}\right) - 0.2443.H \quad (2)$$

where:

Q_n net calorific value (J/g);
Q_s gross calorific value (J/g);
HC hydrogen content, in percentage by mass, of the dry fuel;
OC oxygen content, in percentage by mass, of the dry fuel;
NC nitrogen content, in percentage by mass, of the dry fuel;
H humidity (%);
0.02443 correction factor of the enthalpy of vaporization for water (J/g) per 1 wt% of moisture).

Results of the samples calorific evaluation (gross caloric value) and calculated net calorific value are shown in Table 3.

Table 3. Results of the samples calorific evaluation results and calculated net calorific values

	Sample					Average
	1	2	3	4	5	
	Calorific evaluation results					
Q_s - gross calorific value (J/g)	21 457	22 147	22 164	21 561	21 698	**21 805.40**
Q_n - net calorific value (J/g)	11 323	11 721	11 731	11 383	11 462	**11 524.00**

4.5. Regional MSW production analyses

In Prešov SGR are 18 MSW deposits (landfills). Some of them are near to achieving maximum capacity. For the region is very important to find long term strategy for increasing of waste recovering of the produced municipal waste.

The trend of the amount of landfilled MSW is relatively stable from 2011 on a level between 150 000 to 180 000 tons per year. Prognose for next period is under uncertainty generated by the increasing trend of the amount of the produced MSW and at the same time a growing trend of recycling. The good results of general MSW management in Prešov SGR are shown in Figure 3 – the amount of recovered MSW is higher year by year.

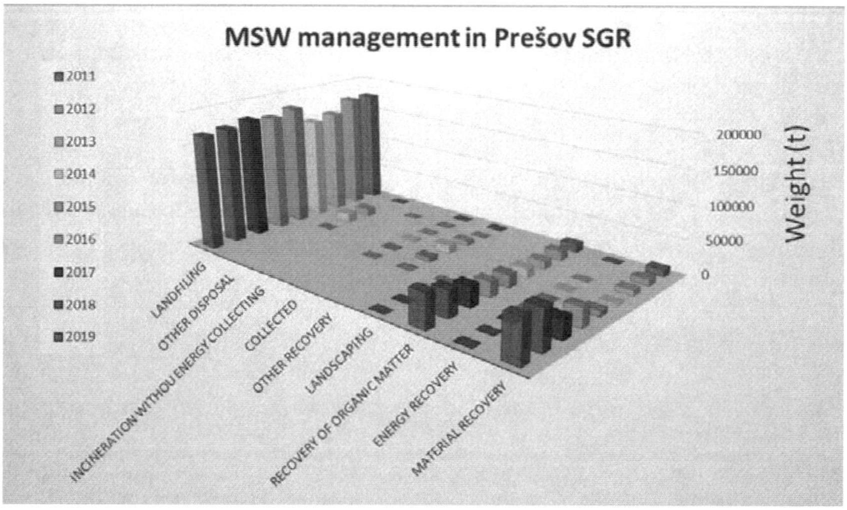

Figure 3. Relative indicators of selected Slovak regions MSW production [12]

Numbers for 2019 presented in Figure 3 are [12]: Total waste production 283 338.00 t/year. In this yearly amount are fractions: Material recovery 61.596 t/year; Energy recovery 927 t/year; Recovery of organic matter 50 862.00 t/year; Landscaping 393 t/year; Other recovery 0 t/year; Collected 0 t/year; Incineration without energy collecting 0 t/year; Other disposals 0 t/year; Landfilling 169.560 t/year.

These same yearly results presented in comparison of total MSW amount and landfilled MSW amount in Figure 4 show the increasing ratio of recovered MSW in time and slowly decreasing ratio for landfilled MSW in time. [13]

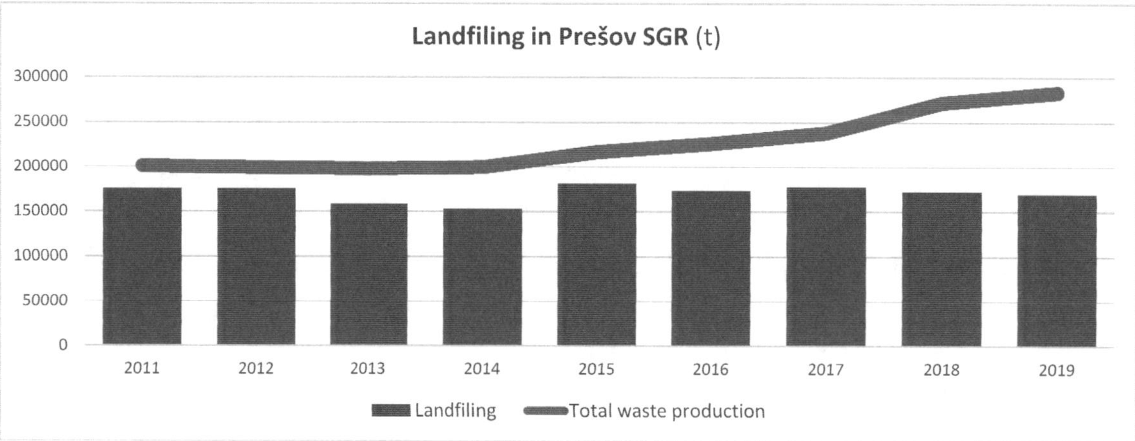

Figure 4. MSW landfilling in Prešov SGR [12]

4.6. Energy recovery potential determination of the region

The official information of the landfilled amount of municipal solid waste 169,560 tons in 2019 was used for calculation of energy potential of the region. This consideration is supported by the long-term development of the amount of waste produced in Prešov SGR in accordance with official statistics presented in Figure 4. [12]

The trend of the generated amount of landfilled MSW is relatively stable in the observed period. Slovakia is also in very good progress to fulfil European Union legislative in decreasing the amount of landfilled MSW.

Next calculations are based on relatively stable production of MSW amount in Prešov SGR for upcoming years. The calculations are divided into two groups – energy-related findings and waste reduction related findings. Both of the calculations are shown in Table 4.

Thermal analysis of the MSW samples was aimed to determine the reduction the MSW by incineration. The analysis was performed in the EnyMSW Laboratory "Small scale system for thermal treatment of municipal solid waste" in Baia Mare, Romania. [14] The result is placed in the table in row 22. The most important of **energy-related** finding is net energy recovery potential of Prešov Self-Governed region in 365 747.14 MWh per year. The most important **landfilling related** finding is the reduction of landfilled MSW from current 169 560.00 tons per year to 31 681.20 ton per year (26 739.61 tons of non-combustible MSW fractions + 4 941.59 tons of incineration rests) produce the decreasing ratio of landfilled MSW volume to 18.68%.

4.7. Environmental impact determination

Thermal analysis to evaluate the production of the emissions by the thermal treatment of MSW was performed in the EnyMSW Laboratory in Baia Mare, Romania, too. [14] The MSW incineration – as every general burning process - is producing air pollution, ash and slag.

Five samples modelling the MSW composition from collected MSW in Prešov SGR (each in weight 5 kg) were thermally treated in the laboratory built by EnyMSW project funding. The produced average amount of ash, slag and air emission readings of the thermal process are shown in Table 5.

Table 4. Figures of analyses results and calculation of energy and waste reduction characteristics

Figures of 2019 for MSW management in Prešov SGR (t/year)		
01	Material recovery	61 596
02	Energy recovery	927
03	Recovery of organic matter	50 863
04	Landscaping	393
05	Other recovery	0
06	Collected	0
07	Incineration without energy collecting	0
08	Other disposal	0
09	Landfiling	169 560
10	Total amount of generated MSW	**283 339**
Figures of EnyMSW analyses in Prešov SGR		
	Energy related findings	
11	Average humidity of MSW samples (%)	21.34
12	Volume of combustible fractures in landfilled MSW (%)	84.23
13	Amount of combustible fractures in landfilled MSW **[r.09*r.12/100]** (t/year)	142 820.39
14	Q_n - net calorific value (MJ/kg)	11.52
16	Energy recovery MSW potential **[r.13*r.14*0.8]** (MJ/year)	1 316 689 721.05
17	Energy recovery MSW potential (MWh/year)	365 747.14
	Landfilling related findings	
21	Volume of landfilled not-combustible MSW **[r.09-r.13]** (t/year)	26 739.61
22	Average MSW amount reduction by incineration (%)	96.54
23	Average rest after reduction by incineration **[r.13*(100-r.22)/100]** (t/year)	4 941.59
24	Total volume for landfilling **[r21+r.23]** (t/year)	**31 681.20**
25	Reduction of landfilling **[r.09/r.24*100]** (%)	**81.32**

Table 5. Environmental impact of the MSW incineration experiments

Incineration by-products average values

Rest in burning chamber	
Ash (%)	3.34%
Slag (%)	0.13%
Air polution	
Dust (mg/m³)	0.0128
O_2 (%)	4.87
CO (mg/m³)	5.732
NO_x (mg/m³)	200.22
NO (mg/m³)	193.072
NO_2 (mg/m³)	1416.242
SO_2 (mg/m³)	1.572

Chemical composition (g/kg)

	Ash	Slag
Ca	82.504	43.875
Mg	1.206	0.188
Na	4.255	1.915
K	52.226	8.361
Li	0.005	0.001
Fe	1.653	2.152
Cu	0.132	0.013
Pb	0.038	0.014
Zn	2.555	2.861
Mn	1.826	2.906
Ni	0.003	0.001
Cr	0.013	0.002
Co	0.012	0.004
Cd	0.001	0.000

The environmental impact of the thermal treatment of MSW is a very controversial topic. They can be observed pros in general: Lowering needed space for landfilling; Decreasing of groundwater contamination level; Energy generation; Carbon footprint lowering. Incineration cons are mainly: High investment to incineration plants; Toxic pollutants emissions; Opportunity costs.

The efficiently controlled waste-to-energy process reaches the reduction of greenhouse gas emission over mechanisms: i) by generating electrical power or steam, waste-to-energy avoids carbon dioxide (CO_2) emissions for fossil fuel-based, ii) the waste-to-energy incineration process effectively avoids methane emissions from landfills and iii) the more energy-efficient recovery of metals than production from raw materials. [15]

By the accounting of the above mechanisms is possible to evaluate the thermal treatment of MSW in a more objective form. Comparison of these mechanisms for the model of MSW divided into two basic groups – 30% of recycled and 70% for landfilling or energy recovery is presented in Figure 5. [16]

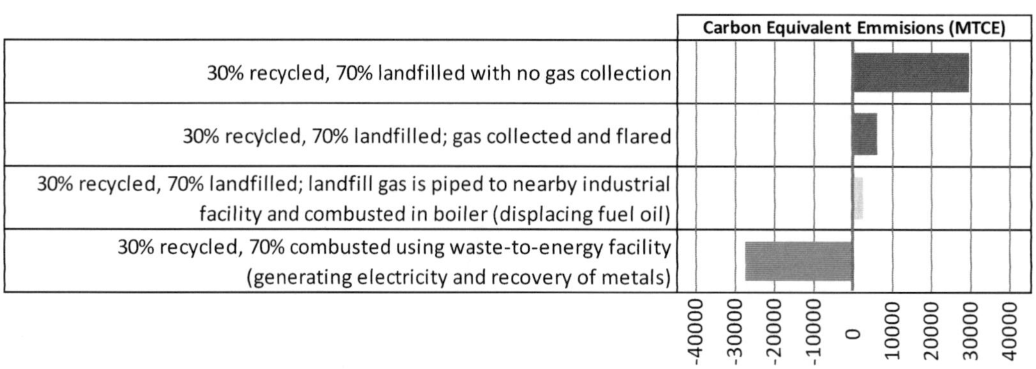

Figure 5. Quantifying the complete accounting of any MSW management options [16]

The methodology used for this evaluation is the USA Decision Support Tool for Materials and Waste Management. The methodology offers relatively good approximation to the real values of the greenhouse gas emission potential of Waste-to-Energy.

Also, the energy produced at waste power plants contributes to climate protection and security of energy supply by replacing the fossil fuels that would be used to produce this energy in conventional power plants. Efficient energy generators heat the domestic waste and thereby reduce methane emissions (a strong greenhouse gas that has 25 times greater climate impact than carbon dioxide) from landfill and CO_2 emissions that would be produced if the amount of energy produced in conventional power plants.

5. Electronic monitoring platform

All research activities were monitored from an information flow point of view. The monitoring resulted in information flow structure design.

The flow structure was optimized for control of data concerning the management of thermal treatment of MSW; for the study of the used methods; and for building application for monitoring all needed consequences. The main required features of the platform were [17], [18]:

For users:

- Easy intuitive use and a good level of reliability;
- Separated access to the application for groups of users;
- To store all information collected in the project;
- To process the collected information in the needed form;
- To present data information in the required form;
- To export stored and processed data in many possible formats and structures;

For administration:

- Flexible data storing and programming environment;
- Data redundancy eliminating and trans actionality;
- Powerful possibilities in connections among different data formats;
- Possibility of simple maintaining of data consistency;
- Simple management of user accounts;
- Ease backup and disaster recovery;
- Very good possibilities to do changes according to a new requirement from application users.

Logical scheme working out following the above requirements can be presented in structure:

- Partners
  - Basic data
    - Countries
    - Regions
    - Partners
    - Stakeholders
    - Reports / Studies
    - Legal documents
  - Research data
    - Waste generation in regions
    - Waste samples evaluation
      - Fractions composition
      - Chemical properties
      - Physical properties
      - Thermal treatment properties
      - Calorific properties
      - Emissions of thermal treatment
      - Ash chemical properties
      - Ash physical properties
      - Slag chemical properties
      - Slag physical properties
    - EnyMSW outputs
      - Studies
      - Project reports

- Stakeholders
    - EnyMSW reports

The application "Electronic Monitoring Platform" (EMP) was built as web-based on SQL technology as an integrated collection of logically related MSW treatment records. This philosophy allows consolidating of separate information into a common pool of data records. The last by not least property of the most important features of this philosophy is: stored data is independent of the application programs using it. This relational approach allows managing physical data storage without affecting access to that data as a logical structure. Operations on database enable EMP to manipulate the data and structures of the research collected information. Logical operations allow for a user to specify the required content; physical operations determine the manner of data access and task processing. [19], [20] Requirements for the platform working mode is off-line registering of data obtained by three geographically distant laboratories in Romania, Ukraine and Slovakia. The main purpose of the platform is to record and process:

1. Basic data
    1.1. Project partners
    1.2. Regions of evaluation
    1.3. Stakeholders involved
    1.4. Waste generation characteristics by regions (generated amount of the MSW per year, the combustible volume of it, energy potential, MSW mass reduction ratio, etc.)
2. Experimental data
    2.1. MSW samples fractions composition (overall moisture, fractions composition - biodegradable, metals, glass, etc.)
    2.2. MSW samples chemical and physical properties (presence and content of circa twenty chemical elements, pH, electroconductivity, etc.)
    2.3. MSW samples calorific properties (Gross calorific value, the content of chemical elements - C, H, N, S and O)
    2.4. MSW samples thermal treatment characteristics (gas consumption, Temperature, duration, the volume of ash and slag)
    2.5. MSW samples thermal treatment emissions (Concentration, dust, O_2, CO_2, NOx, NO, NO_2, SO_2)
    2.6. Thermally treated MSW samples ash physical and chemical properties (presence and content of circa twenty chemical elements, pH, electroconductivity, etc.)
    2.7. Thermally treated MSW samples slag physical and chemical properties (presence and content of circa twenty chemical elements, pH, electroconductivity, etc.)
3. EnyMSW reports
    3.1. Experiments reports
    3.2. EnyMSW studies
    3.3. MWS related legislation documents

All the data are recorded manually in the application because of very wide area of collected data obtained from equipment with hardware and software inconsistency from geographically remote regions. [21]
The application was tested by real data processing and debugged for operability. The programmer team mainly focuses on monitoring of robustness, resilience, deployment, and latency of the application. Robustness was tested by test persons – testers - oriented to use erroneous inputs for the application by more possible hardware and software platforms in the first step (resistance to bad inputs). Next step was testing to the possibility to reach stored data for unauthorized persons (hacker resistance). Platform resilience was tested to recovery ability after network disruption, electricity fails and server crash (by artificial shot down) and followed by an examination of data replication.
Deployment testing and documentation of the platform was performed after every required modification of the data structure or migration to a different server. The deployment procedure was done manually because of a relatively simple application. Latency testing was done by testers to obtain information

about the response of the application by using a combination of the most commonly used devices, operating systems and browsers. Testers focused on the most the vertical (independency concerning "machine") and horizontal (to maintain the flexibility of connected entities) scalability were taken into account as an important condition for the future development of the application, too. The Electronic Monitoring Platform is fully operated in the frame of the EnyMSW project activities.

6. Conclusions

The analysis of the energy potential of MSW of the Prešov Self-Governed Region and the building of the electronic monitoring platform are the subjects of this study. Municipal Solid Waste samples were collected in Prešov SGR, analysed and incinerated within the EnyMSW Laboratory in Baia Mare to find emissions of the thermal process and to determine the reduction the MSW by incineration was performed. Heat treatment the samples was performed in the Inciner I8-10S Incinerator, at high temperatures above 850°C, the recording of emissions during the experiment was done with Flue gas analyser testo 350, weighing of samples and solid residues with Kern ECE 10 K -3N scale. [14] The samples calorific characteristics evaluation was performed in EnyMSW laboratory in Prešov by IKA A10 mill and IKA Calorimeter C1. The most important of energy-related findings is net energy recovery potential of Prešov Self-Governed region in 365 747.14 MWh per year. The most important findings of landfilling related findings are reduction of landfilled MSW from current 169 560. 00 tons per year to 31 681.20 tons per year (26 739.61 tons of non-combustible MSW fractions plus 4 941.59 tons of incineration rests) produce ratio to landfilling reduction 18.68%. Characteristics of statistics of the amount of previously generated MSW and composition of it was used as a basement for forecasting of the future trends of global MSW management in Prešov SGR, shown in Figure 6.

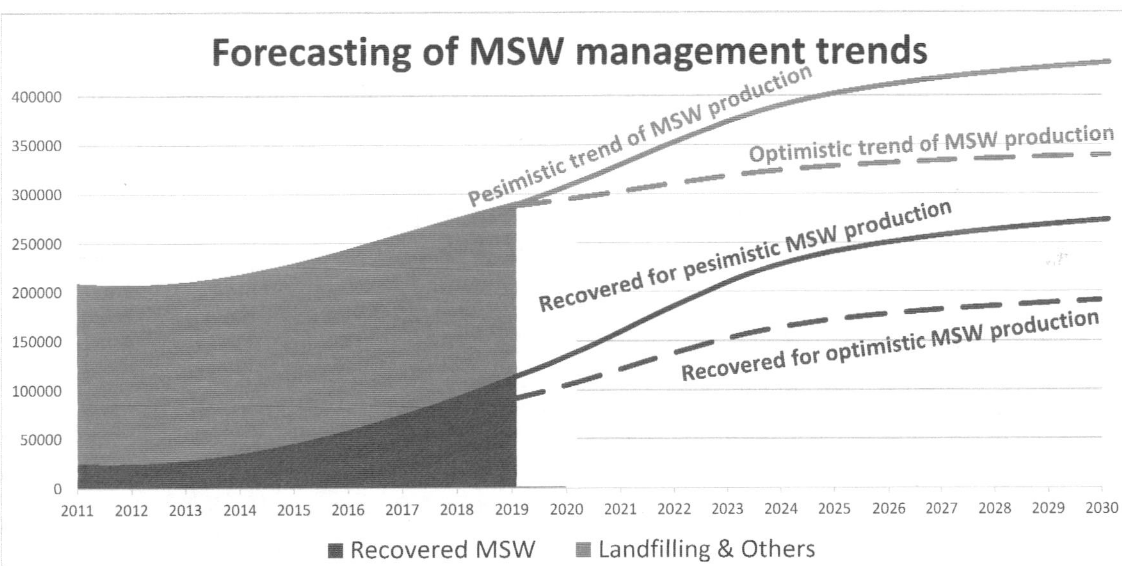

Figure 6. Forecasting of MSW management trends (t/year)

The prediction for the pessimistic trend of the MSW production is based on a hypothetical environmental Kuznets curve for the evaluated region. [22] The Prešov SGR is now under strong development and it is powerfully focusing on improving human resources and the business atmosphere while protecting the environment. The prediction of the optimistic trend of the MSW production is based on the European Union strategy of enviro-protection by legislative and educational tools.

Predictions of optimistic and pessimistic trends of MSW recovering is based on EU and Slovak goals for decreasing of the nature pollutions stated in legislative and previously forecasted trends for MW production. The legislative goal is to recovery 80% of packaging materials and 70% of MSW in 2035.

Acknowledgements
This contribution was prepared with the financial support of the European Union by Hungary-Slovakia-Romania-Ukraine ENI (European Neighbourhood Instrument) Cross-border Cooperation Programme 2014-2020 for the project entitled "Energy Recovery from Municipal Solid Waste by Thermal Conversion Technologies in Cross-border Region - EnyMSW", reference Number: HUSKROUA/1702/6.1/0015.

The thermal analysis to determine the reduction the MSW by incineration and to find emissions of the thermal process was performed in the EnyMSW Laboratory "Small scale system for thermal treatment of municipal solid waste". The laboratory was created at the University of Cluj-Napoca, North University Centre in Baia Mare, Romania under the support of the HUSKROUA/1702/6.1/0015 project.

The project donation serves for EnyMSW laboratory for calorimetric evaluation at the Technical University of Košice, Faculty of Manufacturing Technologies in Prešov. All presented calorimetric results were obtain in the laboratory.

References

[1] Removal and disposal of waste - Separation of waste. Available on https://www.olo.sk/separacia-odpadov/

[2] EU, "Directive 2010/75/EU of the European Parliament and of the Council of 24 November 2010 on industrial emissions (integrated pollution prevention and control)," EUR-Lex, Brussel, 2010.

[3] EC: "Integrated Pollution Prevention and Control, Reference Document on the Best Available Techniques for Waste Incineration," European Commission, Brussel, 2006.

[4] Puna Jaime F. and Teresa M.: Conversion Technologies for Solid Wastes: A New Way to Produce Sustainable Energy. DOI: 10.5772/8461

[5] Greener Slovakia - Strategy of the Environmental Policy of the Slovak Republic until 2030, Bratislava, 2019

[6] Making the Slovak Republic a more resource efficient economy, Country study, OECD environment policy paper No. 7, 2017, ISSN 2309-7841

[7] Hricová R. et al: Selected aspects of municipal solid waste recovery in the Prešov self-governing region (in Slovak language: Vybrané aspekty problematiky zhodnocovania tuhého komunálneho odpadu v Prešovskom samosprávnom kraji), RISE Association, Prague, 2020, ISBN 978-80-87670-25-5

[8] Hricova R. et al: Suggestions for Solving the Problem of Regional Inequality in The Context of Industry 4.0, In: Proceedings of 13th International Technology, Education and Development Conference (INTED), Valencia, 2019

[9] Dorcak D. et al: Total Logistic Plant Solutions, In: Journal-Technology Education Management Informatics, Vol. 5, Issue: 1, 2016

[10] Waste Catalogue - Decree of the Ministry of the Environment of the Slovak Republic no. 365/2015 Coll

[11] Elbersen W. et al: Lignocellulosic Biomass Quality, In: Modeling and Optimization of Biomass Supply Chains, 2017, ISBN 978-0-12-812303-4

[12] DATAcube database of the Slovak Statistical Office. Available on http://datacube.statistics.sk

[13] Hricova, R.et al: Acquisition and Processing of Information from Slovak Publicly Available Data, In: Proceedings of International Conference on Management of Manufacturing Systems, Stary Smokovec, 2017

[14] Ungureanu M. et al: Study on energy recovery from municipal solid waste by thermal conversion technologies in cross-border region Maramures - Ivano Frankivsk - Prešovský, RISOPRINT Cluj-Napoca, 2020, ISBN 978-973-53-2552-7

[15] Michaels T.: Waste not, want not: the facts Behind waste-to-energy, Energy recovery council, 2009
[16] Thorneloe S.A. et al: Application of the U.S. Decision Support Tool for Materials and Waste Management. WM Journal 2006 August.
[17] Monka P., Monkova K.: Design of the Codes Structure for Information System Working on I4.0 Principles, In: Proceedings of the 4th International Conference on the Industry 4.0 Model for Advanced Manufacturing, Industry 4.0 and Internet of Things for Manufacturing (AMP), University of Belgrade, 2019
[18] Monkova K., Monka P: Newly Developed Software Application for Multiple Access Process Planning In: Advances in Mechanical Engineering , 2014
[19] Monka P., Monkova K.: Basic mathematical principles for internal structure of new CAPP software application, In: annals of DAAAM for 2009 & proceedings of the 20th international DAAAM symposium, Volume: 20, 2009
[20] Monka P., Monkova K.: Multi Variant Process Plans Design in Relation to the European Market In: Annals of DAAAM for 2008 & Proceedings of the 19th International DAAAM Symposium: Intelligent Manufacturing & Automation: Focus on Next Generation of Intelligent Systems and Solutions, 2008
[21] Kinik, Daniel et al: On-line monitoring of technological process of material abrasive water jet cutting, In: Tehnicki Vjesnik-Technical Gazette, Volume: 22, Issue: 2, 2015
[22] Monkova K. et al: Simulation of the Chip Formation, In: Proceedings of the International Conference on Advanced Manufacture Technology and Industrial Application (AMTIA), Shanghai, 2016

Thermodynamic analysis and optimization of Transcritical and Supercritical Organic Rankine and Brayton Cycles coupled to parabolic trough collectors.

T Papingiotis[1], I Koronaki[1], N Komninos[1], G Antonakos[1]

[1] Laboratory of Applied Thermodynamics, Thermal Section, School of Mechanical Engineering, National Technical University of Athens

*Corresponding author's e-mail: koronaki@central.ntua.gr

Abstract. The main objective of this work is to investigate the efficiency of three different power cycles coupled with parabolic through collectors in Athens/Greece. The power cycles analyzed where: subcritical ORC with n-Pentane and R245fa as working fluids, trans-critical ORC with R245fa as working fluid and super-critical Brayton cycle with CO_2 and R245fa as working fluid. Mathematical models were developed for both the solar farm and the power cycles in order to optimize the necessary parameters for the optimal operation of the plant. The inlet temperature and pressure of the turbine are the optimization parameters. The overall efficiency of the system (power to solar) for the sub-critical ORC with n-Pentane and R245fa is 14.12% and 10.31% respectively. For the trans-critical ORC with R245fa the efficiency is 15.14% and for the super-critical Brayton cycle for CO_2 and R245fa, 13.23% and 8.87% respectively.

1. Introduction

Solar energy is expected to contribute significantly to the electricity production as more countries diversify their power generation systems. Concentrating solar power (CSP) is one technology used for electricity generation. With the integration of thermal energy storage, this technology has the potential to replace conventional power plants as they have the ability to separate the electrical power generation phase and the solar thermal production [1]. These systems are mainly adopted in large scale plants [2] and the most common cycle used is the Rankine cycle..

One of the most promising application of CSP is the integration of parabolic trough solar collectors with ORC plants for small scale power generation. The organic fluids are more suitable for use with low grade heat sources as the lead to higher efficiency. Several test facilities with PTC have been constructed [1],[3],[4].

This configuration, PTC and subcritical ORC, is largely studied in literature. Design optimizations under design conditions have been performed with different working fluids and different ORC configurations [5]–[8]. In addition, research have been conducted also with transcritical ORC and low heat sources [9],[10]. With the rise of the super-critical cycles, thermodynamic analysis of supercritical closed Brayton cycle integrated with PTC have been conducted with both CO2 and organic fluids as working fluids. [11]–[13]

Up to date, there have been studies undertaken on the efficiency of both Organic Rankine Cycle and supercritical cycles separately. This paper focuses on the optimization of these technologies efficiency related with CSP in Athens, Greece and the comparative analysis of the results. For this purpose, a mathematical model for the optimization of the solar farm with parabolic through collectors and various ORC and supercritical cycles is developed

2. System description

The present study considers a typical solar ORC plant and a simple recuperated super-critical plant. Both could be split into three main functional units which are the solar farm (SF), the thermal energy storage (TES) and the power block (PB). The general plant layout is presented in Figure 1.

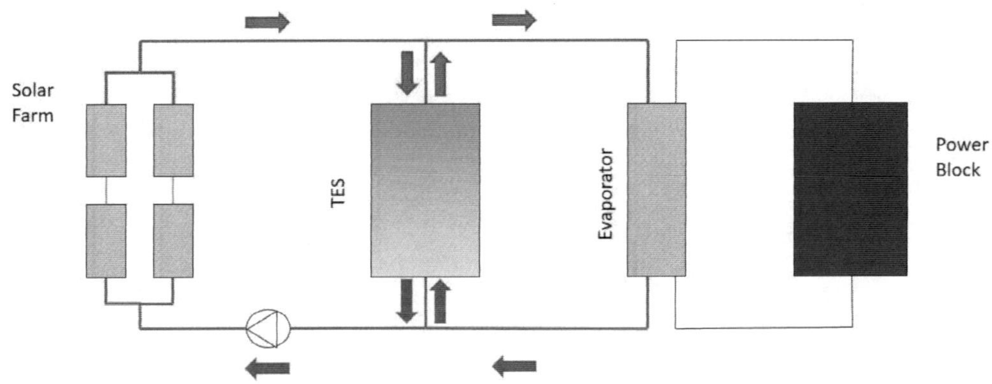

Figure 1 Power plant layout

For the solar field, parabolic trough collectors are used with evacuated tubes and Therminol 66 is used as heat transfer fluid. The central feed configuration, presented at Figure 2, will be adopted for this study as with this structure the piping required is minimized [14]. A thermocline energy storage system will be added to the plant to balance out the solar variations. Regarding the power block components, they depend on the cycle used (ORC or super-critical) and various working fluid would be studied.

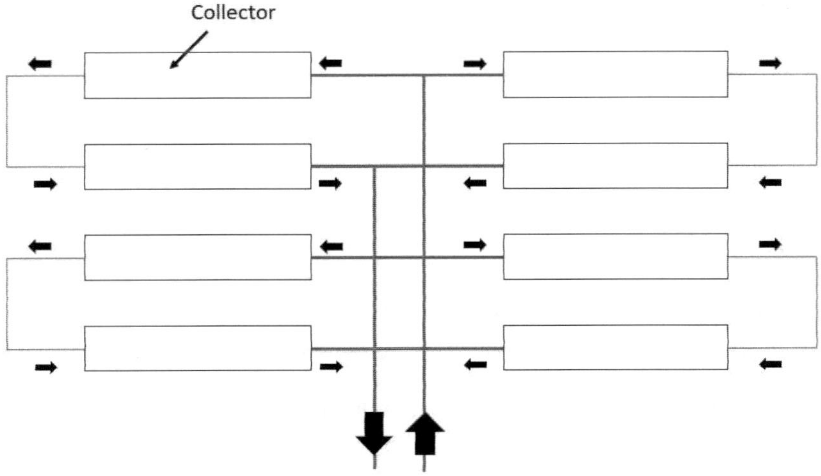

Figure 2 Central feed configuration

3. Mathematical Modelling

The modelling of the systems is discussed in this section. The mathematical model of the parabolic trough collector (PTC) has been implemented and is followed by the modelling of the ORC and super-

IOP Conf. Series: Materials Science and Engineering 1037 (2021) 012044 doi:10.1088/1757-899X/1037/1/012044

critical cycle. Energy analysis of the PTC system has been performed by using the method derived from previous research [15],[16]

The specification for the receiver [17] and the PTC module [18] are presented at Table 1 and Table 2.

Table 1 Receiver data

Receiver data		
Receiver type	Schott PTR 70	
Absorber tube material	Steel type: DIN 1.4541	
Surface roughness	eps_{suf}	0.002 mm
Absorber tube inside diameter	D_{ia}	0.066 m
Absorber tube outside diameter	D_{oa}	0.07 m
Absorber tube thermal conductivity	k_a	$15\frac{W}{m \times K}$
Absorber tube emissivity	ε_α	0.095
Absorber tube absorptance	a_a	0.96
Glass cover material	Borosilicate glass	
Glass cover inside diameter	D_{ic}	0.12 m
Glass cover outside diameter	D_{oc}	0.125 m
Glass cover thermal conductivity	k_c	$1.2\frac{W}{m \times K}$
Glass cover emissivity	ε_c	0.86
Glass cover absorptance	a_c	0.02
Glass cover transmittance	τ_c	0.97

Table 2 PTC module data

PTC module		
Collector	Ultimate Trough	
Collector aperture length	l_{ap}	7.51 m
Length of a single collector module	l_m	24.5 m
Number of modules per SCA	N	10
Length of SCA	L	245 m

3.1. Thermal analysis of the receiver

With the thermal analysis the thermal efficiency of the collector and the useful heat gained is calculated. The incoming solar irradiation is concentrated by the collector to the receiver tube. As a consequence, the temperature at the surface of the receiver is increased and heat is conducted through the wall of the absorber. The useful heat is absorbed by the heat transfer fluid (HTF), passing by the absorber pipe, by force convection and the rest is lost back to the annulus by a combination of natural convention and radiation. The lost heat is then conducted through the glass cover and from the outside surface of the cover to the ambient air by convention and to the reflector and sky by radiation.

The methodology considers a single differential area of a PT collector receiver. Energy balance is applied to the control volume. The useful heat is calculated at each incremental step, and based on that, the mean fluid temperature in determined. The mean temperature is then extrapolated to the solar field and the total heat gained is calculated. The following assumptions will be used for the analysis.

- The heat transfer from the retainers is considered marginal.
- The pipes transporting the HTF outside the PTC are believed to be completely insulated.

To assess the useful heat gained, the thermal resistance method is used.

IOP Conf. Series: Materials Science and Engineering 1037 (2021) 012044 doi:10.1088/1757-899X/1037/1/012044

The solar absorption in the glass cover is determined by:

$$\dot{Q}_{sol,c} = I_{b.n} n_{opt} a_c l_{ap} dx, \tag{1}$$

where the effective optical efficiency at the glass cover, n_{opt}, is:

$$n_{opt} = e_{sh} e_{ge} e_{tr} e_{dm} e_{da} e_{un} \rho_{cl} K_{\theta}, \tag{2}$$

The incidence angle modifier K_{θ} is:

$$K_{\theta} = \cos\theta_i + 0.000884\theta_i - 0.00005369\theta_i^2, \tag{3}$$

The parameters used to calculate the optical efficiency are presented in Table 3. The assumptions that these parameters are independent of temperature can be made. [17]

Table 3 Optical efficiency parameters

Optical efficiency		
Clean mirror reflectance	ρ_{cl}	0.94
Tracking error	e_{tr}	0.994
Geometry error	e_{ge}	1
Dirt on mirror	e_{dm}	0.97
Dirt on receiver	e_{da}	0.995
Receiver shadowing	e_{sh}	0.98
Unaccounted error	e_{un}	0.96

The solar energy accumulated on the outside surface of the receiver tube is calculated by:

$$\dot{Q}_{sol,a} = I_{b.n} n_a a_a l_{ap} dx, \tag{4}$$

Where the effective optical efficiency at the receiver tube, n_a is:

$$n_a = n_{opt}\tau_c, \tag{5}$$

The heat transfer through the absorber tube is determined by:

$$\dot{Q}_{cond,a} = \frac{T_{oa} - T_{ia}}{R_{cond,a}}, \tag{6}$$

Where:

$$R_{cond,a} = \frac{\ln(D_{oa}/D_{ia})}{2\pi k_a \Delta x}, \tag{7}$$

The heat is transferred to the HTF by internal forced convection. The flow is considered turbulent [17] and as a result the Reynolds number is high and the heat transfer can be expressed by the following equations.

$$Nu_{HTF} = \frac{\sqrt{f/8}\, Re_{HTF} Pr_{HTF}}{12.48 Pr_{HTF}^{2/3} - 7.853 Pr_{HTF}^{1/3} + 3.613\ln(Pr_{HTF}) + 5.8 + C}, \tag{8}$$

For $0.2 \le Pr \le 2000$
And $10^4 \le Re \le 5 * 10^6$
Where $C = 2.78\ln\left(\sqrt{f/8}\, Re_{HTF}/90\right)$ and f is the Fanning friction factor, which is defined for smooth pipes as $f = (0.79\ln Re_{HTF} - 1.64)^{-2}$.
The convective heat transfer coefficient on the inside of the absorber tube:

$$h_{HTF} = Nu_{HTF}\frac{k_{HTF}}{D_{ia}}, \tag{9}$$

The thermal convective resistance is:

$$R_{conv.HTF} = \frac{1}{\pi D_{ia} h_{HTF} \Delta x}, \tag{10}$$

The convective heat transfer inside the absorber tube:

$$\dot{Q}_{conv,HTF} = \frac{T_{ia} - T_{bulk}}{R_{conv.HTF}}, \tag{11}$$

Where $T_{bulk} = \frac{T_{m,i-1} + T_{m,i+1}}{2}$

All the HTF properties will be evaluated at the mean film temperature: $T_{flm} = \frac{T_{bulk} - T_{ia}}{2}$

For the heat transfer at the annulus region by thermal radiation:

$$\dot{Q}_{rad,a} = \frac{T_{oa} - T_{ic}}{R_{rad,a}}, \tag{12}$$

Where:

$$R_{rad,a} = \frac{1}{\pi D_{oa} h_{rad,a} \Delta x}, \tag{13}$$

With $h_{rad,a}$ simplified for infinitively long concentric circles as

$$h_{rad,a} = \frac{\sigma (T_{oa} - T_{\iota a})({T_{oa}}^2 + {T_{ic}}^2)}{\frac{1}{\varepsilon_a} + \frac{1-\varepsilon_c}{\varepsilon_c}\left(\frac{D_{oa}}{D_{ic}}\right)}, \tag{14}$$

The heat transfer from the cover to the surroundings occurs with both thermal radiation to the sky and convection, both natural and forced, heat loss to the ambient sky due to varying wind loads. The convection is calculated as shown. The heat transfer coefficient for an isothermal horizontal cylinder in an isothermal environment can be obtained by the following equations.

$$Nu^T = 0.772 Ra_{air}^{1/4}, \tag{15}$$

$$f = 1 - \frac{0.13}{{Nu^T}^{0.16}}, \tag{16}$$

$$Nu_l = \frac{2f}{\ln(1 + 2f/Nu^T)}, \tag{17}$$

$$Nu_t = 0.103 Ra_{air}^{1/4}, \tag{18}$$

$$Nu_N = \left[(Nu_l)^{10} + ({Nu_l}^{10})\right]^{1/10}, \tag{19}$$

$$Nu_F = \alpha Re^n, \tag{20}$$

Using the equality $Nu_N = Nu_F$ and the constants α and n, the value of Re_i is calculated

$$Re_i = (Nu_N/\alpha)^{1/n}, \tag{21}$$

$$Re_{eff} = [(Re_i + Re_{air}\cos\varphi)^2 + (Re_{air}\sin\varphi)^2], \tag{22}$$

$$Nu_{air} = \alpha Re_{eff}^n, \tag{23}$$

The heat transfer coefficient for the mixed convective regime is:

$$h_{air} = Nu_{air}\frac{k_{air}}{D_{oc}}, \tag{24}$$

The thermal convective resistance is:

$$R_{conv,air} = \frac{1}{\pi D_{oc} h_{air} \Delta x}, \tag{25}$$

And:

$$\dot{Q}_{cond,air} = \frac{T_{oc} - T_a}{R_{conv,air}}, \tag{26}$$

For the calculation of the radiation heat transfer the following assumptions were made:
The bottom half of the cover emits radiation to the reflector, which is at local air temperature
The top half of the cover emits radiation to the sky
The heat transfer rates for the cover are:

$$\dot{Q}_{rad,sky} = \frac{T_{oc} - T_s}{R_{rad,sky}}, \tag{27}$$

$$\dot{Q}_{rad,ref} = \frac{T_{oc} - T_a}{R_{rad,a}}, \tag{28}$$

Where:

$$R_{rad,sky} = \frac{1}{\pi (D_{oc}/2) h_{rad,sky} \Delta x}, \tag{29}$$

$$R_{rad,ref} = \frac{1}{\pi (D_{oc}/2) h_{rad,ref} \Delta x}, \tag{30}$$

With $h_{rad,sky}$ and $h_{rad,ref}$ calculated as:

$$h_{rad,sky} = \sigma \varepsilon_c (T_{oc} + T_{sky})({T_{oc}}^2 + {T_{sky}}^2), \tag{31}$$

$$h_{rad,ref} = \sigma \varepsilon_c (T_{oc} + T_a)({T_{oc}}^2 + {T_a}^2), \tag{32}$$

The heat conduction through the glass cover wall:

$$\dot{Q}_{cond,c} = \frac{T_{ic} - T_{oc}}{R_{cond,c}}, \tag{33}$$

Where:

$$R_{cond,c} = \frac{\ln(D_{oc}/D_{ic})}{2\pi k_c \Delta x}, \tag{34}$$

3.2. Thermal analysis of the receiver

The thermal efficiency of the PT field is the ratio of the heat transferred to the HTF to the total solar radiation into the PT receiver and is expressed as:

$$n_{th} = \frac{\dot{Q}_{gain}}{\dot{Q}_{in} n_{opt}}, \tag{35}$$

Where n_{opt} is the effective optical efficiency from equation 2.

$$\dot{Q}_{in} = I_{b.n} A_a, \tag{36}$$

The overall efficiency of the solar field is:

IOP Conf. Series: Materials Science and Engineering 1037 (2021) 012044 doi:10.1088/1757-899X/1037/1/012044

$$n_{SF} = \frac{\dot{Q}_{gain}}{\dot{Q}_{in}}, \tag{37}$$

3.3. PT Collector Efficiency

In order to determine the power requirements for the pumps at the solar field, the pressure drop in the solar field's receiver tube is determined.

$$\Delta P = f\frac{L}{D}\frac{\rho {V_{avg}}^2}{2}, \tag{38}$$

Where f is the Darcy friction factor which is calculated by:

$$\frac{1}{\sqrt{f}} \cong -1.8\log\left[\frac{6.9}{Re} + \left(\frac{\varepsilon/D}{3.7}\right)^{1.11}\right], \tag{39}$$

And the power requirements for the pump is:

$$\dot{W}_{SF,pump} = \frac{\dot{V}\Delta P_L}{n_{SF,pump}}, \tag{40}$$

3.4. Thermal storage

A thermal energy storage system is used in order to adapt the available energy supply to the energy demand as they often not used at the same time. Moreover, it increases the annual hours of operations and the amount of electricity produced [19]. As mentioned in the system description, a thermocline energy storage unit will be incorporated in the system to balance out the solar variations.

Since the focus of this study is the thermodynamic optimization of concentrated parabolic solar collectors integrated with various power cycles under design conditions, the thermal storage systems is assumed to work at steady state conditions and thus the optimal size is not calculated.

3.5. Modeling the Cycle

3.5.1. Subcritical Organic Rankine Cycle. The Organic Rankine Cycle consists of a pump, an evaporator, a turbine, a recuperator and a condenser. The pump delivers the organic fluid to the evaporator where it is evaporated at approximately constant pressure by the heat supplied by the solar farm. The saturated working fluid is expanding in the turbine, which is driving an electric generator. The working fluid leaving the turbine passes through an internal heat exchanger in order to recover the thermal energy at the turbine outlet and preheat the compressed liquid before the entrance to the evaporator, and then is condensed in the condenser. The fluid leaves the condenser as saturated and then the pressure is increased by the feed pump, completing the power cycle. Figure 3 shows the corresponding cycle in the T-s diagram for a typical organic fluid with saturated conditions at the turbine inlet and Figure 4 show a schematic diagram of the ORC.

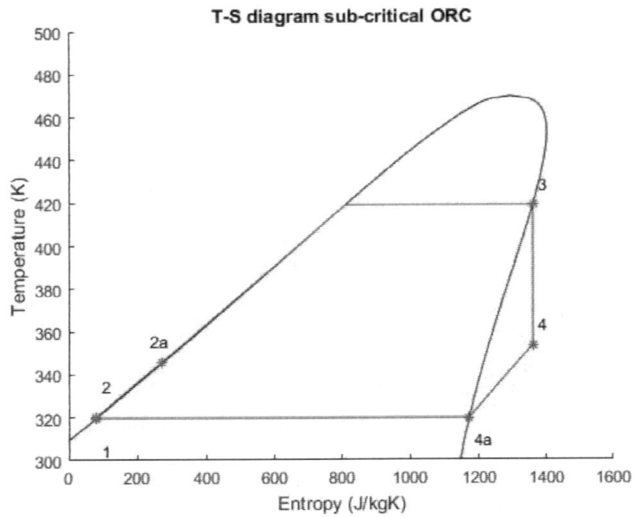

Figure 3 T-S diagram for sub-critical cycle

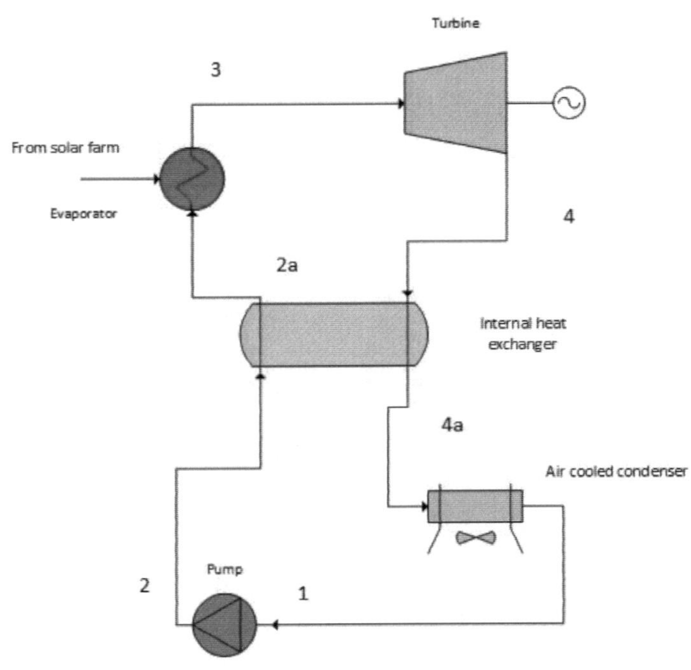

Figure 4 Schematic diagram of the ORC

To simplify the model in this study, the ORC systems are assumed to be in steady state and the heat exchangers are counter flow. A multistage centrifugal pump is chosen by this study and the efficiency is calculated as shown in Macchi & Astolfi [20].

The turbine inlet temperature and turbine inlet pressure are optimized.

The heat and mass balances across the power block are used and the procedure followed is: The mass flow rate of the organic fluid in the power block is given by:

$$\dot{m}_{PB} = \frac{W_{gross}}{(h_3 - h_4)n_T n_{gen}}, \tag{41}$$

The heat transferred to the evaporator is given by:

$$Q_{PB,e} = \dot{m}_{PB}(h_3 - h_{2a}), \tag{42}$$

In this analysis an air cooled condenser (ACC) is used to reduce the maintenance cost. As a result the condensing temperature depends on the air temperature. A typical value for the temperature difference is used. A typical temperature difference is 27 °C [17].

The heat rejected by the condenser is calculated by:

$$Q_{PB,c} = \dot{m}_{PB}(h_{4a} - h_3), \tag{43}$$

In order to determine the pumping power of the pump at the power block, the pressure drops at the turbine, condenser, recuperator and evaporator have to be calculated.

The pressure drop over the turbine is evaluated from the known thermodynamic states by:

$$\Delta P_T = P_3 - P_4, \tag{44}$$

The pressure drops in the evaporator, condenser and recuperator are minor in comparison to the pressure drop over the turbine and according to Macchi & Astolfi [20] it is acceptable to assume typical pressure drops for the above parts of the power block.

The work required for the power block pump is:

$$W_{PB,pump} = \frac{\dot{V}(\Delta P_T + \Delta P_{evap} + \Delta P_{cond} + \Delta P_{rec})}{n_{PB,pump}}, \tag{45}$$

To simplify the process a general relation between the work required by the condenser's fan and the turbine power output is adopted.

$$W_{cond,fan} = 0.02 W_{gross}, \tag{46}$$

The net power output of the plant is then calculated as:

$$W_{net} = W_{gross} - W_{PB,pump} - W_{cond,fan} - W_{SF,pump}, \tag{47}$$

3.5.2. Transcritical Organic Rankine Cycle. For the transcritical ORC, in contrast to subcritical ORC, the evaporation process does not occur and the organic fluid temperature increases continuously during the vapor generation process. Figure 5 shows the corresponding cycle in the T-s diagram.

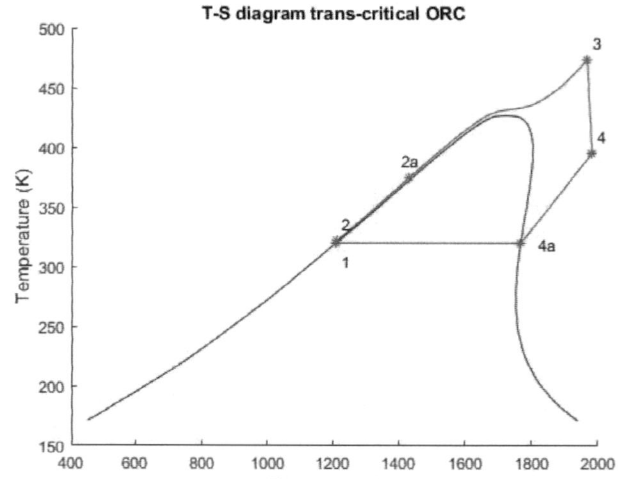

Figure 5 T-S diagram trans-critical ORC

The working fluid is calculated by:

$$\dot{m}_{PB} = \frac{W_{gross}}{(h_3 - h_4) n_T n_{gen}}, \tag{48}$$

The power consumed by the working fluid feed pump is:

$$W_{PB,pump} = \frac{m_{PB}(h_{2s} - h_1)}{n_{PB,pump}}, \tag{49}$$

Where h_{2s} is the organic fluid enthalpy at the pump outlet for an isentropic expansion process.
The power required for the cooling process is calculated from the equation 46.
The net power output of the plant is then calculates as:

$$W_{net} = W_{gross} - W_{PB,pump} - W_{cond,fan} - W_{SF,pump}, \tag{50}$$

3.5.3. Supercritical Cycle. For this analysis a simple recuperated supercritical cycle is considers. This cycle consists of a compressor, recuperator, primary heat exchanger, turbine, and precooler. The working fluid enters the compressor (state 1) at supercritical state in order to increase the pressure of the fluid. Then, the fluid flows out of the compressor (state 2) and into the recuperator where the energy comes from the outlet of the turbine. The fluid subsequently enters the primary heat exchanger (state 3) where the heat from the solar farm is transferred to the working fluid. The fluid leaves the heat exchanger (state 4) and expands through the turbine. The work generated by the turbine is used to produce electricity and spin the compressor that compresses the working fluid. The fluid leaving the turbine passes through the recuperator (state 5) and after that is further cooled to just above the vapor dome. [21]

Figure 6 shows the corresponding cycle in the T-s diagram for a typical organic fluid with saturated conditions at the turbine inlet and Figure 7 shows a schematic diagram of the plant.

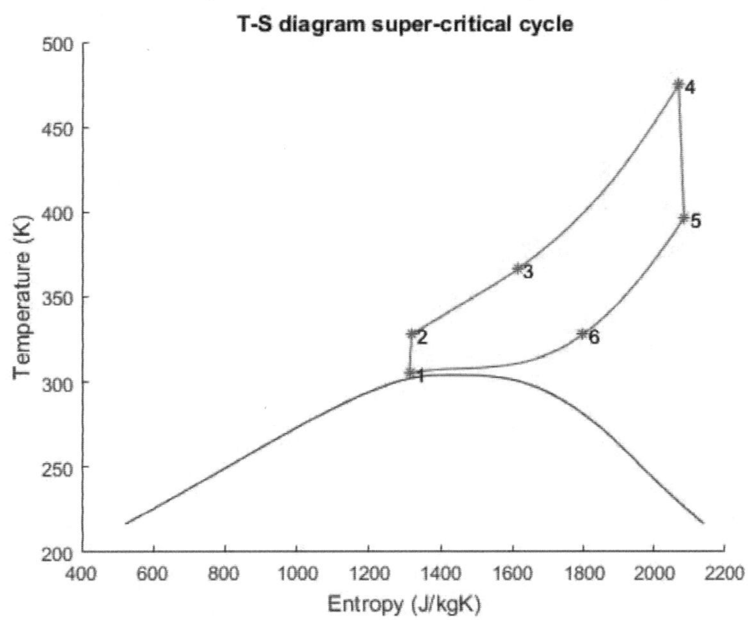

Figure 6 T-S diagram of super-critical cycle

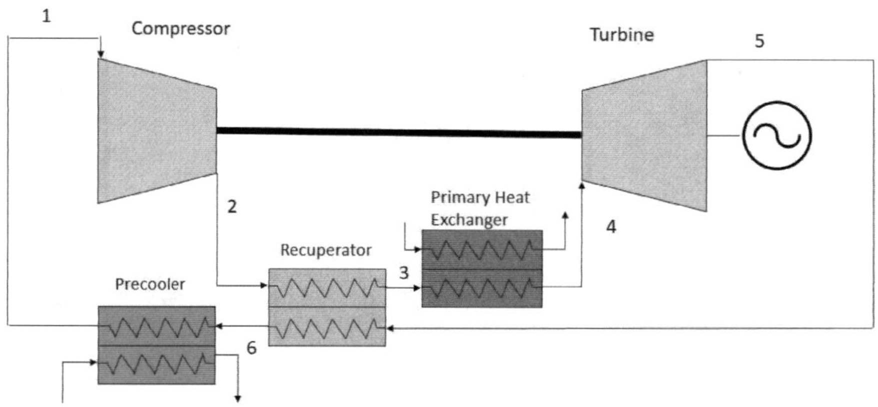

Figure 7 Super-critical plant schematic

For this study, compressor and turbine efficiency are assumed to be 0.9the effectiveness of the heat exchangers is assumed to be 0.9 and the pressure drop through the heat exchangers is negligible.

For state 1, it is assumed that the fluid is at 8 MPa and 32 °C, just above the critical point of the fluid.

The energy balance equations applied to each component of the cycle are shown below. For the compressor:

$$\dot{W}_c = \frac{\dot{m}(h_{2s} - h_1)}{n_c}, \tag{51}$$

For the recuperator:

$$\dot{Q}_{rec} = \dot{m}_{PB}(h_3 - h_2) = \dot{m}(h_5 - h_6), \tag{52}$$

For the turbine:

$$\dot{W}_t = \frac{(h_4 - h_{5s})}{n_t}, \tag{53}$$

Finally the net power output of the power plant is:

$$W_{net} = W_{gross} - W_c - W_{cond,fan} - W_{SF,pump}, \tag{54}$$

4. Optimization

A thermodynamic optimization is performed on the solar farm and the power block where the total plant efficiency is used as objective function. The total plant efficiency is set to the ratio of the net power output and the maximum received thermal heat.

$$n_{plant} = \frac{W_{net}}{Q_{in}}, \tag{55}$$

Where

$$Q_{in} = I_{ba} * A_{sf}, \tag{56}$$

I_{ba} is the solar irradiation at the site of the solar farm.

For the subcritical organic rankine cycle the optimizing variable is the evaporating temperature. The constraints regarding the power cycle depend on the organic fluid used and its properties. The upper limit depends on the working fluid critical temperature.

A design choice is made to set the condensing temperature as a constant value which depends on the air temperature, due to the fact that an air-cooled condenser is used.

For the transcritical cycle, the turbine inlet temperature and the turbine inlet pressure are optimized. The turbine inlet pressure lower limit is set above the critical point at 4 MPa and the upper limit is set as 8 MPa because this value is safe, simple and cost effective to achieve.

For the simple recuperated supercritical cycle, the turbine inlet pressure and temperature are optimized. A design choice was made to set the turbine outlet pressure at 8 MPa and the temperature at the state point 1 at 32 °C.

More detailed information about the constraints and constants are listed in Table 4, Table 5 and

Table 6.

Table 4 Sub-critical cycle

	Parameter	Working Fluid N-Pentane	R245fa
Subcritical ORC	Condensing Temperature	$T_{air} + 27$ °C $= 46.6$°C	$T_{air} + 27$ °C $= 46.6$°C
	Minimum evaporation temperature	$T_{cond} + 2$°C	$T_{cond} + 2$°C
	Maximum Evaporation temperature	197 °C	154 °C
	Evaporator pressure drop	100 kPa	100 kPa
	Condenser working fluid side pressure drop	1 kPa	1 kPa
	Recuperator pressure drop	50 kPa	50 kPa
	Pump efficiency	0.75	0.75
	Generator efficiency	0.96	0.96
	Turbine efficiency	variable	variable

Table 5 Trans-critical cycle

	Parameter	
Transcritical ORC	Working Fluid	R245fa
	Condensing Temperature	$T_{air} + 27$ °C $= 46.6$°C
	Minimum turbine inlet temperature	$T_{crit} + 3$°C $= 157$ °C
	Maximum turbine inlet temperature	250°C
	Minimum turbine inlet pressure	4 MPa
	Maximum turbine inlet pressure	8 MPA
	Pump efficiency	0.75
	Generator efficiency	0.96
	Turbine efficiency	variable

Table 6 Super-critical cycle

Cycle type	Parameter	Working Fluid	
		CO2	R245fa
Supercritical	Compressor inlet temperature (state 1)	32 °C	155 °C
	Compressor inlet pressure (state 1)	8 MPa	4 MPa
	Turbine inlet maximum temperature	250 °C	310 °C
	Turbine inlet minimum pressure	9 MPa	4 MPa
	Turbine inlet maximum pressure	20 MPa	20 MPa
	Compressor efficiency	0.9	0.9
	Generator efficiency	0.96	0.96
	Turbine efficiency	0.9	0.9

The data for the meteorological conditions of Athens where taken from Technical Engineering Handbook of the Engineering Chamber of Greece.

Table 7 Meteorological Data for Athens

Parameter	Value/Description
Site	Athens
Latitude φ	38 °
Solar Irradiation I_{ba}	800 $\frac{W}{m^2}$
Air temperature T_{air}	19.6 °C
Sky temperature T_{sky}	0 °C
Wind speed	3.65 m/s

For the 1 MW gross power output power block, the optimization is split in two parts. Firstly, by differentiating the solar field inlet temperature and heat transfer fluid mass flow rate the operating conditions of a parabolic through collector are defined. The developed algorithm calculates the solar field efficiency, the solar field outlet temperature and the heat gained for every single combination of solar field inlet temperature and mass flow rate. The values are stored and used at the second stage of the optimization process.

Depending on the cycle used at the power block, for each iteration of evaporating temperature and pressure, there are a range of condition of the solar farm (pressure and temperature) that can offer the necessary amount of heat to the power block. Among these conditions the most efficient is chosen for the solar farm.

All model equations are solved in sequence using MatLab as simulating software and the values of the thermodynamics properties of the working fluids are calculated from CoolProp [22] of MatLab. Thus, the most efficient plant design is determined.

5. Results

The operating conditions of a parabolic trough (PT) collector were defined, as part of the first stage of the optimization, by varying the inlet temperature by increments of one degree and the mass flow of the heat transfer fluid by increments of 0.5 kg/s. The impact of the mass flow with 3 different input temperatures of the collector to the solar field efficiency, PT outlet temperature, heat gained and pressure drop are presented in Figure 8, Figure 9, Figure 10 and Figure 11.

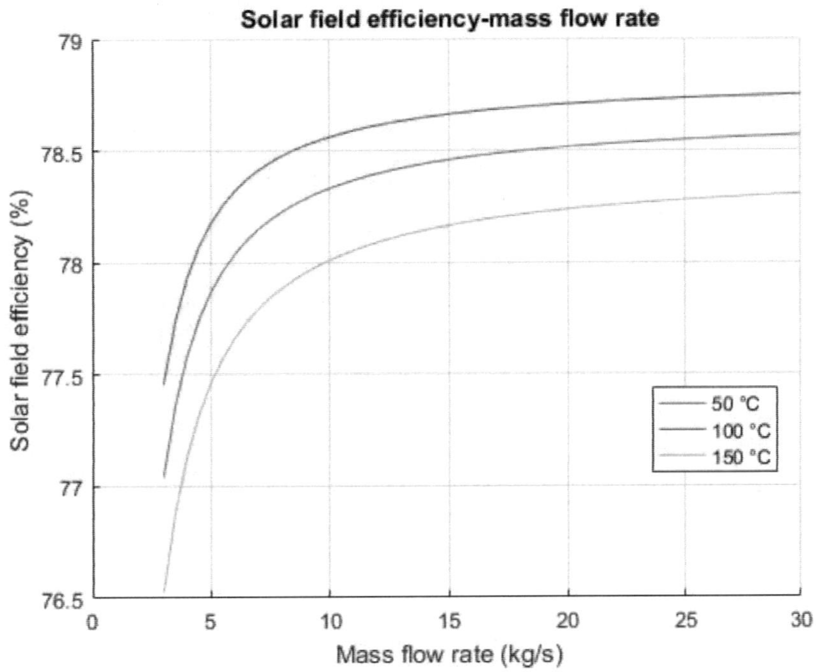

Figure 8 Solar field efficiency-mass flow rate

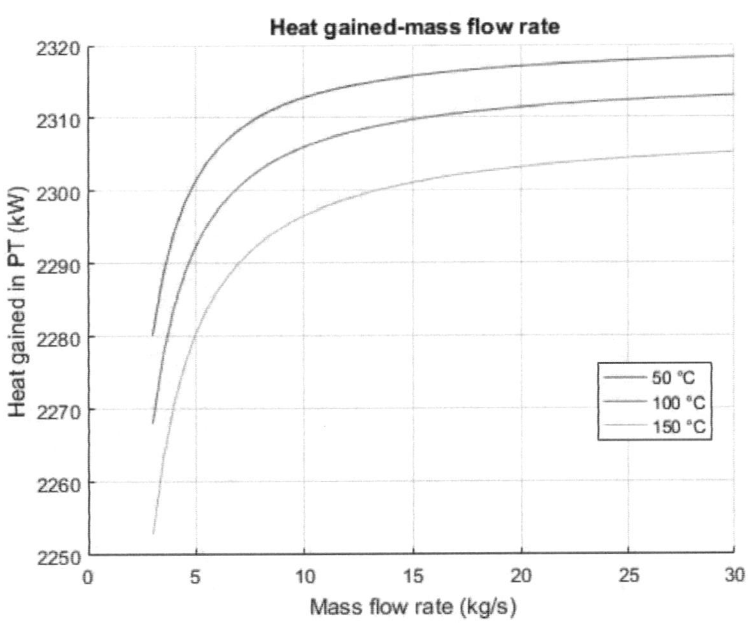

Figure 9 Heat gained-mass flow rate

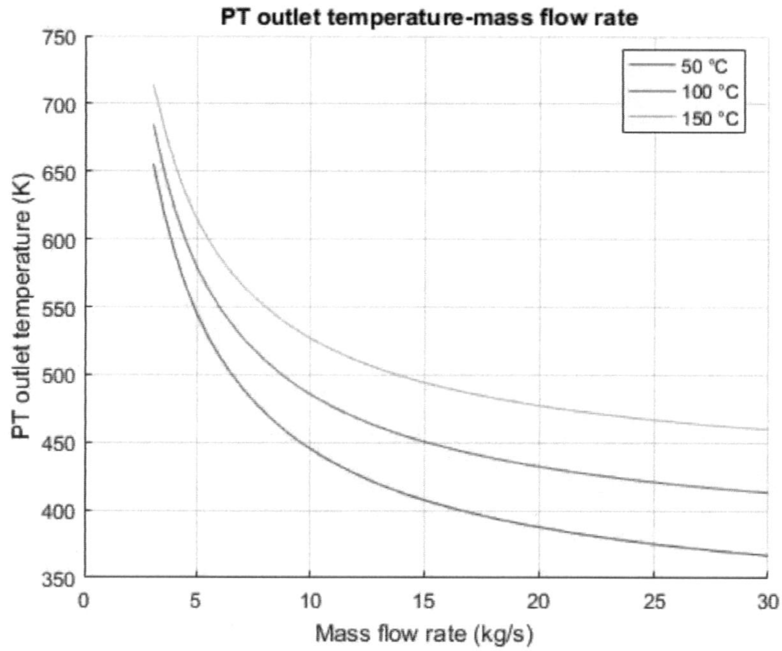

Figure 10 PT outlet temperature-mass flow

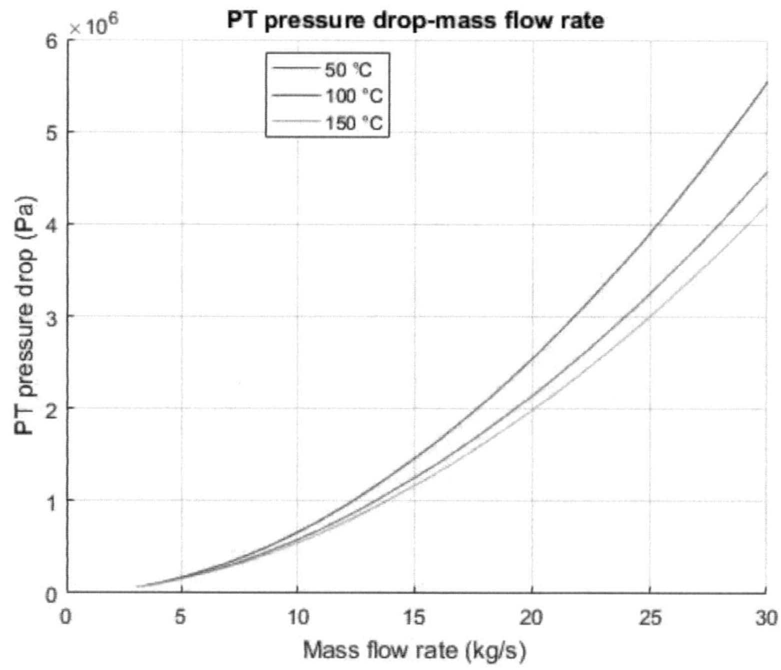

Figure 11 PT pressure drop-mass flow rate

From figures it could be deduced that greater mass flow is required for increased solar field performance as the efficiency and the heat gained rises with the increase in mass flow rate. At the beginning, the rise in both efficiency and heat gained is steep but flattens out promptly. In figure, it could be concluded that the outlet temperature of a PT collector decreases as the mass flow rate increases. At low mass flow rates the outlet temperature is above the constraint of max temperature

IOP Conf. Series: Materials Science and Engineering 1037 (2021) 012044 doi:10.1088/1757-899X/1037/1/012044

allowed for the heat transfer fluid. As a result, the minimum mass flow allowed is monitored during the optimization process. At figure, as the mass flow increases the pressure drop increases too. As a result, the pumping power required to overcome this pressure drop, increases.

In all results, it can be observed that the lowest inlet temperature presents better results regarding the efficiency of the collector and the pressure drop.

From the above observations, higher mass flow rates improves the thermal performance of the solar field although the net electricity production is lower as a result of the higher consumption of the pump used. Taking into considerations the conclusion analyzed before, in order to establish the best overall system performance, the mass flow rate is selected by weighing the overall thermal efficiency and the power consumed by the pump.

After determining the solar field operating conditions, the thermodynamic optimization is performed for each of the power cycle where the total system efficiency is used as the objective function.

For the subcritical ORC, the design point, determined through the optimization, with n-Pentane as the working fluid is achieved at a turbine inlet temperature of 458 K. The mass flow at the power block is 10.86 kg/s. For the solar field, the total mass flow is 15.09 kg/s and the inlet and outlet temperature are 353 K and 512.5 K respectively. For the above conditions, the net power output of the power plant is 897.05 kWe and the overall efficiency of the system is 14.12 %. The effect of the turbine inlet temperature on the efficiency of the plant is presented on the Figure 13.

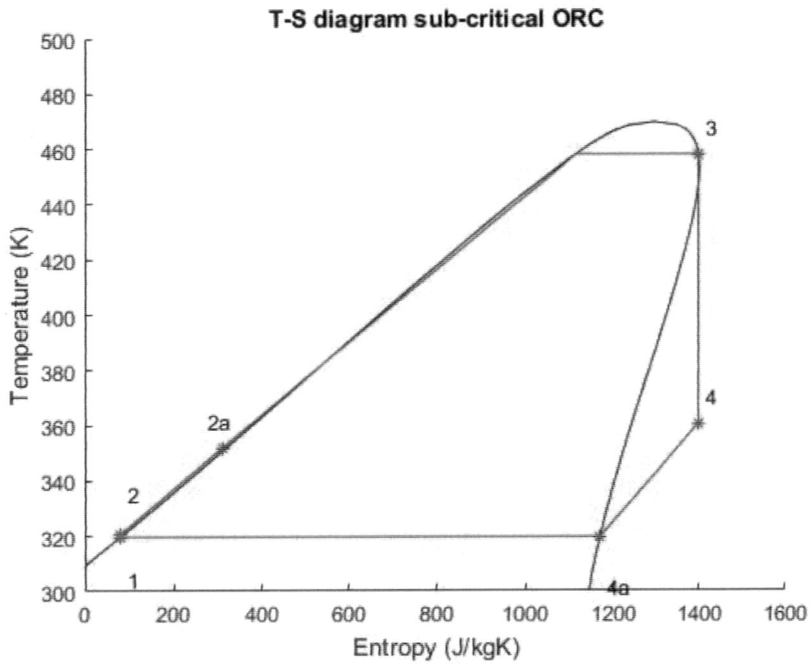

Figure 12 Design Point T-S diagram (n-Pentane)

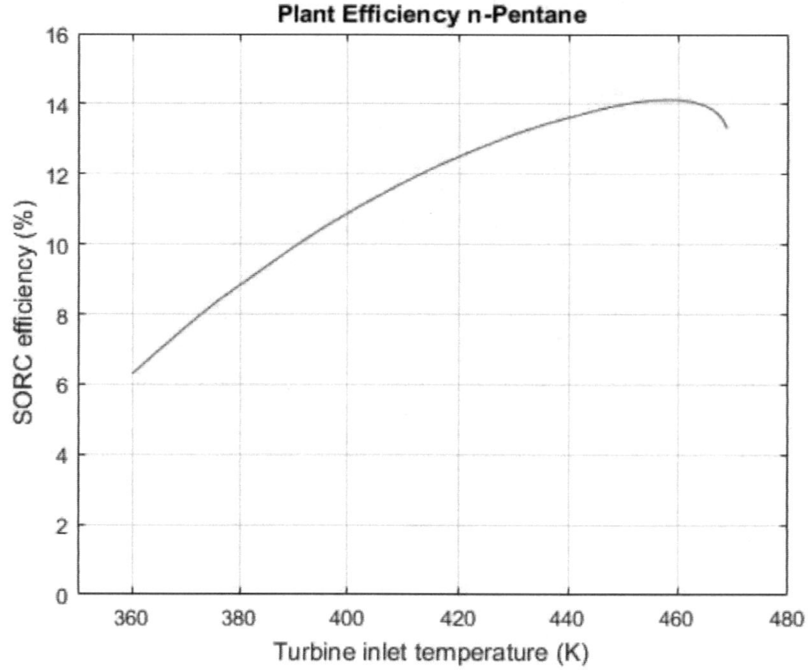

Figure 13 Effect of turbine inlet temperature on plant efficiency

For the subcritical ORC the design point with R245fa as working fluid, is achieved at a turbine inlet temperature of 415 K. The mass flow at the power block is 30.57 kg/s. For the solar field, the total mass flow is 19.95 kg/s and the inlet and outlet temperature are 328 K and 493.8 K respectively. For the above conditions, the net power output of the power plant is 865.54 kWe and the overall efficiency of the system is 10.31 %. The effect of the turbine inlet temperature on the efficiency of the plant is presented in Figure 15.

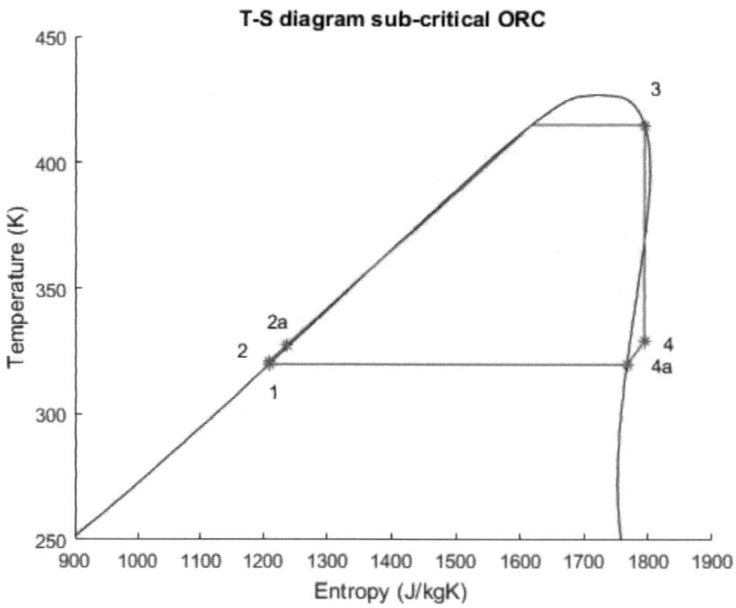

Figure 14 Design Point T-S diagram (R245fa)

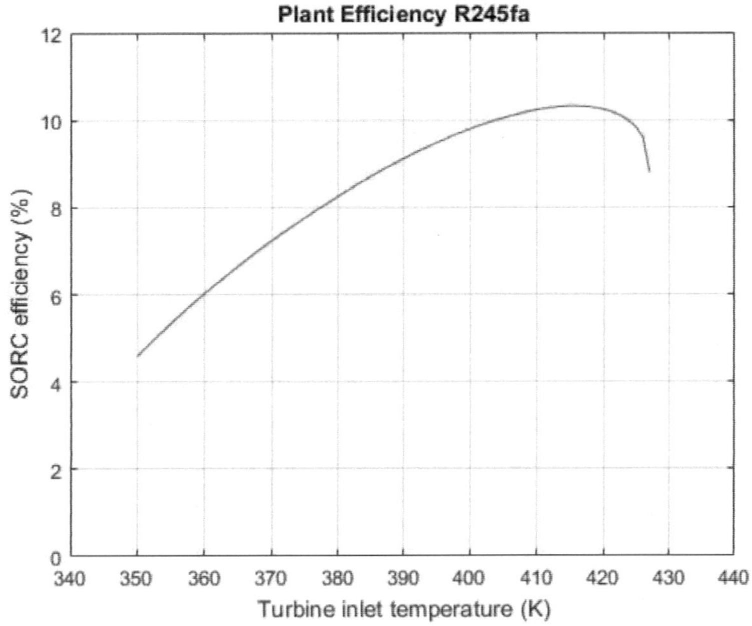

Figure 15 Effect of turbine inlet temperature on plant efficiency (R245fa)

From the above results, the conclusion is that the n-Pentane is the best working fluid choice for the subcritical ORC. This is because n-Pentane has higher critical temperature resulting in a higher temperature difference at the turbine inlet. In addition, from Figure 13 and Figure 15, the conclusion drawn is that at the highest operating temperature of each fluid, the efficiency decreases as the negative effects of the high temperatures of the solar field outperform the positives to the power block.

For the transcritical ORC, with R245fa as the working fluid, the design point is achieved at a turbine inlet temperature of 489 K and pressure of 5 MPa. The mass flow at the power block is 19.91 kg/s. For the solar field, the total mass flow is 13.97 kg/s and the inlet and outlet temperature are 324 K and 490.9 K respectively. For the above conditions, the net power output of the power plant is 889.76 kWe and the overall efficiency of the system is 15.14 %. The effect of the turbine inlet temperature, depending on the turbine pressure, on the efficiency of the plant is presented on the Figure 17.

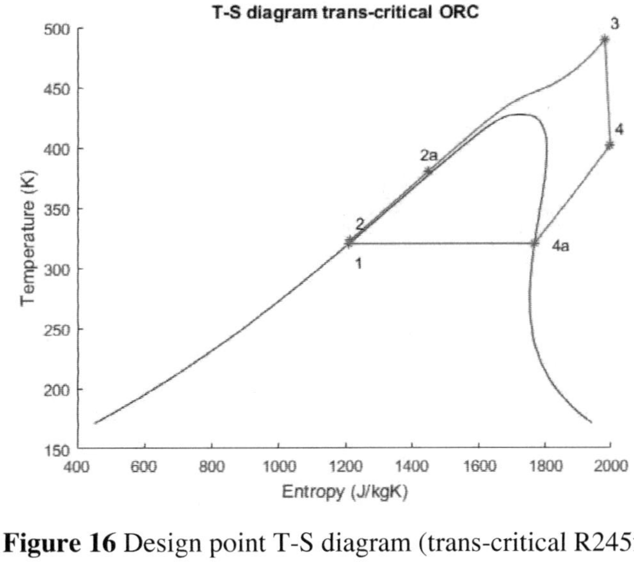

Figure 16 Design point T-S diagram (trans-critical R245fa)

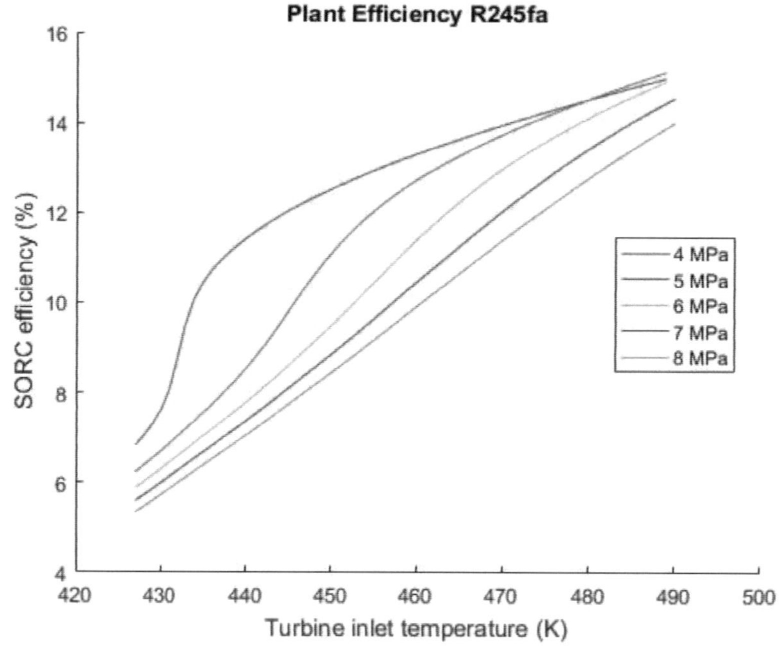

Figure 17 Effect of turbine inlet temperature on plant efficiency (trans-critical R245fa)

For the supercritical cycle with CO2 as working fluid the design point is achieved at a turbine inlet temperature of 493.4 K and pressure of 20 MPa. The mass flow at the power block is 19.11 kg/s. For the solar field, the total mass flow is 11.36 kg/s and the inlet and outlet temperature are 330 K and 495.3 K respectively. For the above conditions, the net power output of the power plant is 631.97 kWe and the overall efficiency of the system is 13.23 %. The effect of the turbine inlet temperature, depending on the turbine pressure, on the efficiency of the plant is presented in Figure 19.

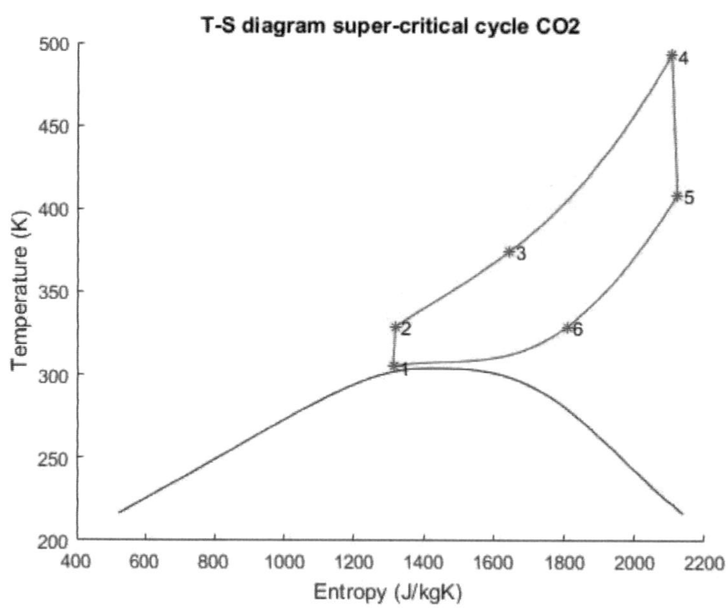

Figure 18 Design point T-S diagram (super-critical CO2)

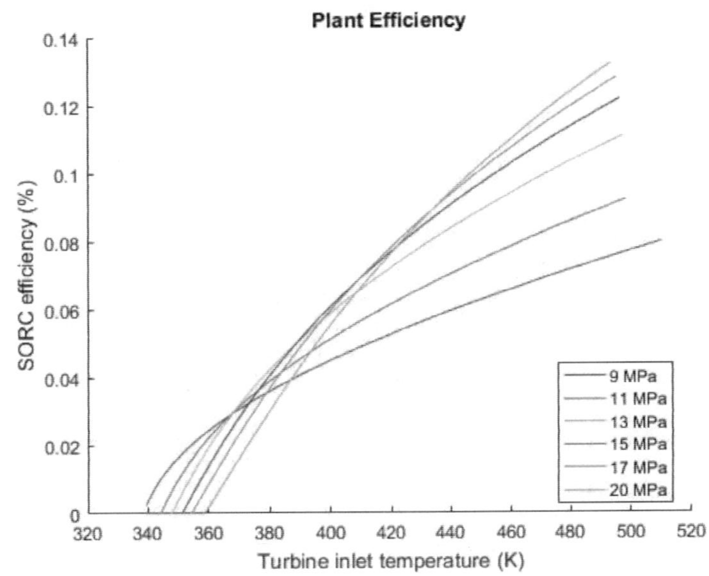

Figure 19 Effect of turbine inlet temperature on plant efficiency (super-critical CO2)

For the supercritical cycle with R245fa as working fluid the design point is achieved at a turbine inlet temperature of 559.36 K and pressure of 10 MPa. The mass flow at the power block is 54.9 kg/s. For the solar field, the total mass flow is 18.02 kg/s and the inlet and outlet temperature are 444 K and 560.5 K respectively. For the above conditions, the net power output of the power plant is 542.06 kWe and the overall efficiency of the system is 8.87 %. The effect of the turbine inlet temperature, depending on the turbine pressure, on the efficiency of the plant is presented on Figure 21.

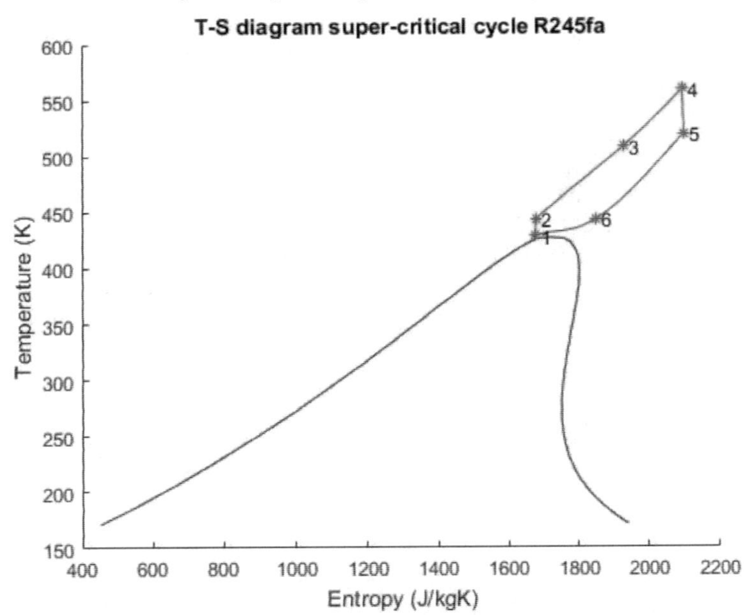

Figure 20 Design point T-S diagram (super-critical R245fa)

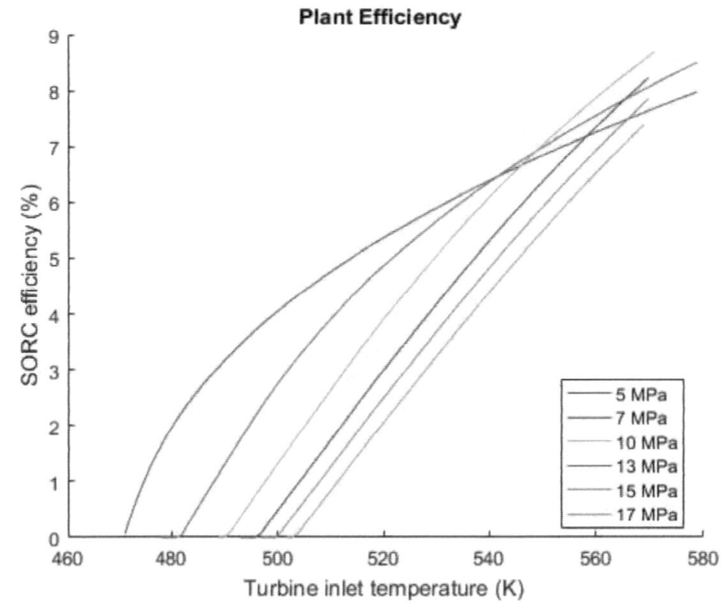

Figure 21 Effect on turbine inlet temperature on plant efficiency (super-critical R245fa)

For the supercritical cycles, the efficiency is lower than expected, as the mass flow of the solar farm, calculated from the optimization process, does not produce high temperature at the outlet.

6. Validation

In order to validate the mathematical model presented at section 3, the results of the overall efficiency for each proposed power plant are compared, for the subcritical ORC with experimental data [23] and for the trans-critical ORC and supercritical cycle with data provided by other papers [24],[11]. The results are presented at Table 8.

Table 8 Validation data

Cycle	Working fluid	Reference	
Sub-critical	n-Pentane	12.1%	14.12%
Trans-critical	Hexamethyldisiloxane (MM) (ref) - R245fa(study)	~21%	15.14%
Super-critical	CO_2	~15%	13.23%
	R245fa	~5%	8.87%

It can be seen that the simulated efficiencies by this paper are in accordance with the values provided by previous research.

7. Conclusions

The aim of this work is to analyze both an ORC and a supercritical cycle integrated with a small-scale parabolic trough collector system. The analysis considered a solar power plant with a gross power output of 1 MWe. The system incorporate the solar farm with Therminol 66 as heat transfer fluid, a thermal energy storage and the power plant. The organic fluids tested are: n-Pentane, R245fa and O_2 .

An optimization procedure was established for both the solar farm and the power plant, based on a mathematical model, in order to maximize the overall efficiency of the system (solar to power). The main conclusion of this study are the following:

- Transcritical ORC with R245fa as a working fluid presents the best overall performance of the cycles studied with solar to power efficiency of 15.14%.
- N-Pentane shows high efficiency as a working fluid for subcritical ORC followed by R245fa with efficiency of 14.12% and 10.31% respectively.
- Supercritical cycles were expected to show higher efficiencies however the low output temperature of the solar collectors, presented by the optimization process of the solar farm, restricts the overall performance of the system.

References

[1] M. Petrollese, G. Cau, and D. Cocco, "The Ottana solar facility: dispatchable power from small-scale CSP plants based on ORC systems," *Renew. Energy*, vol. 147, pp. 2932–2943, 2020, doi: 10.1016/j.renene.2018.07.013.

[2] Z. Aqachmar, A. Allouhi, A. Jamil, B. Gagouch, and T. Kousksou, "Parabolic trough solar thermal power plant Noor I in Morocco," *Energy*, vol. 178, pp. 572–584, 2019, doi: 10.1016/j.energy.2019.04.160.

[3] R. Taccani, J. B. Obi, M. De Lucia, D. Micheli, and G. Toniato, "Development and Experimental Characterization of a Small Scale Solar Powered Organic Rankine Cycle (ORC)," *Energy Procedia*, vol. 101, no. September, pp. 504–511, 2016, doi: 10.1016/j.egypro.2016.11.064.

[4] T. Chambers, J. Raush, and B. Russo, "Installation and operation of parabolic trough organic Rankine cycle solar thermal power plant in South Louisiana," *Energy Procedia*, vol. 49, pp. 1107–1116, 2014, doi: 10.1016/j.egypro.2014.03.120.

[5] L. Karsten, J. Hoffmann, and F. Dinter, "An Organic Rankine Cycle as technology for smaller concentrated solar powered systems," *ISES Sol. World Congr. 2017 - IEA SHC Int. Conf. Sol. Heat. Cool. Build. Ind. 2017, Proc.*, pp. 85–96, 2017, doi: 10.18086/swc.2017.04.05.

[6] N. B. Desai and S. Bandyopadhyay, "Thermo-economic analysis and selection of working fluid for solar organic Rankine cycle," *Appl. Therm. Eng.*, vol. 95, pp. 471–481, 2016, doi: 10.1016/j.applthermaleng.2015.11.018.

[7] Y. L. He, D. H. Mei, W. Q. Tao, W. W. Yang, and H. L. Liu, "Simulation of the parabolic trough solar energy generation system with Organic Rankine Cycle," *Appl. Energy*, vol. 97, pp. 630–641, 2012, doi: 10.1016/j.apenergy.2012.02.047.

[8] U. Caldiño-Herrera, L. Castro, O. A. Jaramillo, J. C. Garcia, G. Urquiza, and F. Flores, "Small Organic Rankine Cycle Coupled to Parabolic Trough Solar Concentrator," *Energy Procedia*, vol. 129, pp. 700–707, 2017, doi: 10.1016/j.egypro.2017.09.097.

[9] P. Morrone, A. Algieri, and T. Castiglione, "Hybridisation of biomass and concentrated solar power systems in transcritical organic Rankine cycles: A micro combined heat and power application," *Energy Convers. Manag.*, vol. 180, no. November 2018, pp. 757–768, 2019, doi: 10.1016/j.enconman.2018.11.029.

[10] J. Li, Q. Liu, Z. Ge, Y. Duan, and Z. Yang, "Thermodynamic performance analyses and optimization of subcritical and transcritical organic Rankine cycles using R1234ze(E) for 100–200 °C heat sources," *Energy Convers. Manag.*, vol. 149, pp. 140–154, 2017, doi: 10.1016/j.enconman.2017.06.060.

[11] A. Erdogan, O. Kizilkan, and C. O. Colpan, "Thermodynamic analysis of a supercritical closed Brayton cycle integrated with parabolic trough solar collectors," *J. Therm. Anal. Calorim.*, no. 0123456789, 2019, doi: 10.1007/s10973-019-09199-0.

[12] H. Singh and R. S. Mishra, "Performance analysis of solar parabolic trough collectors driven combined supercritical CO2 and organic Rankine cycle," *Eng. Sci. Technol. an Int. J.*, vol. 21, no. 3, pp. 451–464, 2018, doi: 10.1016/j.jestch.2018.03.015.

[13] D. Novales, A. Erkoreka, V. De Peña, and B. Herrazti, "Sensitivity analysis of supercritical CO 2 power cycle energy and exergy efficiencies regarding cycle component efficiencies for concentrating solar power," *Energy Convers. Manag.*, vol. 182, no. January, pp. 430–450,

2019, doi: 10.1016/j.enconman.2018.12.016.
[14] K. Lovegrove and W. Stein, *Concentrating Solar Power Technology: Principles, Developments and Applications*. Woodhead Publishing, 2012.
[15] S. A. Kalogirou, "A detailed thermal model of a parabolic trough collector receiver," *Energy*, vol. 48, no. 1, pp. 298–306, 2012, doi: 10.1016/j.energy.2012.06.023.
[16] H. Yılmaz and M. S. Söylemez, "Thermo-mathematical modeling of parabolic trough collector," vol. 88, pp. 768–784, 2014, doi: 10.1016/j.enconman.2014.09.031.
[17] L. Karsten, "The Analysis of an Organic Rankine Cycle for Smaller Concentrated Solar Powered Systems," Stellenbosch University, 2018.
[18] K. Riffelmann, T. Richert, P. Nava, and A. Schweitzer, "Ultimate Trough® - A significant step towards cost-competitive CSP," in *Energy Procedia*, 2014, vol. 49, pp. 1831–1839, doi: 10.1016/j.egypro.2014.03.194.
[19] J. A. Duffie and W. A. Beckman, *Solar Engineering of Thermal Processes: Fourth Edition*. Wiley, 2013.
[20] E. Macchi and M. Astolfi, Eds., *Organic Rankine Cycle (ORC) Power Systems*. Elsevier, 2017.
[21] K. Brun, P. Friedman, and R. Dennis, *Fundamentals and Applications of Supercritical Carbon Dioxide (SCO2) Based Power Cycles*. Woodhead Publishing, 2017.
[22] I. H. Bell, J. Wronski, S. Quoilin, and V. Lemort, "Pure and pseudo-pure fluid thermophysical property evaluation and the open-source thermophysical property library coolprop," *Ind. Eng. Chem. Res.*, vol. 53, no. 6, pp. 2498–2508, 2014, doi: 10.1021/ie4033999.
[23] G. Cohen, R. Cable, D. Brosseau, and H. Price, "Parabolic Trough Organic Rankine Cycle Power Plant," National Renewable Energy Lab.(NREL), Golden, CO (United States), 2005.
[24] C. Li, G. Kosmadakis, D. Manolakos, E. Stefanakos, G. Papadakis, and D. Y. Goswami, "Performance investigation of concentrating solar collectors coupled with a transcritical organic Rankine cycle for power and seawater desalination co-generation," *Desalination*, vol. 318, pp. 107–117, Jun. 2013, doi: 10.1016/j.desal.2013.03.026.

The wastewater using in technologies of bio-oil production from microalgae: CO_2 capture and storage

N I Chernova[1], S V Kiseleva[1]
[1] Lomonosov Moscow State University, Leninskie Gory, 1, Moscow, Russia

*Corresponding author's e-mail: k_sophia_v@mail.ru

Abstract. Bioenergy remains the largest renewable energy industry. At the same time, microalgae (MA) are a promising object of research among other types of biomass and the scale of microalgae using for energy purposes is increasing. Industrial cultivation of microalgae opens up great opportunities for CO_2 utilization and wastewater treatment from organic and mineral pollutants, and also significantly reduces the load on fresh water supplies. To reduce the cost of biofuels, optimization of a whole number of technological stages, including the cultivation of MA, is necessary. The paper presents the results of developing methods for cultivating MA to optimize their growth and absorption of nutrients from wastewater. A culture of microalgae/cyanobacteria *Arthrospira platensis rsemsu P (Bios)* (collection of RSE Laboratory at Lomonosov MSU) which grew well in wastewater was experimentally selected. The results of the wet MA biomass conversion into bio-oil by hydrothermal liquefaction (HTL) technology with the associated production of biochar are presented. Biochar producing can be considered as a method for capturing and storing carbon.

1. Introduction

Nowadays the gradual movement of the energy sector towards the active development of renewable energy sources and the implementation of projects for carbon capture and storage (CCS) has become a global trend. In this regard, bioenergy can be considered as an effective solution of energy problems, as well as a promising approach to improving the environmental situation. Microalgae (MA) are considered a promising feedstock for the production of third and fourth generation biofuels, as they have higher growth rates than land biomass, and at the same time do not compete with food and feed production. Due to the simple structure and simple type of nutrition, MA have a theoretically higher efficiency of photosynthesis (up to 12.6%) compared to terrestrial plants [1].

The lability of the MA biochemical composition allows obtaining biomass with desired properties and, accordingly, increasing the yield of biofuel of a given type (for example, the yield of biodiesel from biomass with an increased lipid content). Like other types of biomass, MA as feedstock for biofuel and energy production are "CO_2-neutral".

Modern technologies for growing MA in open and closed systems can be optimized for the purpose of capturing and storing carbon in MA biomass - both short-term (biofuel production) and long-term (conversion of CO_2 into geologically stable forms of carbon). Approximately 1.83–1.88 kg of CO_2 can be fixed in 1 kg of microalgae biomass [2-3]. The industrial flue gas utilization for microalgae cultivation has been already applied in breweries, cement factories and gas-fired power plants, as these factories generate gases with relatively low SO_x and NO_x concentrations [4]. The coal-derived flue gas has been also utilized for microalgae cultivation when it was diluted with air or added in small amount [4, 5-6]. A mitigation of industrial CO_2 emission from the unfiltered smoke stack of a 4MW coal-fired power plant through biotransformation into microalgal biomass with *Desmodesmus sp.* as the dominant microalgae was demonstrated in [7].

For the cultivation of microalgae, not only gaseous, but also liquid wastes (in particular, sewage) can be used, which makes it possible to reduce the need for water and nutrients by 90% [8]. Integration of biological wastewater treatment and CO2 capture is a promising area of biofuel production from microalgae. The aim of this work is studying the effectiveness of wastewater treatment and CO_2 capture through the cultivation of microalgae and subsequent production of biofuel by the hydrothermal liquefaction of biomass.

2. Materials and methods

A microalgae *A. platensis rsemsu P (Bios)* with a high growth rate and the ability to respond to changing environmental conditions was selected for the study purposes. *Arthrospira platensis* is a well-known culture that is used in the production of many commercial products including healthy food supplements, food for animals, cosmetics and pharmaceutical products. It is grown usually in an open manner in a large scale. *A. platensis* is classified as cyanobacteria by International Code of Bacterial Nomenclature (ICNB), although according to the International Code of Nomenclature for Algae, Fungi and Plants (Melbourne Codex 2012) it is classified as blue-green microalgae. When it is used as biomass feedstock for biofuel production, it is often considered as microalgae [9-11]. *A. platensis* strain was obtained by semi-continuous cultivation in open pond with volume of 1000 l and illumination of 55 ± 5 μE/ (m^2 s) at steady-lighting conditions and temperature equal to T=21°C. The pond is equipped by near-surface mixing. *A. platensis* was cultivated using Zarrouk's medium [12]. *A. platensis* strain has straight trichomes formed due to natural morphological variability during prolonged cultivation for 30 years under laboratory conditions. Microalgae are stably associated with heterotrophic bacteria, which we assigned to the genus *Pseudomonas and Bacillus*. These bacteria are able to assimilate dissolved organic matter and toxic elements and provide efficient municipal wastewater treatment.

The municipal waste waters of Moscow were used for experimental studies. The content of biogenic elements in the waste waters of Moscow, as well as typical and artificial waste waters (data from scientific publications) are presented in Table 2. It can be seen that the waste waters used in the experiments have a typical composition, but with an underestimated COD value, and an overestimated ammonium nitrogen content. The amount of nutrients (nitrogen ammonia, nitrates and orthophosphates) are present in a small amount (110, 2.4 and 15 mg/l, respectively) in comparison with the Zarrouk's medium, which is optimal for the *A. platensis* cultivation. COD (total organic matter in heavily polluted effluents) 140 mgO/l and BOD_5 (readily oxidizable organic matter concentration in effluents) 85 mgO_2/l indicate that the sample contains a small amount of dissolved organic matter. According to the value of the hydrogen index (pH = 8.2), the wastewater is alkaline. The amount of oil products (2.7 mg/l) and total aromatic substances (0.668 mg/l) exceeds the average for this class of wastewater.

The *A. platensis* cultivation was carried out in undiluted and in diluted waste water (diluted by a factor of 2 ... 25) in Erlenmeyer flasks with a volume of 250–500 ml. The biomass growth rate of the same microalgae on the classic nutrient Zarrouk's medium served as a control [12]. The growing duration was 10–47 days. The degree of nutrients utilization from wastewater at different stages of the MA cultivation was controlled by chemical analyzes.

HTL experiments were carried out on the setup previously described at the [15] in collaboration with Joint Institute for High Temperatures of the Russian Academy of Sciences specialists. Hydrothermal liquefaction (HTL) with the production of bio-oil as end product is of most interest. During HTL the biomass is thermally treated under humid conditions at temperatures up to 370 °C and pressures up to 25 MPa. Due to this treatment, the biomass components are passed through hydrolysis and pyrolysis reactions forming a number of liquid hydrocarbons, both soluble and insoluble in water, as well as gaseous and solid reaction products. HTL products are bio-oil, aqueous solution with water soluble organics, gas product and solid residue - biochar.

Table 1. The chemical composition of the municipal wastewater sample from the observation well of Moscow, artificial and typical waste waters

Composition indicator, units	Waste water sources							
	Waste water, Moscow	Artificial waste water [13]					Typical waste waters [14]	
COD (bichromate oxidation), mg O/l	140	749	900	820	798	925	–	–
BOD_5, mgO_2/l	85	–	–	–	–	–	–	–
Potassium, mg/l	15	2,1	8.9	3,9	11,7	17,4	4,5	5,4
Nitrates, mg/l	2,4	3,5	13,8	8,7	7,4	71,1	0,4	0,4
Ammonium, mg/l	110	26,9	53,6	85,2	12,0	14,9	42	56

3. Results and discussion

Due to the toxic compounds (phenols, ammonium, anionic surfactants) presence in the waste water, it was necessary to experimentally determine the degree of its dilution for sustainable growth of microalgae *A. platensis* with heterotrophic bacteria-associates. A steady growth of *A.platensis* was found for 12 days on undiluted municipal waste water but with the addition of 1/2 sodium bicarbonate doze from the composition of the Zarrouk's culture medium. Subsequently, the biomass accumulation lagged behind the control, and after 25 days the growth of arthrospira cells in the wastewater stopped. A 5-fold dilution of waste water was determined as the highest, at which a consortium of *A. platensis* with associate bacteria can effectively grow and assimilate nutrients within 8 days. The sodium bicarbonate addition to the wastewater created more favorable conditions for the MA growth. Under dilution of wastewater by 3-1.5 times with the addition of sodium bicarbonate, the effective growth of this consortium of MA with associate bacteria was observed within 47 days. As a result, the concentration of nitrogen and phosphorus ions decreased to trace values. Moreover, the amount of readily soluble organic matter (BOD_5) decreased from 85 to 53 mgO_2/l, which indicates the ability of MA to switch to mixotrophic growth with the consumption of organic matter.

The wastewater treatment efficiency under microalgae cultivation was determined by analyzing the water chemical composition before and after growing microalgae (Table 2).

Table 2. The wastewater main chemical indicators before and after growing MA (growing time 47 days, dilution 3x-1,5x)

Composition indicator, units	Experimental results		Relative error, %
	Initial concentration	Final concentration	
COD (bichromate oxidation), mgO/l	140	120	15
BOD_5, mgO_2/l	85	53	13
Potassium, mg / l	15	4	10
Nitrates, mg / l	2,4	<0,1	20
Ammonium, mg / l	110	1,5	12

A comparative analysis of the wastewater main chemical indicators before and after growing MA showed that *A. platensis* is capable of efficient growth in slightly diluted wastewater. However, a prerequisite for growth is the creation of alkaline conditions in a nutrient medium by adding $NaHCO_3$ in an amount of at least 6 g/l. *A. platensis* cells completely consumed phosphorus from the wastewater in the form of orthophosphates and nitrogen in the form of ammonium and nitrates, the concentration of which decreased by 4, 73 and 24 times, respectively.

The grown microalgae biomass was subjected to hydrothermal liquefaction at temperatures of 270, 300 and 330°C. Hydrothermal liquefaction (HTL) allows to process wet biomass and to convert all carbon-containing components into fuel, the components of which are bio-oil, synthesis gas and coal (Table 3, Figure1).

Table 3. Conversion products of *A. platensis* biomass by hydrothermal liquefaction at different temperatures

HTL- products	Temperature, °C		
	270°C	300°C	330°C
Bio-oil, %	34,6	38,8	45,7
Syn- gas, %	12,7	14,8	17,5
Biochar, %	28,7	27,4	26,0
Aqua-phase, %	24,0	19,0	10,8

As can be seen from Table 3, during the hydrothermal liquefaction of MA biochar is formed in a significant amount (from 26.0 to 28.7% depending on the process temperature). Consequently, the HTL technology provides not only the conversion of MA into biofuel, but also the removal of CO2 from the atmosphere by converting part of the carbon into a stable form that can be deposited for a long time [16]. Thus, to some extent, CO2 emissions from the cultivation and processing of MA are compensated. The aqueous phase formed after HTL as a by-product can also be partially utilized by microalgae as biogenic elements. Aqua-phase or aqueous solution is a by-product with low energy value and also needs to be processed. Previously we have described the recycling of nutrients from an aqueous solution after HTL [15].

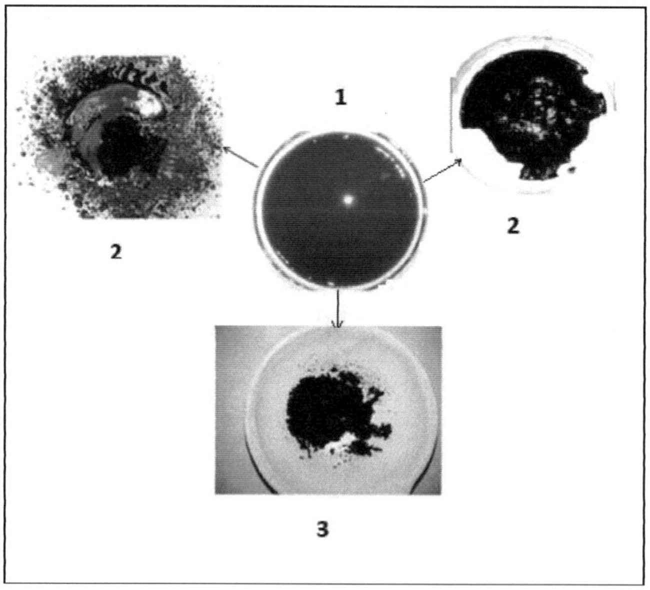

Figure 1. *A. platensis* conversion products by HTL-technology under various experimental conditions.
Initial biomass of *A.platensis* (1), bio-oil (2) and biochar (3).

4. Conclusion

Cultivation of MA using wastewater in combination with biofuel production is promising as the cost of nutrients (fertilizers), CO_2 emissions and pressure on fresh water resources are reduced. Experimentally selected strains of microalgae (*Arthrospira platensis rsemsu P (Bios)*), capable of growing for a long time in municipal wastewater with efficient utilization of nutrients. By means of hydrothermal liquefaction, the *Arthrospira* biomass, grown in wastewater, was transformed into biofuel (bio-oil and synthesis gas) and by-products – biochar and an aqueous phase. It has been proposed that biochar, which is 90% carbon and is a geologically stable form, should be considered as another alternative strategy carbon capture and storage.

Acknowledgments
The work was funded by Lomonosov Moscow State University

References
[1] Zhu XG, Long SP, Ort DR 2008 *J. Curr Opin Biotechnol* **19** 153–9
[2] Cheah W Y, Show P L, Chang J-S, Ling T C, Juan J C 2015 *J. Bioresource Technology* **184** 190-201
[3] Zhao B, Su Y 2014 *J. Renewable and Sustainable Energy Reviews* **31** 121-132
[4] Van Den Hende S, Vervaeren H, Boon N 2012 *J. Biotechnology Advances* **30** 1405-424
[5] Kao C-Y, Chen T-Y, Chang Y-B, Chiu T-W, Lin H-Y, Chen C-D, Chang J-S, Lin C-S 2014 *J.Bioresource Technology* **166** 485-493
[6] Moheimani NR 2016 *J. Applied Phycology*, **28** 2139-2146
[7] Aslam A, Thomas-Hall S R, Mughal T A, Schenk P M 2017 *J. Bioresource Technology* **233** 71-283
[8] Yang J, Xu M, Zhang X, Hu Q, Sommerfeld M, Chen Y 2011 *J. Bioresource Technology* **102(1)** 159–165
[9] Gouveia L, Oliveira A C 2009 *J. Indust. Microbiol & Biotechn* **36** 269–274
[10] Raikova S, Smith-Baedorf H, Bransgrove R, Barlow O, Santomauro F, Wagner J, Allen M J, Bryan C G, Sapsford D, Chuck C J 2016 *J. Fuel Proc. Technol* **142** 219–227
[11] Kostyukevich Y, Vlaskin M, Vladimirov G, Zherebker A, Kononikhin A, Popov I, Nikolaev E 2017 *Europ. J. Mass Spectrometry* **23** 83–88
[12] Zarrouk C 1966 *Influence de diverse facteursphysiques et chimiques sur la croissance et la photosynthese de Spirulina maxima (Setch et Gardner)* Geitler: University of Paris
[13] Jämsä M, Lynch F, Santana-Sánchez A, Laaksonen P, Zaitsev G, Solovchenko A, Allahverdiyev Y 2017 *J Algal Res* **26** 65–73
[14] Hao R, Yu Z, Li J C, Gao M, Ma W L, Zhu Y 2017 *Int J Agric & Biol Eng* **10 (1)** 107-114
[15] Chernova N, Kiseleva S, Vlaskin M, Grigorenko A, Rafikova Y 2019 *IOP Conference Series: Materials Science and Engineering* **564** 012101–1–012101–6
[16] Hielmann SM, Davis HT, Jader LR, Lefebvre PA, Sadowsky MJ, Schendel FJ, von Keitz MG, Valentas KJ 2010 *J. Biomass and Bioenergy* **34** 875–882

Comprehensive analysis of the energy supply potential using local types of renewable energy sources

A A Tulegenova[1], S V Kiseleva[1*]
[1] Lomonosov Moscow State University, Leninskie Gory, 1, Moscow, Russia

[*]Corresponding author's e-mail: k_sophia_v@mail.ru

Abstract. Over the past decade, a steady tendency towards the increasing use of renewable energy sources in the fuel and energy complex of separate countries and regions is observed. Countries that have large hydrocarbon reserves such as the Republic of Kazakhstan are no exception. The main challenges of using renewable energy sources in these countries are related not only to their spatial and temporal variety, but also to the high cost of produced energycomparing to the traditional energy в sources based on hydrocarbon fuel. The solution to the mentioned problems can be achieved by optimizing the use of renewable energy resources, which requires a comprehensive analysis of the territory in terms of physical-geographic conditions and economic factors. The article offers methodological approaches and the results from a spatial analysis of the energy supply potential to consumers in the Republic of Kazakhstan using the cartography tools.It takes into account not only the distribution of resource potential, but also the environmental factors (specially protected natural areas, water conservation zone, etc.) limiting the installation of power plants based on RES, and the energy consumption schedule. Economic factors include the performance of typical power plants based on RES, as well as the present value of potentially generated energy.

Key words: renewable energy resource, Kazakhstan, energy generation plant output, energy demand, spatial analysis, solar energy, wind energy, Geographic Information System (GIS)

1. Introduction

Ecological issues and the climate agenda stimulate the society to develop resource-saving technologies. Renewable energy can be considered as the flagship of this trend.Even with a drastic decrease of prices for the hydrocarbon fuel, the renewable energy industry (REI) will remain relevant, particularly for the remote regions of countries with a large territory, where the central power supply is missing, and the high cost of laying the power lines stimulates an autonomous type of power supply system.

The electric energy production in the Republic of Kazakhstan (RK) is currently based on the use of fossil hydrocarbon fuel.Most of the electric energy (81%) in Kazakhstan is generated by coal power plants [1], which entail a high intensity of air pollution. One the one hand, the low population density in Kazakhstan leads to the presence of seasonal or permanent autonomous energy consumers, while on the other hand, it provides significant areas for the construction of power plants on renewable sources.The generating capacities in the RK are combined into several energy zones, between which the transmission of energy leads to significant losses (in 2018, losses were about 3 GWh/year).The steady growth of losses is caused by the aging of the network infrastructure that was created back in the days of the USSR. At the same time, the increasing energy consumption and the shortage of its own power in the Southern energy zone require a solution to the problem of "North-South" electric energy transit.

The ratification of the Kyoto Protocol has served as a trigger for the formation of institutional conditions for the development of REI in Kazakhstan. The development of this energy industry started in 2009, when the Law №165-IV «On support of the use of renewable energy sources (RES)» has been adopted. In 2013, a center of calculation and analytics was created to ensure the purchase and sale of electric energy from RES; together with the European Reconstruction and Development Bank and the Asian Development Bank, a concept for the transition of the RK to a green economy was developed.The implementation of the concept assumes the fulfillment of several targets with a maximum increase in power capacity of wind (up to 1200 MW) and solar (up to 1100 MW) power plants by 2025. To a certain extent, the renewable energy plants will be used to replace the old thermal power plants (TPP) [2].

As in many countries, to achieve these targets, the RK has adopted mechanisms of state support and incentives for the industry. Until February 2018, a mechanism of fixed tariffs was functioning in the RK to stimulate the use of RES. This mechanism was replaced by the auction trading.
As a result, from 2014 to 2018, a positive dynamics of installed power capacities and energy production from RES is observed, the maximum growth was in wind power plants (WPP), in second and third places are small hydroelectric power plants (HPP) and solar power plants (SPP), respectively (table 1).

Table 1. The volume of generation and the share of electricity production by renewable sources of the total generation in the Republic of Kazakhstan (compiled by the authorsbased on data [2,3]).

Year	The volume of electricity generation from renewable energy plants, million kWh (WPP/SPP/SmallHPP/BioTPP)	The share of electricity generation from RES in the energy balance(%)
2016	271/86/81/-	0.98
2017	335/90/144/-	1.07
2018	398/138/242/1	1.26

Thus, Kazakhstan is a state with atypical conditions for the active development of renewable energy.However, despite the significant reserves of fossil fuels, the country actively stimulates the development of REI, where wind and solar energy are almost equally considered as the leading industries of renewable energy. To assess the possible role of renewable sources in energy supply, it is necessary to assess the potential for energy supply to the regions of Kazakhstan based on REI.The energy supply potential means the share (%) of electricity consumption that can be provided by generating electric energy based on RES.The energy supply potential is determined by: the needs for energy supply in a given area; the productivity of renewable energy plants; the restrictions on the use of stations (installations) for power generation based on RES; competitiveness of energy from RES in a given area.

2. Methodology and initial data

2.1General characteristics of regional electricity consumption in the RK

The electricity industry in Kazakhstan is a combination of the following sectors: production, transmission, supply, consumption and other activities in the field of energy supply. As of 01.01.2019, there are 138 power plants operating in the Republic with a total installed capacity of 21901.9 MW[1]. A significant number of regions are characterized by energy deficiency (table2).The Akmola region is characterized by the maximum negative electricity balance among the regions of Kazakhstan, the deficit of electric energy amounted to 4792 GWh/year.

Table 2. Electric energy balance for the period of 2013-2019 by regions of the RK (compiled by the authors based on data [4]).

Region of the RK	The balance of electric energy production and consumption, GWh/year				
	2014	**2015**	**2016**	**2017**	**2018**
Pavlodar	23528	17808	16594	22706	26194
Karagandy	-1696	-713	-608	-1283	-2160
Almaty	-4222	-3590	-2467	-3043	-3758
Akmola	-4991	-4679	-4653	-4603	-4792
East-Kazakhstan	-807	197	1393	1432	569
Aktobe	-1453	-1750	-1754	-2289	-2522
Mangystau	205	210	230	-27	187
Atyrau	-172	-222	-276	-208	-521
Kostanay	-3800	-3176	-3299	-3627	-3889
Turkestan	-2807	-3166	-3189	-4035	-4104
Zhambyl	-1180	-637	-339	-815	-2014
West-Kazakhstan	-182	-313	-107	134	214
North-Kazakhstan	1029	1171	1527	1499	1416
Kyzylorda	-1178	-1191	-1286	-1312	-1250
Total	2332	2274	1764	4527	3569

The largest amount of the electric energy is generated and consumed in the Pavlodar region, which is the main donor of the electric energy in Kazakhstan. The balance flow to the Russian Federation and Central Asia in 2018 was 3566 GWh and 2.8 GWh, respectively.

2.2Performance assessment of wind and solar power plants

At the assessment of the potential for energy supply based on RES, an analysis of resources and the productivity calculation of the projected plants should be carried out.In order to estimate the gross energy potential, the NASAPOWER climate database [5] was used as the initial data. The NASAPOWER database contains data arrays about the incoming solar radiation on a horizontal surface, temperature, wind speed at different heights and other characteristics significant for solar and wind energy industry on a regular spatial grid (1x1)°.The data are presented in the format of daily sum (for solar radiation), average daily values (for temperature, wind speed, etc.) for each day of the period from 1983 to the present. To obtain the resource characteristics of solar and wind energy on the territory of Kazakhstan, the sequences of data on the arrival of solar radiation and wind speed for 1983-2005 were used.As a result, the average annual and seasonal daily sums of the total solar radiation, the average annual values and the frequency of the wind speed for each cell of the spatial grid (1x1)° were calculated, with the following generalization up to the grid (2x2)°.It was shown that inter-annual variations in characteristics of radiation and average wind speeds range are from 2% to 13%.Further, it was assumed that the accuracy of estimating the stations productivity is determined by inter-annual variations in resource characteristics and lie within the same limits.

In this research, to calculate *the performance of solar and wind power plants*, typical stations of the same power capacity (15 MW) with the following technical characteristics were selected:

–a solar power plant (SP) consisting of the photo-voltaic modules (PVM) with a unit power of 265 W, a surface area of 1.65 m^2, an efficiency of 16.2% and inverters with an efficiency that varies in the range of 89-98% depending on the PVM power;

–wind power plant (WP), consisting of wind turbines (WT) of the Nordex N 27-50 model with a rated power of 150 kW each, a mast height of 50 m and a wind wheel diameter of 27 m.

For the spatial analysis of the productivity of solar and wind power plants (technical potential), the density of the generated energy per unit surface area (1 km^2) (specific productivity) was considered(figure 1,2). In this case, the minimum possible distance between WT/PVM at the station sites was assumed (1 WT/km^2, 143 PVM/km^2).The calculated specific energy generation for the territory of the RK: from 63 to 342 MWh/year/km^2 for the WPP, and from 41 to 68 MWh/year/km^2 for the SPP (Figure 1, 2).

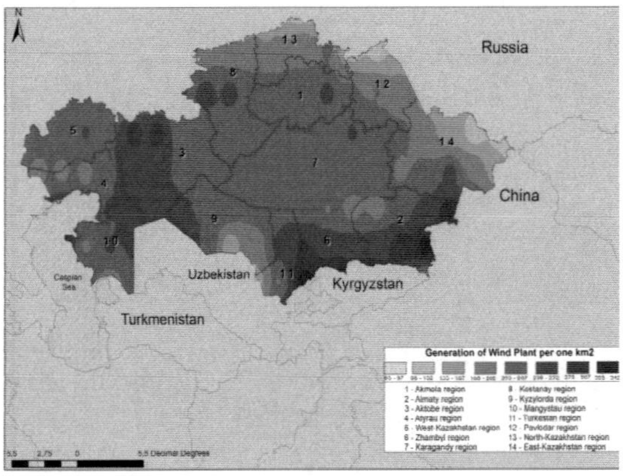

Figure 1. Generation ofWP, MWh/km^2 for 2018 (compiled by the authors).

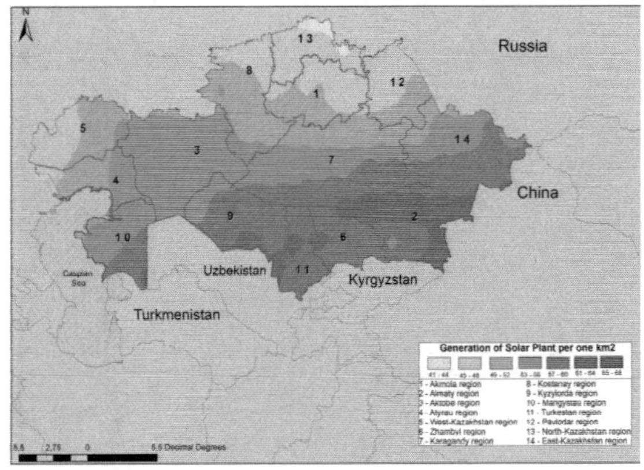

Figure 2. Generation of SP MWh/km^2 for 2018 (compiled by the authors).

It can be observed that the generation of the wind power plants is 1.5-5 times higher than the specific productivity of the solar power plants (for the selected types of SPP and WPP).

2.3Limitingfactors.

In order to assess the potential of energy supply, the influence of limiting (specially protected natural areas, forests, settlement, water bodies, key bird areas, buffer zones of settlements, etc.) andfacilitatingfactors (power lines, road network and electrical substations) has to be taken into account.The tools of spatial analysis allow to exclude the territories, where the construction of power plants is prohibited, from consideration and to calculate the area of available territories.In this study, power lines were considered as a facilitating factor.After the allocation of available territories on the basis of the factors mentioned above, an additional restriction of the territory (no further than 5 km from power lines) was introduced due to the high cost of the power lines construction.As a result, the data on the areas available for construction were obtained.According to the obtained results, significant land resources for the placement of power plants based on RES are held by the following regions: Karagandy - 33 131 km^2 for the WPP and 34 271 km^2 for the SPP, Akmola - 27 445 km^2 for the WPP and 29 357 km^2 for the SPP, and Pavlodar - 22 111 km^2 for the WPP and 22 614 km^2 for the SPP.

2.4Assessment of power supply potential

Subsequently, the value of electricity generation by solar and wind power plants was calculated and compared with the value of electricity consumption in the regions of Kazakhstan, taking into account the calculated values for the available areas. The total potential of the energy supply for the entire Republic is 43% (44975 GWh/year) for the WPP and about 11% (11023 GWh/year) for the SPP, if they occupy all permitted territories (Figure 3).

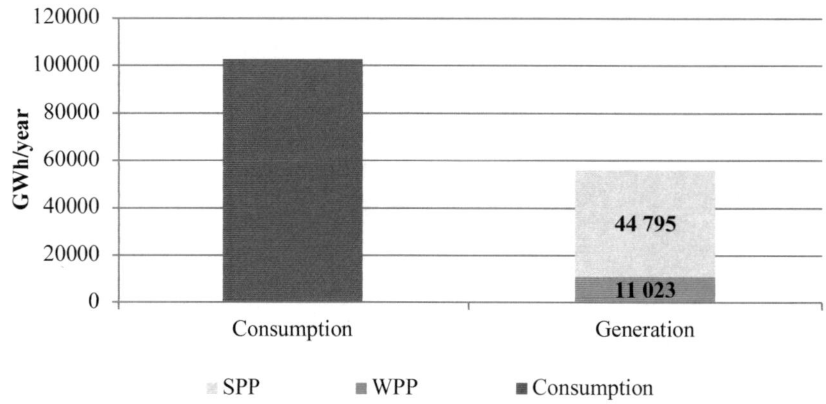

Figure 3. Electricity consumption and productivity of the WPP and the SPP in the RK (compiled by the authors).

If the regional potential of energy supply is considered, then the annual output of the WPP in the permitted territories can potentially satisfy all the needs for electricity in the regions with relatively low total electricity consumption such as North Kazakhstan and Kyzylorda regions.Zhambyl, West Kazakhstan, Kostanay and Akmola regions also have a high potential for energy supply from RES. In these regions more than 65% (1 295-5 898 GWh/year) of consumed electricity can be satisfied by the renewable energy (Figure 4).

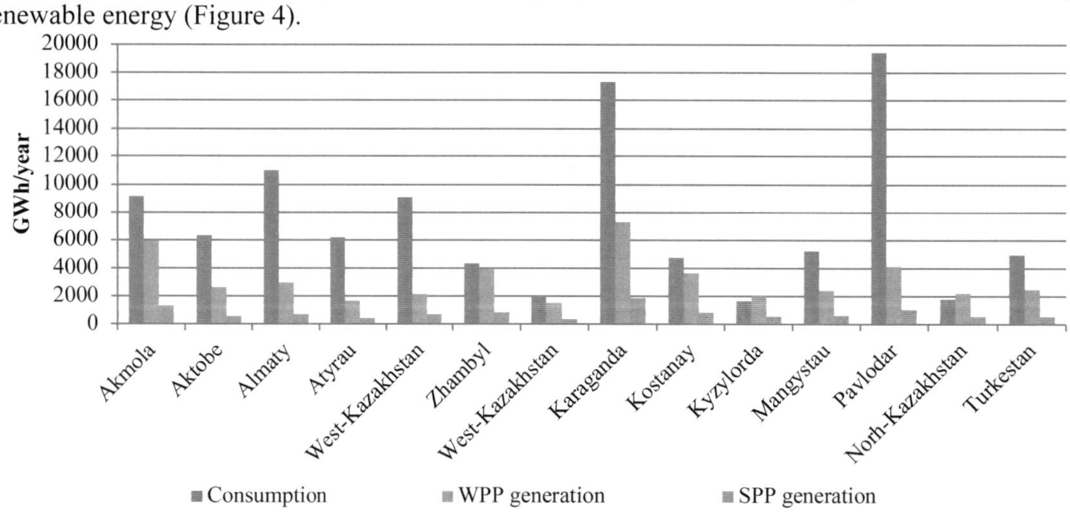

Figure 4. Electricity consumption and productivity of the WPP and the SPP in the RK regions (compiled by the authors)

The most promising for the use of the SPP are Kyzylorda, Zhambyl and North Kazakhstan regions, where the solar power plants can potentially provide from 20 to 32% (544-873 GWh) of consumed electricity.

2.5 Economic assessment

The Levelized Cost of Energy (LCOE) was used as a key characteristic for the cost assessment of electricity generated by solar and wind power plants.The assessment of capital (equipment costs) and operating (equipment delivery, construction, operation of power plants) costs for the SPP was carried out on the basis of estimate documents for the construction of the Kosh-Agach solar power plant (Republic of Altai, the Russian Federation) that was built in 2014, taking into account the inflation and changes in the PVM cost for the past period.

IOP Conf. Series: Materials Science and Engineering 1037 (2021) 012046 doi:10.1088/1757-899X/1037/1/012046

The capital and operating costs for the construction and operation of the wind power plants were taken from the current estimates of international analytical agencies [6]. The discount rate was taken equal to 10%, while the service life was taken to be 20 years.

As a result, the present values for the SPP and the WPP were0.078USD/kWhand0.012 USD/kWh, respectively (The transfer was made according to the rate of the Central Bank of Russia dated 29.09.2020 (1 USD = 78.6713 RUB)).The cost price of electric energy in the Kazakhstan wholesale market of energy and power generated by using the traditional fuel is 0.007-0.034 USD for individuals and 0.009-0.072 USD for legal entities (the minimum - the maximum daily rate). Thus, the cost price of energy generated at solar and wind power plants, taking into account the estimated nature of the calculations, is close to a competitive.

3. Conclusion

The Republic of Kazakhstan actively develops the renewable energy industry, the created institutional conditions, including the developments programs. The legislative framework and methods of supporting renewable energy industries have led to an increase in the installed power capacity of renewable energy power plants.The highest specific productivity in Kazakhstan is characterized by the wind power plants, the energy generation by which, according to the estimates, is several times higher than the energy generation by the solar power plants, which is determined by the spatial distribution of resources for these types of RES.

The potential of energy supply based on solar and wind energies in the regions of Kazakhstan varies in wide ranges (from 5% to 90%).At the same time, the wind power plants in each region of the RK can potentially cover a significant share or full needs in the electricity supply.The solar power plants are characterized by a significantly lower potential of energy supply, and can only be used as an additional source of the energy. The Kyzylorda, North Kazakhstan and Zhambyl regions have the maximum integrated potential of the energy supply based on RES.

The results of the conducted economic assessment of the generated electricity based on wind and solar energy generation indicatetheir competitiveness.

References:

[1] Kazakhstan Electricity Grid Operating Company*Annual report 2018* (Astana) p. 146

[2] Kazakhstan Association of Oil and Gas and Energy Complex Organizations «Kazenergy»*National energy report2019* (Astana) p.213

[3] Financial settlement center of RE *Annual report 2018* (Astana) 82 p.

[4] Kazakhstan electricity and power market operator JSC *Annual report 2018* (Astana) p.190

[5] NASA POWER database Electronic resource (date of appeal - 09/09/2020) (http:// power.larc.nasa.gov)

[6] Stehly T, Beiter P, Heimiller D. and Scott G. 2018.*2017 Cost of Wind Energy Review*. Golden, CO: National Renewable Energy Laboratory 46p.

IOP Conf. Series: Materials Science and Engineering 1037 (2021) 012047 doi:10.1088/1757-899X/1037/1/012047

The potential of wave energy in the coastal region of Romania

M Rus[1], E Mereuta[1,*] and M Novetschi[1]

[1]"Dunarea de Jos" University of Galati, Mechanical Engineering Department, 111 Domneasca Street

[*]Corresponding author's e-mail: elena.mereuta@ugal.ro

Abstract: In this study, the wave energy potential in the coastal region of Romania was analyzed. For this purpose, a reanalysis data set downloaded from ECMWF (European Centre for Medium-Range Weather Forecasts) was used. The reanalysis data contains information for various models. In this analysis the significant height of combined wind waves and swell, the mean wave direction and the mean wave period were analyzed in order to determine the potential of extracting clean energy from waves. The aim of this research is to analyze some performances factors (capacity factor, rated capacity, operating capacity, annul electricity production) of some of the most used wave energy convertors.

1. Introduction

The driving force of a constantly changing society is the energy sector.

The traditional resources (oil, gas, coal, etc.) are in limited quantities, constantly decreasing but also very polluting for the environment. For these reasons, primary resources, the sources (sun, wind, water, biomass, etc.) have captured the attention of many researchers around the world.

Studies in coastal areas of Canada, Australia, Ireland, United Kingdom, France, Portugal, Spain, Greece, Bulgaria, Romania have allowed the assessment of both wind and hydroelectric energy potential [1-4].

The hydroelectric source is considered as a major source worldwide due to the fact that the installed power is high with relatively low capture costs [5].

The wave characterized by height and period can be considered as an important hydroelectric source.

Wave energy is converted into electricity using energy converters (WEC), which have developed a global scale [6,7].

In order to make investments more efficient, it is very important to find the best locations for WEC placement [8].

The seasonal variability of the waves raises a real and important problem in the energy production and that is why it is very important to create maps with their distribution over time [9].

The height and power of the waves are also variable according to the geographical area. Thus, it has been found that the coastal area of the ocean is more recommended in the location of WECs than the maritime areas [10].

According to Eurostat [11], of the 28 EU member states, 11 have already reached the level necessary to achieve the 2020 national target energy: Bulgaria, Czechia, Denmark, Estonia, Croatia, Italy, Lithuania, Hungary, Romania, Finland and Sweden.

Romania ranks 9th in the hierarchy of EU states with 24% of energy consumed in 2020, a percentage that has increased 67.5% compared to 2004 (16.2%).

According to the data recorded in the National Energy System [12], during the last 10 years, it is observed that the average energy production generated by waters represents between 20.45% and

33.03% of the total energy production showing an upward trend, while the average energy production from coal represents between 23.64% and 41.94% of the total production, the trend being downward. A uniform trend is found in nuclear energy whose average production varies between 17.77% and 20.09% of total average production.

Regarding the energy produced from photovoltaic and biomass sources, an increasing trend is observed even if the average production has relatively low values of 2.34% (photovoltaic) and 0.74% (biomass) (figure 1 and 2).

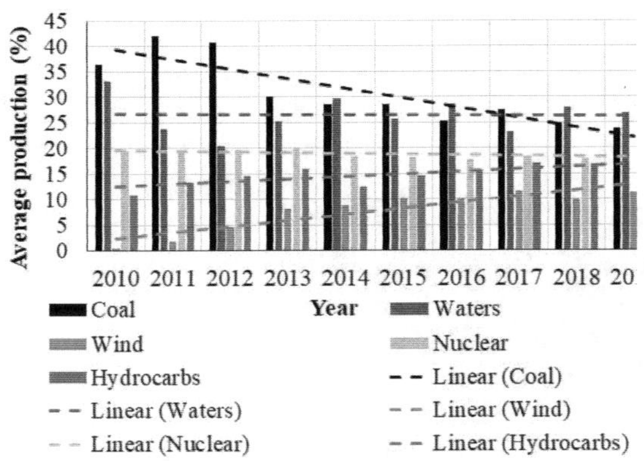

Figure 1. Average production and trend of conventional/unconventional energy

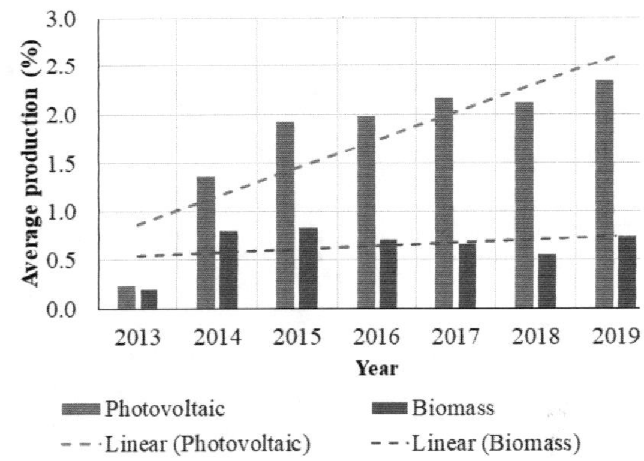

Figure 2. Average production and trend of unconventional energy (photovoltaic and biomass)

The data from figure 3 show that in Romania the production from coal ranges from 50 MW to 2068 MW for 50% of the analyzed time and 50% between 2068 MW and 4408 MW. The smallest variations are found for three of the seven types of energy recorded, namely: biomass, photovoltaic and nuclear. The production of energy from water ranges from 1276 MW to 4728 MW for 75% of the analyzed time and wind energy varies between 821MW and 2820 MW only 25% of the analyzed time (figure 3).

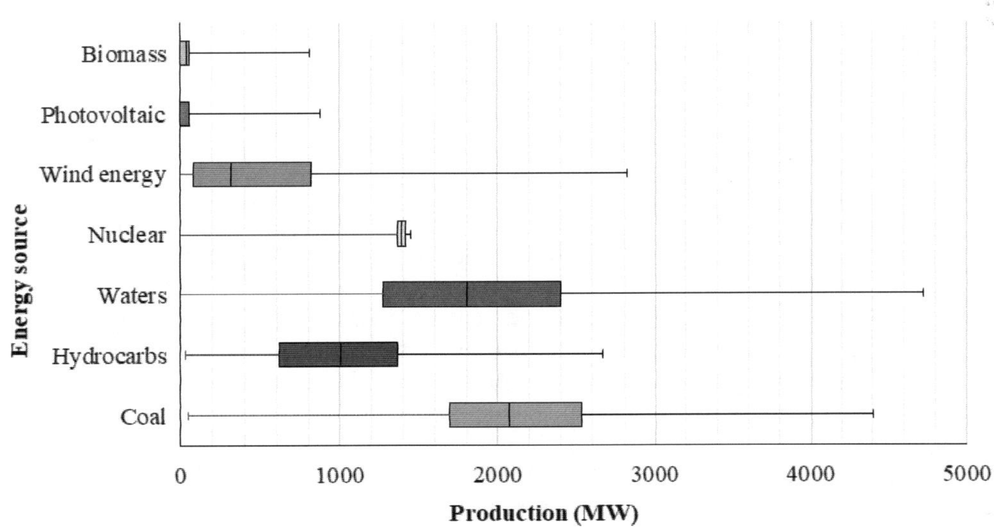

Figure 3. The representation of energy production recorded for 10 years (2010-2019) according to SEN [12]

The purpose of this study is the analysis of the energy potential of the Romanian coastal area by evaluating performance factors capacity factor, rated capacity, operating capacity, annul electricity production) of some of the most used wave energy convertors.
For this purpose, a reanalysis data set downloaded from ECMWF (European Centre for Medium-Range Weather Forecasts) was used.

2 Research methods

In this study, the Black Sea from the coastal area of Romania is analyzed on a distance from the coast between 35 km and 115 km (figure 4).
The three analyzed points are located near the localities Sulina, Navodari and Mangalia at a depth between 114 m and 323 m, whose geographical indications are indicated in table 1. The points were chosen near some cities precisely to facilitate their energy supply.

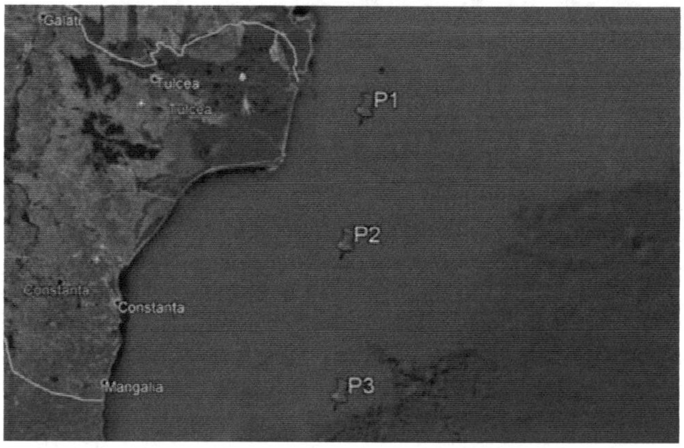

Figure 4. The geographical locations of the 3 reference points considered in the study

Table 1. The considered points coordinates

	Latitude	Longitude	Elevation	Distance to shore [km]
P 1	45.00°	30.05°	- 114	35.5
P 2	44.25°	30.00°	- 205	103.2
P 3	43.45°	30.00°	- 323	115.6

To determine the energy potential from waves in the area shown in figure 3, the following parameters were analyzed: the significant height of combined wind waves and swell (Hs) which represents the average height of the highest third of the surface of the waves generated by wind and swell; the mean wave direction (MWD) which represents the average direction of surface waves calculated as the average of all frequencies and directions of the two-dimensional wave spectrum and the mean wave period (MWP) which represents the average time required between two consecutive wave crests passing through a fixed point. All these parameters are used in coastal applications to assess the state of the sea and swell [13].
The data were downloaded from the ERA5 database, which is the latest and most accurate version of the European Center for Medium-Range Weather Forecasts (ECMWF). ERA5 provides hourly estimates of a large number of climate variables having a temporal resolution of 1 h and a spatial resolution of 31 km and the period covered is 1950 [13].
The analyzed database covers a period of 20 years, namely from 01 January 2000 to 31 December 2019. The data were prevalent every 6 hours (0:00:00, 6:00:00, 12:00:00, 18:00:00) every day for the period mentioned above.

3. Results and discussions

For the analysis and prediction for the next 5 years of the Hs wave height for the 3 points considered, the moving average method was used. The method of moving averages consists in calculating the seasonal components of the chronological series by dividing the trend by the total of the successive values of the series [14].

In figures 5, 6 and 7 is presented the Hs variation for the period 01 January 2000 to 31 December 2019. Also, these figures contains the Hs forecast and the linear trend for the analised data.

For example, the seasonality irregularity was in 2001 6% (for P1), 8% (for P2) and 9% (for P3) above the baseline (CMA), while in 2011 was 5% (for P3), 6% (for P2) and 7% (for P1) below the baseline.

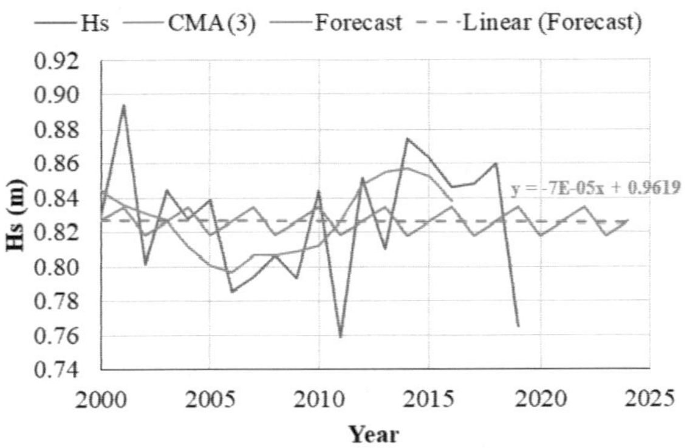

Figure 5. The representation of variation Hs for P1

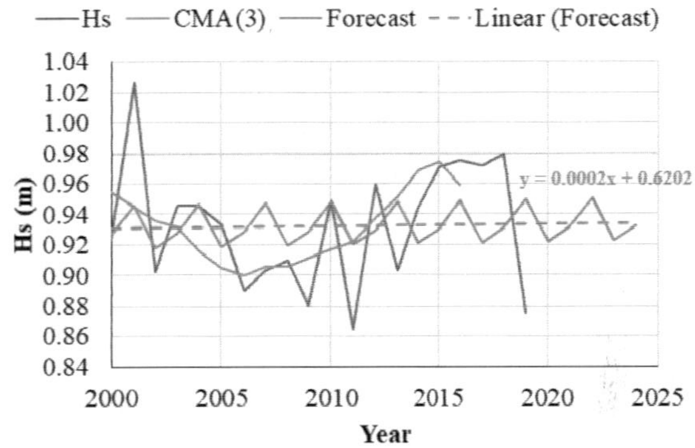

Figure 6. The representation of variation Hs for P2

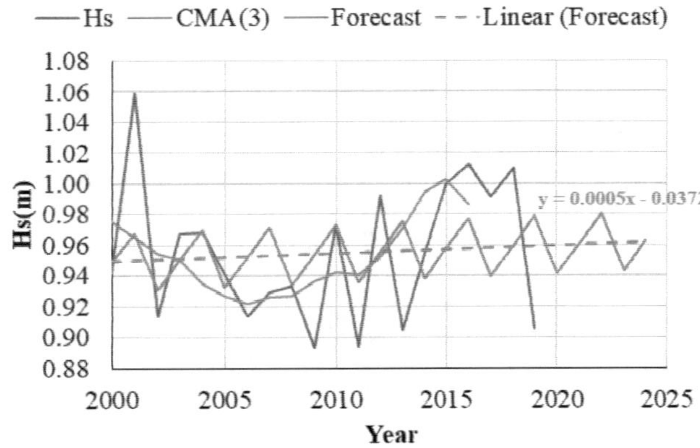

Figure 7. The representation of variation Hs for P3

One of the most important performance factors of converters is the capacity factor (CF). It shows us the percentage of WEC (wave energy convertors) running at full capacity.

Initially 6 WEC where analysed: Wave Dragon, Langlee, OE Bouy, Pontoon, Wavebob, Seabased AB. For each one the following parameters were evaluated: capacity factor CF (%) and annual energy production AEP (kW).

The wave Dragon has a rated power of 5900 kW, the Langlee – 1665 kW, the OE Bouy – 2880 kW, the Pontoon – 3619 kW, the Wavebob – 1000 kW and Seabased AB – 15kW [5, 16].

The calculated CF values for the 3 points considered and the 6 WEC are presented in Table 2.

From the presented data it can be seen that the highest performances are recorded for Seabased AB, followed by Wave Dragon and the weakest for OE Bouy.

The highest values for CF are recorded in point P3 (10.57% for Seabased AB, 6.88% for Wave Dragon and 1.54% for OE Bouy), which is the most profitable.

Table 2. The considered points coordinates

CF%	Wave Dragon	Langlee	OE Bouy	Pontoon	Wavebob	Seabased AB
P 1	4.810867	2.26479	0.936494	2.746353	1.636509	8.428131
P 2	6.228961	2.975793	1.331474	3.352864	2.255537	10.25934
P 3	6.876712	3.234538	1.535112	3.616167	2.547016	10.57002

The power matrices that formed the basis of the CF calculations are presented in figures 8, 9, 10, considering for a more detailed analysis only the 3 types of WEC mentioned above.

Hs(m)	Tm (s)					
	4	5	6	7	8	9
1.0	1282148	813276	129516	2097	0	0
1.5	472976	1326475	353250	41960	0	0
2.0	6210	669438	601596	75492	3706	0
2.5	0	81979	885416	103356	10408	0
3.0	0	3248	342664	294030	13400	0
3.5	0	0	49039	358080	29337	0
4.0	0	0	7076	134444	55352	0
4.5	0	0	0	51909	54660	0
5.0	0	0	0	0	5900	5900
5.5	0	0	0	0	5900	11800
6.0	0	0	0	0	5900	5900

Figure 8. The power matrice for P1 and Wave Dragon

Hs(m)	Tm (s)						
	4	5	6	7	8	9	10
1.0	52232	47175	7020	336	0	0	0
1.5	20162	126516	35502	4608	0	0	0
2.0	270	85422	97632	13430	448	0	0
2.5	0	8208	141622	33250	2450	0	0
3.0	0	0	60512	101878	4032	0	0
3.5	0	0	7968	100032	13034	0	0
4.0	0	0	866	46920	30464	942	0
4.5	0	0	0	10332	31640	0	0
5.0	0	0	0	1060	7000	1470	0
5.5	0	0	0	0	6760	1780	0
6.0	0	0	0	1530	4040	2120	1870
6.5	0	0	0	0	0	0	2190
7.0	0	0	0	0	0	5760	0

Figure 9. The power matrice for P2 and OE Bouy

Hs (m)	Tm (s)						
	4	5	6	7	8	9	10
1.0	7315.2	4225	415.2	25.2	0	0	0
1.5	2308.8	8057.5	1984.9	169.4	11.5	2	0
2.0	30.8	4156	4158.8	457.2	31.5	0	0
2.5	0	306	4206.8	994.5	50.6	0	0
3.0	0	0	1413.7	1897.2	102.6	5.4	0
3.5	0	0	151.2	1730.1	303.6	11.6	0
4.0	0	0	0	576.2	425.6	34	0
4.5	0	0	0	66.5	348	30.4	0
5.0	0	0	0	0	176.4	8.6	0
5.5	0	0	0	0	10.1	17.8	8.1
6.0	0	0	0	0	0	20.2	0
6.5	0	0	0	0	11.6	20.8	0
7.0	0	0	0	0	0	21.8	30

Figure 10. The power matrice for P3 and Seabased AB

4. Conclusions

The energy produced by the world's seas and oceans is a clean and permanent resource. Moreover, humanity has an impressive reservoir, the planetary ocean. Research in the field has intensified in the last decade, thus highlighting the economic and environmental potential of the planet's ocean. Although the technologies that capture the energy produced by this resource are expensive and difficult to implement, mankind is making impressive progress in this direction.

In this study we analyzed the wave data from the ERA5 database for 3 points in the coastal area of the Romanian coast located near Sulina, Navodari and Mangalia. These data cover a period of 20 years, namely 1 January 2000 and 31 December 2019, being prevailed every 6 hours (0:00:00, 6:00:00, 12:00:00, 18:00:00) every day for the period mentioned above.

We analyzed the Hs trend and the forecast for the next 5 years and we noticed that the trend is approximately constant, slightly decreasing for point P1 while for point P3 it is slightly increasing.

In order to determine the efficiency of some WEC technologies in the Black Sea (close to the Romanian coast) we determined the parameters CF (%) and AEP (kW) based on the wave height and the period.

6 WEC were compared: Wave Dragon, Langlee, OE Bouy, Pontoon, Wavebob, Seabased AB. If we analyze the power matrices from manufacturers (in optimal operating conditions) with those resulting from this research, we notice that such technologies have a low economic efficiency but have important implications for environmental protection.

References:

[1] Rusu L, Onea F 2017 The performance of some state-of-the-art wave energy converters in locations with the worldwide highest wave power, *Renewable & Sustainable Energy Reviews* **75**, 1348–1362.

[2] Ganea D, Amortila V, Mereuta E, Rusu E 2017 A joint evaluation of the wind and wave energy resources close to the Greek Islands, *Sustainability*, **9**, 1025.

[3] Ganea D, Mereuta E, Rusu L 2018 Future projection of the wind power potential for the Romania Territorry, SGEM Conference, **18 (43)**, 479-487.

[4] Veresiu S, Rus M, Mereuta E 2019 Wind power prediction in the Dobrogea region of Romania, 19th International Multidisciplinary Scientific GeoConference SGEM 2019, Volume **19**, pp. 619-626, 30 June - 6 July, Albena Bulgaria, DOI: 10.5593/sgem2019/4.1/S17.079 https://www.sgem.org/index.php/elibrary?view=publication&task=show&id=5884

[5] Guilloua N, Chapalaina G 2018 Annual and seasonal variabilities in the performances of wave energy converters. *Energy, Elsevier*, hal-01897275.

[6] Falcao A 2010 Wave energy utilization: a review of the technologies, *Renewable & Sustainable Energy Reviews* **14**, 899–918.

[7] Lindroth S, Leijon M 2011 Offshore wave power measurements - A review, *Renewable & Sustainable Energy Reviews* **15**, 4274–4285.

[8] Ganea D, Rusu L, Mereuta E 2019 Turbines evaluation in the nearshore and offshore of the Black Sea, SGEM Conference, *Renewable Energy Sources and Clean Technologies* https://www.researchgate.net/publication/334290806.

[9] Rus M, Mereuta E, Veresiu S 2019 Wave parameters distribution in coast region of Bulgaria, 19th International Multidisciplinary Scientific GeoConference SGEM 2019, Volume **16**, pp.611-618, 30 June - 6 July, Albena Bulgaria, DOI:10.5593/sgem2019/4.1/S17.078 https://www.sgem.org/index.php/elibrary?view=publication&task=show&id=5883

[10] Rusu E, Onea F 2016 Estimation of the wave energy conversion efficiency in the Atlantic Ocean close to the European Islands, *Renewable Energy*, **85**, 687-703.

[11] Eurostat. Renewable energy in the EU. Available on: https://ec.europa.eu/eurostat/documents/2995521/9571695/8-12022019-AP-EN.pdf/b7d237c1-ccea-4adc-a0ba-45e13602b428 (accessed on 15.04.2020)

[12] Sistemul Energetic National. Available on: http://www.transelectrica.ro/widget/web/tel/sen-grafic/-/SENGrafic_WAR_SENGraficportlet (accesed on 15.04.2020)

[13] Copernicus. Available on: https://cds.climate.copernicus.eu/cdsapp#!/dataset/reanalysis-era5-single-levels?tab=overview (accesed on 20.04.2020)

[14] Veresiu S., Mereuta E., Rus M., Ganea D., Amortila V. 2017 Forecast of a quality indicator in academic activity using statistical methods, IManEE International Conference 2017; www.matec - conferences.org/articles/matecconf/pdf/2017/26/matecconf_imane2017_08008.pdf

[15] Diaconu S., Rusu E. 2013 Evaluation of various WEC devices in the Romanian near shore, Proceedings of Advances in Environment Technologies, Agriculture, Food and Animal Science.

The efficiency of extracting clean energy in the Dobrogea Region of Romania

S Veresiu[1], E Mereuta[1*] and M Novetschi[1]

[1]"Dunarea de Jos" University of Galati, Mechanical Engineering Department, Domneasca Street 111, 800201 – Galati, Romania

[*]Corresponding author's e-mail: elena.mereuta@ugal.ro

Abstract. The aim of this study is to evaluate the efficiency of extracting clean energy in the Dobrogea region of Romania. More precisely was analyzed the wind potential in correspondence with some economical parameters that determines the efficiency of extracting clean energy. For this purpose, a data set that contains wind zonal and meridional components for a period of 20 years was processed. In this study, the data set was downloaded from the European Centre for Medium-Range Weather Forecasts. By analyzing the levelized cost of electricity, the annual energy production, the annual expensive and other parameters it can be easily observed that implementing new project for extracting renewable energy can be efficient.

1. Introduction

Energy is of strategic importance for economic development and is also a basic utility, with a major impact on the quality of life. The continuous evolution of the energy sector has led to a significant increase in the last century of greenhouse gas emissions, responsible for the global warming phenomenon, of acidifying gases - cause of acid precipitation, of other emissions with harmful effect on the environment and human health.

Given the worldwide responsibility has for future generations and the environment, there are clear objectives in terms of energy production from renewable sources (RES) [1-3]. They help to diversify energy sources available and most likely will eventually replace energy sources based on fossil fuels (coal, gas, oil) are steadily declining. Using RES instead of fossil fuels can substantially reduce greenhouse gases, especially CO_2 emissions and other pollutants. Also, the use of renewable sources increases the economic independence of the users who develop such energy sources.

The reality facing any country that wants to implement clean renewable energy extraction technologies is that these technologies are much more expensive than the traditional ones.

One of the most abundant renewable resources is wind energy, regardless of the location of exploitation (onshore, nearshore or offshore). Thus, the problem of finding the best method and location to extract wind energy remains an extremely important one [4,5]. Scientists around the world are making constant efforts to analyze the intermittency of natural resources. Wind power production is facing a major problem because these resources have a variability over time [6-9].

Favorable natural conditions for harnessing wind energy resources have been identified in several areas in Romania. The present analysis aims to characterize the potential in the Dobrogea area by evaluating the performance of some of the most used wind turbines. By the obtained results, the developers of extractive technologies will be able to adopt or manufacture optimized technologies for specific evolution patterns.

2. Data and methods

In this study, data coming from ERA 5 databases has been processed. ERA 5 is the latest climate reanalysis produced by the European Center for Medium-Range Weather Forecasts (ECMWF) [10, 11]. ERA 5 data are available on regular latitude-longitude grids at 0.25 x 0.25 resolution. The time window considered in this paper is 20 years, 01 January 2000 to 31 December 2019. The data set is divided in four intervals for each day, corresponding to 0:00:00, 06:00:00, 12:00:00, 18:00:00. The model analysed in this paper contains data for the zonal (*U*) and the meridional (*V*) wind velocity at 10 meter height (U_{10}). By interpolating these parameters, it can be obtained the wind velocity field for a certain altitude, location or zone.

The target area for this study is Dobrogea region, Romania (fig. 1). More precisely, for this analysis, 8 representative points were selected for the wind potential (Table 1).

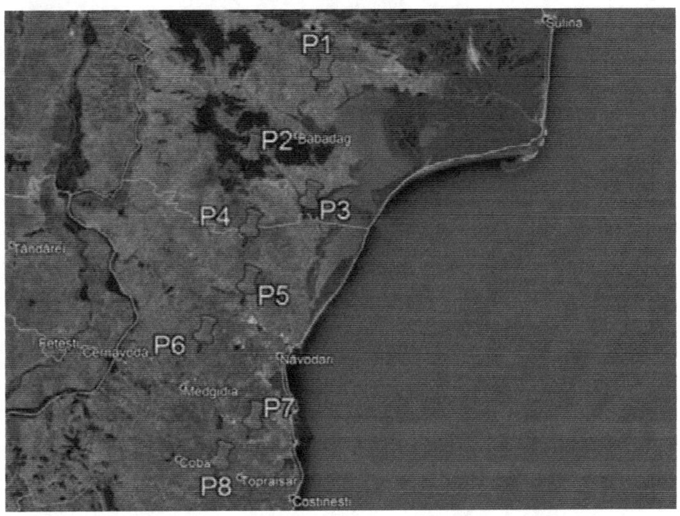

Figure 1. The geographical locations of the reference points

Table 1. The geographical locations of the points

Point	Latitude	Longitude	Elevation (m)	Point	Latitude	Longitude	Elevation [m]
P1	45.08°	28.89°	115	P5	44.63°	28.53°	195
P2	44.86°	28.20°	120	P6	44.42°	28.57°	39
P3	44.94°	28.80°	33	P7	44.28°	28.36°	38
P4	44.59°	28.27°	177	P8	43.94°	28.39°	75

Two wind turbines were considered in this study. Thus were considered a Vestas V126-3.0 turbine with a nominal power of 3 MW and a Gamesa G128-4.5 turbine with a nominal power of 4.5 MW. Several parameters such as: capacity factor (*CF%*), the annual profit (P_a) and the Break Even Point (*BEP*) were processed [12].

The CF% represents the ratio between the produced electrical power (P_{gen}) and the maximum output of a turbine (P_{rat}).

$$CF\% = (P_{gen}/P_{rat}) \cdot 100 \tag{1}$$

The annual profit from a turbine is:

$$P_a = (FCR - P_{cost\ to\ gen}) \cdot AEP \tag{2}$$

where: *FCR* is fixed charge rate, $P_{cost\ to\ gen}$ is the profitability (the cost to generate power with a wind turbine) and *AEP* is the net annual energy production.
The Break Even Point (*BEP*) represents the ratio between the initial capital (*IC*) and the annual profit (P_a) The *BEP* parameter show how long until your investment is paid for.

$$BEP = IC/P_a \tag{3}$$

3. Results

The wind speed components, zonal *U* and meridional *V*, at 10-meter were computed to determine the wind speed at 120-meter, by using a method that assumes neutral conditions:

$$U_{120} = U_{10} \frac{ln\frac{z_{120}}{z_0}}{ln\frac{z_{10}}{z_0}} \tag{4}$$

where U_{120} represent the wind speed at 120-meter height (in m/s), U_{10} is the wind at 10-meter height, z_0 is the surface roughness, while z_{120} and z_{10} are the reference heights at 120 m and 10 m, respectively.
Figure 2 show an analysis of data of the wind speed at 120-m heigh. The data were processed for the entire interval.
Each box in figure 2 contains several elements known as quartiles and whiskers. First quartile (represent the 25% of each data) is the median of the lower half of the dataset. The third quartile (represent the 25% of each data) is the median of the upper half of the dataset. The box corresponding to the middle quartile is equivalent to the 50% of data and the horizontal line drawn in it is the median.

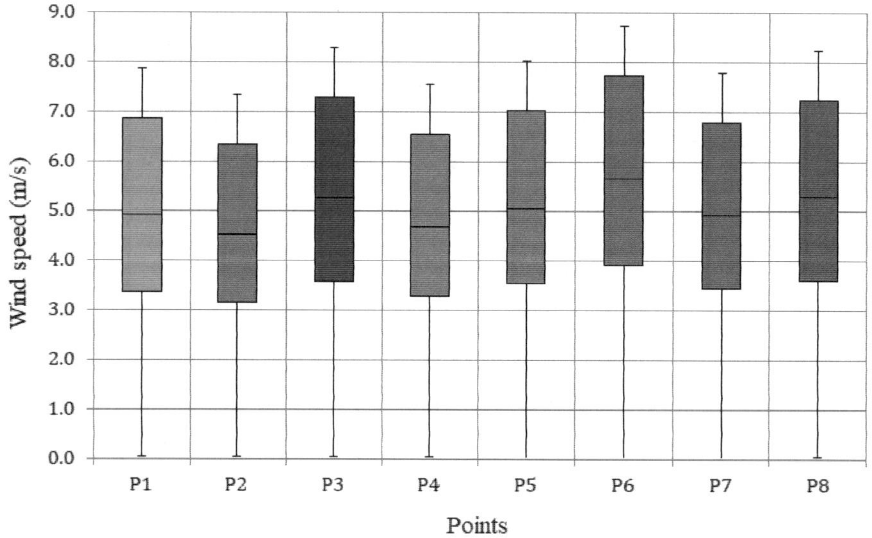

Figure 2. The assessment of the wind speed at 120-meter height for all points corresponding to 20-years interval (01 January 2000 to 31 December 2019)

By analyzing figure 2 it is observed that the wind does not vary significantly in the studied points. However, this small variation of the wind speed ranges generates high variation of the capacity factor and implicitly of the annual energy production.
Capacity factor is the measure of how often a wind turbine runs for a specific period of time. This factor (*CF%*) shows what percentage the wind turbine operates at full capacity throughout the operation period. This factor is closely and directly proportional to wind speed.
Even if the wind turbines have different power curves and capacities, they will have similar capacity factors (figure 3).

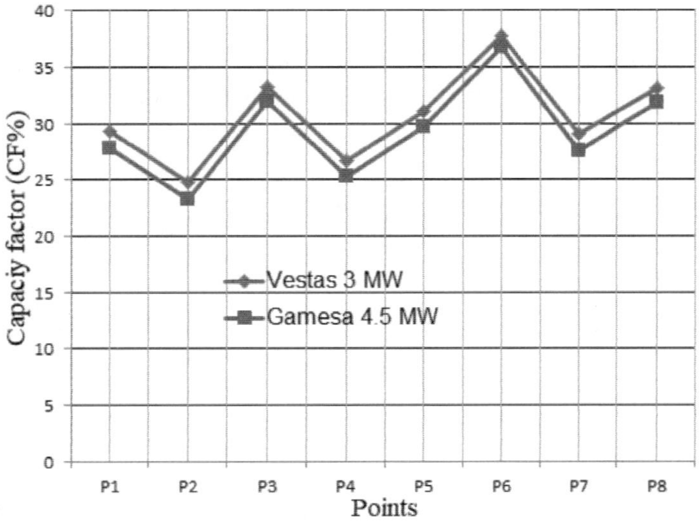

Figure 3. The mean values for the capacity factor (CF%) for all points corresponding to 20-years interval (01 January 2000 to 31 December 2019)

The difference between the two turbines results if we analyze the annual energy produced by them. Because the 4.5 MW turbine has a higher capacity factor, it will produce about 43% more than the 3 MW turbine (figure 4).

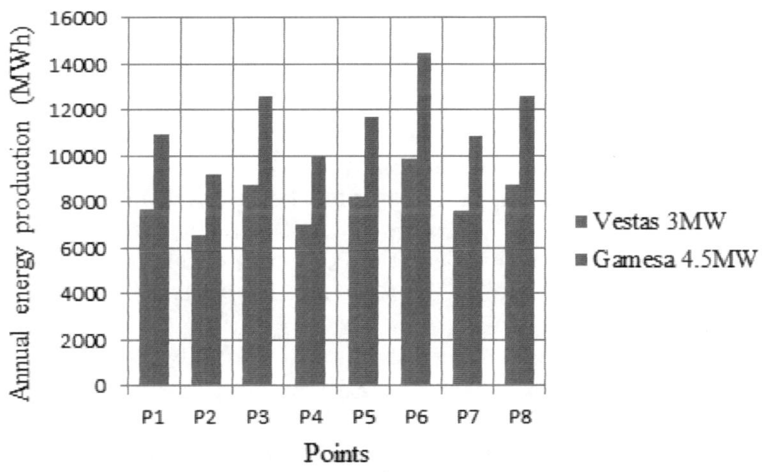

Figure 4. The mean values for the annual energy production (AEP) for all points in MWh, corresponding to 20-years interval (01 January 2000 to 31 December 2019)

In order to decide which of the two turbines is more profitable, the economic factors must be analyzed. It is known that 1 MW of installed turbine costs 1.3 million euros. Thus for the two turbines analyzed, the initial capital (*IC*) required is about 3.900.000 euro (for 3 MW) respectively 5.850.000 euro (for 4.5 MW).

Nowadays turbines are designed for a service life of 20 years, but they can continue to operate after that if properly maintained.

The results from the economic analysis for the most representative points (P3, P5, P6, P8) are presented in table 2. As it can be seen even if the capacity factor does not vary significantly in value, at these points, the break-even point does. According to the results, it can be noticed that for the two analyzed

turbines the investment it is not profitable after a certain point, respectively a capacity factor less than 37%.
In the table 2 *IC* represent the initial capital, *O&M* is the cost for operations and maintenance, T_{ae} is the total annual expense, I_g is the annual gross income, P_a is the annual profit and *BEP* represent the break-even point.

Table 2. The results of the economic analysis for the most representative points

Point	mean CF%		mean AEP (MWh)		IC (mil. €)		O&M (€)		T_{ae} (€)	
	3MW	*4.5MW*	*3MW*	*4.5MW*	*3MW*	*4.5MW*	*3MW*	*4.5MW*	*3MW*	*4.5MW*
P3	33	32	8727	12575	3.9	5.85	50269	72433	534069	794133
P5	31	30	8181	11709	3.9	5.85	47122	67442	530922	789142
P6	38	37	9911	14484	3.9	5.85	57085	83426	540885	805126
P8	33	32	8710	12557	3.9	5.85	50171	72326	533971	794026
Point	**mean CF%**		**mean AEP (MWh)**		**I_g (€)**		**P_a (€)**		**BEP (years)**	
	3MW	*4.5MW*	*3MW*	*4.5MW*	*3MW*	*4.5MW*	*3MW*	*4.5MW*	*3MW*	*4.5MW*
P3	33	32	8727	12575	628358	905413	94289	111280	41	53
P5	31	30	8181	11709	589020	843031	58098	53888	67	109
P6	38	37	9911	14484	713566	1042829	172680	237703	23	25
P8	33	32	8710	12557	627134	904079	93164	110053	42	53

4. Conclusions

Renewable power is booming, as innovation brings down costs and starts to deliver on the promise of a clean energy future. While renewable energy is often thought of as a new technology, harnessing nature's power has long been used for heating, transportation, lighting, and more. This means that renewables are increasingly displacing "dirty" fossil fuels in the power sector, offering the benefit of lower emissions of carbon and other types of pollution.
The wind is an intermittent energy source, which cannot make electricity nor be dispatched on demand. It also gives variable power, which is consistent from year to year but varies greatly over shorter time scales. Therefore, it must be used together with other electric power sources or storage to give a reliable supply. Wind power is one of the fastest-growing renewable energy technologies. Usage is on the rise worldwide, in part because costs are falling.
In this paper was studied the profitability of investments in wind turbines. Thus, 2 turbines of 3 MW and 4.5 MW were chosen in this study. The technical parameters of these turbines were computed along wind speed data coming from ECMWF. More precisely, data coming from ERA 5 databases has been processed. ERA 5 is an ongoing global reanalysis project developed and maintained by the European Center for Medium Range Weather Forecast (ECMWF).
In this study, the ERA 5 data with a spatial resolution of 0.25^0 x 0.25^0 was processed. The time window considered in this paper is 20 years, 01 January 2000 to 31 December 2019. The model analyzed in this research contains data for the zonal (*U*) and the meridional (*V*) wind velocity at 10-meter height (U_{10}). By computing and interpolating these parameters, was obtained the wind velocity field for a certain altitude (120 meter), location or zone. Wind direction patterns were also determined. The result presented in this paper provides a comprehensive picture of the wind speed model and also profitability. For the 2 turbines considered, an economic analysis was made. By analyzing the annual energy production (*AEP*), cost for operations and maintenance (*O&M*), annual expense (T_{ae}), annual gross (I_g), annual profit and others, new zones for implementing wind farm on the Romania territory are revealed. It is especially important that the place of implementation of a wind farm is chosen appropriately. Even if there are small variations in terms of capacity factor and thus wind speed, they can have a negative impact on depreciation.

Regarding this aspect, according to data obtained points P3, P6 and P8 are the most efficient in economic terms.
We believe that such technologies should be implemented in less efficient areas. Even if the cost amortization factor is low, the environmental implications are much greater.
However, there are also a number of negative effects on the environment generated by wind farms, such as noise pollution and effects on soil, subsoil, flora and fauna.
Therefore, before the location of wind turbines, environmental impact studies are needed.

References

[1] Ganea D, Mereuta E, Rusu L 2019 The economic aspects of extracting wind energy in Romania, *SGEM Conference*, **19(4.2)**, p. 375
[2] Jung C and Schindler D 2018 On the inter-annual variability of wind energy generation – A case study from Germany, *Applied Energy*, **230**, p. 845-854.
[3] Ganea D, Mereuta E and Rusu E 2019 An Evaluation of the Wind and Wave Dynamics along the European Coasts, *Journal of Marine Science and Engineering*, **7(2)**, p. 43.
[4] Koletsis I, Kotroni V, Lagouvardos K, & Soukissian T 2016 Assessment of offshore wind speed and power potential over the Mediterranean and Black Seas under future climate changes, *Renewable and Sustainable Energy Reviews*, **60**, p 234-245.
[5] Onea F, Deleanu L, Rusu L, Georgescu C 2016 Evaluation of the wind energy potential along the Mediterranean Sea coasts, *Energy Exploration & Exploitation*, **Vol. 34(5)**, p.766–792.
[6] Egli F, Steffen B, & Schmidt T 2018 A dynamic analysis of financing conditions for renewable technology, *Nature Energy*, **3**.
[7] Rusu L, Bernardino M, Guedes Soares C 2014 Wind and Wave Modelling in the Black Sea, *Journal of Operational Oceanography*, **7**, p. 5-20.
[8] Onea F, Rusu E 2014 Wind Energy Assessments Along the Black Sea Basin, *Meteorological Application*, **21**, p. 316-329.
[9] Ganea D, Mereuta E, Rusu L 2018 Estimation on the Near Future Wind Power Potential in the Black Sea, *Energies*, **11**, p. 3198.
[10] ERA5. Available on: https://www.ecmwf.int/. Acces on 15.05.2020
[11] Olauson J, 2018 ERA5: The new champion of wind power modelling? *Renewable Energy*, **126**, p. 322-331.
[12] Sedaghat A, Alkhatib F, Eilaghi A, Sabati M, Borvayeh L, Mostafaeipour A 2019 A new strategy for wind turbine selection using optimization based on rated wind speed, *Energy Procedia*, **160**, p. 582-589.

Acknowledgments

The ECMWF data used in this study have been obtained from the ECMWF data servers.

IOP Conf. Series: Materials Science and Engineering 1037 (2021) 012049 doi:10.1088/1757-899X/1037/1/012049

Modal analysis of the nozzle guide vane in low pressure turbine system of aircraft engine

R Robak[1,2] **and M Szczepanik**[3]

[1] Silesian University of Technology, Faculty of Energy and Environmental Engineering, Konarskiego 18, 44-100 Gliwice, Poland
[2] AvioPolska Sp. z o.o., Michała Grażyńskiego 141, 43-346 Bielsko-Biała, Poland
[3] Silesian University of Technology, Faculty of Mechanical Engineering, Konarskiego 18A, 44-100 Gliwice, Polska

[*]Corresponding author's e-mail: Rafal.Robak@avioaero.it

Abstract. The aim of the article is to present modal analysis of nozzle guide vane in low pressure turbine system of aircraft engine. Comparison of results obtained by means of finite element analysis with results of experimental one in terms of frequencies and modal forms are presented. Additionally, the article presents consistent results which will be used in further analyses and researches related to the optimization of dynamic characteristics with application of artificial intelligence.

1. Introduction
Modal analysis [1], [2] is fundamental assessment in the design phase of nozzle guide vane in low pressure turbine system. Evaluation is crucial for new concept design but also in case of design modification domain. Technical requirement is to ensure appropriate durability level (number of flight cycles) and reliability of the system. Understanding of dynamic behavior is one of a key element in the high cycle fatigue (HCF) evaluation. Finite element method (FEM) analysis [3] are widely used in new product introduction phase to verify modal characteristics with respect to operating range and engine orders. Verification at early stage of design is providing an opportunity to address dynamic issues before first prototype enters in the engine certification tests.

2. Modal analysis
The result of the analysis is a natural frequency and mode shape. Basing on the modal shape additionally from analysis can be obtained modal strain and stresses. Mode shapes are normalized to the unitary modal mass or unitary displacement therefore obtained strains and stresses are not real and need to be adjusted using dedicated experiment or frequency response analysis. Nevertheless, information about strain and stress patterns are important in order to optimize strain gage position and identify critical location from HCF perspective.

2.1. Analysis theory
Basic system to demonstrate relationship between modal and mechanical characteristics is mass-spring system presented in figure 1.

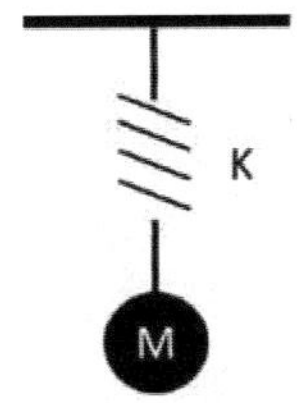

Figure 1. Mass-spring system

Natural frequency of the system is evaluated using equation 1.

$$f = \frac{\omega}{2\pi}\ Hz;\ \omega = \sqrt{\frac{k}{m}}\ \text{rad/s} \tag{1}$$

where: f and ω – natural frequency, k – stiffness, m – mass.

Understanding of basic system and relationship between stiffness, mass and natural frequency is crucial to develop proper strategy to resolve dynamic problems (increase stiffness and/or reduce mass in order to raise a natural frequency).

2.2. Campbell diagram
Information about dynamic characteristic needs to be verified together with engine order and assessed in the operating range against potential resonance points (aligning of natural frequency and excitation source frequency; harmonic function) as presented in figure 2. Identified crossing not necessarily means resonance because structure could still have enough damping to prevent amplitude growth however such condition is not desired from durability standpoint (high dynamic stress or extensive wear problems).

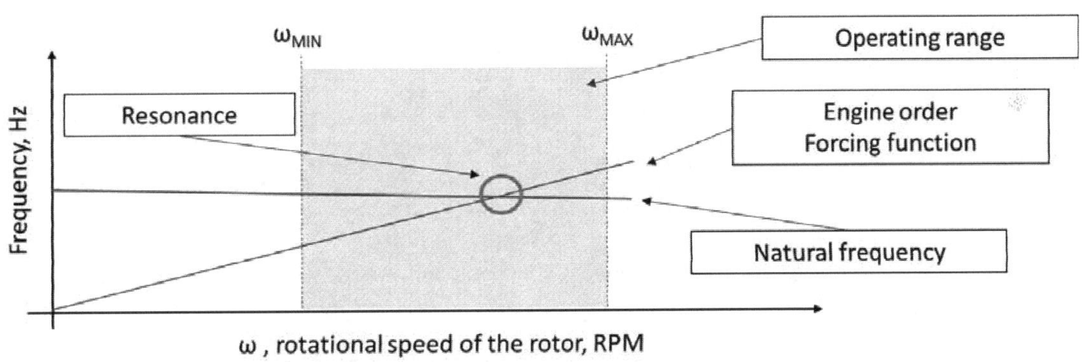

Figure 2. Campbell diagram (scheme)

Identified resonance appears together with assumption of not adequate damping.

Typical engine input in the turbine are mechanical extortion caused by rotor imbalance and aeromechanical due to blade passing phenomena (down and upstream excitation).

2.3. FEM analysis
Finite element model in order to evaluate natural frequency and modal shape is expressed by equation 2.

$$M\ddot{x} + Kx = 0 \tag{2}$$

where: M – mass matrix, K – stiffness matrix, x – displacement

In order to successfully perform analysis by means of finite element method [4] the applied numerical model should be possibly consistent with real device and should include all effects which may affect results. Typically, in the analysis of the nozzle guide vane, the important factors are temperature and presence of adjacent components since are impacting model stiffness.

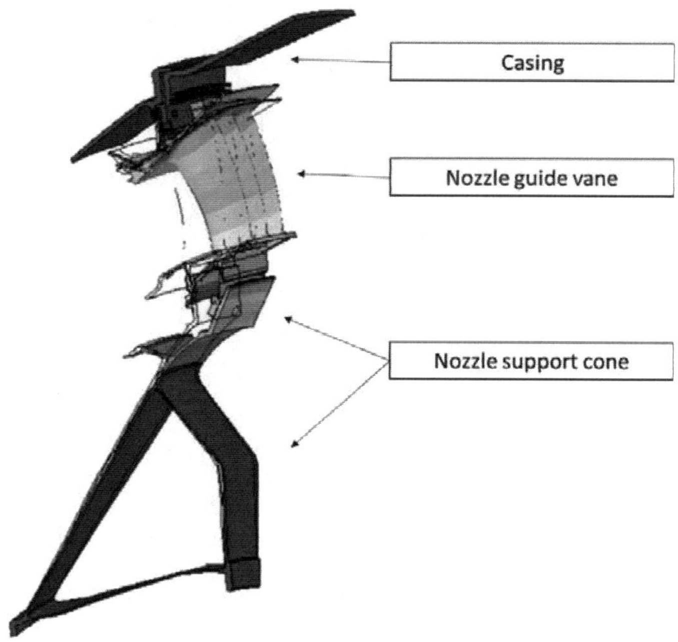

Figure 3. First system mode – pendulum - axial

Figure 3 shows first system mode of nozzle guide vane with pendulum axial mode shape. Making more detailed verification on strain energy distribution above system can be simplified to the basic mass-spring system.

Table 1. Modal strain energy distribution for pendulum mode-shape

Mass-spring system	Strain energy contribution
Nozzle guide vane (mass element)	5%
Nozzle support cone (spring element)	95%

Crossing of natural frequency with engine input can be eliminated by increasing a stiffness associated with high strain energy contribution and reducing a mass on nozzle side.

3. Modal analysis validation

Main purpose of the validation is to benchmark airfoil modal results between FEM model and component test (real hardware) results. This step demonstrates finite element model consistency in terms of frequencies and modal shapes. Experiment is conducted in ambient temperature by use of the special fixture. The device (special grip) is subjected to the harmonic loads with frequencies identified by ping test. FEM model has been constrained by use of the zero displacements in all three directions at outer band forward and rear hook as presented in figure 4.

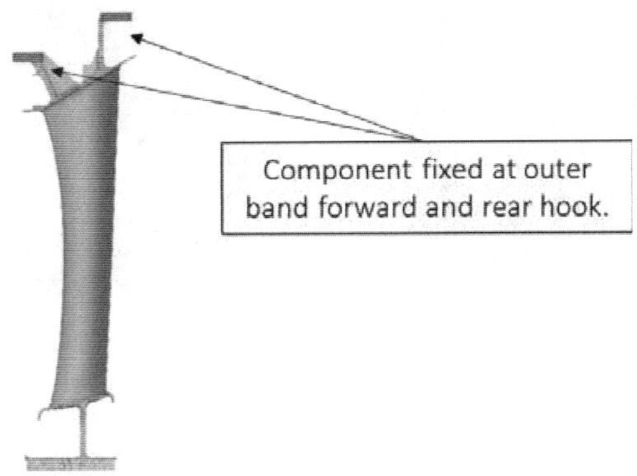

Figure 4. FEM model constrains

Model discretization has been completed with tetrahedral elements considering mid side nodes (quadratic shape function) in order reflect correct model stiffness. Modal analysis uses Block Lanczos eigen solver method with results normalized to the unitary modal mass.

Table 2. Comparison of modal analysis results coming from FEM and experiment tests

Fifth flexural airfoil mode (5F)	
Numerical prediction	Component test results
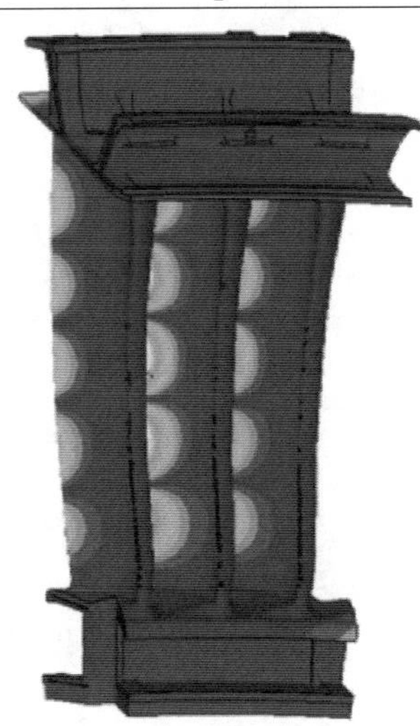	

Table 3. Comparison of modal analysis results coming from FEM and experiment tests

The second flexural airfoil mode (2F)	
Numerical prediction	Component test results

IOP Conf. Series: Materials Science and Engineering 1037 (2021) 012049 doi:10.1088/1757-899X/1037/1/012049

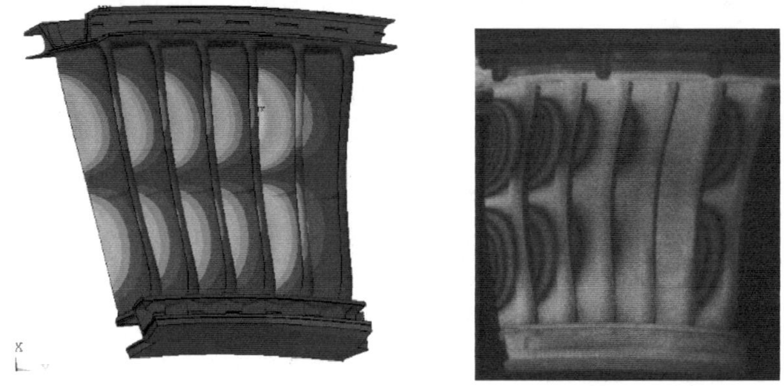

Mode shape for both FEM model and component test results present the same modal family; fifth flexural mode for table 2 results and the second flexural mode reported in table 3. Maximum amplitude occurs at trailing edge section of the aerodynamic profile. Numerical model prediction for presented in table 2 mode shape frequency is 7497 Hz compared to the component test results 7560 Hz. Difference is well below 1% and acceptable from design quality standpoint. Comparison proves good accordance between results obtained by means of FEM analysis and real system tests. All aerodynamic profiles of the nozzle guide vane show the same displacement pattern however in real hardware slight variation can be observed. The main reason is the fact that numerical model by the definition accounts nominal condition of the geometry (design intent) while in real component variation within tolerance can be observed and affects a modal pattern.

4. Summary and conclusions

The purpose of the test was to confirm the correctness of the FEM model in the context of the obtained modal forms and thus also the deformation distributions, which are crucial in the subsequent phases of validating the structure dynamics.

Numerical models take into account not only the complex geometry of the component, but also high-tech materials (directly solidified or single crystal alloys) that exhibit the characteristics of orthotropic materials. Numerical simulations of real system with complex geometry made and materials, gives proper results which are in good accordance with real experiment. Positive outcome from validation is providing confidence in the sequential steps basing on numerical models (e.g. definition of strain gage position and orientation, prediction of HCF critical location).

References

[1] Geradin M and Rixen D J 2015 *Mechanical Vibrations* (Wiley)
[2] Case J., Chilver A. H. 1986 *Strength of materials and structures* (Baltimore: Edward Arnold)
[3] Zienkiewicz O.C. 2013 *The Finite Element Method: Its Basis and Fundamentals*, ELSEVIER
[4] https://www.ansys.com/

IOP Conf. Series: Materials Science and Engineering 1037 (2021) 012050 doi:10.1088/1757-899X/1037/1/012050

The influence of the hybridization factor on vehicle energetic consumption for different e-gearbox configurations

C Rențea[1], M Bățăuș[1], S Maican[2], M Oprean[1] and G Frățilă[1]

[1] Automotive Engineering Department, University Politehnica of Bucharest,
[2] Faculty of Mechanical Engineering and Mechatronics, University Politehnica of Bucharest

*Corresponding author's e-mail: rentea.cristian@yahoo.com

Abstract To reduce the CO_2 emission level for the new CAFE regulations, it is mandatory that automobile producers should choose the hybridization approach.
Relevant published literature show that for a mild hybrid passenger car, increased values of the hybridization factor (HF) have marginal influence over the fuel economy for high speed values if a fixed ratio e-gearbox is used. This paper analyzes the energetic consumption for a Through the Road Hybrid Vehicle (TTR-HEV) when using a two-gear ratio transmission for the electric axle.
To investigate the energetic performances, a complex model developed in a performant simulation environment is used. The energetic consumption is determined for five different test cycles in terms of speed and duration by analyzing the fuel consumption ratio (FCR) for different values of the HF.
In order to compare the test cycles, the FCR parameter is used, after which the correlation with vehicle speed fluctuation (VSF) is explored.
The simulation results obtained for the current application are compared with a hybrid vehicle equipped with a single gear in the rear axle. The results are also expressed in terms of CO_2 emission.

1. Introduction
Compared to the vehicles solely equipped with internal combustion engines, hybrid vehicles improve the fuel consumption as well as the CO_2, given the fact that these two requirements are strongly related to one-another.

According to the traction component types of the powertrain, these vehicles are categorized in three different types: parallel, serial and mixed respectively.

Another classification based on the hybridization factor [1], defined as the ratio between the power of the electric motor and the total power for both thermal and electric sources of the vehicle (equation 1) is presented in table 1.

$$HF = \frac{P_{em}}{P_{em} + P_{ICE}} \quad (1)$$

where P_{em} is the power of the electric motor and P_{ICE} represents the power of the internal combustion engine.

A particular parallel structure is approached for this study, known in the published literature as the Through the Road Hybrid Vehicle (TTR-HEV).

Table 1. Vehicle classifications based on HF values

HF	Hybrid type	Characteristics
0 ... 0.1	Minimal Hybrid	No pure electric mode
0.25 ... 0.5	Mild Hybrid	The energy provided by the purely electric mode cannot support a full standard driving cycle completion
0.5 ... 0.7	Full Hybrid	The electric system fully supports the vehicle to follow a standard cycle

The specific powertrain components for this structure are an internal combustion engine, which is placed on the front axle, and an electric motor that provides power to the rear axle.

The electric motor also functions as a generator, in order to allow the regenerative braking, [2].

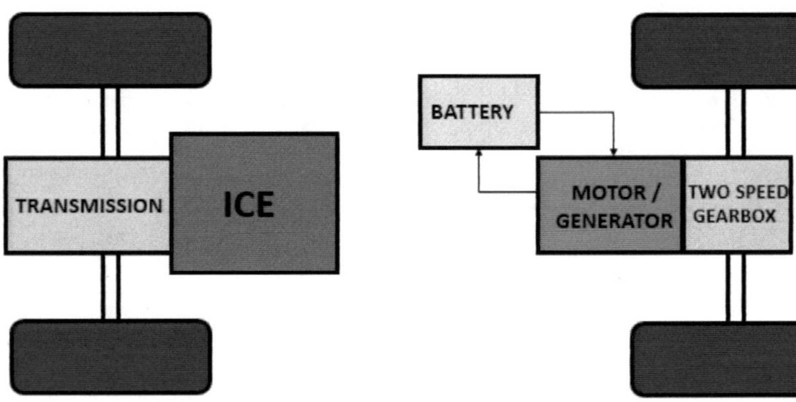

Figure 1. Through the road parallel hybrid vehicle structure

As for the pure electric mode of the TTR-HEV but requiring improvements in terms of vehicle performances such as: driving on steep roads or driving at high speed on a horizontal road, an increased number of gears for the electric axle represents a good solution.

Furthermore, increasing the number of gears results in the cut back of the energetic consumption, therefore imposing the electric motor to function at high efficiency.

On the other hand, this implies more weight for the electric axle, a reduced transmission efficiency, and also an additional gear shifting mechanism needs to be installed, thus making the manufacturing cost high.

Consequently, the optimum number of gear ratios must be chosen such that the adopted solution should reduce the energetic consumption, but in the same time the vehicle's overall cost to be maintained as low as possible.

The purpose of this paper is to analyse the energetic consumption for a middle class vehicle when two gear ratios are used for the electric axle of a TTR-HEV, considering three different values of the hybridization factor from the mild hybrid range. For all HF values, the battery capacity was kept constant at 5 Ah in the hybrid model simulation.

All the parameters will be determined for five driving cycles: US06, Japanese 10-15, HWFET, NEDC and WLTC.

2. Model development

The hybrid vehicle model, figure 2, has been developed based on specialized libraries from the simulation environment LMS Imagine.lab AMESim, [3, 4].

There are 10 main submodels used for the HEV model, all of which have the parameters modified in order to correlate the vehicle speed with the cycle profile and also to simulate all the topologies aforementioned.

The main submodels used for the HEV model are: a driver submodel (DRVDRVA00A-1) necessary to follow the cycle given by the mission profile and ambient data submodel (DRVMP2A-1), electric motor (DRVELMT0A), battery (DRVBAT03), internal combustion engine (DRVICE01E), control unit for a parallel hybrid vehicle (DRVCUPH1A), gearbox (DRVMGC01), e-Gearbox (DRVDCT01), Transmission Control Unit for Automatic Gearbox (DRVAGTCU01), which allows the shifting gears according to a proposed shifting schedule and 1-D vehicle (DRVVEH4A).

For the power electronic converter that includes the standard DRVPET0B submodel, two special supercomponents were added, one having the role of completely interrupting the electric machine circuit when the torque command is null and the other which removes the discontinuities caused by portions of the test cycle when the vehicle is stationary.

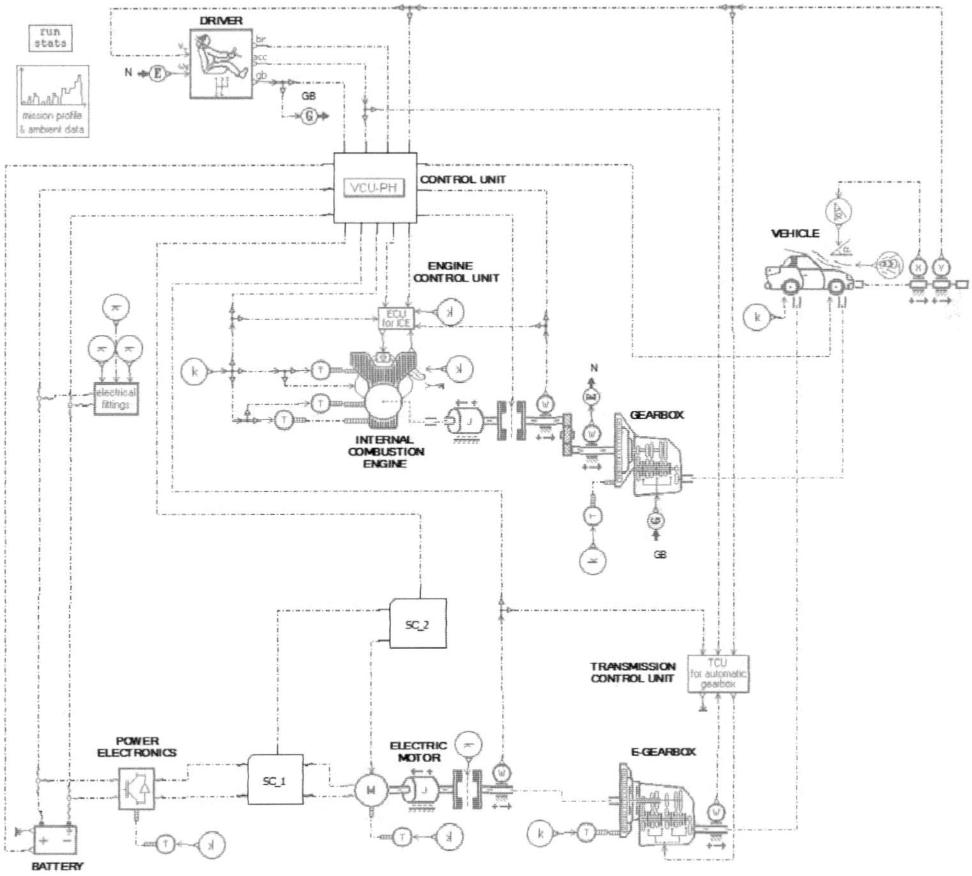

Figure 2. TTR-HEV model for the energetic consumption study

For the front axle driver command, the gearbox ratio control is the signal input, whereas three more signal inputs are used for the mechanical components such as wheel rotary velocity, engine torque and gearbox temperature.

For the rear axle, an automatic gearbox linked to another control unit is preferred when shifting two gears, whereas for the single gear model the latter component is no longer needed.

The most important submodel for a hybrid vehicle is the electronic control unit. It gets information from the driver, the engine, the electric machine, the battery and it receives vehicle speed data.

3. Gear ratios and shifting schedule

For a TTR-HEV that has a two speed gearbox on the electric axle, the first gear is used to gain acceleration at low speeds and to climb inclines, whereas for high speeds the second gear is necessary [5].

For this study, the first and second gear ratios are determined using the equation (2) and equation (3) respectively, [6].

$$i_{t\,max} = \frac{r_r \cdot m \cdot g \cdot (f_r \cdot \cos\alpha + \sin\alpha)}{T_m \cdot \eta_{pt}} \quad (2)$$

where r_r – rolling radius (0.3 m), m – vehicle mass (1600 kg), g – gravity, f_r – rolling resistance coefficient (0.01), α – road incline angle (20°), T_m – maximum torque (208 Nm), η_{pt} – e-gearbox efficiency (0.95).

$$i_{\min speed} = \frac{3.6 \cdot \pi \cdot n_m \cdot r_r}{30 \cdot V_{max}} \quad (3)$$

where n_m – maximum motor speed (6800 rev/min), V_{max} – maximum vehicle speed in pure electric mode (128 km/h).

The obtained values for gear ratio are $i_{t\,max}$ =8.4 and $i_{\min speed}$ = 6, the variation of vehicle acceleration being represented for these two values in figure 3.

The pure electric mode was obtained based on the TTR-HEV model, where the internal combustion engine was set unfunctional, thus making the front axle non-driven. The vehicle acceleration and energy consumption in this mode are presented in figure 3 and figure 4.

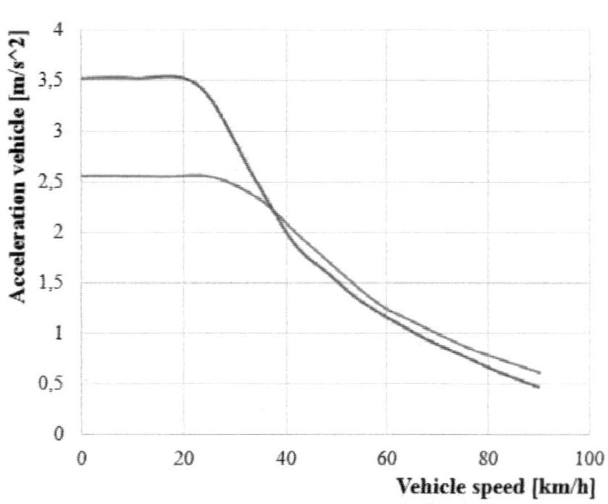

Figure 3. Vehicle acceleration

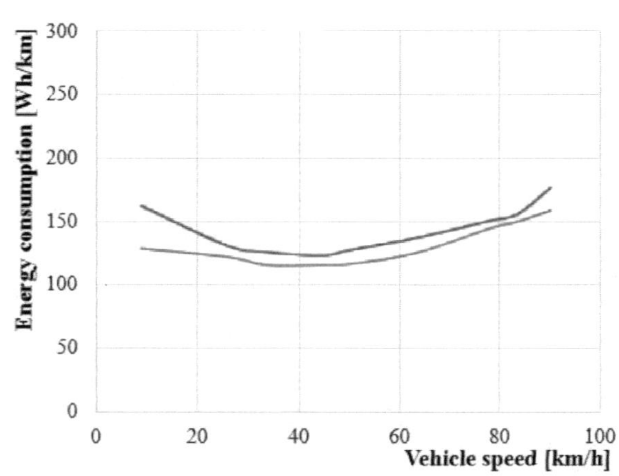

Figure 4. Energy consumption at constant speeds

As the transmission efficiency depends on the gear, the values for the e-gearbox efficiency are 0.95 for an 8.4 ratio and 0.97 for a 6 ratio, respectively.

Figure 5 reflects the shifting schedule used for this study [7], by considering the energy consumption at constant speeds for the two gears (figure 4) and the variation of vehicle acceleration (in the domain of interest).

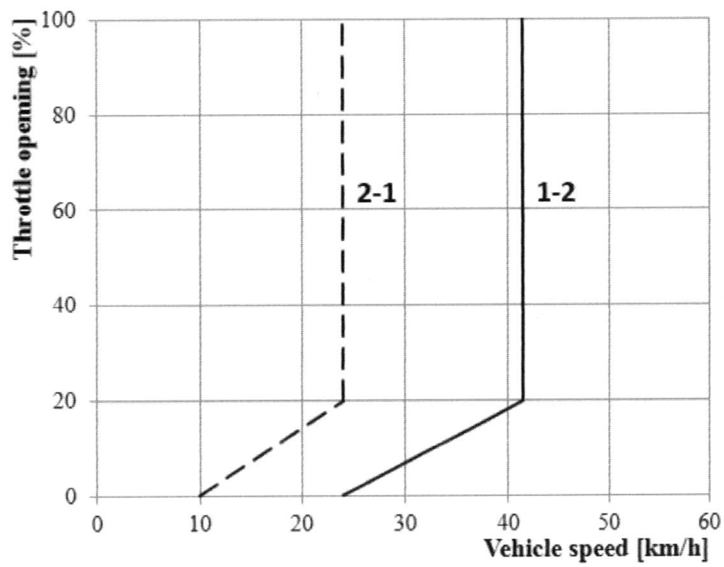

Figure 5. Shifting schedule for two speed transmission

The vehicle performances in pure electric mode resulted within simulations with the chosen gear ratios, are presented in table 2.

Table 2. Vehicle performances in pure electric mode

No.	Parameter	One gear 8.4	One gear 6	Two gears 8.4/6
1	Maximum speed [km/h]	110	128	128
2	Maximum acceleration [m/s]	3.52	2.07	3.48
3	Acceleration time 0-50 km/h [s]	5.4	6.18	5.42
4	Maximum grade [°]	20	14	20
7	Energy consumption at 50 km/h [Wh/km]	127.3	115.8	116.4

4. Results

In the hybrid mode, the fuel consumption for different powertrain solutions is being determined, considering three values for the hybridization factor.
Table 3 shows the fuel consumption classified in two categories, based on the hybridization factor and also considering different e-gearbox configurations (one that has two gear ratios and a single ratio gearbox).

Table 3. TTR-HEV fuel consumption for different HF

No.	Cycle	HF=0.22			HF=0.32			HF=0.42		
		6	Two gears	8.4	6	Two gears	8.4	6	Two gears	8.4
1	NEDC	5.366	5.355	5.3	5.5482	5.535	5.664	5.934	6.062	6.39
2	WLTP	5.786	5.840	5.818	5.902	6.034	6.14	6.257	6.406	6.833
3	10-15	4.797	5.076	5.032	5.458	5.702	5.944	6.338	6.654	6.93
4	US-06	6.28	6.354	6.44	6.342	6.397	6.569	6.343	6.39	6.57
5	HWFET	4.99	5.015	4.896	4.91	4.929	4.908	4.91	4.93	5.03

From table 3, one can observe that as the hybridization factor increases, the difference between the fuel consumption obtained for using a single speed e-gearbox with i=6 and i=8.4 increases for each cycle driven, thus utilizing a 2-speed e-gearbox clearly becomes advantageous. It can be stated that the fuel consumption for using a 2-speed e-gearbox is close to the one for a single gear with i=6 ratio, but with a far superior dynamic performance, as shown in table 2.

It can be noticed that for average speed higher than 80 km/h cycles, US-06 and HWFET, when the hybridization factor is 0.32 and 0.42, the fuel consumption for a 2-speed e-gearbox configuration is similar. Also, from table 3 it can be observed that when driving on HWFET cycle, the three configurations studied have the least influence on fuel consumption.

In figure 6, the fuel consumption obtained from 5 driving cycles considering 3 transmission configurations for HF=0.42 is shown. The data obtained from the gasoline vehicle model at a constant speed equal to the average speed value for each cycle is considered as the reference curve.

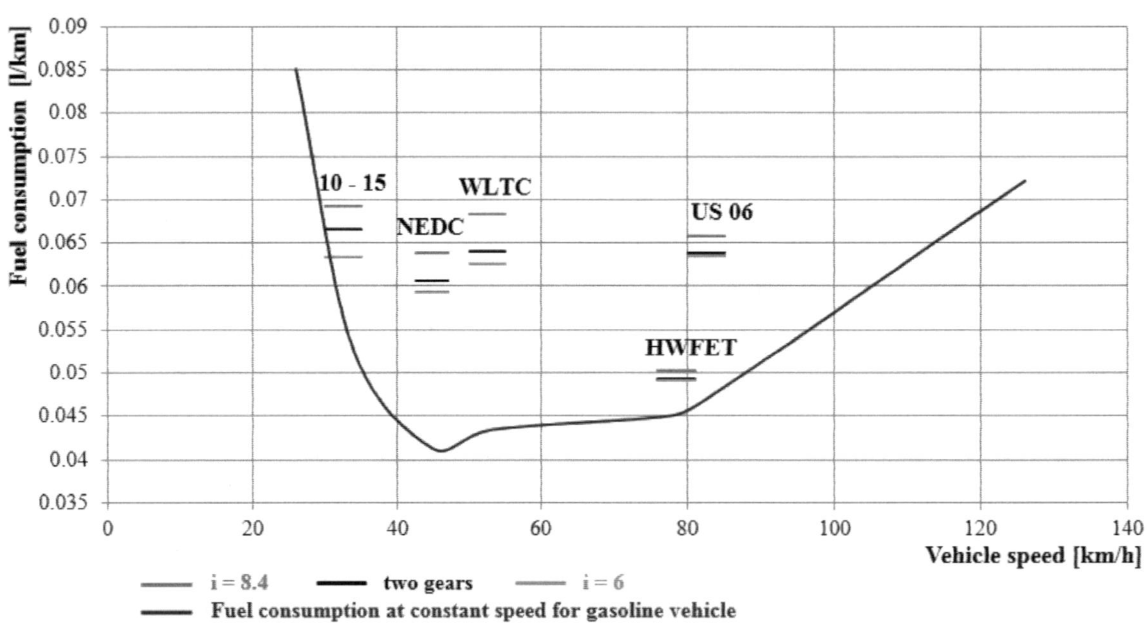

Figure 6. Fuel consumption for different gear ratios at HF=0.42

After completing the simulations, it is observed that using a 2-speed e-gearbox becomes increasingly advantageous as the average speed of the driven cycle is higher.

In figure 7, the correlation between FCR and VSF is shown, [8]. Four regression lines are represented, one for each degree of hybridization for a 2-speed e-gearbox configuration and a regression line for the gasoline vehicle with a point of intercept set at (0,100).

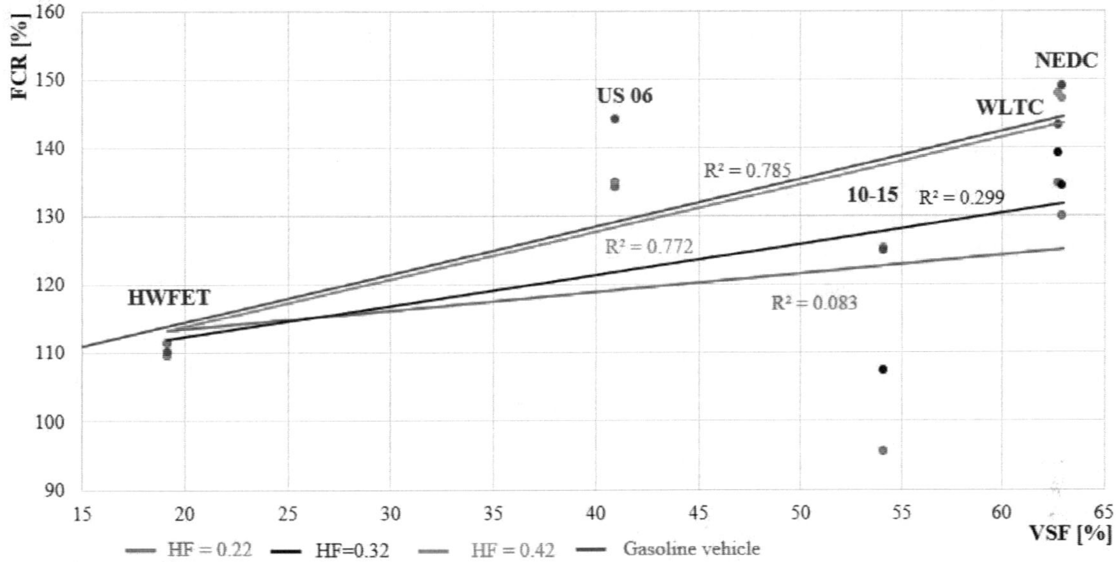

Figure 7. Correlation between FCR and VSF for three degrees of hybridization

Based on the low value of the coefficient of determination R^2, it can be noted that there is no linear correlation between FCR and VSF for the TTR-HEV equipped with a two-speed e-gearbox.

5. Conclusions

As the hybridization factor increases, implementing a 2-speed e-gearbox solution clearly becomes advantageous because the difference between the fuel consumption obtained when using a single speed e-gearbox for two different ratios increases for each test cycle.

The fuel consumption for using a 2-speed e-gearbox is close to the one for a single gear with a ratio of 6, but with a much better dynamic performance.

One can observe that the linear dependency between FCR and VSF, defined by the coefficient of determination, is being reduced as the hybridization factor decreases.

In terms of CO_2 emissions of the two-speed e-gearbox vehicle model, the relative difference from a maximum value, obtained for the Japanese 10-15 cycle, and a minimum value, achieved for HWFET mode, is of 25%. Also, it can be acclaimed that the hybridization degree has a minor influence (less than 3.5%) for cycles with high average speed (US-06 and HWFET cycles).

For the current application, no notable improvement in terms of fuel consumption has been observed, but for the pure electric mode, the dynamic performance and the energetic consumption are far superior. Thus, a two-speed e-gearbox can be very useful to be implemented on a plug-in hybrid vehicle.

References

[1] Aurelio Somà *Trends and Hybridization Factor for Heavy - Duty Working Vehicles*
[2] Pisanti C, Rizzo G, Marano V *Energy Management of Through-The-Road Parallel Hybrid Vehicles*
[3] Siemens Industry Software NV – LMS Imagine.Lab Amesim 2015 *IFP Drive Library 14* (User's guide)

[4] Renţea C, Băţăuş M V, Maican S, Oprean M, Frăţilă G. 2019 *The influence of the hybridization factor on the energetic performance in test cycle* (The XXX-th SIAR International Automotive and Transport Engineering Congress, SMAT 2019, Craiova, România)
[5] Jiageng Ruan, Paul D Walker, Jinglai Wu, Nong Zhang and Bangji Zhang 2018 *Development of continuously variable transmission and multi-speed dual clutch transmission for pure electric vehicle* (Advances of Mechanical Engineering vol. 10(2) 1-15)
[6] Paul D Waker, Salisa Abdul Rahman, Bo Zhu and Nong Zhang 2013 *Modelling, Simulations and Optimisation of Electric Vehicles for Analysis of Transmission Ratio Selection* (Advances of Mechanical Engineering Volume 2013)
[7] Bo Zhu 2015 *Research of Two Speed DCT Electric Power-train and Control System* (University of Technology, Sydney)
[8] Renţea C, Oprean M, Băţăuş M V, Frăţilă G. 2018 *The study of the energetic performance in test cycles considering a variable powertrain thermal regime* (The IVth International Congress of Automotive and Transport Engineering, Technical University of Cluj Napoca, România)

Acknowledgement: This work has been funded by the European Social Fund from the Sectoral Operational Programme Human Capital 2014-2020, through the Financial Agreement with the title "Scholarships for entrepreneurial education among doctoral students and postdoctoral researchers (Be Antreprenor!)", Contract no. 51680/09.07.2019 - SMIS code: 124539.

Comparative study of different electric vehicle configurations in terms of energy consumption

C Rențea[1], M Bățăuș[1], S Maican[2], M Oprean[1] and G Frățilă[1]

[1] Automotive Engineering Department, University Politehnica of Bucharest,
[2] Faculty of Mechanical Engineering and Mechatronics, University Politehnica of Bucharest

*Corresponding author's e-mail: rentea.cristian@yahoo.com

Abstract. In order to respect the new fleet-wide average emission regulations, recent studies show that car manufacturers produce an increased number of electric vehicles. For improving the dynamic performances of an electric vehicle while maintaining a low energy consumption, different solutions are used, such as the multi-speed transmission, or the continuously variable transmission (CVT). Another solution more advantageous than other EV typologies consists of a dual motor two speed direct drive, that implies each motor being connected with the final drive through a separate gear. The aim of this paper is to analyse the energy consumption and the overall transmission performance in different test cycles for a middle class with different configurations. The study was carried out by developing a model in a performant simulation environment. The obtained simulation results from the chosen configuration are compared in different test cycles in terms of energy consumption ratio (ECR) and vehicle speed fluctuation (VSF) with the ones from a direct drive EV and a two-gear transmission EV.

1. Introduction

The electric powertrain solution that consists of one motor and a two-speed transmission is considered a better alternative to the electric vehicles that have a single motor and one gear ratio, with respect to the dynamic performance and the energy efficiency.

Adjusting the operating point of the motor in a high efficiency region can be achieved by using a two-speed transmission and an optimized gear shifting schedule [1].

In addition, by making use of the high and low gear ratios corresponding to the speed regime imposed by the vehicle it is possible to obtain a much better dynamic performance in terms of acceleration and maximum speed.

However, a multi-speed transmission has a disadvantage in terms of drivability because the problem of torque interruption during the gear shifting process persists.

Several studies conducted for electric vehicles on multi-motor and multi-speed transmissions show that these topologies achieve higher efficiency than a single motor and multi-speed transmission solutions [2].

Some papers investigate the powertrain solution which consists of two motors working together but being limited in terms of rotational speed [2, 3], while other sources are not focused on applying a specific limitation for the electric machines.

In this paper, one of the electric machines is being decoupled by a clutch at a certain rotational speed through a functioning strategy that was implemented.

To eliminate the torque interruption inconvenience, each motor needs to be connected to a gear set, thus making the use of a separate shifting device unnecessary. Therefore, for a non-shifting EV powertrain, it is essential that the number of motors to be equal to the number of gears [4].

Figure 1 illustrates the structure of a dual motor and a two-speed direct drive EV powertrain.

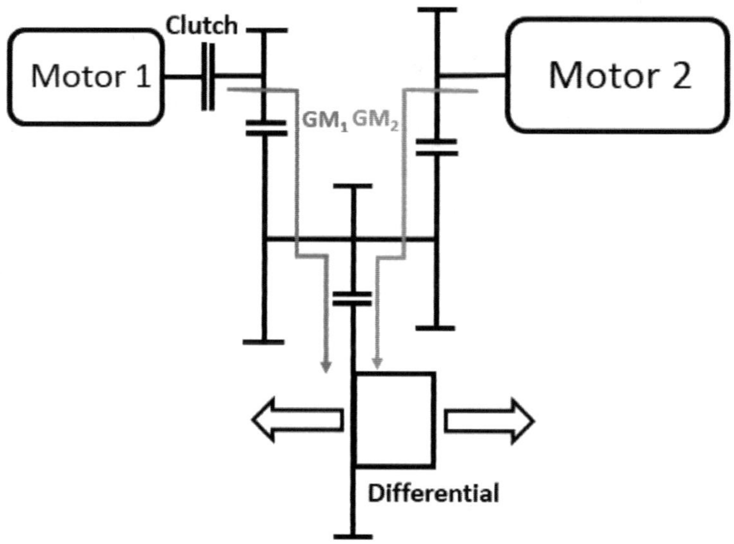

GM – gear set of the motor

Figure 1. Structure of the dual motor two speed direct drive EV powertrain

Each motor is connected to a gear and these gears share a single output shaft. The torque of the output shaft is finally transferred to the driving wheels [1].

The aim of this paper is to analyse the energy consumption and dynamic performance of a middle class vehicle for one motor with one speed, one motor with two-speed transmission and dual-motor with two-speed direct drive.

All the parameters will be determined for five driving cycles: US06, Japanese 10-15, HWFET, NEDC and WLTC.

2. Model development

The simulation model was created with the aid of LMS Imagine Lab AMESim, a specialized software that contains several components organized in libraries [5].

The main submodels were extracted from the IFP Drive library and some of them are listed as follows: a driver submodel (DRVDRVA00A) necessary to follow the cycle given by the mission profile and ambient data submodel (DRVMP2A), two electric motors (DRVELMT0A), two control units (DRVCUE0A), a battery (DRVBAT04), a power electronic converter (DRVPET0B) and 1D vehicle (DRVVEH4A).

Aside from these components, four more blocks were composed as supercomponents, two of which having the role of interrupting the electric machine circuit when the torque command is null, and the other two being used for coupling and decoupling the two motors based of the functioning strategy.

The clutch corresponding to motor 2 is added only based on causality considerations of the simulation model, having no functional role.

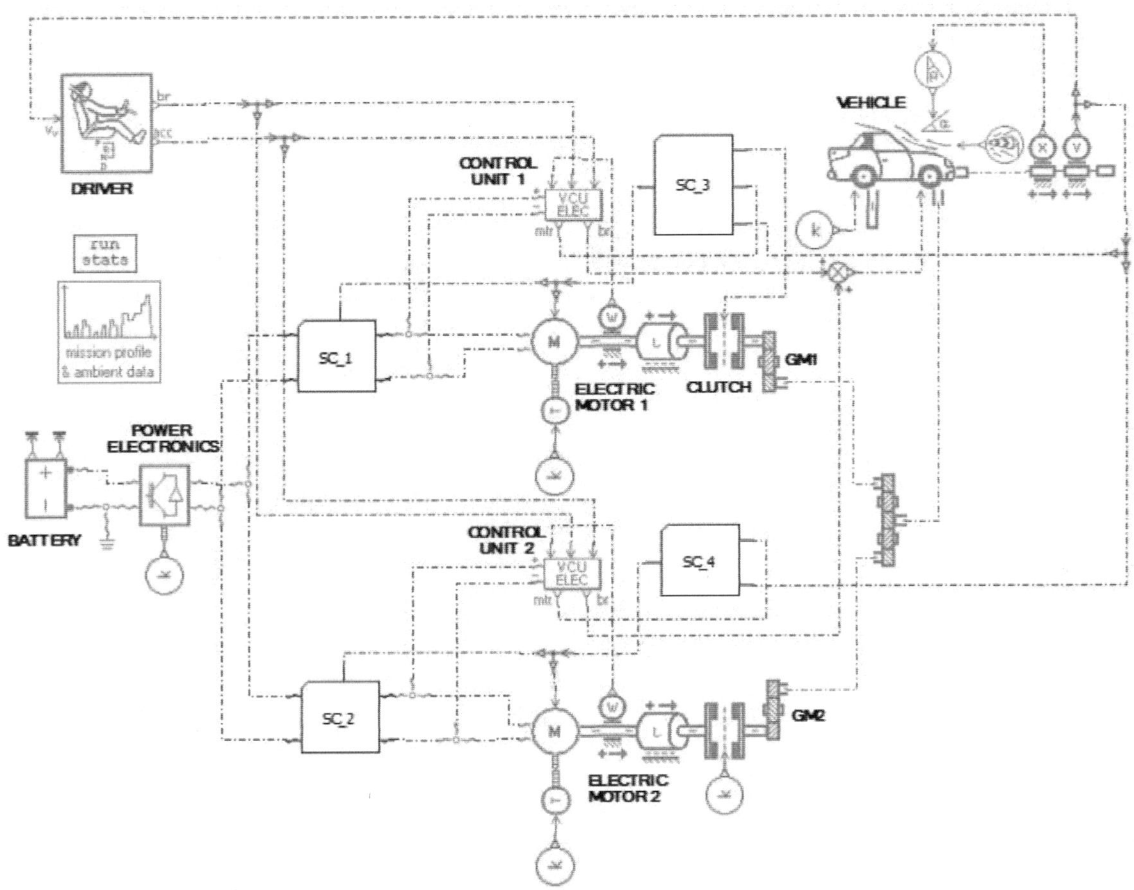

Figure 2. Model design of the dual motor two speed direct drive EV for the energy consumption study

3. Gear ratio determination

Because the first gear ratio is always higher than the second gear ratio ($r_2 \le r_1$), when the vehicle is driving at the maximum speed only the second motor generates the driving power, as the first motor is decoupled (it does not exceed the allowable maximum motor speed).

Since the power gets split between the two motors, thus reaching the maximum reference speed being difficult to achieve, the power of motor 2 is calculated such that the maximum vehicle speed is reduced with 10% compared with the one motor and two speed reference powertrain.

The two motors were rescaled such that the sum of their maximum powers is equal to the power obtained from the reference motor [6] and motor 2 should have the required power to reach the maximum speed.

The power required for obtaining the maximum imposed speed is calculated with equation (1)

$$P = \frac{1}{\eta_t}(f_r \cdot m \cdot g \cdot cos\,\alpha + \frac{1}{2} \cdot \rho \cdot C_d \cdot A \cdot V_{max}{}^2) \cdot \frac{V_{max}}{3600} \quad (1)$$

where: η_t - overall efficiency of the powertrain (0.95), f_r - rolling resistance coefficient (0.01), m - vehicle mass (1966 kg), α - road incline angle (0°), ρ - air density (1.225 kg/m^3), C_d - air drag coefficient (0.3), V- maximum vehicle speed (152 km/h), A - frontal area of the vehicle (2.5 m^2)

The obtained value for the required power is 45 kW.

Based on the required maximum vehicle speed, the gear ratio for motor 2 (r_2) can be expressed as follows:

$$r_2 = \frac{3.6 \cdot \pi \cdot n_P \cdot r_r}{30 \cdot V_{max}} \tag{2}$$

where n_P –motor speed for maximum power (6500 rev/min), r_r – rolling radius (0.31 m), V_{max} - maximum vehicle speed (152 km/h).

The value for r_2 has therefore been set to 5.

The gear ratio for motor 1 (r_1) is calculated such that the same traction force from the reference motor with the maximum torque of 220 Nm and a gear ratio of 9.52 is obtained [6]. This will ensure the same maximum slope that the vehicle can drive on.

$$F_{t\,ref} = F_{t1} + F_{t2} \tag{3}$$

Expressing the traction force as a function of each motor's torque, the below expression is obtained,

$$\frac{T_{ref} \cdot i_{ref} \cdot \eta_{ref}}{r_r} = \frac{T_1 \cdot r_1 \cdot \eta_{t1}}{r_r} + \frac{T_2 \cdot r_2 \cdot \eta_{t2}}{r_r} \tag{4}$$

From eq. 4 results

$$r_1 = \frac{T_{ref} \cdot i_{ref} \cdot \eta_{ref} - T_2 \cdot GM_2 \cdot \eta_{t2}}{T_1 \cdot \eta_{t1}} \tag{5}$$

Thus, it resulted a value of 22 for the gear ratio of motor 1.

The variation of the vehicle acceleration for different powertrain configurations is represented in figure 3.

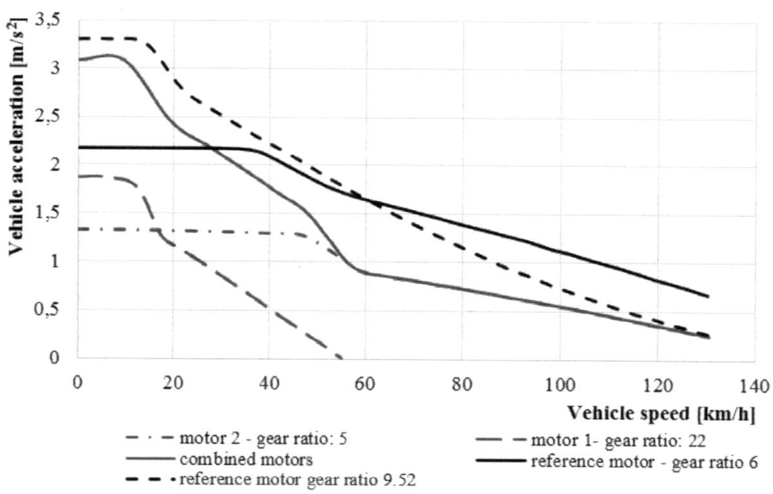

Figure 3. Vehicle acceleration

Therefore, it resulted a smaller maximum acceleration than the reference vehicle acceleration due to the high ratio of GM_1 that amplifies Motor 1 inertia.

The efficiency map of the two motors was obtained by rescaling the power losses and the torque maps of the reference electric machine.

Figure 4 shows the energetic consumption at constant speed values, including the medium speed values of each driving cycle, for different powertrain solutions for every gear.

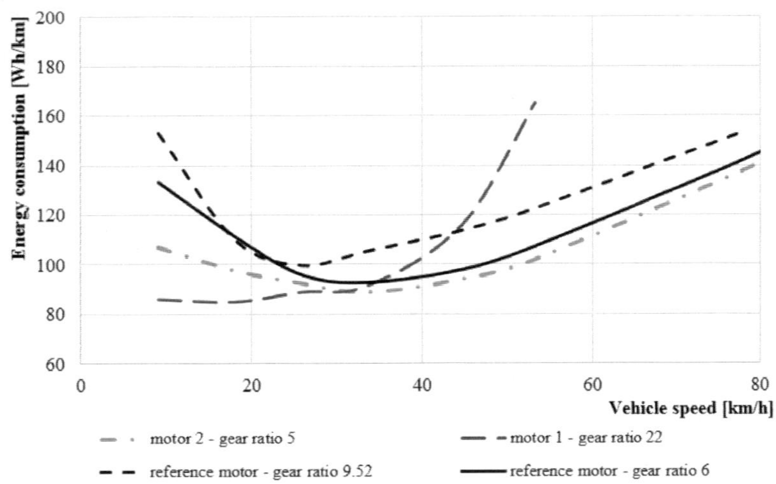

Figure 4. Energy consumption at constant speed

Table 1 presents the dynamic performances of the electric vehicle when its powertrain is consisted of motor 1 and motor 2 individually with their corresponding gear ratios and with their combined functioning. In case of combined motor operation, motor 1 is decoupled when it reaches the maximum speed, the torque being provided only by motor 2.

Also, the last column shows the results when the vehicle is equipped with the reference motor and a two speed gearbox.

Table 1. Electric vehicle performances

No.	Parameter	Motor 1	Motor 2	Combined motors	Reference motor Two gears
1	Maximum speed [km/h]	54.9	152.6	152.6	171
2	Maximum acceleration [m/s]	1.87	1.33	3.1	3.3
3	Acceleration time 0-50 km/h [s]	19.5	10.8	6.5	4.93
4	Acceleration time 0-100 km/h [s]	–	29.1	24.8	14
5	Maximum grade [%]	21.2	13.5	30	30
6	Energy consumption at 50 km/h [Wh/km]	131.4	98.3	98.3	103.5

4. Results

In principle, due to the low values of torque developed by motor 1, the electric vehicle cannot undergo the driving cycles without motor 2, therefore one must establish strategies that can satisfy this condition.

For the vehicle to go through the test cycles, an efficient control based on load and rotational speed has been developed, thus three different strategies have been implemented, as follows.

- S1 – the electric vehicle has only motor 2 set working,
- S2 – both motors act simultaneous, motor 1 being decoupled at 30 km/h. After improving the motor load and rotary velocity control, the minimum consumption for which motor 1 is decoupled has been obtained for the vehicle speed of 30 km/h.
- S3 - The two motors work alternatively. Motor 1 functions with the load appropriate to the minimum energetic consumption, therefore, when the vehicle reaches 30 km/h, it gets decoupled. For vehicle speed values higher than 30 km/h, motor 2 is set functional.

Table 2 shows the energetic consumption for the presented strategies.

Table 2. The energetic consumption for different motor functioning strategies

No.	Cycle	Energy consumption [Wh/km]			
		S1	S2	S3	Reference vehicle
1	NEDC	150.4	133	131.3	152.5
2	WLTC	170.9	154.7	151.7	158.3
3	10-15	135	113.7	106.3	132.5
4	US-06	200.1	192.3	190	206.1
5	HWFET	152.1	146.5	146.3	158.9

The energetic consumption obtained at constant speed values is compared with the energetic consumption for five different driving cycles when the functioning strategy is S2, as shown in figure 5.

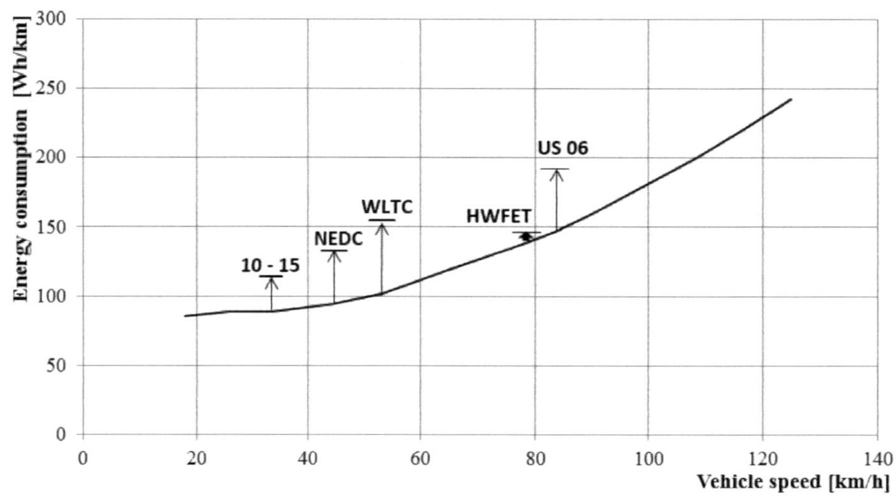

Figure 5. Energy consumption comparison

In figure 6, the correlation between ECR and VSF is presented. Also, two regression lines are illustrated, having set as an imposed point of the constant speed movement with coordinates (0,100).

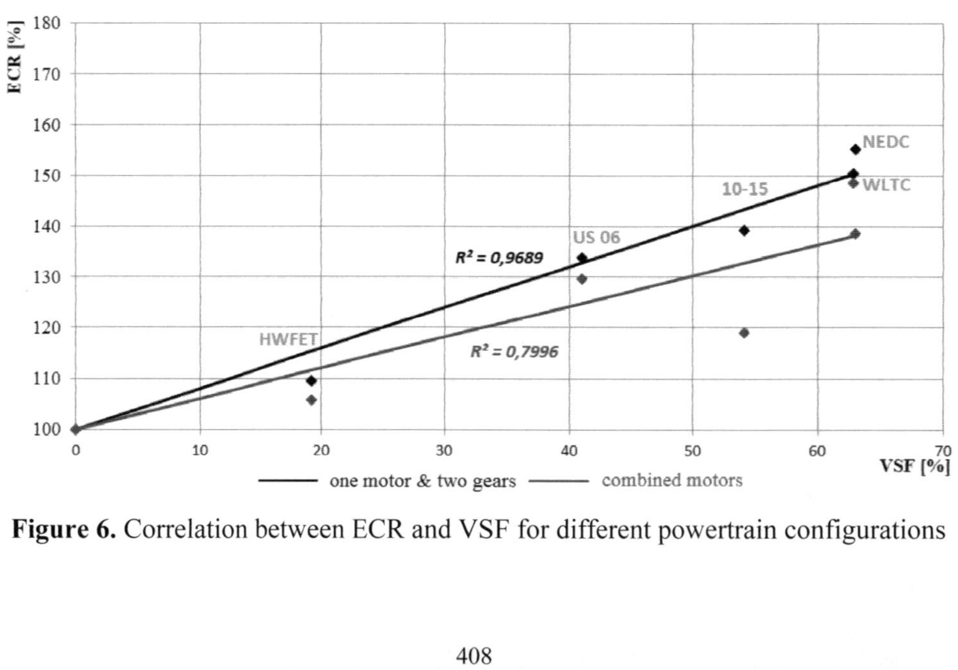

Figure 6. Correlation between ECR and VSF for different powertrain configurations

For these lines, the coefficient of determination is also indicated.

Regarding the above regression lines for the electric vehicle, when strategy S2 is implemented, a less linear correlation between ECR and VSF can be observed.

5. Conclusions

In the current application, when compared with the two gear transmission reference vehicle, the dynamic performances at high velocity are drastically reduced. Consequently, the acceleration time from 0 to 100 km/h is increased with 77.1%.

When applying a functioning strategy that uses an alternative coupling of the two motors (S3), the drivability of the vehicle is less convenient because of the shock induced when the motors are coupled/decoupled, compared with other strategies that allow a smoother torque shift (S2).

The energetic consumption obtained for an alternative motor coupling strategy (S3) is lower than for when the two motors work simultaneously.

When compared with the reference vehicle, a clear improvement of energy consumption was achieved: 14.2% for 10-15 Japanese cycle, 12.8% for NEDC, with the smallest benefit for WLTC of 2.3%.

For the present study, as opposed to the reference vehicle [6], a less linear correlation between ECR and VSF is highlighted.

References

[1] Kihan Kwon, Minsik Seo, Seungjae Min 2019 *Efficient multi-objective optimization of gear ratios and motor torque distribution for electric vehicles with two-motor and two-speed powertrain system*

[2] Jiageng Ruan and Qiang Song 2019 *A Novel Dual-Motor Two-Speed Direct Drive Battery Electric Vehicle Drivetrain*

[3] Mingjie Zhao, Junhui Shi, Cheng Lin 2019 *Optimization of integrated energy management for a dual-motor coaxial coupling propulsion electric city bus*

[4] Zhu B, Zhang N, Walker P, Zhan W, Yueyuan W, Ke N, Zhou X. 2013 *Two motor two speed power-train system research of pure electric vehicle* (SAE Technical Paper)

[5] Siemens Industry Software NV – LMS Imagine.Lab Amesim 2015 IFP Drive Library 14 (User's guide)

[6] Rențea C, Oprean M, Bățăuș M V, Frățilă G. 2019 *The influence of multi-speed transmissions on electric vehicles energy consumption* (The 23rd edition of IManEE 2019 International Conference, Pitești, România)

Acknowledgement: This work has been funded by the European Social Fund from the Sectoral Operational Programme Human Capital 2014-2020, through the Financial Agreement with the title "Scholarships for entrepreneurial education among doctoral students and postdoctoral researchers (Be Antreprenor!)", Contract no. 51680/09.07.2019 - SMIS code: 124539.

IOP Conf. Series: Materials Science and Engineering 1037 (2021) 012052 doi:10.1088/1757-899X/1037/1/012052

Drivability improvements obtained using a mild hybrid electrical system on a vehicle equipped with a manual gearbox

Ștefan Saragea[1*], Gheorghe Frățilă[2], Gabriel Badea[1]

[1] PhD Student, Automotive Engineering Department, Politehnica University of Bucharest, Bucharest RO-060042, Romania

[2] Professor, Automotive Engineering Department, Politehnica University of Bucharest, Bucharest RO-060042, Romania

[*]Corresponding author's e-mail: stefansaragea@gmail.com

Abstract. In order to fulfil the latest and future emission standards, all vehicles will need to be equipped with hybrid powertrains or advanced depollution systems. For low cost and entry level range of vehicles, in order to keep the production costs low, the simplest choice of hybridization is the mild-hybrid system with integrated starter-generator (ISG). This paper contains a model of a mild-hybrid electric vehicle equipped with a manual gearbox with focus on the drivability performance of the powertrain. The goal of this paper is to find out the drivability performances of a hybrid vehicle by simulation and to show the improvements compared to a similar vehicle equipped with a conventional powertrain.

1. Introduction

In order to reduce de CO_2 emissions for the passenger vehicles, the latest European Union regulations [1] (EC regulation 443/2009) sets a target of 95 g CO_2/km for the average emissions of the new car fleet for each car manufacturer starting with the year 2021.

In order to fulfill the emission target, manufacturers are required to reduce the fuel consumption of their vehicles. One of the solution for reducing the fuel consumption is the hybridization of the vehicles. The simplest way of hybridization is the mild-hybrid system with Integrated Starter Generators (ISG).

The current forecast [2] shows that the mild hybrid vehicle global market share will reach 15% by 2030, compared to a decrease of market share for the conventional ICE vehicles to 47% by 2030 (from a market share of 78% in 2020 – figure 1).

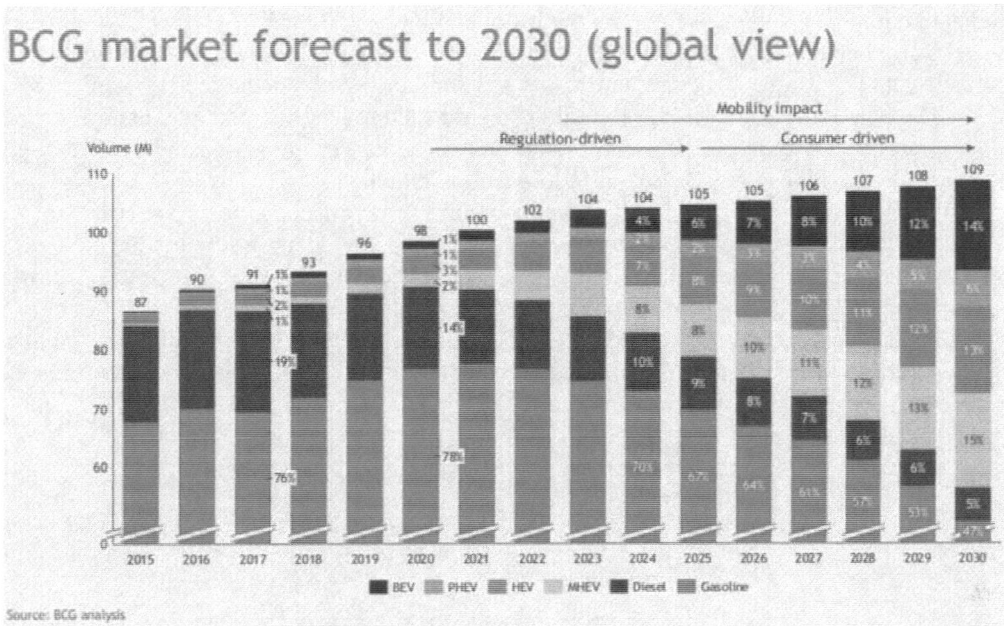

Figure 1. Market share forecast for ICE, hybrid vehicles and electric vehicles by 2030 [2]

2. Simulation model presentation

The model of the 48V mild hybrid electric vehicle was simulated using LMS Imagine.Lab AMESim program [3]. The simulation contains a B-class vehicle equipped with a 6 speed manual gearbox, 48V electric motor (Integrated starter generator) and 1 liter (998 cm^3) petrol internal combustion engine (ICE).

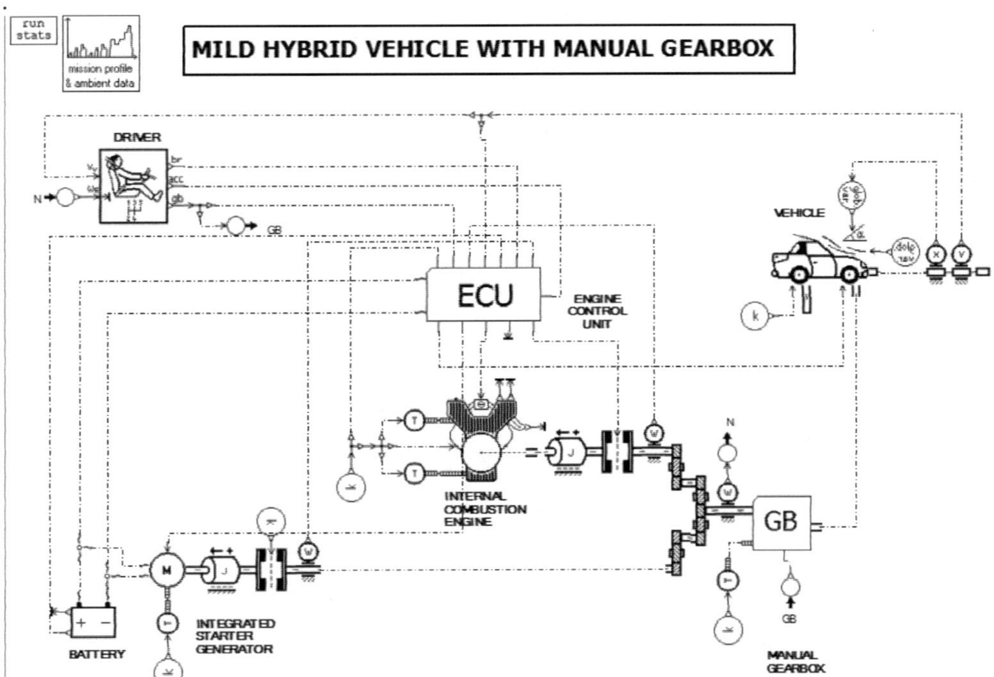

Fig. 2. AMESim model of a mild-hybrid electric vehicle equipped with a 6 speed Manual Transmission

The main components of the model are the following:

- Internal combustion engine (ICE) –The inputs for the engine model are torque structure and the engine load. The torque structure and losses are similar to a turbocharged 998 cm^3 engine with 3 cylinders. The output of the engine are the torque transmitted to gearbox and exhaust gasses.
- Gearbox and clutch (GB) – the clutch and gearbox were modeled in order to simulate the clutch, gearbox and driveshaft torque oscillations during clutch closing.
- Engine control unit (ECU) - the engine control unit is used in order to set the source of the torque distributed to the drive wheels between the ICE and EM. The inputs for the ECU are: driver acceleration, braking and clutch command, battery state of charge, current and voltage, gearbox ratio and ICE speed. The output from the ECU are: ICE and EM load control.
- Electric motor (EM) – the EM model is a simple model with the following inputs: battery voltage, load control from ECU and temperature. The output is the torque produced by the motor. The electric motor parameters are presented in Table 1.

Table 1. Electric motor parameters

Nr.	Item	Value
1	Maximum Output Torque	40 Nm
2	Mechanical power	14 kW
3	Speed	20.000 rpm

- Battery – the battery model is a simple Lithium-Ion battery model with the following outputs: battery current and voltage and state of charge. The battery parameters are presented in Table 2.

Table 2. Battery parameters [4]

Nr.	Item	Value
1	Capacity	6 Ah
2	Operating Voltage	24-54 V
3	Configuration	14 series / 1 parallel
4	Weight	7kg
5	State of charge range	30-80 %
6	Usable energy	138 Wh
7	Dimensions	304 x 92 x 180 mm

The selected battery has compact dimensions, in order to integrate the battery in the body of the car without major modifications of the car body.

- Car body - this sub model is a simple model of a vehicle load used to calculate the longitudinal acceleration, velocity and displacement of the car body. The input parameters for this model are the vehicle mass, the aerodynamic coefficient, the frontal area and the air density.

3. Simulation results

One of the main disadvantage of downsized engines with reduced displacement and 3 cylinders is the need of heavy flywheel in order to reduce the engine vibrations transmitted to vehicle body. Due to the heavy flywheel, the engine inertia is high and the engine speed derivative in no load conditions is low compared to similar performance non-downsized engines.

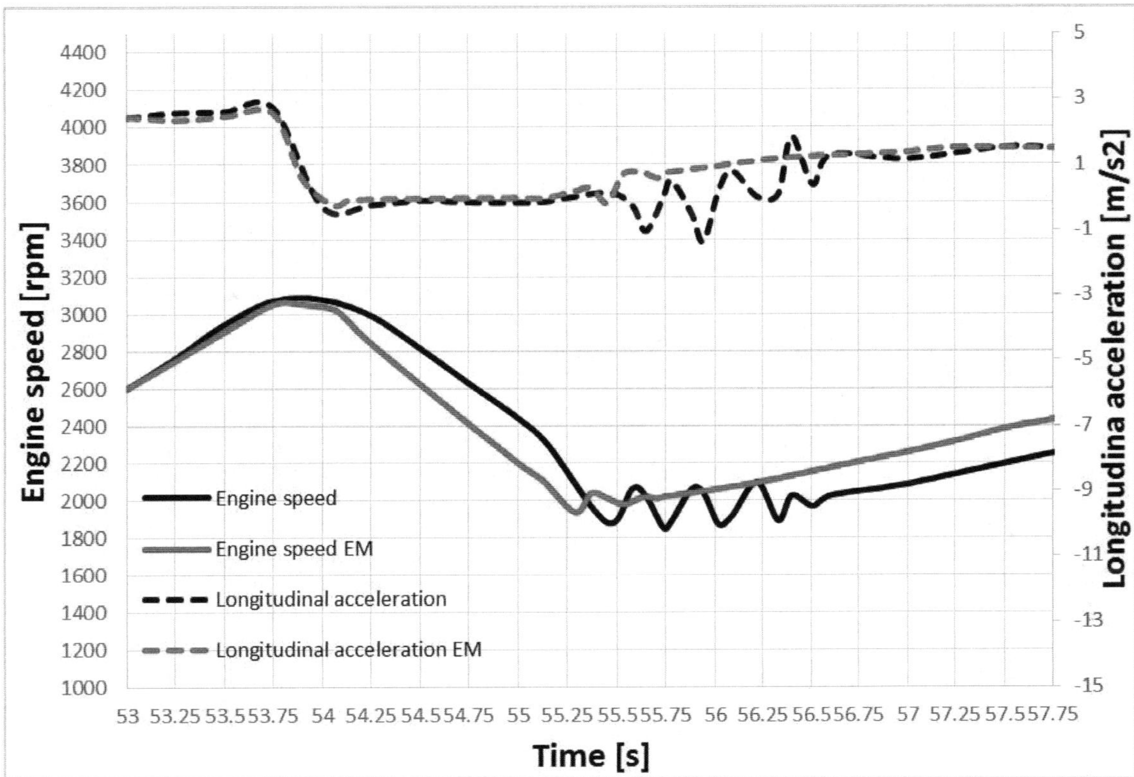

Figure 3. Engine speed and longitudinal acceleration variation during gear change from 2nd to 3rd gear

In figure 3 is presented a gear change from 2nd to 3rd gear with medium engine load. As can be seen, due to engine high inertia, the engine speed derivative is low (600 rpm/s). During the closing phase of the clutch, the engine speed is forced down to the primary shaft speed for the 3rd gear. This behavior causes high engine speed (Engine speed) and torque oscillations in the drivetrain which in turn causes high longitudinal acceleration oscillations to the car body. The maximum longitudinal acceleration oscillation is 2.5 m/s^2 (Longitudinal acceleration), this value causes discomfort for passengers during gear changes.

In case of the mild hybrid vehicles equipped with ISG, during the gear change, the engine speed derivative can be increased to 1000 rpm/s by increasing the engine load with the ISG. In this case, when the clutch is closed, the engine speed (Engine speed EM) is closer to the primary shaft speed and the maximum car body longitudinal oscillation is 0.9 m/s^2 (Longitudinal acceleration), compared to a maximum longitudinal acceleration of 2.5 m/s^2 for the vehicle without ISG.

A shift strategy with internal combustion engine speed control can be implemented in order to improve comfort during gear shifts. This strategy is similar with the strategy presented in a paper before [5].

IOP Conf. Series: Materials Science and Engineering 1037 (2021) 012052 doi:10.1088/1757-899X/1037/1/012052

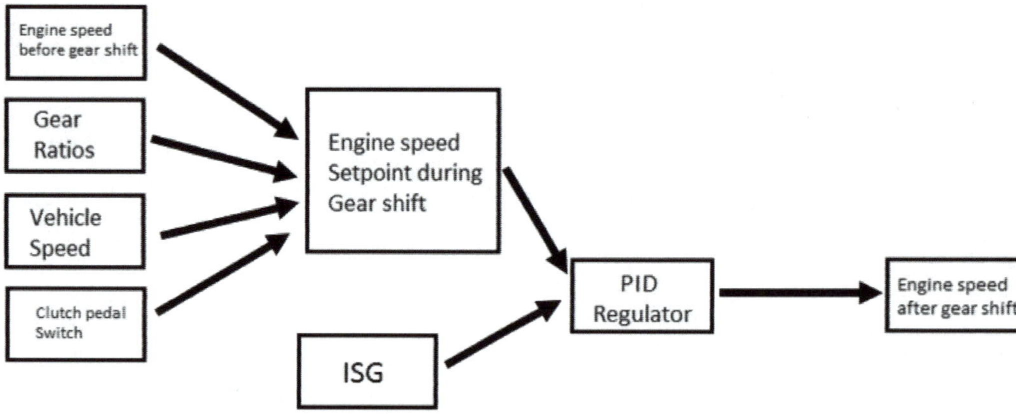

Figure 4. Engine speed control strategy during gear shifts

The input for the engine speed control is the engine speed, the vehicle speed, the gear ratios and the clutch pedal position. With these inputs, the engine speed set point during gear shift is calculated. The calculated engine speed, the fuel injection and the body throttle are inputs for engine speed control using a PID regulator. The output of this strategy is the speed of the ICE which should be equal to the speed of the input shaft of the gearbox when the clutch is closed in order to produce a seamless gear shift.

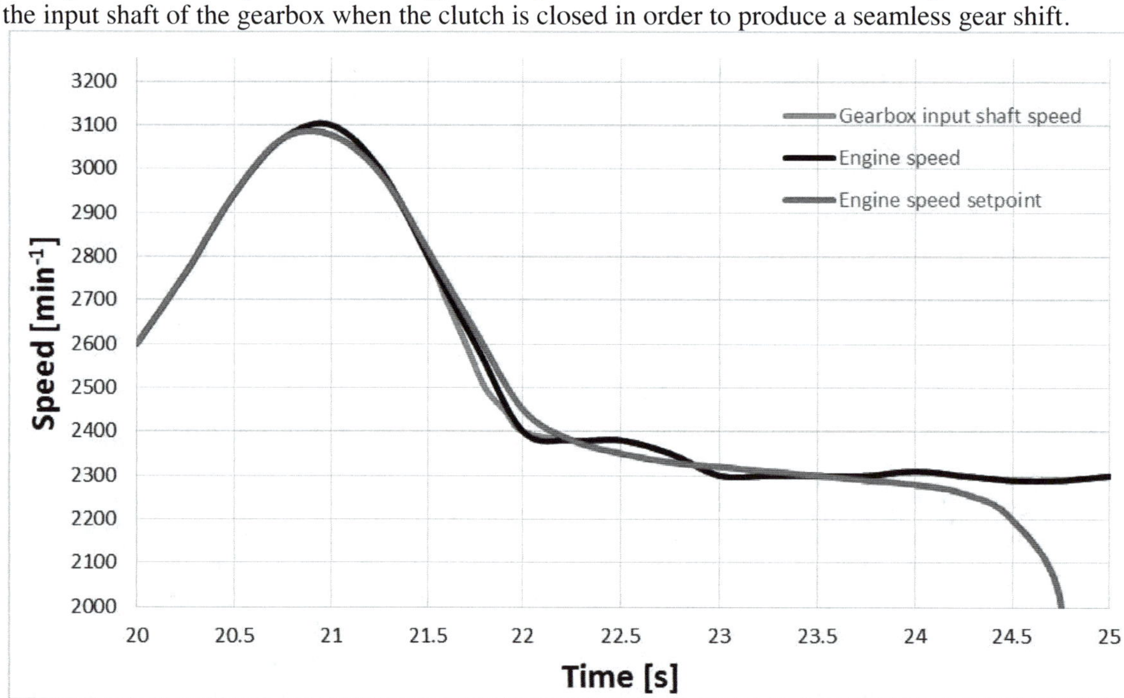

Figure 5. Gear shift strategy with internal combustion engine speed control

In figure 5 is presented the gear shift strategy with the internal combustion engine speed control during gear shift. In order to cover different types of drivers, with different shifting times, the internal combustion engine speed (black signal) is maintained at the speed of the gearbox input shaft (green signal) using a PID regulator in order to follow the engine speed shift set point (red signal). In order to reduce the fuel consumption and to reduce the exhaust gases, during the gear shift, when the clutch is pressed, the engine speed is controlled by the electric motor (ISG).

Another improvement obtained using an ISG on a vehicle is the dynamic performance of the vehicle. The ISG can provide assistance power to the ICE when high engine load is requested by the driver, for example in an overtaking maneuver.

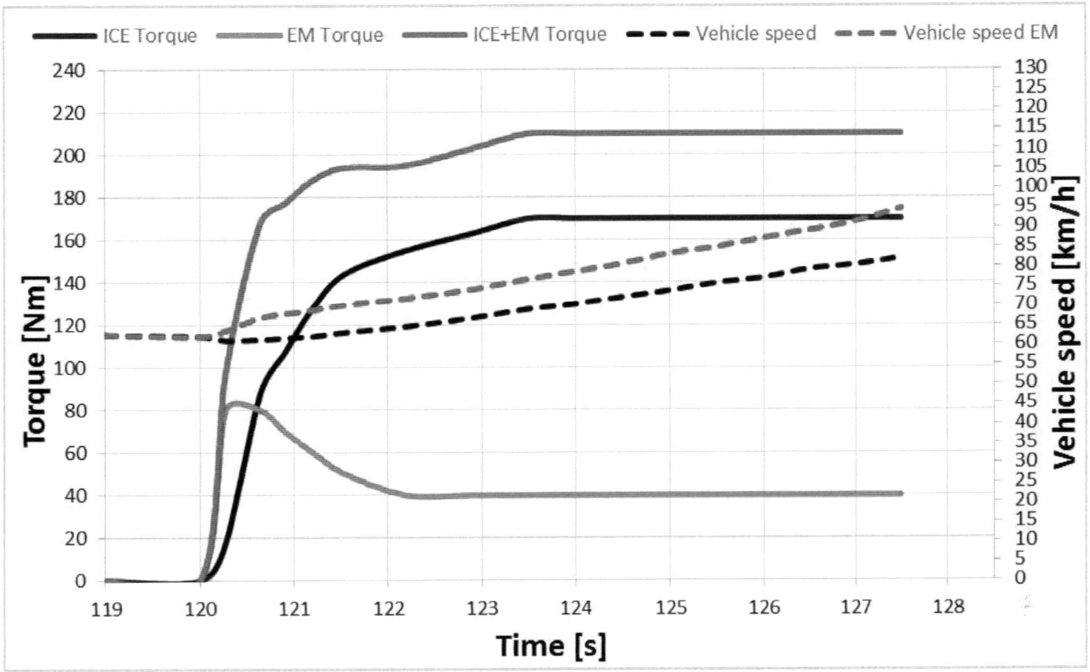

Figure 6. Vehicle acceleration in 4th gear from 60 km/h

In figure 6 is presented an overtaking maneuver, a vehicle acceleration from 60 km/h in 4th gear for a conventional vehicle and a mild-hybrid vehicle equipped with ISG. For the conventional vehicle equipped with only an ICE, the speed (Vehicle speed) of 80 km/h is reached in 7 seconds and the maximum torque (ICE Torque) of 170 Nm is obtained in 3.4 seconds after the acceleration pedal is pressed, due to the delay of increasing pressure in the turbocharger and intake manifold. For the mild hybrid vehicle equipped with ISG, the speed (Vehicle speed EM) of 80 km/h is reached in 4.5 seconds (2.5 seconds faster compared to conventional ICE time) and the ICE maximum torque of 170 Nm is obtained after 0.5 seconds (compared to 3.5 seconds with conventional ICE). This improvement on crankshaft torque is possible due to assistance from the ISG. A maximum of 80 Nm (EM Torque) at crankshaft can be produced by the ISG for a short amount of time (about 1.5 to 2 seconds), after this time, the ISG can still assist the ICE continuously with 40 Nm. The maximum torque produced by the mild hybrid vehicle powertrain (ICE + EM Torque) is 210 Nm (compared to 170 Nm for conventional ICE vehicle), achieved after 3.5 seconds. The response time of the mild hybrid powertrain can be further improved by using an electrically assisted turbocharger (e-turbocharger), which reduces the time needed for increasing pressure in the compressor and intake manifold.

4. Conclusion

In this paper are presented the drivability improvements using a mild hybrid electric vehicle using a 48V Integrated Starter Generator with focus on gear shifts and vehicle performance during acceleration. Using the Integrated Starter Generator, the internal combustion engine speed can be controlled during gear shifts in order to match the internal combustion engine speed with the gearbox input shaft speed. The results shows improvement in longitudinal acceleration oscillations after the clutch is closed and also reduced acceleration times during overtaking maneuvers.

Other improvements obtained using mild hybrid electrical powertrains are reduced vehicle emissions, improved thermal comfort for passenger (possibility to equip vehicle with electrical AC compressor which can provide AC when the engine is stopped in start&stop mode and electrical cabin heaters during cold weather), electrical vacuum pump in order to improve braking performance without ICE torque consumption and engine oil heating in order to achieve maximum ICE performance faster during cold weather.

References

[1] https://ec.europa.eu/clima/policies/transport/vehicles/cars
[2] https://www.greencarcongress.com/2017/11/20171103-bcg.html
[3] LMS. AMESim Help. AMEHelp, 2012
[4] Saragea Ș, Oprean I M, Bățăuș M, Frățilă G 2019 Driveability Improvements obtained using a mild-hybrid electrical system on a vehicle equipped with an automated manual transmission, *Science and management of automotive and transportation engineering 4th edition* (Craiova)
[5] Saragea Ș, Oprean IM, Badea G, Frățilă G 2018 Modeling the Torque Oscillations During Gear Shifts on Automated Manual Transmission. *Proceedings of the 4th International Congress of Automotive and Transport Engineering (AMMA 2018)* ed Burnete N, Varga B (Cluj) pp 634-641

Acknowledgments

This work has been funded by the European Social Fund from the Sectoral Operational Programme Human Capital 2014-2020, through the Financial Agreement with the title "Scholarships for entrepreneurial education among doctoral students and postdoctoral researchers (Be Antreprenor!)", Contract no. 51680/09.07.2019 - SMIS code: 124539.

Fuel consumption and emission improvements obtained using a mild hybrid electrical system on a vehicle equipped with a manual gearbox

Ştefan Saragea[1*], Gheorghe Frăţilă[2], Gabriel Badea[1]

[1] PhD Student, Automotive Engineering Department, Politehnica University of Bucharest, Bucharest RO-060042, Romania
[2] Professor, Automotive Engineering Department, Politehnica University of Bucharest, Bucharest RO-060042, Romania

*Corresponding author's e-mail: stefansaragea@gmail.com

Abstract. The latest emission requirements for vehicles requires major improvement in vehicles fuel consumption and emission performance. For entry level range of vehicles, in order to keep the production costs low, the simplest choice of hybridization is the mild-hybrid system with integrated starter-generator (ISG). This paper contains a model of a mild-hybrid electric vehicle equipped with a manual gearbox and a model of conventional vehicle with focus on the emission performance. The goal of this paper is to find out the fuel consumption and emission of a mild hybrid electric vehicle by simulation and to show the improvements compared to a similar vehicle equipped with a conventional powertrain.

1. Introduction

The latest emission requirements for vehicles requires major improvement in vehicles fuel consumption and emission performance. For example, the latest European Union regulations [1] (EC regulation 443/2009) sets a target of 95 g CO_2/km for the average emissions of the new car 95% of the fleet for each car manufacturer starting with the year 2020 and for all vehicle fleet starting with 2021. If the emission target are not met, major fines are given to manufacturers, up to 95€/each excess CO_2/km emission per vehicle sold.

In order to fulfill the emission target, internal combustion engines (ICE) need to be more efficient or hybrid systems should implemented on vehicles. The easiest and cheapest way of hybridization is the mild-hybrid system with Integrated Starter Generators (ISG).

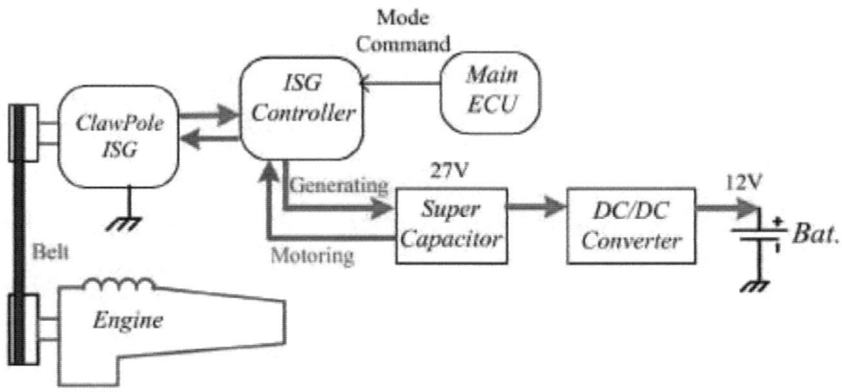

Figure 1. Structure of belt driven Integrated Starter Generator (ISG) system [2]

2. Simulation model presentation

Two models of simulation were made in LMS Imagine.Lab AMESim program [3]: A model of a 48V mild hybrid electric vehicle was simulated The simulation contains a B-class low cost vehicle equipped with a 6 speed manual gearbox, 48V electric motor (Integrated starter generator) and 1.6 liter (1698 cm^3) petrol internal combustion engine (ICE). Another simulation was made for a conventional vehicle equipped with internal combustion engine (ICE).

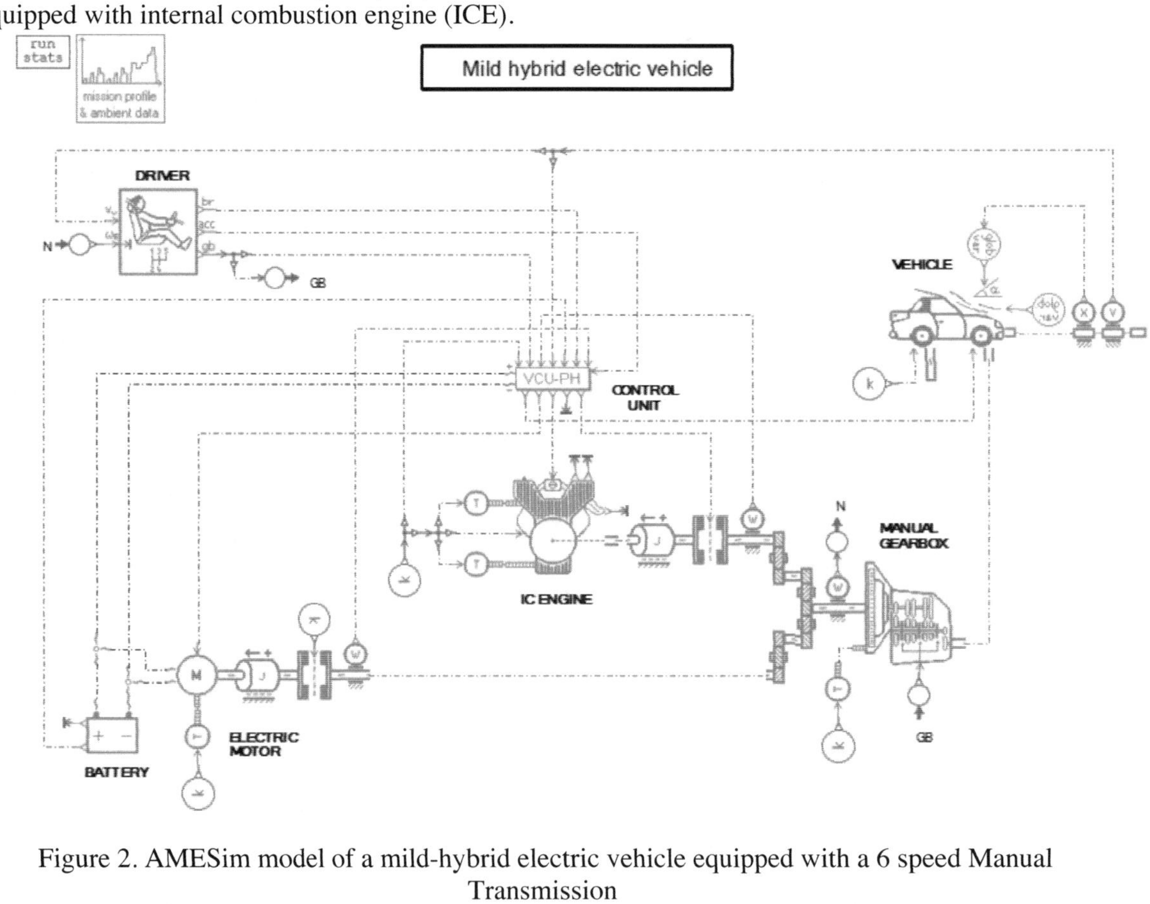

Figure 2. AMESim model of a mild-hybrid electric vehicle equipped with a 6 speed Manual Transmission

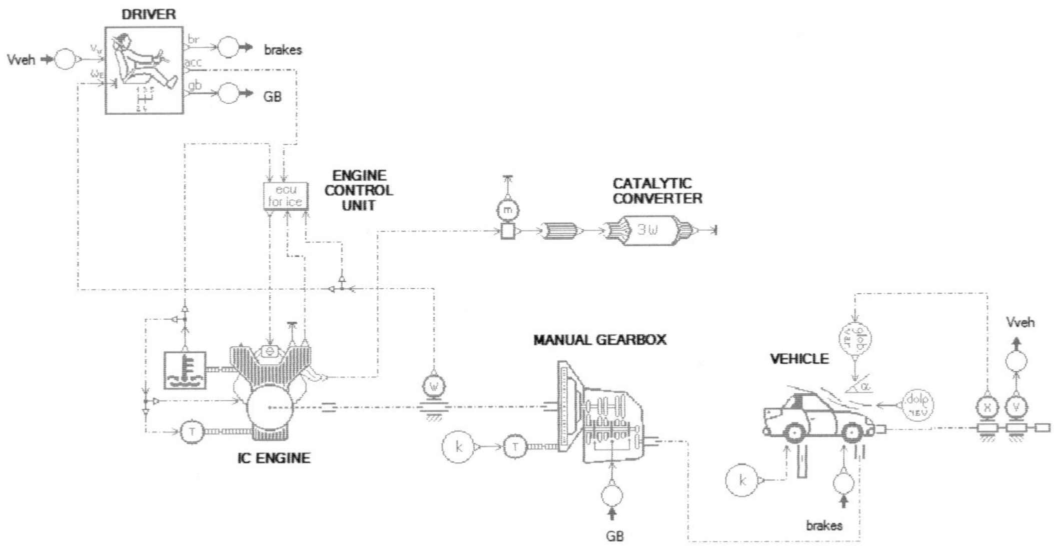

Figure 3. AMESim model of a conventional vehicle equipped with ICE and a 6 speed Manual Transmission

The main components of the models are the following:

- Internal combustion engine (ICE) –The inputs for the engine model are torque structure and the engine load. The torque structure and losses are similar to a naturally aspirated 1698 cm^3 engine with 4 cylinders. The internal parameters are the torque curve, torque losses and thermal losses.
- Catalytic converter –the 3 way catalytic converter inputs are the engine exhaust gases.
- Gearbox and clutch (GB) – the model of the gearbox is a simple model. The inputs for the gearbox is the oil temperature and clutch command. The internal parameters are the gear ratios and the gear efficiency.
- Engine control unit (ECU) - the engine control unit is used in order to set the source of the torque distributed to the drive wheels between the ICE and EM. The inputs for the ECU are driver acceleration, braking and clutch command, battery state of charge, current and voltage, gearbox ratio and ICE speed. The output from the ECU are: ICE and EM load control.
- Electric motor (EM) – the EM model is a simple model with the following inputs: battery voltage, load control from ECU and temperature. The output is the torque produced by the motor.
- Battery – the battery model is a simple Lithium-Ion battery model with the following outputs: battery current and voltage and state of charge. The battery parameters are presented in Table 2.

Table 1. Battery parameters [4]

Nr.	Item	Value
1	Capacity	6 Ah
2	Operating Voltage	24-54 V
3	Configuration	14 series / 1 parallel
4	Weight	7kg
5	State of charge range	30-80 %
6	Usable energy	138 Wh
7	Dimensions	304 x 92 x 180 mm

The selected battery has compact dimensions, in order to integrate the battery in the body of the car without major modifications of the car body in order to keep the development costs low.

- Car body - this sub model is a simple model of a vehicle load used to calculate the longitudinal acceleration, velocity and displacement of the car body.

3. Simulation results

In order to determine the emission and fuel consumption level and compare the results to a similar vehicle equipped with conventional ICE, both vehicles were simulated on NEDC cycle.

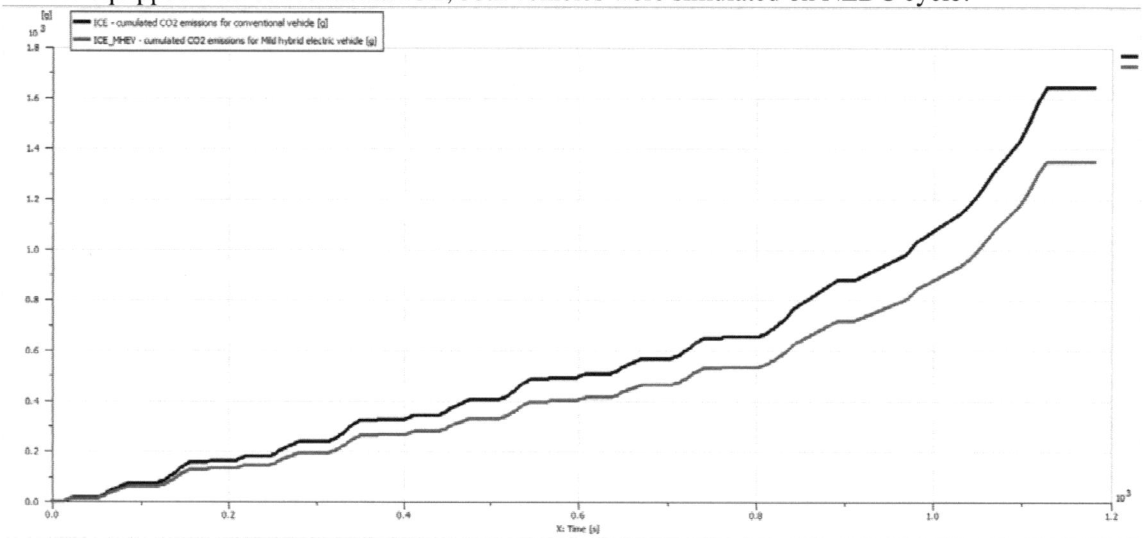

Figure 4. Total CO_2 emissions on NEDC cycle for conventional and MHEV

In figure 4 is presented the total CO_2 emissions for conventional vehicle with internal combustion engine and for mild hybrid electric vehicle (MHEV). For the conventional vehicle, the total amount of CO_2 emission is 1640 g CO_2 which is equivalent to 143 g CO_2/km. For the MHEV, total amount of CO_2 emission is 1349 g/CO_2 which is equivalent to 118 g CO_2/km. which is 20% improvement over conventional vehicle. The result is still above the EU limit of 95 g CO_2/km for 2021 year, but this simulation is applied to entry level and low cost vehicles, the average fleet for each manufacturer can be lowered in order to fulfill the target value with full hybrid or electric vehicles.

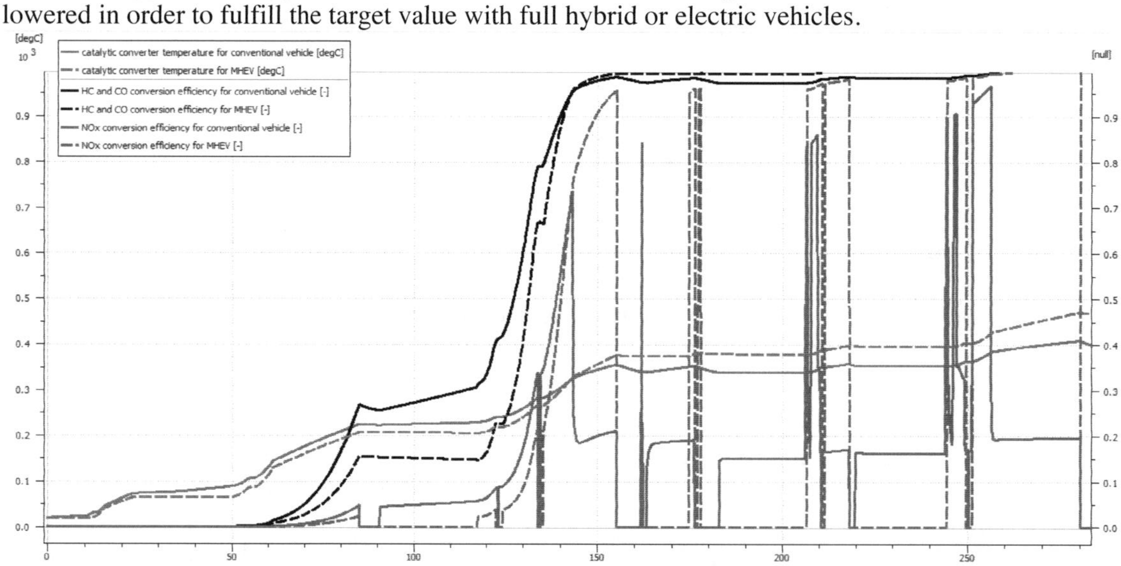

Figure 5. Catalytic converter temperature and efficiency for conventional vehicle and MHEV

In figure 5 is presented the temperature of the catalytic converter for the first 200 seconds of the NEDC cycle (continuous line signals - conventional vehicle, dotted signals – MHEV). In the first 150 seconds, the catalytic converter temperature is higher in the conventional compared to MHEV because the MHEV is equipped with start&stop system and the ICE is turned off when the vehicle is stationary. The catalytic converter temperature for MHEV can be increased higher by improving the simulation model with a strategy to retard the ignition advance of the ICE in order to increase the exhaust valve temperature when the ICE is turned on.

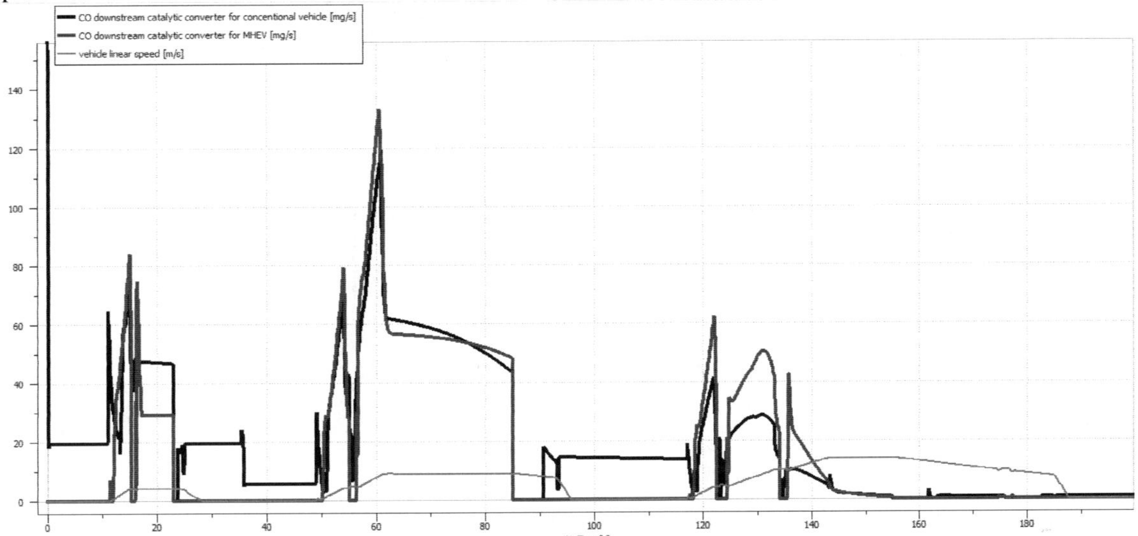

Figure 6. CO emission for conventional ICE vehicle and MHEV

In figure 6 is presented the CO emissions for conventional vehicle with internal combustion engine (black signal) and for mild hybrid electric vehicle (green signal). For the conventional vehicle, the mean value of CO for the first 200 seconds emission is 43 mg/s. For the MHEV, the mean value of CO emission for the first 25 seconds is 28 mg/s which is an improvement of 35% over the performance of conventional vehicle equipped with ICE.

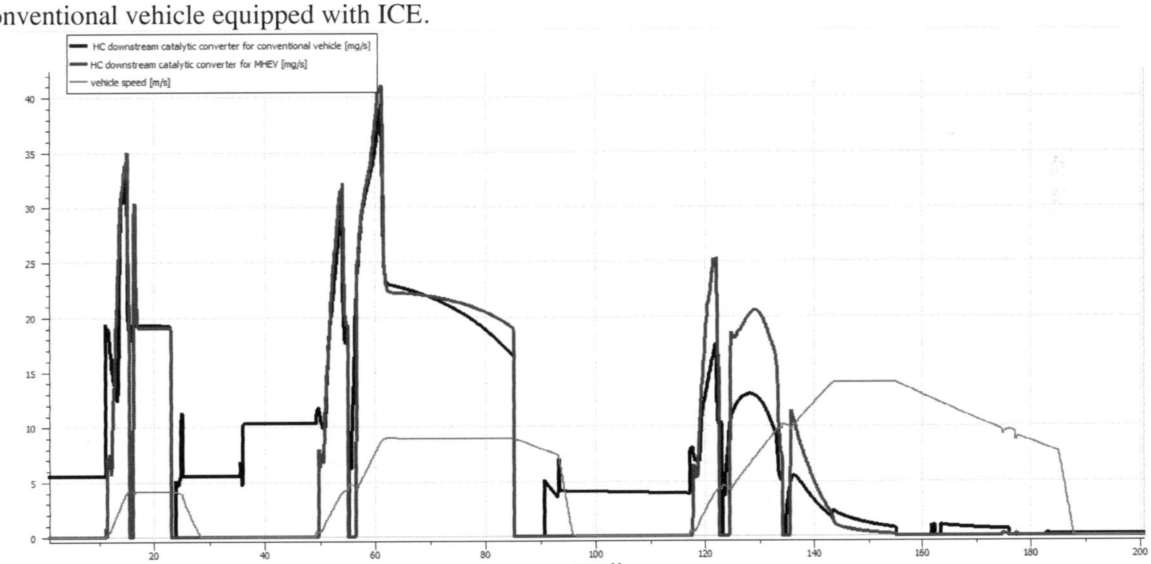

Figure 7. HC emission for conventional ICE vehicle and MHEV

In figure 7 is presented the HC emissions for conventional vehicle with internal combustion engine (black signal) and for mild hybrid electric vehicle (green signal). For the conventional vehicle, the mean value of HC emission for the first 200 seconds is 25 mg/s. For the MHEV, the mean value of CO emission for the first 200 seconds is 13 mg/s which is an improvement of 48% over the performance of conventional vehicle equipped with ICE.

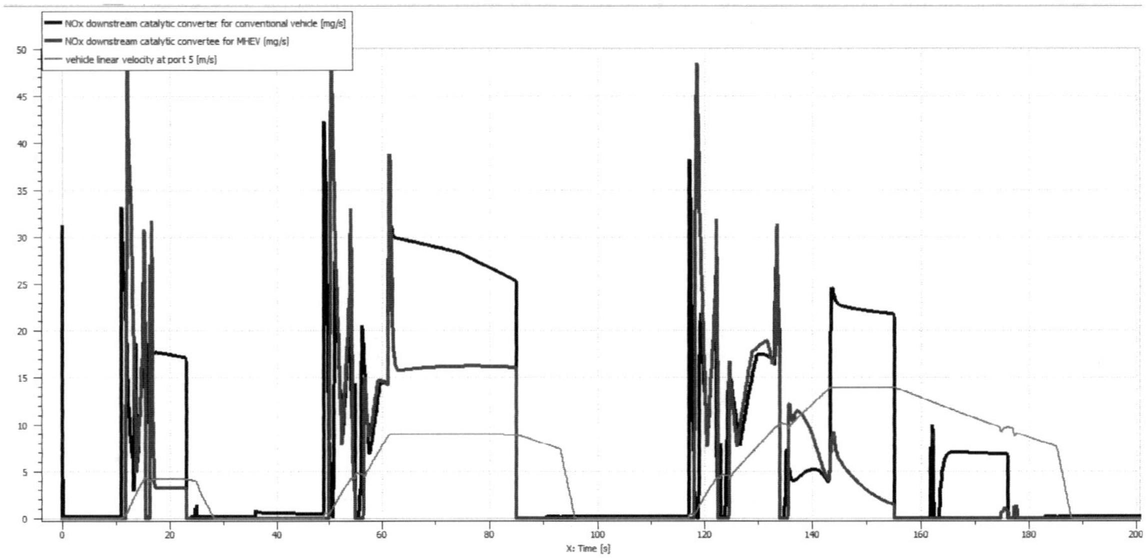

Figure 8. NO_x emission for conventional ICE vehicle and MHEV

In figure 8 is presented the NO_x emissions for conventional vehicle with internal combustion engine (black signal) and for mild hybrid electric vehicle (green signal). For the conventional vehicle, the mean value of NO_x emission for the first 200 seconds is 23 mg/s. For the MHEV, the mean value of NO_x emission for the first 200 seconds is 15 mg/s which is an improvement of 35% over the performance of conventional vehicle equipped with ICE.

4. Conclusion

In this paper are presented the fuel consumption and emission improvements using a mild hybrid electric vehicle using a 48V Integrated Starter Generator on NEDC driving cycle. Using the Integrated Starter Generator, the emission level of the vehicle are lower compared to conventional vehicle especially during urban driving cycle due to the presence of start&stop system. Also, the fuel consumption of the vehicle is lower with the ISG due to start&stop in urban driving cycle and power assistance in highway cycle.

References

[1] https://ec.europa.eu/clima/policies/transport/vehicles/cars
[2] Geun-Ho L, Geo-Seung C, Choi W 2011 Design Considerations for Low Voltage Claw Pole Type Integrated Starter Generator (ISG) Systems, *Jurnal of power electronics*
[3] LMS. AMESim Help. AMEHelp, 2012
[4] Saragea Ş, Oprean I M, Băţăuş M, Frăţilă G 2019 Driveability Improvements obtained using a mild-hybrid electrical system on a vehicle equipped with an automated manual transmission, *Science and management of automotive and transportation engineering 4th edition* (Craiova)
[5] Saragea Ş, Oprean IM, Badea G, Frăţilă G 2018 Modeling the Torque Oscillations During Gear Shifts on Automated Manual Transmission *Proceedings of the 4th International Congress of Automotive and Transport Engineering (AMMA 2018)* ed Burnete N, Varga B (Cluj) pp 634-641

Acknowledgments

This work has been funded by the European Social Fund from the Sectoral Operational Programme Human Capital 2014-2020, through the Financial Agreement with the title "Scholarships for entrepreneurial education among doctoral students and postdoctoral researchers (Be Antreprenor!)", Contract no. 51680/09.07.2019 - SMIS code: 124539.

IOP Conf. Series: Materials Science and Engineering 1037 (2021) 012054 doi:10.1088/1757-899X/1037/1/012054

Optimization of dynamic absorber parameters for protecting the machines against the vibrations transmitted through their base

R Ibănescu[1] and M Ibănescu[2]

[1]Department of Theoretical Mechanics, "Gheorghe Asachi" Technical University of Iași, Bd. Dimitrie Mangeron nr. 67, 700050, Iași, Romania
[2]Department of Structural Mechanics, "Gheorghe Asachi" Technical University of Iași, Bd. Dimitrie Mangeron nr. 67, 700050, Iași, Romania

*Corresponding author's e-mail: ribanesc@yhoo.com

Abstract. The correct functioning of machines could be affected by the vibrations transmitted through their base from other devices. The first possible protection measure is to support these machines on elastic elements with damping. The present work presents a protection solution which consists in the attachment of a dynamic absorber with viscous friction to the considered machine, in order to diminish the amplitudes of the vibrations caused by base movement. The mathematical model is developed for the real case when the base vibrations are not transmitted to the machine, but to the elastic supporting elements. The analytical expression of machine vibration amplitude in steady state is obtained. This amplitude could be diminished by an accurate choice of the three dynamic absorber parameters (mass, elastic constant, coefficient of viscous friction). An optimization program conceived in MATLAB provides the values of these parameters which lead to the minimum amplitude of machine vibrations.

1. Introduction

The normal functioning of machine tools could be affected by the vibrations coming from outer sources. For diminishing these effects and ensuring the machine tools functioning with no perturbations, the amplitude of these vibrations must be reduced, as much as possible. There are several possibilities for reaching this goal. One of the most used procedure is represented by the use of devices for the passive control of vibrations. Between them, the dynamic vibration absorber has been intensively studied, because of its wide range of application in different engineering fields, as civil engineering, automotive, shipbuilding [1], [2], [3], [4]. This wide use is explained by its high effectiveness relative to its constructive simplicity. A lot of recent studies have been focused on the optimization of the dynamic absorber parameters, the absorber being considered not only in the classical variant [5], but also in the new variant, with inverter and negative stiffness [6], [7], [8], [9], [10]. The present paper presents the model of a machine tool subjected to a sinusoidal displacement, transmitted through its support base. For mitigating the vibrations induced to the considered machine, a classical dynamic absorber is attached to it. The parameters of this dynamic absorber are optimized by using a software dedicated to MATLAB facilities. The approach in this paper is a lot more user friendly compared to the theoretical approaches of other works [11], [12], because it answers quickly at the practical problem of finding the absorber optimum parameters for a given vibrating system, by using three very explicit diagrams.

2. The mechanical system model

The model of the mechanical system contains a mass M, which is the mass of the machine tool supported on the ground through a spring, whose stiffness is k and a damper with the damping coefficient c, as shown in figure 1.

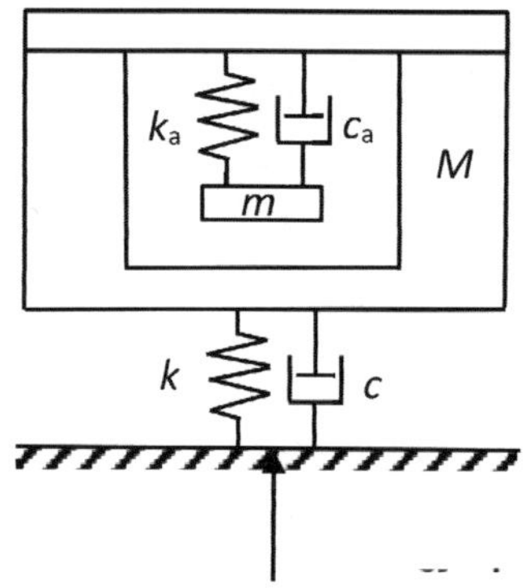

Figure 1. The model of the mechanical device.

It is considered that the supporting surface oscillates according to a sinusoidal law:

$$x = X_0 \sin(\omega_0 t) \tag{1}$$

A classical vibration dynamic absorber (C-DVA) is attached to mass M. This absorber consists of a mass m, a spring of stiffness k_a and a damper having the damping coefficient c_a, as it can be seen in figure 1.

This model differs from those presented in other works because it contains the damper of damping coefficient c_a between mass M and the ground, and that the excitatory vibration is induced by the displacement of the supporting surface, which is transmitted to the spring and damper, on which mass M is supported.

3. Mathematical model

The mechanical system, without the vibration dynamic absorber, is considered. By denoting the displacement of the mass M by x_1, the motion differential equation for this mass is:

$$M\ddot{x}_1 + c\dot{x}_1 + kx_1 = kx + c\dot{x} \tag{2}$$

By substituting equation (1) in equation (2), it results:

$$\ddot{x}_1 + \frac{c}{M}\dot{x}_1 + \frac{k}{M}x_1 = \frac{k}{M}X_0 \sin(\omega_0 t) + \frac{c}{M}\omega_0 X_0 \cos(\omega_0 t) \tag{3}$$

The steady state response will be the particular solution of a non-homogeneous equation, presumed of the following form:

$$x_{1p} = A\sin(\omega_0 t) + B\cos(\omega_0 t) \tag{4}$$

It is imposed the condition for x_{1p} to satisfy equation (3) and so, the values of coefficients A and B are determined. After performing the calculus, by identification, the amplitude of mass M oscillation, denoted by A_M, is determined:

$$A_M = X_0(k^2 + c^2\omega_0^2)^{1/2} \Big/ \left[(k^2 - M\omega_0^2)^2 + c^2\omega_0^2\right]^{1/2} \tag{5}$$

The angular velocity corresponding to the maximum amplitude is obtained by differentiating relation (5) and finding the roots of the derived equation. The root corresponding to the maximum amplitude is:

$$\omega_0 = \left[(Q - Mk^2)/(Mc^2) \right]^{1/2} \tag{6}$$

where

$$Q = (M^2k^4 + 2Mc^2k^3)^{1/2} \tag{7}$$

There are considered the following numerical values: M = 5000kg, X_0 = 0.0001m, c = 2000Ns/m, k = 100000N/m. The resulted maximum amplitude of mass M is 1.124x10^{-3}m, corresponding to an angular velocity ω_0=4.461s^{-1}. The graph which describes the variation of the amplitude in terms of angular velocity, in case of no dynamic absorber, is pictured in figure 2.

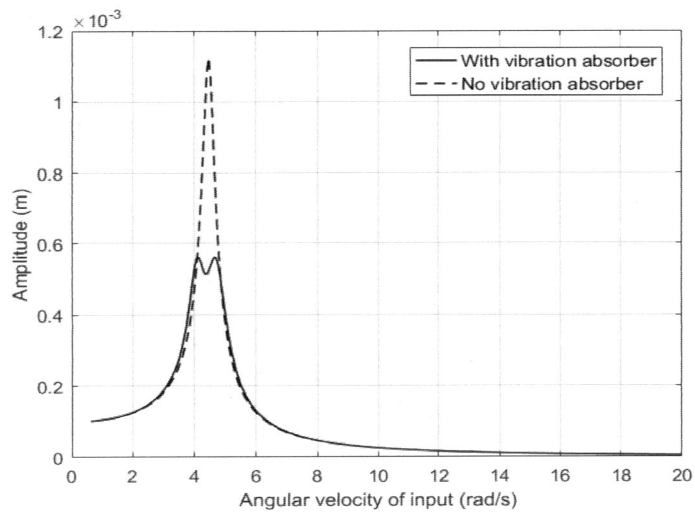

Figure 2. The amplitude of the mass M.

For diminishing the maximum value of the amplitude, a vibration dynamic absorber is attached to mass M. In this new case, denoting the displacement of mass m by x_2, the system of differential equations that describes the dynamics of the mechanical system shown in figure 1 is:

$$M\ddot{x}_1 + k(x_1 - x) + k_a(x_1 - x_2) + c(\dot{x}_1 - \dot{x}) + c_a(\dot{x}_1 - \dot{x}_2) = 0 \tag{8}$$

$$m\ddot{x}_2 - k_a(x_1 - x_2) - c_a(\dot{x}_1 - \dot{x}_2) = 0 \tag{9}$$

The system of state equations is determined based on equations (8) and (9)and it has the shape:

$$\dot{v}_1 + v_1(c + c_a)/M - v_2c_a/M + x_1(k + k_a)/M - x_2k_a/M = xk/M + \dot{x}c_1/M \tag{10}$$

$$\dot{v}_2 - v_1c_a/m + v_2c_2/m - x_1k_2/m + x_2k_2/m = 0 \tag{11}$$

$$\dot{x}_1 = v_1 \tag{12}$$

$$\dot{x}_2 = v_2 \tag{13}$$

where v_1 and v_2 are the velocities of mass M and m, respectively. The steady state response is given by a particular solution, having the form:

$$x_1 = a_1 \sin(\omega_0 t) + b_1 \cos(\omega_0 t) \tag{14}$$

$$x_2 = c_1 \sin(\omega_0 t) + d_1 \cos(\omega_0 t) \tag{15}$$

$$v_1 = e_1 \sin(\omega_0 t) + f_1 \cos(\omega_0 t) \tag{16}$$

$$v_2 = g_1 \sin(\omega_0 t) + h_1 \cos(\omega_0 t) \tag{17}$$

where a_1,b_1, c_1, d_1, e_1, f_1, g_1 and h_1 are unknown coefficients, which result by identification, after replacing equations (14), (15), (16) and (17) and their derivatives in equations (10), (11), (12) and (13).There are eliminated the unknowns e_1, f_1, g_1 and h_1, by expressing them in terms of a_1,b_1, c_1 and d_1. The following system of algebraic equations results:

$$Aa_1 - Bb_1 - Cc_1 + Dd_1 = kX_0 / M \tag{18}$$

$$Ba_1 + Ab_1 - Dc_1 - Cd_1 = c\omega_0 X_0 / M \tag{19}$$

$$-Ea_1 + Fb_1 + Gc_1 - Hd_1 = 0 \tag{20}$$

$$-Fa_1 - Eb_1 + Hc_1 + Gd_1 = 0 \tag{21}$$

where:

$$\begin{array}{ll} A = (k + k_a) / M - \omega_0^2 & E = k_a / m \\ B = (c + c_a)\omega_0 / M & F = c_a \omega_0 / M \\ C = k_a / M & G = k_a / m - \omega_0^2 \\ D = c_a \omega_0 / M & \mathrm{H} = \mathrm{c}_a \omega_0 / m \end{array} \tag{22}$$

The system of equations is solved and the amplitude x_1 of mass M is obtained according to the formula:

$$x_1 = (a_1^2 + b_1^2)^{1/2} \tag{23}$$

4. Optimization of vibration dynamic absorber parameters

This optimization of dynamic absorber parameters consists in the assessment of m, c_a and k_a values, which correspond to the minimum amplitude of mass M.

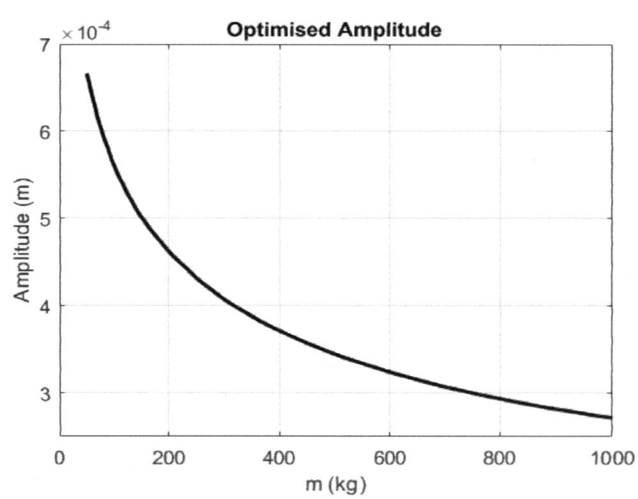

Figure 3. The amplitude with respect to the absorber mass m.

The original values of these parameters are: m=100kg, c_a=1000Ns/m, k_a=100000N/m.

The calculus of the optimum values is performed by using a software which has been created by the authors, based on function *fmincon* from MATLAB.

The optimum values have been computed when $\omega_0 \in [\pi / 5,\ 16\pi]$ for the following intervals of the considered parameters $c_a \in [1,\ 10000]$, Ns/m $k_a \in [100,\ 200000]$N/m and $m \in [1,\ 100]$kg.

The resulted optimum values of absorber parameters are: m=100 kg, c_a=79.1 Ns/m and k_a=1882.2 N/m.

For these values, the maximum amplitude is A=5.635x10^{-4}m, for an angular velocity ω_0 = 4.147s^{-1}.

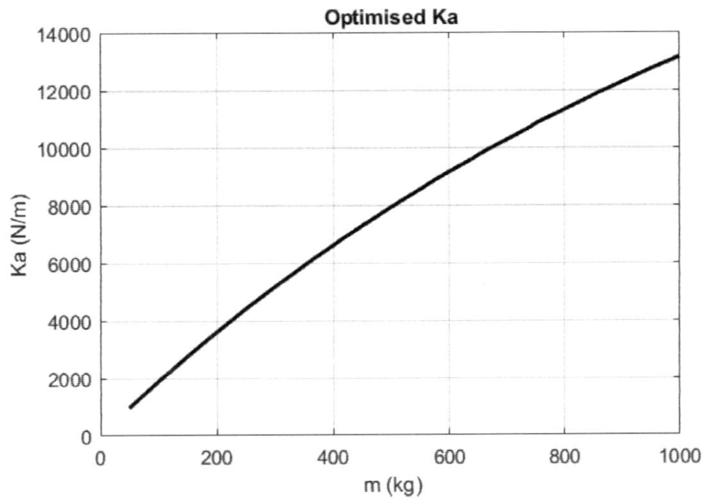

Figure 4. The stiffness k_a with respect to the absorber mass m.

The graph which shows the relation amplitude versus angular velocity is pictured in figure 2. It can be seen that the maximum amplitude decreases with 50% compared with the maximum amplitude of 11.24x10^{-4}m, corresponding to the case without absorber.

It has been noticed that in case of changing the maximum value for mass m in the previously mentioned interval, the software provides as optimum value that maximum value. For this reason, the optimization of this parameter, mass m, has been abandoned.

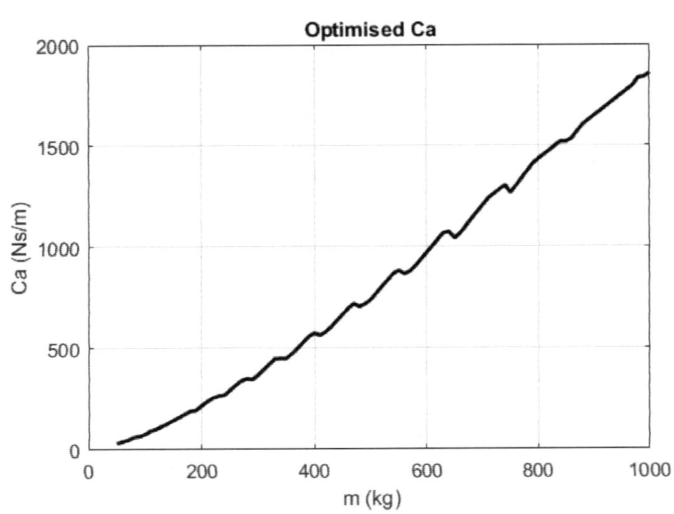

Figure 5. The damping coefficient c_a with respect to the absorber mass m.

In these circumstances, another optimization software has been conceived, which provides the minimum amplitude and the values of parameters c_a and k_a, for each considered value for mass *m*. The results are presented in figure 3, figure 4 and figure 5. From figure 3, the constructive accepted couple mass-minimum amplitude is chosen. The parameter k_a results from figure 4, while c_a results from figure 5, both corresponding to a selected mass *m*.

As an example, *for* m=400kg, the resulted optimum parameters are: the minimum amplitude is $A_{min}=3.715\text{x}10^{-4}$ m, k_a=6600N/m and c_a=574.9Ns/m.

It has been noticed that the amplitude mitigates less and less when the value of mass *m* increases, as shown in figure 3. So, when the mass *m* increases from 100kg to 500kg, the amplitude decreases from $5.635\text{x}10^{-4}$m to $3.446\text{x}10^{-4}$m, which means a decrease of 38.84%. When mass m=1000kg, the corresponding amplitude is $2.717\text{x}10^{-4}$m. So, compared to the value which results for m=500kg, the amplitude decreases with 21.15%.

5. Conclusions

The goal of this paper is to solve the problem of selection for classical dynamic vibration absorber (C-DVA) in terms of their effectiveness for a wide frequency range, around the maximum frequency.

The solution provided by the present work is based on an optimization software created by the authors, which provides the optimum values for the parameters of the dynamic absorber, attached to a vibrating system with damping.

Because the absorber mass has always the maximum value in the considered interval, the conceived software offers the possibility of selecting the best solution for the couple absorber mass–minimum amplitude. Then, for this solution, the optimum corresponding parameters, that is the spring stiffness and the absorber damping coefficient, can be immediately determined.

In this manner, each user can conceive an optimum dynamic absorber, based on the absorber mass and the corresponding optimum amplitude.

6. References

[1] Magdaleno A and Lorenzana A 2020 *Mech. Sist. Sign. Proc.* **135** 106378
[2] Love J S and Taylor Z J 2019 *J. of Wind Eng. & Ind. Aerodyn.* **194** 103977
[3] Huang X, Su Z and Hua H 2018 *Ocean Eng.* **155** pp 131-143
[4] Tehrani G G and Dardel M 2019 *Commun. Nonlin. Sci. Numer. Simulat.* **69** pp 1-30
[5] Shen Y, Chen L, Yang X, Shi D and Yang J 2016 *J. Sound. Vibr.* **361** pp 148-158
[6] Chun S, Lee Y and Kim T H 2015 *J. Sound. Vibr.* **335** pp 55-65
[7] Hu Y and Chen M Z Q 2015 *Int. J. Mech. Sci.* **99** pp 297-307
[8] Zhou S, Jean-Mistral C and Chesne S 2019 *Int. J. Mech. Sci.* **157-158** pp 528-541
[9] Shen Y, Peng H, Li X and Yang S 2017 *Mech. Sist. Sign. Proc.* **85** pp 193-203
[10] Shen Y, Xing Z, Yang S and Sun J 2019 *Mech. Sist. Sign. Proc.* **133** 106282
[11] Brown B and Singh T 2011 *J. Sound. Vibr.* **330** pp 2437-2448
[12] Tigli O F 2012 *J. Sound. Vibr.* **331** pp 3035-3049

IOP Conf. Series: Materials Science and Engineering 1037 (2021) 012055 doi:10.1088/1757-899X/1037/1/012055

Simplified mathematical model for frontal impact study

A M Manea[1], D Iozsa[1] and C Stan[1]

[1]Automotive Engineering Dept., Faculty of Transports, POLITEHNICA University of Bucharest, Splaiul Independentei nr. 313, sector 6, Bucuresti, ROMANIA, RO-060042
E-mail: mannea_annamaria@yahoo.com, daniel_iozsa@yahoo.com, corraista@yahoo.com

Abstract. The structure of the front compartment is built to provide safety to passengers during an impact. Safety is ensured by designing structures capable of taking up much of the impact energy. During an impact, the energy of the impact is consumed by the ability of the body structure to deform. Thus, a large part of the kinetic energy of the impact is transformed into deformation energy. It ensures that decelerations are reduced in a short period of time to the limit that passengers can bear. Two mathematical models are proposed for the study of the frontal impact between two vehicles. In the mathematical model the component elements of the frontal compartmental structures are treated separately. Two simplified numerical models with two degrees of freedom are proposed. The first mathematical model is of the table-spring type, and the second is the mathematical model of the table-spring-shock absorber type. The body elements and the main subassemblies of the vehicle are considered to be two concentrated table bodies. The connection between these bodies will be made by elastic elements and damping elements. Based on Lagrange's generalized equations, the equations of motion necessary for the impact study will be obtained. Furthermore, the solving of the numerical systems of the proposed models will be done with the help of the modelling, simulation and analysis environment offered by the MATLAB Simulink software. In this environment, the main initial parameters and impact conditions between the two bodies will be established and block schemes will be built for both systems of equations. With the help of these mathematical models it is desired to determine the maximum deformation recorded by each vehicle during the frontal impact between the two.

1. Introduction

The structure of the vehicle is designed to provide special areas capable of dissipating kinetic energy by deformation during an impact. In most accidents between two vehicles there are elastic and plastic deformations (elasto-plastic deformations) [1], [2]. In this case, both the resistance structure of the vehicle and the passenger protection systems are thus designed to ensure the protection of the occupants of the vehicle. Also, taking into account the fact that the deformations during the impact are of the elasto-plastic type, it is possible to propose for study mathematical models of mass-spring type or mathematical models of mass-spring-damper type [3]...[5].

In the paper [6], a simplified one-dimensional calculation model with a single degree of freedom was studied. This model solves the impact phenomenon between a vehicle hitting a fixed rigid barrier. The vehicle is considered as a unit mass.

Compared to the study in the paper [6], this paper aims to study the impact phenomenon between two vehicles.

2. Simplified numerical models with two degrees of freedom consisting of two bodies (two vehicles)

2.1. Mathematical model of mass-spring type

The simplified numerical model with two degrees of freedom consists of two bodies (two vehicles) [2]. Based on this simplified two-body model [6], the behavior of two vehicles that are involved in a frontal impact will be determined. The bodies are connected to each other by elastic elements of helical spring type with stiffness k_1, respectively k_2. In the figure 1, presented below you can see the simplified model for studying the impact between two vehicles.

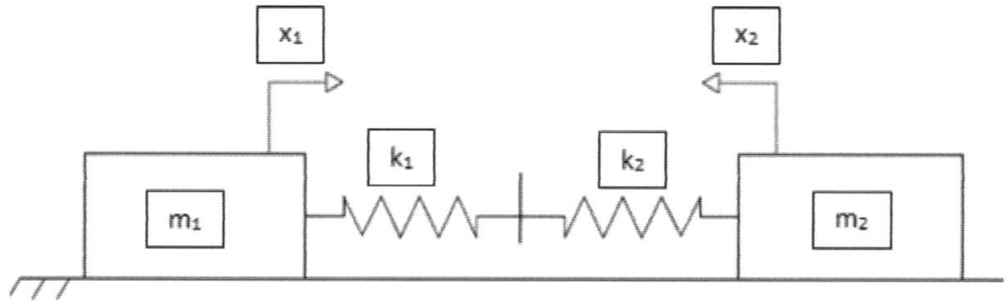

Figure 1. Simplified numerical model of mass-spring type with two degrees of freedom, where: m_1 - mass of vehicle 1; m_2 - mass of vehicle 2; x_1 - movement of vehicle 1; x_2 - movement of vehicle 2; k_1 - elastic constant corresponding to the deformable area of vehicle 1; k_2 - elastic constant corresponding to the area of vehicle 2.

The equations of motion of the vehicles involved in the collision are obtained on the basis of Lagrange's generalized equation [3] [7]:

$$\frac{d}{dt}\frac{\partial E_c}{\partial \dot{x}_i} - \frac{\partial E_c}{\partial x_i} + \frac{\partial V}{\partial x_i} = 0 \tag{1}$$

where: E_c – is the kinetic energy of the system; V – is the potential deformation energy of the system; x_i – are the generalized coordinates (distances of the centers of gravity of the bodies in the longitudinal direction x_1 and x_2).

The kinetic energy of the system formed by the two vehicles is [8]:

$$E_c = \frac{1}{2}m_1\dot{x}_1^2 + \frac{1}{2}m_2\dot{x}_2^2 \tag{2}$$

where: $\dot{x}_1, \dot{x}_2$ - are the speeds corresponding to the two vehicles.

To write the potential deformation energy of the system formed by the two vehicles is considered as the two elastic elements connected in series. Thus, $k_s = \frac{k_1 k_2}{k_1+k_2}$ is considered to be the equivalent elastic constant of the two elastic elements k_1 and k_2 [8].

$$V = \frac{1}{2}k_s(x_1 + x_2)^2 \tag{3}$$

It is considered that only the collision force acts on the vehicles during the impact, the other forces being negligible in relation to it.

The generalized Lagrange equation for the two vehicles will be applied below:

$$\begin{cases} \frac{d}{dt}\frac{\partial E_c}{\partial \dot{x}_1} - \frac{\partial E_c}{\partial x_1} + \frac{\partial V}{\partial x_1} = 0 \\ \frac{d}{dt}\frac{\partial E_c}{\partial \dot{x}_2} - \frac{\partial E_c}{\partial x_2} + \frac{\partial V}{\partial x_2} = 0 \end{cases} \tag{4}$$

Solving this system of equations by replacing with the corresponding expressions the kinetic energy and the potential deformation energy, the equations of motion of the vehicles involved in a frontal impact are obtained.

$$\begin{cases} m_1\ddot{x}_1 + k_s(x_1 + x_2) = 0 \\ m_2\ddot{x}_2 + k_s(x_1 + x_2) = 0 \end{cases} \tag{5}$$

The result is a system of two degree II differential equations in which the accelerations of the two vehicles are unknown:

$$\begin{cases} \ddot{x}_1 = \frac{-k_s(x_1+x_2)}{m_1} \\ \ddot{x}_2 = \frac{-k_s(x_1+x_2)}{m_2} \end{cases} \tag{6}$$

2.2. Mathematical model of mass-spring-damper type

The simplified numerical model with two degrees of freedom consists of two bodies (two vehicles) which are connected to each other by elastic spring-type elements with stiffness k_1 and k_2 and two damping elements having damping constants c_1 and c_2) [2] [4]. In the figure 2, presented below you can see the simplified model for studying the impact between two vehicles.

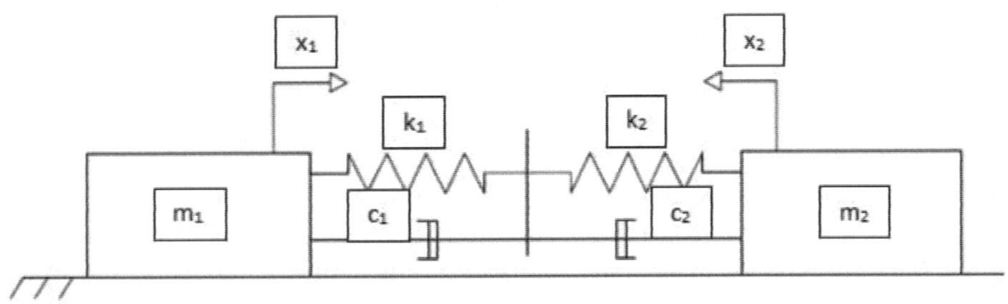

Figure 2. Simplified numerical model of mass-spring-damper type with two degrees of freedom, where: m_1 - mass of vehicle 1; m_2 - mass of vehicle 2; x_1 - movement of vehicle 1; x_2 - movement of vehicle 2; k_1 - elastic constant corresponding to the deformable area of vehicle 1; k_2 - elastic constant corresponding to the area of vehicle 2; c_1 - the damping constant corresponding to the deformable area of the vehicle 1; c_2 - the damping constant corresponding to the deformable area of the vehicle 2.

The equations of motion of the vehicles involved in the collision are obtained on the basis of Lagrange's generalized equation [3] [7]:

$$\frac{d}{dt}\frac{\partial E_c}{\partial \dot{x}_i} - \frac{\partial E_c}{\partial x_i} + \frac{\partial E_a}{\partial \dot{x}_i} + \frac{\partial V}{\partial x_i} = 0 \tag{7}$$

where: E_c – is the kinetic energy of the system; E_a- e is the energy dissipated by the damping elements; $\dot{x}_i$ - are the corresponding speeds of the vehicles; V – is the potential deformation energy of the system; x_i – are the generalized coordinates (distances of the centers of gravity of the bodies in the longitudinal direction x_1 and x_2).

The kinetic energy of the system formed by the two vehicles is [8]:

$$E_c = \frac{1}{2}m_1\dot{x}_1^2 + \frac{1}{2}m_2\dot{x}_2^2 \tag{8}$$

where: $\dot{x}_1, \dot{x}_2$ - are the speeds corresponding to the two vehicles.

To write the potential deformation energy of the system formed by the two vehicles is considered as the two elastic elements connected in series. Thus, $k_s = \frac{k_1 k_2}{k_1+k_2}$ is considered to be the equivalent elastic constant of the two elastic elements k_1 and k_2 [8].

$$V = \frac{1}{2}k_s(x_1 + x_2)^2 \tag{9}$$

In order to write the energy dissipated by the damping elements of the system formed by the two vehicles, it is considered that the two dampers connected in series. Thus, $cc_s = \frac{c_1 c_2}{c_1+c_2}$ is considered to be the equivalent depreciation constant of the two depreciation elements c_1 and c_2, [8].

$$E_a = \frac{1}{2}c_s(\dot{x}_1 + \dot{x}_2)^2 \tag{10}$$

The generalized Lagrange equation for the two vehicles will be applied below:

$$\begin{cases} \frac{d}{dt}\frac{\partial E_c}{\partial \dot{x}_1} - \frac{\partial E_c}{\partial x_1} + \frac{\partial E_a}{\partial \dot{x}_1} + \frac{\partial V}{\partial x_1} = 0 \\ \frac{d}{dt}\frac{\partial E_c}{\partial \dot{x}_2} - \frac{\partial E_c}{\partial x_2} + \frac{\partial E_a}{\partial \dot{x}_2} + \frac{\partial V}{\partial x_2} = 0 \end{cases} \tag{11}$$

Solving this system of equations by replacing with the corresponding expressions the kinetic energy and the potential deformation energy, the equations of motion of the vehicles involved in a frontal impact are obtained.

$$\begin{cases} m_1\ddot{x_1} + c_s(\dot{x_1} + \dot{x_2}) + k_s(x_1 + x_2) = 0 \\ m_2\ddot{x_2} + c_s(\dot{x_1} + \dot{x_2}) + k_s(x_1 + x_2) = 0 \end{cases} \tag{12}$$

The result is a system of two differential equations in which the accelerations of the two vehicles are unknown:

$$\begin{cases} \ddot{x}_1 = \frac{-c_s(\dot{x_1}+\dot{x_2})-k_s(x_1+x_2)}{m_1} \\ \ddot{x}_2 = \frac{-c_s(\dot{x_1}+\dot{x_2})-k_s(x_1+x_2)}{m_2} \end{cases} \tag{13}$$

3. Solving mathematical models in Matlab Simulink

3.1. Solving mathematical model of mass-spring type in Matlab Simulink

The numerical solution of the mass-spring type system and of the mass-spring-damper type system will be performed with the help of the modeling, simulation and analysis environment offered by the Matlab Simulink software, [6].

The solution of the system of differential equations, according to the relations (6), is presented in figure 3 through the block diagram of the mass-spring type. The following initial data are considered to solve this system: m_1= 1200 kg; m_2= 1400 kg; k_1 =170000 N/m; k_2=210000 N/m; k_s= 93947 N/m; v_1= v_2= 13,8 m/s.

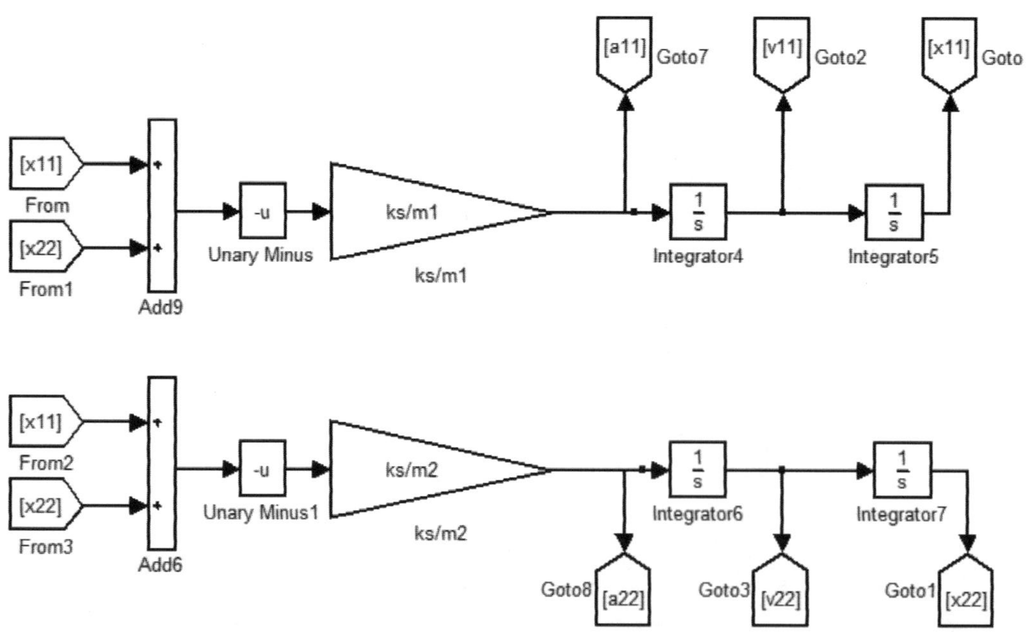

Figure 3. Block diagram of the mass-spring system in Matlab Simulink

IOP Conf. Series: Materials Science and Engineering 1037 (2021) 012055 doi:10.1088/1757-899X/1037/1/012055

x_{11} - movement of vehicle 1; x_{22} - movement of vehicle 2; v_{11} and v_{22}- initial vehicle speeds at the time of impact; a_{11} and a_{22} - vehicles accelerations; k_s - equivalent rigidity of the system with two springs connected in series.

Furthermore, with the help of this scheme, the evolutions in time of the movement, speed and, respectively, of the acceleration of the two vehicles subjected to the theoretical test to a frontal impact will be represented graphically. From the following block diagrams, from figure 4, these evolutions will be represented in the figure 5.

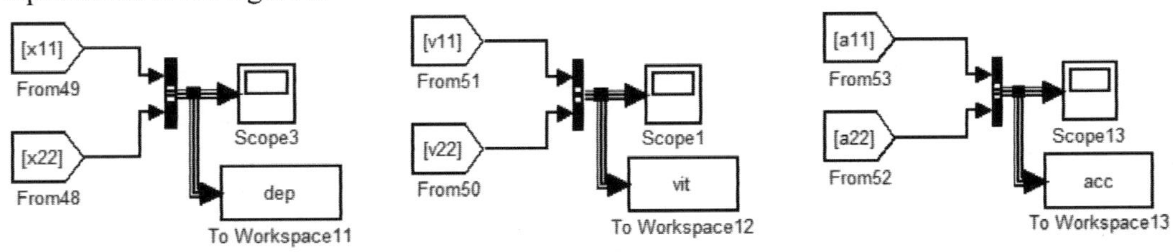

Figure 4. Block diagrams made to determine the displacement, velocity and acceleration

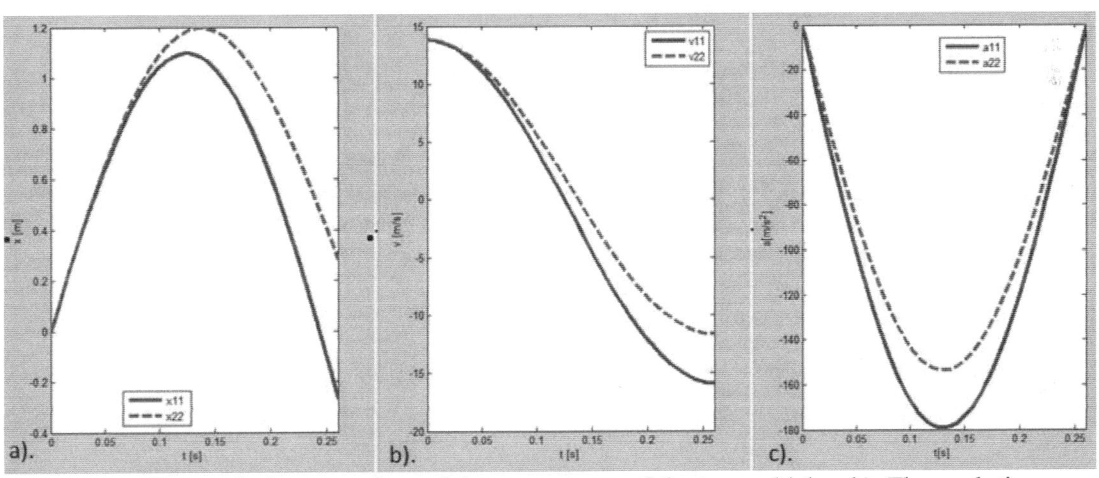

Figure 5.a). The evolution over time of the movements of the two vehicles; b). The evolution over time of the speeds of the two vehicles; c). The evolution over time of the accelerations of the two vehicles

Using the block diagram in Figure 6, the characteristics for relative deformation, relative velocity and relative acceleration will be represented. Due to the order of magnitude, the relative deformation and relative velocity are multiplied by 30 and 10. They are represented in overlapping graphs in Figure 7 to observe the evolution of the impact.

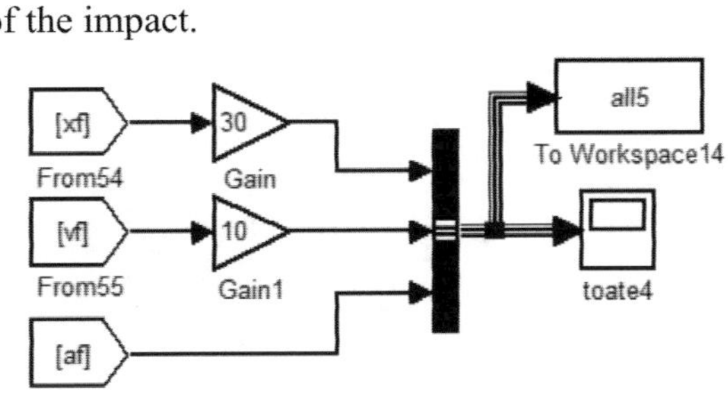

Figure 6. Block diagrams made to determine the evolution over time of relative displacement, relative velocity and relative acceleration

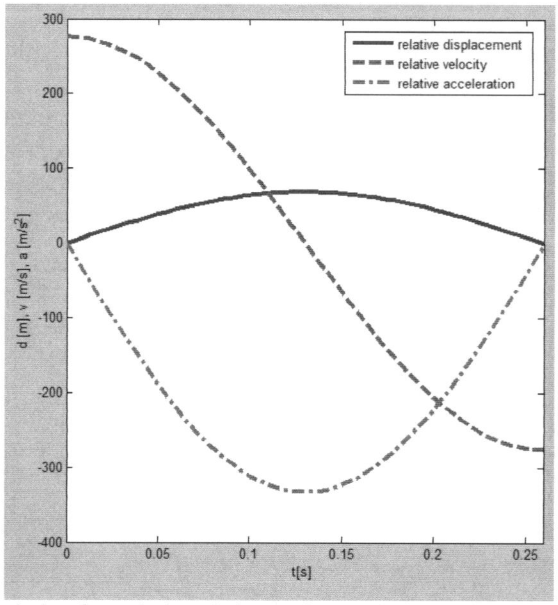

Figure 7. The characteristics for relative deformation, relative velocity and relative acceleration

3.2. Solving mathematical model of mass-spring-damper type in Matlab Simulink

The solution of the system of differential equations, according to the relations (13), is presented in figure 8 through the block diagram of the mass-spring-damper type. The following initial data are considered to solve this system: m_1= 1200 kg; m_2= 1400 kg; k_1 =170000 N/m; k_2=210000 N/m; k_s= 93947 N/m; c_1=1500 Ns/m; c_2=2000 Ns/m; c_s= 857 Ns/m; v_1= v_2= 13,8 m/s.

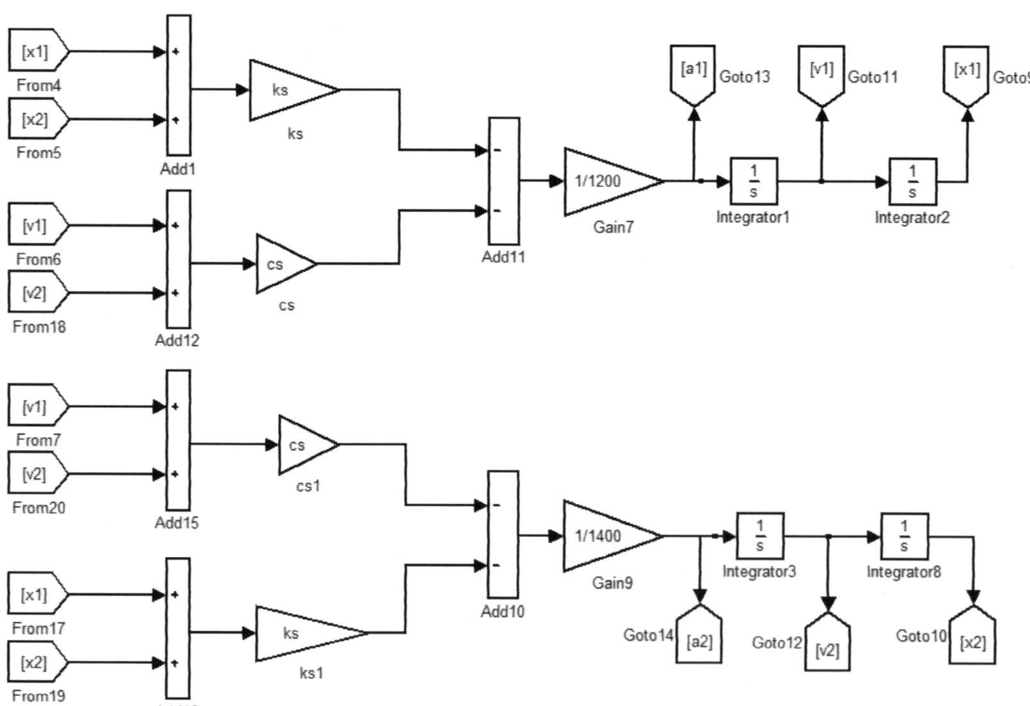

Figure 8. Block diagram of the mass-spring-damper system in Matlab Simulink

where: x_1 - movement of vehicle 1; x_2 - movement of vehicle 2; v_1 and v_2- initial vehicle speeds at the time of impact; a_1 and a_2 - vehicles accelerations; k_s - equivalent rigidity of the system with two springs connected in series; c_s - the equivalent damping constant of the system with two dampers connected in series.

Similar to the mass-arc system, with the help of this scheme will be represented graphically the evolutions in time of the displacement, velocity and acceleration of the two vehicles subjected to the frontal impact test. The following figure 9 will represent the block diagrams, which will present the graphs for the evolution over time of the vehicle parameters according to the simulation of the mass-spring-damper system (figure 10).

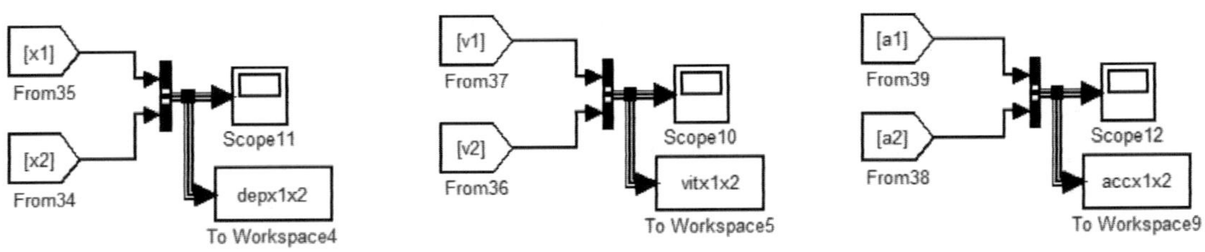

Figure 9. Block diagrams made to determine the displacement, velocity and acceleration

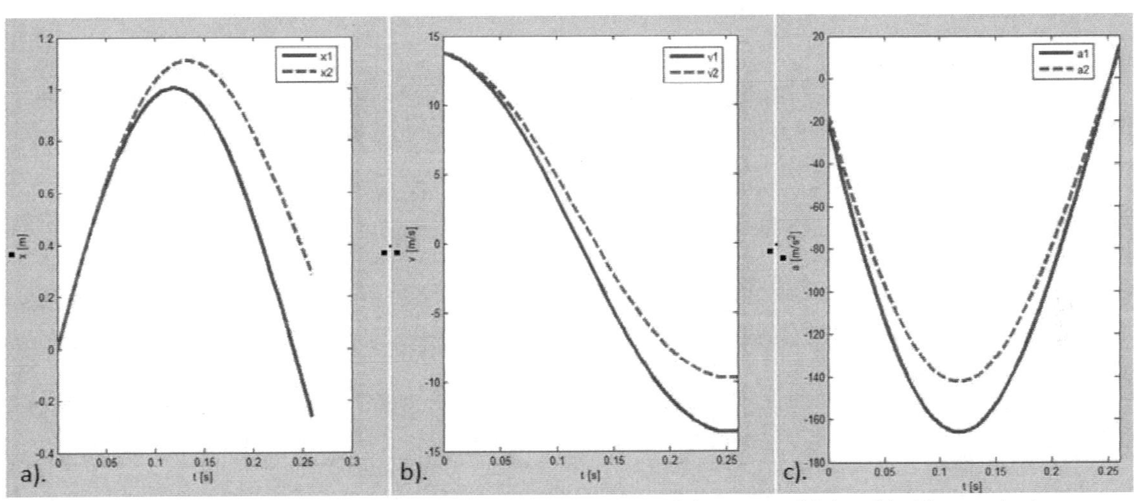

Figure 10 a). The evolution over time of the movements of the two vehicles; b). The evolution over time of the speeds of the two vehicles; c). The evolution over time of the accelerations of the two vehicles

Figure 11 shows the block diagram that will represent the evolution over time for relative deformation, relative velocity and relative acceleration. They are represented in overlapping graphs in Figure 12 to observe the evolution of the impact.

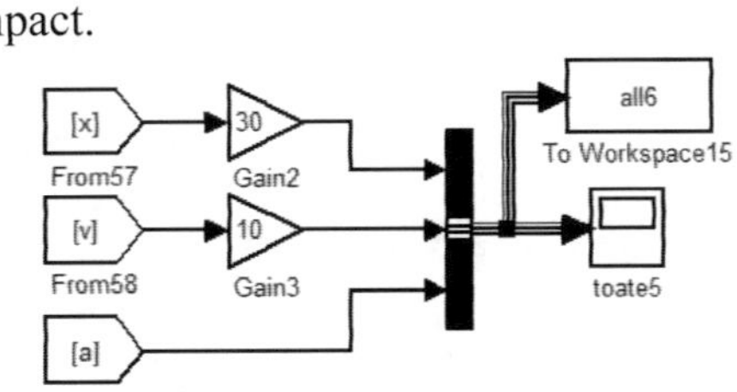

Figure 11. Block diagrams made to determine the evolution over time of relative displacement, relative velocity and relative acceleration

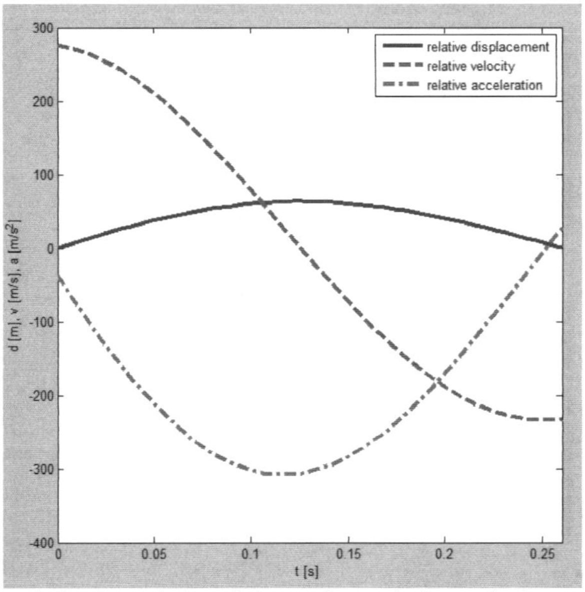

Figure 12. The characteristics for relative deformation, relative velocity and relative acceleration

4. Conclusions

In conclusion, in order to study the phenomena and differences between the two mathematical models during the frontal impact, the relative displacements, relative speeds and relative accelerations will be plotted in parallel, using the block diagrams in figure figura 13. Thus, a parallel analysis of the frontal impact behavior of the two proposed calculation models will be performed. These parameters were obtained using the modeling, simulation and analysis environment offered by the Matlab Simulink program.

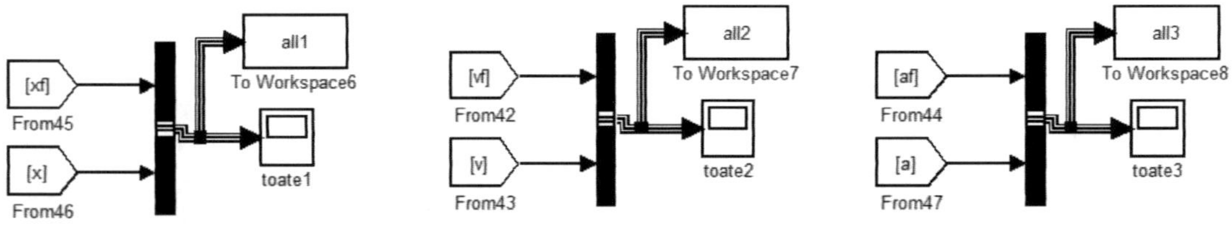

Figure 13. Parallel analysis of the frontal impact behavior of the proposed calculation models

It can be seen from figure 14.a). the fact that the deformations in the case of the mass-arc calculation model are larger than those registered in the case of the mathematical mass-spring-damper model. From this result it can be stated that the mass-spring-damper system has an advantage in terms of the absorption of kinetic energy during an impact. Also, through those physical properties of the damping system, it can be specified that this system makes a considerable contribution to the deformation mode of vehicles and thus, it is produced in a controlled manner. From the evolutions of the other characteristics, the relative speeds and the relative accelerations, it can be observed that also in the case of the mass-spring-damper model the values obtained are lower than those obtained in the case of the mass-spring system.

In conclusion, the mathematical models proposed for the study of the frontal impact between two vehicles have shown that they can help in the development of the evolution of passive safety. Applying fundamental notions from classical mechanics, satisfactory results can be obtained with few

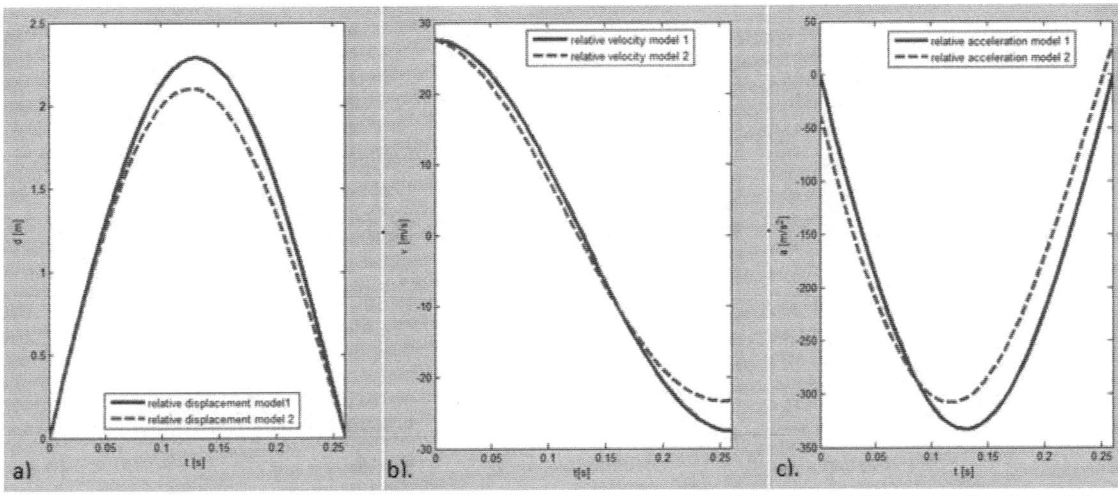

Figure 14. Characteristics obtained from the parallel analysis of the frontal impact behavior of the proposed calculation models: a). relative displacement; b). relative speed; c). relative acceleration

resources. These two models can also be analyzed with the help of structural analysis by the finite element method.

Acknowledgments
This work has been funded by the European Social Fund from the Sectoral Operational Programme Human Capital 2014-2020, through the Financial Agreement with the title *"Scholarships for entrepreneurial education among doctoral students and postdoctoral researchers (Be Antreprenor!)"*, Contract no. 51680/09.07.2019 - SMIS code: 124539.

5. References

[1] Malen D 2011 Fundamentals of Automobile Body Structure Design SAE International USA;
[2] Du Bois P, Chou C C, Fileta B B, Khalil T B, King A I, Mahmood H F, Mertz H J, Wismans J, VEHICLE CRASHWORTHINESS AND OCCUPANT PROTECTION, American Iron and Steel Institute, 2000 Town Center Southfield, Michigan, 2004;
[3] Ilie S, Modeling the impact phenomena of vehicles, Publishing house AGIR, Bucharest, 2012;
[4] Deac S C, Perescu A, Simoiu D, Nyaguly E, Crâştiu I, Bereteul, Modeling and simulation of cars in frontal collision, International Conference on Applied Sciences, IOP Conference Series: Materials Science and Engineering, 2017;
[5] Munyazikwiye B B, Vysochinskiy D, Mikhail Khadyko M, Robbersmyr K G, Prediction of Vehicle Crashworthiness Parameters Using Piecewise Lumped Parameters and Finite Element Models, Designs, 2018, (www.mdpi.com/journal/designs).
[6] Manea A M, Iozsa D, Stan C, Teodorescu C, Frățilă Gh, Mathematical model for study the behavior of the bodywork at impact, he 30th SIAR International Congress of Automotive and Transport Engineering. SMAT 2019. Springer, Cham. https://doi.org/10.1007/978-3-030-32564-0_26
[7] Radeş M, Mechanical vibrations, Publishing house Printech, Bucharest, 2008;
[8] Alecu A, Mechanics Course, Politehnica University of Bucharest, Bucharest, 2014;

IOP Conf. Series: Materials Science and Engineering 1037 (2021) 012056 doi:10.1088/1757-899X/1037/1/012056

Reliability model of freight bogie damper based on real field data

M A Spiroiu[1]

[1]University POLITEHNICA of Bucharest, Department of Railway Rolling Stock, Splaiul Independenţei 313, Bucharest, Romania

[*]Corresponding author's e-mail: marius.spiroiu@upb.ro

Abstract. In the present paper is estimated the reliability probabilistic model of the dry friction damper of the freight bogie Y25. The analysis is made on the basis of real field data, observed in actual operations of freight wagons. The theoretical probability distributions used to model the reliability are exponential and Weibull. The Weibull reliability models are estimated according to the Regression Method and to the Maximum Likelihood Method.

1. Introduction

Taking into account the nature of the service provided, reliability is an essential feature of railway vehicles. Many of the possible failure modes are potential causes of derailments, so it may have important social and / or economic implications and - in the worst case scenario – may lead to significant material damage or even to fatalities [1]. For this reason, the railway vehicles reliability is the subject of quite a lot of research. For example, in [2] are evaluated the failures and the reliability of three types of metro vehicles operating in Poland. The failures were classified and their occurrence in time was analysed. Also, in [3] a reliability analysis of subway vehicles subsystems was developed based on operational failures data. Failure statistics were made, and the optimal failure distribution model of each subsystem was determined. A reliability model of components of bogie is proposed in [4] based on the real data of the CRH2 high speed train.

Since the wheelset is a critical item in terms of safety, a significant number of studies address the issue of its reliability. Of these, can be mentioned the analysis of freight car wheelset failure modes based on field data, a reliability model being deduced using the regression method [5]. Also, in [1] are analysed the railway wheel failure modes and, for the major four failure modes, is carried out an additional analysis regarding the evolution in time of the failure rates, as well as the estimation of the failure probabilistic models, based on the Weibull distribution.

Besides the wheelset, the railway vehicle suspension is also a system with a high impact on safety. The suspension system has to isolate the car body from the vibrations determined by track irregularities, rail discontinuities, wheel defects, etc. It has also to ensure a stable dynamic behaviour of the vehicle in straight line and in curves and to reduce the vehicle-track mutual forces for reasons of traffic safety and passenger comfort or protection of the transported goods [6]. As shown in [7], the bogie suspension and structure failures are one of the main categories of rolling stock failures that have produced derailments in recent years in railway freight transport - both in European Union (second place) and in Romania (first place).

Among the concerns related to the reliability of the suspension, there is a growing interest in the field of condition monitoring and fault detection. For example, in [8,9] are proposed sensor configurations and are developed the methods (i.e. the associated mathematical models), while in [10] is proposed a failure detection method of the primary suspension damper, based on the analysis of cross-correlation of the bogie vertical accelerations.

As for reliability studies presenting systematically the failure modes of the suspension system or of its main components (dampers, springs) or in which is derived a reliability probabilistic model thereof, these are quite hard to find in the scientific literature. It can be mentioned a study investigating the freight vehicles suspension failures is carried out in [11], but the major emphasis was on the springs failure causes, the investigation consisting in failed spring material analysis and in stress analysis using analytical and finite element methods.

Under these circumstances, the present paper aim is to estimate the reliability probabilistic model of freight bogie dampers, the analysis being made based on real field data.

2. Estimation of the reliability probabilistic model

To determine a failure mathematical model, it is necessary to select a theoretical probability distribution and to estimate its parameters so that the obtained reliability model is adequate to the experimental data (i.e. accurately models, in terms of reliability, the real behaviour of the tested system). In this paper are analysed the cases of exponential and Weibull distributions, for the latter being used two distinct estimation methods.

2.1. Estimation of exponential reliability model

The exponential distribution is, chronologically speaking, the first distribution used extensively in reliability theory. Its fundamental feature is the constant failure rate (i.e. independent of time). Due to this peculiarity, the exponential distribution can be used especially when the effect of wear is not important, as is the case of the basic (useful) period of life of systems.

The exponential reliability equation is:

$$R(t) = \exp(-\lambda t), \tag{1}$$

where t is the time and λ is the failure rate. The exponential model is quite simple and its form leads to very simple mathematical expressions of the reliability indicators. On the other hand, this simplicity – given by the presumed constant failure rate - makes the exponential model less versatile, thus leads sometimes to a less accurate approximation of the empirical reliability.

It can be seen in equation (1) that the only parameter to determine is the failure rate λ, its estimator being given by:

$$\hat{\lambda} = \frac{r}{TTT(n,r)}, \tag{2}$$

where r is the number of failures registered during the observation period and $TTT(n,r)$ is the total time on test statistic, defined as the sum of the operating times of all units, regardless of whether they are still in service (units without failure) or not (units that have failed). Considering a sample of n units and given their respective failure times arranged in ascending order, the empirical TTT up to the rth order statistic is:

$$TTT(n,r) = \sum_{i=1}^{r} t_i + (n-r)t_r, \tag{3}$$

Equation (3) gives thus the TTT demonstrated by all n units in the case of incomplete data, assuming that the test ends after the rth failure.

2.2. Estimation of Weibull reliability model using the Regression Method

The Weibull distribution is widely used in the reliability theory because of its versatility. Usually it is used the two-parameter version of Weibull distribution, according to which the evolution in time of the reliability $R(t)$ is given by:

$$R(t) = \exp\left[-\left(\frac{t}{\eta}\right)^{\beta}\right], \tag{4}$$

where t is the time, β and η are the shape and the scale parameters, respectively. The versatility mentioned above is given by the structure of the probabilistic law (i.e. by the two parameters), which makes it suitable to model the failure law of practically any industrial product.

To estimate the Weibull reliability model involves determining the values of the shape and the scale parameters so that the model is appropriate to the experimental data. The most used estimation method is the Regression (Least Squares) Method. It is based on the linearized form of the Weibull law and on the Least Squares principle in order to estimate the parameters of the Weibull distribution. The linear form is obtained by applying twice the natural logarithm in both members of equation (4), resulting:

$$\ln\ln\frac{1}{R(t)} = \beta\ln t - \beta\ln\eta, \tag{5}$$

thus the Weibull probability law can be written under a linear form:

$$Y = a_0 + a_1 X \tag{6}$$

where

$$X = \ln t\,;\; Y = \ln\ln\frac{1}{R(t)}. \tag{7}$$

To estimate the Weibull distribution parameters can then be used the linear regression and the Least Squares Method, which implies to minimize the sum of the squares of the deviations:

$$S = \sum_{i=1}^{n}(y_i - a_0 - a_1 x_i)^2 \tag{8}$$

where

$$x_i = \ln t_i\,;\; y_i = \ln\ln\frac{1}{1-F(i)}, \tag{9}$$

n is the sample size, t_i the failure time of damper i and $F(i)$ an estimation of the failures cumulative distribution function.

The conditions for minimizing the expression in equation (8) are:

$$\frac{\partial S}{\partial a_0} = \frac{\partial S}{\partial a_1} = 0. \tag{10}$$

Solving the corresponding equations system, the expressions for coefficients a_1 and a_0 are:

$$a_1 = \left(\sum_{i=1}^{n} x_i y_i - \bar{y}\sum_{i=1}^{n} x_i\right)\left(\sum_{i=1}^{n} x_i^2 - \bar{x}\sum_{i=1}^{n} x_i\right)^{-1};\; a_0 = \left(\bar{y}\sum_{i=1}^{n} x_i^2 - \bar{x}\sum_{i=1}^{n} x_i y_i\right)\left(\sum_{i=1}^{n} x_i^2 - \bar{x}\sum_{i=1}^{n} x_i\right)^{-1} \tag{11}$$

where

IOP Conf. Series: Materials Science and Engineering 1037 (2021) 012056 doi:10.1088/1757-899X/1037/1/012056

$$\bar{x} = \frac{1}{n}\sum_{i=1}^{n} x_i \,;\; \bar{y} = \frac{1}{n}\sum_{i=1}^{n} y_i \,. \tag{12}$$

Consequently – see equations (5) and (6) - the estimations of parameters β and η are:

$$\hat{\beta} = a_1 \,;\; \hat{\eta} = \exp\left(-\frac{a_0}{a_1}\right). \tag{13}$$

The relationships above apply to the complete data sets (i.e. all n failure times are known). In the case of a censored data set (i.e. only r of the n dampers have failed) the sums in equations (8) and (11) have the upper limit r instead of n and the means in equations (12) are calculated only for the corresponding r terms.

The failures cumulative distribution function in equation (9) can be expressed using either the mean ranks or the median ranks. In most cases is preferred the Median Ranks Regression, situation in which the cumulative distribution function is estimated using Benard's approximation [12]:

$$F(i) = \frac{i - 0.3}{n + 0.4} \,. \tag{14}$$

2.3. Estimation of Weibull reliability model using the Maximum Likelihood Method.
The principle of the method consists in determining the most likely estimators (i.e. that maximize the likelihood of occurrence of the empirical data set) for the unknown parameters of a distribution. Considering a complete set of experimental data $x_1, x_2, \ldots x_n$, the likelihood function is defined as [13]:

$$L(x_1, x_2, \ldots, x_n; \theta) = f(x_1;\theta)\cdot f(x_2;\theta)\cdot \ldots \cdot f(x_n;\theta) = \prod_{i=1}^{n} f(x_i;\theta), \tag{15}$$

where θ is the parameter to be estimated (there may be more than one) and f is the probability density function of the assumed distribution.

The estimator that is most likely to have generated the observed sample is the one that maximizes the likelihood function, so it can be found by solving the equation [13]:

$$\frac{dL(x_1, x_2, \ldots, x_n; \theta)}{d\theta} = 0 \,. \tag{16}$$

In the situation of a censored data set, when for the n observed units only the first r failure times are known ($r<n$), the likelihood function is defined as [13]:

$$L(x_1, x_2, \ldots, x_n; \theta) = \prod_{i=1}^{r} f(x_i;\theta) \prod_{j=1}^{n-r} R(x_j;\theta), \tag{17}$$

where R is the reliability function.

Since the Weibull probability density function is given by:

$$f(t) = \frac{\beta}{\eta}\left(\frac{t}{\eta}\right)^{\beta-1} \exp\left[-\left(\frac{t}{\eta}\right)^{\beta}\right], \tag{18}$$

the corresponding likelihood function for a censored data set of r failure times given by equation (17) is:

$$L(t_1,t_2,...,t_r;\beta,\eta)=\prod_{i=1}^{r}\frac{\beta}{\eta}\left(\frac{t_i}{\eta}\right)^{\beta-1}\exp\left[-\left(\frac{t_i}{\eta}\right)^{\beta}\right]\prod_{j=1}^{n-r}\exp\left[-\left(\frac{t_j}{\eta}\right)^{\beta}\right]. \tag{19}$$

Applying the natural logarithm:

$$\ln L(t_1,t_2,...,t_r;\beta,\eta)=r\ln\beta-r\beta\ln\eta+(\beta-1)\sum_{i=1}^{r}\ln t_i-\sum_{i=1}^{r}\left(\frac{t_i}{\eta}\right)^{\beta}-(n-r)\left(\frac{t_r}{\eta}\right)^{\beta}, \tag{20}$$

where t_r – the last observed failure time – is the censoring time, thus the time assigned to the *(n-r)* units which have survived the time t_r.

The maximum conditions are

$$\frac{\partial\ln L(t_1,t_2,...,t_r;\beta,\eta)}{\partial\beta}=0\quad\frac{\partial\ln L(t_1,t_2,...,t_r;\beta,\eta)}{\partial\eta}=0. \tag{21}$$

By solving the equation system above are obtained the following expressions of the estimators for β and η:

$$\hat{\eta}=\left(\frac{\sum_{i=1}^{r}t_i^{\hat{\beta}}+(n-r)t_r^{\hat{\beta}}}{r}\right)^{1/\hat{\beta}} \tag{22}$$

$$\frac{1}{\hat{\beta}}+\frac{1}{r}\sum_{i=1}^{r}\ln t_i-\frac{\sum_{i=1}^{r}t_i^{\hat{\beta}}\ln t_i+(n-r)t_r^{\hat{\beta}}\ln t_r}{\sum_{i=1}^{r}t_i^{\hat{\beta}}+(n-r)t_r^{\hat{\beta}}}=0. \tag{23}$$

3. Results and discussions

In this section is established the reliability mathematical model of freight bogie dampers, based on the real field data. The reliability model is estimated according to the previously presented probabilistic distributions and methods, so that more variants of this model are be obtained and a comparison is possible.

The analysis is based on empirical data, observed in actual operations of freight bogies in the period of their useful (normal) life. The sample size was n=1802 units, of which there have been observed failures in the case of r = 66 units over the testing period of one year. The studied dampers are of dry friction type, equipping the freight bogies type Y25. They are known as Lenoir dampers, and they have as an important feature the possibility to modify the damping force according to the axle load.

In the case of the exponential model, as the number of failures is known, it is only necessary to calculate the total time on test given by equation (3). Using the set of empirical data, it is obtained a *TTT* of 643214 days, which leads, through relationship (2), to an estimation of the failure rate of $\hat{\lambda}=1.0261\cdot10^{-4}(\text{days}^{-1})$.

For the Weibull distribution, the parameters of the reliability model are estimated using two distinct estimation methods. In the case of the model estimated using the Regression Method (RM) it is enough to simply apply the procedure described above. In the case of the Maximum Likelihood Method (MLM) the two estimates are determined using iterative procedures, by considering an initial value for the form parameter in equation (23) and by adjusting this value until the expression becomes null. As initial value it can be used the previously determined by the Regression Method. Once the form parameter estimation found, it is inserted in equation (22) thus the estimation of the scale parameter can be found.

In table 1 are presented the calculated estimations of the parameters of the Weibull reliability model corresponding to the two methods.

Table 1. Estimated parameters of Weibull distribution.

	$\hat{\beta}$	$\hat{\eta}$ (days)
RM	0.8799	12666
MLM	0.7884	22905

It can be seen that there are significant differences between the results obtained for both parameters by using the two estimation methods. The scale parameter obtained through the Maximum Likelihood Method is significantly higher (almost double) than the one determined using Regression Method. On the other hand, the form parameter is about 10% lower. This confirms the widely accepted opinion according to which the use of the Maximum Likelihood Method leads to more optimistic models of the reliability.

Of the two Weibull parameters, the shape parameter β has a special significance, its value setting the monotony of the failure rate. For $\beta = 1$ the failure rate is constant, while for values higher and lower than 1, the failure rate is increasing and decreasing, respectively. A value close to 1 is an expected (usual) one in the case of a system during its normal (useful) life, as it is the case of the freight bogie dampers. In this regard, it can be seen in Table 1 that the β values are slightly below 1, this indicating a decreasing failure rate, i.e. a negative ageing (NWU – 'new worse than used') of the dampers.

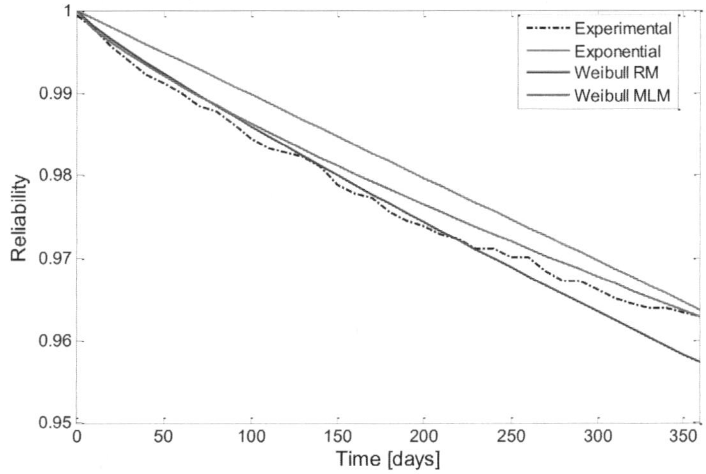

Figure 1. Theoretical and experimental damper reliability.

In figure 1 are shown the damper reliability laws according to the three estimated probabilistic models and the damper empirical reliability (dash-dot line).

It is obvious that the exponential model performs the poorest approximation of the empirical (real) damper reliability. This was somehow to be expected, as long as the exponential distribution, by its very structure, is suitable for modeling failure laws at a constant rate.

Better matching with experimental data is obtained in the case of the Weibull reliability models. Of them, as previously mentioned, the model resulting from the Maximum Likelihood estimation method indicates a higher reliability than the real one.

4. Conclusion

The aim of the present paper was to estimate the reliability probabilistic model of the dry friction dampers of the freight bogie Y25, the study being motivated by the fact that bogie suspension failures are one of the main categories of rolling stock failures that have produced derailments in recent years in railway freight transport.

The analysis was based on real field empirical data, observed for one year in actual operations of freight wagons, in the period of their useful life.

The estimation of the reliability model was made using the exponential distribution and the Weibull distribution. Based on the latter, two models were obtained, following the estimation by two methods, the Regression Method and the Maximum Likelihood Method.

By comparing the obtained reliability models with the empirical reliability, it has been observed that the exponential model has the weakest match with the experimental data.

Regarding the Weibull-based models, although the two estimation methods produced significant differences in the two parameters of the Weibull distribution, the resulting models proved a good accuracy. Of the two models, it turned out that the Maximum Likelihood Method led to a more optimistic model of the damper reliability.

The shape parameter estimations of the two Weibull reliability models were slightly below 1, this indicating a slightly decreasing failure rate, i.e. a negative ageing (NWU – 'new worse than used') of the dampers. The fact that the values of parameter β are close to 1 (situation which corresponds to a constant failure rate) is in line with the general theory regarding the behavior of systems during their normal/useful life - as it is the case of the analysed system.

References

[1] Spiroiu MA and Nicolescu M 2018 *MATEC Web of Conferences* **178** 06005
[2] Melnik R, Koziak S, Sowiński B and Chudzikiewicz A 2019 *Transp. Res. Proc.* **40** 808–814
[3] Yin H, Wang K, Qin Y et al 2017 *J. Wireless Com. Network* **2017** 212
[4] Yun T, Qin Y et al. 2016 *Prognostics and System Health Management Conf. (Chengdu)* (New York: IEEE) p 1
[5] Spiroiu MA 2015 *Appl. Mechan. Mater.* **809-810** 1097-1102
[6] Spiroiu MA 2018 *Mater. Plast.* **55(1)** 24-27
[7] Nicolescu M and Spiroiu MA 2019 *U.P.B. Sci. Bull. Series D* **81(3)** 61-68
[8] Melnik R and Koziak S 2017 J. Vibroeng. **19(1)** 487-501
[9] Wei XK, Jia LM and Liu H 2013 *Vehicle Syst. Dyn.* **51(5)** 700-720
[10] Dumitriu M 2019 *Mech. Ind.* **20(1)** 102
[11] Kumbhalkara MA, Bhopeb DV and Vanalkara AV 2015 *Proc. Mat. Sci.* **10** 331-343
[12] McCool JI 2012 *Using the Weibull Distribution: Reliability, Modeling and Inference* (Hoboken: John Wiley & Sons)
[13] Rausand M and Hsyland A 2004 *System Reliability Theory Models, Statistical Methods and Applications* (Hoboken: John Wiley & Sons)

Load capacity evaluation for single lap joints applied in the railway vehicles

T Zadorozny[1*], M Szczepanik[2]

[1]Alstom Konstal S.A., Engineering Department, Metalowców 9, 41-500 Chorzow, Poland
[2]Silesian University of Technology in Gliwice, Faculty of Mechanical Engineering, Institute of Computational Mechanics and Engineering, Konarskiego 18a, 44-100 Gliwice, Poland

[*]Corresponding author's e-mail: zadorozny.tomasz@gmail.com

Abstract. The article will present a comparison performed for two different approaches based on the application of Finite Element Method used to overlap joints "HUCK BOM" modeling. For the given geometric form of the joint two different numerical models will be presented and their load capacity will be determined for each of the modeling methodologies. The first methodology where a simplified joint model using beam elements and rigid elements of the RBE2 type will be compared with the second where the initial tension and the influence of friction on the joint load capacity are taking into account. Finally, a comparison of the results obtained with reference to the time of connection modeling and computation will be given.

1. Introduction

The method of joining elements using "HUCK BOM" fasteners is very efficient and ensures high repeatability due to the use of dedicated tools. For these reasons, those fasteners are widely used in structures exposed to high static and dynamic loads (bridges, towers, trains, etc.). Such structures (and especially serial-produced ones) require calculations that allow obtaining a compromise between the load capacity of the lap joint and the number of fasteners used. Modeling structures composed of thousands of fasteners requires modeling each fastener, which is an extremely tedious process, which is why the right choice of modeling is a key issue. There are various ways of modeling this type of connection in the FEA environment. The most commons are the use of 1D elements for mapping fasteners and 3D modeling, which is much more time-consuming. This study aims to compare the above-mentioned methods in terms of the obtained results and the time needed to solve the task. This paper is an introduction to further research aimed at carrying out real tests to determine the consistency with numerical calculations, taking into account the influence of various factors on the load capacity of the joint (e.g. the accuracy of making holes). HyperWorks (one of the commercial programs using the Finite Element Method (FEA)) was used to solve the problem defined above.

2. Description of samples

The tested joint (fig. 2) consists of 12mm thick steel sheets and 8mm thick aluminum sheet, "HUCK BOM" fasteners with the R20 designation were used to connect the plates. Table 1 presents the basic dimensions of the fastener (fig.1). Holes are made in steel sheets and then a tensile force is applied evenly on the cylindrical surface [6,7].

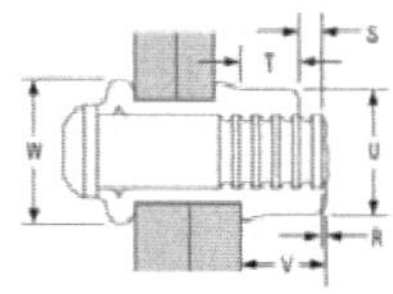

Figure 1. Clamped HUCK BOM fastener [6]

Table 1. Dimensions of the clamped HUCK BOM fastener [6]

R [mm]	S [mm]	T [mm]	U [mm]	V [mm]	W [mm]
2	8	11.1	26.1	18.3	32.2

Table 2 presents material data:

Table 2. Material properties

	Young's modulus [MPa]	Density[g/cm^3]	Poisson's ratio	Yield [MPa]
S355	$2.1 \cdot 10^5$	7.85	0.3	355
RAP6082	$7 \cdot 10^4$	2.6	0.3	260

The load capacity of the joint in a given load case is determined by the surface pressure between the aluminum plate and the fastener on the cylindrical surface of the hole. The analytical capacity of the lap joint is calculated below, implying the lap is perfectly stiff and the influence of friction between the connected elements is nonexistent. Due to surface pressures, the calculations are carried out assuming that the force pressing fastener against the wall of the hole acts on the surface of a projection fastener shaft on a surface perpendicular to the direction of the force [2,3].

Section area:

$$A = D \cdot t_a = 17.5 \cdot 8 = 140mm^2 \tag{1}$$

Permissible stress (safety factor is n=1.15):

$$\sigma_a = \frac{R_e}{n} = \frac{260}{1.15} = 226MPa \tag{2}$$

The maximum value of force extending the lap joint:

$$F \leq \sigma_a \cdot A \cdot 4 = 226 \cdot 140 \cdot 4 = 126.56kN \tag{3}$$

where:

σ_a –Permissible stress for aluminum sheet
D – Diameter of the hole for the fastener HUCK BOM
t_a – Thickness of aluminum sheet
n – Safety factor

3. Determining load capacity of the lap joint using 1D elements

In the first attempt, the connection was modeled using CBAR beam elements that were connected to the steel and aluminum sheets using rigid RBE2 elements. Figure 2 shows the layout of the lap joint and basic dimensions, the steel plate is marked yellow, and the aluminum sheet purple. The model does not take into account the friction contact between the connected elements and the pre-tension, all forces are transferred directly through the 1D elements. The numerical model consists of 66138 elements and 87140 nodes.

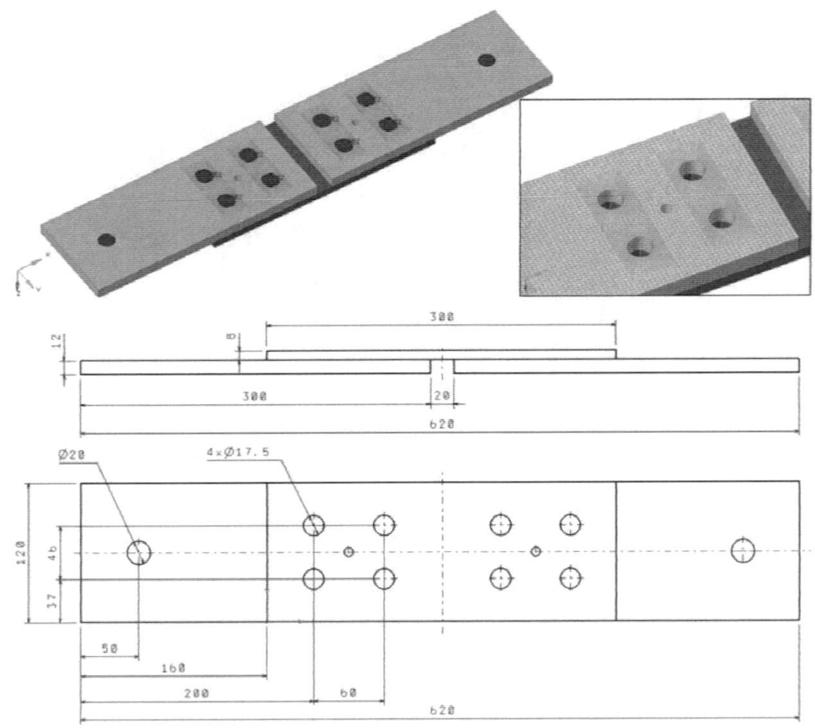

Figure 2. Lap joint with fasteners designed using 1D elements

Figure 3 presents the boundary conditions and the numbering of the connectors. The model was fixed in one of the technological holes and the force determined by the analytical method equal to 126.56kN was applied to the hole on the opposite side, simulating the tensile effect of the sample.

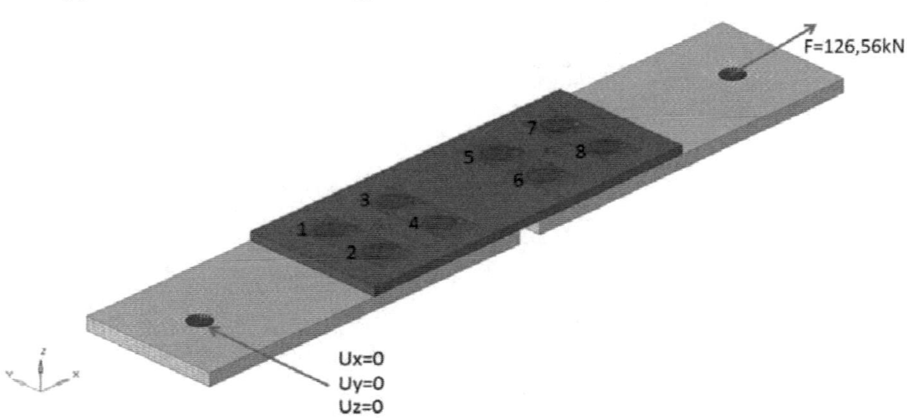

Figure 3. Boundary conditions

IOP Conf. Series: Materials Science and Engineering 1037 (2021) 012057 doi:10.1088/1757-899X/1037/1/012057

Table 3 presents the values of forces in 1D elements representing the fasteners, and the stress values determined on their basis. The stress results obtained from the numerical analysis are similar to the results obtained by the analytical method, differences of less than 2% are caused by the deformation of the joined elements. Figure 4 shows how the force values in the 1D fasteners were read.

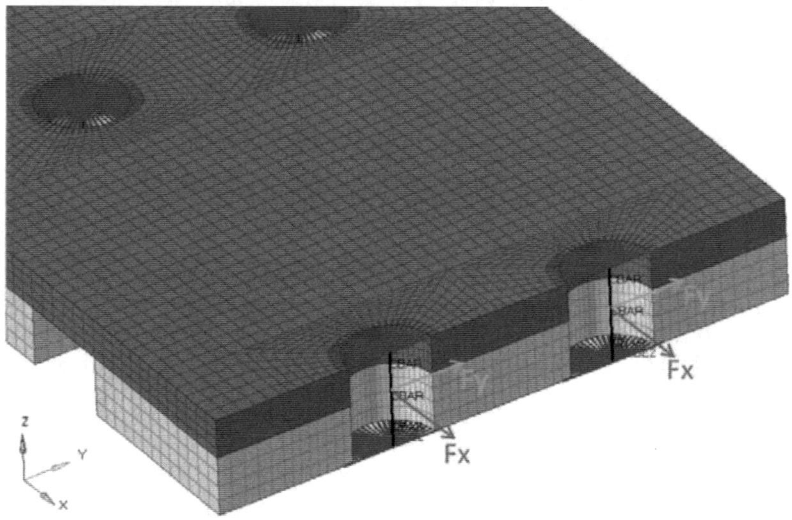

Figure 4. Forces in 1D elements

Table 3. Forces and stress values for HUCK BOM fasteners at a load of 126.56kN

L.p.	Fy [N]	Fx [N]	Fw [N]	σ [Mpa]
1	-948.17	-31117	31 131.4	222.4
2	948.16	-31117	31 131.4	222.4
3	966.55	-32183	32 197.5	230.0
4	-966.54	-32183	32 197.5	230.0
5	966.54	32183	32 197.5	230.0
6	-966.53	32183	32 197.5	230.0
7	-948.17	31117	31 131.4	222.4
8	948.16	31117	31 131.4	222.4

where:

F_y – Force acting in the direction of the Y axis
F_x – Force acting in the direction of the X axis
F_w – Resultant of forces
σ– Maximum stresses determined on the basis of forces in 1D elements

Figure 5 shows the displacement distribution on the deformed model in the 2:1 scale, the maximum displacement value is 8.76 mm.

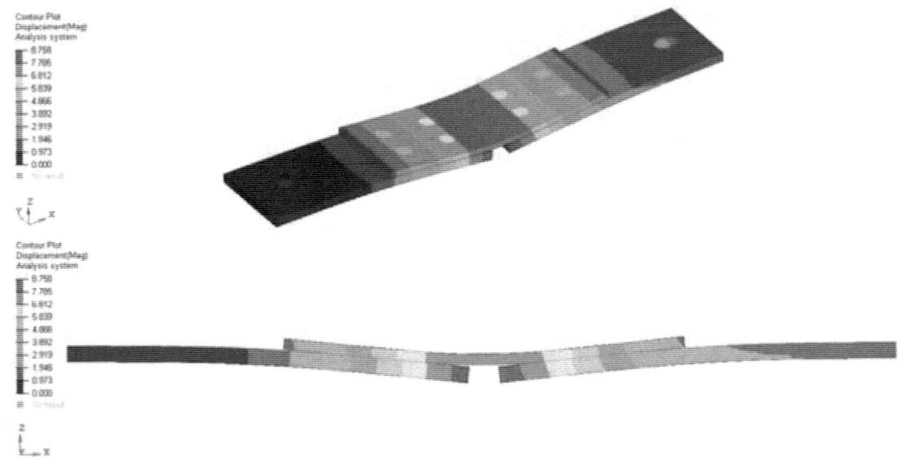

Figure 5. Displacement distribution in the 2:1 scale for a tensile force of 126.56kN [mm]

Knowing the results presented above, the lap joint capacity corresponding to the maximum value of force F_z=124.03kN (fig. 6) was determined so that the maximum stresses do not exceed the permissible stress.

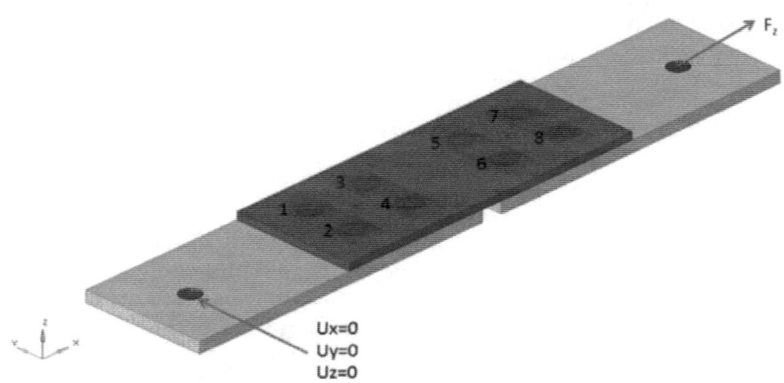

Figure 6. Point of application of Fz force, determining the maximum load capacity of the joint

In order to check numerically, table 4 presents the values of forces in elements 1D for the sample loaded with force F_z=124.03kN. The determined stresses do not exceed the permissible value.

Table 4. Forces and stress values for HUCK BOM fasteners at a load of F_z=124.03kN

L.p.	Fy [N]	Fx [N]	Fw [N]	σ [Mpa]
1	-929.2	-30495	30 509.2	217.9
2	929.19	-30495	30 509.2	217.9
3	947.22	-31539	31 553.2	226.0
4	-947.21	-31539	31 553.2	226.0
5	947.21	31539	31 553.2	226.0
6	-947.2	31539	31 553.2	226.0
7	-929.21	30495	30 509.2	217.9
8	929.19	30495	30 509.2	217.9

4. Determining load capacity of the lap joint using 3D elements modeled for fasteners

In the second test, the load capacity of lap joint with fasteners modeled using 3D elements was examined. Friction contact was modeled between the fasteners and plates on the head surfaces and cylindrical surfaces. The model also includes pretension N_u=24kN. The coefficient of friction used is μ=0.2. The friction contact in this case also prevents the penetration of joined elements. The numerical model consists of 75954 elements and 99804 nodes (fig.7).

The theoretical friction force in the lap joint was analytically determined below:

$$T = 4 \cdot N_u \cdot \mu = 4 \cdot 24 \cdot 0.2 = 19.2kN \tag{4}$$

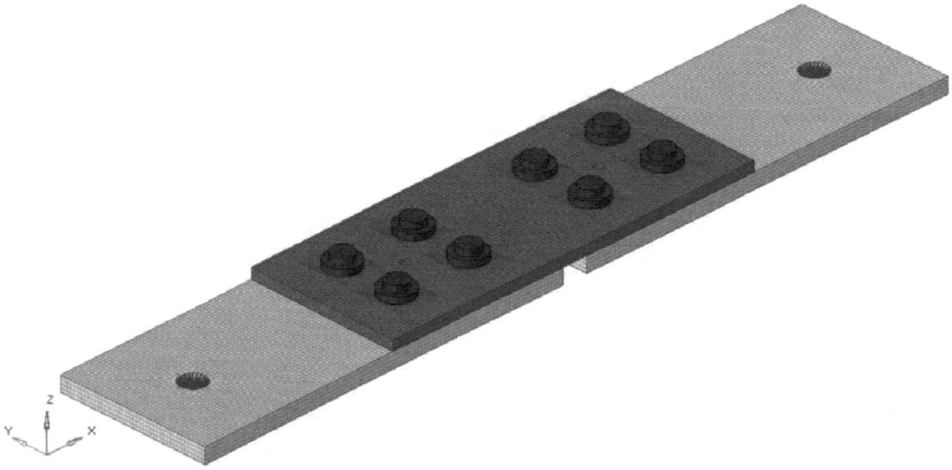

Figure 7. Lap joint with fasteners designed using 3D elements

The model was loaded in the same way as in the first attempt (fig.3). Figure 8 presents the distribution of H-M-H reduced stresses on an aluminum plate in the area of the location of the fasteners. Figure 9 presents a detailed distribution of stresses on the cylindrical part of the hole in contact with the HUCK BOM fastener. The maximum stresses in the contact area are 235.8MPa, however, these are local stress accumulation in the nodes, that are in direct contact with the connectors. The difference of 4% in relation to the analytical calculation results comes from the deformation of the joined elements and the contact used (different contact areas in contact) on the cylindrical surfaces.

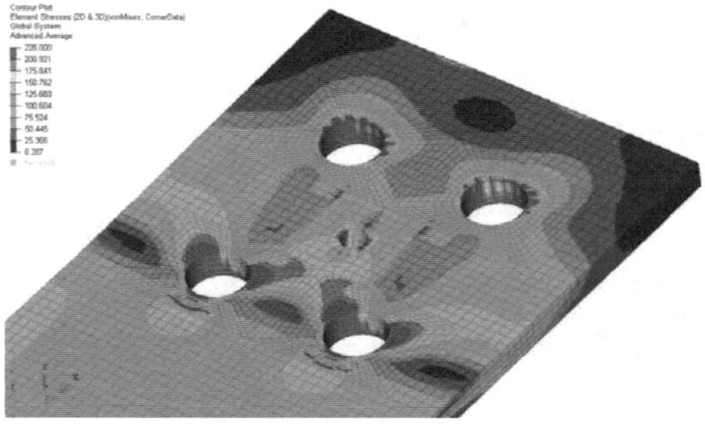

Figure 8. Distribution of H-M-H reduced stress in aluminum plate for tensile force 126.56kN [MPa]

IOP Conf. Series: Materials Science and Engineering 1037 (2021) 012057 doi:10.1088/1757-899X/1037/1/012057

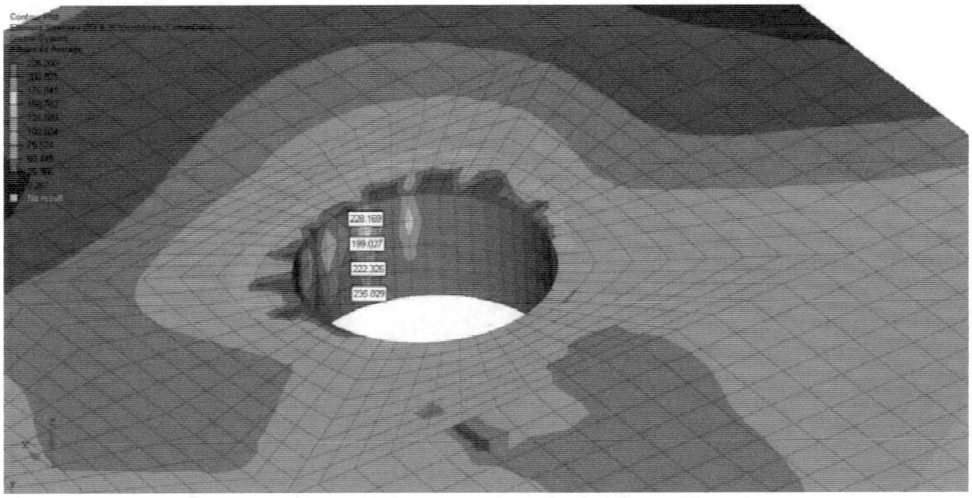

Figure 9. Distribution of H-M-H reduced stress in cylindrical part of the hole for tensile force 126.56kN [MPa]

The real contact surfaces are shown below in figure 10, the red color marks where the cylindrical surface of the fastener closely adheres to the cylindrical surface of the hole, thus transferring the predetermined load.

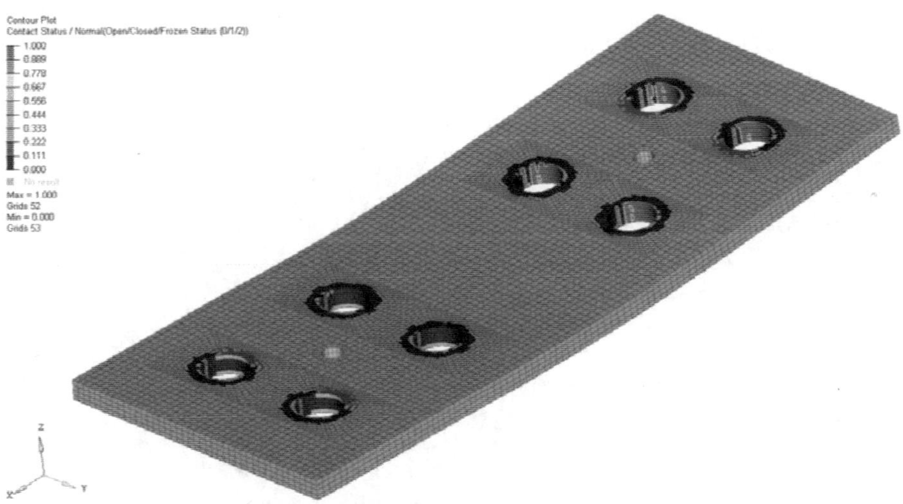

Figure 10. Contact status

Figure 11 shows the displacement distribution on the deformed model in the 2:1 scale, the maximum displacement value is 8.48mm.

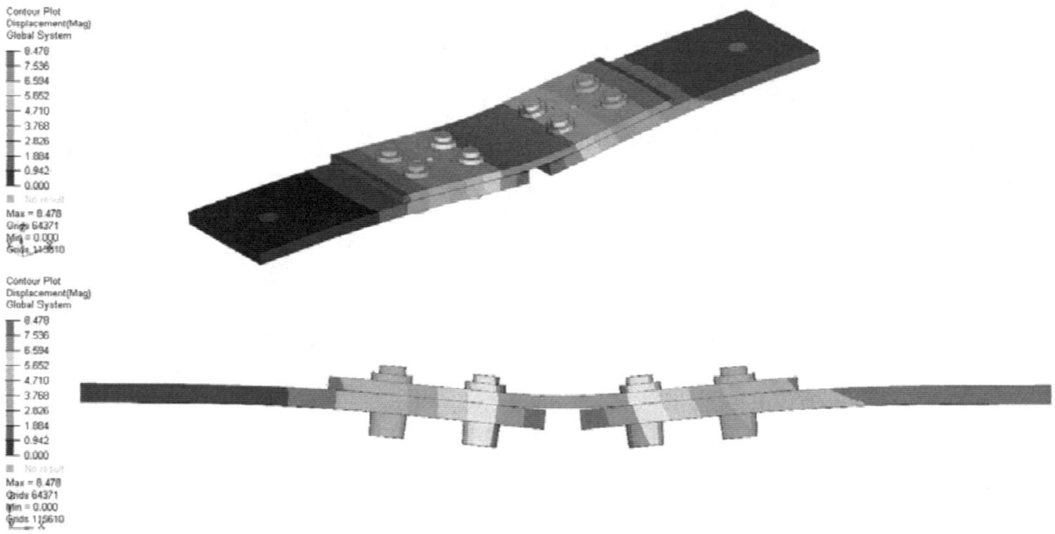

Figure 11. Displacement distribution in the 2:1 scale for a tensile force of 126.56kN [mm]

Like before, knowing the results presented above, the lap joint capacity corresponding to the maximum value of force F_z=120.63kN was determined so that the maximum stresses do not exceed the permissible stress.

The detailed distribution of stresses on the cylindrical part of the hole for the tensile force F_z=120.63kN is presented below. The maximum stresses (225.97MPa) are within the acceptable range.

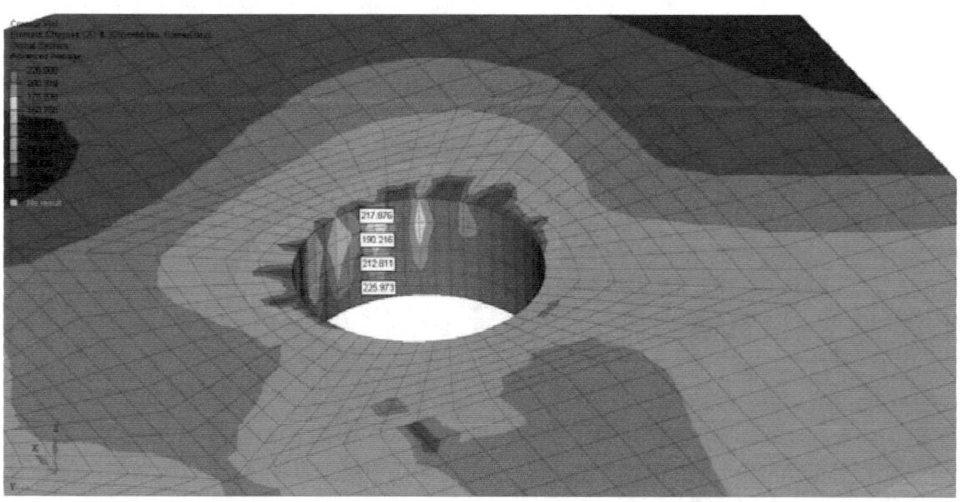

Figure 12. Distribution of H-M-H reduced stress in cylindrical part of hole for tensile force 120.63kN [MPa]

5. Summary

The time needed to solve the problem using 1D beam elements is about two hours. In order to solve the same task by modeling fasteners using 3D elements, one should spend an additional hour modeling connectors that have a relatively complex geometry. In addition, one should create groups of contacts between the combined elements and fasteners, additionally extending the time of solving the task. Table 5 provides a comparison of approximate modeling times and numerical solutions.

Table 5. Comparison of approximate times for modeling and analysis solutions

	1D	3D
Modeling of connected plates	90min	90min
Modeling of fasteners	30min	60min
Modeling of contacts	-	30min
Numerical solution of the task	2min	70min
Summary	122min	250min

The results indicate that the lowest load capacity of 120.63kN was obtained using 3D elements modeling for the fastener, although the numerical model taking into account the phenomenon of friction between the joined elements. This method can be considered the most accurate because it precisely maps the joint system and its stiffness, but it is crucial to remember that the maximum stress value occurs locally, in the nodes where the contact between the aluminum plate and the HUCK BOM fastener occurs. Close attention should be paid to the fact that with the decrease of the friction force, the contact stresses will increase resulting in a reduced load capacity of the single lap joint. Modeling the fasteners using 1D elements, the joint capacity was 124.03kN. Table 6 summarizes the comparison of load capacity for each of the methods described above.

Table 6. Comparison of single lap joint load capacity, depending on the calculation method

	Load capacity of the single lap joint
Analytical method	126.56kN
Fastener 1D	124.03kN
Fastener 3D	120.63kN

Differences in the load capacity of the joint between carried out numerical analyses are insignificant, the time needed to solve the problem by modeling the fastener using 3D elements is more than twice as long. On the basis of the obtained results, it can be concluded that such slight differences indicate that for fasteners with relatively low stiffness, a simplified method of modeling connectors with 1D elements can be successfully applied to take into account the appropriate safety factor. This approach allows shortening the computation time several times, which translates into cost reductions while maintaining the safety margin.

6. References

[1] Zienkiewicz O C 1977 *The Finite Element Method* (McGraw Hill)
[2] Mazanek E 2009 *Basics of calculations from the fundamental of machine design* (WNT Warsaw)
[3] Kurmaz L W, Kurmaz O L 2011 *Basics of design the nodes and machine parts* (Kielce)
[4] Dyląg Z, Jakubowicz A, Orłoś Z 2007 *Strength of material* (WNT Warsaw)
[5] Cebulski L, Boneill C, Drazetic P, Markiewicz E 2003 *Experimental and numerical behavior of railway rivets under quasi-static promptings* (Engineering Computations 20(7):795-809)
[6] Official HUCK website: *https://www.hfsindustrial.com/brands/huck.html* (10.2020)
[7] Altair HyperWorks helpdesk

IOP Conf. Series: Materials Science and Engineering 1037 (2021) 012058 doi:10.1088/1757-899X/1037/1/012058

Aerodynamic shape optimization of racing car front wing

M Kalinowski[1,2], M Szczepanik[2]

[1] Alstom Konstal S.A., Metalowców Street No.9, 41-500 Chorzów, Poland
[2] Silesian University of Technology in Gliwice, Faculty of Mechanical Engineering, Department of Computational Mechanics and Engineering, Konarskiego Street No.18a, 44-100 Gliwice, Poland

*Author's e-mails: marcin.kalinowski@alstomgroup.com; miroslaw.szczepanik@polsl.pl

Abstract. Nowadays, the automotive industry is one of the largest areas of CFD (Computational Fluid Dynamics) simulation applications. The big challenge that engineers dealing with CFD simulations in this industry have to face is shape optimization, which is often very time-consuming and requires a lot of iterations. To reduce the time needed to achieve an optimal shape and make the design process more efficient, more and more CFD analysts are turning to Adjoint Solver (AS) formulas, a new methodology used for optimization in the automotive as well as railway industry. The AS method calculates gradients (directions, quantities) directly by solving conjugate equations, which makes them independent of design variables. The article concerns optimization the shape of the front wing of the racing car in order to obtain the highest possible downforce-to-drag ratio. At the beginning, 2D shape optimization of the front wing was performed. The next stage of work was the preparation of a CAD (3D) model that took into account the change in the shape of the wing, obtained during the optimization process, and then performing a flow analysis for it.

1. Introduction

Nowadays, the automotive industry is one of the largest areas of CFD (Computational Fluid Dynamics) simulation applications. The big challenge that engineers dealing with CFD simulations in this industry have to face is shape optimization, which is often very time-consuming and requires a lot of iterations. To reduce the time needed to achieve an optimal shape and make the design process more efficient, more and more CFD analysts are turning to Adjoint Solver formulas, a new methodology used for optimization in the automotive as well as railway industry. The AS method calculates gradients (directions, quantities) directly by solving conjugate equations, which makes them independent of design variables.

ANSYS Fluent software offers shape optimization capabilities that can automatically adjust the geometrical parameters of a specific design to specific optimization goals. The tool responsible for shape optimization in ANSYS Fluent is the Adjoint Solver module. It provides information on how to modify the geometry to achieve the goals of the design (e.g., minimalization of the drag or lift force). This information is communicated by modifying the mesh to see the effect of the recommended changes. The Adjoint Solver can be used to optimize downforce (F1, sports cars), reduce drag (for passenger cars) and also to reduce overall pressure drop (for ducts and pipes, Figure 1). [1,2]

So far, few articles dealing with the problem of aerodynamic optimization using Adjoint Solver have been written. In article [3] authors describe variant optimization of the open-wheel race car, focusing mainly on the change in the drag coefficient due to the change in radiator cooling channel angle. Optimization method presented in this article consist on choosing the best of the "n" variants under consideration and does not concern the determination of the extremum (local or global) of the objective function. On the other hand authors of the article [4] present shape optimization of racing cars using Computational Fluid Dynamics optimization process, which is based on using the mesh morphing techniques to create new designs for analysis by morphing the CFD mesh of the original design. Article [4] is the most similar to the subject under consideration in this article, but focus only on the optimization of the downforce (not downforce-to-drag ratio as presented in this article) for rear wing of racing car (not front wing as presented in this article).

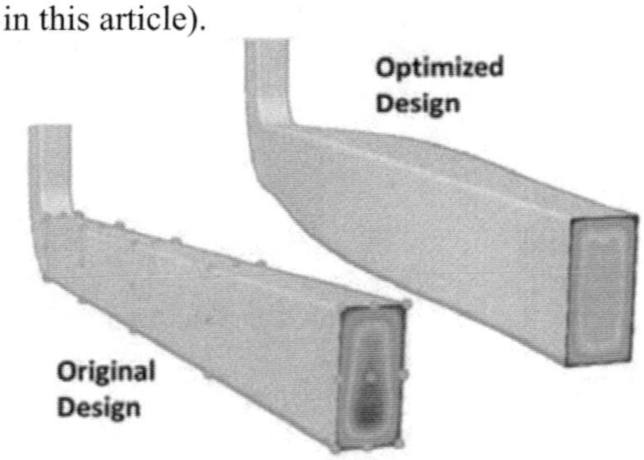

Figure 1. Optimizing the shape of the L-shaped conduit using the Adjoint Solver module [1]

The article is divided into two main parts. In the first part of the work, optimization of the shape of the front wing was carried out. In order to simplify the problem, the analyzed model was a 2D model and contained only a cross-section through main plane and upper flap. The second part of the work was the preparation of a CAD model, which took into account the change in the shape of the wing, obtained during the optimization process, and then performing a flow analysis for it.

Aerodynamic analysis was carried out in the ANSYS Fluent software. The shape optimization, which is the main goal of this article, was made in the "Adjoint Solver" add-on, built into Fluent.

2. The principle of operation of the Adjoint Solver module

During the standard (conventional) aerodynamic calculation process in ANSYS Fluent, the input values are geometric properties, finite element mesh, fluid properties (e.g. air, water). After the aerodynamic analysis, we get maps (e.g. velocity, pressure), graphs, scalar values (forces: resistance, lift) on the output. Based on them, engineer decides to change the shape.

The conventional process of carrying out aerodynamic analyses is presented in the Figure 2 [5,6].

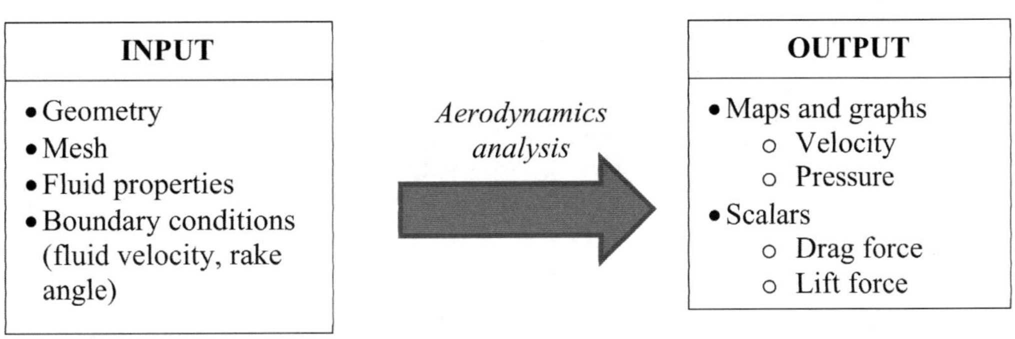

Figure 2. Conventional process for conducting aerodynamic analyses

Thanks to the use of the Adjoint Solver module, we can, based on the results, change the shape and finite elements mesh in the examined flow problem in order to maximize / minimize the observed output value of the aerodynamic analysis (e.g. drag force). In other words, in a conjugate solver, the output values have an effect on the input variables. The process of conducting aerodynamic analyses with the Adjoint Solver is shown in Figure 3 [2].

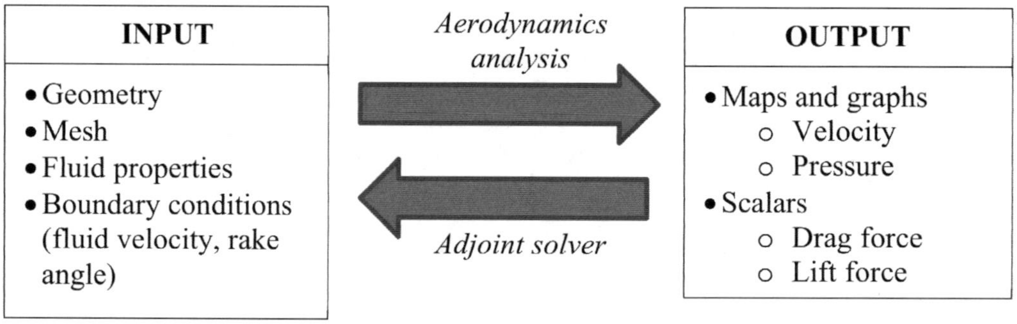

Figure 3. The process of conducting aerodynamic analyses using the Adjoint Solver [2]

A more detailed comparison of the standard process for calculating flow properties and calculations performed with a coupled solver is presented below.
The standard process of calculating flow properties begins with the designer creating a 3D model. The next step is to create a finite element mesh, which may be external (external flows, e.g. a car while moving) or internal (internal flows e.g. cooling fluid flow through a pipe). The production of a high-quality mesh is very important to obtain correct and accurate results. On the other hand, too fine mesh results in a very long computational time.
The next step in the standard flow analysis process is flow analysis from which the output is derived. Only on their basis, the CFD engineer makes decisions about the changes necessary to obtain the optimal flow. These changes are successively presented to the designer and, after obtaining his approval, are introduced into the CAD model. Detailed scheme of the fluid properties calculation process is presented in Figure 4 [5,6].

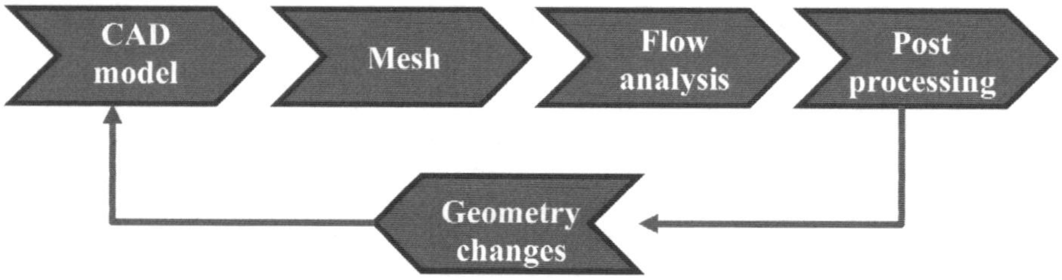

Figure 4. Diagram of the process of calculating the flow properties using a standard method

As mentioned before, this approach is very ineffective as it requires a lot of iterations. Moreover, with the complex geometry of the analysed model and with non-standard flow conditions (e.g. turbulent flow), it is very difficult to indicate the appropriate "direction" of the changes made.
These inconveniences can be solved by using the coupled solver, which is an extension of the conventional CFD solver. The process of performing a flow optimization can be described in the following few steps:

1. **Step I:**
   Standard flow analysis. This step covers the first 3 steps of the standard flow property calculation process (i.e. CAD model building, model discretization, flow analysis).
2. **Step II:**
   Activation of the Adjoint Solver module and determination of the observed values (e.g. drag force, lift force, lift to drag ratio, pressure etc.), which will be arguments of the optimization objective function In this paper, the observed value is the ratio of the downforce to the drag force, which is to achieve the highest possible value.

3. **Step III:**
   Selection of Adjoint Solver parameters and conditions of convergence.
4. **Step IV:**
   Perform a flow analysis consisting of initialization and iteration until convergence.
5. **Step V:**
   Post processing stage. During it, velocity/pressure maps can be generated, convergence graph can be checked and the change in the observed value can be evaluated.
6. **Step VI:**
   Conversion of the mesh using a dedicated tool. This tool defines the dimensions of a quad element (in the case of 2D analyses) or a hexagonal element (in the case of 3D analyses) inside which the mesh morphing is to take place. The actual amount of change within a quad/hexagonal depends on the number of control points and the percentage of modification allowed. Before modifying the shape, the morphing tool reports the predictable effect on the change in the observed value, taking into account the sensitivity data and morphing settings.

Optimization in the Adjoint Solver is gradient-based topology optimization which is combined with a mesh morphing tool. Parameters in optimization task considered in this article are the coordinates of each and every point on the boundary of the region (contour of the main plane and upper flap) thereby allowing efficient shape improvements to be made.
Due to the fact that the optimization process in Adjoint Solver is iterative, after completing the sixth step, solver go back to the fourth step (re-analysis of the flow is carried out in order to evaluate the change made). Figure 5 shows a diagram showing the process of calculating flow properties using the Adjoint Solver module [1,2].

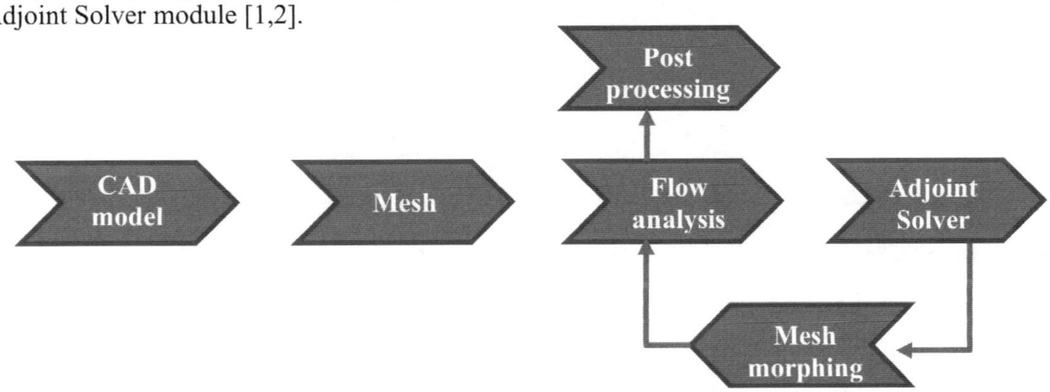

Figure 5. Diagram of the flow properties calculation using the Adjoint Solver module [2]

3. Optimization

3.1. The input model to the optimization process
As mentioned in previous chapters, the input model for the optimization process is based on the front wing CAD model of the racing car.
Due to the fact that the optimization process is usually a multi-iteration process, a very important aspect is the maximum limitation of the duration of one iteration. As mentioned in previous chapters, the finite element mesh (dense mesh - long computation time) has the greatest impact on the calculation time. Therefore, the 3D model has been replaced with a 2D model. As a result, the number of mesh nodes was reduced to 171,000 (previously 750,000), and the number of elements to 60,000 (previously 3,200,000). The input model consists of a cross-section through the main plane and the upper flap. The input model for optimization in the Adjoint Solver module is shown in Figure 6.

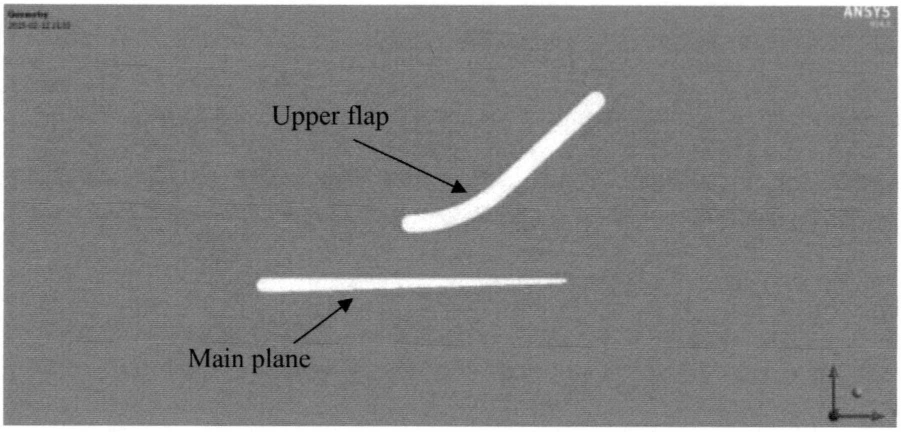

Figure 6. Input CAD model for optimization

3.2. FEM model

The FEM (Finite Element Method) model consists of a rectangle simulating the environment (air) in which the shape of the front wing has been cut out (Figure 7). The FEM model was created in the CAD software through a Boolean operation, consisting in subtracting from the rectangle (environment) the cross-section of the front wing (main plate and upper flap).

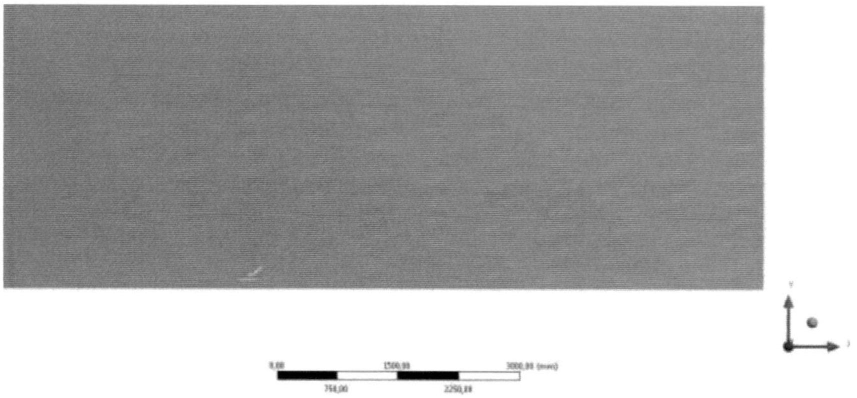

Figure 7. Input CAD model for optimization

The dimensions of the rectangle should be selected depending on the length of the model. Due to the fact that this analysis is two-dimensional and only covers the cross-section of main plate and upper flap, it is not physically possible to determine the total length of the model. Therefore, it was assumed that the front wing model would not be longer than 1000 mm.

For the assumed model length, the rectangle dimensions are as follows:

- length (-X): 3 * L = 3000 mm
- length (+ X): 5 * L = 5000 mm
- height (+ Y): 2 * L = 2000 mm

The above dimensions of the rectangle simulating the environment were selected on the basis of the document [7] and own observations resulting from CFD analyzes. The distance from the ground was set to 70 mm.

The FEM model is a non-deformed (rigid) model. This means that the front wing of the car under analysis does not deform under the influence of the flowing air. The following air properties were assumed:

- density = 1.225 kg/m^3
- viscosity = 1.7894e-5 kg/ms

IOP Conf. Series: Materials Science and Engineering 1037 (2021) 012058 doi:10.1088/1757-899X/1037/1/012058

A computational pressure-based solver was used in the analysis. This solver is well suited for vehicle aerodynamic calculations, as it gives precise solutions in the speed range up to 367 km/h, when the air flow can be treated as incompressible. Temperature is not taken into account. During optimization, only one case was considered, characterized by the air stream speed equal to 100 km/h.

3.3. Mesh

As mentioned earlier, the model was meshed approximately with 171,000 nodes and 60,000 elements (triangle and quadrilateral). The sizes of the elements differ depending on the location in the FEM model. Elements at a considerable distance from the front wing model are large, but while approaching to the front wing model are getting smaller and smaller. The refinement of the mesh near the analysed model allows for more accurate results of the flow analysis (mainly in terms of pressure and velocity values). Further away from the model, the mesh can be coarse as the airflow is almost intact there. The "Sphere of influence" function was used to condensate the mesh (Figure 8).

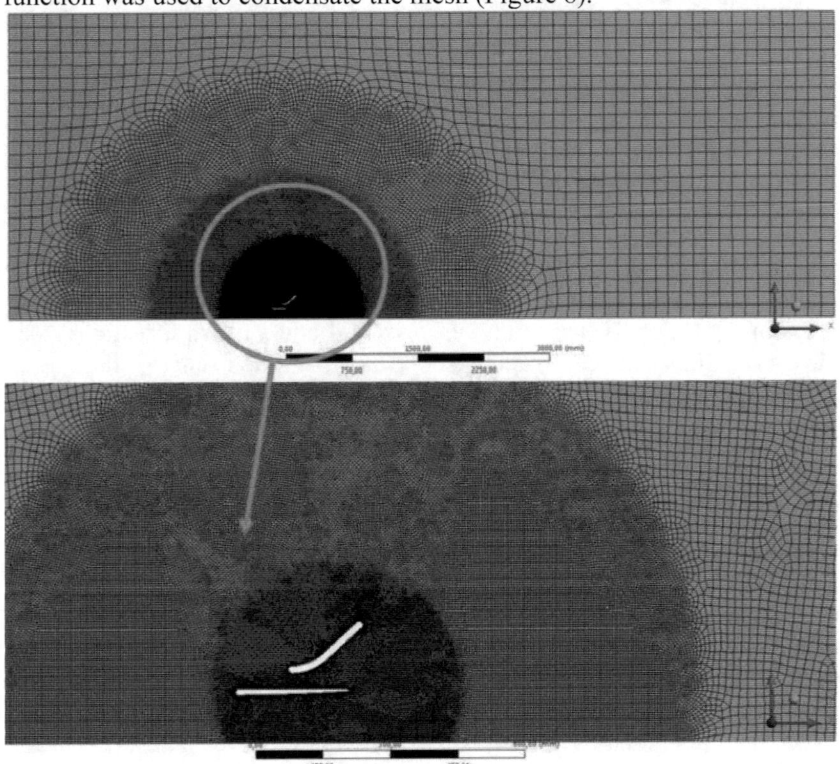

Figure 8. Mesh

3.4. Results

This section presents the results of flow optimization in the Adjoint Solver module. The most important result is a change in the shape of the front wing and thus a change in the downforce-to-drag ratio. The drag and the lift coefficients, respectively, have a decisive influence on the drag force and the downforce. These coefficients directly depend on the shape of the analysed body, so their change will be directly proportional to the change of the corresponding forces. Forces (drag and downforce) also depend on the frontal area, which will also change, but compared to the change in the coefficients (drag and lift), this change is negligible.

Figure 9 shows the changes in the coefficients of the drag force and the downforce as a function of the number of iterations. Drag force coefficient decreases, while the downforce coefficient increases (in absolute value). It follows that the ratio of the downforce to the drag force tends to infinity. It is worth mentioning here that the negative sign at the value of the downforce coefficient only informs about the direction.

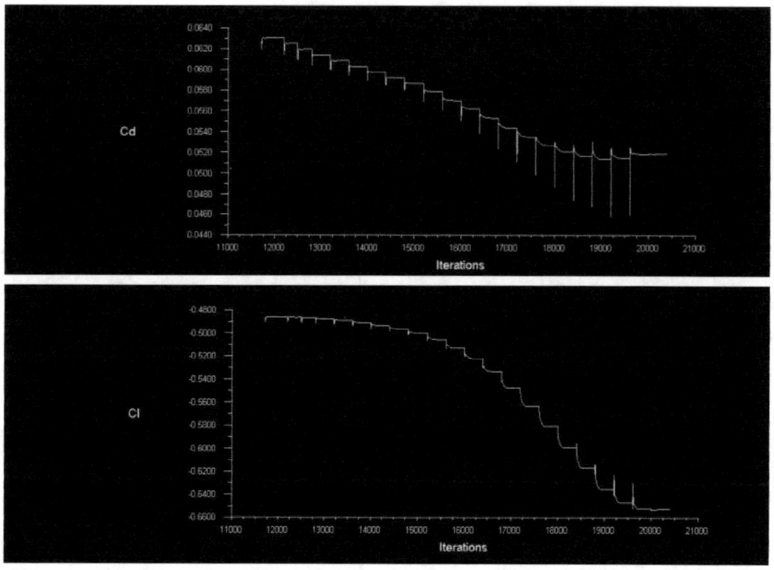

Figure 9. Change of the drag force (top) and downforce (bottom) coefficient as a function of the number of iterations

Optimization was interrupted when the change in the value of both coefficients became negligible. In order to check the correctness of the calculations, the model convergence should be checked by assessing the residual values of continuity, velocity in the horizontal and vertical directions (x-velocity and y-velocity), kinetic energy of turbulence (k) and the coefficient of kinetic energy dissipation of turbulence (ε). In a correctly conducted CFD analysis, the values of these values should decrease until reaching a steady state. For multi-stage and multi-iterative analysis, the convergence plot should take the shape of a comb. With each successive step of the analysis, forced by the change of the mesh, the graphs should in the first step of the new step increase very rapidly to the level that was reached at the beginning of the previous step, and then gradually decline until they reach a steady state. The result of such behaviour of the function course is the "comb" shape of the graph [7].

As a result of the optimization carried out, the shape of the front wing was obtained, as shown in Figure 10. For the resulting wing shape, velocity and pressure maps were generated and compared with the aerodynamic analysis performed for the input geometry (Figure 11).

Pressure and stream velocity maps for the optimized shape are much better aerodynamically than for the input geometry. By analysing only the area of the medium under the main plane for the optimized front wing, it can be seen that the pressure there is much lower (there is even a negative pressure) compared to the same area for the input shape. Moreover (according to Bernoulli's law), under the modified main plane, one can observe a much higher air stream velocity than it was for the input shape.

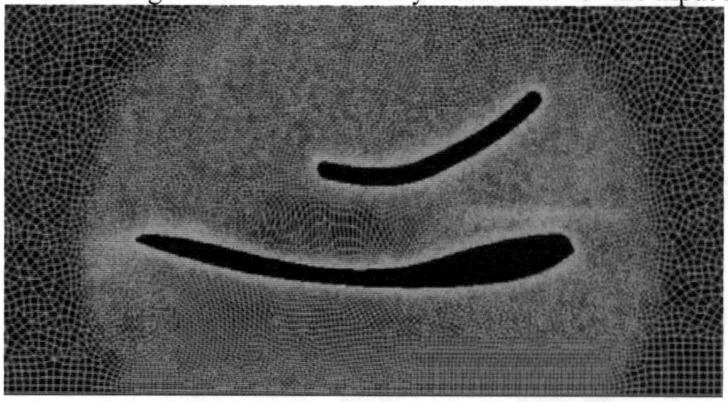

Figure 10. Optimized shape of the front wing

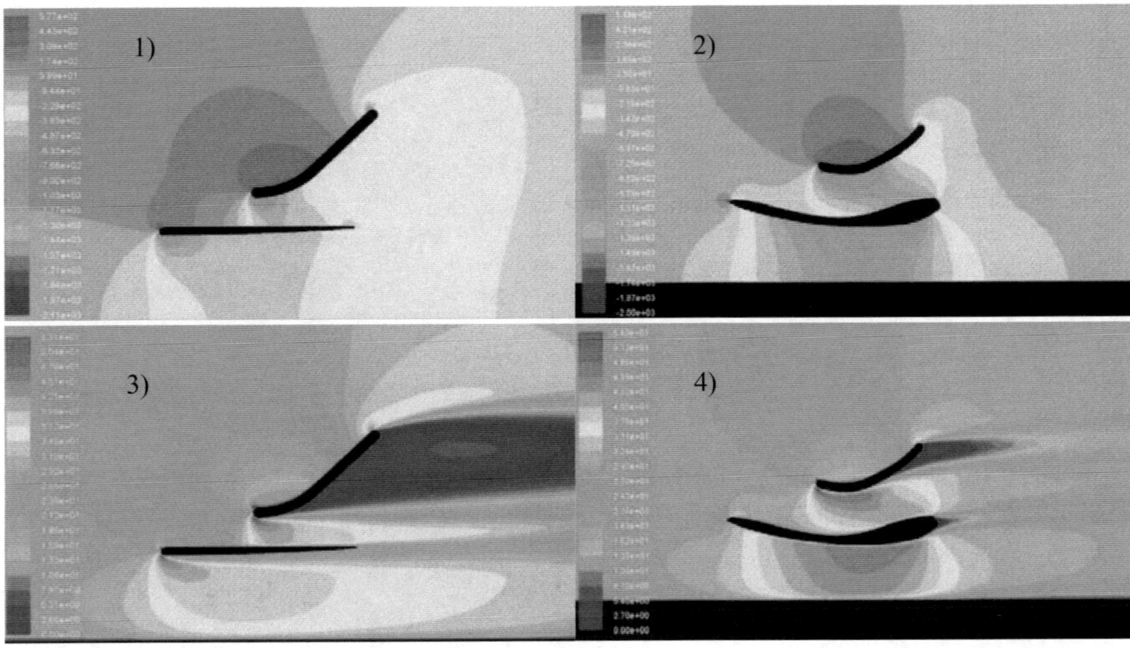

Figure 11. Pressure maps for input (1) and optimized (2) model; velocity maps for input (3) and optimized (4) model

The downforce / drag ratio for the optimized shape of the front wing is 12.65. It is worth remembering that this coefficient is calculated for a two-dimensional model with a large part of the unaccounted for components such as the nose of the car or end plates. In addition, elements omitted from optimization in the three-dimensional analysis (presented in the next chapter) may cause vortices adversely affecting the drag and pressure force (increasing and decreasing respectively). In fact, it should be expected that the value of this coefficient will be lower.

4. 3D flow analysis of the optimized front wing

The 3D model was made in the SolidWorks program based on the shape obtained in the optimization process.

Due to the fact that the front wing model is symmetrical in the YZ plane, only half of the model was analyzed. Thanks to this, the computation time was significantly shortened and the model could be discretized with greater accuracy (finer mesh). The FEM model consists of a cuboid simulating the environment (air) in which the shape of one half of the front wing has been cut.

Model was meshed approximately with 1,100,000 nodes and 5,000,000 second order tetras elements. The sizes of the elements differ depending on the location in the FEM model. Elements at a considerable distance from the front wing model are large, but while approaching to the front wing model are getting smaller and smaller. The refinement of the mesh near the analysed model allows for more accurate results of the flow analysis (mainly in terms of pressure and velocity values).

During 3D flow analysis, only one case was considered, characterized by the air stream speed equal to 100 km/h.

CAD model, FEM model and detailed mesh view were presented in the Figure 12 and results from 3D flow analysis were presented in the Figure 13.

The values of the aerodynamic forces (determined in the 3D analysis) acting on the front wing are presented in the Table 1.

Table 1. Forces acting on the front wing

Drag force [N]	Downforce [N]	Downforce-to-drag ratio	Drag coefficient	Downforce coefficient
34.16	-152.78	4.48	0.388	-1.748

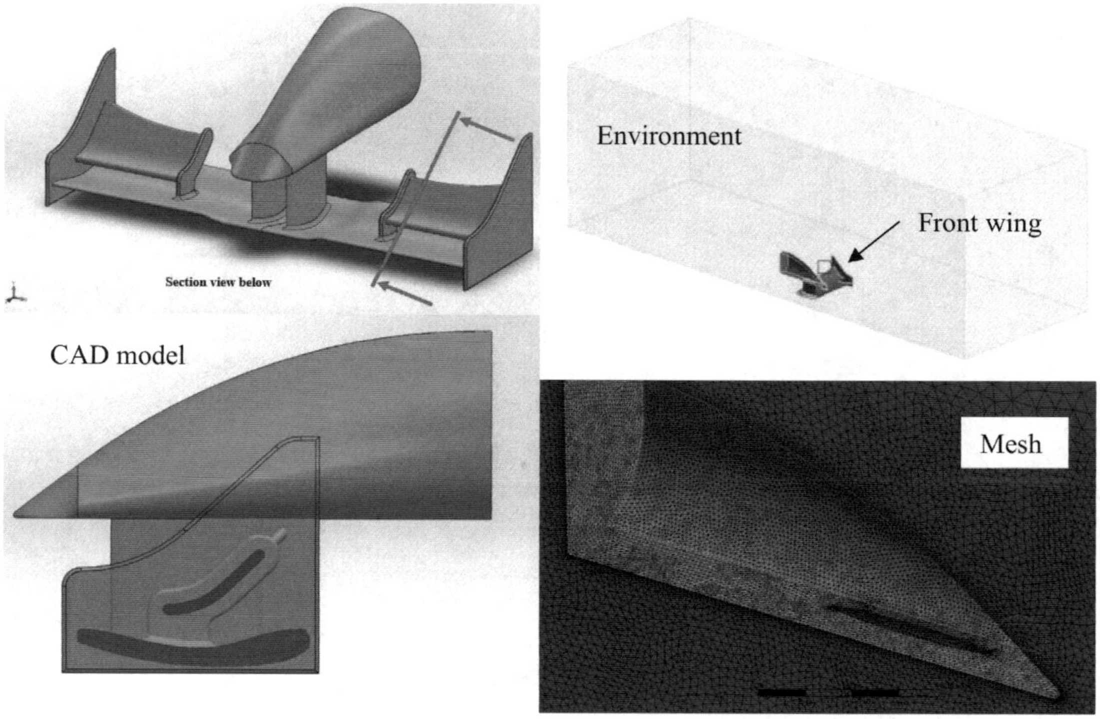

Figure 12. CAD model, FEM model and detailed mesh view

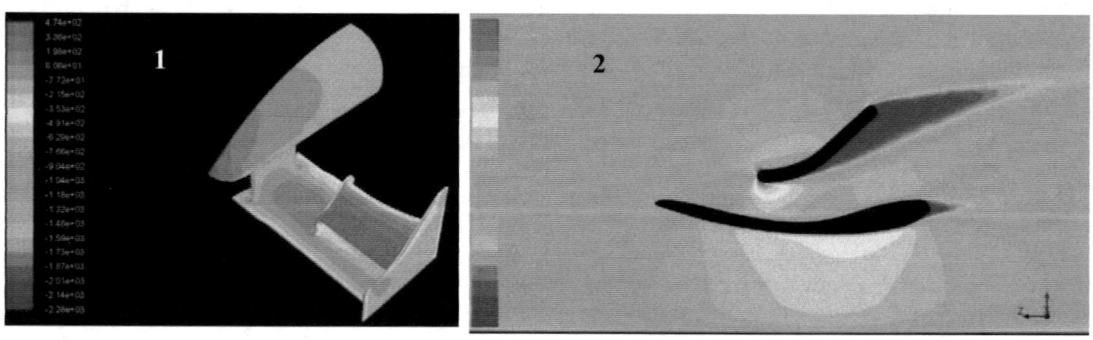

Figure 13. Pressure (1) and velocity (2) maps

5. Summary

Within the framework of this article, the shape of the front wing of the racing car was optimized in terms of maximizing the downforce / drag ratio, thereby achieving the goal of the study. The front wing was designed without relying on existing solutions (due to the lack of precise design data), it fulfils its primary function, i.e. generating the downforce. The downforce-to-drag ratio is equal 4.48, what is twice the size of the non-optimized front wing.

It is true that the resulting value of the downforce-to-drag ratio is more than twice lower than the value determined in the two-dimensional analysis (12.65), but it should be remembered that the two-dimensional analysis did not take into account, among others the prow of the car and endplates. The direction of further research may be a comparative analysis of the gradient optimization method with evolutionary algorithms or with swarm algorithms based on the aerodynamic optimization of the front wing of a racing car. The gradient optimization method has a solid mathematical basis, gives a strict solution and is characterized by high speed of operation, but its biggest disadvantage is the low resistance of the algorithm (there is a high risk of the algorithm converging to the local optimum). In the case of evolution and swarming algorithms, despite the longer waiting for a solution, the risk of the algorithm converging in the local optimum is much lower compared to the gradient method.

6. Acknowledgement
The research is co-financed under the program of the Ministry of Science and Higher Education "Implementation Doctorate" in accordance with the contract RJO15/SDW/002-04 concluded on 29.06.2020

REFERENCES
[1] http://www.ansys.com/
[2] Ansys fluent adjoint solver user's guide
[3] Al-Obaidi A Sh, Sun L C, *Calculation and optimization of the aerodynamic drag of an open-wheel race car*. Journal of Engineering Science and Technology, EURECA 2013 Special Issue August, 1-15, 2014.
[4] Singh R, Golsch, K, *A Downforce Optimization Study for a Racing Car Shape*, SAE Technical Paper 2005-01-0545,
[5] Anderson J, *Computational fluid dynamics*. McGraw-Hill, New York, 1995.
[6] Malasekera W, Versteeg H K, *An introduction to computational fluid dynamics*. Pearson Education Limited, New Jersey, 2007.
[7] Lanfrit M, *Best practice guidelines for handling automotive external aerodynamics with fluent*. Fluent Deutschland GmbH, Darmstadt, 2005.

Global technology trends through patent data analysis

T O Sobolieva[1*], N G Holionko[1], L P Batenko[1] and T I Reshetniak[1]

[1]Kyiv National Economic University named after Vadym Hetman, Management Department, 54/1 Prospect Peremogy, Kyiv, Ukraine

[*]Corresponding author's e-mail: tetiana.sobolieva@kneu.ua

Abstract. The future success of companies directly depends on their innovative activity, research and development of newly technologies. The article examines global technological trends by comparing the number of patent applications in different fields and the degree of development of prominent business areas. These areas are also identified by the highest-ranking positions of the companies in the annual research of innovation in the world. The study highlights the most patentable technological fields, as well as those with the highest average growth of this indicator. The emphasis is made on the regional context of technology patenting and the most active companies in patenting.

1. Introduction

The development and competitiveness of companies, and as a consequence, of national economies, are constantly influenced by shifts in the global economy, political processes, environmental and climatic dynamics, replacement and dominance of scientific and technological paradigms. Moreover, the scientific and technological factors in recent decades have formed an absolutely new vector of development of entire industries and sciences, which, in turn, radically transform the key factors of competitiveness. The leading role among them plays the development and active implementation of innovations, the main place among which is given to radical innovations. The impact for the emergence and already noticeable spread of radical innovations are new exponential technologies that lead to the emergence and disappearance of big corporations and even entire industries.

2. Theoretical base of the research

The global technological trends contribute to the expansion of innovations and the strengthening of competition, therefore, the ability of an organization to use technology to its advantage will determine its effectiveness in the future. Modern technologies define the strategic context and become an apparent priority of many businesses.

2.1. Overview of emerging technology trends

Leading consulting firms and analysts identify technology trends and determine their greatest global impact over the next few years. Consultants from Deloitte and Forbes predict some groups of technological changes that form innovations and enabling organizations to transform [1, 2]. Among them are the following.

1. Artificial intelligence (AI) –machines and mechanisms, that have human abilities and can learn from the own experience in order to adapt to specified parameters. At the same time, AI is the trigger technology for the future innovations.
2. Internet of Things (IoT) -Internet-connected "smart" devices and items designed to collect and transmit significant data. This technology is considered as the foundation for innovative business models, platforms and capabilities.
3. The technology of digital reality, including augmented reality (AR) and virtual reality (VR) - systems that combine in real-time three-dimension image, location, sound, kinesthetic, and other features. It could be used in the wide range of areas - from military to art, from manufacturing companies using Digital Twins to training and trade. The goal of this technology is seen as a natural, intuitive and understandable interaction with core business technologies.
4. Blockchain - the most secure technology of registration and verification of transactions, record keeping and data protection. The rapid development of this technology by fintech companies can revolutionize many business models in the health care, mass media and telecommunications sectors.
5. The quantum computing - processing the large volumes of information on universal quantum machines that use quantum-mechanical phenomena and are able to perform new types of calculations. Such technologies can stimulate the development of innovative discoveries in science, machine learning methods for the quick diagnosis of diseases, life-saving therapies, etc.
6. The exponential intelligence - data interpretation techniques, that use a number of innovative tendencies (analytics, big data, artificial intelligence, the Internet of Things, machine learning, etc.) that contribute to speed up the meaning-making processes by machines or mechanisms.

These groups of technology particularly and many other innovations will generate, in the opinion of the consultants, the greatest influence on the business landscape and development of the Industry 4.0 all over the world. As evidenced by scientific research on the development of technology trends in different countries, such macro-technological forces remain the basis for innovations and business transformation. Thus, the authors [3] underline that the focus of research and characteristics of innovations (for example, usage of the Internet of Things technology) are different in many countries. Consequently, the dynamics and coverage of such technology in patent documents, for example, in Korea and the United States, are also differ.

It is important, in our opinion, to compare innovative technological trends followed by leading companies during recent decades. It is well known that at the beginning of the third millennium, manufacturing companies such as General Electric, Ford, BMW and others were prevailed in the ranking of the largest companies all over the world. However, in the first decade of the twenty-first century tech Apple, Facebook, Amazon, eBay and others began to move more convincingly to the first positions in the ratings. The second and especially forecasts for the third decade of the twenty-first century foreshadow a significant preponderance of innovative tech companies in the ranking. Moreover, in pursuit of the goals to increase efficiency, speed and adaptability to work with customers and partners, such mega companies as Walmart are also digitized. Since 2015, this company has been modernizing its core systems by introducing mobile technologies, cloud storage, digital and augmented reality technologies and personalizing customer relationships.

Researchers call technological foresight as a modern methodology for identifying new technologies [4]. The authors conduct a patent analysis of the telecommunications equipment sector in order to find areas that may be influenced by such scientific and applied investigations. As a result of the study of patents over the eight-year period up to 2010, the authors managed to classify technology areas into four categories, depending on their relevance and growth rate.

2.2. Global technological shifts in the modern economy

The history of Kodak is an example of such dramatic changes. In 1998, the company produced 85% of the world's volume of paper for cameras and there were 170 thousands employees. However, three years later digital cameras have replaced the company with its market and it became bankrupt. In addition to

IOP Conf. Series: Materials Science and Engineering 1037 (2021) 012059 doi:10.1088/1757-899X/1037/1/012059

digital technologies, exponential technologies include artificial intelligence, nanotechnology, three-dimensional printing, virtual reality, electric vehicles, as well as self-driving cars and some others. These technologies already change or will significantly change our lives in the next 5-10 years. Computer software radically changes an architecture and logic of business processes. Uber is the largest taxi company in the world, but it is practically a computer program and does not physically own the cars that provide transportation. The automotive industry will also undergo dramatic changes. In addition to electric vehicles, autonomic (self-piloted) transport will be extended. It will solve the exploitation problem of the traditional personal cars, because the purchase costs, maintenance, refueling or charging, parking will be not necessary, moreover, taking into account the fact that the majority of car owners often use their machines not more than 15% of the daytime.

One of the features that unite these technologies is their high science intensity, which arises as a result of the high-level intellectual activities of researchers and scientists. Created objects of intellectual property (IP) provide the overwhelming added value for the new products and the higher capitalization of companies. Among such objects, the greatest importance have inventions, industrial designs, trademarks and brands. On average, intellectual property can form 40 to 80% of a company's value. Its value at the most valuable companies in the world (Table 1) [5] indicate the importance of a brand as an intellectual asset of a company.

It has sense, in our opinion that in the top of the rating there are high-tech companies. They demonstrate the dominance (according to Forbes) in this rating, for the most of them the impetus for development were given by exponential technologies.

The power of influence and importance of innovations in building successful business is illustrated by comparison rankings of the most expensive brands and the most innovative companies in the world, provided by the one of the world-renowned consulting companies – BCG (Boston Consulting Group) [7]. In 2019, the list of the most innovative companies includes Alphabet/Google, Amazon, Apple, Microsoft, Samsung, Netflix, IBM, Facebook, Tesla, Adidas and others.

Table 1. Top 10 Most Valuable Brands 2019

Rank 2020	Brand	Brand Value	1-Yr Value Change,%	Brand revenue	Industry
1	Apple	$ 241.2 B	17	$ 260.2 B	Technology
2	Google	$ 207.5 B	24	$ 145.6 B	Technology
3	Microsoft	$ 162.9 B	30	$ 125.8 B	Technology
4	Amazon	$ 135.4 B	40	$ 260.5 B	Technology
5	Facebook	$ 70.3 B	-21	$ 49.7 B	Technology
6	Coca-Cola	$ 64.4 B	9	$ 25.2 B	Beverages
7	Disney	$ 61.3 B	18	$ 38.7 B	Leisure
8	Samsung	$ 50.4 B	-5	$ 209.5 B	Technology
9	Louis Vuitton	$ 47.2 B	20	$ 15 B	Luxury
10	McDonald's	$ 46.1 B	5	$ 100.2 B	Restaurants

Duplication of many companies in these rankings confirms, in our opinion, the relationship and the key role of intensive research and implementation of innovations. The belonging of these companies to new rapidly developing industries and technologies demonstrates the direction of changes in technological trends and the formation of a new paradigm for the development of society. In 2018, about nineteen out of fifty companies in this rating initially built their business on digital technologies, and two more such companies were recently included in this rating - the Chinese Alibaba Group and Uber. This technological change is particularly noticeable when the brand rankings are compared among the companies that peaked in their value before the last decade. Most of those rating companies have built their businesses in the traditional industries, including oil industry. Innovative companies that operate in digital, nanotechnology and artificial intelligence spheres have pushed now all of those companies out of this ranking.

3. Results of the study

As was mentioned above, intellectual property is the source of development and main asset of innovative companies. Changes and dynamics of indicators in this area are in direct relationship with global economic trends. At the same time, the dynamics of indicators in the field of intellectual property is an indicator of global technological changes. They reveal technologies and industries in which active patenting processes are taking place and indicate the most dynamic and rapidly growing technological trends (figures 1-6) [6]. Thus, in the Figure1 the global amount of the published patent applications in electrical engineering is presented. Some of them demonstrate high growth rate over ten year period from 2007 to 2017. They are electrical machinery, apparatus, energy (average growth 6.8%), digital communication (8.5 %), computer technology (6.2 %), IT methods for management (10.1 %). Most of all patents were published in 2017 in the field of computer technology - almost 230 thousands (229269 patents). It is the maximum figure not only in the electrical engineering field, but also among the all patented technologies in the world. 197645 patents were published in electrical, apparatus, energy machinery and 144669 – in digital communication. The decline in patenting activity in audio – visual technology and telecommunications is understandable. For the analyzed period, the average growth of these industries is negative (-1.2% and -1.3%, respectively).

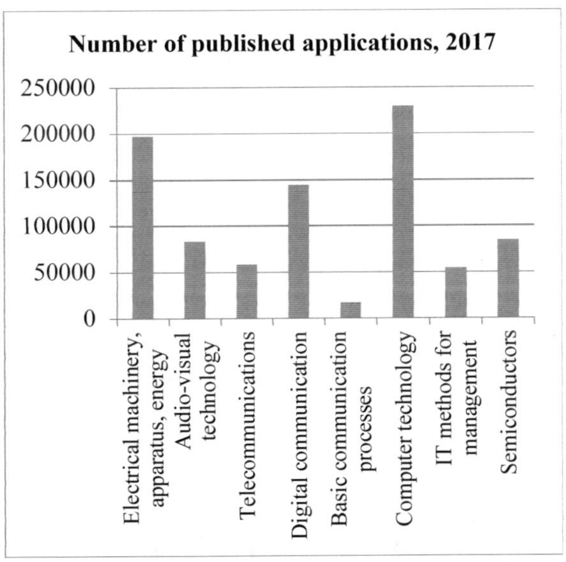

Figure 1. Published applications for the period 2007-2017 in electrical engineering field

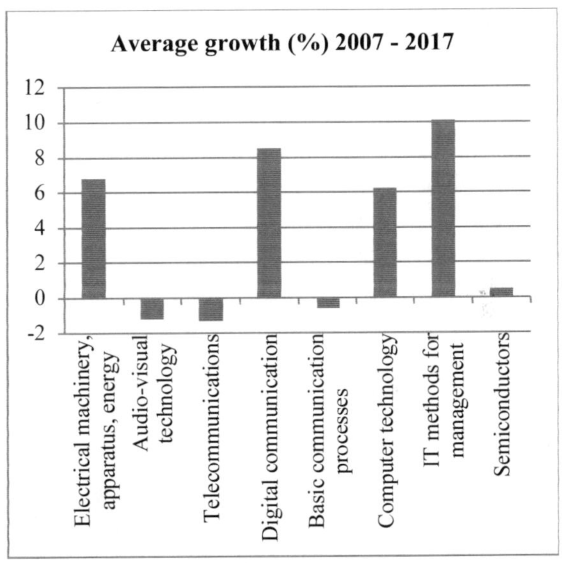

Figure 2. Average growth for the period 2007-2017 in electrical engineering field

Among chemistry technologies (figure 3,4), patents were published the most actively in organic fine chemistry with 132,863 patents (average growth in 2007-2017 was 5.8%), pharmaceuticals with 106312 published patents (3.5% average growth), basic materials chemistry (95776 patents and 9.2% average growth), as well as chemical engineering (80378 patents and 9.0% average growth). The highest average growth in the chemistry field demonstrate food chemistry (13.4%), environmental technology (9.8%), and basic materials chemistry (9.2%). In general, in this field of technology, all areas have a positive average growth of patent applications. On average, the number of applications increase every year by the range from 3.5% (pharmaceuticals) to 13.4% (food chemistry). The micro-structural and nanotechnology fields have the smallest number of applications (5294) within chemistry technologies,

IOP Conf. Series: Materials Science and Engineering 1037 (2021) 012059 doi:10.1088/1757-899X/1037/1/012059

but there is an active increase in their growth - 7.4%. It can be explained most likely by the insufficiently accumulated knowledge base (including expensive equipment and materials) for a technological breakthrough in the use of nanomaterials.

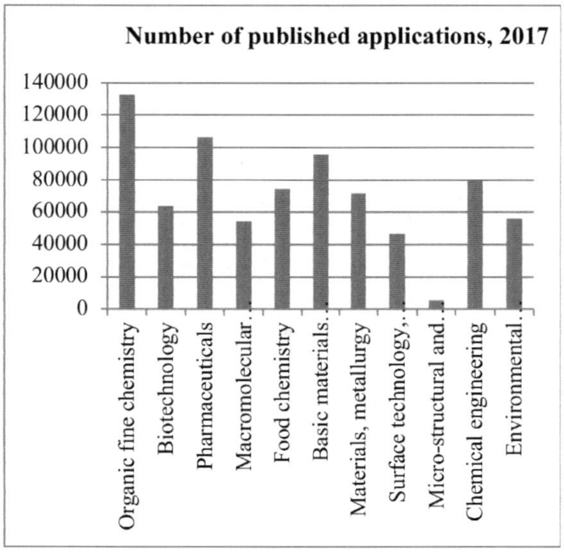

Figure 3. Published applications for the period 2007-2017 in chemistry field

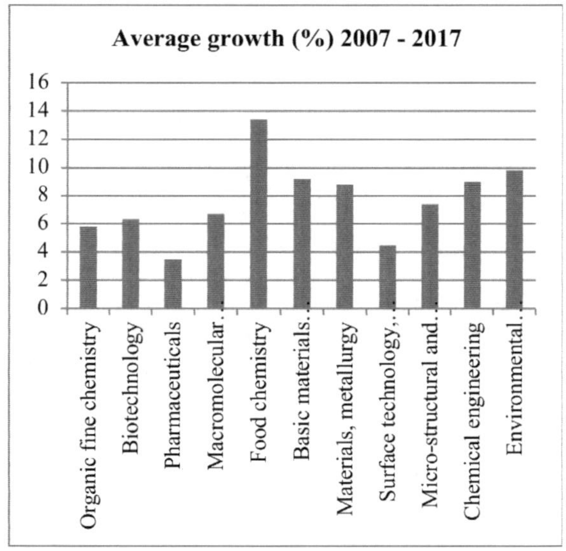

Figure 4. Average growth for the period 2007-2017 in chemistry field

In the mechanical engineering field (figure 5, 6), the largest number of patents were published in transport (124203 applications), other special machines (117901). The latter direction also shows the greatest average growth for the analyzed period (10.1%). There is also a high growth rate (9.2%) and 89742 published applications in machine tools field in 2017.

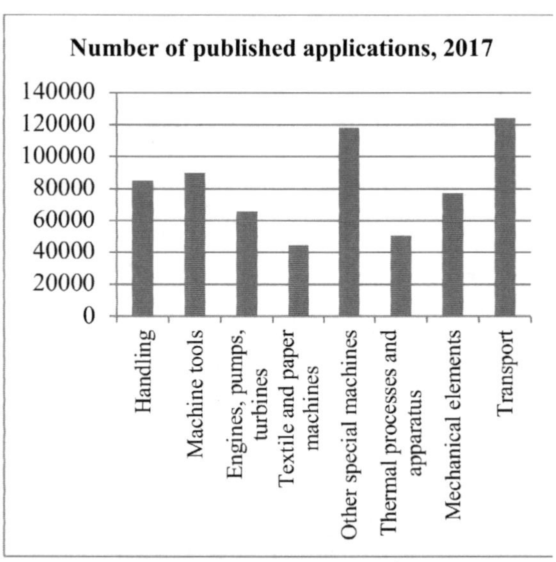

Figure 5. Published applications for the period 2007-2017 in mechanical engineering field

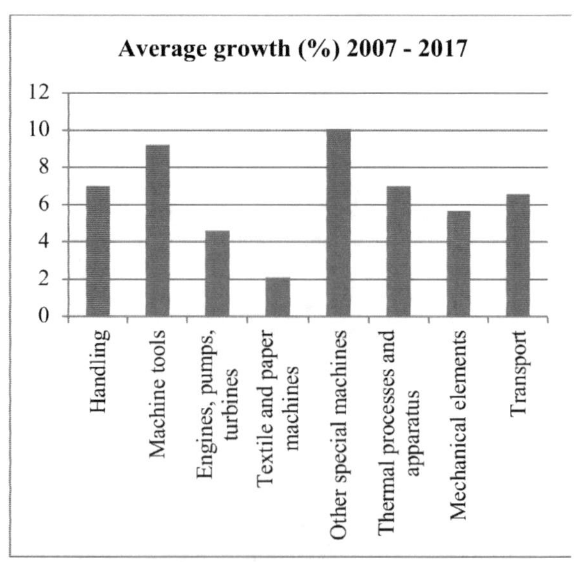

Figure 6. Average growth for the period 2007-2017 in mechanical engineering field

2,950,605 applications were published in 2017 in all areas of technology all over the world; the average growth between 2007 and 2017 was 5.6%. Thus, food chemistry, other special machines and IT

IOP Conf. Series: Materials Science and Engineering 1037 (2021) 012059 doi:10.1088/1757-899X/1037/1/012059

methods for management turned out to be the most innovatively active (actively patentable) with an average growth rate of more than 10%. Electrical machinery, apparatus, energy have been the most patented in Japan, Republic of Korea and Germany. The largest applicants in this area of technology are State Grid Corp of China, Mitsubishi Electric Corp, Toyota Jidosha KK and Robert Bosch Gmbh. The USA and China were more active in computer technology, Switzerland in pharmaceuticals and Russian Federation in food chemistry. IBM is actively filing patents in digital communication, computer technology and semiconductors. Huawei Technologies and Samsung Electronics specialize in digital communication and computer technology, LG Electronics Inc emphasizes research on digital communication and telecommunications.

4. Discussion and conclusion

The conducted research confirms the existence of the relationship between the number of patents in a certain technological area of business and global trends in technological development in the world. In this regard, an increase in patenting activity is observed simultaneously with market success (brand value) and innovative activity of companies in the same technological fields. Some of these fields (computer technology, for example) form the basis for emerging technology trends. Regional displacement of innovation activity in developing technology fields towards Asian countries is also diagnosed from analysis of research data.

In our opinion, the conclusion is also confirmed about the connection between the strengthening of patenting activity, for example, in some fields of electrical engineering, which is happening today, with these fields' possible upcoming development and stimulation of related areas, machine learning methods or diagnostic technologies.

Exponential technology trends rapidly change the business landscape around the world, which will obviously increase the number of patent applications in connected fields. We continue to study with interest new trends and results in technological development and activity on their patenting.

References

[1] Briggs B, Buchholz S and Sharma S K 2020 Macro technology forces at work Available from https://www2.deloitte.com/us/en/insights/focus/tech-trends/2019/macro-technology-trends-forces-at-work.html

[2] Bernard M 2020 These 25 Technology Trends Will Define The Next Decade Available from https://www.forbes.com/sites/bernardmarr/2020/04/20/these-25-technology-trends-will-define-the-next-decade/#7c76be5029e3

[3] Xiang L, Chanil P and Kexin B 2020 *Technology Analysis & Strategic Management* **32** (1) 104-118

[4] Michelino F, Cammarano A, Lamberti E and Caputo M 2016 *International Conf. on Engineering, Technology and Innovation/IEEE International Technology Management Conference (ICE/ITMC)* 1-9

[5] The World's Most Valuable Brands 2020 Available from https://www.forbes.com/the-worlds-most-valuable-brands/#e169d53119c0

[6] WIPO 2019 *World Intellectual Property Indicators 2019* World Intellectual Property Organization (Geneva: WIPO Economics & Statistics Series) p 228

[7] The Most Innovative Companies of 2019 According to BCG Available from https://www.forbes.com/sites/louiscolumbus/2019/03/24/the-most-innovative-companies-of-2019-according-to-bcg/#555e4ba9486d

IOP Conf. Series: Materials Science and Engineering 1037 (2021) 012060 doi:10.1088/1757-899X/1037/1/012060

The influence of combustion properties of fuel materials upon human evacuation in case of fire

Z C Grigoras[1] and M Ibanescu[1]

[1] "Gheorghe Asachi" Technical University of Iași, Bd. Dimitrie Mangeron nr. 67, 700050, Iași, Romania

*Corresponding author's e-mail: zeno.grigoras@gmail.com

Abstract. The purpose of this article is to analyze the variation of the tenability criteria (convected heat, radiant heat, toxic gases and smoke visibility) on the main evacuation paths inside of an educational building, in case of fire. The analysis is performed by numerical simulation for three different fire scenarios. The fire load density of fire space (an office type room) is the same, but the main combustible material is changed: wood, plastic and polyurethane foam. The analyzed building can house a great number of people (over 600) and its inner atrium could facilitate the spread of smoke and hot gases in case of a real fire. According to the numerical analysis, it has been concluded that the convected heat variation is not a significant one (within the engineering limits), but the radiant heat, toxic gases and smoke visibility are highly dependent on the type of considered combustible materials.

1. Introduction

The case study consists in the numerical simulation of the development of a fire in an enclosed space – educational building, considering three different fire scenarios. The fire load density is kept the same, but the main combustible materials are changed: wood, plastic and polyurethane foam. The study also aims to monitor the parameters considered relevant in influencing the tenability criteria for human evacuation in the event of a fire.

The applicable part of this paper consists of:

- identification of a building that houses a large number of people, the architecture of the building favouring the spread of smoke and hot gases;
- identification of a rational design fire scenario;
- establishing the design fire corresponding to the fire scenario variants;
- establishing the relevant parameters for the evacuation of persons in the fire situation, in order to monitor: air temperature, gas toxicity, visibility through the smoke layer, thermal radiation on the escape routes, at a height of 2.00 m above the floor level [1], [2], [3];
- construction of the discrete virtual model with finite volumes for the analyzed space, in the FDS software;
- performing the comparative analysis based on monitoring the relevant parameters in the sections of interest and presenting the conclusion.

CFD (Computational Fluid Dynamics) simulation of fire development used in the case of human evacuation is the current design procedure for complex buildings housing large crowded spaces (like airports, shopping mall, subway stations or skyscrapers) or buildings housing people which cannot evacuate themselves (like hospitals, nurseries or retirement homes).

Simulating the real tenability criteria on evacuation paths is essential for the assessment of occupant safety in a building under fire conditions. When hazardous conditions set in before human evacuation, then different measures must be considered for increasing the available safe egress time (ASET), like: replacing the combustible finishes and furniture inside the building, shortening the length of the escape paths, using a automatic fire extinguishing system or using a smoke and heat exhaust system.

2. The model

2.1. The analyzed space

The analysed space is an educational building which belongs to the Faculty of Civil Engineering and Building Services, from "Gheorge Asachi" Tehnical University of Iaşi, Romania (Figure 1).

Figure 1 Perspective (left) and interior atrium (right)

The building, intended for higher education, has the following height regime:

- semi-basement: it houses technical spaces;
- ground floor: it houses an exhibition space and the two entrances of the building;
- mezzanine: it houses the hall of the Romanian Academy of Scientists, it is not intended for educational activities;
- $1^{st} \div 4^{th}$ floor: it houses spaces for educational activities;
- 5th floor: it houses technical spaces.

2.2. The design fire scenario

Studying the characteristics of the building, it was identified as a reasonable (unfavourable) scenario the initiation of a fire in a meeting room located at the mezzanine of the building.

This space houses combustible materials with high thermal loads, which, by releasing a large amount of toxic gases and smoke, can endanger the safety of building users in a probable evacuation process.

An important aspect in establishing this fire scenarios is that the location of this room facilitates the rapid movement of smoke into the atrium.

The design fire scenarios assume that all the doors of the analysed building are closed, except for the access doors of the buildings, the door of the burned space and the doors to the stairwells located on each floor (the self-closing mechanisms of the doors are supposed to be faulty).

All windows of the building are closed, except for the windows of the burning space (the self-closing mechanisms of the doors are supposed to be faulty). The walls are made of bricks with mortar plaster and the slabs are made of reinforced concrete with mortar plaster, materials with common thermal properties [4] [5].

2.3. Variants of fire scenarios and design fires

The combustible materials existing in the analyzed meeting room are presented in table 1.

Table 1. Thermal load corresponding to the combustible materials in the analyzed space

Object	*No.*	*Combustible materiale*	*Total mass [kg]*	H_u *[MJ/kg]*	*Thermal load [MJ]*
Desk	1	Wood	36	17,5	630
		Coton	4	20	80
Chair	36	Plastic	90	40	3600
		Polyurethane	7	25	175
Drapes	1	Polyester	5	30	150
Carpet	1	Polypropylene	45	40	1801
Hanger	1	Wood	55	17.5	963
Smart board	1	PVC	30	20	600
Carpentry	1	PVC	70	20	1400
Door	1	Wood	30	17.5	525

Note: Quantities were calculated by estimating the volume and mass of each object.

Based on the thermal loads (fuel energy transformed into heat), table 1, the design value of the thermal load density is determined: 254 [MJ/m^2]; the value is much lower than the statistical value for office space (it can be deduced that the design values specified in the European regulations [6] are coverage in most cases).

The percentage in the thermal load of the types of combustible materials inside the burned space are presented in table 2. It can be seen that the predominant group of combustible materials is pure plastic hydrocarbons (PE-polyester, PS-polystyrene, PP-polypropylene), followed by the group of wood and PVC.

Table 2 Percentage in thermal load of each type of combustible material

Combustible material	H_u *[MJ/kg]*	*Caracteristic thermal load [MJ]*	*Percentage [%]*
Pure plastic hydrocarbons (PE, PS, PP)	40	5401	54
Celulozic (wood+coton)	17,5	2198	22
PVC	20	2000	20
Polyurethane foam	25	175	2
Polyester	30	150	2

Within the numerical simulations, three variants were assumed for the design fire scenario, *Vi*, generated by the predominant types of combustible materials:

- for V1: cellulose (wood), with the generic name *Wood*, the most common combustible material in office spaces;
- for V2: pure plastic hydrocarbons (PE, PS, PP), with the generic name *Plastic*, predominant for the situation of the analyzed space;
- for V3: polyurethane, generic name *Polyurethane foam*, large generator of smoke and toxic gases.

The design fires corresponding to the variants of the design fire scenario were established, based on the procedures [6], considering that each of them is produced, in turn, by one of the three types of combustible materials: wood, plastic and polyurethane foam. The parameters of each design fire are presented for comparison in table 3.

Table 3 Design fire parameters corresponding to fire scenario variants

Parameter	*Combustible material*		
	V1 *Wood*	*V2* *Plastic*	*V3* *Polyurethane foam*
Characteristic thermal load density [MJ/m²] $q_{f,k}$	420	420	420
Burning coeficient [-] m	0.74	0.82	0.72
Risk of fire due to compartment size [-] δ_{q1}	1.125	1.125	1.125
Risk of fire due to compartment destination [-] δ_{q2}	1	1	1
Active protection measures [-] δ_n	1	1	1
Design thermal loade density [MJ/m²] $q_{f,d} = m \cdot q_{f,k} \cdot \delta_{q1} \cdot \delta_{q2} \cdot \delta_n$	350	387	340
Reference space area [m²] A	39.15	39.15	39.15
Total energy [MJ] $E_t = q_{f,d} \cdot A$	13690	15170	13320
Fuel-air ratio [-] $1/r$	4.95	14.22	9.54
Lower calorific value of the combustible material [MJ/kg] $\mathrm{H_u}$	17.50	40.00	25.00
The area of the surface through which the ventilation of the burned space occurs [m²] A_v	10.89	10.89	10.89
The average height of openings throught wich the ventilation of the burned space occurs [m] h_{eq}	1.36	1.36	1.36
Peak heat realease rate [MW] $Q_{\max} = 0,52 \cdot \frac{1}{r} \cdot m \cdot \mathrm{H_u} \cdot A_v \cdot \sqrt{h_{eq}}$	17.28	15.23	12.46
The development speed of the fire correlated with the destination of the analyzed space [s] t_α	300	300	300
Duration of the fire growth phase [s] $t_g = t_1 = t_\alpha \cdot \sqrt{Q_{\max}}$	1247	1171	1059

Parameter	*Combustible material* *V1* *Wood*	*V2* *Plastic*	*V3* *Polyurethane foam*
Energy released during the growth phase of the fire [MJ] $E_g = \frac{t_1 \cdot Q_{\max}}{3}$	7181	5945	4398
Energy released during the fully developed phase of the fire [MJ] $E_{fd} = 0.7 \cdot E_t - E_g$	2401	4673	4925
Duration of the fully developed phase of the fire [s] $t_{fd} = \frac{E_{fd}}{Q_{\max}}$	139	307	395
The time measured at the end of the fully developed phase of the fire [s] $t_2 = t_g + t_{fd}$	1386	1478	1454
Energy released during the decay phase of the fire [MJ] $E_d = 0.3 \cdot E_t$	4107	4551	3996
Duration of the decay phase [s] $t_d = \frac{2 \cdot E_d}{Q_{\max}}$	475	597	641
Time measured at the end of the decay phase [s] $t_3 = t_g + t_{fd} + t_d$	1861	2075	2096

The variation of the heat release rate over time for each variant of the design fire scenario is shown in Figure 2. The flashover occurs within about 15÷20 minutes from the fire initiation.

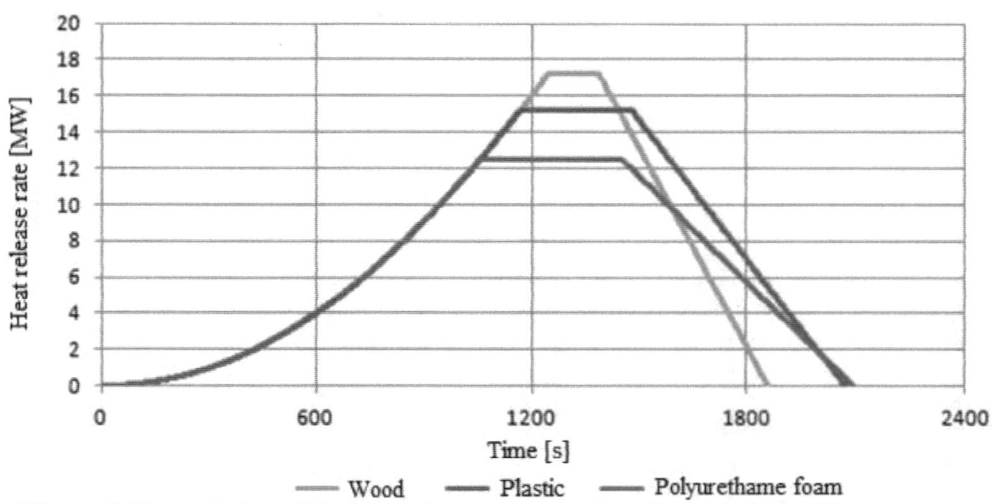

Figure 2 The variation of the heat release rate over time for each design fire scenario

Table 4 specifies the values of the parameters that describes each fuel (declared as input data in FDS software), for the case of simple combustion model [7].

Table 4 Parameters that describes each fuel (declared as input data in FDS software)

Parametru	*Fuel*		
	V_1-Wood	*V_2-Plastic*	*V_3-Polyurethane foam*
C atoms	1	1	1
H atoms	3.584	1	1.7400
O atoms	1.550	0	0.3230
N atoms	0	0	0.0698
Lower calorific value [kJ/kg]	17500	40000	20000
CO released by fuel after combustion [g/g]	0.004	0.024	0.02775
Soot released by fuel after combustion [g/g]	0.015	0.059	0.1875
H released by fuel after combustion [g/g]	0.1	0.1	0.1

2.4. Fire safety relevant parameters

In order to monitor the effects of combustion on the human evacuation in case of fire, the following parameters were considered relevant: *air temperature, visibility through the smoke layer, gas toxicity, thermal radiation*; these parameters are correlated with the tenability criteria imposed by the fire safety engineering regarding the evacuation of people from buildings [1], [2], [3].

The parameters were monitored with virtual devices, defined in the FDS software (MS for the Main Staircase and SS for the Secondary Staircase), located in the entrance of each staircase – considered a point of interest – at 2.00 m above the floor.

The critical value of each monitored parameter at a height of 2.00 m above the floor is [1], [2], [3]:
-100[°C] for the maximum air temperature;
- 10 [m], for the minimum visibility through the smoke layer, the case of large rooms;
- 1.0 [-] for maximum FED toxicity (*Fractional Effective Dose*), in the case of adults;
- 2.5 [kW/m^2] for the maximum radiant heat flux.

2.5. Discretization of the computational domain

The FDS software user manual [8] recommends that a cell size / discretization volume be in the range [D* / 16; D* / 4], where D* is the characteristic diameter of the fire. Applying the user manual [8] recomandation for the design fires, corresponding to the three variants of fire scenario it was decided to use a medium-coarse discretization cell of 0.5 m × 0.5 m × 0.5 m, with which the geometry of the building can be easily described and reduces the running time or hardware resources required for numerical simulations.

The discrete virtual model with finite volumes is the one in Figure 3.

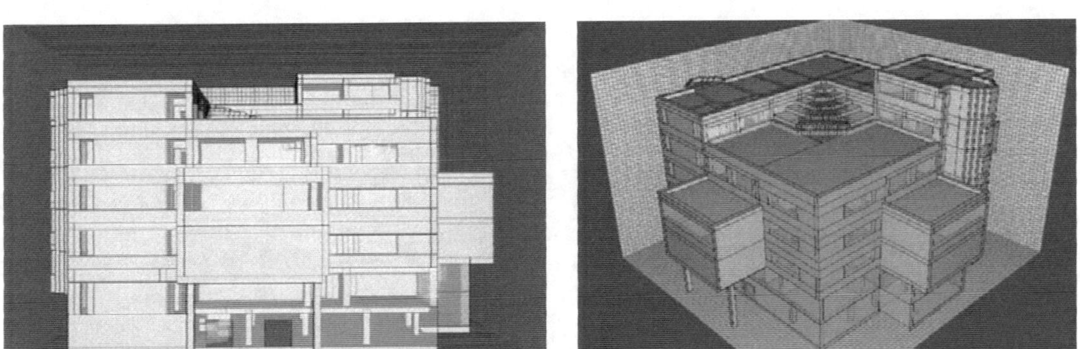

Figure 3 Discreet virtual model with finite volumes - main facade and perspective view

IOP Conf. Series: Materials Science and Engineering 1037 (2021) 012060 doi:10.1088/1757-899X/1037/1/012060

3. Analysis of results

Next, the comparative graphs are presented (in green for V1-Wood, in red for V2-Plastic and in blue for V3-Polyurethane foam; the horizontal red line mark the critical value), on each monitored parameter at the ground floor of the building (where each person must pass) for the MS – main staircase exit and SS – secondary staircase exit.

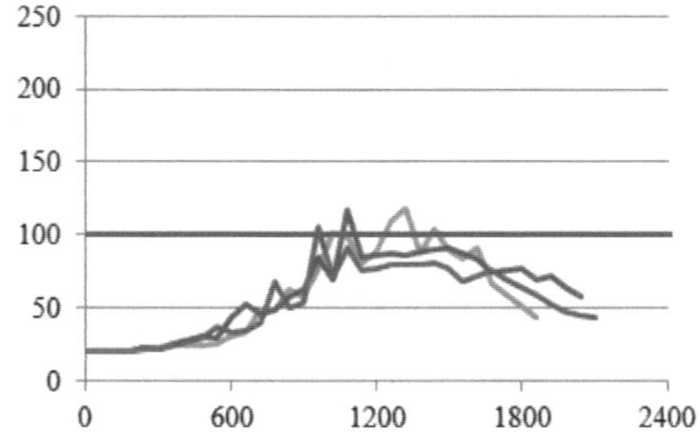

***Figure* 4.1** Air temperature MS [°C]

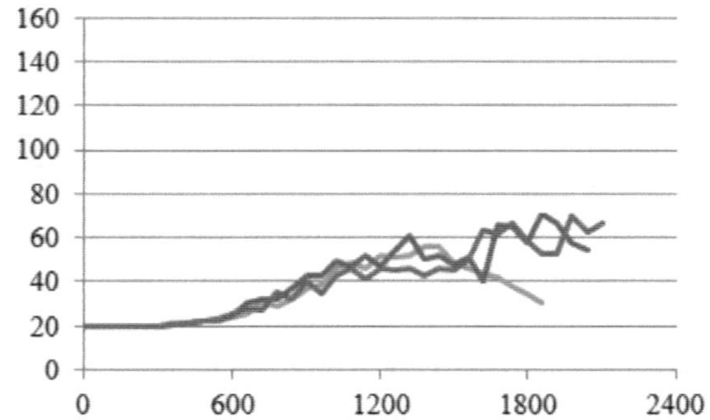

Figure 4.2 Air temperature SS [°C]

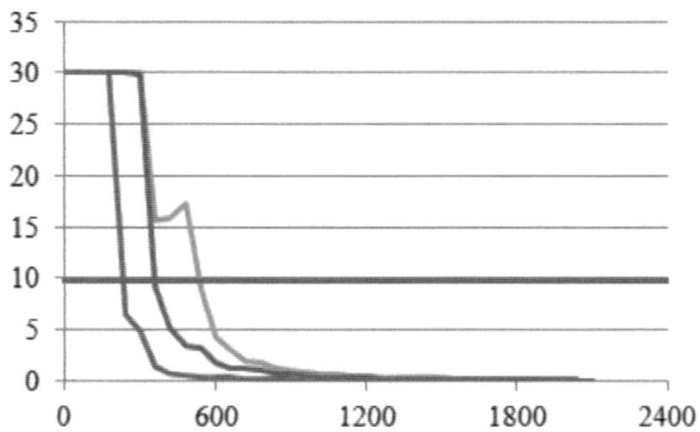

Figure 4.3 Visibility MS [m]

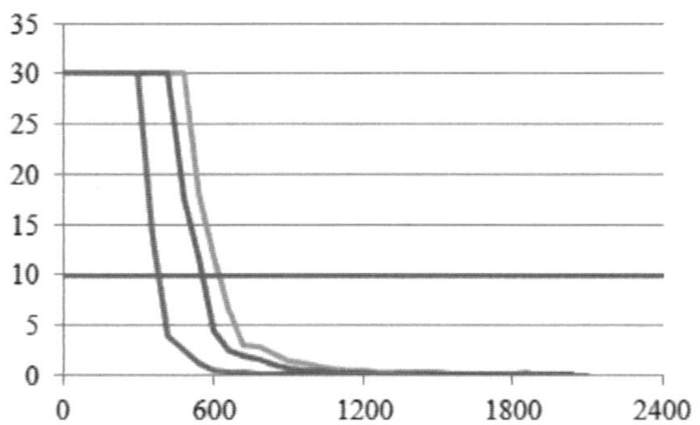

Figure 4.4 Visibility SS [m]

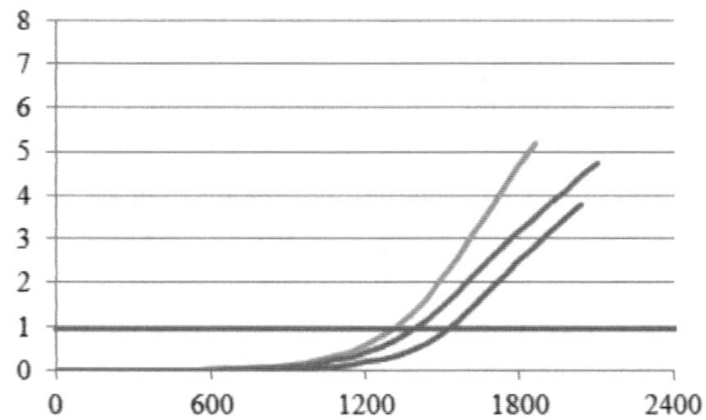

Figure 4.5 FED MS [-]

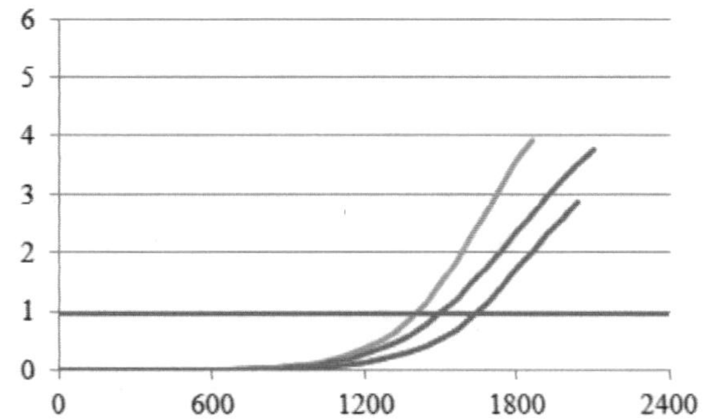

Figure 4.6 FED MS [-]

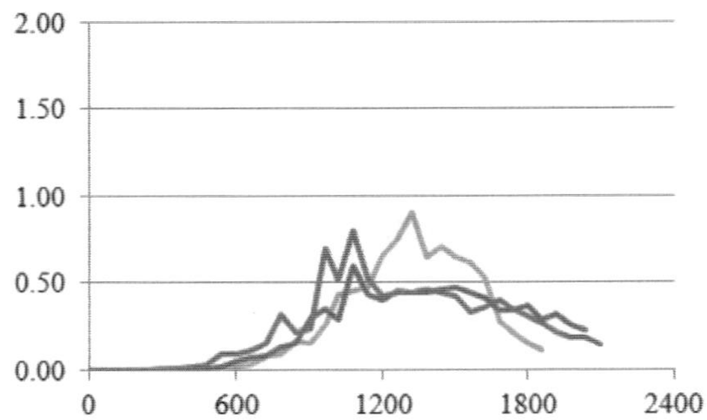

Figure 4.7 Thermal radiation MS [kW/m^2]

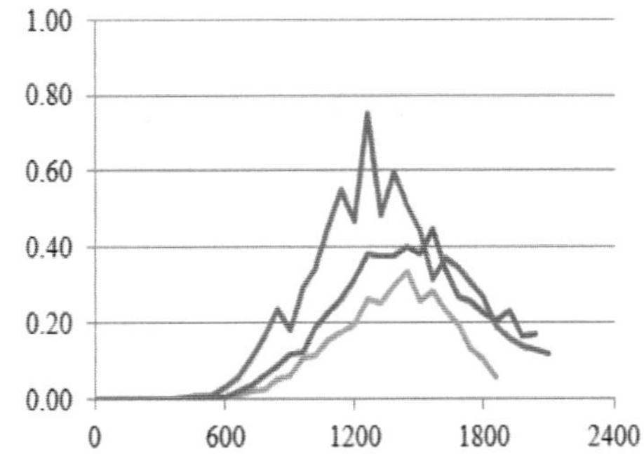

Figure 4.8 Thermal radiation SS [kW/m^2]

4. Conclusions

Based on the present study, the following conclusions could be drawn:

1. The variation in time of the analyzed parameters presents different values for each type of fuel, but these variations have a similar development.
2. The variation of heat release rate in time is different for each type of fuel, but the parameter values are of the same order of magnitude.
3. The air temperature and visibility through the smoke layer, measured at a height of 2.00 m above floor level for each level at the entrance to the stairwells, shows similar values.
4. Gas toxicity, measured at a height of 2,00 m above the floor level, for each level at the entrance to the stairwells, presents different values.
5. The level of thermal radiation of the air, measured at a height of 2.00 m above the floor, for each level at the entrance to the stairwells, has different values. The highest radiation level is provided by Polyurethane Foam fuel and the lowest radiation level is provided by wood type fuel.
6. The temperature on the surface of the burned elements has different values. The highest temperatures have been obtained by using wood fuel and the lowest values have been by using Polyurethane Foam fuel. The type of combustible material significantly influences the evolution of a fire developed indoors. The accuracy in defining the calculation fire by choosing the correct fuel material is very important, because it ensures a rational and close to reality fire scenario for the numerical simulations.

References

[1] Purser D 2009 Institution of FireEngineers, AGM Conference and exhibition (Glasgow UK)
[2] Poh W 2011 *eColibRiuM*, p 32-37
[3] Grigoras Z C 2014 *The Bulletin of the Polytechnic Institute of Jassy, Construction. Architecture Section, Tomul LX (LXIV), Fasc. 2*, p 155-166
[4] ***, *C107-1-2005: Design code regarding the calculation of the global thermal insulation coefficient at residential buildings.*
[5] ***, *www.infrared-thermography.com*
[6] ***, *EN 1991-1-2 Eurocode 1: Actions on structures – Part 1-2: General actions – Actions on structures exposed to fire*
[7] DiNenno P Drysdale D Beyler C Douglas Walton W Custer R Hall J Watts J 2002 *SFPE Handbook of Fire Protection Engineering*, Third Edition, Published by the National Fire Protection Association, Massachusetts
[8] McGrattan K McDermott R Weinschenk C Hostikka S Floyd J 2014 *Fire Dynamics Simulator – User's Guide*, National Institute of Standards and Technology Special Publication 1019, USA

IOP Conf. Series: Materials Science and Engineering 1037 (2021) 012061 doi:10.1088/1757-899X/1037/1/012061

Distribution of resources for equipping handling fronts in rail consolidation terminals

A Ş Stere[1*], E A Roman[1]

[1] Transport, Traffic and Logistics Department, Faculty of Transport, University POLITEHNICA of Bucharest, Spl. Independenţei, 313, Bucharest, Romania

*Corresponding author's e-mail: armand_serban.stere@upb.ro

Abstract. An indispensable factor of the socio-economic life, the transports realize the connection between the economic zones, which facilitates the attraction of all the companies to the economic circuit, to the capitalization of the available resources and to their overall development. The influence of transport volumes is clearly manifested on the capacity of the terminals, the number of means of transport necessary to satisfy the demand, the degree of their use, the economy of the goods handling installations, the capacity of the handling fronts and labour productivity. Therefore, it is of a great importance to design the constructions and equipments of the terminals according to a criteria based on the optimal structure of the technical endowment subject to variations in the volume of transport activities. The present paper aims to present a theoretical framework for the distribution of resources for equipping handling fronts of the rail terminals. A mathematical model and a algorithm based on the projected gradient method are developed to solve this problem.

1. Introduction

Freight transport is a vital component of the national economy. It is the basis of production, trade, consumer activities, ensuring the movement and availability of raw materials and products [1]. The analysis of freight transport sector represents a subject characterized by a high degree of uncertainty. The complexity of this sector is given by the variety of commodities transported and by the here are real constraints due to the location of goods production activities. Because it is a constantly evolving sector, the phenomena related to the transport of different type of commodities are very important in all fields of study that deal with territorial structures and their transformations. In recent years, there has been an increase in the demand for services in the transport of goods in the Member States of the European Union, in close correlation with the increase in the level of consumption of goods and services in society [4].

Variations in transport volumes affect all aspects of the operation of road, rail, river, sea and air transport. A correct assessment of the size of the capacity required, for example, traffic sections and technical and shunting yards, their processing capacity, the capacity of the cargo handling equipment at the terminals is the basis for all necessary construction and installations. Consequently, the volume of investments and the terms of their recovery depend on it.

Therefore, the problem of the appropriate distribution of the resources available for the endowment of the handling furnaces within the consolidation terminals arises as one of great importance. This is a sizing problem of the constructions and installations that are subject to variable operating demands. These requests are undesirable because they involve maintaining additional capacity during peak hours. If the installation is designed according to the maximum load then most of the time it will not be used in full [5]. From this we can deduce that it is not indicated the design of the installations corresponding to the demand during the rush hours because the investments and expenses corresponding to the additional technical endowment will not be recovered.

Also, it is not recommended to design the constructions and installations from the endowment of the consolidation terminals to the minimum volume because, in this case, it is practically impossible to accomplish the operation process. Therefore, the aim will be to identify that economic criterion that can allow the determination of the optimal structure of the technical endowment in the case of machinery and installations subject to variations in the volume of transport activity.

2. Mathematical formulation of the problem

This category of problems generated by variations in the volume of transport activity also includes the problem of allocating available resources in order to endow the handling fronts of the consolidation terminals. This is an example of a non-linear programming problem and involves developing a perspective plan that can distribute these resources over a m period of years to equip a n number of machine handling fronts. The plan must be in direct and close accordance with the requirement that the level of expenditure on the handling of goods must be minimized. Therefore, it can be stated that the problem consists of the distribution of the investment fund on each individual handling front for each year of the planning period.

The following matrix corresponding to the distribution of the volume of goods at each of the n handling fronts during the m years of the planning period shall be considered [5]:

$$\mathrm{Q} = \begin{pmatrix} Q_{11} & Q_{12} & \cdots & Q_{1n} \\ Q_{21} & Q_{22} & \cdots & Q_{2n} \\ \cdots & \cdots & \cdots & \cdots \\ Q_{m1} & Q_{m2} & \cdots & Q_{mn} \end{pmatrix} \tag{1}$$

while the distribution of the fleet of equipment used to handle goods on the n fronts during the m years is given by a matrix of the form:

$$\mathrm{X} = \begin{pmatrix} X_{11} & X_{12} & \cdots & X_{1n} \\ X_{21} & X_{22} & \cdots & X_{2n} \\ \cdots & \cdots & \cdots & \cdots \\ X_{m1} & X_{m2} & \cdots & X_{mn} \end{pmatrix} \tag{2}$$

The number of machines in year I of the period from front k can be calculated with a relation of the form $\alpha_{ik} = \frac{I_{ik}}{V_k}$ in which I_{ik} represents the value of the existing resources in year I to the handling front k while V_k represents the value of a machine with which front k is equipped.

The main objective is to obtain an expression of the optimization criterion C_{ik}, taking into account all the elements of expenditure that are dependent on the value of existing resources in year I at the handling front k. This category includes the depreciation rate and the expenses related to the repair of the handling equipment and those of stationary of the wagons in order to carry out the loading-unloading operations of the goods. Therefore, the value of the optimization criterion C_{ik} is given by the following formula [5] :

$$C_{ik}(I_{ik}) = a_k I_{ik} + \frac{e_k \cdot Q_{ik}^2 \cdot V_k}{I_{ik} \cdot r_{ik} \cdot \eta_{ik}} \tag{3}$$

where :

- a_k - the annual rate of return on investment for equipment of type k ;
- e_k - the value equivalent of a wagon – stationary hour which is related to the ton of goods loaded in the wagon;
- r_{ik} - the daily number of rounds in which the wagons from year I are introduced at front k;
- η_{ik} - the operating productivity achieved in year I by the machines from front k.

Next we will consider a simplifying hypothesis according to which a number of two manipulation fronts will be endowed. Therefore, the Eq. (3) of the optimization criterion can be written as follows [5] :

$$C_{ik}(I_{ik})=\sum_{i=1}^{m}\sum_{k=1}^{2}C_{ik}(I_{ik})=\sum_{i=1}^{m}\sum_{k=1}^{2}a_kI_{ik}+\sum_{i=1}^{m}\sum_{k=1}^{2}\frac{e_k\cdot Q_{ik}^2\cdot V_k}{I_{ik}\cdot r_{ik}\cdot\eta_{ik}} \tag{4}$$

It must be taken into account the condition that, for each of the years of the investment planning period, the following identity exists [5] : $\sum_{i=1}^{m}I_i = I_{i1}+I_{i2}$ (where I_i represents the investment made in year i while terms I_{i1} and I_{i2} define the existing resources on the two fronts in year i).

To obtain a mathematical expression of the optimization criterion, the sum given by Eq. (4) will be divided into two amounts. By developing the two sums, we obtain that:

$$S_1=\sum_{i=1}^{m}\sum_{k=1}^{2}a_kI_{ik}=\sum_{i=1}^{m}\left(a_1I_{i1}+a_2I_{i2}\right)=a_1\sum_{i=1}^{m}I_{i1}+a_2\sum_{i=1}^{m}I_{i2} \tag{5}$$

If we consider $\sum_{i=1}^{m}I_i = I_{11}+I_{21}+\ldots+I_{m1}+I_{12}+I_{22}+\ldots+I_{m2}$ we will obtain the following result : $mI_1 + (m-1)I_2 + \ldots + I_m = I_{11}+I_{21}+\ldots+I_{m1}+I_{12}+I_{22}+\ldots+I_{m2}$

By numerical substitution in the expression of the first sum, we get that : $S_1 = (a_1-a_2)(I_{11}+I_{21}+\ldots+I_{m1})+a_2[mI_1 + (m-1)I_2 + \ldots + I_m]$.

In the case of the second sum, we have : $S_2=\sum_{i=1}^{m}\sum_{k=1}^{2}\frac{e_k\cdot Q_{ik}^2\cdot V_k}{I_{ik}\cdot r_{ik}\cdot\eta_{ik}} = \sum_{i=1}^{m}\left(\frac{e_1\cdot Q_{i1}^2\cdot V_1}{I_{i1}\cdot r_{i1}\cdot\eta_{i1}}+\frac{e_2\cdot Q_{i2}^2\cdot V_2}{I_{i2}\cdot r_{i2}\cdot\eta_{i2}}\right)=e_1\cdot V_1\left(\frac{Q_{11}^2}{I_{11}\cdot r_{11}\cdot\eta_{11}}+\frac{Q_{21}^2}{I_{21}\cdot r_{21}\cdot\eta_{21}}+\ldots+\frac{Q_{m1}^2}{I_{m1}\cdot r_{m1}\cdot\eta_{m1}}\right)+e_2\cdot V_2\left(\frac{Q_{12}^2}{I_{12}\cdot r_{12}\cdot\eta_{12}}+\frac{Q_{22}^2}{I_{22}\cdot r_{22}\cdot\eta_{22}}+\ldots+\frac{Q_{m2}^2}{I_{m2}\cdot r_{m2}\cdot\eta_{m2}}\right)$.

By making the necessary substitutions on the terms that quantify the value of the existing resources at the two fronts of manipulation in each of the m years of the planning period, we obtain that:

$$S_2 = e_1 \cdot V_1 \cdot \left(\frac{Q_{11}^2}{I_{11} \cdot r_{11} \cdot \eta_{11}} + \frac{Q_{21}^2}{I_{21} \cdot r_{21} \cdot \eta_{21}} + \ldots + \frac{Q_{m1}^2}{I_{m1} \cdot r_{m1} \cdot \eta_{m1}} \right) + e_2 \cdot V_2 \cdot \left[\frac{Q_{12}^2}{(I_1 - I_{11}) \cdot r_{12} \cdot \eta_{12}} + \right.$$

$$\left. \frac{Q_{22}^2}{(I_1 + I_2 - I_{21}) \cdot r_{22} \cdot \eta_{22}} + \ldots + \frac{Q_{m2}^2}{\left(\sum_{i=1}^{m} I_i - I_{m1} \right) \cdot r_{m2} \cdot \eta_{m2}} \right] \tag{6}$$

Therefore, the optimization criterion takes the following form:

$$C_{ik}(I_{ik}) = \sum_{i=1}^{m} \sum_{k=1}^{2} C_{ik}(I_i, I_{ik}) = (a_1 - a_2)(I_{11} + I_{21} + \ldots + I_{m1}) + a_2[mI_1 + (m-1)I_2 + \ldots + I_m] +$$

$$e_1 \cdot V_1 \cdot \left(\frac{Q_{11}^2}{I_{11} \cdot r_{11} \cdot \eta_{11}} + \frac{Q_{21}^2}{I_{21} \cdot r_{21} \cdot \eta_{21}} + \ldots + \frac{Q_{m1}^2}{I_{m1} \cdot r_{m1} \cdot \eta_{m1}} \right) + e_2 \cdot V_2 \cdot \left[\frac{Q_{12}^2}{(I_1 - I_{11}) \cdot r_{12} \cdot \eta_{12}} + \right.$$

$$\left. \frac{Q_{22}^2}{(I_1 + I_2 - I_{21}) \cdot r_{22} \cdot \eta_{22}} + \ldots + \frac{Q_{m2}^2}{\left(\sum_{i=1}^{m} I_i - I_{m1} \right) \cdot r_{m2} \cdot \eta_{m2}} \right] \tag{7}$$

The conditions that must be imposed on the problem are the following:

$$\sum_{i=1}^{m} I_i \leq I \tag{8a}$$

$$I_{0i1} \leq I_{i1} \leq \sum_{i=1}^{m} I_i - I_{0i2} \tag{8b}$$

$$\begin{cases} I_{011} + I_{012} \leq I_1 \leq I \\ 0 \leq I_2 \leq I - (I_{011} + I_{012}) \\ \ldots\ldots\ldots\ldots\ldots\ldots\ldots\ldots\ldots\ldots \\ 0 \leq I_i \leq I - (I_{0(i-1)1} + I_{0(i-1)2}) \end{cases} \tag{8c}$$

where :

- I – the total value of the investment allocated for the m years for the k fronts ;
- I_{0ik} – the minimum amount of resources at its disposal in year I for handling the quantity of commodities Q_{ik}.

Condition (8a) requires that the investments made in each of these years not exceed the total value I of the investment allocated to those k handling fronts. The restriction (8b) takes into account the need to ensure mechanized handling of the volumes of goods from each of the k handling fronts. The set of restrictions (8c) aims to limit the investments that are made each year. The value of parameter I_{0ik} can be determined by using the bellow formula [5] :

$$I_{0ik} = \frac{Q_{ik} \cdot V_k}{365 \cdot n_s \cdot t_s \cdot \eta_{ik} \cdot \tau_0} \tag{9}$$

in which n_s represents the number of shifts needed to work at the front, t_s quantify the duration of a shift (in hours) and τ_0 is the working time utilization coefficient. As a result, the problem will be to establish the minimum value of the objective function given by relationship (7) that satisfies the constraints (8a) - (8c).

To identify the minimum expression of the optimization criterion given by Eq. (7) (which can be seen to be nonlinear), the projected gradient method (or Rosen method) can be used for solving. This method is used to obtain the solution of minimizing a convex function that is subject to linear constraints. It is an iterative method that projects the opposite direction to the gradient on active constraints [2]. In essence, the projected gradient method involves following three steps [3] that suppose to determine an initial solution x* of the problem either by transforming the set of inequations (8a) – (8c) into equations or by using Simplex algorithm, to calculate the gradient of the functions $\nabla C(x^*)$şi $P_r\nabla C(x^*)$ and check the optimality conditions $P_r\nabla C(x^*) = 0$ and $P_r = I - A_r'(A_r A_r')^{-1} A_r \geq 0$. In case both optimality conditions are satisfied, the algorithm stops and the initial solution is the optimal one.

On the contrary, it is necessary to go to the next step by improving the initial solution x* with another solution x^k. If x^k does not represent a solution after following a number of k steps, we will try to find a new x^{k+1} as follows :

- If $P_r \nabla f(x^*) \neq 0$, we will consider $s = P_r \nabla f(x^*)$
- If $P_r \nabla f(x^*) = 0$ but $A_r'(A_r A_r')^{-1} A_r \nabla f(x^*) \neq 0$, we will consider $s = P_{r-1} \nabla f(x^*)$

With the new s solution found, we will determine [3] :

$$\beta = \min_{\beta_j \geq 0} \left\{ \beta_j \middle| \beta_j = -\frac{g_j(x^*)}{\alpha_j s}, j = r+1, r+2, ..., m \right\} \tag{10 a}$$

and we will note $z = x^k + \beta$. If $s' \nabla f(x^*) \geq 0$ we shall consider $x^{k+1} = z$. In the contrary case ($s' \nabla f(x^*) \leq 0$) we will consider:

$$x^{k+1} = \frac{s'\nabla f(x^k)}{s'\nabla f(x^k) - s'\nabla f(z)} z + \left(1 - \frac{s'\nabla f(x^k)}{s'\nabla f(x^k) - s'\nabla f(z)}\right) x^k \tag{10 b}$$

With the new x^{k+1} we shall return to the second step. The logical scheme of the algorithm for the optimal distribution of resources for endowment of the handling fronts in the consolidation terminals applying the projected gradient method is represented in figure 1.

3. Numerical example

In this chapter we will provide a numerical example of the mathematical model of resource allocation for the endowment of handling fronts in the consolidation terminals. We consider the following complex machining process of two handling fronts that extends over a period of 4 years. The available funds are worth I = 4200000 €. The following variables are also known :

- annual depreciation rate for equipment type k $a_1 = 0.21$ and $a_2 = 0.15$;
- the equivalent value of a wagon - stationary hour in relation to the ton of goods loaded in the wagon $e_1 = 0.06$ € and $e_2 = 0.04$ € ;
- the value of a machine of the type with which the handling front k is equipped : $V_1 = 600000$ € and $V_2 = 900000$ € ;
- the operating productivity that is achieved in year i by the existing equipment at front k $q_{11} = q_{21} = q_{31} = q_{41} = 55$ tfh and $q_{12} = q_{22} = q_{32} = q_{42} = 65$ tfh ;
- working time utilization coefficient : $\lambda_0 = 0.55$.

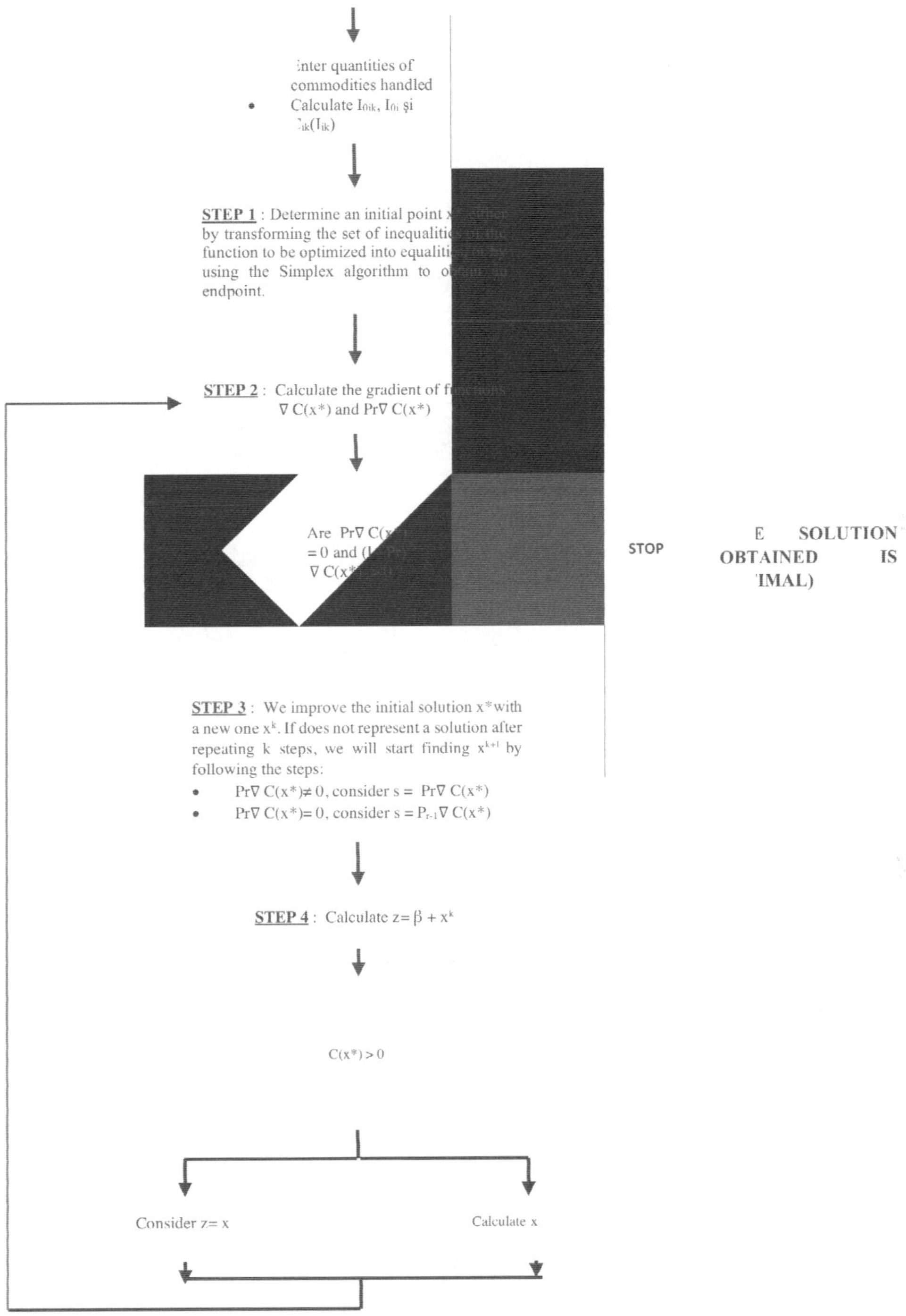

Figure 1. Projected gradient method to solve the distribution of resources for equipping handling fronts in consolidation terminals

The quantities to be handled on two fronts are centralized in the following table. The terms and indicate the quantities of goods handled during a recovery on each of the two handling fronts.

Based on the values in Table 1, the value of I_{0ik}, which represents the minimum level of resources available to one of the two handling fronts in year i, must be determined in order to be able to handle the quantities of goods.

Table 1. Quantities of commodities handled and minimum values of resources available on the two handling fronts

Year i	Front I		Front II	
	$Q_{i1} \times 10^{-3}$	$Q^r_{i1} = Q_{i1}/r_1$	$Q_{i2} \times 10^{-3}$	$Q^r_{i2} = Q_{i2}/r_2$
1	610	428	590	401
2	690	479	620	414
3	740	510	700	483
4	780	534	720	532

By applying Eq. (9), the bellow results will be achieved :

$$I_{011} = \frac{610000 \cdot 6 \cdot 10^5}{365 \cdot 3 \cdot 8 \cdot 55 \cdot 0.55} = 13.81 \cdot 10^5 €$$

$$I_{031} = \frac{740000 \cdot 6 \cdot 10^5}{365 \cdot 3 \cdot 8 \cdot 55 \cdot 0.55} = 16.76 \cdot 10^5 €$$

$$I_{012} = \frac{590000 \cdot 9 \cdot 10^5}{365 \cdot 3 \cdot 8 \cdot 65 \cdot 0.55} = 16.96 \cdot 10^5 €$$

$$I_{032} = \frac{700000 \cdot 9 \cdot 10^5}{365 \cdot 3 \cdot 8 \cdot 65 \cdot 0.55} = 20.12 \cdot 10^5 €$$

$$I_{021} = \frac{690000 \cdot 6 \cdot 10^5}{365 \cdot 3 \cdot 8 \cdot 55 \cdot 0.55} = 15.62 \cdot 10^5 €$$

$$I_{041} = \frac{780000 \cdot 6 \cdot 10^5}{365 \cdot 3 \cdot 8 \cdot 55 \cdot 0.55} = 17.67 \cdot 10^5 €$$

$$I_{022} = \frac{620000 \cdot 9 \cdot 10^5}{365 \cdot 3 \cdot 8 \cdot 65 \cdot 0.55} = 17.82 \cdot 10^5 €$$

$$I_{042} = \frac{770000 \cdot 9 \cdot 10^5}{365 \cdot 3 \cdot 8 \cdot 65 \cdot 0.55} = 22.13 \cdot 10^5 €$$

Therefore, the minimum resources required in each of the four years are:

$$I_{01} = I_{011} + I_{012} = 30.78 \cdot 10^5 €$$

$$I_{03} = I_{031} + I_{032} = 36.88 \cdot 10^5 €$$

$$I_{02} = I_{021} + I_{022} = 33.44 \cdot 10^5 €$$

$$I_{04} = I_{041} + I_{042} = 39.80 \cdot 10^5 €$$

It can be seen that the allocated investment fund ensures the mechanized handling of the entire quantity of goods ($I_{04} < I$). Therefore, the expression of the optimization criterion for the two handling fronts considered and for a four-year investment duration is:

$$C = 0.06\left(I_{11} + I_{21} + I_{31} + I_{41}\right) + 0.15\left(4I_1 + 3I_2 + 2I_3 + I_4\right) + \frac{121}{I_{11}} + \frac{156}{I_{21}} + \frac{179}{I_{31}} + \frac{200}{I_{41}} +$$

$$\frac{96}{I_1 - I_{11}} + \frac{107}{I_1 + I_2 - I_{21}} + \frac{136}{I_1 + I_2 + I_3 - I_{31}} + \frac{165}{I_1 + I_2 + I_3 + I_4 - I_{41}} \quad (11)$$

We are interested in obtaining a minimum value of the optimization criterion with the following restrictions:

$$\begin{cases} I_1+I_2+I_3+I_4\le 42 \\ 30.78\le I_1\le 42 \\ 0\le I_2\le 11.22 \\ 0\le I_3\le 8.56 \\ 0\le I_4\le 5.12 \end{cases} \quad \begin{array}{l} 13.82\le I_{11}\le I_1-16.96 \\ 15.62\le I_{21}\le I_1+I_{24}-17.82 \\ 16.76\le I_{31}\le I_1+I_2+I_{34}-20.12 \\ 17.67\le I_{41}\le I_1+I_2+I_3+I_4-22.13 \end{array} \tag{12}$$

To find the minimum of the expression C (which is nonlinear with respect to the variables I_i and I_{ik}), we can use for solving the projected gradient method:

$$\begin{cases} \dfrac{\partial C}{\partial I_1}=0.6-\dfrac{96}{(I_1+-I_{11})^2}-\dfrac{107}{(I_1+I_2-I_{21})^2}-\dfrac{136}{(I_1+I_2+I_3-I_{31})^2}-\dfrac{165}{(I_1+I_2+I_3+I_4-I_{41})^2} \\ \dfrac{\partial C}{\partial I_2}=0.45-\dfrac{96}{(I_1+-I_{11})^2}-\dfrac{107}{(I_1+I_2-I_{21})^2}-\dfrac{136}{(I_1+I_2+I_3-I_{31})^2}-\dfrac{165}{(I_1+I_2+I_3+I_4-I_{41})^2} \\ \dfrac{\partial C}{\partial I_3}=0.3-\dfrac{96}{(I_1+-I_{11})^2}-\dfrac{107}{(I_1+I_2-I_{21})^2}-\dfrac{136}{(I_1+I_2+I_3-I_{31})^2}-\dfrac{165}{(I_1+I_2+I_3+I_4-I_{41})^2} \\ \dfrac{\partial C}{\partial I_4}=0.155-\dfrac{96}{(I_1+-I_{11})^2}-\dfrac{107}{(I_1+I_2-I_{21})^2}-\dfrac{136}{(I_1+I_2+I_3-I_{31})^2}-\dfrac{165}{(I_1+I_2+I_3+I_4-I_{41})^2} \end{cases}$$

It can be noticed the fact that:

$$\frac{\partial^2 C}{\partial I_1^2}=\frac{\partial^2 C}{\partial I_2^2}=\frac{\partial^2 C}{\partial I_3^2}=\frac{\partial^2 C}{\partial I_4^2}=\frac{192}{(I_1+-I_{11})^3}+\frac{214}{(I_1+I_2-I_{21})^3}+\frac{272}{(I_1+I_2+I_3-I_{31})^3}+\frac{320}{(I_1+I_2+I_3+I_4-I_{41})^3}$$

It is necessary to specify the restrictions of the problem (corresponding to the investments made in each of the four years):
$$\begin{cases} I_1+I_2+I_3+I_4-42\le 0 \\ I_1-30.78\le 11.22 \\ I_2-11.22\le 0 \\ I_3-8.56\le 0 \\ I_4-5.12\le 0 \end{cases}$$

The system of restrictions can be written as a matrix that has the rank r = 4: $\begin{pmatrix} 1 & 1 & 1 & 1 \\ 1 & 0 & 0 & 0 \\ 0 & 1 & 0 & 0 \\ 0 & 0 & 1 & 0 \\ 0 & 0 & 0 & 1 \end{pmatrix}$.

The gradient of the objective function C is:

$$\nabla C=\begin{bmatrix} \dfrac{\partial C}{\partial I_1} \\ \dfrac{\partial C}{\partial I_2} \\ \dfrac{\partial C}{\partial I_3} \\ \dfrac{\partial C}{\partial I_4} \end{bmatrix}=\begin{bmatrix} 0.6-\dfrac{96}{(I_1+-I_{11})^2}-\dfrac{107}{(I_1+I_2-I_{21})^2}-\dfrac{136}{(I_1+I_2+I_3-I_{31})^2}-\dfrac{165}{(I_1+I_2+I_3+I_4-I_{41})^2} \\ 0.45-\dfrac{96}{(I_1+-I_{11})^2}-\dfrac{107}{(I_1+I_2-I_{21})^2}-\dfrac{136}{(I_1+I_2+I_3-I_{31})^2}-\dfrac{165}{(I_1+I_2+I_3+I_4-I_{41})^2} \\ 0.3-\dfrac{96}{(I_1+-I_{11})^2}-\dfrac{107}{(I_1+I_2-I_{21})^2}-\dfrac{136}{(I_1+I_2+I_3-I_{31})^2}-\dfrac{165}{(I_1+I_2+I_3+I_4-I_{41})^2} \\ 0.155-\dfrac{96}{(I_1+-I_{11})^2}-\dfrac{107}{(I_1+I_2-I_{21})^2}-\dfrac{136}{(I_1+I_2+I_3-I_{31})^2}-\dfrac{165}{(I_1+I_2+I_3+I_4-I_{41})^2} \end{bmatrix} \tag{13}$$

A point $M_1(I_1,I_2,I_3,I_4)$ must be chosen so that the conditions are met at the limit, namely: $M_1 = \begin{pmatrix} 30.78 \\ 3.74 \\ 3.74 \\ 3.74 \end{pmatrix}$. We must also keep in mind that compliance with the condition is also necessary: $P_4 \cdot \nabla C(M_1) = 0$. Because $A_5 = \begin{pmatrix} 1 & 1 & 1 & 1 \\ 1 & 0 & 0 & 0 \\ 0 & 1 & 0 & 0 \\ 0 & 0 & 1 & 0 \\ 0 & 0 & 0 & 1 \end{pmatrix}$ and the matrix transpose is $A_5' = \begin{pmatrix} 1 & 1 & 0 & 0 & 0 \\ 1 & 0 & 1 & 0 & 0 \\ 1 & 0 & 0 & 1 & 0 \\ 1 & 0 & 0 & 0 & 1 \end{pmatrix}$, we will obtain that $A_5 \cdot A_5' = \begin{pmatrix} 4 & 1 & 1 & 1 & 1 \\ 1 & 1 & 0 & 0 & 0 \\ 1 & 0 & 1 & 0 & 0 \\ 1 & 0 & 0 & 1 & 0 \\ 1 & 0 & 0 & 0 & 1 \end{pmatrix} \Rightarrow (A_5 \cdot A_5')^{-1} = \begin{pmatrix} 1 & -1 & -1 & -1 & -1 \\ -1 & 2 & 1 & 1 & 0 \\ -1 & 1 & 2 & 1 & 0 \\ -1 & 1 & 0 & 2 & 0 \\ -1 & 1 & 1 & 1 & 1 \end{pmatrix}$.

From this we can deduce that: : $(A_5 \cdot A_5')(A_5 \cdot A_5')^{-1} = \begin{pmatrix} 1 & 0 & 0 & 0 \\ 0 & 1 & 0 & 0 \\ 0 & 0 & 1 & 0 \\ 0 & 0 & 0 & 1 \end{pmatrix} = I_4 \Rightarrow P_4 = I_4 - (A_5 \cdot A_5')(A_5 \cdot A_5')^{-1} = I_4 - I_4 = 0 \Rightarrow P_4 \cdot \nabla C(M_1) = 0$, which means that the first condition is met. The second optimality condition that must be met is that the following product has all the positive components:

$$(A_5 \cdot A_5')^{-1} A_5 \cdot \nabla C(M_1) = \begin{pmatrix} 0 & 0 & 0 & 0 \\ 1 & 0 & 0 & -1 \\ 0 & 1 & 0 & -1 \\ 0 & 0 & 1 & -1 \\ 0 & 0 & 0 & 0 \end{pmatrix} \cdot \begin{pmatrix} -0.743 \\ -0.893 \\ -1.043 \\ -1.188 \end{pmatrix} = \begin{pmatrix} 0 \\ 0.445 \\ 0.295 \\ 0.145 \\ 0 \end{pmatrix} \quad (14)$$

Because all values $u_j \geq 0$ (where j=1,...,5), the algorithm stops at this step, which means that our basic initial solution $M_1 = \begin{pmatrix} 30.78 \\ 3.74 \\ 3.74 \\ 3.74 \end{pmatrix}$ is also the optimal one. By numerical substitution in equation (11), we obtain that the expression of the optimization criterion for the two handling fronts considered and for an investment duration of four years is worth $C_{optim} = 93{,}130 * 10^5$ €.

4. Conclusions

Due to trends in today's transport sector that requires an adequate planning of sustainable logistics processes in order to achieve successful business strategies, a correct evaluation of of the size of the elements underlying the design of all constructions and installations in the endowment of the terminals is necessary. In this paper, a mathematical framework for solving the problem of the distribution of resources for equipping handling fronts in rail consolidation terminals was developed.

The main objective of the problem is to solve the optimization criterion that takes into account the depreciation rate and the expenses related to the repair of the handling equipment and those of

IOP Conf. Series: Materials Science and Engineering 1037 (2021) 012061 doi:10.1088/1757-899X/1037/1/012061

stationary of the wagons in order to carry out the loading-unloading operations of the commodities. To find the minimum of the optimization criterion, the projected gradient method was used starting from an initial point. The coordinates were chosen so as to correspond to a set of restrictions based on the values of level of resources available to one of the three handling fronts in year i for handling a certain volume of commodity and the total value of the investment allocated for the m year for each of the fronts. The number of the iterations of the algorithm depends on the scale coefficients adopted.

Acknowledgment

<< This work has been funded by the European Social Fund from the Sectoral Operational Programme Human Capital 2014-2020, through the Financial Agreement with the title "*Scholarships for entrepreneurial education among doctoral students and postdoctoral researchers (Be Antreprenor!)*", Contract no. 51680/09.07.2019 - SMIS code: 124539.>>

References

[1] Costescu, D. 2012, *Asupra modelării reţelelor de transport intermodal*, Buletinul AGIR, Supliment 1/2012, pp. 22 – 27

[2] Florian, M., I. Constantin, and D. Florian 2009, *A new look at projected gradient method for equilibrium assignment*. Transportation Research Record: Journal of the Transportation Research Board, No. 2090, pp. 10–16

[3] Masgras, V. 2007, *Programare neliniară. Programare dinamică* (Bucharest, Fair Partners)

[4] Nowakowska – Grunt, J., Strzelczyk, M. 2019, *The current situation and the directions of changes in road freight transportation in the European Union*, Transportation Research Procedia 39, pp.350 – 359

[5] Raicu, Ş., Maşală, Ghe. 1981, *Transport feroviar. Funcţionare. Dezvoltare. Eficienţă* (Bucharest, Romania, Editura Ştiinţifică şi Enciclopedică) 225 – 230

IOP Conf. Series: Materials Science and Engineering 1037 (2021) 012062 doi:10.1088/1757-899X/1037/1/012062

The use of the genetic algorithms for optimizing public transport schedules in congested urban areas

A. Şerban[1*]

[1] Transport Engineering Department, University Politehnica of Bucharest, 313 Spl. Independentei, Bucharest, Romania

[*]Corresponding author's e-mail: serban.alin06@gmail.com

Abstract Transit operators face with travel time variability issues related to the design of new transit lines or related to the improvement of the performances of existing lines, especially when transit vehicles without exclusive line. Nowadays, the availability of large data through automated monitoring allows more in-depth this phenomenon of variability in timetables to be pointed out with recorded data. The lack of schedule stability confuses the traveling public and thus the degree of dissatisfaction increases. This leads to a decrease of the public transportation users, especially because of those users having travel alternatives to public transportation. The reliability of the travel time and the vehicle operation according to a stable timetable are the most relevant attributes for the users fidelisation. At the same time, it is well known that the demand for public transportation in congested networks has a large variability over space and time. In this paper, we analyse the deviations from the planned (declared and publicly) schedule for a tram line in Bucharest city in different stations, in diferrent times of the day and in different days of the week, in order to build an optimisation model for needed adjustments to the planned schedule. The data are collected with the Automatic Vehicle Location (AVL) system installed on trams' board. The number of the adjustments solutions is very large and the genetic algorithm is engaged for the optimisation model solving.

1. Introduction

A major impediment to scientific progress in many fields is the inability to make sense of the huge amounts of data that have been collected via experiment or computer simulation.

In public transportation the lack of schedule stability confuses the traveling public and thus the degree of dissatisfaction increases, leading in some cases to a decrease in the number of users of urban public transport, especially when they are not fully dependent on this transport system.

A traffic schedule must be correlated with traffic conditions. Starting from this principle, we analyzed the traffic schedule of tram line no. 10, during the peak periods of the day, in 4 days of the week (Monday, Wednesday, Friday and Sunday) between April 2019 and May 2020, to identify the size deviations and to propose solutions for improving the schedule for at least certain stations, where the highest values appear.

Reliable long-term travel time forecasts availability is one of the most relevant attributes for transit operators, to be used for improving transit service planning in terms of optimization of timetable and vehicle scheduling. In urban areas the demand for public transport is unevenly distributed over space and time [12].

Traffic congestion is one of the most annoying problems nowdays. It expresses the side effect for the many benefits obtained from urban agglomeration and economic activities. Due to the fact that the supply of land is limited, and the construction of road infrastructure is not necessarily cheap, it would be expensive to invest in order to reach the capacity to permanently ensure traffic regimes close to the free one. Even if at present, assuming that, through the development of infrastructure, this regime would have achieved, over time, the stimulating effects on demand will certainly again lead to congestion [12].

Public passenger transport must take over the most important flows of systematic travel to/from work, and the attractiveness of public transport is represented, in particular, by the observance of the displayed/planned traffic schedule, in each station within the transport network.

Strathman et al. point out that by changing traffic patterns, delays caused by traffic congestion in the case of public transport can be integrated into business plans, thus reducing the impact on deviations and delays. On the other hand, there are factors that need to be addressed in real time due to the unpredictability and impact on the service. This group of factors is more difficult to manage, even in terms of the operator's ability to operate quickly [15].

According to several authors (Pignataro, 1973; Gazis, 1974; Bell, 1997), congestion is defined by prolonging the travel time in relation to free flow conditions. The maximum level of congestion occurs when this delay exceeds the threshold of accepted limits (norms) [12].

The conditions in which this type of congestion occurs must be analyzed differently depending on the economic and natural space that determines different behaviors and perceptions of travelers.

The travel times used to determine the degree of congestion can be observed (relevant) or estimated, if they cannot be observed. Of course, the observed data are those that allow assessments of the dynamics of congestion on a certain portion, the spatial determination of blockages or the efficiency of traffic management systems and measures to reduce polluting effects [12].

The traffic schedule for each transmission line must meet two requirements: to ensure the necessary transport capacity and to ensure a minimum frequency of service (for the maximum tracking interval) necessary in terms of the level of service.

The information required to carry out a program for a transport line is given by the estimated volume of passengers and their distribution in time and space.[1]

The paper presents a solution for optimizing public transport schedules that is working with genetic algorithms. GA is a local search algorithm, which works starting from an initial collectionof strings (or a population) representing possible solutions of the problem. Each string of the population is called a *chromosome*, and has associated a value called *fitness function* that contributes in the generation of new populations by means of geneticoperators (denoted *reproduction*, *crossover* and *mutation*, respectively). At each generation, the algorithm uses the fitness function values to evaluate the survival capacity of each string *i* of the population using simple operators in order to create a new population which try to improve on the current fitness function values by using pieces of the oldest ones [10].

2. Literature review

There are many genetic algorithms applied to various parts of transportation research since it was introduced in 1975, and it is expanding more and more its boundary. Several utilizations of GA in transport and traffic have been reviewed and the most important ones synthesized as follows.

The travel times considering the degree of congestion can be observed or estimated. Of course, the observed data are those that allow the assessment of the dynamics of congestion on a certain part of road, the location of traffic jams or the efficiency of traffic management systems and measures to reduce polluting effects [12].

[1]Dinu, O.A., *Proiectarea programelor de circulaţie ale transportului public în condiţiile neuniformităţilor spaţiale şi temporale ale cererii*, Buletinul AGIR, 2009

The accurate prediction of travel time in congestion conditions is difficult to provide, especially due to the very dynamically changing of the traffic flow.

In such cases, exhaustive and random search techniques are useful. The biggest disadvantage of these methods is that they often require thousands of function evaluations, even for the simplest functions to reach an optimal level. The genetic algorithms proposed by J. Holland are based on exhaustive and random search techniques, which prove to be robust for optimizing nonlinear and non-convex functions [13].

Most recent studies using GAs in *transit network problem can be found for optimization of route network design* in which bus scheduling is usually a part of route optimization function.
Development of optimal schedules is an extremely difficult task even for small transit network.The difficulty arises because of large number of variables and constraints. The discrete nature of variables and non-linearity involved in the objective function and the constraints further increase the complexity and computational burden. Chakroborty et al. highlighted similar type of problem. Even after linearizing the problem the complexity remains very large. The benefit obtained through linearization is offset by the increase in the number of variables and constraints [13].

Park S.J has focused to develop new computerized optimization models for the bus scheduling problem in a coordinated bus network, namely, new models to find optimized *bus headways and slack times*. In the new models bus headways are optimized first under the deterministic arrival process, and then slack times are introduced to optimization problem for the stochastic arrival processes with a dispatching strategy that a bus arriving at transfer center earlier than its departure time leaves on time, regardless of the arrival time of passengers transferring from other routes [10].

3. Methodology of research

3.1. Study area

For the study presented in this paper, there has been selected a tram line (line no. 10) which make the connection of the southern area of Bucharest with the western area, the section with the ends of Romprim (station 101**1**) and Banu Manta Blvd (station 101**24**) (figure 1). The trams are monitored on the entire route, when passing through the stations, where they must make stops to pick up passengers.

Line 10 takes over 34000 passengers on average every day, along the route being transfer points to other urban surface lines, but also to subway lines.

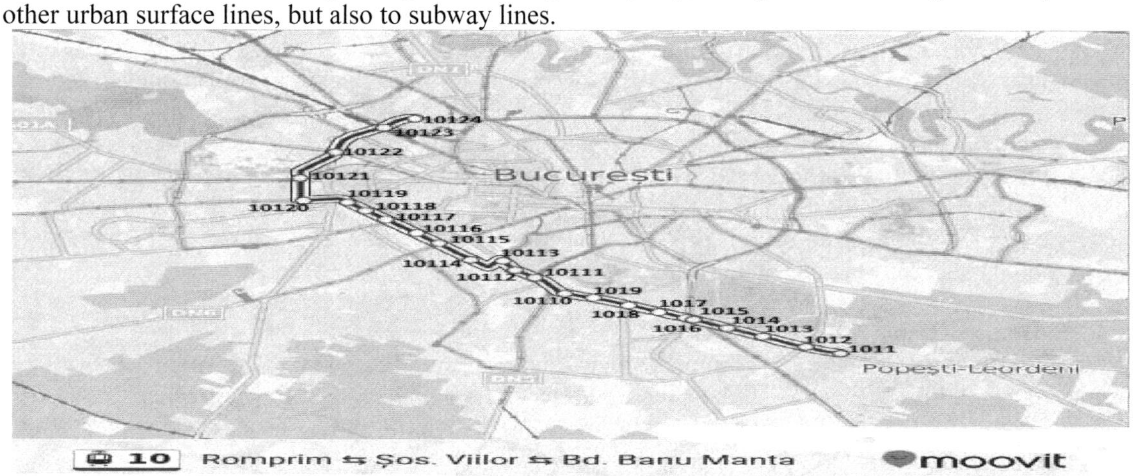

Figure 1. Line no. 10 route (Source: www.moovitapp.com)

Characteristic data of the selected line

Table 1. The time required to cover the 23 stations of line no 10 for a typical working day/for a weekend day

Current number	Name of stations/code	Distance [m]	Average travel time between stations for outward journey/return journey[min]	Average travel time between stations for outward journey/return journey during peak periods [min]
1	Romprim-Şoseaua Vitan Bârzeşti/1011-1012	520	2.15/2.09	2.35/2.29
2	Şoseaua Vitan Bârzeşti-Stadion/1012-1013	691	2.86/2.78	3.12/3.04
3	Stadion-Lunca Bârzeşti/1013-1014	520	2.15/2.09	2.35/2.29
4	Lunca Bârzeşti-Piaţa Sudului/1014-1015	531	2.20/2.14	2.40/2.34
5	Piaţa Sudului-Piaţa Sudului/1015-1016	127	0.53/0.51	0.57/0.56
6	Piaţa Sudului-Huşi/1016-1017	386	1.60/1.55	1.74/1.70
7	Huşi-Bd. Constantin Brâncoveanu/1017-1018	462	1.91/1.86	2.08/2.03
8	Bd. Constantin Brâncoveanu-Opriş Ilie/1018-1019	515	2.13/2.07	2.32/2.27
9	Opriş Ilie-Cimitirul Şerban Vodă/1019-10110	379	1.57/1.52	1.71/1.67
10	Cimitirul Şerban Vodă-Piaţa Eroii Revoluţiei/10110-10111	712	2.94/2.86	3.21/3.13
11	Piaţa Eroii Revoluţiei-Ştefan Hepites/10111-10112	388	1.60/1.56	1.75/1.71
12	Ştefan Hepites-Dr. Constantin Istrati/10112-10113	354	1.46/1.42	1.60/1.56
13	Dr. Constantin Istrati-Spătarul Preda/10113-10114	664	2.75/2.67	3.00/2.92
14	Spătarul Preda-Calea Rahovei/10114-10115	711	2.94/2.86	3.21/3.13
15	Calea Rahovei-Inox/10115-10116	414	1.71/1.66	1.87/1.82
16	Inox-Calea 13 Septembrie/10116-10117	608	2.51/2.45	2.74/2.67
17	Calea 13 Septembrie-Cartier Panduri/10117-10118	368	1.52/1.48	1.66/1.62
18	Cartier Panduri-Piaţa Danny Huwe/10118-10119	481	1.99/1.93	2.17/2.12
19	Piaţa Danny Huwe-Bd. Vasile Milea/10119-10120	579	2.39/2.33	2.61/2.55
20	Bd. Vasile Milea-Bd. Iuliu Maniu/10120-10121	756	3.13/3.04	3.41/3.33
21	Bd. Iuliu Maniu-Pod Grozăveşti/10121-10122	1011	4.18/4.07	4.56/4.45
22	Pod Grozăveşti-Pasajul Basarab/10122-10123	1115	4.61/4.48	5.03/4.90
23	Pasajul Basarab-Bd. Manu Manta/10123-10124	515	2.13/2.07	2.32/2.27

-the *commercial speed* on line no 10, in a *working day*, is 14.51 km/h and the speed considered in the peak periods is 13.30 km/h, so a decrease of 8.4% when the degree of congestion increases.
-the *frequency of scheduled departures*, from the station 1011 to 10124, in the time interval 7^{00}-8^{00} am, has a value of 6.62 minutes and the scheduled duration (for the morning peak) to go through the 23 stations mentioned in table 1 is 53 minutes.
-the commercial speed on line 10, in a weekend day, is 14.92 km/h and the speed considered in the peak periods is 13.64 km/h, so a decrease of 8.6% when the degree of congestion increases.

Table 2. Line no 10 indicators (Source: the public transport operator STB SA)

	Working day	Weekend
Average speed	14.51 km/h	14.92 km/h
Average speed in peak periods	13.30 km/h	13.64 km/h
Average used transport capacity	1791 passengers/h	1531 passengers/h
Maximum capacity	2241 passengers/h	1915 passengers/h

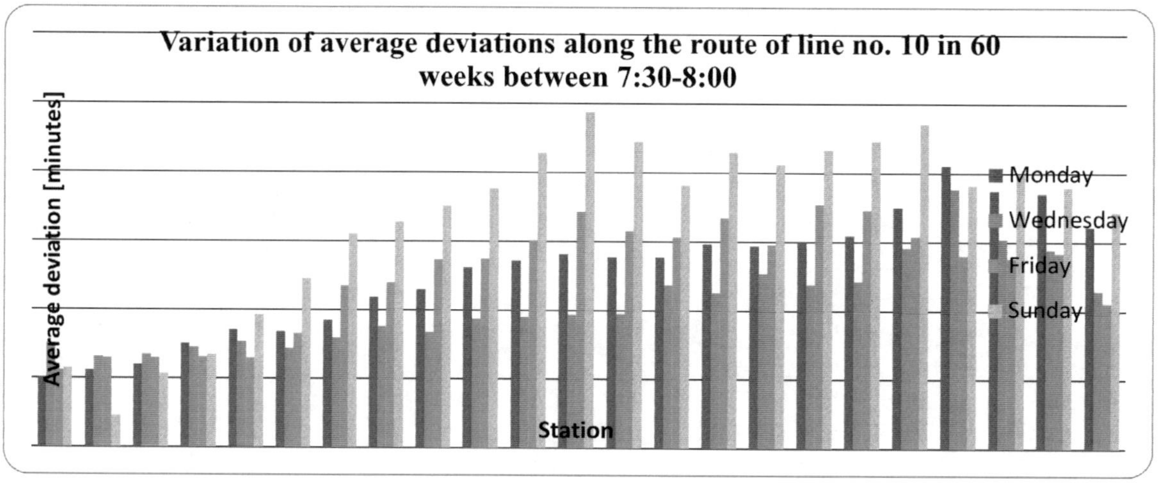

Figure 2. Evolution of average deviations along the line no. 10 in 60 weeks (Source: Author)

The highest values of deviations are recorded in the connection stations (Dr. Constantin Istrati – station 101**12**, Cartier Panduri – station 101**17**, Piața Danny Huwe – station 101**18** – these stations are located at the edge of heavily trafficked intersections, where frequent blockages occur during peak hours). In addition, it can be noticed that the biggest deviations are recorded on Sundays, because due to a smaller number of users and a lighter road traffic, the vehicles travel in advance along the entire route.

3.2. Adjustments of planned schedule with GA

A genetic algorithm performs specific operations in a reproductive process governed by genetic operators. The new solutions are created by selecting and recombining existing chromosomes, in order to optimize an evaluation function specific to each problem. The significance of this function is not relevant to the algorithm, what matters is only its value.

They generally start from a randomly generated population of chromosomes, and each new population generated by reproduction partially or completely replaces the previous generation. The global evaluation function is moving towards the optimal and offers better and better solutions to the problem. The process is analogous to the neo-Darwinian theory of biological evolution, which states that organisms continuously adapted to environmental changes have the highest chances of survival [7], [11].

To find the optimal solution to a problem, a genetic algorithm starts with a given or randomly generated set of potential solutions and improves it by going through a number of iterations (generations), the performance of each solution is evaluated through the fitness function [3].
The population of chromosomes will be:

$C_1 = x_{1,j_1}, x_{1,j_2}, \dots, x_{1,j_{25}}$

$C_2 = x_{2,j_1}, x_{2,j_2}, \dots, x_{2,j_{25}}$

....

$C_{10} = x_{10,j_1}, x_{10,j_2}, \dots, x_{10,j_{25}}$

$C_{11} = x_{11,j_1}, x_{11,j_2}, \dots, x_{11,j_{25}}$

....

$C_n = x_{n,j_1}, x_{n,j_2}, \dots, x_{n,j_{25}}$

(1, n) = (1; 60) = (1.04.2019; 22.05.2020)

Fitness function f(x) = $X_{planned}$ - X_{real}, X = arrival time in the station

$f(x_i) = \Delta X_i$

We establish the following:
-the chromosomes used have a constant length;
-the number of chromosomes is constant.

Each chromosome (vector element) represents a potential solution to the problem and it is implemented in the form of an S data structure. Each solution is evaluated as a measure of "fitness" or life expectancy. This fitness represents the quality of the chromosome. Usually, the more promising the chromosome, the higher the fitness. In the case of our problem, fitness function must be minimized in the idea of optimizing the traffic schedule for line no. 10.

A new population (iteration t+1) is formed by selecting several individual matches, choosing the most promising chromosomes (selection step) from the current population (from iteration t). Some of the newly formed members of the population undergo transformations (the step of modification) through the "genetic" operation of the new solutions.

The number of iterations is essential in achieving the optimization mechanism. This is justified by the fact that only after a certain number of iterations the values of the fitness function tends to a constant, which means that the best solution of the latest generation is very close to the optimal solution. [14]

The following describes how the experiment is performed and the results obtained for a run are presented.

The parameters of the genetic algorithm used are as follows: population sizePop_{size}= 60 (same for each type of day).

Objective function:

minimize {M $[\sum_{n=1}^{60} |x_{n,j_1}|,|x_{n,j_2}|,\ldots,|x_{n,j_{23}}|]$; W$[\sum_{n=1}^{60} |x_{n,j_1}|,|x_{n,j_2}|,\ldots,|x_{n,j_{23}}|]$; F$[\sum_{n=1}^{60} |x_{n,j_1}|,|x_{n,j_2}|,\ldots,|x_{n,j_{23}}|]$}

x_{n,j_i} – deviation in n day and in station j_i; $i = \overline{1,\ 23}$, $n = \overline{1,\ 60}$

M – Monday, W – Wednesday, F - Friday

The initial population was created as follows: a chromosome was considered as the string formed by the values obtained by the difference between the planned time and the actual time of passage through each station, being a total of 23 stations, meaning the string will have a length of 23 bits represented by real positive and negative values.

The population comprises 60 chromosomes, each corresponding to a day of Monday/Wednesday/Friday/Sunday, during 60 weeks, starting with 01.04.2019.

Following the evaluation, the best individual was determined as C_5, and the weakest as C_{33} - for Mondays;C_{21} - the best and C_{34} the weakest for Wednesdays; C_{35} – the best and C_{30} the weakest for Fridays; C_{46} – the best and C_{23} the weakest for Sundays.

$P_{t+1} = \{C_1, C_2, \ldots, C_{60}\}$

Table 3. The initial population of deviations obtained for Monday

	S_1	S_2	S_3	S_4	S_5	S_6	S_7	S_8	S_9	S_{10}	S_{11}	S_{12}	S_{13}	S_{14}	S_{15}	S_{16}	S_{17}	S_{18}	S_{19}	S_{20}	S_{21}	S_{22}	S_{23}
C_1 F(x)=109.06	-1.41	-0.68	-0.14	-2.37	-3.84	-3.57	-3.97	-5.26	-5.12	-5.74	-5.56	-5.22	-5.42	-6.28	-7.05	-7.88	-8.12	-8.17	-8.71	-9.70	-8.18	-8.03	-5.22
C_2 F(x)=49.75	-1.38	-2.21	-1.76	-2.14	-3.44	-3.02	-2.62	-2.13	-2.67	-1.96	-2.06	-1.64	-2.57	-2.76	-2.66	-2.02	-1.85	-1.79	-1.12	-4.07	-2.42	-2.57	-2.05
C_3 F(x)=111.33	-1.59	-1.20	-0.71	-2.74	-5.58	-5.04	-4.48	-5.01	-5.14	-6.56	-6.48	-7.07	-6.90	-7.41	-7.40	-7.20	-7.37	-7.14	-7.67	-7.77	-6.30	-5.15	-4.31
⋮	⋮	⋮	⋮	⋮	⋮	⋮	⋮	⋮	⋮	⋮	⋮	⋮	⋮	⋮	⋮	⋮	⋮	⋮	⋮	⋮	⋮	⋮	

C_{58} F(x)=20.38	0.02	0.47	1.19	1.63	0.06	0.34	0.68	0.51	0.60	0.36	0.57	0.96	1.20	1.49	1.47	1.72	1.80	1.80	1.34	-2.13	-0.33	1.75	1.01
C_{59} F(x)=23.31	-2.09	-1.81	-0.97	-0.30	-1.21	-0.96	-0.38	0.07	0.36	-0.36	-0.03	0.49	0.73	0.07	0.47	0.50	0.61	0.81	1.31	1.98	3.38	3.28	2.91
C_{60} F(x)=15.78	-1.08	-0.63	-0.19	0.61	0.12	0.48	0.85	0.12	0.03	0.58	0.69	0.99	0.50	1.14	1.39	1.05	1.18	1.30	-1.46	-2.05	-0.87	0.58	0.41

For Mondays, given the values of the fitness function, we will establish as a condition that value of f(x) is not greater than 66, so that 14 chromosomes will be removed before the crossover operation.

The blue color indicates the minimum value of deviation in the respective chromosome and the yellow color indicates the maximum value.

The value of the fitness function for each chromosome in the initial population is calculated as follows:

Each chromosome is made up of a series of values of the deviations in each station, from the respective day with the departure time, initially established. These values were summed, the negative ones being taken in the module:

F(x) for C_1 is $\sum |x_{1,j_1}|, |x_{1,j_2}|, \ldots, |x_{1,j_{23}}| = 1.41 + 0.68 + \ldots + 5.22 = 109.6$

F(x) for C_{60} is $\sum |x_{60,j_1}|, |x_{60,j_2}|, \ldots, |x_{60,j_{23}}| = 1.08 + 0.63 + \ldots + 0.41 = 15.78$

Table 4. The initial population of deviations obtained for Wednesday

	S_1	S_2	S_3	S_4	S_5	S_6	S_7	S_8	S_9	S_{10}	S_{11}	S_{12}	S_{13}	S_{14}	S_{15}	S_{16}	S_{17}	S_{18}	S_{19}	S_{20}	S_{21}	S_{22}	S_{23}
C_1 F(x)=36.96	-0.74	-0.41	-0.02	-0.79	-1.14	-0.67	-0.35	-0.19	-0.30	-1.56	-1.35	-0.97	-1.02	-2.79	-2.68	-2.38	-2.25	-2.00	-3.24	-4.72	-3.32	-2.35	-2.00
C_2 F(x)=23.17	-1.11	-0.26	-0.09	-1.19	-2.19	-1.87	-1.70	-1.58	-1.72	-1.66	-1.66	-1.22	-0.50	-0.63	-0.51	0.12	0.06	-0.05	-1.09	-2.45	-1.13	0.12	0.25
C_3 F(x)=38.62	-0.24	-0.31	0.13	-2.84	-2.94	-2.72	-2.38	-2.78	-2.45	-2.16	-1.80	-1.32	-1.05	-2.23	-2.30	-2.10	-1.79	-1.80	-1.47	-0.90	-0.17	1.63	1.11
⋮	⋮	⋮	⋮	⋮	⋮	⋮	⋮	⋮	⋮	⋮	⋮	⋮	⋮	⋮	⋮	⋮	⋮	⋮	⋮	⋮	⋮	⋮	⋮
C_{58} F(x)=104.59	-4.59	-5.08	-4.86	-5.04	-6.03	-5.67	-5.40	-4.86	-4.97	-5.36	-5.23	-4.91	-4.82	-5.26	-5.00	-4.30	-4.22	-4.35	-4.06	-4.73	-3.00	-1.42	-1.45
C_{59} F(x)=67.55	-3.83	-2.98	-2.47	-1.99	-2.81	-2.39	-2.07	-1.94	-1.84	-2.82	-2.70	-2.39	-2.77	-3.61	-3.40	-3.25	-3.27	-3.19	-4.32	-4.92	-3.42	-2.82	-2.37
C_{60} F(x)=75.31	-3.63	-3.01	-2.42	-2.24	-3.53	-3.16	-3.37	-3.46	-3.34	-3.99	-3.86	-3.57	-3.12	-3.88	-3.60	-3.00	-2.87	-2.84	-3.01	-5.38	-3.77	-2.30	-1.98

Table 5. The initial population of deviations obtained for Friday

	S_1	S_2	S_3	S_4	S_5	S_6	S_7	S_8	S_9	S_{10}	S_{11}	S_{12}	S_{13}	S_{14}	S_{15}	S_{16}	S_{17}	S_{18}	S_{19}	S_{20}	S_{21}	S_{22}	S_{23}
C_1 F(x)=36.05	-0.26	0.60	0.98	0.16	-1.03	-0.49	-0.20	-0.58	-0.30	-1.26	-1.31	-1.69	-2.37	-2.94	-3.28	-2.48	-2.54	-2.52	-3.64	-3.72	-2.38	-0.85	-0.77
C_2 F(x)=47.62	-2.18	-1.56	-1.21	-1.35	-2.34	-2.07	-1.97	-2.13	-2.04	-1.96	-2.00	-1.51	-2.25	-3.04	-2.81	-2.88	-2.60	-2.34	-2.74	-3.22	-2.10	-0.62	-0.71
C_3 F(x)=16.08	-1.69	-0.91	-0.22	0.50	0.11	0.48	0.92	0.36	0.61	0.99	1.14	1.48	0.66	-0.18	0.00	0.62	0.83	1.05	0.64	-1.00	0.03	1.32	0.35
⋮	⋮	⋮	⋮	⋮	⋮	⋮	⋮	⋮	⋮	⋮	⋮	⋮	⋮	⋮	⋮	⋮	⋮	⋮	⋮	⋮	⋮	⋮	⋮
C_{58} F(x)=56.13	0.76	1.62	2.31	2.96	1.66	2.33	2.68	2.91	3.30	2.98	3.25	3.59	3.40	3.41	3.52	2.12	2.51	2.61	0.79	0.50	2.33	3.37	1.22
C_{59} F(x)=83.18	0.84	1.60	2.33	2.81	2.07	2.53	2.85	3.14	3.50	3.43	3.55	3.93	4.03	3.72	4.09	4.72	5.00	4.98	4.68	4.57	5.12	5.60	4.11
C_{60} F(x)=59	-0.57	-0.43	0.24	1.26	1.04	1.46	2.28	2.89	3.36	3.09	3.25	3.66	3.28	3.47	3.57	3.68	3.78	3.91	2.46	1.25	2.98	3.87	3.19

Table 6. The initial population of deviations obtained for Sunday

	S_1	S_2	S_3	S_4	S_5	S_6	S_7	S_8	S_9	S_{10}	S_{11}	S_{12}	S_{13}	S_{14}	S_{15}	S_{16}	S_{17}	S_{18}	S_{19}	S_{20}	S_{21}	S_{22}	S_{23}
C_1 F(x)=151.82	2.49	3.47	4.18	4.48	4.47	4.83	5.50	6.06	6.53	6.83	7.10	7.54	8.10	8.06	8.59	9.17	9.20	9.00	7.26	5.63	6.88	8.95	7.50
C_2 F(x)=58.94	0.56	1.20	1.74	1.83	1.47	1.84	2.33	2.19	1.81	1.84	1.95	2.06	1.90	2.52	2.72	2.98	2.91	3.05	3.98	2.97	4.82	5.70	4.55
C_3 F(x)=152.07	1.40	1.93	2.77	2.83	4.53	5.35	6.10	6.35	6.93	7.50	8.27	9.02	7.78	6.25	7.05	8.28	9.05	9.28	10.58	9.67	9.02	6.72	5.40
⋮	⋮	⋮	⋮	⋮	⋮	⋮	⋮	⋮	⋮	⋮	⋮	⋮	⋮	⋮	⋮	⋮	⋮	⋮	⋮	⋮	⋮	⋮	⋮
C_{58} F(x)=89.49	-1.07	-0.87	-0.12	0.22	2.02	3.00	3.75	4.45	5.58	4.77	5.55	6.55	6.32	4.23	4.97	5.97	6.65	5.98	7.52	3.73	2.42	2.02	1.76
C_{59} F(x)=113.56	0.33	0.52	1.08	0.78	3.10	4.08	5.05	5.35	5.78	6.63	7.45	8.38	7.73	6.57	7.32	4.98	5.83	6.03	6.98	5.70	5.47	4.58	3.81
C_{60} F(x)=102.24	-0.10	0.05	0.72	1.17	3.58	4.37	4.98	5.13	6.02	5.33	6.00	7.02	5.53	4.60	5.18	5.52	6.33	6.58	6.60	5.35	4.48	4.10	3.49

The procedure is generative. It makes use of three main operators; reproduction, crossover and mutation. Each generation of a genetic algorithm consists of a new population produced from the previous generation.

For Mondays, given the values of the fitness function, we will establish as a condition that F(x) is not greater than 66, so that 14 chromosomes will be removed before the crossover operation.

For Wednesdays, given the values of the fitness function, we will establish as a condition that F(x) is not greater than 66, so that 13 chromosomes will be removed before the crossover operation.

For Fridays, given the values of the fitness function, we will establish as a condition that value of F(x) is not greater than 66, so that 21 chromosomes will be removed before the crossover operation.

For Sundays, given the values of the fitness function, we will establish as a condition that value of F(x) is not greater than 88, so that 29 chromosomes will be removed before the crossover operation.

In the table below I have centralized (for working days) the results obtained by crossing selected chromosomes (two by two) from the initial population of solutions generated by the function: f(x) = $X_{planned} - X_{real}$

Table 7. The results obtained after the first crossing operation (workday)

R_{1M} F(x)=26	-1.38	-2.21	-1.54	-1.34	-1.73	-1.29	-0.73	-0.54	-0.19	-0.71	-0.48	-0.12	0.40	0.19	0.60	1.00	1.20	1.36	-1.12	1.45	-2.42	-2.57	-1.43
R_{1W} F(x)=15.61	-0.74	-0.26	-0.26	-0.79	-1.14	-0.67	-0.35	-0.19	-0.30	-1.56	-1.35	-0.97	-0.50	-0.63	-0.51	0.12	0.06	-0.05	-1.09	-2.45	-1.13	0.12	0.37
R_{1F} F(x)=33.17	-0.26	0.60	0.98	0.16	-1.03	-0.49	-0.20	-0.58	-0.30	-1.26	-1.31	-1.51	-2.25	-2.94	-2.81	-2.48	-2.54	-2.34	-2.74	-3.22	-2.10	-0.62	-0.45
⋮	…	…	…	…	…	…	…	…	…	…	…	…	…	…	…	…	…	…	…	…	…	…	…
R_{22M} F(x)=12.08	-1.08	-0.63	-0.19	-0.30	0.12	0.48	-0.38	0.12	0.03	-0.36	-0.03	0.49	0.50	0.07	0.47	0.50	0.61	0.81	1.31	1.98	-0.87	0.58	0.17
R_{22W} F(x)=15.29	-1.61	-0.65	0.01	0.60	-0.14	0.29	0.98	0.86	0.86	-0.64	-0.64	0.11	0.78	1.37	1.80	0.27	0.58	0.75	-0.26	-1.00	0.17	0.53	0.39
R_{19F} F(x)=24.40	0.76	1.62	-2.04	-1.87	1.66	2.33	-1.82	-1.49	-1.07	-0.94	-0.70	-0.36	0.20	0.41	0.69	0.48	0.65	0.65	-0.31	0.50	0.05	2.25	1.55

The fitness value of each result obtained can be found in the first column, and depending on this value, the following chromosomes will be chosen for the crossing operation.

The crossover operation was performed as follows: the elements of the parent chromosomes were compared and the smallest bit was selected. (if $|x_{1,j_1}|<|x_{2,j_1}|$, choose x_{1,j_1})

4. Conclusions

Considering the characteristic data of the selected tram line, as well as the traffic analysis in the period 01.04.2019 – 22.05.2020, the conclusions are as follows:

the time needed to go through each station was calculated both during peak periods and in the rest of the day. The resulting values are often exceeded by the values of the deviations from the planned program.

The planned program for this selected line, displayed on the official website of the public transport operator of Bucharest (STB SA) was over 2 years old, which leads us to think that this program was no longer correlated with current traffic conditions, where the degree of congestion increased by over 4% compared to 2018 [18].

The largest deviations are generally recorded in stations 101**16**-101**20**, and some values may decrease towards the last station 101**24**, others increase to the end and in this case it is not possible for the respective tram to recover the time lost on the route.

Related to the GA involving for the adjustments of the planned schedules, the main conclusions are as follows:

Regarding the use of genetic algorithms in solving the studied problem, the following can be said:

- there are many solutions to the problem, and identifying one needs to be proven as overall optimal in future research. Therefore, up to now the involved genetic algorithm are used to obtain at least an optimal solution;
- a wide range of genetic algorithms can be designed to solve a problem. There are thus difficulties in choosing the most appropriate genetic algorithm, the literature providing too little information in this regard.

There are no strict rules to identify situations in which genetic algorithms are the most appropriate option to solve a problem, but we chose this method to optimize the traffic schedule of a public transport

IOP Conf. Series: Materials Science and Engineering 1037 (2021) 012062 doi:10.1088/1757-899X/1037/1/012062

line because: the search space for solutions is large enough that an exhaustive search would be virtually impossible; it is not necessary to determine the global optimum, but to find a good enough solution in a relatively short time.

References

[1] Chambers, L.D., 1999, *"Practical Handbook of Genetic Algorithms"*.
[2] Dediu, A.H., 2001, *Genetic Algorithms* (in Romanian: Algoritmi genetici), Editura Printech București.
[3] Dumitrescu, D., 2000, *Genetic algorithms and evolutionary strategies – Applications in artificial intelligence and related fields* (in Romanian: Algoritmi genetici și strategii evolutive – Aplicații în inteligența artificială și în domenii conexe), Editura Albastră, Cluj-Napoca.
[4] Goldberg, D.E., 1989, *Genetic Algorithms in Search, Optimization and Machine Learning*, Addison-Wesley, Reading, MA.
[5] Laszlo, D., 2000, *Neural networks and Fuzzy logic in automation* (in Romanian: Rețele neuronale și logica Fuzzy în automatizări), Editura Universității "Petru Maior", Târgu-Mureș.
[6] Luchian, H., Breabăn, M., 2007, *Genetic Algorithms* (in Romanian: Algoritmi genetici), Universitatea "Alexandru Ioan Cuza".
[7] Michalewicz, Z., 1996, "*Genetic Algorithms + Data Structures = Evolution Programs*", Springer-Verlag.
[8] Mitchell, M., 1997, "*An introduction to Genetic Algorithms*", MIT Press.
[9] Oltean, M., *Design and implementation of algorithms* (in Romanian: Proiectarea și implementarea algoritmilor), Computer Libris Agora, Cluj-Napoca, 2000.
[10] Park, S.J., 2005, *Bus Network Scheduling with Genetic Algorithms and Simulation*, Faculty of the Graduate School of the University of Maryland, College Park.
[11] Popa, R., 1998, *Research on the development of new methods for testing numerical structures* (in Romanian: "Cercetări privind elaborarea unor noi metode de testare a structurilor numerice"), Teza de doctorat, Universitatea "Dunărea de Jos" din Galați.
[12] Raicu S, 2004, *Transport systems* (in Romanian: Sisteme de transport), Editura AGIR.
[13] Shrivastava, P., Dhingra, S.L., Gundaliya, P.J., *Application of Genetic Algorithm for Scheduling and Schedule Coordination Problems*, Journal of Advanced Transportation, Vol.36 No. I , pp. 23-41
[14] Silva, C., Sousa, J., Runkler, T., 2008, *Rescheduling and Optimization of Logistics Process Using GA and ACO*, Journal of Engineering Application of Artificial Intelligence, Vol.21, No.3.
[15] Strathman, J. G., Dueker, K. J., Kimpel, T. J., Gerhart, R.L., Turner, K., Taylor, P., 1999, *Automated Bus Dispatching,Operations Control, and Service Reliability*, Transportation Research Record, Issue 1666, pag. 28-36.
[16] Xiong, Y.E. and Schneider, J.B., 1993, *Transportation network design using a cumulative genetic algorithm and neural network*. Transportation Research Record, 1364, 37-44.
[17] Zaharia, C.N., 2002, *Genetic algorithms and neural networks* (in Romanian: Algoritmi genetici și rețele neuronale), Editura Academiei Române, București.
[18] https://www.tomtom.com/en_gb/ accessed in 10.08.2020

Acknowledgment: This work has been funded by the European Social Fund from the Sectoral Operational Programme Human Capital 2014-2020, through the Financial Agreement with the title "Scholarships for entrepreneurial education among doctoral students and postdoctoral researchers (Be Antreprenor!)", Contract no. 51680/09.07.2019 – SMIS code: 124539

Software application for uniform quantitative evaluation of social economy projects

B Florea[1], D Marcu[1*], A Ioana[1], A Semenescu[1], R Solea[1], G Iacob[1] and F Niculescu[1]
[1] University Politehnica of Bucharest, Engineering and Management of Metallic Material Department, Splaiul Independenţei Street No. 313, Romania

[*]Corresponding author's e-mail: dragos.marcu@arasnet.ro

Abstract. The qualitative impact calculation for the implementation of a social economy project can be done using dedicated computing software. In this paper we have defined PQS which is an application on JAVA support that allows use irrespective of the existing support platform on the computer. The created application has the advantage that it can run both on computers and on other devices (tablets, phones, etc.). The input data are represented by the values assumed and realized in the project and are stored in a MySQL database.

1. Introduction

The advantage of using the software created and presented in this article is that it allows the automatic loading of data taken from different databases by automatically calculating a complex indicator.
The evaluation of the degree of accomplishment of the different projects, from different fields of activity can be done in this way in a much simpler and at the same time transparent way.
The programming language used in the development of the application allows the integration of the application in the management system of the Structural and Cohesion Funds, thus being able to become a valuable tool for real-time verification of the project realization stage.
Such an application can be useful both for the managing authorities and for the beneficiaries and project implementers.
The great advantage is the possibility of application in heterogeneous fields of activity, allowing comparisons even between projects from totally independent fields.
The managing authorities in Romania, at present, manually and non-uniformly evaluate the implementation of the projects. By achieving a complex indicator and digitizing the calculation mechanism, a much more objective and correct evaluation is allowed, both static and dynamic, of the implementation stage.
Given the realities in the field of project implementation, discussions with both the beneficiaries and the managing authority, we considered it useful to carry out a software to automate the previously defined indicator.

IOP Conf. Series: Materials Science and Engineering 1037 (2021) 012063 doi:10.1088/1757-899X/1037/1/012063

Starting from Project Quality Score, we can create a computerized system to evaluate social economy projects, using the JAVA language for application development.
Project quality indicator can be defined as:

$$P_{QS} = (\sum_{a=1}^{n} t_a \times p_{at}) \times pt + (\sum_{a=1}^{m} pi_a \times p_{ai}) \times pi \qquad (1)$$

with
P_{QS} = Project Quality Score,
t_a= target group value,
p_{at} = importance of the element,
pt = importance of target group in indicator,
pi_a = project indicators value,
p_{ai} = importance of the element,
p_i = importance of project indicators value.

2. Working Methodology

2.1.Logical scheme

For the achievement of this applicatioı lating the values of the indicator. Sitetic. The logical diagı

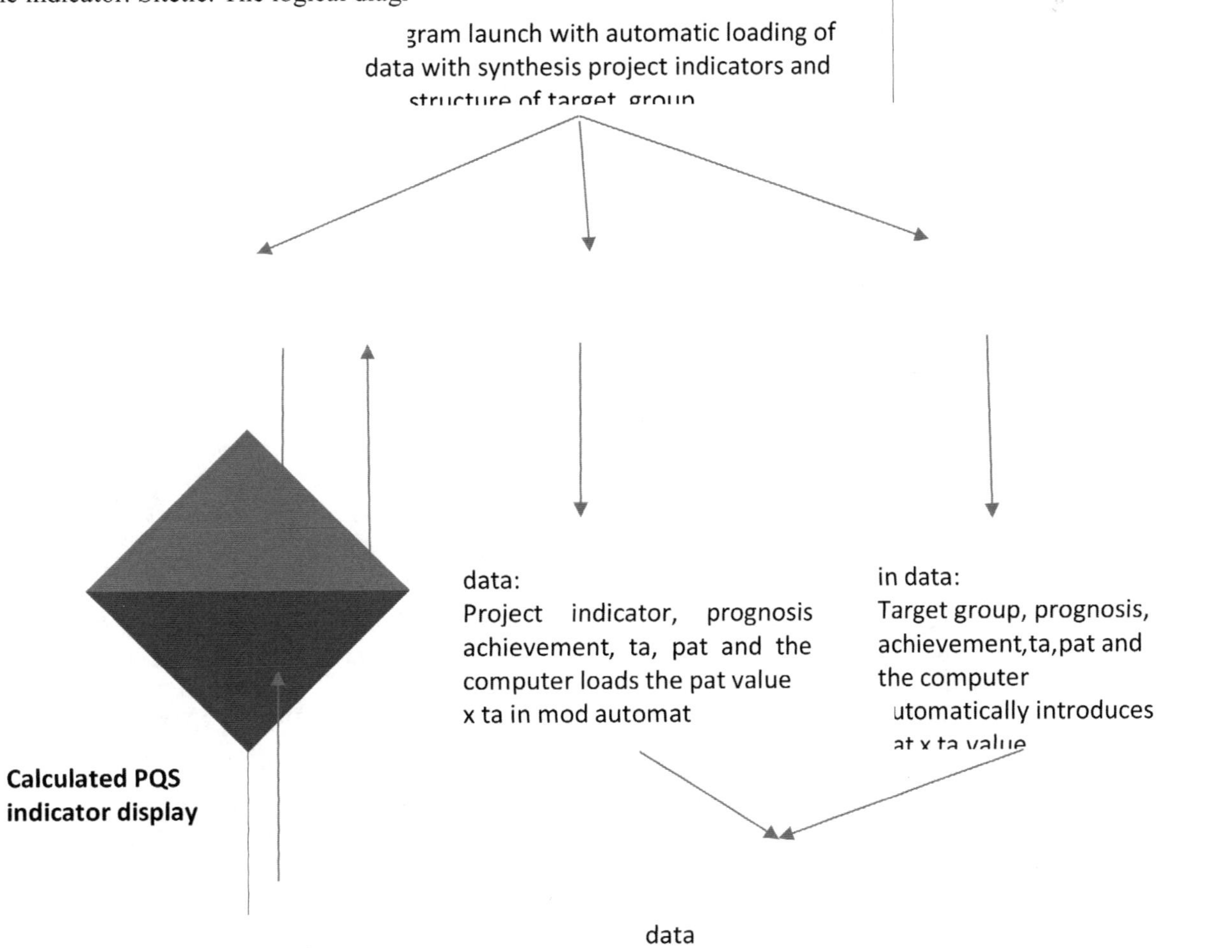

Figure 1. Logical diagram

2.2.Discussions

The input values are represented by the values assumed and realized in the project, according to the table below

Table 1. Synthesis of project indicators

Indicators	Prognosis	Achievement
Immediate production indicators (output)		
Number of incorporated social economy structures	11	11
Number of people benefiting from guidance / counselling - social economy	125	147
Number of participants in training - social economy	136	135
Number of ESF participants - women	15	19
Number of ESF participants - Roma persons	25	32
Number of ESF participants - people with disabilities	15	31
Number of ESF participants - other vulnerable groups	70	65
Number of communication and promotion events - social economy	2	0
Result indicators		-
Number of jobs created by social economy structures	46	47
Number of certified training participants - social economy	125	135
Number of jobs maintained by social economy structures	46	47

Table 2. Structure of target group

Target group	Prognosis		Achievement	
	Women	Men	Women	Men
Families having more than 2 children	4	6	8	4
Mono-parental families	4	1	7	1
Women	15	0	20	0
Managers of the social enterprises	4	7	4	7
People who left school early	3	7	4	6
People living from the guaranteed minimum income	20	20	9	15
Persons with disabilities	8	7	14	17
Roma people	10	15	13	19
Young people over 18 leaving the institutionalized child protection system	3	2	0	0

At launch time the software application is loaded with the data from the two tables, taking over from the created database, as in the figure Figure 2.

IOP Conf. Series: Materials Science and Engineering 1037 (2021) 012063 doi:10.1088/1757-899X/1037/1/012063

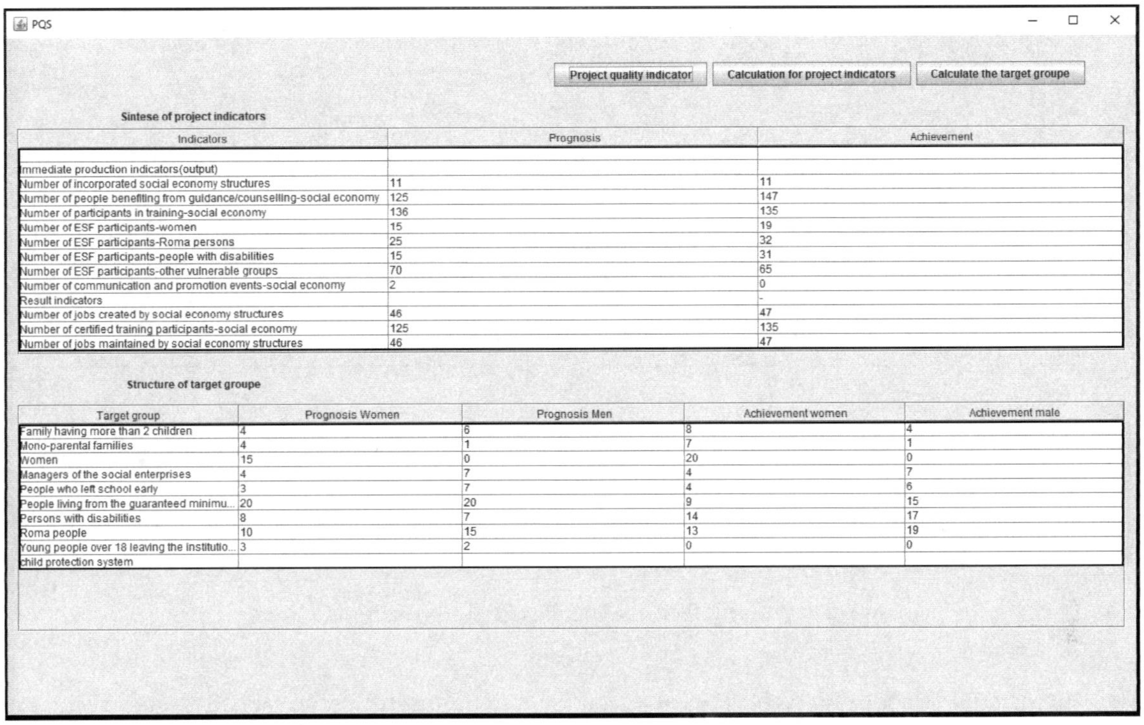

Figure 2. Launching PQS software application

Records downloaded automatically can be validated and manually modified in the application as are presented in figure 3 and figure 4.

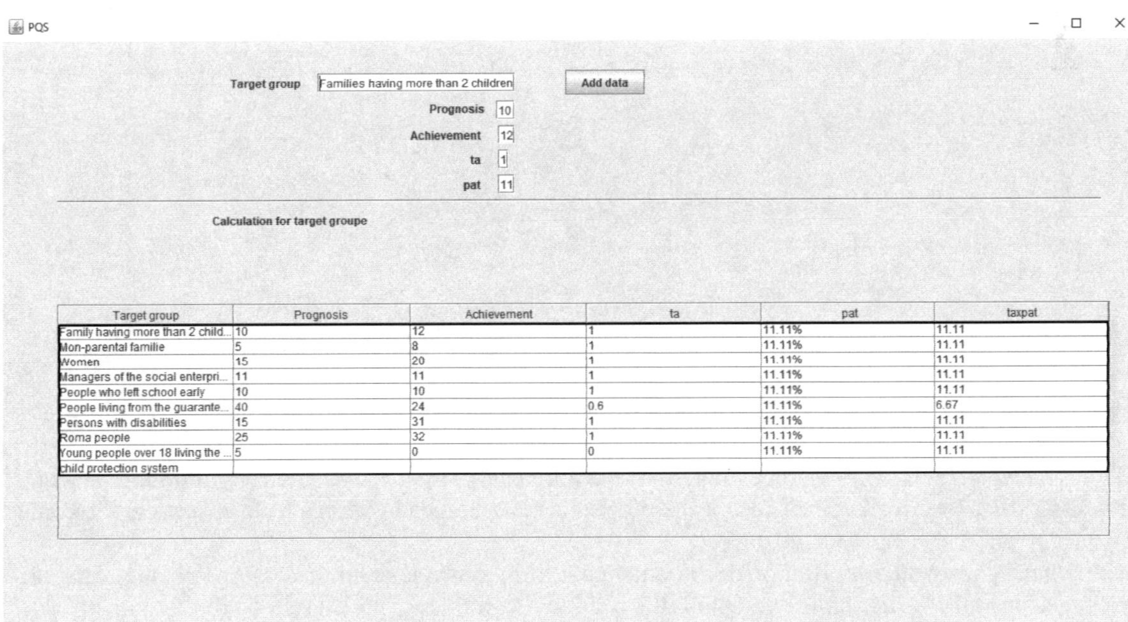

Figure 3. Adding records in database for target group

IOP Conf. Series: Materials Science and Engineering 1037 (2021) 012063 doi:10.1088/1757-899X/1037/1/012063

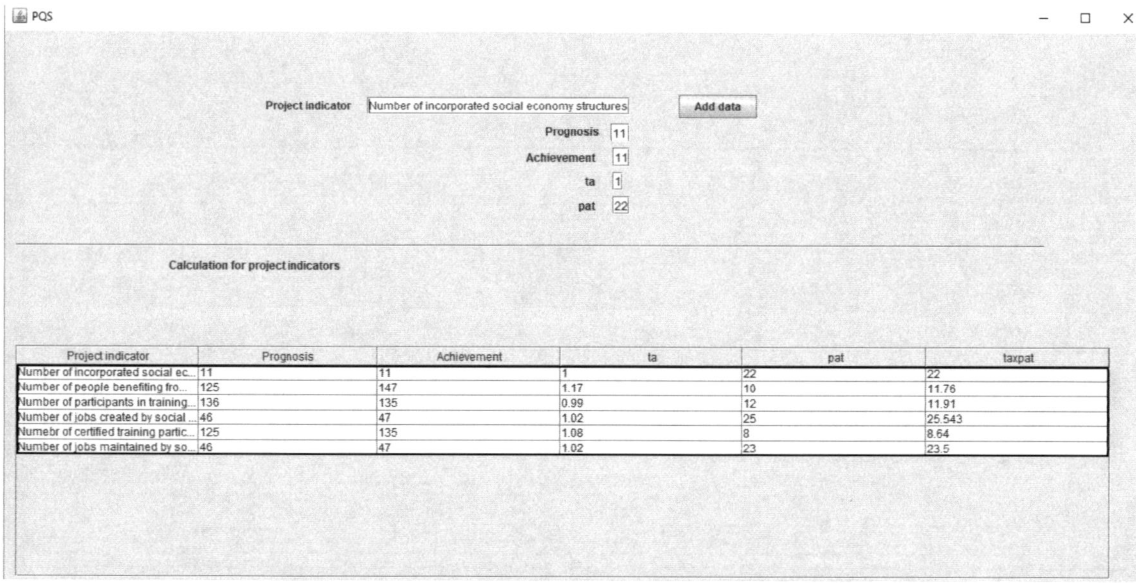

Figure 4. Adding records in database for project indicators

The intermediate calculation values of the complex indicator of project quality are calculated automatically and are offered as values available for further analysis of project efficiency and effectiveness.

Based on these data, the application calculates the general indicator according to the formula (1).The result is presented in figure 5.

$$P_{GQS} = \left(\sum_{a=1}^{n} t_a \times p_{at}\right) \times pt + \left(\sum_{a=1}^{m} pi_a \times p_a\right) \times pi = \mathbf{0.9087} \quad (2)$$

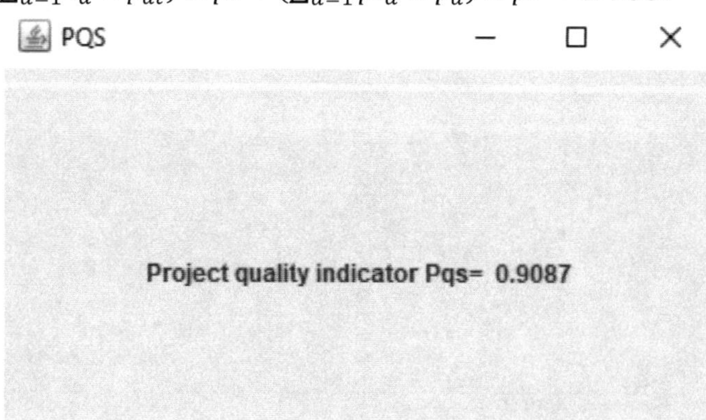

Figure 5. Calculated result

The Java language used in the development of the application is a very secure programming language, providing strict security mechanisms for the programs materialized by: dynamic verification of the code for the detection of dangerous sequences.

Also, when choosing the application development platform, it was taken into account that it is a language independent of the work platform, without the need to recompile it on different operating systems such as Windows, Linux, Mac OS, Solaris, which brings substantial savings to application development companies as well as the possibility of running this software in the cloud.

The application ensures a management of the resources of the hardware equipment on which it runs, being optimized to work even on mobile media such as phones, tablets.

3. Conclusions

The application allows the storage of intermediate values in the database, so that later the results can be analyzed, providing graphs with the evolutions of the degree of project accomplishment.
At the same time, the application can be a facility for management bodies in order to automatically measure the fulfillment of the indicator.
Among the advantages of the application we mention:
- Operation regardless of the operating system
- Does not require licensing costs, being made using the java support (open source)
- The application can be scaled to the level of multiple sectors of activity in the national economy, considering the general character of the application
- In the general conditions of company computerization, as well as in the current, particular conditions generated by the COVID pandemic, we consider that this instrument can constitute a valuable basis for ensuring a secure project management.

As it can be seen in the study, such software allows a real-time analysis of the degree of realization of the project indicators and the target group.
The analysis is qualitative, but with comparable results in terms of both the evolution over time and between projects in non-unitary fields of activity.
In the current period, the vast majority of projects that are implemented at European level are based on the assumption that two objectives are met: A well-defined and well-aimed at target group and measurable and traceable project indicators.
The main advantage of the complex indicator of the quality of projects, especially as a result of the realization of a software that automatically processes the data received from both the beneficiaries and the managing authorities, is given precisely by the possibility of achieving this in an automatic and objective manner.

References

[1] Marcu D, Costoiu M, Semenescu A, Ioana A, Tufeanu D, 2019, *Quantitative measurement of the uality of a social economy project*, 23rd International Conference on INNOVATIVE MANUFACTURING ENGINEERING AND ENERGY
[2] Frijns P, Van Leeuwen F, Bierwolf R ,2017 *Project management-a more balanced approach* , Technology & Engineering Management Conference (TEMSCON), IEEE , pp 234-238
[3] Harrison F, Lock D ,2016 *Advanced project management: a structured approach.* , ISBN 9781138270633, Ed. Routledge,p 336
[4] Harris E, 2009, *Strategic project risk appraisal and management*, ISBN 9780566088483, Ed. Routledge , p 126
[5] Tanasa S, Andrei S, Olaru C,2007, *Java de la 0 la expert*, ISBN:9789734624058, Ed Polirom, , Bucuresti, p 560
[6] Stanciu I, Paraianu E , Schileru I,1998, *Calimetrie I.,* ISBN 973-9264-34-4, Ed Oscar Print Bucuresti ,p 321
[7] Paraschivescu O A, 2006, *Managementul calitatii, ediția a II-a revizuita si adaugita,* Ed Tehnopress , p 248

IOP Conf. Series: Materials Science and Engineering 1037 (2021) 012064 doi:10.1088/1757-899X/1037/1/012064

Software application for optimizing raw material costs, including materials recovered from batteries and accumulators

B Florea[1], D Marcu[1*], A Ioana[1], A Semenescu[1], R Solea[1], G Iacob[1] and F Niculescu[1]

[1] University Politehnica of Bucharest, Engineering and Management of Metallic Material Department, Splaiul Independenţei Street No. 313, Romania

[*]Corresponding author's e-mail: dragos.marcu@arasnet.ro

Abstract. In this article we aim to create a software application aimed at highlighting the importance of the technical and economic estimate in making the bushings. This software was created as a web platform, available at: https://www.bucsa-sim.com , in order to maximize the accessibility, speed and efficiency of the complex process that underlies the construction of parts, in our case, bushings. The idea of developing this software started from the necessity generated by the fact that currently there is only one website on the Romanian market available at: https://bucsebronz.ro/calculator-piese/, which does not allow users to calculate the technical-economic estimate in real time, but only to send a simple form to the company. The use of this application is part of the new concept of Industrialization 4.0 because it allows operation in the cloud with immediate access to the display of the price of a bushing made of elements recovered from waste. Taking into consideration that materials such as copper and lead can come from recycling of batteries and accumulators, this application contributes to a better integration of the processing of these materials in the priciples or the circular economy.

1.Introduction

Industrialization 4.0 is an organizational concept that consists of four basic organizational principles. These principles support companies in identifying and implementing Industrialization 4.0 scenarios

• Interconnection: machines, devices, sensors, and people can be networked and can communicate over the car internet or over the Internet between people.

• Transparency of information: the data delivered by the sensors extend the digital information systems of manufacturing models to create a virtual image of the real world.

• Technical assistance: assistance systems support people with the help of collected, visualized and easy to understand information. Thus, clear decisions can be made quickly if problems arise. In addition, people who perform excessive or dangerous physical effort are being supported.

•Decentralized decisions: cyber systems are able to make decisions and solve problems autonomously. Only in exceptional cases, for example, in the event of interference or conflicts of interest, signals are transmitted at a higher level.
ERP (Enterprise Resource Planning) is an integrated system of software applications, which help manage functional processes in an organization / enterprise. ERP software applications use a common database and are called integrated because they incorporate all the relevant information circulating in a company.
The expansion of MRP II functionalities led to the emergence of the first ERP systems in the late 1980s, as software packages that support all functional areas: planning, production, sales, marketing, distribution, accounting, finance, human resources, project management, inventory, service, maintenance, logistics, e-business.
Taking into account the features of the program developed and presented in the paper, this can be included in the ERP concept in the marketing package.
The novelty of such a soft compared to classical methods of market study is the fact that it allows taking over in real time the trading prices from different available sources regarding the price of raw materials.
At the same time, the application allows the uniformization of values by converting available heterogeneous data into a predetermined currency.
The software is part of the new trend represented by the analysis of "big data", allowing the correlation of different sources of information.
It allows the takeover from the waste suppliers of the quantities available at any time so that they can supply, depending on the customer's need and the purchase price of the product.

2. Software functionality

For maximum accessibility and ease of use, the software has been created as a web interface, which can be accessed from any device connected to the Internet (with the help of any browser), anywhere and anytime, at the next web address: www.bucsa-sim.com
The creation of the website consisted of the use of 3 programming languages HTML, CSS and Java, reaching over 1000 code lines, these being created in the coding program "Brackets". Also, as secondary programs for graphics and design, the following programs were used:
Adobe Photoshop CC 2019 image creation and editing;
Adobe Premier PRO - video editing;
Autocad 2019 - creation and editing of 2D and 3D models;
The first step in creating the website was to create 4 main files:
Index.html - HTML interface that allows navigation, data entry for calculation and display of figures, pictures, videos and 3D model
Resultss.html - HTML interface that allows navigation and display of results such as tables and graphs
Style.css - this file written in CSS, allows you to change the elements in the two HTML pages, such as: the size of an element, the positioning of an element, the color, etc.
Script.JS - we can say that this file created in JavaScript is the "brain" behind the website, where all the data entered is being processed.
Once the web page is loaded, the interface in figure 1 will appear on the computer or mobile phone screen.

Figure 1. Header web interface displayed online

2.1. Application functionality

In order to show how the application works, we extracted from a client an example of taking data and providing the price for certain quotas of a bushing. The figure below (fig.2) shows the software interface in which the customer enters the dimensions of the desired product, respectively the outer diameter, the inner diameter and the height of the piece.

The created application allows setting the time period in which you want to display the price while providing a tool for analyzing the evolution of raw materials on the world market and the composition of the bronze alloy from which you want to achieve it.

Figure 2 shows the web interface for entering data for the "Quantity of Parts" section for the purpose of processing these data.

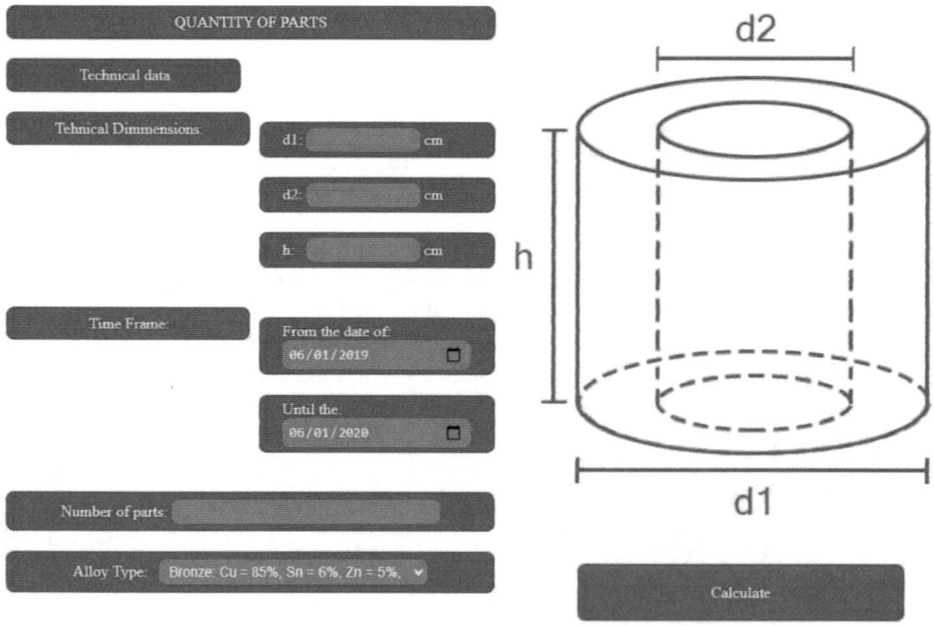

Figure 2. Web interface displayed for data entry

IOP Conf. Series: Materials Science and Engineering 1037 (2021) 012064 doi:10.1088/1757-899X/1037/1/012064

Using the Java programming language, the script.js (JavaScript) file was created which takes all the data entered in the previous steps and processes it.
To calculate material costs, the code written by us in the script.js file takes the statistical data provided by the www.QUADL.com website using an API or Application Programming Interface, and processes them according to the data entered in the previous steps (figure 2). "Quadl provides data on commodity prices for nearly 100 commodities, including gold prices, silver prices and copper prices from several sources."
The article presents a case study obtained by using the application created and used in real time through the website https://www.bucsa-sim.com.
Figure 3 shows the amount of material required for the number of pieces chosen and for a certain bronze composition.

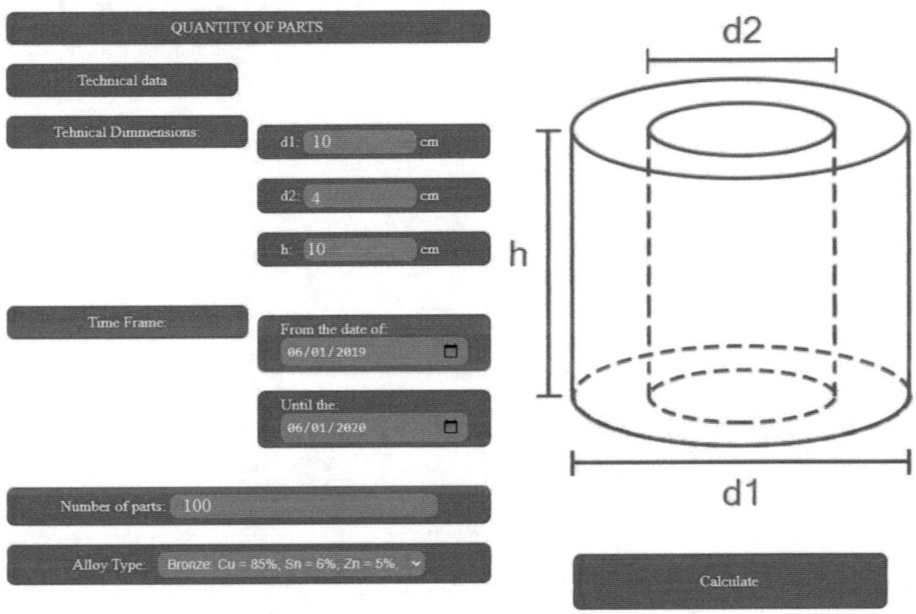

Figure 3. With the data introduced online on the website

By pressing the calculate button, the ava program returns a page with the necessary quantity to make 100 products as well as a table with the evolution of the price for several months depending on the desired period figure 4.

IOP Conf. Series: Materials Science and Engineering 1037 (2021) 012064 doi:10.1088/1757-899X/1037/1/012064

Outer Diameter (d1):	10 cm
Inner Diameter (d2):	4 cm
Height (h):	10 cm
Alloy Density:	8.75 grame/cm3
Track Volume:	659.73 cm3
Part Weight:	5772.64 grams
Total Material:	577264.00 grams
Number of parts:	100
Amount of Copper:	469373.36 grams
Quantity of Tin:	17202.47 grams
Amount of Lead:	22686.48 grams

The below Prices are calculate at the following rate: $1.00= RON 4.50

Data	**2020-05-31**	**2020-04-30**	**2020-03-31**	**2020-02-29**	**2020-01-31**	**2019-12-31**
Copper	10944.49	10564.64	10825.02	11880.07	12597.46	12693.23
Zinc	597.75	575.97	576.05	639.48	712.43	687.69
Plumb	164.19	167.34	175.10	189.04	194.23	191.87
Tin	1179.03	1144.65	1170.53	1261.58	1303.60	1312.16
Total	**12885.46 RON**	**12452.60 RON**	**12746.70 RON**	**13970.17 RON**	**14807.72 RON**	**14884.95 RON**

Figure 4. Results returned by the soft

3. Discussions and conclusions

The online price simulation has the advantage that it allows a customer to identify the alloy solution he can choose and the price difference between them. It can provide both the beneficiary and the manufacturer with information about the evolution of the price market and thus a much cheaper value from recoveries than from pure materials can be presented.

It has the advantage that it can transmit the cost of the product through the modules within the software package related to the ERP concept, thus automatically recording the stocks of raw materials, issuing the consumption receipt for production leading to optimizing the production management of the company that uses it. Such software which streamlines the use of recovered materials, contributes to an economy based on the principles of sustainable development.

The advantage of the app is that it allows the trend analysis of the part price, depending on the evolution of both the prices and the exchange rate.

The app allows the automatic export of the request for offers to the partners from the company's database in order to later make the parts.

It allows the comparison of the received offers both with an average price resulting from the specific calculation and with a trend of it on the market.

The calculation relationships on which the application is based can be harmonized depending on the calculation requirement.

References

[1] Kagermann H, Lukas W D, Wahlster D, *Industrie 4.0: Mit dem Internet der Dinge auf dem Weg zur 4.industriellen Revolution*, http://www.wolfgang-wahlster.de/wordpress/wpcontent/uploads/Industrie_4_0_Mit_dem_Internet_der_Dinge_auf_dem_Weg_zur_vierten_industriellen_Revolution_2.pdf , [accesed 1.09.2020].

[2] Shaul L, Tauber D, 2010, *Hierarchical examination of success factors across ERP life cycle,* MCIS 2010 Proceedings.: 79
[3] Lequeux J L, 2008, *Manager avec les ERP, Architecture Orientée Services (SOA) (in French Edition),* ISBN 978-2-212-54094-9, Ed Eyrolles Group, p 380
[4] Sergi B S , Popkova E G, Bogoviz A V, Litvinova T N, 2019, *Understanding Industry 4.0:AI Internet of things and the future of work,* Ed Emerald Publishing Limited, p 322
[5] Wagner R, Wyke R A, 2002, *Javascript* ISBN 973-20-0127-5, Ed Teora, p 815
[6] Mircea M, 2009, *Managementul afacerilor in economia digitala,* ISBN 978-973-709-438-4, Ed Economica, p 254

Acknowledgments
The work has been funded by the Operational Programme Human Capital of the Ministry of European Funds through the Financial Agreement 51668/09.07.2019, SMIS code 124705.

AUTHOR INDEX